FUNDAMENTALS OF MOMENTUM, HEAT, AND MASS TRANSFER

FUNDAMENTALS OF MOMENTUM, HEAT, AND MASS TRANSFER

THIRD EDITION

JAMES R. WELTY

Professor and Head
Department of Mechanical Engineering
Oregon State University

CHARLES E. WICKS

Professor and Head
Department of Chemical Engineering
Oregon State University

ROBERT E. WILSON

Professor of Mechanical Engineering
Oregon State University

JOHN WILEY & SONS

New York • Chichester • Brisbane • Toronto • Singapore

Library of Congress Cataloging in Publication Data:

Welty, James R.
    Fundamentals of momentum, heat, and mass transfer.

Includes indexes.
1. Fluid mechanics. 2. Heat—Transmission.
3. Mass transfer. I. Wicks, Charles E. II. Wilson,
Robert E. (Robert Elliott) III. Title.
TA357.W45 1984   532   83-17065
ISBN 0-471-87497-3

Printed in the United States of America

20   19   18   17   16

PREFACE

This third edition of *Fundamentals of Momentum, Heat, and Mass Transfer* is, in style, much like the second. We have retained a mixture of SI and English units, introduced new material, deleted some of the old, and altered and rearranged a good deal of the rest. Our objectives have been those of retaining the strengths of earlier editions while bringing in new developments in the subject matter and its treatment.

Some additional detail has been introduced concerning numerical methods for solving transport problems. Although we recognize that a capability will exist for engineers to employ computers to solve problems in this area in their jobs, the ability to solve problems is neither the principal objective of this text nor, presumably, of the courses for which it is used. We continue to believe that a solid understanding of fundamental transport processes must precede a treatment of problem solution techniques. Therefore, numerical methods are discussed but do not represent a central theme of our treatment.

It is quite possible—and we do so in our own course—to assign problems requiring numerical solution using digital computers in courses using this textbook. Such problems are considered to be an application of the material included in this text using techniques learned elsewhere in their execution.

The critical comments of students and teachers, both at Oregon State University and elsewhere, have been extremely helpful in preparing this third edition. We acknowledge all with our thanks. The continued fine cooperation and support of Merrill Floyd and others associated with John Wiley is also greatly appreciated.

Corvallis, Oregon
June 1983

J. R. Welty
C. E. Wicks
R. E. Wilson

PREFACE
TO THE SECOND EDITION

The basic objectives of this edition, as expressed in the preface to the first edition, remain unaltered. The transport process, a single unified subject, is still the basic subject being discussed and for which this book provides the study tool.

In this edition we update the material, bringing in applications of current-day technology. We have also amended the treatment of the first edition to include more detail and additional discussion in those areas where we have found students to have greatest difficulty. We believe—and certainly hope—that this edition will retain the strengths that so many people have commented on regarding the first, with the additional improvement as noted.

Certainly the most obvious change in this edition is the incorporation of SI units. We introduce a balanced treatment of SI and English units, both in example problems and in assigned problems at the end of each chapter. Tables of physical properties have also been amended to include SI data for solids and gases. A good compilation of liquid properties in SI units does not, in our experience, exist; for this reason, it is still necessary for the instructor and student to make appropriate conversions for liquids when properties are needed in SI units. In each example problem, whether it is worked in English or in SI units, the corresponding value in the alternate system is added parenthetically following the final result. This practice attempts to give a feel for orders of magnitude of numbers in each system.

We remain committed to the belief that a good understanding and facility in problem solving in the area of transport processes is an absolute necessity for a competent engineer, regardless of his or her fundamental discipline. The course for which the first edition has served as text for the past six years at Oregon State University has been increasingly well accepted by all engineering disciplines on this campus. We hope that a unified treatment of the transport processes will increase in popularity in many other institutions as well.

The assistance and critical comments of numerous students over the past several years have been most helpful in the preparation of this edition. We particularly appreciate the encouragement and advice of several of our colleagues who have taught using the first edition. Hopefully, we have incorporated those things which, in the opinions of many, will strengthen the text.

Corvallis, Oregon   **J. R. Welty**
September 1976   **C. E. Wicks**
                                                                                R. E. Wilson

PREFACE
TO THE FIRST EDITION

Traditionally, engineering curricula have included courses in momentum transfer of fluid mechanics, usually in Departments of Civil or Mechanical Engineering; energy or heat transfer courses have been parts of Chemical and Mechanical Engineering curricula; and the subject of mass transfer or diffusion has been almost the exclusive domain of the chemical engineers. When studied in this fragmented form, the similarities in both qualitative and quantitative descriptions between each subject are often ignored or thought to be coincidental.

In 1960, with the publication of *Transport Phenomena*, R. B. Bird, W. E. Stewart, and E. N. Lightfoot of the University of Wisconsin united these three previously fragmented subjects into a single volume with a unified approach to the *transfer process*. Thus, students are able to consider a single discipline rather than three and can use the similarities of description and calculation to strengthen their understanding of individual transfer processes. An additional reason for the popularity of the unified approach is the increased interest in situations where two or sometimes all three kinds of transfer are involved in a single process. A systematic, fundamental description of the transport process is invaluable in this regard.

A gradual evolution of engineering curricula to include more core areas of basic subject matter has led many institutions to offer courses in momentum, heat, and mass transfer. In such cases, the transfer process is considered just as fundamental as mechanics, thermodynamics, materials science, and basic electricity and magnetism to the background of the engineering student. It was in this context that the present text has evolved. This material has been developed and used in part by classes in a junior-level course, entitled Transfer and Rate Processes, at Oregon State University since 1963. This book is the result of class notes which have been revised and rewritten at least once during each of the past five years. The opinions and critical comments of students and fellow instructors in this sequence have been a great help to the authors.

Certain compromises are necessary in putting together a textbook of this nature. The authors have been interested primarily in writing a basic textbook to increase a student's understanding of momentum, energy, and mass transfer. Specific applications of this material are kept at a minimum; it is expected that laboratory or design courses to be encountered subsequent to the use of this text will dwell on specific applications and techniques of problem solutions. Three "applications" chapters are included in this text (Chapters 14, 22, and 31). These chapters appear near the end of each section to provide some knowledge of equipment and to indicate the sort of problems that may be approached with the material introduced in this text. These chapters have been included to motivate the student while still giving a minimum of applications to those for whom this text will be a terminal study of momentum, energy, and mass transfer.

This book has been written at a level for junior-year engineering students. The student is assumed to have taken courses in mechanics and mathematics through differential equations, as well as introductory courses in chemistry and physics. Additionally, it would be helpful if a course in thermodynamics be taken before or concurrent with the use of this textbook.

The mathematical level of this textbook has been of much concern to the authors. Vector notation has been employed, primarily in the development of fundamental equations. The compactness, generality, and neatness of vector notation are considered sufficient to overcome objections of those who have suggested that this treatment is too sophisticated. Still others, although fewer in number, have suggested that more general tensor notation and operations would be best. The choice made has been a compromise, deemed best by the authors. A knowledge of differential equations through the solution of second-order equations is necessary. Three example problems involving the solution to partial differential equations by the method of separation of variables are included; however, these may be omitted without any serious loss in comprehension by the student.

Two general approaches to this subject material may be employed. These are depicted diagrammatically opposite. The text is organized in a "vertical" manner; the subjects of momentum transfer, energy transfer, and mass transfer are presented in that order. The alternative "horizontal" approach is indicated on the diagram. This approach involves a coverage of similar subject matter for all three types of transfer, considering one transfer mechanism at a time. The authors are aware that significant numbers of faculty wish to treat the material in each way. The text has been organized to accommodate both schools of thought.

The first three chapters may or may not be studied depending on the particular instructor. The material included is probably covered in previously taken courses but may help bring students of various backgrounds to the same level of understanding before they study transfer processes.

Chapters 4, 5, and 6 are fundamental for the entire text. These chapters must be considered in depth and should be well understood before proceeding. The control volume approach, introduced at this point, is basic to all subsequent chapters. This approach to transport processes is one of the main differences between this text and that of Bird, Stewart, and Lightfoot.

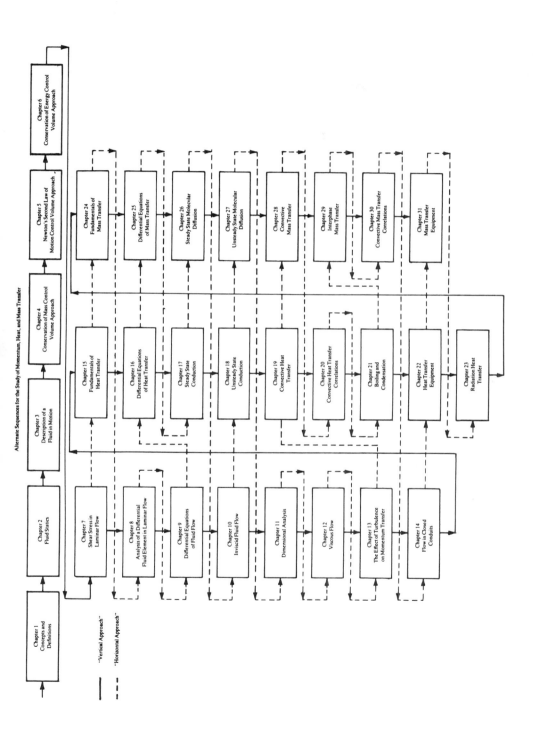

Chapters 7 through 14 deal exclusively with momentum transfer, Chapters 15 through 23 consider energy transfer, and Chapters 24 through 31 involve mass transfer. All may be considered in this sequence or in the "horizontal" manner as mentioned before. The only portion standing alone is Chapter 23, dealing with radiant heat transfer, which has no counterpart in either momentum or mass transfer.

The authors are firmly committed to the consideration of rate processes as fundamental for engineering studies. We believe that the lack of a well-accepted textbook has hindered the adoption of this approach at numerous institutions. Hopefully this text will influence some schools to treat momentum, energy, and mass transfer as a basic part of their curricula, thereby providing an additional, vital background for their graduates.

Corvallis, Oregon   **J. R. Welty**
April 1969   **C. E. Wicks**
                                                                         R. E. Wilson

CONTENTS

1
CONCEPTS
AND DEFINITIONS

Momentum transfer in a fluid involves the study of the motion of fluids and the forces that produce these motions. From Newton's second law of motion it is known that force is directly related to the time rate of change of momentum of a system. Excluding action-at-a-distance forces such as gravity, the forces acting on a fluid, such as those resulting from pressure and shear stress, may be shown to be the result of microscopic (molecular) transfer of momentum. Thus the subject under consideration which is historically fluid mechanics may equally be termed momentum transfer.

The history of fluid mechanics shows the skillful blending of the 19th- and 20th-century analytical work in hydrodynamics with the empirical knowledge in hydraulics that man has collected over the ages. The mating of these separately developed disciplines was started by Ludwig Prandtl in 1904 with his boundary-layer theory, which was verified by experiment. Modern fluid mechanics, or momentum transfer, is both analytical and experimental.

Each area of study has its phraseology and nomenclature. Momentum transfer being typical, the basic definitions and concepts will be introduced in order to provide a basis for communication.

1.1   FLUIDS AND THE CONTINUUM

A fluid is defined as a substance which deforms continuously under the action of a shear stress. An important consequence of this definition is that when a fluid is at rest, there can be no shear stresses. Both liquids and gases are fluids. Some substances such as glass are technically classified as fluids. However, the rate of deformation in glass at normal temperatures is so small as to make its consideration as a fluid impractical.

Concept of a Continuum. Fluids, like all matter, are composed of molecules whose numbers stagger the imagination. In a cubic inch of air at room conditions

1

there are some 10^{20} molecules. Any theory which would predict the individual motions of this many molecules would be extremely complex, far beyond our present abilities. While both the kinetic theory of gases and statistical mechanics treat the motions of molecules, this is done in terms of statistical groups rather than in terms of individual molecules.

Most engineering work is concerned with the macroscopic or bulk behavior of a fluid rather than with the microscopic or molecular behavior. In most cases it is convenient to think of a fluid as a continuous distribution of matter or a *continuum*. There are, of course, certain instances in which the concept of a continuum is not valid. Consider, for example, the number of molecules in a small volume of a gas at rest. If the volume were taken small enough, the number of molecules per unit volume would be time-dependent for the microscopic volume even though the macroscopic volume had a constant number of molecules in it. The concept of a continuum would be valid only for the latter case. The validity of the continuum approach is seen to be dependent upon the type of information desired rather than the nature of the fluid. The treatment of fluids as continua is valid whenever the smallest fluid volume of interest contains a sufficient number of molecules to make statistical averages meaningful. The macroscopic properties of a continuum are considered to vary smoothly (continuously) from point to point in the fluid. Our immediate task is to define these properties at a point.

1.2   PROPERTIES AT A POINT

When a fluid is in motion the quantities associated with the state and the motion of the fluid will vary from point to point. The definition of some fluid variables at a point is presented below.

Density at a Point. The density of a fluid is defined as the mass per unit volume. Under flow conditions, particularly in gases, the density may vary greatly throughout the fluid. The density, ρ, at a particular point in the fluid is defined as

$$\rho = \lim_{\Delta V \to \delta V} \frac{\Delta m}{\Delta V}$$

where Δm is the mass contained in a volume ΔV, and δV is the smallest volume surrounding the point for which statistical averages are meaningful. The limit is shown in Figure 1.1.

The concept of the density at a mathematical point, that is, at $\Delta V = 0$ is seen to be fictitious; however, taking $\rho = \lim_{\Delta V \to 0} (\Delta m / \Delta V)$ is extremely useful, as it allows us to describe fluid flow in terms of continuous functions. The density, in general, may vary from point to point in a fluid and may also vary with respect to time as in a punctured automobile tire.

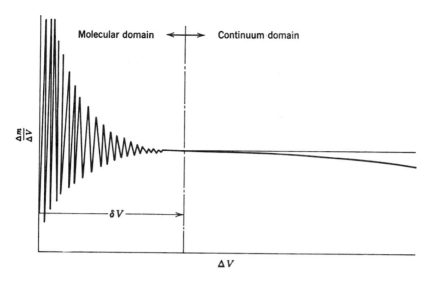

Figure 1.1   Density at a point.

Fluid Properties and Flow Properties. Some fluids, particularly liquids, have densities which remain almost constant over wide ranges of pressure and temperature. Fluids which exhibit this quality are usually treated as being incompressible. The effects of compressibility, however, are more a property of the situation than of the fluid itself. For example, the flow of air at low velocities is described by exactly the same equations that describe the flow of water. From a static viewpoint air is a compressible fluid and water incompressible. Instead of being classified according to the fluid, compressibility effects are considered a property of the flow. A distinction, often subtle, is made between the properties of the fluid and the properties of the flow, and the student is hereby alerted to the importance of this concept.

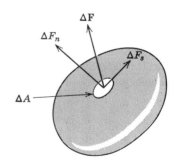

Figure 1.2   Force on an element of fluid.

Stress at a Point. Consider the force $\Delta \mathbf{F}$ acting on an element ΔA of the body shown in Figure 1.2. The force $\Delta \mathbf{F}$ is resolved into components normal and parallel to the surface of the element. The force per unit area or stress at a point is defined as the limit of $\Delta \mathbf{F}/\Delta A$ as $\Delta A \to \delta A$, where δA is the smallest area for which statistical averages are meaningful:

$$\lim_{\Delta A \to \delta A} \frac{\Delta F_n}{\Delta A} = \sigma_{ii} \qquad \lim_{\Delta A \to \delta A} \frac{\Delta F_s}{\Delta A} = \tau_{ij}$$

Here σ_{ii} is called the normal stress and τ_{ij} the shear stress. In this text the double-subscript stress notation as used in solid mechanics will be employed. The student will recall that normal stress is positive in tension. The limiting process for the normal stress is illustrated in Figure 1.3.

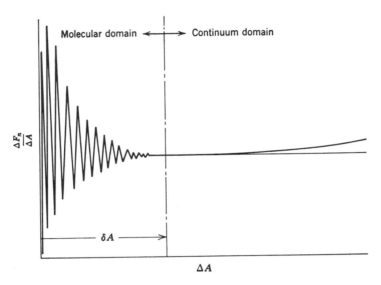

Figure 1.3   Normal stress at a point.

Forces acting on a fluid are divided into two general groups: body forces and surface forces. Body forces are those which act without physical contact, for example, gravity and electrostatic forces. On the other hand, pressure and frictional forces require physical contact for transmission. Since a surface is required for the action of these forces they are called surface forces. Stress is therefore a surface force per unit area.*

Pressure at a Point in a Static Fluid. For a static fluid, the normal stress at a point may be determined from the application of Newton's laws to a fluid element as the fluid element approaches zero size. It may be recalled that there can be no shearing stress in a static fluid. Thus the only surface forces present will be those due to normal stresses. Consider the element shown in Figure 1.4. This element, while at rest, is acted upon by gravity and normal stresses. The weight of the fluid element is $\rho g(\Delta x\,\Delta y\,\Delta z/2)$.

For a body at rest, $\sum \mathbf{F} = 0$. In the x direction,

$$\Delta F_x - \Delta F_s \sin\theta = 0$$

---

* Mathematically, stress is classed as a tensor of second order, since it requires magnitude, direction, and orientation with respect to a plane for its determination.

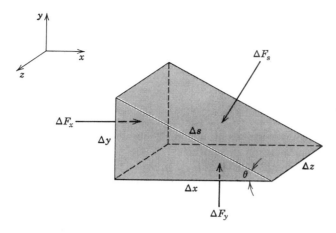

Figure 1.4   Element in a static fluid.

Since $\sin \theta = \Delta y / \Delta s$, the above equation becomes

$$\Delta F_x - \Delta F_s \frac{\Delta y}{\Delta s} = 0$$

Dividing through by $\Delta y \, \Delta z$ and taking the limit as the volume of the element approaches zero, we obtain

$$\lim_{\Delta V \to 0} \left[\frac{\Delta F_x}{\Delta y \, \Delta z} - \frac{\Delta F_s}{\Delta s \, \Delta z} \right] = 0$$

Recalling that normal stress is positive in tension, we obtain, by evaluating the above equation,

$$\sigma_{xx} = \sigma_{ss} \tag{1-1}$$

In the y direction, applying $\sum \mathbf{F} = 0$ yields

$$\Delta F_y - \Delta F_s \cos \theta - \rho g \frac{\Delta x \, \Delta y \, \Delta z}{2} = 0$$

Since $\cos \theta = \Delta x / \Delta s$, one has

$$\Delta F_y - \Delta F_s \frac{\Delta x}{\Delta s} - \rho g \frac{\Delta x \, \Delta y \, \Delta z}{2} = 0$$

Dividing through by $\Delta x \, \Delta z$ and taking the limit as before, we obtain

$$\lim_{\Delta V \to 0} \left[\frac{\Delta F_y}{\Delta x \, \Delta z} - \frac{\Delta F_s}{\Delta s \, \Delta z} - \frac{\rho g \, \Delta y}{2} \right] = 0$$

which becomes

$$-\sigma_{yy} + \sigma_{ss} - \frac{\rho g}{2}(0) = 0$$

or

$$\sigma_{yy} = \sigma_{ss} \tag{1-2}$$

It may be noted that the angle θ does not appear in equation (1-1) or (1-2), thus the normal stress at a point in a static fluid is independent of direction, and is therefore a scalar quantity.

Since the element is at rest, the only surface forces acting are those due to the normal stress. If we were to measure the force per unit area acting on a submerged element, we would observe that it acts inward or to place the element in compression. The quantity measured is, of course, pressure, which in light of the preceding development, must be the negative of the normal stress. This important simplification, the reduction of stress, a tensor, to pressure, a scalar, may also be shown for the case of zero shear stress in a flowing fluid. When shearing stresses are present, the normal stress components at a point may not be equal; however, the pressure is still equal to the average normal stress; that is,

$$P = -\tfrac{1}{3}(\sigma_{xx} + \sigma_{yy} + \sigma_{zz})$$

with very few exceptions, one being flow in shock waves.

Now that certain properties at a point have been discussed, let us investigate the manner in which fluid properties vary from point to point.

1.3 POINT-TO-POINT VARIATION OF PROPERTIES IN A FLUID

In the continuum approach to momentum transfer, use will be made of pressure, temperature, density, velocity, and stress fields. In previous studies the concept of a gravitational field has been introduced. Gravity, of course, is a vector, and thus a gravitational field is a vector field. In this book, vectors will be written in boldfaced type. Weather maps illustrating the pressure variation over this country are published daily in our newspapers, Since pressure is a scalar quantity, such maps are an illustration of a scalar field. Scalars in this book will be set in regular type.

In Figure 1.5 the lines drawn are the loci of points of equal pressure. The pressure, of course varies continuously throughout the region, and one may observe the pressure levels and infer the manner in which the pressure varies by examining such a map.

Of specific interest in momentum transfer is the description of the point-to-point variation in the pressure. Denoting the directions east and north in Figure 1.5 by x and y, respectively, we may represent the pressure throughout the region by the general function $P(x, y)$.

The change in P, written as dP, between two points in the region separated by the distances dx and dy is given by the total differential

$$dP = \frac{\partial P}{\partial x}\, dx + \frac{\partial P}{\partial y}\, dy \tag{1-3}$$

Figure 1.5   Weather map—an example of a scalar field.

In equation (1-3) the partial derivatives represent the manner in which P changes along the x and y axes, respectively.

Along an arbitrary path s in the xy plane the total derivative is

$$\frac{dP}{ds} = \frac{\partial P}{\partial x}\frac{dx}{ds} + \frac{\partial P}{\partial y}\frac{dy}{ds} \qquad (1\text{-}4)$$

In equation (1-4) the term dP/ds is the directional derivative, and its functional relation describes the rate of change of P in the s direction.

A small portion of the pressure field depicted in Figure 1.5 is shown in Figure 1.6. The arbitrary path s is shown, and it is easily seen that the terms dx/ds and

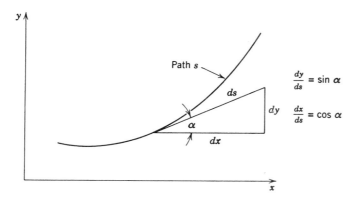

Figure 1.6   Path s in the xy plane.

dy/ds are the cosine and sine of the path angle, α, with respect to the x axis. The directional derivative, therefore, may be written

$$\frac{dP}{ds} = \frac{\partial P}{\partial x} \cos \alpha + \frac{\partial P}{\partial y} \sin \alpha \qquad (1\text{-}5)$$

There are an infinite number of paths to choose from in the xy plane; however, two particular paths are of special interest: the path for which dP/ds is zero and that for which dP/ds is maximum.

The path for which the directional derivative is zero is quite simple to find. Setting dP/ds equal to zero, we have

$$\left. \frac{\sin \alpha}{\cos \alpha} \right|_{dP/ds=0} = \tan \alpha \Big|_{dP/ds=0} = -\frac{\partial P/\partial x}{\partial P/\partial y}$$

or, since $\tan \alpha = dy/dx$, we have

$$\left. \frac{dy}{dx} \right|_{dP/ds=0} = -\frac{\partial P/\partial x}{\partial P/\partial y} \qquad (1\text{-}6)$$

Along the path whose slope is defined by equation (1-6), we have $dP = 0$, and thus P is constant. Paths along which a scalar is constant are called *isolines*.

In order to find the direction for which dP/ds is a maximum, we must have the derivative $(d/d\alpha)(dP/ds)$ equal to zero, or

$$\frac{d}{d\alpha} \frac{dP}{ds} = -\sin \alpha \frac{\partial P}{\partial x} + \cos \alpha \frac{\partial P}{\partial y} = 0$$

or

$$\tan \alpha \Big|_{dP/ds \text{ is max}} = \frac{\partial P/\partial y}{\partial P/\partial x} \qquad (1\text{-}7)$$

Comparing relation (1-6) and (1-7), we see that the two directions defined by these equations are perpendicular. The magnitude of the directional derivative when the directional derivative is maximum is

$$\left. \frac{dP}{ds} \right|_{max} = \frac{\partial P}{\partial x} \cos \alpha + \frac{\partial P}{\partial y} \sin \alpha$$

where $\cos \alpha$ and $\sin \alpha$ are evaluated along the path given by equation (1-7). Since the cosine is related to the tangent by

$$\cos \alpha = \frac{1}{\sqrt{1 + \tan^2 \alpha}}$$

we have

$$\cos \alpha \Big|_{dP/ds \text{ is max}} = \frac{\partial P/\partial x}{\sqrt{(\partial P/\partial x)^2 + (\partial P/\partial y)^2}}$$

Evauluating $\sin \alpha$ in a similar manner gives

$$\frac{dP}{ds}\bigg|_{max} = \frac{(\partial P/\partial x)^2 + (\partial P/\partial y)^2}{\sqrt{(\partial P/\partial x)^2 + (\partial P/\partial y)^2}} = \sqrt{\left(\frac{\partial P}{\partial x}\right)^2 + \left(\frac{\partial P}{\partial y}\right)^2} \qquad (1\text{-}8)$$

Equations (1-7) and (1-8) suggest that the maximum directional derivative is a vector of the form

$$\frac{\partial P}{\partial x}\mathbf{e}_x + \frac{\partial P}{\partial y}\mathbf{e}_y$$

where $\mathbf{e}_x$ and $\mathbf{e}_y$ are unit vectors in the x and y directions, respectively.

The directional derivative along the path of maximum value is frequently encountered in the analysis of transfer processes and is given a special name, the *gradient*. Thus the gradient of P, grad P, is

$$\text{grad } P = \frac{\partial P}{\partial x}\mathbf{e}_x + \frac{\partial P}{\partial y}\mathbf{e}_y$$

where $P = P(x, y)$. This concept can be extended to cases in which $P = P(x, y, z)$. For this more general case,

$$\text{grad } P = \frac{\partial P}{\partial x}\mathbf{e}_x + \frac{\partial P}{\partial y}\mathbf{e}_y + \frac{\partial P}{\partial z}\mathbf{e}_z \qquad (1\text{-}9)$$

Equation (1-9) may be written in more compact form by use of the operator ∇ (pronounced "del"), giving

$$\nabla P = \frac{\partial P}{\partial x}\mathbf{e}_x + \frac{\partial P}{\partial y}\mathbf{e}_y + \frac{\partial P}{\partial z}\mathbf{e}_z$$

where

$$\nabla = \frac{\partial}{\partial x}\mathbf{e}_x + \frac{\partial}{\partial y}\mathbf{e}_y + \frac{\partial}{\partial z}\mathbf{e}_z \qquad (1\text{-}10)$$

Equation (1-10) is the defining relationship for the ∇ operator in cartesian coordinates. This symbol indicates that differentiation is to be performed in a prescribed manner. In other coordinate systems, such as cylindrical and spherical coordinates, the gradient takes on a different form.* However, the geometric meaning of the gradient remains the same; it is a vector having the direction and magnitude of the maximum rate of change of the dependent variable with respect to distance.

1.4 UNITS

In addition to the International Standard (SI) system of units, there are two different English systems of units commonly used in engineering. These systems

---

* The forms of the gradient operator in rectangular, cylindrical, and spherical coordinate systems are listed in Appendix B.

have their roots in Newton's second law of motion: force is equal to the time rate of change of momentum. In defining each term of this law, a direct relationship has been established between the four basic physical quantities used in mechanics: force, mass, length, and time. Through the arbitrary choice of fundamental dimensions, some confusion has occurred in the use of the English systems of units. Adoption of the SI system of units as a world standard will overcome these difficulties.

The relationship between force and mass may be expressed by the following statement of Newton's second law of motion:

$$\mathbf{F} = \frac{m\mathbf{a}}{g_c}$$

where g_c is a conversion factor which is included to make the equation dimensionally consistent.

In the SI system, mass, length and time are taken as basic units. The basic units are mass in kilograms, (kg), length in meters, (m), and time in seconds, (s). The corresponding unit of force is the newton, (N). One newton is the force required to accelerate a mass of one kilogram at a rate of one meter per second per second (1 m/s^2). The conversion factor, g_c, is then equal to one kilogram meter per newton per second per second $(1 \text{ kg} \cdot \text{m/N} \cdot \text{s}^2)$.

In engineering practice, force, length, and time have been frequently chosen as defining fundamental units. With this system, force is expressed in pounds force (lb_f), length in feet, and time in seconds. The corresponding unit of mass will be that which will be accelerated at the rate of 1 ft/(s)^2 by 1 lb_f.

This unit of mass having the dimensions of $(\text{lb}_f)(\text{s})^2/(\text{ft})$ is called the *slug*. The conversion factor, g_c, is then a multiplying factor to convert slugs into $(\text{lb}_f)(\text{s})^2/(\text{ft})$, and its value is $1 \text{ (slug)(ft)}/(\text{lb}_f)(\text{s})^2$.

A third system encountered in engineering practice involves all four fundamental units. The unit of force is 1 lb_f, the unit of mass is 1 lb_m; length and time are given in units of feet and seconds, respectively. When 1 lb_m at sea level is allowed to fall under the influence of gravity, its acceleration will be $32.174 \text{ (ft)/(s)}^2$. The force exerted by gravity on 1 lb_m at sea level is defined as 1 lb_f. Therefore the conversion factor, g_c, for this system is $32.174 \text{ (lb}_m)(\text{ft})/(\text{lb}_f)(\text{s})^2$.*

A summary of the values of g_c is given in Table 1.1 for these three English systems of engineering units, along with the units of length, time, force, and mass.

Since all three systems are in current use in the technical literature, the student should be able to use formulas given in any particular situation. Careful check for dimensional consistency will be required in *all* calculations. The conversion factor, g_c, will correctly relate the units corresponding to a system. There will be no attempt by the authors to incorporate the conversion factor in any equations; instead, it will be the reader's responsibility to use units which are consistent with every term in the equation.

---

* In subsequent calculations in this book g_c will be rounded off to a value of $32.2 \text{ lb}_m \text{ ft/s}^2 \text{ lb}_f$.

TABLE 1.1

System	Length	Time	Force	Mass	g_c
1	meter	second	newton	kilogram	$1\dfrac{\text{kg} \cdot \text{m}}{\text{N} \cdot \text{s}^2}$
2	foot	second	lb_f	slug	$\dfrac{1\,(\text{slug})(\text{ft})}{(\text{lb}_f)(\text{s})^2}$
3	foot	second	lb_f	lb_m	$\dfrac{32.174\,(\text{lb}_m)(\text{ft})}{(\text{lb}_f)(\text{s})^2}$

PROBLEMS

1.1 The number of molecules crossing a unit area per unit time in one direction is given by

$$N = \tfrac{1}{4}n\bar{v}$$

where n is the number of molecules per unit volume and $\bar{v}$ is the average molecular velocity. As the average molecular velocity is approximately equal to the speed of sound in a perfect gas, estimate the number of molecules crossing a circular hole 10^{-3} in. in diameter. Assume that the gas is at standard conditions. At standard conditions, there are 4×10^{20} molecules per in.^{3}

1.2 Find the pressure gradient at point (a, b) when the pressure field is given by

$$P = \rho_\infty v_\infty{}^2 \left(\sin\frac{x}{a} \sin\frac{y}{b} + 2\frac{x}{a} \right)$$

where ρ_∞, v_∞, a, and b are constants.

1.3 Find the temperature gradient at point (a, b) at time $t = (L^2/\alpha)\ln e$ when the temperature field is given by

$$T = T_0 e^{-\alpha t/4L^2} \sin\frac{x}{a} \cosh\frac{y}{b}$$

where T_0, α, a, and b are constants.

1.4 Are the fields described in problems 1.2 and 1.3 dimensionally homogeneous? What must the units of ρ_∞ be in order that the pressure be in pounds per square foot when v_∞ is given in feet per second (problem 1.2)?

1.5 Which of the quantities listed below are flow properties and which are fluid properties?

pressure	temperature	velocity
density	stress	speed of sound
specific heat	pressure gradient	

1.6 Show that the unit vectors $\mathbf{e}_r$ and $\mathbf{e}_\theta$ in a cylindrical coordinate system are related to the unit vectors $\mathbf{e}_x$ and $\mathbf{e}_y$ by

$$\mathbf{e}_r = \mathbf{e}_x \cos \theta + \mathbf{e}_y \sin \theta$$

and

$$\mathbf{e}_\theta = -\mathbf{e}_x \sin \theta + \mathbf{e}_y \cos \theta$$

1.7 Using the results of problem 1.6, show that $d\mathbf{e}_r/d\theta = \mathbf{e}_\theta$ and $d\mathbf{e}_\theta/d\theta = -\mathbf{e}_r$.

1.8 Using the geometric relations given below and the chain rule for differentiation, show that

$$\frac{\partial}{\partial x} = -\frac{\sin \theta}{r}\frac{\partial}{\partial \theta} + \cos \theta \frac{\partial}{\partial r}$$

and

$$\frac{\partial}{\partial y} = \frac{\cos \theta}{r}\frac{\partial}{\partial \theta} + \sin \theta \frac{\partial}{\partial r}$$

when $r^2 = x^2 + y^2$ and $\tan \theta = y/x$.

1.9 Transform the operator ∇ to cylindrical coordinates (r, θ, z), using the results of problems 1.6 and 1.8.

1.10 For a fluid of density ρ in which solid particles of density ρ_s are uniformly dispersed, show that if x is the mass fraction of solid in the mixture, the density is given by

$$\rho_{\text{mixture}} = \frac{\rho_s \rho}{\rho x + \rho_s(1 - x)}$$

1.11 A scalar field is given by the function $\phi = 3x^2 y + 4y^2$.
(a) Find $\nabla\phi$ at the point $(3, 5)$.
(b) Find the component of $\nabla\phi$ that makes a $-60°$ angle with the x axis at the point $(3, 5)$.

1.12 If the fluid of density ρ in problem 1.10 obeys the perfect gas law, obtain the equation of state of the mixture, that is, $P = f(\rho_s, (RT/M), \rho_m, x)$. Will this result be valid if a liquid is present instead of a solid?

1.13 Using the expression for the gradient in polar coordinates (Appendix A), find the gradient of $\psi(r, \theta)$ when

$$\psi = Ar \sin \theta \left(1 - \frac{a^2}{r^2}\right)$$

Where is the gradient maximum? The terms A and a are constant.

1.14 Given the following expression for the pressure field where x, y, and z are space coordinates, t is time and P_0, ρ, V_∞, and L are constants. Find the pressure gradient.

$$P = P_0 + \tfrac{1}{2}\rho V_\infty^2 \left[2\frac{xyz}{L^3} + 3\left(\frac{x}{L}\right)^2 + \frac{V_\infty t}{L}\right]$$

2
FLUID STATICS

The definition of a fluid variable at a point was considered in Chapter 1. In this chapter the point-to-point variation of a particular variable, pressure, will be considered for the special case of a fluid at rest.

A frequently encountered static situation exists for a fluid which is stationary on the Earth's surface. Although the Earth has some motion of its own, we are well within normal limits of accuracy to neglect the absolute acceleration of the coordinate system which, in this situation, would be fixed with reference to the Earth. Such a coordinate system is said to be an *inertial reference*. If, on the other hand, a fluid is stationary with respect to a coordinate system which has some significant absolute acceleration of its own, the reference is said to be *noninertial*. An example of this latter situation would be the fluid in a railroad tank car as it travels around a curved portion of track.

The application of Newton's second law of motion to a fixed mass of fluid reduces to the expression that the sum of the external forces is equal to the product of the mass and its acceleration. In the case of an inertial reference we would naturally have the relation $\sum \mathbf{F} = 0$; while the more general statement $\sum \mathbf{F} = m\mathbf{a}$ must be used for the noninertial case.

2.1 PRESSURE VARIATION IN A STATIC FLUID

From the definition of a fluid it is known that there can be no shear stress in a fluid at rest. This means that the only forces acting on the fluid are those due to gravity and pressure. As the sum of the forces must equal zero throughout the fluid, Newton's law may be satisfied by applying it to an arbitrary free body of fluid of differential size. The free body selected, shown in Fig. 2.1, is the element of fluid $\Delta x\, \Delta y\, \Delta z$ with a corner at the point xyz. The coordinate system xyz is an inertial coordinate system.

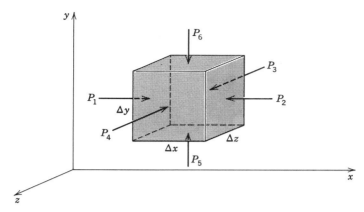

Figure 2.1   Pressure forces on a static fluid element.

The pressures which act on the various faces of the element are numbered 1 through 6. To find the sum of the forces on the element the pressure on each face must first be evaluated.

We shall designate the pressure according to the face of the element upon which the pressure acts. For example, $P_1 = P|_x$, $P_2 = P|_{x+\Delta x}$, and so on. Evaluating the forces acting on each face, along with the force due to gravity acting on the element $\rho \mathbf{g} \, \Delta x \, \Delta y \, \Delta z$, we find that the sum of the forces is

$$\rho \mathbf{g}(\Delta x \, \Delta y \, \Delta z) + (P|_x - P|_{x+\Delta x}) \, \Delta y \, \Delta z \, \mathbf{e}_x$$
$$+ (P|_y - P|_{y+\Delta y}) \, \Delta x \, \Delta z \, \mathbf{e}_y + (P|_z - P|_{z+\Delta z}) \, \Delta x \, \Delta y \, \mathbf{e}_z = 0$$

Dividing by the volume of the element $\Delta x \, \Delta y \, \Delta z$, we see that the above equation becomes

$$\rho \mathbf{g} - \frac{P|_{x+\Delta x} - P|_x}{\Delta x} \mathbf{e}_x - \frac{P|_{y+\Delta y} - P|_y}{\Delta y} \mathbf{e}_y - \frac{P|_{z+\Delta z} - P|_z}{\Delta z} \mathbf{e}_z = 0$$

where the order of the pressure terms has been reversed. As the size of the element approaches zero, Δx, Δy and Δz approach zero and the element approaches the point (x, y, z). In the limit

$$\rho \mathbf{g} = \lim_{\Delta x, \Delta y, \Delta z \to 0} \left[\frac{P|_{x+\Delta x} - P|_x}{\Delta x} \mathbf{e}_x + \frac{P|_{y+\Delta y} - P|_y}{\Delta y} \mathbf{e}_y + \frac{P|_{z+\Delta z} - P|_z}{\Delta z} \mathbf{e}_z \right]$$

or

$$\rho \mathbf{g} = \frac{\partial P}{\partial x} \mathbf{e}_x + \frac{\partial P}{\partial y} \mathbf{e}_y + \frac{\partial P}{\partial z} \mathbf{e}_z \tag{2-1}$$

Recalling the form of the gradient, we may write equation (2-1) as

$$\rho \mathbf{g} = \nabla P \tag{2-2}$$

Equation (2-2) is the basic equation of fluid statics and states that the maximum rate of change of pressure occurs in the direction of the gravitational vector. In addition, since isolines are perpendicular to the gradient, constant pressure lines are perpendicular to the gravitational vector. The point-to-point variation in pressure may be obtained by integrating equation (2-2).

EXAMPLE 1
The manometer, a pressure measuring device, may be analyzed from the previous discussion. The simplest type of manometer is the U-tube manometer shown in Figure 2.2.

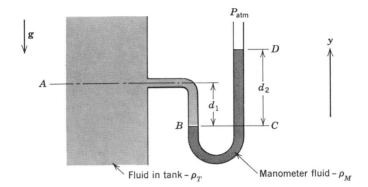

Figure 2.2 A U-tube manometer.

The pressure in the tank at point A is to be measured. The fluid in the tank extends into the manometer to point B. Choosing the y axis in the direction shown, we see that equation (2-2) becomes

$$\frac{dP}{dy}\mathbf{e}_y = -\rho g \mathbf{e}_y$$

Integrating between C and D in the manometer fluid, we have

$$P_{\text{atm}} - P_C = -\rho_m g d_2$$

and then integrating between B and A in the tank fluid, we obtain

$$P_A - P_B = -\rho_T g d_1$$

Since Pascal's principle requires that the pressure in a fluid at rest be the same at all points with the same elevation, the above equations may be combined to give

$$P_A - P_{\text{atm}} = \rho_m g d_2 - \rho_T g d_1$$

The U-tube manometer measures the difference between the absolute pressure and the atmospheric pressure. This difference is called the *gage pressure* and is frequently used in pressure measurement.

EXAMPLE 2

In the fluid statics of gases, a relation between the pressure and density is required to integrate equation (2-2). The simplest case is that of the isothermal perfect gas, where $P = \rho RT/M$. Here R is the universal gas constant, M is the molecular weight of the gas, and T is the temperature, which is constant for this case. Selecting the y axis parallel to $\mathbf{g}$, we see that equation (2-2) becomes

$$\frac{dP}{dy} = -\rho g = -\frac{PMg}{RT}$$

Separating variables, we see that the above differential equation becomes

$$\frac{dP}{P} = -\frac{Mg}{RT}dy$$

Integration between $y = 0$ (where $P = P_{atm}$) and $y = y$ (where the pressure is P) yields

$$\ln \frac{P}{P_{atm}} = -\frac{Mgy}{RT}$$

or

$$\frac{P}{P_{atm}} = \exp\left\{-\frac{Mgy}{RT}\right\}$$

In the above examples, the atmospheric pressure and a model of pressure variation with elevation have appeared in the results. Since performance of aircraft, rockets, and many types of industrial machinery varies with ambient pressure, temperature, and density, a standard atmosphere has been established in order to evaluate performance. At sea level standard atmospheric conditions are

$$P = 29.92 \text{ in. Hg} = 2116.2 \text{ lb}_f/\text{ft}^2 = 14.696 \text{ lb}_f/\text{in.}^2 = 101\,325 \text{ N/m}^2$$

$$T = 519°R = 59°F = 288 \text{ K}$$

$$\rho = 0.07651 \text{ lb}_m/\text{ft}^3 = 0.002378 \text{ slug/ft}^3 = 1.226 \text{ kg/m}^3$$

A table of the standard atmospheric properties as a function of altitude is given in Appendix G.*

2.2   UNIFORM RECTILINEAR ACCELERATION

For the case in which the coordinate system xyz in Figure 2.1 is not an inertial coordinate system, equation (2-2) does not apply. In the case of uniform rectilinear acceleration, however, the fluid will be at rest with respect to the accelerating coordinate system. With a constant acceleration we may apply the

---

* These performance standard sea-level conditions should not be confused with gas-law standard conditions of $P = 29.92$ in. Hg $= 14.696$ lb/in.$^2 = 101\,325$ Pa, $T = 492°R = 32°F = 273$ K.

same analysis as in the case of the inertial coordinate system except that $\sum \mathbf{F} = m\mathbf{a} = \rho \, \Delta x \, \Delta y \, \Delta z \mathbf{a}$, as required by Newton's second law of motion. The result is

$$\nabla P = \rho(\mathbf{g} - \mathbf{a}) \tag{2-3}$$

The maximum rate of change of pressure is now in the $\mathbf{g} - \mathbf{a}$ direction, and lines of constant pressure are perpendicular to $\mathbf{g} - \mathbf{a}$.

The point-to-point variation in pressure is obtained from integration of equation (2-3).

EXAMPLE 3

A fuel tank is shown in Figure 2.3. If the tank is given a uniform acceleration to the right, what will be the pressure at point B? From equation (2-3) the pressure gradient is in the $\mathbf{g} - \mathbf{a}$ direction, therefore the surface of the fluid will be perpendicular to this direction.

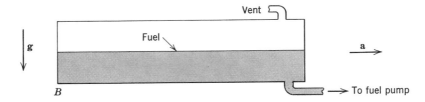

Figure 2.3   Fuel tank at rest.

Choosing the y axis parallel to $\mathbf{g} - \mathbf{a}$, we find that equation (2-3) may be integrated between point B and the surface. The pressure gradient becomes $dP/dy \, \mathbf{e}_y$ with the selection of the y axis parallel to $\mathbf{g} - \mathbf{a}$ as shown in Figure 2.4. Thus

$$\frac{dP}{dy}\mathbf{e}_y = -\rho |\mathbf{g} - \mathbf{a}| \mathbf{e}_y = -\rho \sqrt{g^2 + a^2}\,\mathbf{e}_y$$

Integrating between $y = 0$ and $y = d$ yields

$$P_{\text{atm}} - P_B = \rho \sqrt{g^2 + a^2}(-d)$$

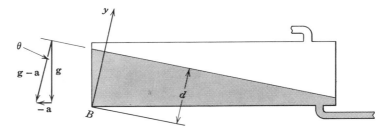

Figure 2.4   Uniformly accelerated fuel tank.

or

$$P_B - P_{atm} = \rho \sqrt{g^2 + a^2}(d)$$

The depth of the fluid, d, at point B is determined from the tank geometry and the angle θ.

2.3 FORCES ON SUBMERGED SURFACES

Determination of the force on submerged surfaces is done frequently in fluid statics. Since these forces are due to pressure, use will be made of the relations describing the point-to-point variation in pressure which have been developed in the previous sections. The plane surface illustrated in Figure 2.5 is inclined at an angle α to the surface of the fluid. The area of the inclined plane is A, and the density of the fluid is ρ.

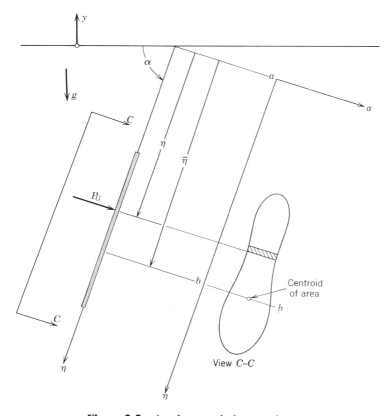

Figure 2.5   A submerged plane surface.

The magnitude of the force on the element dA is $P_G dA$, where P_G is the gage pressure; $P_G = -\rho g y = \rho g \eta \sin \alpha$, giving

$$dF = \rho g \eta \sin \alpha \, dA$$

Integration over the surface of the plate yields

$$F = \rho g \sin \alpha \int_A \eta \, dA$$

The definition of the centroid of the area is

$$\bar{\eta} \equiv \frac{1}{A} \int_A \eta \, dA$$

and thus

$$F = \rho g \sin \alpha \, \bar{\eta} A \qquad (2\text{-}4)$$

Thus the force due to the pressure is equal to the pressure evaluated at the centroid of the submerged area multiplied by the submerged area. The point at which this force acts (the center of pressure) is not the centroid of the area. In order to find the center of pressure, we must find the point at which the total force on the plate must be concentrated in order to produce the same moment as the distributed pressure, or

$$F\eta_{\text{c.p.}} = \int_A \eta P_G dA$$

Substitution for the pressure yields

$$F\eta_{\text{c.p.}} = \int_A \rho g \sin \alpha \, \eta^2 \, dA$$

and since $F = \rho g \sin \alpha \, \bar{\eta} A$, we have

$$\eta_{\text{c.p.}} = \frac{1}{A\bar{\eta}} \int_A \eta^2 \, dA = \frac{I_{aa}}{A\bar{\eta}} \qquad (2\text{-}5)$$

The moment of the area about the surface may be translated from an axis aa located at the fluid surface to an axis bb through the centroid by

$$I_{aa} = I_{bb} + \bar{\eta}^2 A$$

and thus

$$\eta_{\text{c.p.}} - \bar{\eta} = \frac{I_{bb}}{A\bar{\eta}} \qquad (2\text{-}6)$$

The center of pressure is located below the centroid a distance $I_{bb}/A\bar{\eta}$.

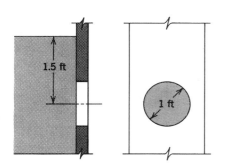

Figure 2.6 Submerged window.

EXAMPLE 4

A circular viewing port is to be located 1.5 ft below the surface of a tank as shown in Figure 2.6. Find the magnitude and location of the force acting on the window.

SOLUTION

The force on the window is

$$F = \rho g \sin \alpha \, A\bar{\eta}$$

where

$$\alpha = \pi/2 \quad \text{and} \quad \bar{\eta} = 1.5 \text{ ft};$$

the force is

$$F = \rho g A\bar{\eta} = \frac{(62.4 \text{ lb}_m/\text{ft}^3)(32.2 \text{ ft/s}^2)(\pi/4 \text{ ft}^2)(1.5 \text{ ft})}{32.2 \text{ lb}_m\text{ft/s}^2 \text{ lb}_f}$$

$$= 73.5 \text{ lb}_f \, (327 \text{ N})$$

The force F acts at $\bar{\eta} + \dfrac{I_{\text{centroid}}}{A\bar{\eta}}$. For a circular area, $I_{\text{centroid}} = \pi R^4/2$, so we obtain

$$\eta_{\text{c.p.}} = 1.5 + \frac{\pi R^4}{2\pi R^2 1.5} \text{ ft} = 1.583 \text{ ft}$$

EXAMPLE 5

Rainwater collects behind the concrete retaining wall shown in Figure 2.7. If the water-saturated soil (specific gravity = 2.2) acts as a fluid, determine the force and center of pressure on a one-meter width of the wall.

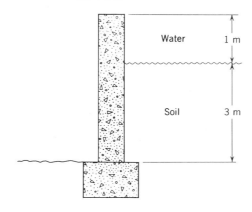

Figure 2.7 Retaining wall.

SOLUTION

The force on a unit width of the wall is obtained by integrating the pressure difference between the right and left sides of the wall. Taking the origin at the top of the wall and measuring y downward, the force due to pressure is

$$F = \int (P - P_{atm})(1)\, dy$$

The pressure difference in the region in contact with the water is

$$P - P_{atm} = \rho_{H_2O} g y$$

and the pressure difference in the region in contact with the soil is

$$P - P_{atm} = \rho_{H_2O} g (1) + 2.2 \rho_{H_2O} g (y - 1)$$

The force F is

$$F = \rho_{H_2O} g \int_0^1 y\, dy + \rho_{H_2O} g \int_1^4 [1 + 2.2(y-1)]\, dy$$

$$F = (1000 \text{ kg/m}^3)(9.807 \text{ m/s}^2)(1 \text{ m})(13.4 \text{ m}^2) = 131\,414 \text{ N}(29\,546 \text{ lb}_f)$$

The center of pressure of the force on the wall is obtained by taking moments about the top of the wall.

$$F y_{c.p.} = \rho_{H_2O} g \left\{ \int_0^1 y^2\, dy + \int_1^4 y[1 + 2.2(y-1)]\, dy \right\}$$

$$y_{c.p.} = \frac{1}{(131\,414 \text{ N})} (1000 \text{ kg/m}^3)(9.807 \text{ m/s}^2)(1 \text{ m})(37.53 \text{ m}^3) = 2.80 \text{ m}(9.19 \text{ ft})$$

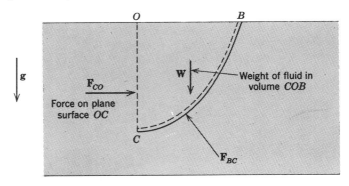

Figure 2.8 Submerged curved surface.

The force on a submerged curved surface may be obtained from knowledge of the force on a plane surface and the laws of statics. Consider the curved surface BC illustrated in Figure 2.8.

By consideration of the equilibrium of the fictitious body BCO, the force of the curved plate on body BCO may be evaluated. Since $\sum \mathbf{F} = 0$, we have

$$\mathbf{F}_{CB} = -\mathbf{W} - \mathbf{F}_{CO} \tag{2-7}$$

The force of the fluid on the curved plate is the negative of this or $\mathbf{W} + \mathbf{F}_{CO}$. Thus the force on a curved submerged surface may be obtained from the weight on the volume BCO and the force on a submerged plane surface.

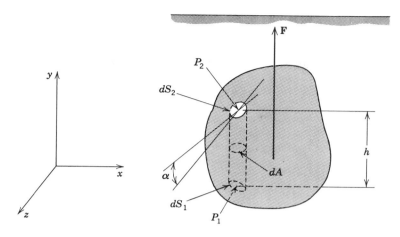

Figure 2.9 Forces on submerged volume.

2.4. BUOYANCY

The body shown in Figure 2.9 is submerged in a fluid with density ρ. The resultant force **F** holds the body in equilibrium.

The element of volume $h\,dA$ has gravity and pressure forces acting on it. The y component of the force due to pressure on the top of the element is $-P_2\,dS_2\cos\alpha\,\mathbf{e}_y$, where α is the angle between the plane of the element dS_2 and the xz plane. The product $dS_2\cos\alpha$ then is the projection of dS_2 onto the xz plane, or simply dA. The net pressure force on the element is $(P_1-P_2)\,dA\,\mathbf{e}_y$, and the resultant force on the element is

$$dF = (P_1 - P_2)\,dA\,\mathbf{e}_y - \rho_B g h\,dA\,\mathbf{e}_y$$

where ρ_B is the density of the body. The difference in pressure $P_1 - P_2$ may be expressed as $\rho g h$, so

$$dF = (\rho - \rho_B) g h\,dA\,\mathbf{e}_y$$

Integration over the volume of the body, assuming constant densities, yields

$$\mathbf{F} = (\rho - \rho_B) g V \mathbf{e}_y \qquad (2\text{-}8)$$

where V is the volume of the body. The resultant force **F** is composed of two parts, the weight $-\rho_B g V \mathbf{e}_y$ and the buoyant force, $\rho g V \mathbf{e}_y$. The body experiences an upward force equal to the weight of the displaced fluid. This is the well-known principle of Archimedes. When $\rho > \rho_B$, the resultant force will cause the body to float on the surface. In the case of a floating body, the buoyant force is $\rho g V_s \mathbf{e}_y$, where V_s is the submerged volume.

EXAMPLE 6

A cube measuring 1 ft on a side is submerged so that its top face is 10 ft below the free surface of water. Determine the magnitude and direction of the applied force necessary to hold the cube in this position if it is made of

(a) cork ($\rho = 10$ lb$_m$/ft^3)
(b) steel ($\rho = 490$ lb$_m$/ft^3)

The pressure forces on all lateral surfaces of the cube cancel. Those on the top and bottom do not, since they are at different depths.

Summing forces on the vertical direction, we obtain

$$\sum F_y = -W + P(1)\big|_{\text{bottom}} - P(1)\big|_{\text{top}} + F_y = 0$$

where F_y is the additional force required to hold the cube in position.

Expressing each of the pressures as $P_{\text{atm}} + \rho_w g h$, and W as $\rho_c g V$, we obtain, for our force balance,

$$-\rho_c g V + \rho_w g \, (11 \text{ ft})(1 \text{ ft}^2) - \rho_w g \, (10 \text{ ft})(1 \text{ ft}^2) + F_y = 0$$

Solving for F_y, we have

$$F_y = -\rho_w g[(11)(1) - 10(1)] + \rho_c g V = -\rho_w g V + \rho_c g V$$

The first term is seen to be a buoyant force, equal to the weight of displaced water.

Finally, solving for F_y, we obtain

(a) $\rho_c = 10$ lb$_m$/ft^3

$$F_y = -\frac{(62.4 \text{ lb}_m/\text{ft}^3)(32.2 \text{ ft/s}^2)(1 \text{ ft}^3)}{32.2 \text{ lb}_m\text{ft/s}^2 \text{ lb}_f} + \frac{(10 \text{ lb}_m\text{ft}^3)(32.2 \text{ ft/s}^2)(1 \text{ ft}^3)}{32.2 \text{ lb}_m \text{ ft/s}^2 \text{ lb}_f}$$

$$= -52.4 \text{ lb}_f \text{ } (downward) \text{ } (-233 \text{ N})$$

(b) $\rho_c = 490$ lb$_m$/ft^3

$$F_y = -\frac{(62.4 \text{ lb}_m/\text{ft}^3)(32.2 \text{ ft/s}^2)(1 \text{ ft}^3)}{32.2 \text{ lb}_m \text{ ft/s}^2 \text{ lb}_f} + \frac{(490 \text{ lb}_m/\text{ft}^3)(32.2 \text{ ft/s}^2)(1 \text{ ft}^3)}{32.2 \text{ lb}_m \text{ ft/s}^2 \text{ lb}_f}$$

$$= +427.6 \text{ lb}_f \text{ } (upward) \text{ } (1902 \text{ N})$$

In case (a) the buoyant force exceeded the weight of the cube, thus to keep it submerged 10 ft below the surface, a downward force of over 52 lb was required. In the second case, the weight exceeded the buoyant force, and an upward force was required.

2.5 CLOSURE

In this chapter the behavior of static fluids has been examined. The application of Newton's laws of motion led to the description of the point-to-point variation in fluid pressure, from which force relations were developed. Specific applications have been considered, including manometry, forces on plane and curved submerged surfaces, and the buoyancy of floating objects.

The static analyses that have been considered will later be seen as special cases of more general relations governing fluid behavior. Our next task is to examine the behavior of fluids in motion and to describe the effect of that motion. Fundamental laws other than Newton's laws of motion will be necessary for this analysis.

PROBLEMS

2.1 What would the height of the atmosphere be if it were incompressible? Use standard conditions to determine the density of air.

2.2 The bulk modulus, β of a substance is given by $\beta = dP/(d\rho/\rho)$. Determine β for a perfect gas.

2.3 In water the modulus, β, defined in problem 2.2 is nearly constant and has a value of 300 000 psi. Determine the percentage volume change in water due to a pressure of 3000 psi.

2.4 Find the pressure at point A.

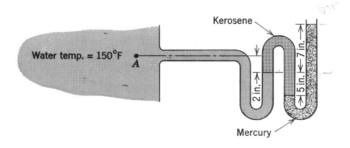

2.5 The car shown in the figure is accelerated to the right at a uniform rate. What way will the balloon go relative to the car?

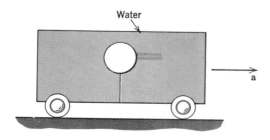

2.6 The tank is accelerated upward at a uniform rate. Does the manometer level go up or down?

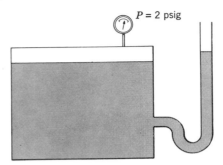

2.7 Glass viewing windows are to be installed in an aquarium. Each window is to be 0.6 m in diameter and to be centered 2 m below the surface level. Find the force and location of the force acting on the window.

2.8 On a certain day the barometric pressure at sea level is 30.1 in. Hg, and the temperature is 70°F. The pressure gage in an airplane in flight indicates a pressure of 10.6 psia, and the temperature gage shows the air temperature to be 46°F. Estimate as accurately as possible the altitude of the airplane above sea level.

2.9 A differential manometer is used to measure the pressure change caused by a flow constriction in a piping system as shown. Determine the pressure difference between points A and B in pounds per square inch. Which section has the higher pressure?

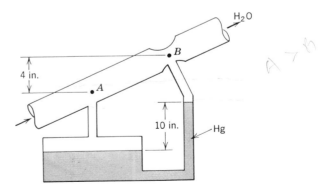

2.10 The open end of a cylindrical tank 2 ft in diameter and 3 ft high is submerged in water as shown. If the tank weighs 250 lb, to what depth h will the tank submerge? The local barometric pressure is 14.7 psia. The thickness of tank wall may be neglected. What additional force is required to bring the top of the tank flush with the water surface?

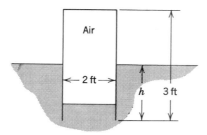

2.11 In problem 2.10 above, find the depth at which the net force on the tank is zero.

2.12 Find the minimum value of h for which the gate shown will rotate counter-clockwise if the gate cross section is (a) rectangular, 4 ft $\times$ 4 ft; (b) triangular, 4 ft at the base $\times$ 4 ft high. Neglect bearing friction.

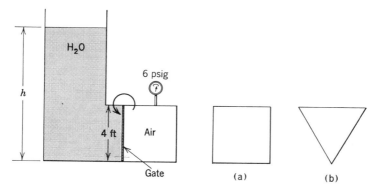

2.13 A cubical piece of wood with an edge L in length floats in the water. The specific gravity of the wood is 0.90. What moment M is required to hold the cube in the position shown? The right-hand edge of the block is flush with the water.

2.14 A circular log of radius r is to be used as a barrier as shown in the figure below. If the point of contact is at O, determine the required density of the log.

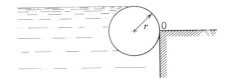

2.15 A rectangular block of concrete 3 ft × 3 ft × 6 in. has its 6-in. side half buried at the bottom of a lake 23 ft deep. What force is required to lift the block free of the bottom? What force is required to maintain the block in this position? (Concrete weighs 150 lb/ft^3.)

2.16 A dam spillway gate holds back water of depth h. The gate weighs 500 lb/ft and is hinged at A. At what depth of water will the gate raise up and permit water to flow under it?

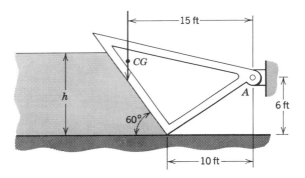

2.17 It is desired to use a 0.75-m diameter beach ball to stop a drain in a swimming pool. Obtain an expression that relates the drain diameter D and the minimum water depth h for which the ball will remain in place.

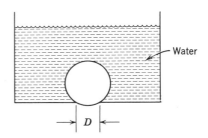

2.18 If the density of sea water is approximated by the equation of state $\rho = \rho_0 \exp{[(p - p_{atm})/\beta)]}$, where β is the compressibility, determine the pressure and density at a point 32 000 ft below the surface of the sea. Assume $\beta = 300\,000$ psi.

2.19 The change in density due to temperature causes the take-off and landing speeds of heavier-than-air craft to increase as the square root of the temperature. What effect do temperature-induced density changes have on the lifting power of rigid lighter-than-air craft?

2.20 The practical depth limit for a suited diver is about 185 m. What is the gage pressure in seawater at that depth? The specific gravity of seawater is 1.025.

2.21 Matter is attracted to the center of the earth with a force proportional to the radial distance from the center. Using the known value of g at the surface where the radius is 6330 km, compute the pressure at the earth's center, assuming the material behaves like a liquid, and that the mean specific gravity is 5.67. (A tube of constant diameter rather than a spherical segment may be considered for convenience.) Obtain first a formula in symbols before numerical values are inserted.

2.22 A watertight bulkhead 22 ft high forms a temporary dam for some construction work. The top 12 ft behind the bulkhead consist of sea water with a density of 2 slugs/ft^3, but the bottom 10 ft being a mixture of mud and water can be considered a fluid of density 4 slugs/ft^3. Calculate the total horizontal load per unit width and the location of the center of pressure measured from the bottom.

2.23 Concrete is used to cast a road divider. The cross section is shown below. Determine the external forces F_1 and F_2 that are required to hold the forms in place. Assume that concrete acts as a fluid and neglect the weight of the forms. At what height does the force F_2 act? Express your answers in terms of the density of concrete, ρ_c, the acceleration of gravity, g, and the dimensions H, a, and r. Assume a unit length for the divider.

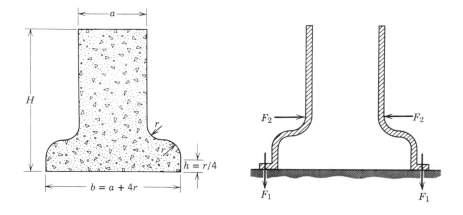

3
DESCRIPTION OF A FLUID IN MOTION

The development of an analytical description of fluid flow is based upon the expression of the physical laws related to fluid flow in a suitable mathematical form. Accordingly, we shall present the pertinent physical laws and discuss the methods used to describe a fluid in motion.

3.1 FUNDAMENTAL PHYSICAL LAWS

There are three fundamental physical laws which, with the exception of relativistic and nuclear phenomena, apply to each and every flow independently of the nature of the fluid under consideration. These laws are listed below with the designations of their mathematical formulations.

Law	Equation
1. The law of conservation of mass	continuity equation
2. Newton's second law of motion	momentum theorem
3. The first law of thermodynamics	energy equation

The next three chapters will be devoted exclusively to the development of a suitable working form of these laws.*

In addition to the above laws, certain auxiliary or subsidiary relations are employed in describing a fluid. These relations depend upon the nature of the fluid under consideration. Unfortunately most of these auxiliary relations have also been termed "laws." Already in our previous studies, Hooke's law, the perfect gas law, and others have been encountered. However accurate these "laws" may be over a restricted range, their validity is entirely dependent upon the nature of the material under consideration. Thus, while some of the auxiliary relations which

---

* The second law of thermodynamics is also fundamental to fluid-flow analysis. An analytic consideration of the second law is beyond the scope of the present treatment.

will be used will be called laws, the student will be responsible for noting the difference in scope between the fundamental physical laws and the auxiliary relations.

3.2 FLUID FLOW FIELDS: LAGRANGIAN AND EULERIAN REPRESENTATIONS

The term *field* refers to a quantity defined as a function of position and time throughout a given region. There are two different forms of representation for fields in fluid mechanics, *Lagrange's* form and *Euler's* form. The difference between these approaches lies in the manner in which the position in the field is identified.

In the Lagrangian approach the physical variables are described for a particular element of fluid as it traverses the flow. This is the familiar approach of particle and rigid-body dynamics. The coordinates (x, y, z) are the coordinates of the element of fluid and, as such, are functions of time. The coordinates (x, y, z) are therefore dependent variables in the Lagrangian form. The fluid element is identified by its position in the field at some arbitrary time, usually $t = 0$. The velocity field in this case is written in functional form as

$$\mathbf{v} = \mathbf{v}(a, b, c, t) \tag{3-1}$$

where the coordinates (a, b, c) refer to the *initial* position of the fluid element. The other fluid-flow variables, being functions of the same coordinates, may be represented in a similar manner. The Lagrangian approach is seldom used in fluid mechanics, since the type of information desired is usually the value of a particular fluid variable at a fixed point in the flow rather than the value of a fluid variable experienced by an element of fluid along its trajectory. For example, the determination of the force on a stationary body in a flow field requires that we know the pressure and shear stress at every point on the body. The Eulerian form provides us with this type of information.

The Eulerian approach gives the value of a fluid variable at a given point at a given time. In functional form the velocity field is written as

$$\mathbf{v} = \mathbf{v}(x, y, z, t) \tag{3-2}$$

where x, y, z, and t are *all* independent variables. For a particular point (x_1, y_1, z_1) and t_1, equation (3-2) gives the velocity of the fluid at that location at time t_1. In this text the Eulerian approach will be used exclusively.

3.3 STEADY AND UNSTEADY FLOWS

In adopting the Eulerian approach we note that the fluid flow will, in general, be a function of the four independent variables (x, y, z, t). If the flow at every point

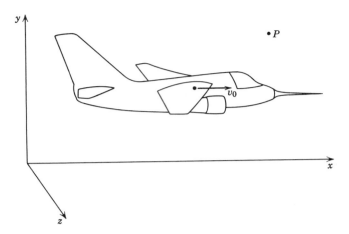

Figure 3.1 Unsteady flow with respect to a fixed-coordinate system.

in the fluid is independent of time, the flow is termed *steady*. If the flow at a point varies with time, the flow is termed *unsteady*. It is possible in certain cases to reduce an unsteady flow to a steady flow by changing the frame of reference. Consider an airplane flying through the air at constant speed v_0, as shown in Figure 3.1. When observed from the stationary x, y, z coordinate system, the flow pattern is unsteady. The flow at the point P illustrated, for example, will vary as the vehicle approaches it.

Now consider the same situation when observed from the x', y', z' coordinate system which is moving at constant velocity v_0 as illustrated in Figure 3.2.

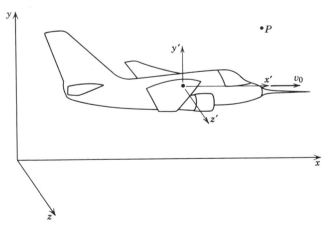

Figure 3.2 Steady flow with respect to a moving-coordinate system.

The flow conditions are now independent of time at every point in the flow field, and thus the flow is steady when viewed from the moving coordinate system. Whenever a body moves through a fluid with a constant velocity, the flow field may be transformed from an unsteady flow to a steady flow by selecting a coordinate system which is fixed with respect to the moving body.

In the wind-tunnel testing of models, use is made of this concept. Data obtained for a static model in a moving stream will be the same as the data obtained for a model moving through a static stream. The physical as well as the analytical simplifications afforded by this transformation are considerable. We shall make use of this transformation whenever applicable.

3.4   STREAMLINES

A useful concept in the description of a flowing fluid is that of a *streamline*. A streamline is defined as the line drawn tangent to the velocity vector at each point in a flow field. Figure 3.3 shows the streamline pattern for ideal flow past a

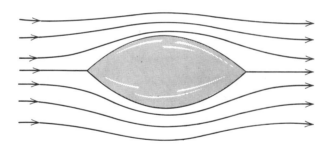

Figure 3.3   Illustration of streamlines.

football-like object. In steady flow, since all velocity vectors are invariant with time, the path of a fluid particle follows a streamline, hence a streamline is the trajectory of an element of fluid in such a situation. In unsteady flow, streamline patterns change from instant to instant. Thus the trajectory of a fluid element will be different from a streamline at any particular time. The actual trajectory of a fluid element as it traverses the flow is designated as a *path line*. It is obvious that path lines and streamlines are coincident only in steady flow.

The streamline is useful in relating the fluid velocity components to the geometry of the flow field. For two-dimensional flow this relation is

$$\frac{v_y}{v_x} = \frac{dy}{dx} \tag{3-3}$$

since the streamline is tangent to the velocity vector having x and y components v_x and v_y. In three dimensions this becomes

$$\frac{dx}{v_x} = \frac{dy}{v_y} = \frac{dz}{v_z} \qquad (3\text{-}4)$$

The utility of the above relations is in obtaining an analytical relation between velocity components and the streamline pattern.

3.5 SYSTEMS AND CONTROL VOLUMES

The three basic physical laws listed in section 3.1 are all stated in terms of a *system*. A system is defined as a collection of matter of fixed identity. The basic laws give the interaction of a system with its surroundings. The selection of the system for the application of these laws is quite flexible and is, in many cases, a complex problem. Any analysis utilizing a fundamental law must follow the designation of a specific system, and the difficulty of solution varies greatly depending on the choice made.

As an illustration, consider Newton's second law, $\mathbf{F} = m\mathbf{a}$. The terms represented are as follows:

$\mathbf{F} =$ the resultant force exerted by the surroundings on the system.

$m =$ the mass of the system.

$\mathbf{a} =$ the acceleration of the center of mass of the system.

In the piston-and-cylinder arrangement shown in Figure 3.4, a convenient system to analyze, readily identified by virtue of its confinement, is the mass of material enclosed within the cylinder by the piston.

In the case of the nozzle shown in Figure 3.5 the fluid occupying the nozzle changes from instant to instant. Thus different systems occupy the nozzle at different times.

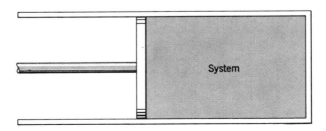

Figure 3.4 An easily identifiable system.

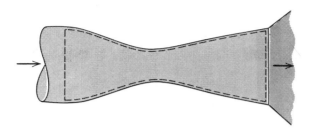

Figure 3.5 Control volume for analysis of flow through a nozzle.

A more convenient method of analysis of the nozzle would be to consider the region bounded by the dotted line. Such a region is a *control volume*. A control volume is a region in space through which fluid flows.*

The extreme mobility of a fluid makes the identification of a particular system a tedious task. By developing the fundamental physical laws in a form which applies to a control volume (where the system changes from instant to instant), the analysis of fluid flow is greatly simplified. The control-volume approach circumvents the difficulty in system identification. Succeeding chapters will convert the fundamental physical laws from the system approach to a control-volume approach. The control volume selected may be either finite or infinitesimal; in fact, we shall obtain the differential equations of fluid flow by the application of the fundamental laws, using infinitesimal control volumes.

---

* A control volume may be fixed or moving uniformly (inertial), or it may be accelerating (noninertial). Primary consideration here will be given to inertial control volumes.

4
CONSERVATION OF MASS: CONTROL VOLUME APPROACH

The initial application of the fundamental laws of fluid mechanics involves the law of conservation of mass. In this chapter we shall develop an integral relationship which expresses the law of conservation of mass for a general control volume. The integral relation thus developed will be applied to some often-encountered fluid-flow situations.

4.1 INTEGRAL RELATION

The law of conservation of mass states that mass may be neither created nor destroyed. With respect to a control volume, the law of conservation of mass may be simply stated as

$$\left\{ \begin{array}{c} \text{Rate of mass} \\ \text{efflux from} \\ \text{control} \\ \text{volume} \end{array} \right\} - \left\{ \begin{array}{c} \text{rate of mass} \\ \text{flow into con-} \\ \text{trol volume} \end{array} \right\} + \left\{ \begin{array}{c} \text{rate of accumu-} \\ \text{lation of mass} \\ \text{within control} \\ \text{volume} \end{array} \right\} = 0$$

Consider now the general control volume located in a fluid flow field, as shown in Figure 4.1.

For the small element of area dA on the control surface, the rate of mass efflux $= (\rho v)(dA \cos \theta)$, where $dA \cos \theta$ is the projection of the area dA in a plane normal to the velocity vector, $\mathbf{v}$, and θ is the angle between the velocity vector, $\mathbf{v}$, and the *outward* directed unit normal vector, $\mathbf{n}$, to dA. From vector algebra we recognize the product

$$\rho v \, dA \cos \theta = \rho \, dA |\mathbf{v}| \, |\mathbf{n}| \cos \theta$$

as the "scalar" or "dot" product

$$\rho (\mathbf{v} \cdot \mathbf{n}) \, dA$$

which is the form we shall use to designate the rate of mass efflux through dA. The

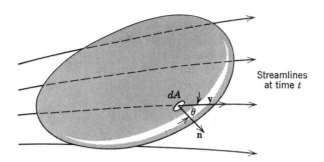

Figure 4.1 Fluid flow through a control volume.

product ρv is the mass flux, often called the mass velocity, G. Physically this product represents the amount of mass flowing through a unit cross-sectional area per unit time.

If we now integrate this quantity over the entire control surface, we have

$$\iint_{c.s.} \rho(\mathbf{v} \cdot \hat{\mathbf{n}}) \, dA$$

which is the net outward flow of mass across the control surface, or the *net mass efflux* from the control volume.

Note that if mass is entering the control volume, that is, flowing inward across the control surface, the product $\mathbf{v} \cdot \mathbf{n} = |\mathbf{v}| \, |\mathbf{n}| \cos \theta$ is negative, since $\theta > 90°$, and $\cos \theta$ is therefore negative. Thus, if the integral is

positive, there is a net efflux of mass;
negative, there is a net influx of mass;
zero, the mass within the control volume is constant.

The rate of accumulation of mass within the control volume may be expressed as

$$\frac{\partial}{\partial t} \iiint_{c.v.} \rho \, dV$$

and the integral expression for the mass balance over a general control volume becomes

$$\iint_{c.s.} \rho(\mathbf{v} \cdot \mathbf{n}) \, dA + \frac{\partial}{\partial t} \iiint_{c.v.} \rho \, dV = 0 \tag{4-1}$$

4.2 SPECIFIC FORMS OF THE INTEGRAL EXPRESSION

Equation (4-1) gives the mass balance in its most general form. We now consider some frequently encountered situations where equation (4-1) may be applied.

If flow is steady relative to coordinates fixed to the control volume, the accumulation term, $\partial/\partial t \iiint_{c.v.} \rho \, dV$, will be zero. This is readily seen when one recalls that, by the definition of steady flow, the properties of a flow field are invariant with time, hence the partial derivative with respect to time is zero. Thus for this situation the applicable form of the continuity expression is

$$\iint_{c.s.} \rho(\mathbf{v} \cdot \mathbf{n}) \, dA = 0 \qquad (4\text{-}2)$$

Another important case is that of an incompressible flow with fluid filling the control volume. For incompressible flow the density, ρ, is constant, hence the accumulation term involving the partial derivative with respect to time is again zero. Additionally, the density term in the surface integral may be canceled. The conservation-of-mass expression for incompressible flow of this nature thus becomes

$$\iint_{c.s.} (\mathbf{v} \cdot \mathbf{n}) \, dA = 0 \qquad (4\text{-}3)$$

The following examples illustrate the application of equation (4-1) to some cases that recur frequently in momentum transfer.

EXAMPLE 1

As our first example, let us consider the common situation of a control volume for which mass efflux and influx are steady and one-dimensional. Specifically, consider the control volume indicated by dashed lines in Figure 4.2.

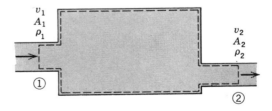

Figure 4.2 Steady one-dimensional flow into and out of a control volume.

Equation (4-2) applies. Since mass crosses the control surface at positions (1) and (2) only, our expression is

$$\iint_{c.s.} \rho(\mathbf{v} \cdot \mathbf{n}) \, dA = \iint_{A_1} \rho(\mathbf{v} \cdot \mathbf{n}) \, dA + \iint_{A_2} \rho(\mathbf{v} \cdot \mathbf{n}) \, dA = 0$$

The absolute value of the scalar product, $(\mathbf{v} \cdot \mathbf{n})$ is equal to the magnitude of the velocity in each integral, since the velocity vectors and outwardly directed normal vectors are collinear both at (1) and (2). At (2) these vectors have the same sense, thus this product is positive, as it should for an efflux of mass. At (1), where mass flows into the control volume, the two vectors are opposite in sense, hence the sign is negative. We may now express the continuity equation in scalar form:

$$\iint_{c.s.} \rho(\mathbf{v} \cdot \mathbf{n})\, dA = -\iint_{A_1} \rho v\, dA + \iint_{A_2} \rho v\, dA = 0$$

Integration gives the familar result

$$\rho_1 v_1 A_1 = \rho_2 v_2 A_2 \tag{4-4}$$

In obtaining equation (4-4) it is noted that the flow situation inside the control volume was unspecified. In fact, this is the beauty of the control-volume approach; the flow inside the control volume can be analyzed from information (measurements) obtained on the surface of the control volume. The box-shaped control volume illustrated in Figure 4.2 is defined for analytical purposes; the actual system contained in this box could be as simple as a pipe or as complex as a propulsion system or a distillation tower.

In solving example 1 we assumed a constant velocity at sections (1) and (2). This situation may be approached physically, but a more general case is one in which the velocity varies over the cross-sectional area.

EXAMPLE 2

Let us consider the case of an incompressible flow, for which the flow area is circular and the velocity profile is parabolic (see Figure 4.3), varying according to the expression

$$v = v_{max}\left[1 - \left(\frac{r}{R}\right)^2\right]$$

where v_{max} is the maximum velocity which exists at the center of the circular passage (i.e., at $r = 0$), and R is the radial distance to the inside surface of the circular area considered.

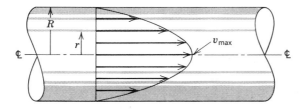

Figure 4.3   A parabolic velocity profile in a circular flow passage.

The above velocity-profile expression may be obtained experimentally. It will also be derived theoretically in Chapter 8 for the case of laminar flow in a circular conduit. This expression represents the velocity at a radial distance, r, from the center of the flow section. Since the average velocity is of particular interest in engineering problems, we will now consider the means of obtaining the average velocity from this expression.

At the station where this velocity profile exists, the mass rate of flow is

$$(\rho v)_{\text{avg}} A = \iint_A \rho v \, dA$$

For the present case of incompressible flow the density is constant. Solving for the average velocity, we have

$$v_{\text{avg}} = \frac{1}{A} \iint_A v \, dA$$

$$= \frac{1}{\pi R^2} \int_0^{2\pi} \int_0^R v_{\text{max}} \left[1 - \left(\frac{r}{R} \right)^2 \right] r \, dr \, d\theta$$

$$= \frac{v_{\text{max}}}{2}$$

In the previous examples we were not concerned with the composition of the fluid streams. Equation (4-1) applies to fluid streams containing more than one constituent as well as to the individual constituents alone. This type application is common to chemical processes in particular. Our final example will use the law of conservation of mass for both the total mass and for a particular species, in this case, salt.

EXAMPLE 3

Let us now examine the situation illustrated in Figure 4.4. A tank initially contains 1000 kg of brine containing 10% salt by mass. An inlet stream of brine containing 20% salt by mass flows into the tank at a rate of 20 kg/min. The mixture in the tank is kept uniform by stirring. Brine is removed from the tank via an outlet pipe at a rate of 10 kg/min. Find

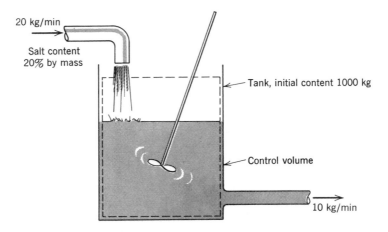

20 kg/min

Salt content
20% by mass

Tank, initial content 1000 kg

Control volume

10 kg/min

Figure 4.4   A mixing process.

the amount of salt in the tank at any time t, and the elapsed time when the amount of salt in the tank is 200 kg.

We first apply equation (4-1) to express the total amount of brine in the tank as a function of time. For the control volume shown,

$$\iint_{c.s.} \rho(\mathbf{v} \cdot \mathbf{n}) \, dA = 10 - 20 = -10 \, \text{kg/min}$$

$$\frac{\partial}{\partial t} \iiint_{c.v.} \rho \, dV = \frac{d}{dt} \int_{1000}^{M} dM = \frac{d}{dt}(M - 1000)$$

where M is the total mass of brine in the tank at any time. Writing the complete expression, we have

$$\iint_{c.s.} \rho(\mathbf{v} \cdot \mathbf{n}) \, dA + \frac{\partial}{\partial t} \iiint_{c.v.} \rho \, dV = -10 + \frac{d}{dt}(M - 1000) = 0$$

Separating variables and solving for M gives

$$M = 1000 + 10t \qquad (\text{kg})$$

We now let S be the amount of salt in the tank at any time. The concentration by weight of salt may be expressed as

$$\frac{S}{M} = \frac{S}{1000 + 10t} \qquad \frac{\text{kg salt}}{\text{kg brine}}$$

Using this definition, we may now apply equation (4-1) to the salt, obtaining

$$\iint_{c.s.} \rho(\mathbf{v} \cdot \mathbf{n}) \, dA = \frac{10S}{1000 + 10t} - (0.2)(20) \qquad \frac{\text{kg salt}}{\text{min}}$$

and

$$\frac{\partial}{\partial t} \iiint_{c.v.} \rho \, dV = \frac{d}{dt} \int_{S_0}^{S} dS = \frac{dS}{dt} \qquad \frac{\text{kg salt}}{\text{min}}$$

The complete expression is now

$$\iint_{c.s.} \rho(\mathbf{v} \cdot \mathbf{n}) \, dA + \frac{\partial}{\partial t} \iiint_{c.v.} \rho \, dV = \frac{S}{100 + t} - 4 + \frac{dS}{dt} = 0$$

This equation may be written in the form

$$\frac{dS}{dt} + \frac{S}{100 + t} = 4$$

which we observe to be a first-order linear differential equation. The general solution is

$$S = \frac{2t(200 + t)}{100 + t} + \frac{C}{100 + t}$$

The constant of integration may be evaluated, using the initial condition that $S = 100$ at $t = 0$ to give $C = 10\,000$. Thus the first part of the answer, expressing the amount of salt present as a function of time, is

$$S = \frac{10\,000 + 400t + 2t^2}{100 + t}$$

The elapsed time necessary for S to equal $200\,\text{kg}$ may be evaluated to give $t = 36.6$ min.

4.3 CLOSURE

In this chapter we have considered the first of the fundamental laws of fluid flow: conservation of mass. The integral expression developed for this case was seen to be quite general in its form and use.

Similar integral expressions for conservation of energy and of momentum for a general control volume will be developed and used in subsequent chapters. The student should now develop the habit of *always* starting with the applicable integral expression and evaluating each term for a particular problem. There will be a strong temptation simply to write down an equation without considering each term in detail. Such temptations should be overcome. This approach may seem needlessly tedious at the outset, but it will always insure a complete analysis of a problem and circumvent any errors which may otherwise result from a too-hasty consideration.

PROBLEMS

4.1 The velocity vector in a two-dimensional flow is given by the expression $\mathbf{v} = 10\mathbf{e}_x + 2x\mathbf{e}_y$ m/s when x is measured in meters. Determine the component of the velocity that makes a $-30°$ angle with the x axis at the point $(2, 2)$.

4.2 Using the velocity vector of the previous problem, determine (a) the equation of the streamline passing through the point $(2, 1)$; (b) the volume of flow that crosses a plane surface connecting the points $(1, 0)$ and $(2, 2)$.

4.3 Water is flowing through a circular conduit with a velocity profile given by the equation $v = 9(1 - r^2/16)$ fps. What is the average water velocity in the 1.5-ft pipe?

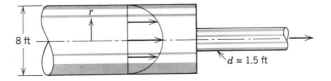

4.4 Water enters a 4-in. square channel as shown at a velocity of 10 fps. The channel converges to a 2-in. square configuration as shown at the discharge end. The outlet section is cut at 30° to the vertical as shown, but the mean velocity of the discharging water remains horizontal. Find the exiting water's average velocity and total rate of flow.

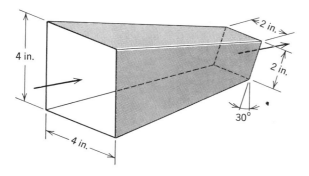

4.5 Water enters one end of a perforated pipe 0.2 m in diameter with a velocity of 6 m/s. The discharge through the pipe wall is approximated by a linear profile. If the flow is steady, find the discharge velocity.

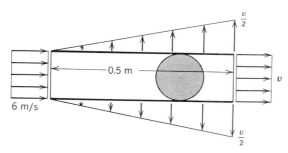

4.6 The velocities in a circular duct of 20-in. diameter were measured as follows:

Distance From Center (in.)	Velocity (fps)	Distance From Center (in.)	Velocity (fps)
0	7.5	7.75	5.47
3.16	7.10	8.37	5.10
4.45	6.75	8.94	4.50
5.48	6.42	9.49	3.82
6.33	6.15	10.00	2.40
7.07	5.81	. . .	. . .

Find (a) the average velocity; (b) the flow rate in cubic feet per second.

4.7 Into a 100-gal tank, initially filled with fresh water, salt water containing 1.92 lb/gal of salt flows at a fixed rate of 2 gal/min. The density of the incoming solution is 71.8 lb/ft³. The solution, kept uniform by stirring, flows out at a fixed rate of 19.2 lb/min.

(a) How many pounds of salt will there be in the tank at the end of 1 hr and 40 min?

(b) What is the upper limit for the number of pounds of salt in the tank if the process continues indefinitely?

(c) How much time will elapse while the quantity of salt in the tank changes from 100 lb to 150 lb?

4.8 In the piston and cylinder arrangement shown below, the large piston has a velocity of 2 fps and an acceleration of 5 fps². Determine the velocity and acceleration of the smaller piston.

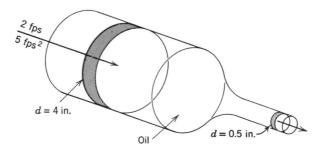

4.9 Show that in a one-dimensional steady flow the following equation is valid:

$$\frac{dA}{A} + \frac{dv}{v} + \frac{d\rho}{\rho} = 0$$

4.10 Using the symbol M for the mass in the control volume, show that equation (4-1) may be written

$$\frac{\partial M}{\partial t} + \iint_{c.s.} d\dot{m} = 0$$

4.11 A shock wave moves down a pipe as shown below. The fluid properties change across the shock, but they are not functions of time. The velocity of

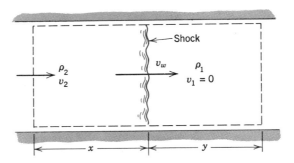

the shock is v_w. Write the continuity equation and obtain the relation between ρ_2, ρ_1, v_2, and v_w. The mass in the control volume at any time is $M = \rho_2 Ax + \rho_1 Ay$.

4.12 The velocity profile in circular pipe is given by $v = v_{max}(1 - r/R)^{1/7}$, where R is the radius of the pipe. Find the average velocity in the pipe.

4.13 In the figure below, the x-direction velocity profiles are shown for a control volume surrounding a cylinder. If the flow is incompressible, what must the rate of flow be across the horizontal control-volume surface?

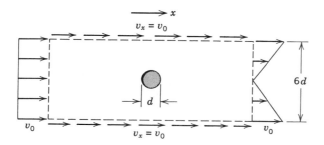

4.14 Two very long parallel plates of length $2L$ are separated a distance b. The upper plate moves downward at a constant rate V. A fluid fills the space between the plates. Fluid is squeezed out between the plates. Determine the mass flow rate and maximum velocity:
(a) If the exit velocity is uniform.
(b) If the exit velocity is parabolic.

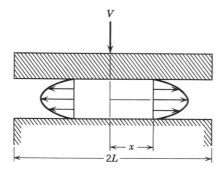

5
NEWTON'S SECOND LAW OF MOTION: CONTROL VOLUME APPROACH

The second of the fundamental physical laws upon which fluid-flow analyses are based is Newton's second law of motion. Starting with Newton's second law, we shall develop integral relations for linear and angular momentum. Applications of these expressions to physical situations will be considered.

5.1 INTEGRAL RELATION FOR LINEAR MOMENTUM

Newton's second law of motion may be stated as follows:

The time rate of change of momentum of a system is equal to the net force acting on the system and takes place in the direction of the net force.

We note at the outset two very important parts of this statement: first, that this law refers to a specific system, and second, that it includes direction as well as magnitude and is therefore a vector expression. In order to use this law it will be necessary to recast its statement into a form applicable to a control volume which contains different fluid particles (i.e., a different system) when examined at different times.

In Figure 5.1 observe the control volume located in a fluid-flow field. The system considered is the material occupying the control volume at time t, and its position is shown both at time t and at time $t + \Delta t$.

Referring to the figure, we see that:

Region I is occupied by the system only at time t.
Region II is occupied by the system at $t + \Delta t$.
Region III is common to the system both at t and at $t + \Delta t$.

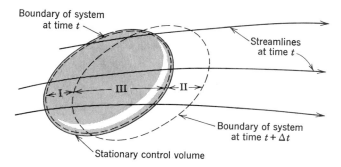

Boundary of system at time t

Streamlines at time t

Boundary of system at time $t + \Delta t$

Stationary control volume

Figure 5.1 Relation between a system and a control volume in a fluid-flow field.

Writing Newton's second law for such a situation, we have

$$\sum \mathbf{F} = \frac{d}{dt}(m\mathbf{v}) = \frac{d}{dt}\mathbf{P} \tag{5-1}$$

where the symbols $\mathbf{F}$, m, and $\mathbf{v}$ have their usual meanings and $\mathbf{P}$ represents the total linear momentum of the system.

At time $t + \Delta t$ the linear momentum of the system which now occupies regions II and III may be expressed as

$$\mathbf{P}|_{t+\Delta t} = \mathbf{P}_{\text{II}}|_{t+\Delta t} + \mathbf{P}_{\text{III}}|_{t+\Delta t}$$

and at time t we have

$$\mathbf{P}|_{t} = \mathbf{P}_{\text{I}}|_{t} + \mathbf{P}_{\text{III}}|_{t}$$

Subtracting the second of these expressions from the first and dividing by the time interval, Δt, between the evaluation gives

$$\frac{\mathbf{P}|_{t+\Delta t} - \mathbf{P}|_{t}}{\Delta t} = \frac{\mathbf{P}_{\text{II}}|_{t+\Delta t} + \mathbf{P}_{\text{III}}|_{t+\Delta t} - \mathbf{P}_{\text{I}}|_{t} - \mathbf{P}_{\text{III}}|_{t}}{\Delta t}$$

We may rearrange the right-hand side of this expression and take the limit of the resulting equation to get

$$\lim_{\Delta t \to 0} \frac{\mathbf{P}|_{t+\Delta t} - \mathbf{P}|_{t}}{\Delta t} = \lim_{\Delta t \to 0} \frac{\mathbf{P}_{\text{III}}|_{t+\Delta t} - \mathbf{P}_{\text{III}}|_{t}}{\Delta t} + \lim_{\Delta t \to 0} \frac{\mathbf{P}_{\text{II}}|_{t+\Delta t} - \mathbf{P}_{\text{I}}|_{t}}{\Delta t} \tag{5-2}$$

Considering each of the limiting processes separately, we have, for the left-hand side,

$$\lim_{\Delta t \to 0} \frac{\mathbf{P}|_{t+\Delta t} - \mathbf{P}|_{t}}{\Delta t} = \frac{d}{dt}\mathbf{P}$$

which is the form specified in the statement of Newton's second law, equation (5-1).

The first limit on the right-hand side of equation (5-2) may be evaluated as

$$\lim_{\Delta t \to 0} \frac{\mathbf{P}_{III}|_{t+\Delta t} - \mathbf{P}_{III}|_t}{\Delta t} = \frac{d}{dt}\mathbf{P}_{III}$$

This we see to be the rate of change of linear momentum of the control volume itself, since, as $\Delta t \to 0$, region III becomes the control volume.

The next limiting process,

$$\lim_{\Delta t \to 0} \frac{\mathbf{P}_{II}|_{t+\Delta t} - \mathbf{P}_{I}|_t}{\Delta t}$$

expresses the net rate of momentum efflux across the control surface during the time interval Δt. As Δt approaches zero, regions II and I become coincident with the control-volume surface.

Considering the physical meaning of each of the limits in equation (5-2) and Newton's second law, equation (5-1), we may write the following word equation for the conservation of linear momentum with respect to a control volume:

$$\left\{\begin{array}{c}\text{Sum of forces} \\ \text{acting on con-} \\ \text{trol volume}\end{array}\right\} = \underbrace{\left\{\begin{array}{c}\text{rate of momen-} \\ \text{tum out of con-} \\ \text{trol volume}\end{array}\right\} - \left\{\begin{array}{c}\text{rate of momen-} \\ \text{tum into con-} \\ \text{trol volume}\end{array}\right\}}_{\begin{array}{c}\text{net rate of momentum efflux from} \\ \text{control volume}\end{array}} + \left\{\begin{array}{c}\text{rate of accumulation} \\ \text{of momentum with-} \\ \text{in control volume}\end{array}\right\}$$

$$(5\text{-}3)$$

We shall now apply equation (5-3) to a general control volume located in a fluid-flow field as shown in Figure 5.2 and evaluate the various terms.

The total force acting on the control volume consists both of surface forces due to interactions between the control-volume fluid, and its surroundings through direct contact, and of body forces resulting from the location of the control volume in a force field. The gravitational field and its resultant force are the most common example of this latter type. We will designate the total force acting on the control volume as $\sum \mathbf{F}$.

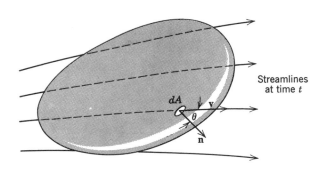

Figure 5.2   Fluid flow through a control volume.

If the small area dA on the control surface is considered, we may write

$$\text{Rate of momentum efflux} = \mathbf{v}(\rho v)(dA \cos \theta)$$

Observe that the product $(\rho v)(dA \cos \theta)$ is the rate of mass efflux from the control volume through dA, as discussed in Chapter 4. Recall further that $dA \cos \theta$ is the area, dA, projected in a direction normal to the velocity vector, $\mathbf{v}$, where θ is the angle between $\mathbf{v}$ and the outwardly directed normal vector, $\mathbf{n}$. We may then multiply the rate of mass efflux by $\mathbf{v}$ to give the rate of momentum efflux through dA. From vector algebra this product may be written as

$$\mathbf{v}(\rho v)(dA \cos \theta) = \mathbf{v}(\rho \, dA)[|\mathbf{v}| \, |\mathbf{n}| \cos \theta]$$

The term in square brackets is the scalar or dot product, $\mathbf{v} \cdot \mathbf{n}$, and the momentum efflux term becomes

$$\rho \mathbf{v}(\mathbf{v} \cdot \mathbf{n}) \, dA$$

Integrating this quantity over the entire control surface, we have

$$\iint_{\text{c.s.}} \mathbf{v}\rho(\mathbf{v} \cdot \mathbf{n}) \, dA$$

which is the *net momentum efflux* from the control volume.

In its integral form the momentum flux term stated above includes the rate of momentum entering the control volume as well as that leaving. If mass is entering the control volume, the sign of the product $\mathbf{v} \cdot \mathbf{n}$ is negative, and the associated momentum flux is an input. Conversely, a positive sign of the product $\mathbf{v} \cdot \mathbf{n}$ is associated with a momentum efflux from the control volume. Thus the first two terms on the right-hand side of equation (5-3) may be written

$$\begin{Bmatrix} \text{rate of momen-} \\ \text{tum out of con-} \\ \text{trol volume} \end{Bmatrix} - \begin{Bmatrix} \text{rate of momen-} \\ \text{tum into con-} \\ \text{trol volume} \end{Bmatrix} = \iint_{\text{c.s.}} \mathbf{v}\rho(\mathbf{v} \cdot \mathbf{n}) \, dA$$

The rate of accumulation of linear momentum within the control volume may be expressed as

$$\frac{\partial}{\partial t} \iiint_{\text{c.v.}} \mathbf{v}\rho \, dV$$

and the overall linear-momentum balance for a control volume becomes

$$\sum \mathbf{F} = \iint_{\text{c.s.}} \mathbf{v}\rho(\mathbf{v} \cdot \mathbf{n}) \, dA + \frac{\partial}{\partial t} \iiint_{\text{c.v.}} \rho \mathbf{v} \, dV \qquad (5\text{-}4)$$

This extremely important relation is often referred to in fluid mechanics as the *momentum theorem*. Note the great similarity between (5-4) and (4-1) in the form of the integral terms; observe, however, that equation (5-4) is a vector expression as opposed to the scalar form of the overall mass balance considered in Chapter 4. In rectangular coordinates the single vector equation, (5-4), may be written as three scalar equations:

$$\sum F_x = \int\int_{c.s.} v_x\rho(\mathbf{v}\cdot\mathbf{n})\,dA + \frac{\partial}{\partial t}\int\int\int_{c.v.}\rho v_x\,dV \qquad (5\text{-}5a)$$

$$\sum F_y = \int\int_{c.s.} v_y\rho(\mathbf{v}\cdot\mathbf{n})\,dA + \frac{\partial}{\partial t}\int\int\int_{c.v.}\rho v_y\,dV \qquad (5\text{-}5b)$$

$$\sum F_z = \int\int_{c.s.} v_z\rho(\mathbf{v}\cdot\mathbf{n})\,dA + \frac{\partial}{\partial t}\int\int\int_{c.v.}\rho v_z\,dV \qquad (5\text{-}5c)$$

When applying any or all of the above equations, it must be remembered that each term has a sign with respect to the positively defined x, y, and z directions. The determination of the sign of the surface integral should be considered with special care, since both the velocity component (v_x) and the scalar product $(\mathbf{v}\cdot\mathbf{n})$ have signs. The combination of the proper sign associated with each of these terms will give the correct sense to the integral. It should also be remembered that since equations (5-5) are written for the fluid in the control volume, *the forces to be employed in these equations are those acting on the fluid.*

A detailed study of the example problems to follow should aid in the understanding of, and afford facility in using, the overall momentum balance.

5.2 APPLICATIONS OF THE INTEGRAL EXPRESSION FOR LINEAR MOMENTUM

In applying equation (5-4) it is first necessary to define the control volume which will make possible the simplest and most direct solution to the problem at hand. There are no general rules to aid in this definition, but experience in handling problems of this type will enable such a choice to be made readily.

EXAMPLE 1

Consider first the problem of finding the force exerted on a reducing pipe bend resulting from a steady flow of fluid in it. A diagram of the pipe bend and the quantities significant to its analysis are shown in Figure 5.3.

The first step is the definition of the control volume. One choice for the control volume, of the several available, is all fluid in the pipe at a given time. The control volume chosen in this manner is designated in Figure 5.4, showing the external forces imposed upon it. The external forces imposed on the fluid include the pressure forces at sections (1) and (2), the body force due to the weight of fluid in the control volume, and the forces due to pressure and shear stress, P_w and τ_w, exerted on the fluid by the pipe wall. The resultant force on the fluid (due to P_w and τ_w) by the pipe will be designated $\mathbf{B}$, and its x and y components symbolized as B_x and B_y, respectively.

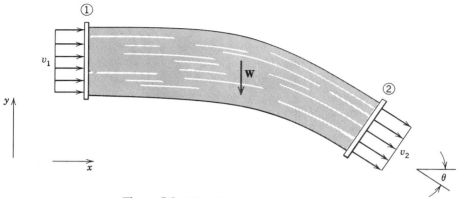

Figure 5.3 Flow in a reducing pipe bend.

Considering the x- and y-directional component equations, (5-5a) and (5-5b), of the overall momentum balance, the external forces acting on the fluid in the control volume are

$$\sum F_x = P_1 A_1 - P_2 A_2 \cos\theta + B_x$$

and

$$\sum F_y = P_2 A_2 \sin\theta - W + B_y$$

Each component of the unknown force, **B**, is assumed to have a positive sense. The actual signs for these components, when a solution is obtained, will indicate whether or not this assumption is correct.

Evaluating the surface integral in both the x and y directions, we have

$$\iint_{c.s.} v_x \rho(\mathbf{v} \cdot \mathbf{n})\, dA = (v_2 \cos\theta)(\rho_2 v_2 A_2) + (v_1)(-\rho_1 v_1 A_1)$$

$$\iint_{c.s.} v_y \rho(\mathbf{v} \cdot \mathbf{n})\, dA = (-v_2 \sin\theta)(\rho_2 v_2 A_2)$$

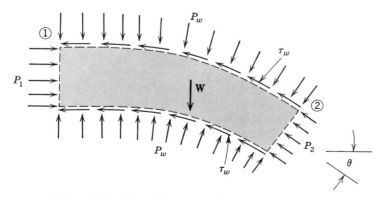

Figure 5.4 Control volume defined by pipe surface.

The accumulation term is zero in both equations, since, for the problem considered, flow is steady.

The complete momentum expressions in the x and y directions are

$$B_x + P_1 A_1 - P_2 A_2 \cos \theta = (v_2 \cos \theta)(\rho_2 v_2 A_2) + v_1(-\rho_1 v_1 A_1)$$

and

$$B_y + P_2 A_2 \sin \theta - W = (-v_2 \sin \theta)(\rho_2 v_2 A_2)$$

Solving for the unknown force components B_x and B_y, we have

$$B_x = v_2{}^2 \rho_2 A_2 \cos \theta - v_1{}^2 \rho_1 A_1 - P_1 A_1 + P_2 A_2 \cos \theta$$

and

$$B_y = -v_2{}^2 \rho_2 A_2 \sin \theta - P_2 A_2 \sin \theta + W$$

Recall that we were to evaluate the force exerted on the pipe rather than that on the fluid. The force sought is the reaction to **B** and has components equal in magnitude and opposite in sense to B_x and B_y. The components of the reaction force, **R**, exerted on the pipe are

$$R_x = -v_2{}^2 \rho_2 A_2 \cos \theta + v_1{}^2 \rho_1 A_1 + P_1 A_1 - P_2 A_2 \cos \theta$$

and

$$R_y = v_2{}^2 \rho_2 A_2 \sin \theta + P_2 A_2 \sin \theta - W$$

Some simplification in form may be achieved if the flow is steady. Applying equation (4-3), we have

$$\rho_1 v_1 A_1 = \rho_2 v_2 A_2 = \dot{m}$$

where $\dot{m}$ is the mass flow rate.

The final solution for the components of **R** may now be written as

$$R_x = \dot{m}(v_1 - v_2 \cos \theta) + P_1 A_1 - P_2 A_2 \cos \theta$$
$$R_y = \dot{m} v_2 \sin \theta + P_2 A_2 \sin \theta - W$$

The control volume shown in Figure 5.4 for which the above solution was obtained represents only one possible choice. Another is depicted in Figure 5.5. This control volume

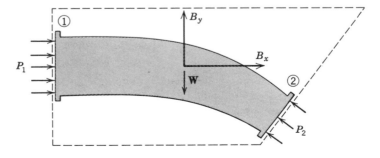

Figure 5.5   Control volume including fluid and pipe.

is bounded simply by the straight planes cutting through the pipe at sections (1) and (2). The fact that a control volume such as this can be used indicates the versatility of this approach, that is, that the results of complicated processes occurring internally may be analyzed quite simply by considering only those quantities of transfer across the control surface.

For this control volume the x- and y-directional momentum equations are

$$B_x + P_1 A_1 - P_2 A_2 \cos \theta = (v_2 \cos \theta)(\rho_2 v_2 A_2) + v_1(-v_1 \rho_1 A_1)$$

and

$$B_y + P_2 A_2 \sin \theta - W = (-v_2 \sin \theta)(\rho_2 v_2 A_2)$$

where the force having components B_x and B_y is that exerted on the control volume by the section of pipe cut through at sections (1) and (2). The pressures at (1) and (2) in the above equations are gage pressures, since the atmospheric pressures acting on all surfaces cancel.

Note that the resulting equations for this control volume are identical to those obtained for the one defined previously. Thus a correct solution may be obtained for each of several chosen control volumes so long as they are analyzed carefully and completely.

EXAMPLE 2

As our second example of the application of the control-volume expression for linear momentum (the momentum theorem) consider the steam locomotive tender schematically illustrated in Figure 5.6, which obtains water from a trough by means of a scoop. The force on the train due to the water is to be obtained.

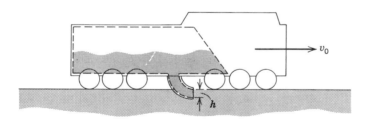

Figure 5.6 Schematic of locomotive tender scooping water from a trough.

The logical choice for a control volume in this case is the water-tank/scoop combination. Our control-volume boundary will be selected as the *interior* of the tank and scoop. Since the train is moving with a uniform velocity, there are two possible choices of coordinate systems. We may select a coordinate system either fixed in space or moving* with the velocity of the train, v_0. Let us first analyze the system by using a moving coordinate system.

---

* Recall that a uniformly translating coordinate system is an inertial coordinate system, hence Newton's second law and the momentum theorem may be employed directly.

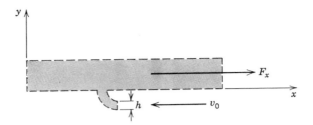

Figure 5.7 Moving coordinate system and control volume.

The moving control volume is shown in Figure 5.7 with the xy coordinate system moving at velocity v_0. All velocities are determined with respect to the x and y axes.

The applicable expression is equation (5-5a),

$$\sum F_x = \iint_{c.s.} v_x\rho(\mathbf{v} \cdot \mathbf{n})\, dA + \frac{\partial}{\partial t} \iiint_{c.v.} v_x\rho\, dV$$

In Figure 5.7 $\sum F_x$ is represented as F_x and is shown in the positive sense. Since the forces due to pressure and shear are to be neglected, F_x is the total force exerted on the fluid by the train and scoop. The momentum flux term is

$$\iint_{c.s.} v_x\rho(\mathbf{v} \cdot \mathbf{n})\, dA = \rho(-v_0)(-1)(v_0)(h) \qquad \text{(per unit length)}$$

and the rate of change of momentum within the control volume is zero, since the fluid in the control volume has zero velocity in the x direction.

Thus

$$F_x = \rho v_0^2 h$$

This is the force exerted by the train on the fluid. The force exerted by the fluid on the train is the opposite of this, or $-\rho v_0^2 h$.

Now let us consider the same problem with a stationary coordinate system (see Figure 5.8). Employing once again the control-volume relation for linear momentum,

$$\sum F_x = \iint_{c.s.} v_x\rho(\mathbf{v} \cdot \mathbf{n})\, dA + \frac{\partial}{\partial t} \iiint_{c.v.} v_x\rho\, dV$$

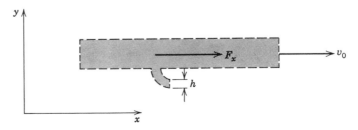

Figure 5.8 Stationary coordinate system and moving control volume.

we obtain

$$F_x = 0 + \frac{\partial}{\partial t} \iiint_{c.v.} v_x \rho \, dV$$

where the momentum flux is zero, since the entering fluid has zero velocity. There is, of course, no fluid leaving the control volume. The terms $\partial/\partial t \iiint_{c.v.} v_x \rho \, dV$, as the velocity, $v_x = v_0 = $ constant, may be written as $v_0 \, \partial/\partial t \iiint_{c.v.} \rho \, dV$ or $v_0(\partial m/\partial t)$, where m is the mass of fluid entering the control volume at the rate $\partial m/\partial t = \rho v_0 h$ so that $F_x = \rho v_0^2 h$ as before.

The student should note that, in the case of a stationary coordinate system and a moving control volume, care must be exercised in the interpretation of the momentum flux

$$\iint_{c.s.} v\rho(\mathbf{v} \cdot \mathbf{n}) \, dA$$

Regrouping the terms, we obtain

$$\iint_{c.s.} \mathbf{v}\rho(\mathbf{v} \cdot \mathbf{n}) \, dA \equiv \iint_{c.s.} \mathbf{v} \, d\dot{m}$$

Thus it is obvious that while $\mathbf{v}$ is the velocity relative to fixed coordinates, $\mathbf{v} \cdot \mathbf{n}$ is the velocity relative to the control-volume boundary.

EXAMPLE 3

The use of the momentum theorem to predict local changes in fluid and flow properties can be illustrated at this point by consideration of a shock wave. A shock wave is a region of very rapid change of flow and fluid properties. For the purposes of engineering calculations, these changes may be considered to occur discontinuously as shown in Figure 5.9.

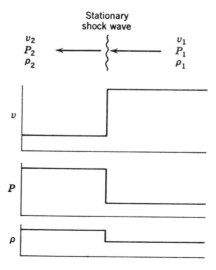

Figure 5.9 Changes in fluid and flow properties across a stationary shock wave. Flow is from right to left.

Shock waves occur in a compressible fluid only. Although liquids are usually considered incompressible, sufficient compressibility exists to produce shock waves, as evidenced by the phenomenon of "water hammer" in pipes. The pressure change across a moving shock wave may be determined from the following analysis. The shock wave illustrated in Figure 5.10 is moving to the right at velocity v_w into a stationary fluid. After

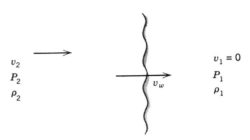

Figure 5.10   Moving shock wave.

the wave passes a given point, the velocity is changed from 0 to v_2, the pressure is changed from P_1 to P_2, and the density is changed from ρ_1 to ρ_2. The illustrated shock wave may be analyzed by considering either a stationary control volume (see problems 4.11 and 5.13) or by using a control volume moving with the speed of the wave, v_w. We shall use the latter approach here. For an observer moving at velocity v_w, the shock wave appears stationary, the fluid in region 1 appears to be flowing from right to left with velocity v_w, and the fluid in region 2 also flows from right to left with velocity $v_w - v_2$. The values of the pressure and density in regions 1 and 2 are unchanged so that the moving shock wave appears as illustrated in Figure 5.11.

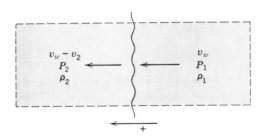

Figure 5.11   Moving shock wave as seen by observer moving with the wave.

Employing the conservation of mass for a control volume, we obtain

$$\iiint_{c.v.} \frac{\partial \rho}{\partial t}^{\,0} \, dV + \iint_{c.s.} \rho(\mathbf{v} \cdot \mathbf{n}) \, dA = 0$$

which becomes, upon substitution and after canceling of the areas,

$$\rho_2(v_w - v_2) - \rho_1 v_w = 0$$

If we designate the positive direction as shown in the figure and use the momentum theorem,

$$\Sigma \mathbf{F} = \iint_{\text{c.s.}} \mathbf{v}\rho(\mathbf{v} \cdot \mathbf{n}) \, dA + \frac{\partial}{\partial t} \iiint_{\text{c.v.}} \mathbf{v}\rho \, dV^{\,0}$$

substitution yields

$$(P_1 - P_2)A = \rho_2(v_w - v_2)^2 A - \rho_1 v_w^2 A$$

The areas may be canceled and the conservation of mass relation, $\rho_2(v_w - v_2) = \rho_1 v_w$, employed, and this yields

$$P_2 - P_1 = \rho_1 v_w v_2$$

The pressure change across the shock wave is seen to be equal to the momentum change across the shock wave, which may be simply expressed as $\rho_1 v_w v_2$. If the wave velocity is approximated by the speed of sound in the fluid under consideration, a useful estimate of the velocity change can be made.

5.3 INTEGRAL RELATION FOR MOMENT OF MOMENTUM

The integral relation for the moment of momentum of a control volume is an extension of the considerations just made for linear momentum.

Starting with equation (5-1), which is a mathematical expression of Newton's second law of motion applied to a system of particles (Figure 5.12),

$$\Sigma \mathbf{F} = \frac{d}{dt}(m\mathbf{v}) = \frac{d}{dt}\mathbf{P} \tag{5-1}$$

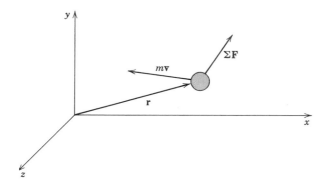

Figure 5.12 A system and its displacement vector r.

we take the vector or "cross" product of a position vector, $\mathbf{r}$, with each term and get

$$\mathbf{r} \times \sum \mathbf{F} = \mathbf{r} \times \frac{d}{dt}(m\mathbf{v}) = \mathbf{r} \times \frac{d}{dt}\mathbf{P} \qquad (5\text{-}6)$$

The quantity on the left-hand side of equation (5-6), $\mathbf{r} \times \sum \mathbf{F}$, is the resultant moment, $\sum \mathbf{M}$, about the origin as shown in Figure 5.12, due to all forces applied to the system. Clearly we may write

$$\mathbf{r} \times \sum \mathbf{F} = \sum \mathbf{r} \times \mathbf{F} = \sum \mathbf{M}$$

where $\sum \mathbf{M}$ is, again, the total moment about the origin of all forces acting on the system.

The right-hand side of equation (5-6) is the moment of the time rate of change of linear momentum. This we can write as

$$\mathbf{r} \times \frac{d}{dt}m\mathbf{v} = \frac{d}{dt}(\mathbf{r} \times m\mathbf{v}) = \frac{d}{dt}(\mathbf{r} \times \mathbf{P}) = \frac{d}{dt}\mathbf{H}$$

Thus this term is also the time rate of change of the moment of momentum of the system. We shall use the symbol $\mathbf{H}$ to designate moment of momentum. The complete expression is now

$$\sum \mathbf{M} = \frac{d}{dt}\mathbf{H} \qquad (5\text{-}7)$$

As with its analogous expression for linear momentum, equation (5-1), equation (5-7) applies to a specific system. By the same limit process as that used for linear momentum we may recast this expression into a form applicable to a control volume and achieve a word equation

$$\begin{Bmatrix} \text{Sum of moments} \\ \text{acting on control} \\ \text{volume} \end{Bmatrix} = \underbrace{\begin{Bmatrix} \text{rate of moment} \\ \text{of momentum out} \\ \text{of control volume} \end{Bmatrix} - \begin{Bmatrix} \text{rate of moment} \\ \text{of momentum in-} \\ \text{to control volume} \end{Bmatrix}}_{\substack{\text{net rate of efflux of moment of momentum} \\ \text{from control volume}}} + \begin{Bmatrix} \text{rate of ac-} \\ \text{cumulation} \\ \text{of moment} \\ \text{of momen-} \\ \text{tum within} \\ \text{control} \\ \text{volume} \end{Bmatrix}$$

$$(5\text{-}8)$$

Equation (5-8) may be applied to a general control volume to yield the following equation:

$$\sum \mathbf{M} = \iint_{\text{c.s.}} (\mathbf{r} \times \mathbf{v})\rho(\mathbf{v} \cdot \mathbf{n})\, dA + \frac{\partial}{\partial t} \iiint_{\text{c.v.}} (\mathbf{r} \times \mathbf{v})\rho\, dV \qquad (5\text{-}9)$$

The term on the left-hand side of equation (5-9) is the total moment of all forces acting on the control volume. The terms on the right-hand side represent the net rate of efflux of moment of momentum through the control surface and the rate of accumulation of moment of momentum within the control volume, respectively.

This single vector equation may be expressed as three scalar equations for the orthogonal inertial coordinate directions x, y, and z as

$$\sum M_x = \iint_{c.s.} (\mathbf{r} \times \mathbf{v})_x \rho (\mathbf{v} \cdot \mathbf{n}) \, dA + \frac{\partial}{\partial t} \iiint_{c.v.} (\mathbf{r} \times \mathbf{v})_x \rho \, dV \qquad (5\text{-}10a)$$

$$\sum M_y = \iint_{c.s.} (\mathbf{r} \times \mathbf{v})_y \rho (\mathbf{v} \cdot \mathbf{n}) \, dA + \frac{\partial}{\partial t} \iiint_{c.v.} (\mathbf{r} \times \mathbf{v})_y \rho \, dV \qquad (5\text{-}10b)$$

and

$$\sum M_z = \iint_{c.s.} (\mathbf{r} \times \mathbf{v})_z \rho (\mathbf{v} \cdot \mathbf{n}) \, dA + \frac{\partial}{\partial t} \iiint_{c.v.} (\mathbf{r} \times \mathbf{v})_z \rho \, dV \qquad (5\text{-}10c)$$

The directions associated with M_x and $(\mathbf{r} \times \mathbf{v})$ are those considered in mechanics in which the right-hand rule is used to determine the orientation of quantities having rotational sense.

5.4 APPLICATIONS TO PUMPS AND TURBINES

The moment-of-momentum expression is particularly applicable to two types of devices, generally classified as pumps and turbines. We shall, in this section, consider those having rotary motion only. If energy is derived from a fluid acting on a rotating device, it is designated a turbine, while a pump adds energy to a fluid. The rotating part of a turbine is called a runner and that of a pump an impeller.

EXAMPLE 4

Let us first direct our attention to a type of turbine known as the Pelton wheel. Such a device is represented in Figure 5.13. In this turbine a jet of fluid, usually water, is directed

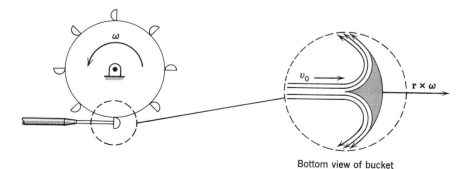

Bottom view of bucket

Figure 5.13 Pelton wheel.

from a nozzle striking a system of buckets on the periphery of the runner. The buckets are shaped so that the water is diverted in such a way as to exert a force on the runner which will, in turn, cause rotation. Using the moment-of-momentum relation, we may determine the torque resulting from such a situation.

We must initially define our control volume. The dashed lines in Figures 5.14 illustrate the control volume chosen. It encloses the entire runner and cuts the jet of water with velocity v_0 as shown. The control surface also cuts the shaft on both sides of the runner.

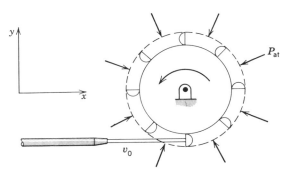

Figure 5.14 Control volume for analysis of Pelton wheel.

The applicable scalar form of the general moment-of-momentum expression is equation (5-10c) written for the z direction. All rotation is in the xy plane, and—according to the right-hand rule—the vector representation of a quantity having angular motion, or a tendency to produce angular motion, has a sense normal to the xy plane, that is, the z direction. Recall that a positive angular sense is that conforming to the direction in which the thumb on the right hand will point when the fingers of the right hand are aligned with the direction of counterclockwise angular motion.

$$\sum M_z = \iint_{\text{c.s.}} (\mathbf{r} \times \mathbf{v})_z \rho(\mathbf{v} \cdot \mathbf{n}) \, dA + \frac{\partial}{\partial t} \iiint_{\text{c.v.}} \rho(\mathbf{r} \times \mathbf{v})_z \, dV$$

Evaluating each term separately, we have, for the external moment,

$$\sum M_z = M_{\text{shaft}}$$

where M_{shaft}, the moment applied to the runner by the shaft, is the only such moment acting on the control volume.

The surface integral,

$$\iint_{\text{c.s.}} (\mathbf{r} \times \mathbf{v})_z \rho(\mathbf{v} \cdot \mathbf{n}) \, dA$$

is the net rate of efflux of moment of momentum. The fluid leaving the control volume is illustrated in Figure 5.15. The x-direction component of the fluid leaving the control volume is

$$\{r\omega - (v_0 - r\omega) \cos \theta\} \mathbf{e}_x$$

Here it is assumed that the z components of the velocity are equal and opposite. The

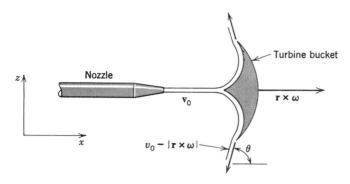

Figure 5.15 Velocity vectors for turbine bucket.

leaving velocity is the vector sum of the velocity of the turbine bucket, $r\omega$, and that of the exiting fluid relative to the bucket and leaving at an angle θ to the direction of motion of the bucket, $(v_0 - r\omega) \cos \theta$. These velocity vectors are shown in the figure. The final expression for the surface integral is now

$$\iint_{\text{c.s.}} (\mathbf{r} \times \mathbf{v})_z \rho(\mathbf{v} \cdot \mathbf{n}) \, dA = r[r\omega - (v_0 - r\omega) \cos \theta]\rho Q - r v_0 \rho Q$$

The last term, $r v_0 \rho Q$, is the moment of momentum of the incoming fluid stream of velocity v_0 and density ρ, with a volumetric flow rate Q.

Since the problem under consideration is one in which the angular velocity, ω, of the wheel is constant, the term expressing the time derivative of moment of momentum of the control volume, $\partial/\partial t \iiint_{\text{c.v.}} (\mathbf{r} \times \mathbf{v})_z \rho \, dV = 0$. Replacing each term in the complete expression by its equivalent, we have

$$\sum M_z = M_{\text{shaft}} = \iint_{\text{c.s.}} (\mathbf{r} \times \mathbf{v})_z \rho(\mathbf{v} \cdot \mathbf{n}) \, dA + \frac{\partial}{\partial t} \iiint_{\text{c.v.}} \rho(\mathbf{r} \times \mathbf{v})_z \, dV$$

$$= r[r\omega - (v_0 - r\omega) \cos \theta]\rho Q - r v_0 \rho Q$$

$$= -r(v_0 - r\omega)(1 + \cos \theta)\rho Q$$

The torque applied to the shaft is equal in magnitude and opposite in sense to M_{shaft}. Thus our final result is

$$\text{Torque} = -M_{\text{shaft}} = r(v_0 - r\omega)(1 + \cos \theta)\rho Q$$

EXAMPLE 5

The radial-flow turbine illustrated in Figure 5.16 may be analyzed with the aid of the moment-of-momentum expression. In this device, the fluid (usually water) enters the guide vanes, which impart a tangential velocity and hence angular momentum to the fluid before it enters the revolving runner which reduces the angular momentum of the fluid while delivering torque to the runner.

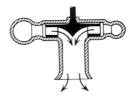

Figure 5.16   Radial-flow turbine.

The control volume to be used is illustrated below in Figure 5.17. The outer boundary of the control volume is at radius r_1, and the inner boundary is at r_2. The width of the runner is h.

We will use equation (5-9) in order to determine the torque. For steady flow this equation becomes

$$\sum \mathbf{M} = \iint_{\text{c.s.}} (\mathbf{r} \times \mathbf{v}) \rho (\mathbf{v} \cdot \mathbf{n}) \, dA$$

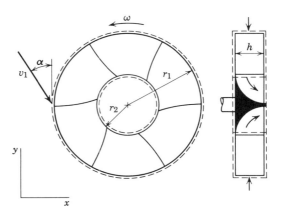

Figure 5.17   Radial-flow turbine-runner control volume.

Evaluating each term separately, we have, for the external moment of the runner on the fluid,

$$\sum \mathbf{M} = M_{fluid}\mathbf{e}_z = -T\mathbf{e}_z$$

where T is the shaft torque. The surface integral requires the evaluation of the vector product $(\mathbf{r} \times \mathbf{v})$ at the outer boundary r_1 and at the inner boundary r_2. If we express the velocity of the water in polar coordinates $\mathbf{v} = v_r\mathbf{e}_r + v_\theta\mathbf{e}_\theta$, so that $(\mathbf{r} \times \mathbf{v}) = r\mathbf{e}_r \times (v_r\mathbf{e}_r + v_\theta\mathbf{e}_\theta) = rv_\theta\mathbf{e}_z$. Thus the surface integral, assuming uniform velocity distribution, is given by

$$\iint_{c.s} (\mathbf{r} \times \mathbf{v})\rho(\mathbf{v} \cdot \mathbf{n}) \, dA = \{r_1 v_{\theta_1}\rho(-v_{r_1})2\pi r_1 h + r_2 v_{\theta_2}\rho v_{r_2}2\pi r_2 h\}\mathbf{e}_z$$

The general result is

$$-T\mathbf{e}_z = (-\rho v_{r_1} v_{\theta_1}2\pi r_1^2 h + \rho v_{r_2} v_{\theta_2}2\pi r_2^2 h)\mathbf{e}_z$$

The law of conservation of mass may be used,

$$\rho v_{r_1}2\pi r_1 h = \dot{m} = \rho v_{r_2}2\pi r_2 h$$

so that the torque is given by

$$T = \dot{m}(r_1 v_{\theta_1} - r_2 v_{\theta_2})$$

The velocity components v_{θ_1} and v_{θ_2} have been expressed in the xyz coordinate system, which is stationary. The velocity at r_1 is seen from Figures 5.16 and 5.17 to be determined by the flow rate and the guide vane angle α. The velocity at r_2, however, requires knowledge of flow conditions on the runner.

The velocity at r_2 may be determined by the following analysis. In Figure 5.18, the flow conditions at the outlet of the runner are sketched. The velocity of the water v_2 is the vector sum of the velocity with respect to the runner v_2' and the runner velocity $r_2\omega$.

The velocity v_{θ_2}, the tangential velocity of the water leaving the runner, is given by

$$v_{\theta_2} = r_2\omega - v_2' \sin \beta$$

where β is the blade angle as shown. The fluid is assumed to flow in the same direction as the

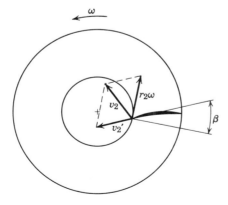

Figure 5.18 Velocity at runner exit (only one blade is shown).

blade. The radial component of the flow may be determined from conservation of mass

$$v_{r_2} = v_2' \cos \beta = \frac{\dot{m}}{2\pi \rho r_2 h}$$

Thus

$$T = \dot{m}\left(r_1 v_{\theta_1} - r_2 \left[r_2 \omega - \frac{\dot{m} \tan \beta}{2\pi \rho r_2 h} \right] \right)$$

In practice, the guide vanes are adjustable to make the relative velocity at the runner entrance tangent to the blades.

5.5 CLOSURE

In this chapter the basic relation involved has been Newton's second law of motion. This law, as written for a system, was recast so that it could apply to a control volume. The result of a consideration of a general control volume led to the integral equations for linear momentum, equation (5-4), and moment of momentum, equation (5-9). Equation (5-4) is often referred to as the momentum theorem of fluid mechanics. This equation is one of the most powerful and often-used expressions in this field.

The student is again urged to start always with the complete integral expression when working a problem. A term-by-term analysis from this basis will allow a correct solution, whereas in a hasty consideration certain terms might be evaluated incorrectly or neglected completely. As a final remark, it should be noted that the momentum theorem expression, as developed, applies to an inertial control volume only.

PROBLEMS

5.1 A two-dimensional object is placed in a 4-ft-wide water tunnel as shown. The upstream velocity, v_1, is uniform across the cross section. For the downstream velocity profile as shown, find the value of v_2.

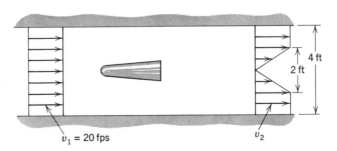

5.2 A stationary jet engine is shown. Air with density of 0.0805 lb_m/ft^3 enters as shown. The inlet and outlet cross-sectional areas are both 10.8 ft^2. The mass of fuel consumed is 1% of the mass of air entering the test section. For these conditions calculate the thrust developed by the engine tested.

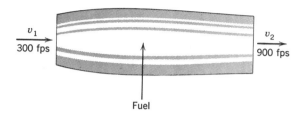

5.3 If, in the system for problem 5.1, the total drag on the object is measured to be 600 N/m of length normal to the direction of flow, and frictional forces at the walls are neglected, find the pressure difference between inlet and outlet sections.

5.4 (a) Determine the magnitude of the x and y components of the force exerted on the fixed blade shown by a 2-ft^3/s jet of water flowing at 25 fps.

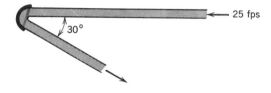

(b) If the blade is moving to the left at 15 fps, find the magnitude and velocity of the water jet leaving the blade.

5.5 An open tank car as shown travels to the right at a uniform velocity of 4.5 m/s. At the instant shown the car passes under a jet of water issuing from a stationary 0.1 m diameter pipe with a velocity of 10 m/s. What force is exerted on the tank by the water jet?

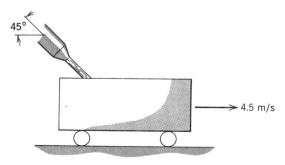

5.6 The pump in the boat shown pumps 8 ft^3/s of water through the submerged water passage, which has an area of 0.25 ft^2 at the bow of the boat and 0.15 ft^2 at the stern. Determine the tension in the restraining rope, assuming that the inlet and exit pressures are equal.

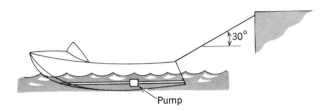

5.7 Oil (sp. gr. = 0.8) flows smoothly through the circular reducing section shown at 2.9 ft^3/s. If the entering and leaving velocity profiles are uniform, estimate the force which must be applied to the reducer to hold it in place.

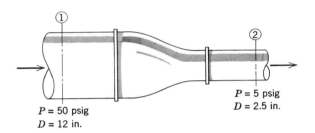

5.8 At the end of a water pipe of 3-in. diameter is a nozzle which discharges a jet having a diameter of $1\frac{1}{2}$ in. into the open atmosphere. The pressure in the pipe is 60 psig (pounds per square inch gage), and the rate of discharge is 400 gal/min. What are the magnitude and direction of the force necessary to hold the nozzle to the pipe?

5.9 A water jet pump has an area $A_j = 0.06$ ft^2 and a jet velocity $v_j = 90$ fps, which entrains a secondary stream of water having a velocity $v_s = 10$ fps in a constant-area pipe of total area $A = 0.6$ ft^2. At section 2, the water is

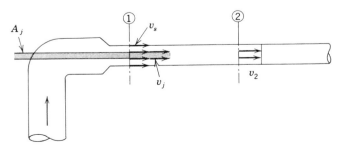

thoroughly mixed. Assuming one-dimensional flow and neglecting wall shear,

(a) find the average velocity of mixed flow at section 2;

(b) find the pressure rise $(P_2 - P_1)$, assuming the pressure of the jet and secondary stream to be the same at section 1.

5.10 If the plate shown is inclined at an angle as shown, what are the forces F_x and F_y necessary to maintain its position? The flow is frictionless.

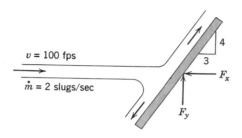

5.11 A plate moves perpendicularly toward a discharging jet at the rate of 5 fps. The jet discharges water at the rate of 2 ft^3/s and a speed of 30 fps. Find the force of the fluid on the plate and compare it with what it would be if the plate were stationary. Assume frictionless flow.

5.12 The illustration below shows a vane with a turning angle θ which moves with a steady speed v_c. The vane receives a jet which leaves a fixed nozzle with speed v.

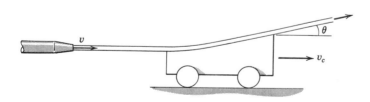

(a) Assume that the vane is mounted on rails as shown in the sketch. Show that the power transmitted to the cart is maximum when $v_c/v = \frac{1}{3}$.

(b) Assuming that there are a large number of such vanes attached to a rotating wheel with peripheral speed, v_c, show that the power transmitted is maximum when $v_c/v = \frac{1}{2}$.

5.13 The shock wave illustrated below is moving to the right at v_w fps. The properties in front and in back of the shock are not a function of time. By using the illustrated control volume, show that the pressure difference

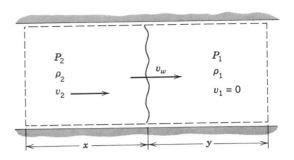

across the shock is

$$P_2 - P_1 = \rho_1 v_w v_2$$

5.14 Consider the differential control volume shown below. By applying the conservation of mass and the momentum theorem, show that

$$dP + \rho v\, dv + g\, dy = 0$$

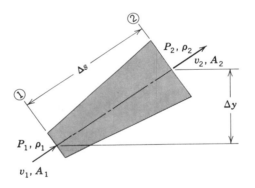

5.15 Water flows steadily through the horizontal 30° pipe bend shown below. At station 1 the diameter is 0.3 m, the velocity is 12 m/s, and the pressure is 128 kPa gage. At station 2 the diameter is 0.38 m and the pressure is 145 kPa gage. Determine the forces F_x and F_z necessary to hold the pipe bend stationary.

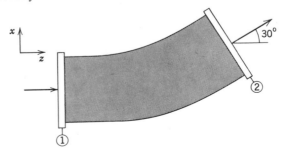

5.16 The rocket nozzle shown below consists of three welded sections. Determine the axial stress at junctions 1 and 2 when the rocket is operating at sea level. The mass flow rate is 770 lb_m/s.

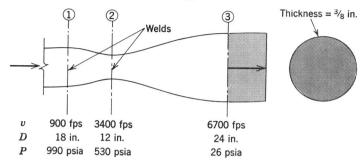

v	900 fps	3400 fps	6700 fps	
D	18 in.	12 in.	24 in.	
P	990 psia	530 psia	26 psia	

5.17 The pressure on the control volume illustrated below is constant. The x components of velocity are as illustrated. Determine the force exerted on the cylinder by the fluid. Assume incompressible flow.

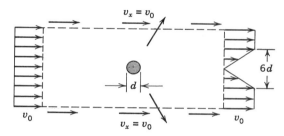

5.18 If the shock-wave velocity in problem 5.13 is approximated by the speed of sound, determine the pressure change causing a velocity change of 10 fps in
(a) air at standard conditions;
(b) water.

5.19 Water flows in a pipe at 4 m/s. A valve at the end of the pipe is suddenly closed. Determine the pressure rise in the pipe.

5.20 Sea water, $\rho = 64$ lb_m/ft^3, flows through the impeller of a centrifugal pump at the rate of 800 gal/min. Determine the torque exerted on the impeller by the fluid and the power required to drive the pump. Assume that the absolute velocity of the water entering the impeller is radial. The dimensions are as follows:

$$\omega = 1180 \text{ rpm} \qquad t_2 = 0.6 \text{ in.}$$
$$r_1 = 2 \text{ in.} \qquad \theta_2 = 135°$$
$$r_2 = 8 \text{ in.} \qquad t_1 = 0.8 \text{ in.}$$

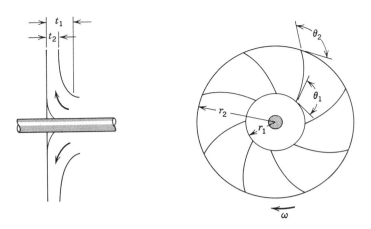

5.21 In problem 5.20 determine:
(a) the angle θ_1 such that the entering flow is parallel to the vanes;
(b) the axial load on the shaft if the shaft diameter is 1 in. and the pressure at the pump inlet is atmospheric.

5.22 A water sprinkler consists of two $\frac{3}{8}$-in. diameter jets at the ends of a rotating hollow rod as shown. If the water leaves at 20 fps, what torque would be necessary to hold the sprinkler in place?

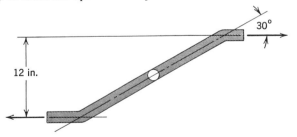

5.23 A lawn sprinkler consists of two sections of curved pipe rotating about a vertical axis as shown. The sprinkler rotates with an angular velocity ω, and the effective discharge area is A, thus the water is discharged at a rate $Q = 2v_rA$, where v_r is the velocity of the water relative to the rotating pipe.

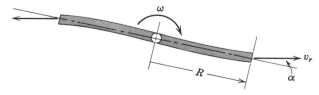

A constant friction torque M_f resists the motion of the sprinkler. Find an expression for the speed of the sprinkler in terms of the significant variables.

5.24 The pipe shown below has a slit of thickness $\frac{1}{4}$ in. so shaped that a sheet of water of uniform thickness $\frac{1}{4}$ in. issues out radially from the pipe. The velocity is constant along the pipe as shown and a flow rate of $8\,\text{ft}^3/\text{s}$ enters at the top. Find the moment on the tube about the axis BB from the flow of water inside the pipe system.

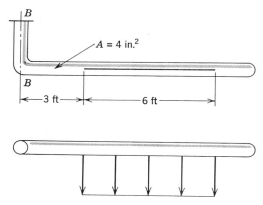

5.25 An open tank L feet long as shown below travels to the right at a velocity v_c fps. A jet of area A_j exhausts fluid of density ρ at a velocity v_j fps relative to the car. The tank car, at the same time, collects fluid from an overhead

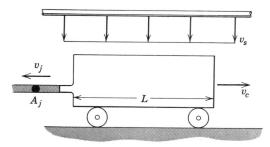

sprinkler which directs fluid downward with velocity v_s. Assuming that the sprinkler flow is uniform over the car area, A_c, determine the net force of the fluid on the tank car.

5.26 A steady, incompressible, frictionless, two-dimensional jet of fluid with breadth h, velocity V, and unit width impinges on a flat plate held at an angle α to its axis. Gravitational forces are to be neglected.
(a) Determine the total force on the plate, and the breadths a, b of the two branches.
(b) Determine the distance l to the center of pressure (c.p.) along the plate from the point 0. (The center of pressure is the point at which the plate can be balanced without requiring an additional moment.)

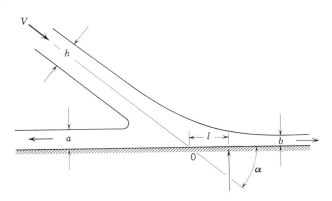

5.27 A dam discharges into a channel of constant width as shown. It is observed that a region of still water backs up behind the jet to a height H. The velocity and height of the flow in the channel are given as V and h, respectively, and the density of the water is ρ. Using the momentum theorem and the control surface indicated, determine H. Neglect the horizontal momentum of the flow that is entering the control volume from above and assume friction to be negligible. The air pressure in the cavity below the crest of falling water is to be taken as atmospheric.

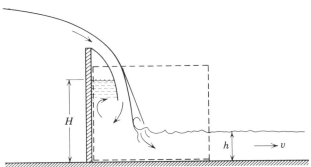

5.28 A liquid of density ρ flows through a sluice gate as shown. The upstream and downstream flows are uniform and parallel, so that the pressure variations at stations 1 and 2 may be considered hydrostatic.
(a) Determine the velocity at station 2.
(b) Determine the force per unit width, R, necessary to hold the sluice gate in place.

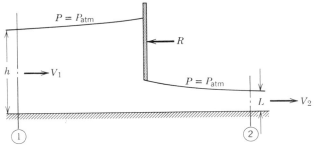

6
CONSERVATION OF ENERGY: CONTROL VOLUME APPROACH

The third fundamental law to be applied to fluid-flow analyses is the first law of thermodynamics. An integral expression for the conservation of energy applied to a control volume will be developed from the first law of thermodynamics, and examples of the application of the integral expression will be shown.

6.1 INTEGRAL RELATION FOR THE CONSERVATION OF ENERGY

The first law of thermodynamics may be stated as follows:

If a system is carried through a cycle, the total heat added to the system from its surroundings is proportional to the work done by the system on its surroundings.

Note that this law is written for a specific group of particles—those comprising the defined system. The procedure will then be similar to that used in Chapter 5, that is, recasting this statement into a form applicable to a control volume which contains different fluid particles at different times. The statement of the first law of thermodynamics involves only scalar quantities however, and thus, unlike the momentum equations considered in Chapter 5, the equations resulting from the first law of thermodynamics will be scalar in form.

The statement of the first law given above may be written in equation form as

$$\oint \delta Q = \frac{1}{J} \oint \delta W \qquad (6\text{-}1)$$

where the symbol $\oint$ refers to a "cyclic integral" or the integral of the quantity evaluated over a cycle. The symbols δQ and δW represent differential heat transfer and work done, respectively. The differential operator, δ, is used, since both heat transfer and work are path functions and the evaluation of integrals of

73

this type requires a knowledge of the path. The more familiar differential operator, d, is used with a "point" function. Thermodynamic properties are, by definition, point functions, and the integrals of such functions may be evaluated without a knowledge of the path by which the change in the property occurs between the initial and final states.* The quantity J is the so-called "mechanical equivalent of heat," numerically equal to 778.17 ft lb/Btu in engineering units. In the SI system, $J = 1$ newton-meter/joule. This factor will not be written henceforth, and the student is reminded that all equations must be dimensionally homogeneous. The responsibility for dimensional consistency remains that of the student.

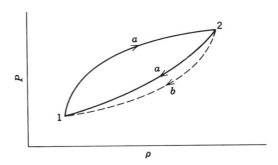

Figure 6.1   Reversible and irreversible
thermodynamic cycles.

We now consider a general thermodynamic cycle, as shown in Figure 6.1. The cycle a occurs between points 1 and 2 by the paths indicated. Utilizing equation (6-1), we may write, for cycle a,

$$\int_{1a}^{2} \delta Q + \int_{2a}^{1} \delta Q = \int_{1a}^{2} \delta W + \int_{2a}^{1} \delta W \qquad (6\text{-}2a)$$

A new cycle between points 1 and 2 is postulated as follows: the path between point 1 and point 2 is identical to that considered previously; however, the cycle is completed by path b between point 2 and point 1, which is any path other than a between these points. Again equation (6-1) allows us to write

$$\int_{1a}^{2} \delta Q + \int_{2b}^{1} \delta Q = \int_{1a}^{2} \delta W + \int_{2b}^{1} \delta W \qquad (6\text{-}2b)$$

Subtracting equation (6-2b) from equation (6-2a) gives

$$\int_{2a}^{1} \delta Q - \int_{2b}^{1} \delta Q = \int_{2a}^{1} \delta W - \int_{2b}^{1} \delta W$$

---

* For a more complete discussion of properties, point functions, and path functions, the reader is referred to G. N. Hatsopoulos and J. H. Keenan, *Principles of General Thermodynamics*. Wiley, New York, 1965, p. 14.

which may be written

$$\int_{2a}^{1} (\delta Q - \delta W) = \int_{2b}^{1} (\delta Q - \delta W) \tag{6-3}$$

Since each side of equation (6-3) represents the integrand evaluated between the same two points but along different paths, it follows that the quantity, $\delta Q - \delta W$, is equal to a point function or a property. This property is designated dE, the total energy of the system. An alternate expression for the first law of thermodynamics may be written:

$$\delta Q - \delta W = dE \tag{6-4}$$

The signs of δQ and δW were specified in the original statement of the first law; δQ is positive when heat is added to the system, δW is positive when work is done by the system.

For a system undergoing a process occurring in time interval dt, equation (6-4) may be written as

$$\frac{\delta Q}{dt} - \frac{\delta W}{dt} = \frac{dE}{dt} \tag{6-5}$$

Consider now, as in Chapter 5, a general control volume fixed in inertial space located in a fluid-flow field, as shown in Figure 6.2. The system under consideration, designated by dashed lines, occupies the control volume at time t, and its position is also shown after a period of time Δt has elapsed.

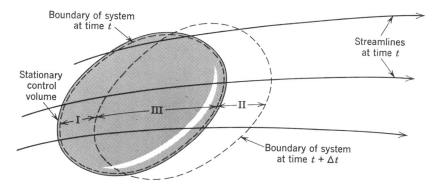

Figure 6.2 Relation between a system and a control volume in a fluid-flow field.

In this figure region I is occupied by the system at time t, region II is occupied by the system at $t + \Delta t$, and region III is common to the system both at t and at $t + \Delta t$.

At time $t + \Delta t$ the total energy of the system may be expressed as

$$E|_{t+\Delta t} = E_{II}|_{t+\Delta t} + E_{III}|_{t+\Delta t}$$

and at time t,

$$E|_t = E_I|_t + E_{III}|_t$$

Subtracting the second expression from the first and dividing by the elapsed time interval, Δt, we have

$$\frac{E|_{t+\Delta t} - E|_t}{\Delta t} = \frac{E_{III}|_{t+\Delta t} + E_{II}|_{t+\Delta t} - E_{III}|_t - E_I|_t}{\Delta t}$$

Rearranging and taking the limit as $\Delta t \to 0$ gives

$$\lim_{\Delta t \to 0} \frac{E|_{t+\Delta t} - E|_t}{\Delta t} = \lim_{\Delta t \to 0} \frac{E_{III}|_{t+\Delta t} - E_{III}|_t}{\Delta t} + \lim_{\Delta t \to 0} \frac{E_{II}|_{t+\Delta t} - E_I|_t}{\Delta t} \qquad (6\text{-}6)$$

Evaluating the limit of the left-hand side, we have

$$\lim_{\Delta t \to 0} \frac{E|_{t+\Delta t} - E|_t}{\Delta t} = \frac{dE}{dt}$$

which corresponds to the right-hand side of the first-law expression, equation (6-5). On the right-hand side of equation (6-6) the first limit becomes

$$\lim_{\Delta t \to 0} \frac{E_{III}|_{t+\Delta t} - E_{III}|_t}{\Delta t} = \frac{dE_{III}}{dt}$$

which is the rate of change of the total energy of the system, since the volume occupied by the system as $\Delta t \to 0$ is the control volume under consideration.

The second limit on the right of equation (6-6),

$$\lim_{\Delta t \to 0} \frac{E_{II}|_{t+\Delta t} - E_I|_t}{\Delta t}$$

represents the net rate of energy leaving across the control surface in the time interval Δt.

Having given physical meaning to each of the terms in equation (6-6), we may now recast the first law of thermodynamics into a form applicable to a control volume expressed by the following word equation:

$$\begin{Bmatrix} \text{Rate of addition} \\ \text{of heat to con-} \\ \text{trol volume from} \\ \text{its surroundings} \end{Bmatrix} - \begin{Bmatrix} \text{rate of work done} \\ \text{by control volume} \\ \text{on its surroundings} \end{Bmatrix} = \begin{Bmatrix} \text{rate of energy} \\ \text{out of control} \\ \text{volume due to} \\ \text{fluid flow} \end{Bmatrix}$$

$$- \begin{Bmatrix} \text{rate of energy into} \\ \text{control volume due} \\ \text{to fluid flow} \end{Bmatrix} + \begin{Bmatrix} \text{rate of accumulation} \\ \text{of energy within} \\ \text{control volume} \end{Bmatrix} \qquad (6\text{-}7)$$

Equation (6-7) will now be applied to the general control volume shown in Figure 6.3.

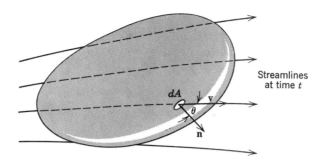

Figure 6.3   Fluid flow through a control volume.

The rates of heat addition to and work done by the control volume will be expressed as $\delta Q/dt$ and $\delta W/dt$.

Consider now the small area dA on the control surface. The rate of energy leaving the control volume through dA may be expressed as

$$\text{rate of energy efflux} = e(\rho v)(dA \cos \theta)$$

The product $(\rho v)(dA \cos \theta)$ is the rate of mass efflux from the control volume through dA, as discussed in the previous chapters. The quantity e is the specific energy or the energy per unit mass. The specific energy includes the potential energy, gy, due to the position of the fluid continuum in the gravitational field; the kinetic energy of the fluid, $v^2/2$, due to its velocity; and the internal energy, u, of the fluid due to its thermal state.

The quantity $dA \cos \theta$ represents the area, dA, projected normal to the velocity vector, $\mathbf{v}$. Theta (θ) is the angle between $\mathbf{v}$ and the outwardly directed normal vector, $\mathbf{n}$. We may now write

$$e(\rho v)(dA \cos \theta) = e\rho \, dA[|\mathbf{v}| \, |\mathbf{n}|] \cos \theta = e\rho(\mathbf{v} \cdot \mathbf{n}) \, dA$$

which we observe to be similar in form to the expressions previously obtained for mass and momentum. The integral of this quantity over the control surface,

$$\iint_{\text{c.s.}} e\rho(\mathbf{v} \cdot \mathbf{n}) \, dA$$

represents the *net efflux of energy* from the control volume. The sign of the scalar product, $\mathbf{v} \cdot \mathbf{n}$, accounts both for efflux and for influx of mass across the control surface as considered previously. Thus the first two terms on the right-hand side of equation (6-7) may be evaluated as

$$\begin{Bmatrix} \text{rate of energy} \\ \text{out of control} \\ \text{volume} \end{Bmatrix} - \begin{Bmatrix} \text{rate of energy} \\ \text{into control} \\ \text{volume} \end{Bmatrix} = \iint_{\text{c.s.}} e\rho(\mathbf{v} \cdot \mathbf{n}) \, dA$$

The rate of accumulation of energy within the control volume may be expressed as

$$\frac{\partial}{\partial t} \iiint_{c.v.} e\rho \, dV$$

Equation (6-7) may now be written as

$$\frac{\delta Q}{dt} - \frac{\delta W}{dt} = \iint_{c.s.} e\rho(\mathbf{v} \cdot \mathbf{n}) \, dA + \frac{\partial}{\partial t} \iiint_{c.v.} e\rho \, dV \qquad (6\text{-}8)$$

A final form for the first-law expression may be obtained after further consideration of the work-rate or power term, $\delta W/dt$.

There are three types of work included in the work-rate term. The first is the shaft work, W_s, which is that done by the control volume on its surroundings that could cause a shaft to rotate or accomplish the raising of a weight through a distance. A second kind of work done is flow work, W_σ, which is that done on the surroundings to overcome normal stresses on the control surface where there is fluid flow. The third type of work is designated shear work, W_τ, which is performed on the surroundings to overcome shear stresses at the control surface.

Examining our control volume for flow and shear work rates, we have, as shown on Figure 6.4, another effect on the elemental portion of control surface, dA. Vector $\mathbf{S}$ is the force intensity (stress) having components σ_{ii} and τ_{ij} in the directions normal and tangential to the surface, respectively. In terms of $\mathbf{S}$ the force on dA is $\mathbf{S} \, dA$, and the rate of work done by the fluid flowing through dA is $\mathbf{S} \, dA \cdot \mathbf{v}$.

The net rate of work done by the control volume on its surroundings due to the presence of $\mathbf{S}$ is

$$-\iint_{c.s.} \mathbf{v} \cdot \mathbf{S} \, dA$$

where the negative sign arises from the fact that the force per unit area *on the surroundings* is $-\mathbf{S}$.

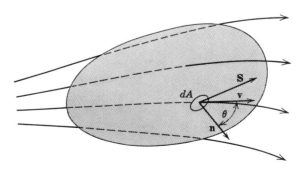

Figure 6.4 Flow and shear work for a general control volume.

The first-law expression, equation (6-8), may now be written as

$$\frac{\delta Q}{dt} - \frac{\delta W_s}{dt} + \iint_{c.s.} \mathbf{v} \cdot \mathbf{S} \, dA = \iint_{c.s.} e\rho(\mathbf{v} \cdot \mathbf{n}) \, dA + \frac{\partial}{\partial t} \iiint_{c.v.} e\rho \, dV \quad (6\text{-}9)$$

where $\delta W_s/dt$ is the shaft work rate.

Writing the normal stress components of $\mathbf{S}$ as $\sigma_{ii}\mathbf{n}$, we obtain, for the net rate of work done in overcoming normal stress,

$$\left(\iint_{c.s.} \mathbf{v} \cdot \mathbf{S} \, dA \right)_{\text{normal}} = \iint_{c.s.} \mathbf{v} \cdot \sigma_{ii}\mathbf{n} \, dA = \iint_{c.s.} \sigma_{ii}(\mathbf{v} \cdot \mathbf{n}) \, dA$$

The remaining part of the work to be evaluated is the part necessary to overcome shearing stresses. This portion of the required work rate, $\delta W_\tau/dt$, is transformed into a form which is unavailable to do mechanical work. This term, representing a loss of mechanical energy, will be included in the derivative form given above and its analysis will be included in example 3, to follow. The work rate now becomes

$$\frac{\delta W}{dt} = \frac{\delta W_s}{dt} + \frac{\delta W_\sigma}{dt} + \frac{\delta W_\tau}{dt}$$

$$= \frac{\delta W_s}{dt} - \iint_{c.s.} \sigma_{ii}(\mathbf{v} \cdot \mathbf{n}) \, dA + \frac{\delta W_\tau}{dt}$$

Substituting into equation (6-9), we have

$$\frac{\delta Q}{dt} - \frac{\delta W_s}{dt} + \iint_{c.s.} \sigma_{ii}(\mathbf{v} \cdot \mathbf{n}) \, dA - \frac{\delta W_\tau}{dt} = \iint_{c.s.} e\rho(\mathbf{v} \cdot \mathbf{n}) \, dA + \frac{\partial}{\partial t} \iiint_{c.v.} e\rho \, dV$$

The term involving normal stress must now be presented in a more usable form. A complete expression for σ_{ii} will be stated in Chapter 9. For the present we may say simply that the normal stress term is the sum of pressure effects and viscous effects. Just as with shear work, the work done to overcome the viscous portion of the normal stress is unavailable to do mechanical work. We shall thus combine the work associated with the viscous portion of the normal stress with the shear work to give a single term, $\delta W_\mu/dt$, the work rate accomplished in overcoming viscous effects at the control surface. The subscript, μ, is used to make this distinction.

The remaining part of the normal stress term, that associated with pressure, may be written in slightly different form if we recall that the bulk stress, $\bar{\sigma}_{ii}$, is the negative of the thermodynamic pressure, P. The shear and flow work terms may now be written as follows:

$$\iint_{c.s.} \sigma_{ii}(\mathbf{v} \cdot \mathbf{n}) \, dA - \frac{\delta W_\tau}{dt} = -\iint_{c.s.} P(\mathbf{v} \cdot \mathbf{n}) \, dA - \frac{\delta W_\mu}{dt}$$

Combining this equation with the one written previously and rearranging slightly will yield the final form of the first-law expression:

$$\frac{\delta Q}{dt} - \frac{\delta W_s}{dt} = \iint_{c.s.} \left(e + \frac{P}{\rho}\right)\rho(\mathbf{v} \cdot \mathbf{n})\, dA + \frac{\partial}{\partial t}\iiint_{c.v.} e\rho\, dV + \frac{\delta W_\mu}{dt} \qquad (6\text{-}10)$$

Equation (6-10), equation (4-1), and equation (5-4) constitute the basic relations for the analysis of fluid flow via the control-volume approach. A thorough understanding of these three equations and a mastery of their application places at the disposal of the student very powerful means of analyzing many commonly encountered problems in fluid flow.

The use of the overall energy balance will be illustrated in the following example problems.

6.2 APPLICATIONS OF THE INTEGRAL EXPRESSION

EXAMPLE 1

As a first example let us choose a control volume as shown in Figure 6.5 under the conditions of steady fluid flow and no frictional losses.

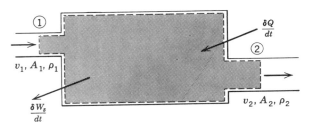

Figure 6.5 Control volume with one-dimensional flow across boundaries.

For the specified conditions the overall energy expression, equation (6-10), becomes

$$\frac{\delta Q}{dt} - \frac{\delta W_s}{dt} = \iint_{c.s.} \rho\left(e + \frac{P}{\rho}\right)(\mathbf{v} \cdot \mathbf{n})\, dA + \underbrace{\frac{\partial}{\partial t}\iiint_{c.v.} e\rho\, dV}_{0\text{--steady flow}} + \underbrace{\frac{\delta W_\mu}{dt}}_{0}$$

Considering now the surface integral, we recognize the product $\rho(\mathbf{v} \cdot \mathbf{n})\, dA$ to be the mass flow rate with the sign of this product indicating whether mass flow is into or out of the control volume, dependent upon the sense of $\mathbf{v} \cdot \mathbf{n}$. The factor by which the mass flow rate is multiplied, $e + P/\rho$, represents the types of energy that may enter or leave the control volume per mass of fluid. The specific total energy, e, may be expanded to include the kinetic, potential, and internal energy contributions, so that

$$e + \frac{P}{\rho} = gy + \frac{v^2}{2} + u + \frac{P}{\rho}$$

Since mass enters the control volume only at section (1) and leaves at section (2), the surface integral becomes

$$\iint_{c.s.} \rho\left(e+\frac{P}{\rho}\right)(\mathbf{v}\cdot\mathbf{n})\,dA = \left[\frac{v_2{}^2}{2}+gy_2+u_2+\frac{P_2}{\rho_2}\right](\rho_2 v_2 A_2)$$

$$-\left[\frac{v_1{}^2}{2}+gy_1+u_1+\frac{P_1}{\rho_1}\right](\rho_1 v_1 A_1)$$

The energy expression for this example now becomes

$$\frac{\delta Q}{dt}-\frac{\delta W_s}{dt} = \left[\frac{v_2{}^2}{2}+gy_2+u_2+\frac{P_2}{\rho_2}\right](\rho_2 v_2 A_2)$$

$$-\left[\frac{v_1{}^2}{2}+gy_1+u_1+\frac{P_1}{\rho_1}\right](\rho_1 v_1 A_1)$$

In Chapter 4 the mass balance for this same situation was found to be

$$\rho_1 v_1 A_1 = \rho_2 v_2 A_2$$

If each term in the above expression is now divided by the mass flow rate, we have

$$\frac{q-\dot W_s}{\dot m} = \left[\frac{v_2{}^2}{2}+gy_2+u_2+\frac{P_2}{\rho_2}\right]-\left[\frac{v_1{}^2}{2}+gy_1+u_1+\frac{P_1}{\rho_1}\right]$$

or, in more familiar form,

$$\frac{v_1{}^2}{2}+gy_1+h_1+\frac{q}{\dot m} = \frac{v_2{}^2}{2}+gy_2+h_2+\frac{\dot W_s}{\dot m}$$

where the sum of the internal energy and flow energy, $u+P/\rho$, has been replaced by the enthalpy, h, which is equal to the sum of these quantities by definition $h \equiv u+P/\rho$.

EXAMPLE 2

As a second example, consider the situation shown in Figure 6.6. If water flows under steady conditions in which the pump delivers 3 horsepower to the fluid, find the mass flow rate if frictional losses may be neglected.

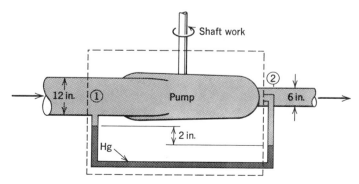

Figure 6.6   A control volume for pump analysis.

Defining the control volume as shown by the dashed lines, we may evaluate equation (6-10) term by term as follows:

$$\frac{\delta Q}{dt} = 0$$

$$\frac{\delta W_s}{dt} = (3 \text{ hp})(2545 \text{ Btu/hp-hr})(778 \text{ ft-lb}_f/\text{Btu})(\text{hr}/3600 \text{ s})$$

$$= 1650 \text{ ft lb}_f/\text{s}$$

$$\iint_{c.s.} \left(e + \frac{P}{\rho}\right)\rho(\mathbf{v} \cdot \mathbf{n})\, dA = \iint_{A_2} \left(e + \frac{P}{\rho}\right)\rho(\mathbf{v} \cdot \mathbf{n})\, dA + \iint_{A_1} \left(e + \frac{P}{\rho}\right)\rho(\mathbf{v} \cdot \mathbf{n})\, dA$$

$$= \left(\frac{v_2^2}{2} + gy_2 + u_2 + \frac{P_2}{\rho_2}\right)(\rho_2 v_2 A_2)$$

$$- \left(\frac{v_1^2}{2} + gy_1 + u_1 + \frac{P_1}{\rho_1}\right)(\rho_1 v_1 A_1)$$

$$= \left[\frac{v_2^2 - v_1^2}{2} + g(y_2 - y_1) + (u_2 - u_1)\right.$$

$$\left. + \left(\frac{P_2}{\rho_2} - \frac{P_1}{\rho_1}\right)\right](\rho v A)$$

Here it may be noted that the pressure measured at station (1) is the static pressure while the pressure measured at station (2) is the stagnation pressure defined by $P_{\text{stagnation}} \equiv P_0 \equiv P_{\text{static}} + \frac{1}{2}\rho V^2$ for an incompressible fluid—hence we may rewrite the energy flux term as

$$\iint_{c.s.} \left(e + \frac{P}{\rho}\right)\rho(\mathbf{v} \cdot \mathbf{n})\, dA = \left(\frac{P_{0_2} - P_1}{\rho} - \frac{v_1^2}{2}\right)(\rho v A)$$

$$= \left\{\frac{2(1 - 1/13.6) \text{ in. Hg}(14.7 \text{ lb/in.}^2)(144 \text{ in.}^2/\text{ft}^2)}{(62.4 \text{ lb}_m/\text{ft}^3)(29.92 \text{ in. Hg})}\right.$$

$$\left. - \frac{v_1^2}{64.4 \,(\text{lb}_m \text{ ft/s}^2 \text{ lb}_f)}\right\} \{(62.4 \text{ lb}_m/\text{ft}^3)(v_1)(\pi/4 \text{ ft}^2)\}$$

$$= \left(2.21 - \frac{v_1^2}{6.64}\right)(49v_1) \text{ ft lb}_f/\text{s}$$

$$\frac{\partial}{\partial t} \iiint_{c.v.} e\rho\, dV = 0$$

$$\frac{\delta W_\mu}{dt} = 0$$

In the evaluation of the surface integral the choice of the control volume coincided with the location of the pressure taps at sections (1) and (2). The pressure sensed at section (1) is the static pressure, since the manometer opening is parallel to the fluid-flow direction. At section (2), however, the manometer opening is normal to the flowing fluid stream. The pressure measured by such an arrangement includes both the static fluid pressure and the pressure resulting as a fluid flowing with velocity v_2 is brought to rest. The sum of these two quantities is known as the impact or stagnation pressure.

The potential energy change is zero between sections (1) and (2) and since we consider the flow to be isothermal, the variation in internal energy is also zero. Hence the surface integral reduces to the simple form indicated.

The flow rate of water necessary for the stated conditions to exist is achieved by solving the resulting cubic equation. The solution is

$$v_1 = 16.65 \text{ fps } (5.075 \text{ m/s})$$

$$\dot{m} = \rho A v = 815 \text{ lb}_m/\text{s } (370 \text{ kg/s})$$

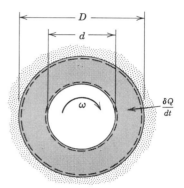

EXAMPLE 3

A shaft is rotating at constant angular velocity ω in the bearing shown in Figure 6.7. The shaft diameter is d and the shear stress acting on the shaft is τ. Find the rate at which energy must be removed from the bearing in order that the lubricating oil between the rotating shaft and the stationary bearing surface remains at constant temperature. The shaft is assumed to be lightly loaded and concentric with the journal.

Figure 6.7 Bearing and control volume for bearing analysis.

The control volume selected consists of a unit length of the fluid surrounding the shaft as shown in Figure 6.7. The first law of thermodynamics for the control volume is

$$\frac{\delta Q}{dt} - \frac{\delta W_s}{dt} = \iint_{\text{c.s.}} \rho\left(e + \frac{P}{\rho}\right)(\mathbf{v} \cdot \mathbf{n}) \, dA + \frac{\partial}{\partial t}\iiint_{\text{c.v.}} \rho e \, dV + \frac{\delta W_\mu}{dt}$$

From the figure we may observe the following:

1. No fluid crosses the control surface.
2. No shaft work crosses the control surface.
3. The flow is steady.

Thus $\delta Q/dt = \delta W_\mu/dt = \delta W_\tau/dt$. The viscous work rate must be determined. In this case all of the viscous work is done to overcome shearing stresses; thus the viscous work is $\iint_{\text{c.s.}} \tau(\mathbf{v} \cdot \mathbf{e}_t) \, dA$. At the outer boundary, $v = 0$ and at the inner boundary $\iint_{\text{c.s.}} \tau(\mathbf{v} \cdot \mathbf{e}_t) \, dA = -\tau(\omega d/2)A$, where $\mathbf{e}_t$ indicates the sense of the shear stress, τ, on the surroundings. The resulting sign is consistent with the concept of work being positive when done by a system on its surroundings. Thus

$$\frac{\delta Q}{dt} = -\tau\frac{\omega d^2 \pi}{2}$$

which is the heat transfer rate required to maintain the oil at a constant temperature.

If energy is not removed from the system then $\delta Q/dt = 0$, and

$$\frac{\partial}{\partial t}\iiint_{\text{c.v.}} e\rho \, dV = -\frac{\delta W_\mu}{dt}$$

As only the internal energy of the oil will increase with respect to time,

$$\frac{\partial}{\partial t}\iiint_{\text{c.v.}} e\rho \, dV = \rho\pi\left(\frac{D^2 - d^2}{4}\right)\frac{du}{dt} = -\frac{\delta W_\mu}{dt} = \omega\frac{d^2\pi}{2}\tau$$

or, with constant specific heat c,

$$c\frac{dT}{dt} = \frac{2\tau\omega\, d^2}{\rho(D^2 - d^2)}$$

where D is the outer bearing diameter.

In this example the use of the viscous-work term has been illustrated. Note that

1. The viscous-work term involves only quantities on the surface of the control volume.
2. When the velocity on the surface of the control volume is zero, the viscous-work term is zero.

6.3   THE BERNOULLI EQUATION

Under certain flow conditions the expression of the first law of thermodynamics applied to a control volume reduces to an extremely useful relation known as the Bernoulli equation.

If equation (6-10) is applied to a control volume as shown in Figure 6.8, in which flow is steady, incompressible, and inviscid, and in which no heat transfer or change in internal energy occurs, a term-by-term evaluation of equation (6-10) gives the following:

$$\frac{\delta Q}{dt} = 0$$

$$\frac{\delta W_s}{dt} = 0$$

$$\iint_{c.s.} \rho\left(e + \frac{P}{\rho}\right)(\mathbf{v} \cdot \mathbf{n})\, dA = \iint_{A_1} \rho\left(e + \frac{P}{\rho}\right)(\mathbf{v} \cdot \mathbf{n})\, dA$$

$$+ \iint_{A_2} \rho\left(e + \frac{P}{\rho}\right)(\mathbf{v} \cdot \mathbf{n})\, dA$$

$$= \left(gy_1 + \frac{v_1^2}{2} + \frac{P_1}{\rho_1}\right)(-\rho_1 v_1 A_1)$$

$$+ \left(gy_2 + \frac{v_2^2}{2} + \frac{P_2}{\rho_2}\right)(\rho_2 v_2 A_2)$$

$$\frac{\partial}{\partial t} \iiint_{c.v.} e\rho\, dV = 0$$

The first-law expression now becomes

$$0 = \left(gy_2 + \frac{v_2^2}{2} + \frac{P_2}{\rho}\right)(\rho v_2 A_2) - \left(gy_1 + \frac{v_1^2}{2} + \frac{P_1}{\rho}\right)(\rho v_1 A_1)$$

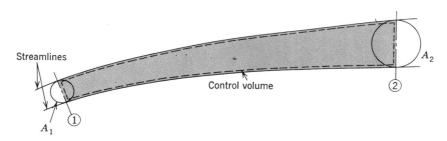

Figure 6.8 Control volume for steady, incompressible, inviscid, isothermal flow.

Since flow is steady, the continuity equation gives

$$\rho_1 v_1 A_1 = \rho_2 v_2 A_2$$

which may be divided through to give

$$gy_1 + \frac{v_1^2}{2} + \frac{P_1}{\rho} = gy_2 + \frac{v_2^2}{2} + \frac{P_2}{\rho} \qquad (6\text{-}11\text{a})$$

Dividing through by g, we have

$$y_1 + \frac{v_1^2}{2g} + \frac{P_1}{\rho g} = y_2 + \frac{v_2^2}{2g} + \frac{P_2}{\rho g} \qquad (6\text{-}11\text{b})$$

Either of the above expressions is designated the Bernoulli equation.

Note that each term in equation (6-11b) has the unit of length. The quantities are often designated "heads" due to elevation, velocity, and pressure, respectively. These terms, both individually and collectively, indicate the quantities which may be directly converted to produce mechanical energy.

Equation (6-11) may be interpreted physically to mean that the total mechanical energy is conserved for a control volume satisfying the conditions upon which this relation is based, that is, steady, incompressible, inviscid, isothermal flow, with no heat transfer or work done. These conditions may seem overly restrictive, but they are met, or approached, in many physical systems. One such situation of practical value is for flow into and out of a stream tube. Since stream tubes may vary in size, the Bernoulli equation can actually describe the variation in elevation, velocity, and pressure head from point-to-point in a fluid-flow field. Such a consideration might also be expected to follow from a differential approach to the general laws of fluid flow, and this is indeed the case. This result will be shown in Chapter 9.

A classic example of the application of the Bernoulli equation is depicted in Figure 6.9, in which it is desired to find the velocity of the fluid exiting the tank as shown.

The control volume is defined as shown by dashed lines in the figure. The upper boundary of the control volume is just below the fluid surface, and thus can be considered to be at the same height as the fluid. There is fluid flow across this

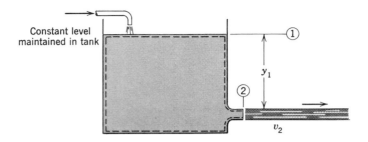

Figure 6.9   Control volume for Bernoulli equation analysis.

surface, but the surface area is large enough that the velocity of this flowing fluid may be considered negligible.

Under these conditions the proper form of the first law of thermodynamics is equation (6-11), the Bernoulli equation. Applying equation (6-11), we have

$$y_1 + \frac{P_{atm}}{\rho g} = \frac{v_2{}^2}{2g} + \frac{P_{atm}}{\rho g}$$

from which the exiting velocity may be expressed in the familiar form

$$v_2 = \sqrt{2gy}$$

As a final illustration of the use of the control volume relations, an example using all three expressions is presented below.

EXAMPLE 4

In the sudden enlargement shown below in Figure 6.10, the pressure acting at section (1) is considered uniform with value P_1. Find the change in internal energy between stations (1) and (2) for steady, incompressible flow. Neglect shear stress at the walls and express $u_2 - u_1$ in terms of v_1, A_1, and A_2. The control volume selected is indicated by the dotted line.

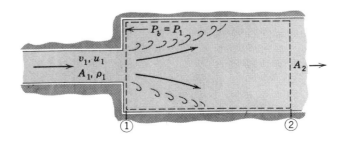

Figure 6.10   Flow through a sudden enlargement.

Conservation of mass

$$\iint_{c.s.} \rho(\mathbf{v} \cdot \mathbf{n}) \, dA + \frac{\partial}{\partial t} \iiint_{c.v.} \rho \, dV = 0$$

If we select station (2) a considerable distance downstream from the sudden enlargement, the continuity expression, for steady, incompressible flow, becomes

$$\rho_1 v_1 A_1 = \rho_2 v_2 A_2$$

or

$$v_2 = v_1 \frac{A_1}{A_2} \qquad (6\text{-}12)$$

Momentum

$$\sum \mathbf{F} = \iint_{c.s.} \rho \mathbf{v}(\mathbf{v} \cdot \mathbf{n}) \, dA + \frac{\partial}{\partial t} \iiint_{c.v.} \rho \mathbf{v} \, dV$$

and thus

$$P_1 A_2 - P_2 A_2 = \rho v_2^2 A_2 - \rho v_1^2 A_1$$

or

$$\frac{P_1 - P_2}{\rho} = v_2^2 - v_1^2 \left(\frac{A_1}{A_2} \right) \qquad (6\text{-}13)$$

Energy

$$\frac{\delta Q}{\partial t} - \frac{\delta W_s}{dt} = \iint_{c.s.} \rho \left(e + \frac{P}{\rho} \right)(\mathbf{v} \cdot \mathbf{n}) \, dA + \frac{\partial}{\partial t} \iiint_{c.v.} \rho e \, dV + \frac{\delta W_\mu}{dt}$$

Thus

$$\left(e_1 + \frac{P_1}{\rho} \right)(\rho v_1 A_1) = \left(e_2 + \frac{P_2}{\rho} \right)(\rho v_2 A_2)$$

or, since $\rho v_1 A_1 = \rho v_2 A_2$,

$$e_1 + \frac{P_1}{\rho} = e_2 + \frac{P_2}{\rho}$$

The specific energy is

$$e = \frac{v^2}{2} + gy + u$$

Thus our energy expression becomes

$$\frac{v_1^2}{2} + gy_1 + u_1 + \frac{P_1}{\rho} = \frac{v_2^2}{2} + gy_2 + u_2 + \frac{P_2}{\rho} \qquad (6\text{-}14)$$

The three control volume expressions may now be combined to evaluate $u_2 - u_1$. From (6-14), we have

$$u_2 - u_1 = \frac{P_1 - P_2}{\rho} + \frac{v_1^2 - v_2^2}{2} + g(y_1 - y_2) \qquad (6\text{-}14a)$$

Substituting (6-13) for $(P_1 - P_2)/\rho$ and (6-12) for v_2 and noting that $y_1 = y_2$, we obtain

$$u_2 - u_1 = v_1{}^2 \left(\frac{A_1}{A_2}\right)^2 - v_1{}^2 \frac{A_1}{A_2} + \frac{v_1{}^2}{2} - \frac{v_1{}^2}{2}\left(\frac{A_1}{A_2}\right)^2$$

$$= \frac{v_1{}^2}{2}\left[1 - 2\frac{A_1}{A_2} + \left(\frac{A_1}{A_2}\right)^2\right] = \frac{v_1{}^2}{2}\left[1 - \frac{A_1}{A_2}\right]^2 \qquad (6\text{-}15)$$

Equation (6-15) shows that the internal energy increases in a sudden enlargement. The temperature change corresponding to this change in internal energy is insignificant, but from (6-14a) it can be seen that the change in total head,

$$\left(\frac{P_1}{\rho} + \frac{v_1{}^2}{2} + gy_1\right) - \left(\frac{P_2}{\rho} + \frac{v_2{}^2}{2} + gy_2\right)$$

is equal to the internal energy change. Accordingly the internal energy change in an incompressible flow is designated as the head loss, h_L, and the energy equation for steady, adiabatic, incompressible flow in a stream tube is written as

$$\frac{P_1}{\rho g} + \frac{v_1{}^2}{2g} + y_1 = h_L + \frac{P_2}{\rho g} + \frac{v_2{}^2}{2g} + y_2 \qquad (6\text{-}16)$$

Note the similarity to equation (6-11).

6.4 CLOSURE

In this chapter the first law of thermodynamics, the third of the fundamental relations upon which fluid-flow analyses are based, has been used to develop an integral expression for the conservation of energy with respect to a control volume. The resulting expression, equation (6-10), is, in conjunction with equations (4-1) and (5-4), one of the fundamental expressions for the control-volume analysis of fluid-flow problems.

A special case of the integral expression for the conservation of energy is the Bernoulli equation, equation (6-11). Although simple in form and use, this expression has broad application to physical situations.

PROBLEMS

6.1 Sea water, $\rho = 1025 \text{ kg/m}^3$, flows through a pump at $0.14 \text{ m}^3/\text{s}$. The pump inlet is 0.25 m in diameter. At the inlet the pressure is -0.15 m of mercury. The pump outlet, 0.152 m in diameter, is 1.8 m above the inlet. The outlet pressure is 175 kPa. If the inlet and exit temperature are equal, how much power does the pump add to the fluid?

6.2 A liquid is heated in a vertical tube of constant diameter, 15 m long. The flow is upward. At the entrance the average velocity is 1 m/s, the pressure 340 000 Pa, and the density is 1001 kg/m^3. If the increase in internal energy is 200 000 J/kg, find the heat added to the fluid.

6.3 Air at 70°F flows into a 10-ft^3 reservoir at a velocity of 90 fps. If the reservoir pressure is 14 psig and the reservoir temperature 70°F, find the rate of temperature increase in the reservoir. Assume the incoming air is at reservoir pressure and flows through a 8-in.-diameter pipe.

6.4 Water flows through a 2-in.-diameter horizontal pipe at a flow rate of 35 gal/min. The heat transfer to the pipe can be neglected, and frictional forces cause a pressure drop of 10 psi. What is the temperature change of the water?

6.5 During the flow of 200 ft^3/s of water through the hydraulic turbine shown, the pressure indicated by gage A is 12 psig. What should gage B read if the turbine is delivering 600 hp at 82% efficiency? Gage B is designed to measure the total pressure, that is, $P + \rho v^2 / 2$ for an incompressible fluid.

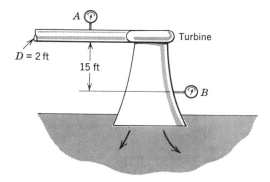

6.6 During the test of a centrifugal pump, a Bourdon pressure gage just outside the casing of the 12-in.-diameter suction pipe reads −6 psig (i.e., vacuum). On the 10-in.-diameter discharge pipe another gage reads 40 psig. The discharge pipe is 5 ft above the suction pipe. The discharge of water through the pump is measured to be 4 ft^3/s. Compute the horsepower input of the test pump.

6.7 A fan draws air from the atmosphere through a 0.30-m-diameter round duct that has a smoothly rounded entrance. A differential manometer connected to an opening in the wall of the duct shows a vacuum pressure of 2.5 cm of water. The density of air is 1.22 kg/m^3. Determine the volume rate of air

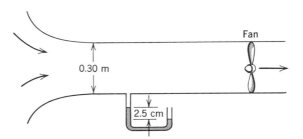

flow in the duct in cubic feet per second. What is the horsepower output of the fan?

6.8 Find the change in temperature between stations (1) and (2) in terms of the quantities A_1, A_3, v_1, v_3, c_v, and θ. The internal energy is given by $c_v T$. The fluid is water and $T_1 = T_3$, $P_1 = P_3$.

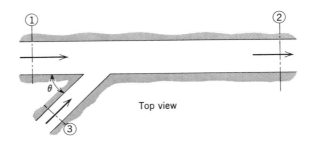

Top view

6.9 A liquid flows from A to B in horizontal pipe line shown at a rate of 2 ft³/s with a friction loss of 0.45 ft of flowing fluid. For a pressure head at B of 24 in., what will be the pressure head at A?

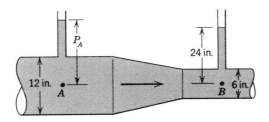

6.10 Water flows steadily up the vertical pipe and is then deflected to flow outward with a uniform radial velocity. If friction is neglected, what is the flow rate of water through the pipe if the pressure at A is 10 psig?

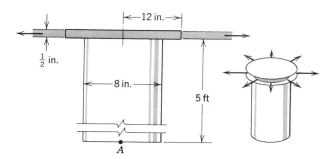

6.11 In problem 6.10 compute the upward force on the device from water and air. Use the results of problem 6.10 as well as any other data given in that problem that you may need. Explain why you cannot profitably use Bernoulli's equation here for a force calculation.

6.12 A Venturi meter with an inlet diameter of 0.6 m is designed to handle 6 m^3/s of standard air. What is the required throat diameter if this flow is to give a reading of 0.10 m of alcohol in a differential manometer connected to the inlet and the throat? The specific gravity of alcohol may be taken as 0.8.

6.13 Water flows through the pipe contraction shown at a rate of 1 ft^3/s. Calculate the differential manometer reading in inches of mercury, assuming no energy loss in the flow. Be sure to give the correct direction of the manometer reading.

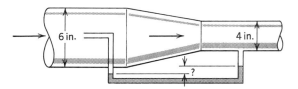

6.14 The figure illustrates the operation of an air lift pump. Compressed air is forced into a perforated chamber to mix with the water so that the specific

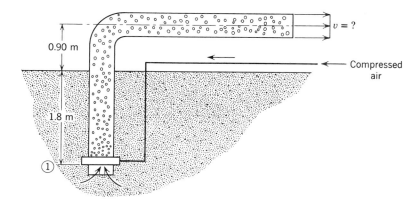

gravity of the air–water mixture above the air inlet is 0.5. Neglecting any pressure drop across section (1), compute the discharge velocity v of the air-water mixture. Can Bernoulli's equation be used across section (1)?

6.15 The pressurized tank shown has a circular cross section of 6 ft in diameter. Oil is drained through a nozzle 2 in. in diameter in the side of the tank. Assuming that the air pressure is maintained constant, how long does it take to lower the oil surface in the tank by 2 ft? The specific gravity of the oil in the tank is 0.85 and that of mercury is 13.6.

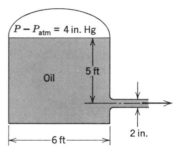

6.16 Air of density 1.21 kg/m^3 is flowing as shown. If $v = 15$ m/s, determine the readings on manometers a and b in the figures below.

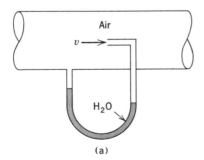
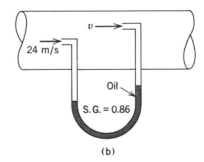

6.17 Referring to the figure, assume the flow to be frictionless in the siphon. Find the rate of discharge in cubic feet per second, and the pressure head at B if the pipe has a uniform diameter of 1 in. How long will it take for the water level to decrease by 3 ft? The tank diameter is 10 ft.

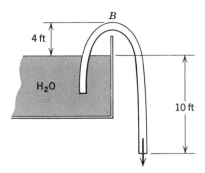

6.18 An automobile is driving into a 45-mph headwind at 45 mph. If the barometer reads 29 in. Hg and the temperature is 40°F, what is the pressure at a point on the auto where the wind velocity is 120 fps with respect to the auto?

6.19 Water is discharged from a 1.27-cm-diameter nozzle which is inclined at a 30° angle above the horizontal. If the jet strikes the ground at a horizontal distance of 3.66 m and a vertical distance of 0.6 m from the nozzle as shown, what is the rate of flow in cubic meters per second? What is the total head of the jet? (See equation 6-11b.)

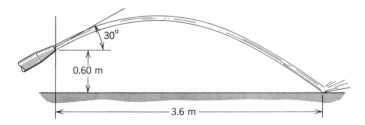

6.20 Assume that the level of water in the tank remains the same and that there is no friction loss in the pipe, entrance, or nozzle. Determine:
(a) the volumetric discharge rate from the nozzle;
(b) the pressure and velocity at points A, B, C, and D.

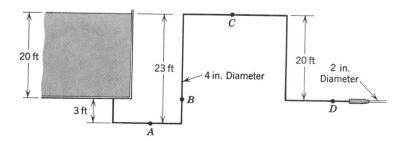

6.21 The pump shown in the figure delivers water at 59°F at a rate of 600 gal/min. The inlet pipe has an inside diameter of 5.95 in. and it is 10 ft long. The inlet pipe is submerged 6 ft into the water and is vertical. Estimate the pressure inside the pipe at the pump inlet.

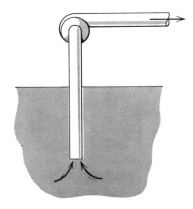

6.22 In the previous problem, determine the flow rate at which the pump inlet pressure is equal to the vapor pressure of the water. Assume that friction causes a head loss of 4 ft. The vapor pressure of water at 59°F is 0.247 psi.

6.23 Using the data of problem 5.20, determine the velocity head of the fluid leaving the impeller. What pressure rise would result from such a velocity head?

6.24 In order to maneuver a large ship while docking, pumps are used to issue a jet of water perpendicular to the bow of the ship as shown in the figure. The

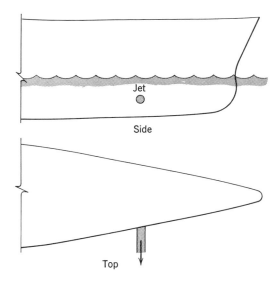

pump inlet is located far enough away from the outlet that the inlet and outlet do not interact. The inlet is also vertical so that the net thrust of the jets on the ship is independent of the inlet velocity and pressure. Determine the pump horsepower required per pound of thrust. Assume that the inlet and outlet are at the same depth. Which will produce more thrust per horsepower, a low-volume, high-pressure pump or a high-volume, low-pressure pump?

6.25 Determine the head loss between stations (1) and (2) in problem 5.7.

6.26 Rework problem 6.14 with the assumption that the momentum of the incoming air at section (1) is zero. Determine the exit velocity v and the magnitude of the pressure drop at section (1).

6.27 Water in an open cylindrical tank 10 ft in diameter discharges into the atmosphere through a nozzle 2 in. in diameter. Neglecting friction and the unsteadiness of the flow, find the time required for the water in the tank to drop from a level of 28 ft above the nozzle to the 4-ft level.

6.28 A fluid of density ρ_1 enters a chamber where the fluid is heated so that the density decreases to ρ_2. The fluid then escapes through a vertical chimney that has a height L. Neglecting friction and treating the flow processes as incompressible except for the heating, determine the velocity, V, in the stack. The fluid velocity entering the heating chamber may be neglected and the chimney is immersed in fluid of density ρ_1.

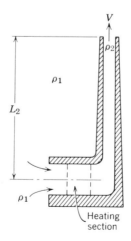

6.29 Repeat the previous problem without the assumption that the velocity in the heating section is negligible. The ratio of the flow area of the heating section to the chimney flow area is R.

7
SHEAR STRESS
IN LAMINAR FLOW

In the analysis of fluid flow thus far, shear stress has been mentioned, but it has not been related to the fluid or flow properties. We shall now investigate this relation for laminar flow. The shear stress acting on a fluid depends upon the type of flow that exists. In so-called laminar flow, the fluid flows in smooth layers or lamina, and the shear stress is the result of the (non-observable) microscopic action of the molecules. Turbulent flow is characterized by the large scale, observable fluctuations in fluid and flow properties, and the shear stress is the result of these fluctuations. The criteria for laminar and turbulent flows will be discussed in Chapters 12 and 13. The shear stress in turbulent flow will be discussed in Chapter 13.

7.1 NEWTON'S VISCOSITY RELATION

In a solid, the resistance to *deformation* is the modulus of elasticity. The shear modulus of an elastic solid is given by

$$\text{shear modulus} = \frac{\text{shear stress}}{\text{shear strain}} \tag{7-1}$$

Just as the shear modulus of an elastic solid is a property of the solid relating shear stress and shear strain, there exists a relation similar to (7-1) which relates the shear stress in a parallel, laminar flow to a property of the fluid. This relation is Newton's law of viscosity,

$$\text{viscosity} = \frac{\text{shear stress}}{\text{rate of shear strain}} \tag{7-2}$$

Thus the viscosity is the property of a fluid to resist the *rate* at which deformation takes place when the fluid is acted upon by shear forces. As a property of the fluid,

96

the viscosity depends upon the temperature, composition, and pressure of the fluid, but is independent of the rate of shear strain.

The rate of deformation in a simple flow is illustrated in Figure 7.1. The flow parallel to the x axis will deform the element if the velocity at the top of the element is different than the velocity at the bottom.

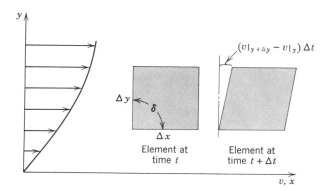

Figure 7.1   Deformation of a fluid element.

The rate of shear strain at a point is defined as $-d\delta/dt$. From Figure 7.1, it may be seen that

$$-\frac{d\delta}{dt} = -\lim_{\Delta x, \Delta y, \Delta t \to 0} \frac{\delta|_{t+\Delta t} - \delta|_t}{\Delta t}$$

$$= \lim_{\Delta x, \Delta y, \Delta t \to 0} \left(\frac{\{\pi/2 - \arctan\left[(v|_{y+\Delta y} - v|_y)\,\Delta t/\Delta y\right]\} - \pi/2}{\Delta t} \right)$$

In the limit, $-d\delta/dt = dv/dy =$ rate of shear strain   (7-3)

Combining equations (7-2) and (7-3) and denoting the viscosity by μ, we may write Newton's law of viscosity as

$$\tau = \mu \frac{dv}{dy} \tag{7-4}$$

The velocity profile and shear stress variation in a fluid flowing between two parallel plates is illustrated in Figure 7.2. The velocity profile* in this case is parabolic; as the shear stress is proportional to the derivative of the velocity, the shear stress varies in a linear manner.

---

* The derivation of velocity profiles is discussed in Chapter 8.

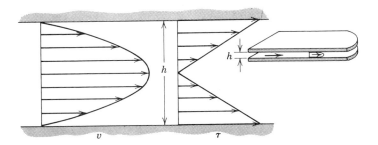

Figure 7.2 Velocity and shear stress profiles for flow between two parallel plates.

7.2 NON-NEWTONIAN FLUIDS

Newton's law of viscosity does not predict the shear stress in all fluids. Fluids are classified as newtonian or non-newtonian, depending upon the relation between shear stress and the rate of shearing strain. In newtonian fluids the relation is linear, as shown in Figure 7.3.

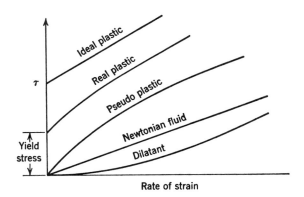

Figure 7.3 Stress rate-of-strain relation for newtonian and non-newtonian fluids.

In non-newtonian fluids, the shear stress depends upon the rate of shear strain. While fluids deform continuously under the action of shear stress, plastics will sustain a shear stress before deformation occurs. The "ideal plastic" has a linear stress rate-of-strain relation for stresses greater than the yield stress. *Thixotropic* substances such as printer's ink have a resistance to deformation that depends upon deformation rate and time.

THE NO-SLIP CONDITION

While the substances above differ in their stress rate-of-strain relations, they are similar in their action at a boundary. In both newtonian and non-newtonian fluids, the layer of fluid adjacent to the boundary has zero velocity relative to the boundary. When the boundary is a stationary wall, the layer of fluid next to the wall is at rest. If the boundary or wall is moving, the layer of fluid moves at the velocity of the boundary, hence the name no-slip (boundary) condition. The no-slip condition is the result of experimental observation and fails when the fluid no longer can be treated as a continuum.

The no-slip condition is a result of the viscous nature of the fluid. In flow situations in which the viscous effects are neglected—so called inviscid flows— only the component of the velocity normal to the boundary is zero.

7.3 VISCOSITY

The viscosity of a fluid is a measure of its resistance to deformation rate. Tar and molasses are examples of highly viscous fluids; air and water, which are the subject of frequent engineering interest, are examples of fluids with relatively low viscosities. An understanding of the existence of the viscosity requires an examination of the motion of fluid on a molecular basis.

The molecular motion of gases can be described more simply than that of liquids. The mechanism by which a gas resists deformation may be illustrated by examination of the motion of the molecules on a microscopic basis. Consider the control volume shown in Figure 7.4.

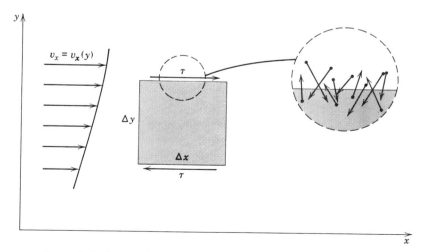

Figure 7.4 Molecular motion at the surface of a control volume.

The top of the control volume is enlarged to show that, even though the top of the element is a streamline of the flow, individual molecules cross this plane. The paths of the molecules between collisions are represented by the random arrows. Because the top of the control volume is a streamline, the net molecular flux across this surface must be zero, hence the upward molecular flux must equal the downward molecular flux. The molecules that cross the control surface in an upward direction have average velocities in the x direction corresponding to their points of origin. Denoting the y coordinate of the top of the control surface as y_0, we shall write the x-directional average velocity of the upward molecular flux as $v_x|_{y-}$, where the minus sign signifies that the average velocity is evaluated at some point below y_0. The x-directional momentum carried across the top of the control surface is then $mv_x|_{y-}$ per molecule, where m is the mass of the molecule. If Z molecules cross the plane per unit time, then the net x-directional momentum flux will be

$$\sum_{n=1}^{Z} m_n(v_x|_{y-} - v_x|_{y+}) \tag{7-5}$$

The flux of x-directional momentum on a molecular scale appears as a shear stress when the fluid is observed on a macroscopic scale. The relation between the molecular momentum flux and the shear stress may be seen from the control-volume expression for linear momentum:

$$\sum \mathbf{F} = \iint_{c.s.} \rho \mathbf{v}(\mathbf{v} \cdot \mathbf{n}) \, dA + \frac{\partial}{\partial t} \iiint_{c.v.} \rho \mathbf{v} \, dV \tag{5-4}$$

The first term on the right-hand side of equation (5-4) is the momentum flux. When a control volume is analyzed on a molecular basis, this term includes both the macroscopic and molecular momentum fluxes. If the molecular portion of the total momentum flux is to be treated as a force, it must be placed on the left-hand side of equation (5-4). Thus the molecular momentum flux term changes sign. Denoting the negative of the molecular momentum flux per unit area as τ, we have

$$\tau = -\sum_{n=1}^{Z} m_n(v_x|_- - v_x|_{y+}) \tag{7-6}$$

We shall treat shear stress exclusively as a force per unit area.

The bracketed term, $(v_x|_{y-} - v_x|_{y+})$ in equation (7-6) may be evaluated by noting that $v_x|_{y-} = v_x|_{yo} - (dv_x/dy|_{yo})\delta$, where $y - = y_0 - \delta$. Using a similar expression for $y+$, we obtain, for the shear stress,

$$\tau = 2 \sum_{n=1}^{Z} m_n \frac{dv_x}{dy}\bigg|_{yo} \delta_n$$

In the above expression δ is the y component of the distance between molecular collisions. Borrowing from the kinetic theory of gases the concept of the mean free path, λ, as the average distance between collisions, and also noting from the same source that $\delta = \frac{2}{3}\lambda$, we obtain, for a pure gas,

$$\tau = \frac{4}{3}m\lambda Z\frac{dv_x}{dy}\bigg|_{y_0} \tag{7-7}$$

as the shear stress.

Comparing equation (7-7) with Newton's law of viscosity, we see that

$$\mu = \frac{4}{3}m\lambda Z \tag{7-8}$$

The kinetic theory gives $Z = N\bar{C}/4$, where

$$N = \text{molecules per unit volume}$$

$$\bar{C} = \text{average random molecular velocity}$$

and thus

$$\mu = \frac{1}{3}Nm\lambda\bar{C} = \frac{\rho\lambda\bar{C}}{3}$$

or, using*

$$\lambda = \frac{1}{\sqrt{2}\pi Nd^2} \quad \text{and} \quad \bar{C} = \sqrt{\frac{8\kappa T}{\pi m}}$$

where d is the molecular diameter and κ is the Boltzmann constant, we have

$$\mu = \frac{2}{3\pi^{3/2}}\frac{\sqrt{m\kappa T}}{d^2} \tag{7-9}$$

Equation (7-9) indicates that μ is independent of pressure for a gas. This has been shown, experimentally, to be essentially true for pressures up to approximately 10 atmospheres. Experimental evidence indicates that at low temperatures the viscosity varies more rapidly than $\sqrt{T}$. The constant-diameter rigid-sphere model for the gas molecule is responsible for the less-than-adequate viscosity-temperature relation. Even though the preceding development was somewhat crude in that an indefinite property, the molecular diameter, was introduced, the interpretation of the viscosity of a gas being due to the microscopic momentum flux is a valuable result and should not be overlooked. It is also important to note that equation (7-9) expresses the viscosity entirely in terms of fluid properties.

A more realistic molecular model utilizing a force field rather than the rigid-sphere approach will yield a viscosity-temperature relationship much more consistent with experimental data than the $\sqrt{T}$ result. The most acceptable expression for nonpolar molecules is based upon the Lennard-Jones potential energy function. This function and the development leading to the viscosity

---

* In order of increasing complexity, the expressions for mean free path are presented in R. Resnick and D. Halliday, *Physics*, Part I, Wiley, New York, 1966, Chap. 24, and E. H. Kennard, *Kinetic Theory of Gases*, McGraw-Hill Book Company, New York, 1938, Chap. 2.

expression will not be included here. The interested reader may refer to Hirschfelder, Curtiss, and Bird* for the details of this approach. The expression for viscosity of a pure monatomic gas that results is

$$\mu = 2.6693 \times 10^{-6} \frac{\sqrt{MT}}{\sigma^2 \Omega_\mu} \tag{7-10}$$

where μ is the viscosity, in pascal-seconds; T is absolute temperature, in K; M is the molecular weight; σ is the "collision diameter," a Lennard-Jones parameter, in Å (Angstroms); Ω_μ is the "collision integral," a Lennard-Jones parameter which varies in a relatively slow manner with the dimensionless temperature $\kappa T/\epsilon$; κ is the Boltzmann constant, $1.38 \cdot 10^{-16}$ ergs/K; and ϵ is the characteristic energy of interaction between molecules. Values of σ and ϵ for various gases are given in Appendix K, and a table of Ω_μ versus $\kappa T/\epsilon$ is also included in Appendix K.

For multicomponent gas mixtures at low density, Wilke† has proposed this empirical formula for the viscosity of the mixture:

$$\mu_{\text{mixture}} = \sum_{i=1}^{n} \frac{x_i \mu_i}{\sum x_j \phi_{ij}} \tag{7-11}$$

where x_i, x_j are mole-fractions of species i and j in the mixture, and

$$\phi_{ij} = \frac{1}{\sqrt{8}} \left(1 + \frac{M_i}{M_j}\right)^{-1/2} \left[1 + \left(\frac{\mu_i}{\mu_j}\right)^{1/2} \left(\frac{M_j}{M_i}\right)^{1/4}\right]^2 \tag{7-12}$$

where M_i, M_j are the molecular weights of species i and j, and μ_i, μ_j are the viscosities of species i and j. Note that when $i = j$, we have $\phi_{ij} = 1$.

Equations (7-10), (7-11), and (7-12) are for nonpolar gases and gas mixtures at low density. For polar molecules the preceding relation must be modified.*

While the kinetic theory of gases is well developed, and the more sophisticated models of molecular interaction accurately predict viscosity in a gas, the molecular theory of liquids is much less advanced. Hence the major source of knowledge concerning the viscosity of liquids is experiment. The difficulties in the analytical treatment of a liquid are largely inherent in the nature of the liquid itself. Whereas in gases the distance between molecules is so great that we consider gas molecules as interacting or colliding in pairs, the close spacing of molecules in a liquid results in the interaction of several molecules simultaneously. This situation is somewhat akin to an N-body gravitational problem. In spite of these difficulties, an approximate theory has been developed by Eyring which illustrates the relation of the intermolecular forces to viscosity.‡ The viscosity of a liquid can be considered due to the restraint caused by intermolecu-

---

* J. O. Hirschfelder, C. F. Curtiss, and R. B. Bird, *Molecular Theory of Gases and Liquids*, Wiley, New York, 1954.

† C. R. Wilke, *J. Chem. Phys.*, **18**, 517–519 (1950).

‡ For a description of Eyring's theory, see R. B. Bird, W. E. Stewart, and E. N. Lightfoot, *Transport Phenomena*, Wiley, New York, 1960, Chap. 1.

lar forces. As a liquid heats up, the molecules become more mobile. This results in less restraint from intermolecular forces. Experimental evidence for the viscosity of liquids shows that the viscosity decreases with temperature in agreement with the concept of intermolecular adhesive forces being the controlling factor.

UNITS OF VISCOSITY

The dimensions of viscosity may be obtained from Newton's viscosity relation,

$$\mu = \frac{\tau}{dv/dy}$$

or, in dimensional form,

$$\frac{F/L^2}{(L/t)(1/L)} = \frac{Ft}{L^2}$$

where F = force, L = length, t = time.

Using Newton's second law of motion to relate force and mass ($F = ML/t^2$), we find that the dimensions of viscosity in the mass-length-time system become M/Lt.

The ratio of the viscosity to the density occurs frequently in engineering problems. This ratio, μ/ρ, is given the name kinematic viscosity and is denoted by the symbol ν. The origin of the name kinematic viscosity may be seen from the dimensions of ν:

$$\nu \equiv \frac{\mu}{\rho} \sim \frac{M/Lt}{M/L^3} = \frac{L^2}{t}$$

The dimensions of ν are those of kinematics: length and time. Either of two names, absolute viscosity or dynamic viscosity, is frequently employed to distinguish μ from the kinematic viscosity, ν.

In the SI system, dynamic viscosity is expressed in pascal-seconds (1 pascal-second$=1\text{N} \cdot \text{s/m}^2 = 10$ poise $= 0.02089$ slugs/ft $\cdot$ s $= 0.02089$ lb$_f$ $\cdot$ s/ft$^2 = 0.6720$ lb$_m$/ft $\cdot$ s). Kinematic viscosity in the metric system is expressed in square meters per second ($1\text{m}^2/\text{s} = 10^4$ stokes $= 10.76$ ft^2/s).

Absolute and kinematic viscosities as functions of temperature are shown in Figure 7.5. A more extensive listing is contained in Appendix I.

7.4 SHEAR STRESS IN MULTIDIMENSIONAL LAMINAR FLOWS OF A NEWTONIAN FLUID

Newton's viscosity relation, discussed previously, is valid for only parallel, laminar flows. Stokes extended the concept of viscosity to three-dimensional

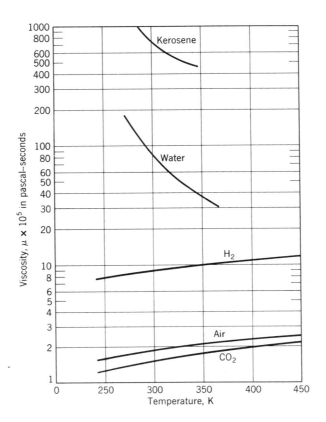

Figure 7.5 Viscosity-temperature variation for some liquids and gases.

laminar flow. The basis of Stokes' viscosity relation is equation (7-2),

$$\text{viscosity} = \frac{\text{shear stress}}{\text{rate of shear strain}} \tag{7-2}$$

where the shear stress and rate of shear strain are those of a three-dimensional element. Accordingly, we must examine shear stress and strain rate for a three-dimensional body.

SHEAR STRESS

The shear stress is a tensor quantity requiring magnitude, direction, and orientation with respect to a plane for identification. The usual method of identification of the shear stress involves a double subscript, such as τ_{xy}. The tensor component, τ_{ij}, is identified as follows:

$$\tau = \text{magnitude}$$

First subscript = direction of axis to which plane of action of shear stress
is normal

Second subscript = direction of action of the shear stress

Thus τ_{xy} acts on a plane normal to the x axis (the yz plane) and acts in the y direction. In addition to the double subscript, a sense is required. The shear stresses acting on an element $\Delta x \, \Delta y \, \Delta z$, illustrated in Figure 7.6, are indicated in the positive sense. The definition of positive shear stress can be generalized for use in other coordinate systems. A shear stress component is positive when both the vector normal to the surface of action and the shear stress act in the same direction (both positive or both negative).

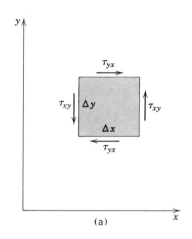

(a)

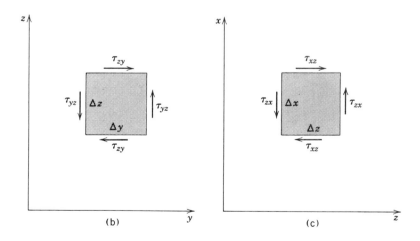

(b)   (c)

Figure 7.6   Shear stress acting in a positive sense.

For example, in Figure 7.6a, the shear stress τ_{yx} at the top of the element acts on surface $\Delta x\ \Delta z$. The vector normal to this area is in the positive y direction. The stress τ_{yx} acts in the positive x direction, hence τ_{yx} as illustrated in Figure 7.6a is positive. The student may apply similar reasoning to τ_{yx} acting on the bottom of the element and conclude that τ_{yx} is also positive as illustrated.

As in the mechanics of solids, $\tau_{ij} = \tau_{ji}$ (see Appendix C).

RATE OF SHEAR STRAIN

The rate of shear strain for a three-dimensional element may be evaluated by determining the shear strain rate in the xy, yz, and xz planes. In the xy plane illustrated in Figure 7.7, the shear strain rate is again $-d\delta/dt$, however, the element may deform in both the x and the y directions.

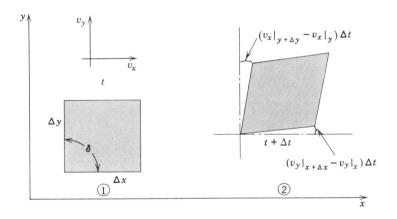

Figure 7.7   Shear strain in the xy plane.

Hence, as the element moves from position 1 to position 2 in time Δt,

$$-\frac{d\delta}{dt} = \lim_{\Delta x,\Delta y,\Delta t\to 0} \frac{\delta|_{t+\Delta t} - \delta|_t}{\Delta t}$$

$$= -\lim_{\Delta x,\Delta y,\Delta t\to 0} \left\{ \frac{\pi/2 - \arctan\{[(v_x|_{y+\Delta y} - v_x|_y)\,\Delta t]/\Delta y\}}{\Delta t} \right.$$

$$\left. -\frac{\arctan\{[(v_y|_{x+\Delta x} - v_y|_x)\,\Delta t]/\Delta x\} - \pi/2}{\Delta t} \right\}$$

Since the shear strain evaluated above is in the xy plane, it will be subscripted xy. In the limit, $-d\delta_{xy}/dt = \partial v_x/\partial y + \partial v_y/\partial x$. In a similar manner, the shear strain rates in the yz and xz planes may be evaluated as

$$-\frac{d\delta_{yz}}{dt} = \frac{\partial v_y}{\partial z} + \frac{\partial v_z}{\partial y}$$

$$-\frac{d\delta_{xz}}{dt} = \frac{\partial v_x}{\partial z} + \frac{\partial v_z}{\partial x}$$

STOKES' VISCOSITY RELATION

A. Shear Stress. Stokes' viscosity relation for the shear-stress components in laminar flow may now be stated with the aid of the preceding developments for rate of shear strain. Using equation (7-2), we have, for the shear stresses written in rectangular coordinate form,

$$\tau_{xy} = \tau_{yx} = \mu\left(\frac{\partial v_x}{\partial y} + \frac{\partial v_y}{\partial x}\right) \tag{7-13a}$$

$$\tau_{yz} = \tau_{zy} = \mu\left(\frac{\partial v_y}{\partial z} + \frac{\partial v_z}{\partial y}\right) \tag{7-13b}$$

and

$$\tau_{zx} = \tau_{xz} = \mu\left(\frac{\partial v_z}{\partial x} + \frac{\partial v_x}{\partial z}\right) \tag{7-13c}$$

B. Normal Stress. The normal stress may also be determined from a stress rate-of-strain relation; the strain rate, however, is more difficult to express than in the case of shear strain. For this reason the development of normal stress, on the basis of a generalized Hooke's law for an elastic medium is included in detail in Appendix D, with only the result expressed below in equations (7-14a), (7-14b), and (7-14c).

The normal stress in rectangular coordinates written for a newtonian fluid is given by

$$\sigma_{xx} = \mu\left(2\frac{\partial v_x}{\partial x} - \tfrac{2}{3}\nabla \cdot \mathbf{v}\right) - P \tag{7-14a}$$

$$\sigma_{yy} = \mu\left(2\frac{\partial v_y}{\partial y} - \tfrac{2}{3}\nabla \cdot \mathbf{v}\right) - P \tag{7-14b}$$

and

$$\sigma_{zz} = \mu\left(2\frac{\partial v_z}{\partial z} - \tfrac{2}{3}\nabla \cdot \mathbf{v}\right) - P \tag{7-14c}$$

It is to be noted that the sum of these three equations yields the previously mentioned result: the bulk stress, $\bar{\sigma} = (\sigma_{xx} + \sigma_{yy} + \sigma_{zz})/3$, is the negative of the pressure, P.

7.5 CLOSURE

The shear stress in laminar flow and its dependence upon the viscosity and kinematic derivatives has been presented for a cartesian coordinate system. The shear stress in other coordinate systems, of course, will occur frequently, and it is to be noted that equation (7-2) forms the general relation between shear stress, viscosity, and the rate of shear strain. The shear stress in other coordinate systems may be obtained from evaluating the shear-strain rate in the associated coordinate systems. Several problems of this nature are included at the end of this chapter.

In conclusion, the student is again reminded that Chapter 7 covers only laminar flow.

PROBLEMS

7.1 An auto lift consists of a 35.56-cm-diameter ram which slides in a 35.58-cm-diameter cylinder. The annular region is filled with oil having a kinematic viscosity of 0.00037 m²/s and a specific gravity of 0.85. If the rate of travel of the ram is 0.15 m/s, estimate the frictional resistance when 2.44 m of the ram is engaged in the cylinder.

7.2 If the ram and auto rack in the previous problem together have a mass of 680 kg, estimate the maximum sinking speed of the ram and rack when gravity and viscous friction are the only forces acting. Assume 2.44 m of the ram engaged.

7.3 The conical pivot shown in the figure has angular velocity ω and rests on an oil film of uniform thickness h. Determine the frictional moment as a

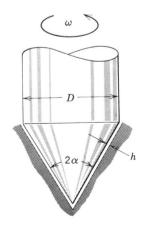

function of the angle α, the viscosity, the angular velocity, the gap distance, and the shaft diameter.

7.4 Determine the number of air molecules crossing a plane of unit area in air under standard conditions. The ratio κ/m is equal to the gas constant R.

7.5 For water flowing in a 0.1-in.-diameter tube the velocity distribution is parabolic (see example 2, chapter 4). If the average velocity is 2 fps, determine the magnitude of the shear stress at the tube wall.

7.6 The rate of shear work per unit volume is given by the product τv. For a parabolic velocity profile in a circular tube (see example 2, chapter 4), determine the distance from the wall at which the shear work is maximum.

7.7 An automobile crankshaft is 3.175 cm in diameter. A bearing on the shaft is 3.183 cm in diameter and 2.8 cm long. The bearing is lubricated with SAE 30 oil at a temperature of 365 K. Assuming that the shaft is centrally located in the bearing, determine how much heat must be removed to maintain the bearing at constant temperature. The shaft is rotating at 1700 rpm, and the viscosity of the oil is 0.01 Pa·s.

7.8 If the speed of the shaft is doubled in problem 7.7, what will be the percentage increase in the heat transferred from the bearing? Assume that the bearing remains at constant temperature.

7.9 Estimate the viscosity of nitrogen, at 175 K using equation (7-10).

7.10 Two ships are traveling parallel to each other and are connected by flexible hoses. Fluid is transferred from one ship to the other for processing and then returned. If the fluid is flowing at 100 kg/s, and at a given instant the first ship is making 4 m/s while the second ship is making 3.1 m/s, what is the net force on ship one when the above velocities exist?

7.11 Sketch the deformation of a fluid element for the following cases:
(a) $\partial v_x/\partial y$ is much larger than $\partial v_y/\partial x$;
(b) $\partial v_y/\partial x$ is much larger than $\partial v_x/\partial y$.

7.12 For a two-dimensional, incompressible flow with velocity $v_x = v_x(y)$, sketch a three-dimensional fluid element and illustrate the magnitude, direction, and surface of action of each stress component.

7.13 Show that the axial strain rate in a one-dimensional flow, $v_x = v_x(x)$, is given by $\partial v_x/\partial x$. What is the rate of volume change? Generalize for a three-dimensional element, and determine the rate of volume change.

7.14 Using a cylindrical element show that Stokes' viscosity relation yields the following shear stress components.

$$\tau_{r\theta} = \tau_{\theta r} = \mu \left[r \frac{\partial}{\partial r} \left(\frac{v_\theta}{r} \right) + \frac{1}{r} \frac{\partial v_r}{\partial \theta} \right]$$

$$\tau_{\theta z} = \tau_{z\theta} = \mu \left[\frac{\partial v_\theta}{\partial z} + \frac{1}{r} \frac{\partial v_z}{\partial \theta} \right]$$

$$\tau_{zr} = \tau_{rz} = \mu \left[\frac{\partial v_z}{\partial r} + \frac{\partial v_r}{\partial z} \right]$$

7.15 The device in the schematic diagram below is a viscosity pump. It consists of a rotating drum inside of a stationary case. The case and the drum are concentric. Fluid enters at A, flows through the annulus between the case and the drum, and leaves at B. The pressure at B is higher than that at A, the difference being Δp. The length of the annulus is L. The width of the annulus h is very small compared to the diameter of the drum, so that the flow in the annulus is equivalent to the flow between two flat plates. Assume the flow to be laminar. Find the pressure rise and efficiency as a function of the flow rate per unit depth.

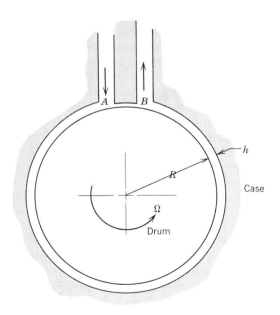

8
ANALYSIS OF A DIFFERENTIAL FLUID ELEMENT IN LAMINAR FLOW

The analysis of a fluid-flow situation may follow two different paths. One type of analysis has been discussed at length in Chapters 4, 5, and 6 in which the region of interest has been a definite volume, the macroscopic control volume. In analyzing a problem from the standpoint of a macroscopic control volume, one is concerned only with gross quantities of mass, momentum, and energy crossing the control surface and the total change in these quantities exhibited by the material under consideration. Changes occurring within the control volume by each differential element of fluid cannot be obtained from this type of overall analysis.

In this chapter we shall direct our attention to elements of fluid as they approach differential size. Our goal is the estimation and description of fluid behavior from a differential point of view; the resulting expressions from such analyses will be differential equations. The solution to these differential equations will give flow information of a different nature than that achieved from a macroscopic examination. Such information may be of less interest to the engineer needing overall design information, but it can give much greater insight into the mechanisms of mass, momentum, and energy transfer.

It is possible to change from one form of analysis to the other, that is, from a differential analysis to an integral analysis by integration and vice versa, rather easily.*

A complete solution to the differential equations of fluid flow is possible only if the flow is laminar; for this reason only laminar-flow situations will be examined in this chapter. A more general differential approach will be discussed in Chapter 9.

---

* This transformation may be accomplished by a variety of methods, among which are the methods of vector calculus. We shall use a limiting process in this text.

8.1   FULLY DEVELOPED LAMINAR FLOW IN A CIRCULAR CONDUIT OF CONSTANT CROSS SECTION

Engineers are often confronted with flow of fluids inside circular conduits or pipes. We shall now analyze this situation for the case of incompressible laminar flow. In Figure 8.1 we have a section of pipe in which the flow is laminar and fully developed; that is, it is not influenced by entrance effects and represents a steady-flow situation. *Fully developed flow* is defined as that for which the velocity profile does not vary along the axis of flow.

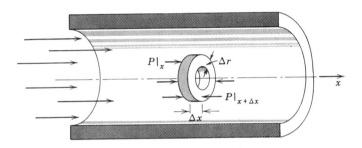

Figure 8.1   Control volume for flow in a circular conduit.

We now consider the cylindrical control volume of fluid having an inside radius r, thickness Δr, and length Δx. Applying Newton's second law to this control volume, we may evaluate the appropriate force and momentum terms for the x direction. Starting with the control-volume expression for linear momentum in the x direction,

$$\Sigma F_x = \iint_{\text{c.s.}} \rho v_x (\mathbf{v} \cdot \mathbf{n})\, dA + \frac{\partial}{\partial t} \iiint_{\text{c.v.}} \rho v_x\, dV \qquad (5\text{-}5a)$$

and evaluating each term as it applies to the control volume shown, we have

$$\Sigma F_x = P(2\pi r\, \Delta r)|_x - P(2\pi r\, \Delta r)|_{x+\Delta x}$$
$$+ \tau_{rx}(2\pi r\, \Delta x)|_{r+\Delta r} - \tau_{rx}(2\pi r\, \Delta x)|_r$$

$$\iint_{\text{c.s.}} v_x \rho (\mathbf{v} \cdot \mathbf{n})\, dA = (\rho v_x)(2\pi r\, \Delta r v_x)|_{x+\Delta x} - (\rho v_x)(2\pi r\, \Delta r v_x)|_x$$

and

$$\frac{\partial}{\partial t} \iiint_{\text{c.v.}} v_x \rho\, dV = 0$$

in steady flow.

The convective momentum flux

$$(\rho v_x)(2\pi r\, \Delta r v_x)|_{x+\Delta x} - (\rho v_x)(2\pi\, \Delta r v_x)|_x$$

is equal to zero, since, by the original stipulation that flow is fully developed, all terms are independent of x. Substitution of the remaining terms into equation (5-5a) gives

$$-[P(2\pi r\, \Delta r)|_{x+\Delta x} - P(2\pi r\, \Delta r)|_x] + \tau_{rx}(2\pi r\, \Delta x)|_{r+\Delta r} - \tau_{rx}(2\pi r\, \Delta x)|_r = 0$$

Canceling terms where possible and rearranging, we find that this expression reduces to the form

$$-r\frac{P|_{x+\Delta x} - P|_x}{\Delta x} + \frac{(r\tau_{rx})|_{r+\Delta r} - (r\tau_{rx})|_r}{\Delta r} = 0$$

Evaluating this expression in the limit as the control volume approaches differential size, that is, as Δx and Δr approach zero, we have

$$-r\frac{dP}{dx} + \frac{d}{dr}(r\tau_{rx}) = 0 \qquad (8\text{-}1)$$

Note that the pressure and shear stress are functions only of x and r, respectively, and thus the derivatives formed are total rather than partial derivatives. In a region of fully developed flow the pressure gradient, dP/dx, is constant.

The variables in equation (8-1) may be separated and integrated to give

$$\tau_{rx} = \left(\frac{dP}{dx}\right)\frac{r}{2} + \frac{C_1}{r}$$

The constant of integration C_1 may be evaluated by knowing a value of τ_{rx} at some r. Such a condition is known at the center of the conduit, $r = 0$, where for any finite value of C_1 the shear stress, τ_{rx}, will be infinite. Since this is physically impossible, the only realistic value for C_1 is zero. Thus the shear-stress distribution for the conditions and geometry specified is

$$\tau_{rx} = \left(\frac{dP}{dx}\right)\frac{r}{2} \qquad (8\text{-}2)$$

We observe that the shear stress varies linearly across the conduit from a value of 0 at $r = 0$, to a maximum at $r = R$, the inside surface of the conduit.

Further information may be obtained if we substitute the newtonian viscosity relationship, that is, assuming the fluid to be newtonian and recalling that the flow is laminar,

$$\tau_{rx} = \mu\frac{dv_x}{dr} \qquad (8\text{-}3)$$

Substituting this relation into equation (8-2) gives

$$\mu\frac{dv_x}{dr} = \left(\frac{dP}{dx}\right)\frac{r}{2}$$

which becomes, upon integration,

$$v_x = \left(\frac{dP}{dx}\right)\frac{r^2}{4\mu} + C_2$$

The second constant of integration, C_2, may be evaluated, using the boundary condition that the velocity, v_x, is zero at the conduit surface (the no-slip condition), $r = R$. Thus

$$C_2 = -\left(\frac{dP}{dx}\right)\frac{R^2}{4\mu}$$

and the velocity distribution becomes

$$v_x = -\left(\frac{dP}{dx}\right)\frac{1}{4\mu}(R^2 - r^2) \tag{8-4}$$

or

$$v_x = -\left(\frac{dP}{dx}\right)\frac{R^2}{4\mu}\left[1 - \left(\frac{r}{R}\right)^2\right] \tag{8-5}$$

Equations (8-4) and (8-5) indicate that the velocity profile is parabolic and that the maximum velocity occurs at the center of the circular conduit where $r = 0$. Thus

$$v_{max} = -\left(\frac{dP}{dx}\right)\frac{R^2}{4\mu} \tag{8-6}$$

and equation (8-5) may be written in the form

$$v_x = v_{max}\left[1 - \left(\frac{r}{R}\right)^2\right] \tag{8-7}$$

Note that the velocity profile written in the form of equation (8-7) is identical to that used in example 4.2. We may therefore use the result obtained in example 4.2 that

$$v_{avg} = \frac{v_{max}}{2} = -\left(\frac{dP}{dx}\right)\frac{R^2}{8\mu} \tag{8-8}$$

Equation (8-8) may be rearranged to express the pressure gradient, $-dP/dx$, in terms of v_{avg}

$$-\frac{dP}{dx} = \frac{8\mu v_{avg}}{\cdot R^2} = \frac{32\mu v_{avg}}{D^2} \tag{8-9}$$

Equation (8-9) is known as the Hagen-Poiseuille equation, in honor of the two men credited with its original derivation. This expression may be integrated over a given length of conduit to find the pressure drop and associated drag force on the conduit resulting from the flow of a viscous fluid.

The conditions for which the preceding equations were derived and apply should be remembered and understood. They are as follows:

1. The fluid (a) is newtonian,
   (b) behaves as a continuum.
2. The flow is (a) laminar,
   (b) steady,
   (c) fully developed,
   (d) incompressible.

8.2 LAMINAR FLOW OF A NEWTONIAN FLUID DOWN AN INCLINED PLANE SURFACE

The approach used in section 8.1 will now be applied to a slightly different situation, that of a newtonian fluid in laminar flow down an inclined plane surface. This configuration and associated nomenclature are depicted in Figure 8.2.

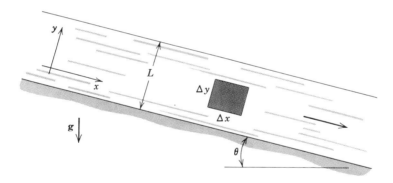

Figure 8.2   Laminar flow down an inclined plane surface.

The analysis again involves the application of the control-volume expression for linear momentum in the x direction, which is

$$\Sigma F_x = \iint_{\text{c.s.}} v_x \rho (\mathbf{v} \cdot \mathbf{n}) \, dA + \frac{\partial}{\partial t} \iiint_{\text{c.v.}} \rho v_x \, dV \qquad (5\text{-}5a)$$

Evaluating each term in this expression for the fluid element of volume $(\Delta x)(\Delta y)(1)$ as shown in the figure, we have

$$\Sigma F_x = P \, \Delta y|_x - P \, \Delta y|_{x+\Delta x} + \tau_{yx} \, \Delta x|_{y+\Delta y} - \tau_{yx} \, \Delta x|_y + \rho g \, \Delta x \, \Delta y \sin \theta$$

$$\iint_{\text{c.s.}} \rho v_x (\mathbf{v} \cdot \mathbf{n}) \, dA = \rho v_x^2 \, \Delta y|_{x+\Delta x} - \rho v_x^2 \, \Delta y|_x$$

and

$$\frac{\partial}{\partial t} \iiint_{\text{c.v.}} \rho v_x \, dV = 0$$

Noting that the convective momentum terms cancel for fully developed flow and that the pressure force terms also cancel because of the presence of a free liquid surface, we see that the equation resulting from the substitution of these terms into equation (5-5a) becomes

$$\tau_{yx} \, \Delta x|_{y+\Delta y} - \tau_{yx} \, \Delta x|_y + \rho g \, \Delta x \, \Delta y \sin \theta = 0$$

Dividing by $(\Delta x)(\Delta y)(1)$, the volume of the element considered, gives

$$\frac{\tau_{yx}|_{y+\Delta y} - \tau_{yx}|_y}{\Delta y} + \rho g \sin \theta = 0$$

In the limit as $\Delta y \to 0$ we get the applicable differential equation

$$\frac{d}{dy} \tau_{yx} + \rho g \sin \theta = 0 \tag{8-10}$$

Separating the variables in this simple equation and integrating we obtain for the shear stress,

$$\tau_{yx} = -\rho g \sin \theta y + C_1$$

The integration constant, C_1, may be evaluated by using the boundary condition that the shear stress, τ_{yx}, is zero at the free surface, $y = L$. Thus the shear stress variation becomes

$$\tau_{yx} = \rho g L \sin \theta \left[1 - \frac{y}{L} \right] \tag{8-11}$$

The consideration of a newtonian fluid in laminar flow enables the substitution of $\mu(dv_x/dy)$, to be made for τ_{yx}, yielding

$$\frac{dv_x}{dy} = \frac{\rho g L \sin \theta}{\mu} \left[1 - \frac{y}{L} \right]$$

which, upon separation of variables and integration, becomes

$$v_x = \frac{\rho g L \sin \theta}{\mu} \left[y - \frac{y^2}{2L} \right] + C_2$$

Using the no-slip boundary condition, that is, $v_x = 0$ at $y = 0$, the constant of integration, C_2 is seen to be zero. The final expression for the velocity profile may now be written as

$$v_x = \frac{\rho g L^2 \sin \theta}{\mu} \left[\frac{y}{L} - \frac{1}{2} \left(\frac{y}{L} \right)^2 \right] \tag{8-12}$$

The form of this solution indicates the velocity variation to be parabolic, reaching the maximum value

$$v_{\max} = \frac{\rho g L^2 \sin \theta}{2\mu} \tag{8-13}$$

at the free surface, $y = L$.

Additional calculations may be performed to determine the average velocity as was indicated in section 8.1. Note that there will be no counterpart in this case to the Hagen-Poiseuille relation, equation (8-9), for the pressure gradient. The reason for this is the presence of a free liquid surface along which the pressure is constant. Thus, for our present case, flow is not the result of a pressure gradient but rather the manifestation of the gravitational acceleration upon a fluid.

8.3 CLOSURE

The method of analysis employed in this chapter, that of applying the basic relation for linear momentum to a small control volume and allowing the control volume to shrink to differential size, enables one to find information of a sort different from that obtained previously. Velocity and shear-stress profiles are examples of this type of information. The behavior of a fluid element of differential size can give considerable insight into a given transfer process and provide an understanding available in no other type of analysis.

This method has direct counterparts in heat and mass transfer, where the element may be subjected to an energy or a mass balance.

In Chapter 9 the methods introduced in this chapter will be used to derive differential equations of fluid flow for a general control volume.

PROBLEMS

8.1 Express equation (8-9) in terms of the flow rate and the pipe diameter. If the pipe diameter is doubled at constant pressure drop, what percentage change will occur in the flow rate?

8.2 A 32-km-long pipeline delivers petroleum at a rate of 5000 barrels per day. The resulting pressure drop is 3.45×10^6 Pa. If a parallel line of the same size is laid along the last 18 km of the line, what will be the new capacity of this network? Flow in both cases is laminar and the pressure drop remains 3.45×10^6 Pa.

8.3 Derive the expressions for the velocity distribution and for the pressure drop for a newtonian fluid in laminar flow in the annular space between two horizontal, concentric pipes. Apply the momentum theorem to an annular fluid shell of thickness Δr and show that the analysis of such a control volume leads to

$$\frac{d}{dr}(r\tau) = r\frac{\Delta P}{L}$$

The desired expressions may then be obtained by the substitution of Newton's viscosity law and two integrations.

8.4 A thin rod of diameter d is pulled at constant velocity through a pipe of diameter D. If the wire is at the center of the pipe, find the drag per unit length of wire. The fluid density is ρ, and the viscosity is μ.

8.5 The viscosity of heavy liquids such as oils is frequently measured with a device that consists of a rotating cylinder inside a large cylinder. The annular region between these cylinders is filled with liquid and the torque required to rotate the inner cylinder at constant speed is computed, a linear velocity profile being assumed. For what ratio of cylinder diameters is the assumption of a linear profile accurate within 1% of the true profile? (See Problem 7.14.)

8.6 A 0.635-cm hydraulic line suddenly ruptures 8 m from a reservoir with gage pressure of 207 kPa. Compare the laminar and inviscid flow rates from the ruptured line, in cubic meters per second.

8.7 Two immiscible fluids of different density and viscosity are flowing between two parallel plates. Express the boundary conditions at the interface between the two fluids.

8.8 Determine the velocity profile for fluid flowing between two parallel plates separated by a distance $2h$. The pressure drop is constant.

8.9 Fluid flows between two parallel plates, a distance h apart. The upper plate moves at velocity v_0; the lower plate is stationary. For what value of pressure gradient will the shear stress at the lower wall be zero?

8.10 Derive the equation of motion for a one-dimensional, inviscid, unsteady compressible flow in a pipe of constant cross-sectional area; neglect gravity.

8.11 A common type of viscosimeter for liquids consists of a relatively large reservoir with a very slender outlet tube, the rate of outflow being determined by timing the fall in the surface level. If oil of constant density flows out of the viscosimeter shown at the rate of $0.273 \text{ cm}^3/\text{s}$, what is the kinematic viscosity of the fluid? The tube diameter is 0.18 cm.

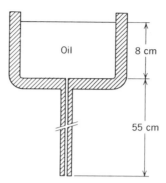

8.12 A continuous belt passes upward through a chemical bath at velocity V_0 and picks up a film of liquid of thickness h, density ρ, and viscosity μ. Gravity tends to make the liquid drain down, but the movement of the belt keeps the fluid from running off completely. Assume that the flow is a well-developed laminar flow with zero pressure gradient, and that the atmosphere produces no shear at the outer surface of the film.
(a) State clearly the boundary conditions at $y = 0$ and $y = h$ to be satisfied by the velocity.
(b) Calculate the velocity profile.
(c) Determine the rate at which fluid is being dragged up with the belt in terms of μ, ρ, h, V_0.

8.13 The element of fluid occupying the illustrated control volume has a velocity $\mathbf{v} = v_\theta(r)\mathbf{e}_\theta$. Show that the time rate of change of the velocity vector along a streamline is $d\mathbf{v}/dt = -(v_\theta^2/r)\mathbf{e}_r$.

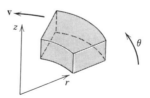

8.14 Sketch the control volume from Problem 8.13 and show the forces and momentum terms. Assume constant density and that gravity acts in the minus z direction.

8.15 Consider the laminar flow of a newtonian fluid down a plane surface inclined at an angle θ to the horizontal. If at the free surface an amount of fluid W is added to the flow per unit area of free surface, find the thickness of liquid film as a function of the distance x measured along the plane. Generate an expression for the velocity profile in the liquid film. Flow velocities normal to the plane surface are to be neglected.

9
DIFFERENTIAL EQUATIONS
OF FLUID FLOW

The fundamental laws of fluid flow, which have been expressed in mathematical form for an arbitrary control volume in Chapters 4, 5, and 6, may also be expressed in mathematical form for a special type of control volume, the differential element. These differential equations of fluid flow provide a means of determining the point-to-point variation of fluid properties. Chapter 8 involved the differential equations associated with some one-dimensional, steady, incompressible laminar flows. In Chapter 9 we shall express the law of conservation of mass and Newton's second law of motion in differential form for more general cases. The basic tools used to derive these differential equations will be the control-volume developments of Chapters 4 and 5.

9.1   THE DIFFERENTIAL CONTINUITY EQUATION

The continuity equation to be developed in this section is the law of conservation of mass expressed in differential form. Consider the control volume $\Delta x \, \Delta y \, \Delta z$ shown in Figure 9.1.

The control volume expression for the convervation of mass is

$$\iint \rho(\mathbf{v} \cdot \mathbf{n}) \, dA + \frac{\partial}{\partial t} \iiint \rho \, dV = 0 \tag{4-1}$$

which states that

$$\left\{ \begin{array}{c} \text{Net rate of mass} \\ \text{flux out of con-} \\ \text{trol volume} \end{array} \right\} + \left\{ \begin{array}{c} \text{rate of accumulation of} \\ \text{mass within control} \\ \text{volume} \end{array} \right\} = 0$$

The mass flux $\rho(\mathbf{v} \cdot \mathbf{n})$ at each face of the control volume is illustrated in Figure 9.1. The mass within the control volume is $\rho \, \Delta x \, \Delta y \, \Delta z$, and thus the time rate of

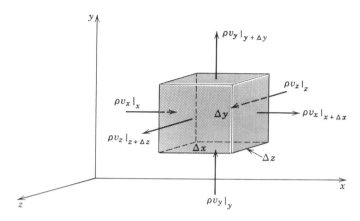

Figure 9.1 Mass flux through a differential control volume.

change of mass within the control volume is

$$\frac{\partial}{\partial t}(\rho\,\Delta x\,\Delta y\,\Delta z)$$

The student is reminded that the density in general can vary from point to point, that is, $\rho = \rho(x, y, z, t)$.

The net mass flux out of the control volume in the x direction is

$$(\rho v_x|_{x+\Delta x} - \rho v_x|_x)\Delta y\,\Delta z$$

in the y direction,

$$(\rho v_y|_{y+\Delta y} - \rho v_y|_y)\Delta x\,\Delta z$$

and in the z direction,

$$(\rho v_z|_{z+\Delta z} - \rho v_z|_z)\Delta x\,\Delta y$$

The total net mass flux is the sum of the above three terms. Substituting into (4-1) yields

$$(\rho v_x|_{x+\Delta x} - \rho v_x|_x)\Delta y\,\Delta z + (\rho v_y|_{y+\Delta y} - \rho v_y|_y)\Delta x\,\Delta z$$

$$+ (\rho v_z|_{z+\Delta z} - \rho v_z|_z)\Delta x\,\Delta y + \frac{\partial}{\partial t}(\rho\,\Delta x\,\Delta y\,\Delta z) = 0$$

The volume does not change with time, so we may divide both sides of equation (9-1) by $\Delta x\,\Delta y\,\Delta z$. In the limit as Δx, Δy, and Δz approach zero, we obtain

$$\frac{\partial}{\partial x}(\rho v_x) + \frac{\partial}{\partial y}(\rho v_y) + \frac{\partial}{\partial z}(\rho v_z) + \frac{\partial\rho}{\partial t} = 0 \qquad (9\text{-}1)$$

The first three terms comprise the divergence of the vector $\rho\mathbf{v}$. The divergence of a vector is the dot product with ∇,

$$\mathrm{div}\,\mathbf{A} \equiv \nabla \cdot \mathbf{A}$$

The student may verify that the first three terms in equation (9-1) may be written as $\nabla \cdot \rho \mathbf{v}$, and thus a more compact statement of the continuity equation becomes

$$\nabla \cdot \rho \mathbf{v} + \frac{\partial \rho}{\partial t} = 0 \qquad (9\text{-}2)$$

The continuity equation above applies to unsteady, three-dimensional flow. It is apparent that, when flow is incompressible, this equation reduces to

$$\nabla \cdot \mathbf{v} = 0 \qquad (9\text{-}3)$$

whether the flow is unsteady or not.

Equation (9-2) may be arranged in a slightly different form to illustrate the use of the *substantial derivative*. Carrying out the differentiation indicated in (9-1), we have

$$\frac{\partial \rho}{\partial t} + v_x \frac{\partial \rho}{\partial x} + v_y \frac{\partial \rho}{\partial y} + v_z \frac{\partial \rho}{\partial z} + \rho \left(\frac{\partial v_x}{\partial x} + \frac{\partial v_y}{\partial y} + \frac{\partial v_z}{\partial z} \right) = 0$$

The first four terms of the above equation comprise the substantial derivative of the density, symbolized as $D\rho/Dt$, where

$$\frac{D}{Dt} = \frac{\partial}{\partial t} + v_x \frac{\partial}{\partial x} + v_y \frac{\partial}{\partial y} + v_z \frac{\partial}{\partial z} \qquad (9\text{-}4)$$

in cartesian coordinates. The substantial derivative is the derivative following the motion of the fluid. The continuity equation may be written as

$$\frac{D\rho}{Dt} + \rho \nabla \cdot \mathbf{v} = 0 \qquad (9\text{-}5)$$

When considering the total differential of a quantity, three different approaches may be taken. If, for instance, we wish to evaluate the change in atmospheric pressure, P, the total differential written in rectangular coordinates is

$$dP = \frac{\partial P}{\partial t} dt + \frac{\partial P}{\partial x} dx + \frac{\partial P}{\partial y} dy + \frac{\partial P}{\partial z} dz$$

where dx, dy, and dz are arbitrary displacements in the x, y, and z directions. The rate of pressure change is obtained by dividing through by dt, giving

$$\frac{dP}{dt} = \frac{\partial P}{\partial t} + \frac{dx}{dt} \frac{\partial P}{\partial x} + \frac{dy}{dt} \frac{\partial P}{\partial y} + \frac{dz}{dt} \frac{\partial P}{\partial z} \qquad (9\text{-}6)$$

As a first approach the instrument to measure pressure is located in a weather station, which is, of course, fixed on the earth's surface. Thus the coefficients dx/dt, dy/dt, dz/dt are all zero, and for a fixed point of observation the total derivative, dP/dt, is equal to the local derivative with respect to time, $\partial P/\partial t$.

A second approach would involve the pressure-measuring instrument housed in an aircraft which, at the pilot's discretion, can be made to climb or descend, or fly in any chosen x, y, z direction. In this case the coefficients dx/dt,

dy/dt, dz/dt are the x, y, and z velocities of the aircraft, and they are arbitrarily chosen, bearing only coincidental relationship to the air currents.

The third situation is one in which the pressure indicator is in a balloon which rises, falls, and drifts as influenced by the flow of air in which it is suspended. Here the coefficients dx/dt, dy/dt, dz/dt are those of the *flow* and they may be designated v_x, v_y, and v_z, respectively. This latter situation corresponds to the substantial derivative and the terms may be grouped as designated below:

$$\frac{dP}{dt} = \frac{DP}{Dt} = \underbrace{\frac{\partial P}{\partial t}}_{\substack{\text{local} \\ \text{rate of} \\ \text{change of} \\ \text{pressure}}} + \underbrace{v_x\frac{\partial P}{\partial x} + v_y\frac{\partial P}{\partial y} + v_z\frac{\partial P}{\partial z}}_{\substack{\text{rate of change} \\ \text{of pressure} \\ \text{due to motion}}} \qquad (9\text{-}7)$$

The substantial derivative is seen to be the derivative following the motion of the fluid. The derivative D/Dt may be interpreted as the time rate of change of a fluid or flow variable along the path of a fluid element. The substantial derivative will be applied to both scalar and vector variables in subsequent sections.

9.2   NAVIER-STOKES EQUATIONS

The Navier-Stokes equations are the differential form of Newton's second law of motion. Consider the differential control volume illustrated in Figure 9.1.

The basic tool we shall use in developing the Navier-Stokes equations is Newton's second law of motion for an arbitrary control volume as given in Chapter 5,

$$\sum \mathbf{F} = \iint_{\text{c.s.}} \rho\mathbf{v}(\mathbf{v}\cdot\mathbf{n})\, dA + \frac{\partial}{\partial t}\iiint_{\text{c.v.}} \rho\mathbf{v}\, dV \qquad (5\text{-}4)$$

which states that

$$\left\{\begin{array}{c} \text{Sum of the external} \\ \text{forces acting on the} \\ \text{c.v.} \end{array}\right\} = \left\{\begin{array}{c} \text{net rate of linear} \\ \text{momentum efflux} \end{array}\right\} + \left\{\begin{array}{c} \text{time rate of change} \\ \text{of linear momentum} \\ \text{within the c.v.} \end{array}\right\}$$

Since the mathematical expression for each of the above terms is rather lengthy, each will be evaluated separately and then substituted into equation (5-4).

The development may be further simplified by recalling that we have, in the prior case, divided by the volume of the control volume and taken the limit as the dimensions approach zero. Equation (5.4) can also be written

$$\lim_{\Delta x, \Delta y, \Delta z \to 0} \frac{\sum \mathbf{F}}{\Delta x\, \Delta y\, \Delta z} = \lim_{\Delta x, \Delta y, \Delta z \to 0} \frac{\iint \rho\mathbf{v}(\mathbf{v}\cdot\mathbf{n})\, dA}{\Delta x\, \Delta y\, \Delta z} + \lim_{\Delta x, \Delta y, \Delta z \to 0} \frac{\partial/\partial t \iiint \rho\mathbf{v}\, dV}{\Delta x\, \Delta y\, \Delta z}$$

$$\underset{①}{} \qquad\qquad \underset{②}{} \qquad\qquad \underset{③}{} \qquad (9\text{-}8)$$

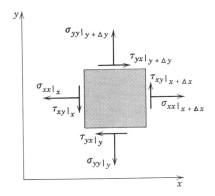

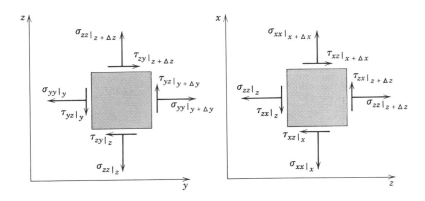

Figure 9.2 Forces acting on a differential control volume.

① *Sum of the External Forces.* The forces acting on the control volume are those due to the normal stress and to the shear stress, and body forces such as that due to gravity. Figure 9.2 illustrates the various forces acting on the control volume. Summing the forces in the x direction, we obtain

$$\sum F_x = (\sigma_{xx}|_{x+\Delta x} - \sigma_{xx}|_x)\Delta y\ \Delta z + (\tau_{yx}|_{y+\Delta y} - \tau_{yx}|_y)\Delta x\ \Delta z$$
$$+ (\tau_{zx}|_{z+\Delta z} - \tau_{zx}|_z)\Delta x\ \Delta y + g_x\rho\ \Delta x\ \Delta y\ \Delta z$$

where g_x is the component of the gravitational acceleration in the x direction. In the limit as the dimensions of the element approach zero this becomes

$$\lim_{\Delta x,\Delta y,\Delta z \to 0} \frac{\sum F_x}{\Delta x\ \Delta y\ \Delta z} = \frac{\partial \sigma_{xx}}{\partial x} + \frac{\partial \tau_{yx}}{\partial y} + \frac{\partial \tau_{zx}}{\partial z} + \rho g_x \tag{9-9}$$

Similar expressions are obtained for the force summations in the y and z

directions:

$$\lim_{\Delta x, \Delta y, \Delta z \to 0} \frac{\sum F_y}{\Delta x \, \Delta y \, \Delta z} = \frac{\partial \tau_{xy}}{\partial x} + \frac{\partial \sigma_{yy}}{\partial y} + \frac{\partial \tau_{zy}}{\partial z} + \rho g_y \qquad (9\text{-}10)$$

$$\lim_{\Delta x, \Delta y, \Delta z \to 0} \frac{\sum F_z}{\Delta x \, \Delta y \, \Delta z} = \frac{\partial \tau_{xz}}{\partial x} + \frac{\partial \tau_{yz}}{\partial y} + \frac{\partial \sigma_{zz}}{\partial z} + \rho g_z \qquad (9\text{-}11)$$

② *Net Momentum Flux Through the Control Volume.* The net momentum flux through the control volume illustrated in Figure 9.3 is

$$\lim_{\Delta x, \Delta y, \Delta z \to 0} \frac{\iint \rho \mathbf{v}(\mathbf{v} \cdot \mathbf{n}) \, dA}{\Delta x \, \Delta y \, \Delta z} = \lim_{\Delta x, \Delta y, \Delta z \to 0} \left[\frac{(\rho \mathbf{v} v_x|_{x+\Delta x} - \rho \mathbf{v} v_x|_x) \Delta y \, \Delta z}{\Delta x \, \Delta y \, \Delta z} \right.$$

$$+ \frac{(\rho \mathbf{v} v_y|_{y+\Delta y} - \rho \mathbf{v} v_y|_y) \Delta x \, \Delta z}{\Delta x \, \Delta y \, \Delta z}$$

$$\left. + \frac{(\rho \mathbf{v} v_z|_{z+\Delta z} - \rho \mathbf{v} v_z|_z) \Delta x \, \Delta y}{\Delta x \, \Delta y \, \Delta z} \right]$$

$$= \frac{\partial}{\partial x}(\rho \mathbf{v} v_x) + \frac{\partial}{\partial y}(\rho \mathbf{v} v_y) + \frac{\partial}{\partial z}(\rho \mathbf{v} v_z) \qquad (9\text{-}12)$$

Performing the indicated differentiation of the right-hand side of equation (9-12) yields

$$\lim_{\Delta x, \Delta y, \Delta z \to 0} \frac{\iint \rho \mathbf{v}(\mathbf{v} \cdot \mathbf{n}) \, dA}{\Delta x \, \Delta y \, \Delta z} = \mathbf{v} \left[\frac{\partial}{\partial x}(\rho v_x) + \frac{\partial}{\partial y}(\rho v_y) + \frac{\partial}{\partial z}(\rho v_z) \right]$$

$$+ \rho \left[v_x \frac{\partial \mathbf{v}}{\partial x} + v_y \frac{\partial \mathbf{v}}{\partial y} + v_z \frac{\partial \mathbf{v}}{\partial z} \right]$$

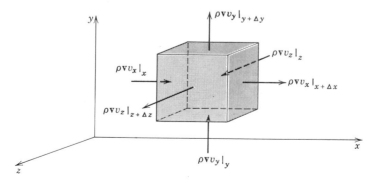

Figure 9.3 Momentum flux through a differential control volume.

The above term may be simplified with the aid of the continuity equation,

$$\frac{\partial \rho}{\partial t} + \frac{\partial}{\partial x}(\rho v_x) + \frac{\partial}{\partial y}(\rho v_y) + \frac{\partial}{\partial z}(\rho v_z) = 0 \tag{9-1}$$

which, upon substitution, yields

$$\lim_{\Delta x, \Delta y, \Delta z \to 0} \frac{\iint \rho \mathbf{v}(\mathbf{v} \cdot \mathbf{n}) \, dA}{\Delta x \, \Delta y \, \Delta z} = -\mathbf{v}\frac{\partial \rho}{\partial t} + \rho \left[v_x \frac{\partial \mathbf{v}}{\partial x} + v_y \frac{\partial \mathbf{v}}{\partial y} + v_z \frac{\partial \mathbf{v}}{\partial z} \right] \tag{9-13}$$

③ *Time Rate of Change of Momentum Within the Control Volume.* The time rate of change of momentum within the control volume may be evaluated directly:

$$\lim_{\Delta x, \Delta y, \Delta z \to 0} \frac{\partial/\partial t \iiint \mathbf{v}\rho \, dV}{\Delta x \, \Delta y \, \Delta z} = \frac{(\partial/\partial t)\rho \mathbf{v} \, \Delta x \, \Delta y \, \Delta z}{\Delta x \, \Delta y \, \Delta z} = \frac{\partial}{\partial t}\rho \mathbf{v} = \rho \frac{\partial \mathbf{v}}{\partial t} + \mathbf{v}\frac{\partial \rho}{\partial t} \tag{9-14}$$

We have now evaluated all terms in equation (9-8):

$$\left(\frac{\partial \sigma_{xx}}{\partial x} + \frac{\partial \tau_{yx}}{\partial y} + \frac{\partial \tau_{zx}}{\partial z} + \rho g_x\right)\mathbf{e}_x \tag{9-9}$$

$$① \lim_{\Delta x, \Delta y, \Delta z \to 0} \frac{\sum \mathbf{F}}{\Delta x \, \Delta y \, \Delta z} = \left\{ \left(\frac{\partial \tau_{xy}}{\partial x} + \frac{\partial \sigma_{yy}}{\partial y} + \frac{\partial \tau_{zy}}{\partial z} + \rho g_y\right)\mathbf{e}_y \right. \tag{9-10}$$

$$\left. \left(\frac{\partial \tau_{xz}}{\partial x} + \frac{\partial \tau_{yz}}{\partial y} + \frac{\partial \sigma_{zz}}{\partial z} + \rho g_z\right)\mathbf{e}_z \right. \tag{9-11}$$

$$② \lim_{\Delta x, \Delta y, \Delta z \to 0} \frac{\iint \rho \mathbf{v}(\mathbf{v} \cdot \mathbf{n}) \, dA}{\Delta x \, \Delta y \, \Delta z} = -\mathbf{v}\frac{\partial \rho}{\partial t} + \rho \left(v_x \frac{\partial \mathbf{v}}{\partial x} + v_y \frac{\partial \mathbf{v}}{\partial y} + v_z \frac{\partial \mathbf{v}}{\partial z} \right) \tag{9-13}$$

$$③ \lim_{\Delta x, \Delta y, \Delta z \to 0} \frac{\partial/\partial t \iiint \rho \mathbf{v} \, dV}{\Delta x \, \Delta y \, \Delta z} = \rho \frac{\partial \mathbf{v}}{\partial t} + \mathbf{v}\frac{\partial \rho}{\partial t} \tag{9-14}$$

It can be seen that the forces are expressed in components, while the rate-of-change-of-momentum terms are expressed as vectors. When the momentum terms are expressed as components, we obtain three differential equations which are the statements of Newton's second law in the x, y, and z directions:

$$\rho\left(\frac{\partial v_x}{\partial t} + v_x \frac{\partial v_x}{\partial x} + v_y \frac{\partial v_x}{\partial y} + v_z \frac{\partial v_x}{\partial z}\right) = \rho g_x + \frac{\partial \sigma_{xx}}{\partial x} + \frac{\partial \tau_{yx}}{\partial y} + \frac{\partial \tau_{zx}}{\partial z} \tag{9-15a}$$

$$\rho\left(\frac{\partial v_y}{\partial t} + v_x \frac{\partial v_y}{\partial x} + v_y \frac{\partial v_y}{\partial y} + v_z \frac{\partial v_y}{\partial z}\right) = \rho g_y + \frac{\partial \tau_{xy}}{\partial x} + \frac{\partial \sigma_{yy}}{\partial y} + \frac{\partial \tau_{zy}}{\partial z} \tag{9-15b}$$

$$\rho\left(\frac{\partial v_z}{\partial t} + v_x \frac{\partial v_z}{\partial x} + v_y \frac{\partial v_z}{\partial y} + v_z \frac{\partial v_z}{\partial z}\right) = \rho g_z + \frac{\partial \tau_{xz}}{\partial x} + \frac{\partial \tau_{yz}}{\partial y} + \frac{\partial \sigma_{zz}}{\partial z} \tag{9-15c}$$

It will be noted that in equations (9-15) above, the terms on the left-hand side represent the time-rate of change of momentum, and the terms on the right-hand side represent the forces. Focusing our attention on the left-hand terms in equation (9-15a), we see that

$$\underbrace{\frac{\partial v_x}{\partial t}}_{} + \underbrace{v_x\frac{\partial v_x}{\partial x} + v_y\frac{\partial v_x}{\partial y} + v_z\frac{\partial v_x}{\partial z}}_{} = \left(\frac{\partial}{\partial t} + v_x\frac{\partial}{\partial x} + v_y\frac{\partial}{\partial y} + v_z\frac{\partial}{\partial z}\right)v_x$$

local rate   rate of change in
of change v_x due to motion
of v_x

The first term, $\partial v_x/\partial t$, involves the time rate of change of v_x at a point, and is called the *local acceleration*. The remaining terms involve the velocity change from point to point, that is, the *convective acceleration*. The sum of these two bracketed terms is the total acceleration. The reader may verify that the terms on the left-hand side of equations (9-15) are all of the form

$$\left(\frac{\partial}{\partial t} + v_x\frac{\partial}{\partial x} + v_y\frac{\partial}{\partial y} + v_z\frac{\partial}{\partial z}\right)v_i$$

where $v_i = v_x$, v_y, or v_z. The above term is the substantial derivative of v_i.

When the substantial derivative notation is used, equations (9-15) become

$$\rho\frac{Dv_x}{Dt} = \rho g_x + \frac{\partial \sigma_{xx}}{\partial x} + \frac{\partial \tau_{yx}}{\partial y} + \frac{\partial \tau_{zx}}{\partial z} \qquad (9\text{-}16a)$$

$$\rho\frac{Dv_y}{Dt} = \rho g_y + \frac{\partial \tau_{xy}}{\partial x} + \frac{\partial \sigma_{yy}}{\partial y} + \frac{\partial \tau_{zy}}{\partial z} \qquad (9\text{-}16b)$$

and

$$\rho\frac{Dv_z}{Dt} = \rho g_z + \frac{\partial \tau_{xz}}{\partial x} + \frac{\partial \tau_{yz}}{\partial y} + \frac{\partial \sigma_{zz}}{\partial z} \qquad (9\text{-}16c)$$

Equations (9-16) are valid for any type of fluid, regardless of the nature of the stress rate-of-strain relation. If Stokes' viscosity relations, equations (7-13) and (7-14), are used for the stress components, equations (9-16) become

$$\rho\frac{Dv_x}{Dt} = \rho g_x - \frac{\partial P}{\partial x} - \frac{\partial}{\partial x}\left(\tfrac{2}{3}\mu\nabla\cdot\mathbf{v}\right) + \nabla\cdot\left(\mu\frac{\partial\mathbf{v}}{\partial x}\right) + \nabla\cdot(\mu\nabla v_x) \qquad (9\text{-}17a)$$

$$\rho\frac{Dv_y}{Dt} = \rho g_y - \frac{\partial P}{\partial y} - \frac{\partial}{\partial y}\left(\tfrac{2}{3}\mu\nabla\cdot\mathbf{v}\right) + \nabla\cdot\left(\mu\frac{\partial\mathbf{v}}{\partial y}\right) + \nabla\cdot(\mu\nabla v_y) \qquad (9\text{-}17b)$$

and

$$\rho\frac{Dv_z}{Dt} = \rho g_z - \frac{\partial P}{\partial z} - \frac{\partial}{\partial z}\left(\tfrac{2}{3}\mu\nabla\cdot\mathbf{v}\right) + \nabla\cdot\left(\mu\frac{\partial\mathbf{v}}{\partial z}\right) + \nabla\cdot(\mu\nabla v_z) \qquad (9\text{-}17c)$$

The above equations are called the Navier-Stokes* equations and are the

---

* L. M. H. Navier, Mémoire sur les Lois du Mouvements des Fluides, *Mem. de l'Acad. d. Sci.*, **6**, 398 (1822); C. G. Stokes, "On the Theories of the Internal Friction of Fluids in Motion," *Trans. Cambridge Phys. Soc.*, **8** (1845).

differential expressions of Newton's second law of motion for a newtonian fluid. As no assumptions relating to the compressibility of the fluid have been made, these equations are valid for both compressible and incompressible flows. In our study of momentum transfer we shall restrict our attention to incompressible flow with constant viscosity. In an incompressible flow, $\nabla \cdot \mathbf{v} = 0$. Equations (9-17) thus become

$$\rho \frac{Dv_x}{Dt} = \rho g_x - \frac{\partial P}{\partial x} + \mu \left(\frac{\partial^2 v_x}{\partial x^2} + \frac{\partial^2 v_x}{\partial y^2} + \frac{\partial^2 v_x}{\partial z^2} \right) \tag{9-18a}$$

$$\rho \frac{Dv_y}{Dt} = \rho g_y - \frac{\partial P}{\partial y} + \mu \left(\frac{\partial^2 v_y}{\partial x^2} + \frac{\partial^2 v_y}{\partial y^2} + \frac{\partial^2 v_y}{\partial z^2} \right) \tag{9-18b}$$

$$\rho \frac{Dv_z}{Dt} = \rho g_z - \frac{\partial P}{\partial z} + \mu \left(\frac{\partial^2 v_z}{\partial x^2} + \frac{\partial^2 v_z}{\partial y^2} + \frac{\partial^2 v_z}{\partial z^2} \right) \tag{9-18c}$$

These equations may be expressed in a more compact form in the single vector equation

$$\rho \frac{D\mathbf{v}}{Dt} = \rho \mathbf{g} - \nabla P + \mu \nabla^2 \mathbf{v} \tag{9-19}$$

The above equation is the Navier-Stokes equation for an incompressible flow. The Navier-Stokes equations are written in cartesian, cylindrical, and spherical coordinate form in Appendix E. As the development has been lengthy, let us review the assumptions and therefore the limitations of equation (9.19). The assumptions are:

1. Incompressible flow.
2. Constant viscosity.
3. Laminar flow.*

All of the above assumptions are associated with the use of the Stokes viscosity relation. If the flow is inviscid ($\mu = 0$), the Navier-Stokes equation becomes

$$\rho \frac{D\mathbf{v}}{Dt} = \rho \mathbf{g} - \nabla P \tag{9-20}$$

which is known as Euler's equation. Euler's equation has only one limitation, that being inviscid flow.

EXAMPLE 1
   Equation (9-19) may be applied to numerous flow systems to provide information regarding velocity variation, pressure gradients, and other information of the type achieved in Chapter 8. Many situations are of sufficient complexity to make the solution extremely

---

* Strictly speaking, equation (9-19) is valid for turbulent flow, as the turbulent stress is included in the momentum flux term. This will be illustrated in Chapter 13.

difficult and are beyond the scope of this text. A situation for which a solution can be obtained is illustrated in Figure 9.4.

Figure 9.4 shows the situation of an incompressible fluid confined between two parallel, vertical surfaces. One surface, shown to the left, is stationary, while the other is moving upward at a constant velocity, v_0. If we consider the fluid newtonian and the flow laminar, the governing equation of motion is the Navier-Stokes equation in the form given by equation (9-19). The reduction of each term in the vector equation into its applicable form is shown below:

$$\rho\frac{D\mathbf{v}}{Dt} = \rho\left\{\frac{\partial\mathbf{v}}{\partial t} + v_x\frac{\partial\mathbf{v}}{\partial x} + v_y\frac{\partial\mathbf{v}}{\partial y} + v_z\frac{\partial\mathbf{v}}{\partial z}\right\} = 0$$

$$\rho\mathbf{g} = -\rho g\mathbf{e}_y$$

$$\nabla P = \frac{dP}{dy}\mathbf{e}_y$$

where dP/dy is constant, and

$$\mu\nabla^2\mathbf{v} = \mu\frac{d^2 v_y}{dx^2}\mathbf{e}_y$$

The resulting equation to be solved is

$$0 = -\rho g - \frac{dP}{dy} + \mu\frac{d^2 v_y}{dx^2}$$

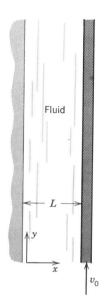

Figure 9.4 Fluid between two vertical plates with the one on the left stationary and the one on the right moving vertically upward with velocity v_0.

This differential equation is separable. The first integration yields

$$\frac{dv_y}{dx} + \frac{x}{\mu}\left\{-\rho g - \frac{dP}{dy}\right\} = C_1$$

Integrating once more, we obtain

$$v_y + \frac{x^2}{2\mu}\left\{-\rho g - \frac{dP}{dy}\right\} = C_1 x + C_2$$

The integration constants may be evaluated, using the boundary conditions that $v_y = 0$ at $x = 0$, and $v_y = v_0$ at $x = L$. The constants thus become

$$C_1 = \frac{v_0}{L} + \frac{L}{2\mu}\left\{-\rho g - \frac{dP}{dy}\right\} \qquad \text{and} \qquad C_2 = 0$$

The velocity profile may now be expressed as

$$v_y = \underbrace{\frac{1}{2\mu}\left\{-\rho g - \frac{dP}{dy}\right\}\{Lx - x^2\}}_{①} + \underbrace{v_0\frac{x}{L}}_{②} \qquad (9\text{-}21)$$

It is interesting to note, in equation (9-21), the effect of the terms labeled ① and ②, which are added. The first term is the equation for a symmetric parabola, the second for a straight line. Equation (9-21) is valid whether v_0 is upward, downward, or zero. In each case the terms may be added to yield the complete velocity profile. These results are indicated in Figure 9.5. The resulting velocity profile obtained by superposing the two parts is shown in each case.

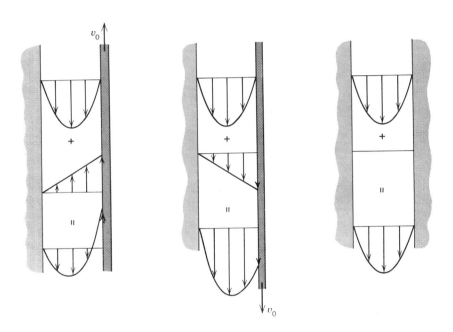

Figure 9.5   Velocity profiles for one surface moving upward, downward, or stationary.

Euler's equation may also be solved to determine velocity profiles, as will be shown in Chapter 10. The vector properties of Euler's equation are illustrated by the example below, in which the form of the velocity profile is given.

EXAMPLE 2

A rotating shaft, as illustrated in Figure 9.6, causes the fluid to move in circular streamlines with a velocity which is inversely proportional to the distance from the shaft. Find the shape of the free surface if the fluid can be considered inviscid.

As the pressure along the free surface will be constant, we may observe that the free surface is perpendicular to the pressure gradient. Determination of the pressure gradient therefore will enable us to evaluate the slope of the free surface.

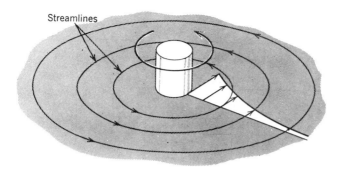

Figure 9.6 Rotating shaft in a fluid.

Rearranging equation (9-20), we have

$$\nabla P = \rho \mathbf{g} - \rho \frac{D\mathbf{v}}{Dt} \qquad (9\text{-}20)$$

The velocity $\mathbf{v} = A\mathbf{e}_\theta/r$, where A is a constant, when using the coordinate system shown in Figure 9.7. Assuming that there is no slip between the fluid and the shaft at the surface of the shaft, we have

$$v(R) = \omega R = \frac{A}{R}$$

and thus $A = \omega R^2$ and

$$\mathbf{v} = \frac{\omega R^2}{r} \mathbf{e}_\theta$$

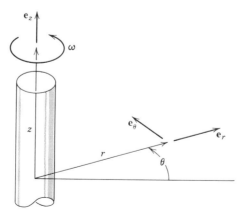

Figure 9.7 Cylindrical coordinate system for rotating shaft and fluid.

The substantial derivative Dv/Dt may be evaluated by taking the total derivative,

$$\frac{d\mathbf{v}}{dt} = -\frac{\omega R^2}{r^2}\mathbf{e}_\theta \dot{r} + \frac{\omega R^2}{r}\frac{d\mathbf{e}_\theta}{dt}$$

where $d\mathbf{e}_\theta/dt = -\dot{\theta}\mathbf{e}_r$. The total derivative becomes

$$\frac{d\mathbf{v}}{dt} = -\frac{\omega R^2}{r^2}\dot{r}\mathbf{e}_\theta - \frac{\omega R^2}{r}\dot{\theta}\mathbf{e}_r$$

Now the fluid velocity in the r direction is zero, and $\dot{\theta}$ for the fluid is v/r, so

$$\left(\frac{d\mathbf{v}}{dt}\right)_{\text{fluid}} = \frac{D\mathbf{v}}{Dt} = -\frac{\omega R^2}{r^2}v\mathbf{e}_r = -\frac{\omega^2 R^4}{r^3}\mathbf{e}_r$$

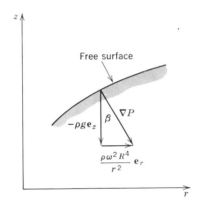

This result could have been obtained in a more direct manner by observing that $D\mathbf{v}/Dt$ is the local fluid acceleration, which for this case is $-v^2\mathbf{e}_r/r$. The pressure gradient becomes

$$\nabla P = -\rho g\mathbf{e}_z + \rho\frac{\omega^2 R^4 \mathbf{e}_r}{r^3}$$

From Figure 9.8 it can be seen that the free surface makes an angle β with the r axis so that

$$\tan\beta = \frac{\rho\omega^2 R^4}{r^3\rho g}$$

$$= \frac{\omega^2 R^4}{gr^3}$$

Figure 9.8 Free-surface slope.

9.3 BERNOULLI'S EQUATION

Euler's equation may be integrated directly for a particular case, flow along a streamline. In integrating Euler's equation the use of streamline coordinates is extremely helpful. Streamline coordinates s and n are illustrated in Figure 9.9. The s direction is parallel to the streamline and the n direction is perpendicular to the steamline, directed away from the instantaneous center of curvature. The flow and fluid properties are functions of position and time. Thus $\mathbf{v} = \mathbf{v}(s, n, t)$, and $P = P(s, n, t)$. The substantial derivatives of the velocity and pressure gradients in equation (9-20) must be expressed in terms of streamline coordinates so that equation (9-20) may be integrated.

Following the form used in equations (9-6) to obtain the substantial derivative, we have

$$\frac{d\mathbf{v}}{dt} = \frac{\partial\mathbf{v}}{\partial t} + \dot{s}\frac{\partial\mathbf{v}}{\partial s} + \dot{n}\frac{\partial\mathbf{v}}{\partial n}$$

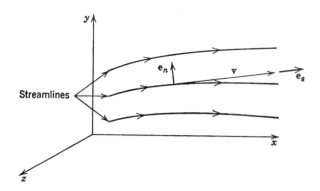

Figure 9.9 Streamline coordinates.

As the velocity of the fluid element has components $\dot{s} = v$, $\dot{n} = 0$ the substantial derivative of the velocity in streamline coordinates is

$$\frac{D\mathbf{v}}{Dt} = \frac{\partial \mathbf{v}}{\partial t} + v\frac{\partial \mathbf{v}}{\partial s} \tag{9-22}$$

The pressure gradient in streamline coordinates may be written as

$$\boldsymbol{\nabla} P = \frac{\partial P}{\partial s}\mathbf{e}_s + \frac{\partial P}{\partial n}\mathbf{e}_n \tag{9-23}$$

Taking the dot product of equation (9-20) with $\mathbf{e}_s\,ds$, and using equations (9-22) and (9-23), we obtain

$$\rho\left(\frac{\partial \mathbf{v}}{\partial t}\cdot \mathbf{e}_s\,ds + v\frac{\partial \mathbf{v}}{\partial s}\cdot \mathbf{e}_s\,ds\right) = \rho\mathbf{g}\cdot\mathbf{e}_s\,ds - \left(\frac{\partial P}{\partial s}\mathbf{e}_s + \frac{\partial P}{\partial n}\mathbf{e}_n\right)\cdot\mathbf{e}_s\,ds$$

or, as $\partial \mathbf{v}/\partial s \cdot \mathbf{e}_s = \partial/\partial s(\mathbf{v}\cdot\mathbf{e}_s) = \partial v/\partial s$, we have

$$\rho\left(\frac{\partial \mathbf{v}}{\partial t}\cdot\mathbf{e}_s\,ds + \frac{\partial}{\partial s}\left\{\frac{v^2}{2}\right\}ds\right) = \rho\mathbf{g}\cdot\mathbf{e}_s\,ds - \frac{\partial P}{\partial s}\,ds \tag{9-24}$$

Selecting $\mathbf{g}$ to act in the $-\mathbf{y}$ direction, we have $\mathbf{g}\cdot\mathbf{e}_s\,ds = -g\,dy$. For *steady incompressible* flow, equation (9-24) may be integrated to yield

$$\frac{v^2}{2} + gy + \frac{P}{\rho} = \text{constant} \tag{9-25}$$

which is known as Bernoulli's equation. The limitations are:

1. Inviscid flow.
2. Steady flow.
3. Incompressible flow.
4. The equation applies along a streamline.

Limitation 4 will be relaxed for certain conditions to be investigated in Chapter 10.

Bernoulli's equation was also developed in Chapter 6 from energy considerations for steady incompressible flow with constant internal energy. It is interesting to note that the constant internal energy assumption and the inviscid flow assumption must be equivalent, since the other assumptions were the same. We may note, therefore, that the viscosity in some way will effect a change in internal energy.

9.4 CLOSURE

We have developed the differential equations for the conservation of mass and Newton's second law of motion. These equations may be subdivided into two special groups:

$$\frac{\partial \rho}{\partial t} + \nabla \cdot \rho \mathbf{v} = 0 \tag{9-26}$$

(continuity equation)

Inviscid flow

$$\rho \frac{D\mathbf{v}}{Dt} = \rho \mathbf{g} - \nabla P \tag{9-27}$$

(Euler's equation)

Incompressible, viscous flow

$$\nabla \cdot \mathbf{v} = 0 \tag{9-28}$$

(continuity equation)

$$\rho \frac{D\mathbf{v}}{Dt} = \rho \mathbf{g} - \nabla P + \mu \nabla^2 \mathbf{v} \tag{9-29}$$

(Navier-Stokes equation
for incompressible flow)

In addition, the student should note the physical meaning of the substantial derivative and appreciate the compactness of the vector representation. In component form, for example, equation (9-29) comprises some 27 terms in cartesian coordinates.

PROBLEMS

9.1 Apply the law of conservation of mass to an element in a polar coordinate system and obtain the continuity equation for a steady, two-dimensional, incompressible flow.

9.2 In cartesian coordinates, show that

$$v_x \frac{\partial}{\partial x} + v_y \frac{\partial}{\partial y} + v_z \frac{\partial}{\partial z}$$

may be written $(\mathbf{v} \cdot \nabla)$. What is the physical meaning of the term $(\mathbf{v} \cdot \nabla)$?

9.3 In an incompressible flow, the volume of the fluid is constant. Using the continuity equation, $\nabla \cdot \mathbf{v} = 0$, show that the fluid volume change is zero.

9.4 Find $D\mathbf{v}/Dt$ in polar coordinates by taking the derivative of the velocity. (Hint: $\mathbf{v} = v_r(r, \theta, t)\mathbf{e}_r + v_\theta(r, \theta, t)\mathbf{e}_\theta$. Remember that the unit vectors have derivatives.)

9.5 For flow at very low speeds and with large viscosity (so-called creeping flows) such as occur in lubrication, it is possible to delete the inertia terms, $D\mathbf{v}/Dt$, from the Navier-Stokes equation. For flows at high velocity and small viscosity it is not proper to delete the viscous term $\nu\nabla^2\mathbf{v}$. Explain this.

9.6 Using the Navier-Stokes equations and the continuity equation, obtain an expression for the velocity profile between two flat, parallel plates.

9.7 Does the velocity distribution in example 2 satisfy continuity?

9.8 The atmospheric density may be approximated by the relation $\rho = \rho_0 \exp(-y/\beta)$, where $\beta = 22\,000$ ft. Determine the rate at which the density changes with respect to body falling at v fps. If $v = 20\,000$ fps at $100\,000$ ft, evaluate the rate of density change.

9.9 In a velocity field where $\mathbf{v} = 400[(y/L)^2\mathbf{e}_x + (x/L)^2\mathbf{e}_y]$ fps, determine the pressure gradient at the point $(L, 2L)$. The y axis is vertical, the density is 64.4 lb$_m$/ft^3 and the flow may be considered inviscid.

9.10 Write equations (9-17) in component form for cartesian coordinates.

9.11 Derive equation (2-3) from equation (9-27).

9.12 In polar coordinates, the continuity equation is

$$\frac{1}{r}\frac{\partial}{\partial r}(rv_r) + \frac{1}{r}\frac{\partial v_\theta}{\partial \theta} = 0$$

Show that
   (a) if $v_\theta = 0$, then $v_r = F(\theta)/r$;
   (b) if $v_r = 0$, then $v_\theta = f(r)$.

9.13 Using the laws for the addition of vectors and equation (9-19), show that in the absence of gravity,
(a) the fluid acceleration, pressure force, and viscous force all lie in the same plane;
(b) in the absence of viscous forces the fluid accelerates in the direction of decreasing pressure;
(c) a static fluid will always start to move in the direction of decreasing pressure.

9.14 Obtain the equations for a one-dimensional steady, viscous, compressible flow in the x direction from the Navier-Stokes equations. (These equations, together with an equation of state and the energy equation, may be solved for the case of weak shock waves.)

9.15 Obtain the equations for one-dimensional inviscid, unsteady, compressible flow.

10
INVISCID FLUID FLOW

An important area in momentum transfer is inviscid flow, in which, by virtue of the absence of shear stress, analytical solutions to the differential equations of fluid flow are possible.

The subject of inviscid flow has particular application in aerodynamics and hydrodynamics and general application to flow about bodies—so called external flows. In this chapter we shall introduce the fundamentals of inviscid flow analysis.

10.1 FLUID ROTATION AT A POINT

Consider the element of fluid shown in Figure 10.1. In time Δt the element will move in the xy plane as shown. In addition to translation, the element may

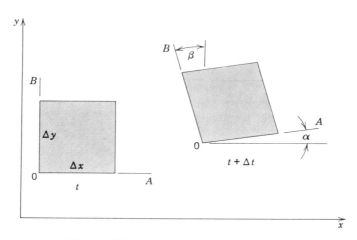

Figure 10.1 Rotation of a fluid element.

also deform and rotate. We have discussed the deformation previously in Chapter 7. Now let us focus our attention on the rotation of the element. Since the element may deform, the orientation will be given by the average rotation of the line segments $\overline{OB}$ and $\overline{OA}$ or by denoting the rotation by

$$\omega_z = \frac{d}{dt}\left(\frac{\alpha+\beta}{2}\right)$$

where the counterclockwise sense is positive. From Figure 10.1 we see that

$$\omega_z = \lim_{\Delta x, \Delta y, \Delta z, \Delta t \to 0} \frac{1}{2}\left(\frac{\arctan\{[(v_y|_{x+\Delta x} - v_y|_x)\,\Delta t]/\Delta x\}}{\Delta t}\right.$$
$$\left. + \frac{\arctan\{-[(v_x|_{y+\Delta y} - v_x|_y)\,\Delta t]/\Delta y\}}{\Delta t}\right)$$

which becomes, in the limit,

$$\omega_z = \frac{1}{2}\left(\frac{\partial v_y}{\partial x} - \frac{\partial v_x}{\partial y}\right) \tag{10-1}$$

The subscript z indicates that the rotation is about the z axis.

In the xz and yz planes the rotation at a point is given by

$$\omega_y = \frac{1}{2}\left(\frac{\partial v_x}{\partial z} - \frac{\partial v_z}{\partial x}\right) \tag{10-2}$$

and

$$\omega_x = \frac{1}{2}\left(\frac{\partial v_z}{\partial y} - \frac{\partial v_y}{\partial z}\right) \tag{10-3}$$

The rotation at a point is related to the vector cross product of the velocity. As the student may verify,

$$\nabla \times \mathbf{v} = \left(\frac{\partial v_z}{\partial y} - \frac{\partial v_y}{\partial z}\right)\mathbf{e}_x + \left(\frac{\partial v_x}{\partial z} - \frac{\partial v_z}{\partial x}\right)\mathbf{e}_y + \left(\frac{\partial v_y}{\partial x} - \frac{\partial v_x}{\partial y}\right)\mathbf{e}_z$$

and thus

$$\nabla \times \mathbf{v} = 2\boldsymbol{\omega} \tag{10-4}$$

The vector $\nabla \times \mathbf{v}$ is also known as the *vorticity*. When the rotation at a point is zero the flow is said to be *irrotational*. For irrotational flow $\nabla \times \mathbf{v} = 0$, as can be seen from (10-4). The significance of fluid rotation at a point may be examined by a different approach. The Navier-Stokes equation for incompressible flow, equation 9-29, may also be written in the form

$$\rho\frac{D\mathbf{v}}{Dt} = -\nabla P + \rho\mathbf{g} - \mu[\nabla \times (\nabla \times \mathbf{v})] \tag{9-29}$$

It may be observed from the above equation that if viscous forces act on a fluid, the flow must be rotational. Viscous forces act on the fluid through the vorticity. Furthermore, there is thermodynamic significance to irrotational flow. For a stady, adiabatic flow, either compressible or incompressible, an irrotational flow is an isentropic flow.

The kinematic condition $\nabla \times \mathbf{v} = 0$ is not the first time we have encountered a kinematic relation that satisfies one of the fundamental physical laws of fluid mechanics. The law of conservation of mass for an incompressible flow, $\nabla \cdot \mathbf{v} = 0$, is also expressed as a kinematic relation. The use of this relation is the subject of the next section.

10.2 THE STREAM FUNCTION

For a two-dimensional, incompressible flow, the continuity equation is

$$\nabla \cdot \mathbf{v} = \frac{\partial v_x}{\partial x} + \frac{\partial v_y}{\partial y} = 0 \tag{9-3}$$

Equation (9-3) indicates that v_x and v_y are related in some way so that $\partial v_x / \partial x = -(\partial v_y / \partial y)$. Perhaps the easiest way to express this relation is by having v_x and v_y both related to the same function. Consider the function $F(x, y)$; if $v_x = F(x, y)$, then

$$\frac{\partial v_y}{\partial y} = -\frac{\partial F}{\partial x} \qquad \text{or} \qquad v_y = -\int \frac{\partial F}{\partial x} \, dy$$

Unfortunately, the selection of $v_x = F(x, y)$ results in an integral for v_y. We can easily remove the integral sign if we make the original $F(x, y)$ equal to the derivative of some function with respect to y. For example, if $F(x, y) = (\partial \Psi(x, y)/\partial y]$, then

$$v_x = \frac{\partial \Psi}{\partial y}$$

As $\partial v_x / \partial x = -(\partial v_y / \partial y)$, we may write

$$\frac{\partial v_y}{\partial y} = -\frac{\partial}{\partial x} \left(\frac{\partial \Psi}{\partial y} \right) \qquad \text{or} \qquad \frac{\partial}{\partial y} \left(v_y + \frac{\partial \Psi}{\partial x} \right) = 0$$

for this to be true in general

$$v_y = -\frac{\partial \Psi}{\partial x}$$

Instead of having two unknowns, v_x and v_y, we now have only one unknown, Ψ. The unknown, Ψ, is called the *stream function*. The physical significance of Ψ can be seen from the following considerations. Since $\Psi = \Psi(x, y)$, the total derivative is

$$d\Psi = \frac{\partial \Psi}{\partial x} \, dx + \frac{\partial \Psi}{\partial y} \, dy$$

Also,

$$\frac{\partial \Psi}{\partial x} = -v_y \qquad \text{and} \qquad \frac{\partial \Psi}{\partial y} = v_x$$

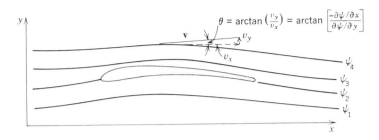

Figure 10.2   Streamlines and the stream function.

and thus

$$d\Psi = -v_y\, dx + v_x\, dy \tag{10-5}$$

Consider a path in the xy plane such that $\Psi =$ constant. Along this path, $d\Psi = 0$, and thus equation (10-5) becomes

$$\left.\frac{dy}{dx}\right|_{\Psi=\text{constant}} = \frac{v_y}{v_x} \tag{10-6}$$

The slope of the path $\Psi =$ constant is seen to be the same as the slope of a streamline as discussed in Chapter 3. The function $\Psi(x, y)$ thus represents the streamlines. Figure 10.2 illustrates the streamlines and velocity components for flow about an airfoil.

The differential equation which governs Ψ is obtained by consideration of the fluid rotation, ω, at a point. In a two-dimensional flow, $\omega_z = \frac{1}{2}[(\partial v_y/\partial x) - (\partial v_x/\partial y)]$, and thus if the velocity components v_y and v_x are expressed in terms of the stream function Ψ, we obtain, for an incompressible, steady flow,

$$-2\omega_z = \frac{\partial^2\Psi}{\partial x^2} + \frac{\partial^2\Psi}{\partial y^2} \tag{10-7}$$

When the flow is irrotational, equation (10-7) becomes Laplace's equation,

$$\nabla^2\Psi = \frac{\partial^2\Psi}{\partial x^2} + \frac{\partial^2\Psi}{\partial y^2} = 0 \tag{10-8}$$

10.3   INVISCID, IRROTATIONAL FLOW ABOUT AN INFINITE CYLINDER

In order to illustrate the use of the stream function, the inviscid, irrotational flow pattern about a cylinder of infinite length will be obtained by solving equation (10-8). The physical situation is illustrated in Figure 10.3. A stationary circular cylinder of radius a is situated in uniform, parallel flow in the x direction.

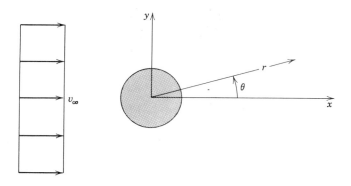

Figure 10.3 Cylinder in a uniform flow.

Making use of the cylindrical symmetry, we shall employ polar coordinates. In polar coordinates,* equation (10-8) becomes

$$\frac{\partial^2 \Psi}{\partial r^2} + \frac{1}{r}\frac{\partial \Psi}{\partial r} + \frac{1}{r^2}\frac{\partial^2 \Psi}{\partial \theta^2} = 0 \tag{10-9}$$

where the velocity components v_r and v_θ are given by

$$v_r = \frac{1}{r}\frac{\partial \Psi}{\partial \theta} \qquad v_\theta = -\frac{\partial \Psi}{\partial r} \tag{10-10}$$

In order to solve equation (10-9), four boundary conditions are required. These are as follows:

(1) The circle $r = a$ must be a streamline. Since the velocity normal to a streamline is zero, $v_r|_{r=a} = 0$ or $\partial \Psi/\partial \theta|_{r=a} = 0$.
(2) From symmetry, the line $\theta = 0$ must also be a streamline. Hence $v_\theta|_{\theta=0} = 0$ or $\partial \Psi/\partial r|_{\theta=0} = 0$.
(3) As $r \to \infty$ the velocity must be finite.
(4) The magnitude of the velocity as $r \to \infty$ is v_∞, a constant.

The solution of equation (10-9) for this case may be obtained using the method of separation of variables. Assuming that a solution exists in the form $\Psi(r, \theta) = F(rt)G(\theta)$, substitution into equation (10-9) yields

$$r^2\frac{F''(r)}{F(r)} + r\frac{F'(r)}{F(r)} = -\frac{G''(\theta)}{G(\theta)} \tag{10-11}$$

As the left-hand side of equation (10-11) is a function of r and the right-hand side

---

* The operator ∇^2 in cylindrical coordinates is developed in Appendix A.

is a function of θ, each side must be constant for the equality to exist over all values of r and θ. Thus equation (10-11) becomes two equations:

$$G''(\theta) + \lambda^2 G(\theta) = 0 \tag{10-12}$$

and

$$r^2 F''(r) + rF'(r) - \lambda^2 F(r) = 0 \tag{10-13}$$

Equation (10-12) is a simple second-order linear differential equation with a solution

$$G(\theta) = A \sin \lambda\theta + B \cos \lambda\theta \tag{10-14}$$

Equation (10-13) is known as an Euler equation* and has the solution

$$F(r) = Cr^\lambda + Dr^{-\lambda} \tag{10-15}$$

The boundary conditions listed above will determine the constants. From boundary condition 1 we have

$$\left.\frac{\partial \Psi}{\partial \theta}\right|_{r=a} = (Ca^\lambda + Da^{-\lambda})\lambda(A \cos \lambda\theta - B \sin \lambda\theta) = 0$$

and thus

$$D = -Ca^{2\lambda}$$

hence

$$\Psi(r,\theta) = (A' \sin \lambda\theta + B' \cos \lambda\theta)\left(r^\lambda - \frac{a^{2\lambda}}{r^\lambda}\right)$$

where

$$A' = AC \qquad \text{and} \qquad B' = BC$$

Boundary condition 2 states that at $\theta = 0$, we have $\partial \Psi/\partial r = 0$. As $\sin \theta = 0$, the only way this requirement can be met is to have $B' = 0$, yielding

$$\Psi(r,\theta) = A'(\sin \lambda\theta)\left(r^\lambda - \frac{a^{2\lambda}}{r^\lambda}\right)$$

Finally, conditions 3 and 4 require that the limit $(v_r^2 + v_\theta^2) = v_\infty^2$. As

$$v_r^2 + v_\theta^2 = A'^2 \frac{\lambda^2 \cos^2 \lambda\theta}{r^2}\left(r^\lambda - \frac{a^{2\lambda}}{r^\lambda}\right)^2 + A'^2\lambda^2 \sin^2 \lambda\theta\left(r^{\lambda-1} + \frac{a^2}{r^{\lambda+1}}\right)^2$$

$$v_r^2 + v_\theta^2 = A'^2\lambda^2\left[(\cos^2 \lambda\theta)\left(r^{\lambda-1} - \frac{a^{2\lambda}}{r^{\lambda+1}}\right)^2 + (\sin^2 \lambda\theta)\left(r^{\lambda-1} + \frac{a^{2\lambda}}{r^{\lambda+1}}\right)^2\right]$$

---

* The differential equation (10-13) is of a type investigated by Euler. This is not the same equation designated as Euler's equation in Chapter 9.

the only value of λ for which the velocity will be finite as $r \to \infty$ is unity. Using $\lambda = 1$ requires $A' = v_\infty$, and the stream function becomes

$$\Psi(r, \theta) = v_\infty r \sin \theta \left[1 - \frac{a^2}{r^2} \right] \tag{10-16}$$

The velocity components v_r and v_θ are calculated from equation (10-10),

$$v_r = \frac{1}{r} \frac{\partial \Psi}{\partial \theta} = v_\infty \cos \theta \left[1 - \frac{a^2}{r^2} \right] \tag{10-17}$$

and

$$v_\theta = -\frac{\partial \Psi}{\partial r} = -v_\infty \sin \theta \left[1 + \frac{a^2}{r^2} \right] \tag{10-18}$$

By setting $r = a$ in the above equations, the velocity at the surface of the cylinder may be determined. This results in

$$v_r = 0$$

and

$$v_\theta = -2v_\infty \sin \theta \tag{10-19}$$

The velocity in the radial direction is, of course, zero, as the cylinder surface is a streamline. The velocity along the surface is seen to be zero at $\theta = 0$ and $\theta = 180°$. These points of zero velocity are known as *stagnation points*. The forward stagnation point is at $\theta = 180°$, and the aft or rearward stagnation point is at $\theta = 0°$.

10.4 IRROTATIONAL FLOW, THE VELOCITY POTENTIAL

In a two-dimensional irrotational flow $\nabla \times \mathbf{v} = 0$, and thus $\partial v_x/\partial y = \partial v_y/\partial x$. The similarity of this equation to the continuity equation suggests that the type of relation used to obtain the stream function may be used again. Note, however, that the order of differentiation is reversed from the continuity equation. If we let $v_x = \partial \phi(x, y)/\partial x$, we observe that

$$\frac{\partial v_x}{\partial y} = \frac{\partial^2 \phi}{\partial x \, \partial y} = \frac{\partial v_y}{\partial x}$$

or

$$\frac{\partial}{\partial x} \left(\frac{\partial \phi}{\partial y} - v_y \right) = 0$$

and for the general case

$$v_y = \frac{\partial \phi}{\partial y}$$

The function ϕ is called the *velocity potential*. In order for ϕ to exist, the flow must be irrotational. As the condition of irrotationality is the only condition required, the velocity potential can also exist for compressible, unsteady flows. The velocity potential is commonly used in compressible flow analysis. Additionally, the velocity potential, ϕ, exists for three-dimensional flows, whereas the stream function does not.

The velocity vector is given by

$$\mathbf{v} = v_x \mathbf{e}_x + v_y \mathbf{e}_y + v_z \mathbf{e}_z = \frac{\partial \phi}{\partial x} \mathbf{e}_x + \frac{\partial \phi}{\partial y} \mathbf{e}_y + \frac{\partial \phi}{\partial z} \mathbf{e}_z$$

and thus, in vector notation,

$$\mathbf{v} = \nabla \phi \tag{10-20}$$

The differential equation defining ϕ is obtained from the continuity equation. Considering a steady incompressible flow, we have $\nabla \cdot \mathbf{v} = 0$; thus, using equation (10-20) for $\mathbf{v}$, we obtain

$$\nabla \cdot \nabla \phi = \nabla^2 \phi = 0 \tag{10-21}$$

which is again Laplace's equation; this time the dependent variable is ϕ. Clearly, Ψ and ϕ must be related. This relation may be illustrated by a consideration of isolines of Ψ and ϕ. An isoline of Ψ is, of course, a streamline. Along the isolines

$$d\Psi = \frac{\partial \Psi}{\partial x} dx + \frac{\partial \Psi}{\partial y} dy$$

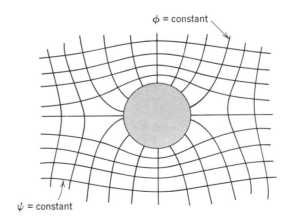

Figure 10.4 Streamlines and constant velocity potential lines for steady, incompressible, irrotational, inviscid flow about a cylinder.

or

$$\frac{dy}{dx}\bigg|_{\Psi=\text{constant}} = \frac{v_y}{v_x}$$

and

$$d\phi = \frac{\partial \phi}{\partial x} dx + \frac{\partial \phi}{\partial y} dy \qquad \frac{dy}{dx}\bigg|_{d\phi=0} = -\frac{v_x}{v_y}$$

accordingly

$$dy/dx\big|_{\phi=\text{constant}} = -\frac{1}{dy/dx}\bigg|_{\Psi=\text{constant}} \tag{10-22}$$

and thus Ψ and ϕ are orthogonal. The orthogonality of the stream function and the velocity potential is a useful property, particularly when graphical solutions to equations (10-8) and (10-21) are employed.

Figure 10.4 illustrates the inviscid, irrotational, steady incompressible flow about an infinite circular cylinder. Both the streamlines and constant-velocity potential lines are shown.

10.5 TOTAL HEAD IN IRROTATIONAL FLOW

The condition of irrotationality has been shown to be of great aid in obtaining analytical solutions in fluid flow. The physical meaning of irrotational flow can be illustrated by the relation between the rotation or vorticity, $\nabla \times \mathbf{v}$, and the total head, $P/\rho + v^2/2 + gy$. For an inviscid flow we may write

$$\frac{D\mathbf{v}}{Dt} = \mathbf{g} - \frac{\nabla P}{\rho} \qquad \text{(Euler's equation)}$$

and

$$\frac{D\mathbf{v}}{Dt} \equiv \frac{\partial \mathbf{v}}{\partial t} + \nabla\left(\frac{v^2}{2}\right) - \mathbf{v} \times (\nabla \times \mathbf{v}) \qquad \text{(Vector identity)}$$

As the gradient of the potential energy is $-\mathbf{g}$, Euler's equation becomes, for incompressible flow,

$$\nabla\left\{\frac{P}{\rho} + \frac{v^2}{2} + gy\right\} = \mathbf{v} \times (\nabla \times \mathbf{v}) - \frac{\partial \mathbf{v}}{\partial t} \tag{10-23}$$

If the flow is steady, it is seen from equation (10-23) that the gradient of the total head depends upon the vorticity, $\nabla \times \mathbf{v}$. The vector $(\nabla \times \mathbf{v})$ is perpendicular to the velocity vector; hence, the gradient of the total head has no component along a streamline. Thus, along a streamline in an incompressible, inviscid, steady flow,

$$\frac{P}{\rho} + \frac{v^2}{2} + gy = \text{constant} \tag{10-24}$$

This is, of course, Bernoulli's equation, which was discussed in Chapters 6 and 9. If the flow is irrotational and steady, equation (10-23) yields the result that Bernoulli's equation is valid throughout the flow field. An irrotational, steady, incompressible flow, therefore, has a constant total head throughout the flow field.*

10.6   UTILIZATION OF POTENTIAL FLOW

Potential flow has great utility in engineering for the prediction of pressure fields, forces, and flow rates. In the field of aerodynamics, for example, potential flow solutions are used to predict force and moment distributions on wings and other bodies.

An illustration of the determination of the pressure distribution from a potential flow solution may be obtained from the solution for the flow about a circular cylinder presented in section 10.3. From the Bernoulli equation

$$\frac{P}{\rho} + \frac{v^2}{2} = \text{constant} \tag{10-25}$$

We have deleted the potential energy term in accordance with the original assumption of uniform velocity in the x direction. At a great distance from the

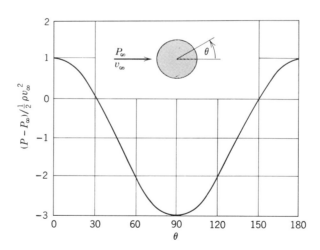

Figure 10.5   Pressure distribution on a cylinder in an inviscid, incompressible, steady flow.

* A more general result, Crocco's theorem, relates the vorticity to the entropy. Thus it can be shown that a steady, inviscid, irrotational flow, either compressible or incompressible, is isentropic.

cylinder the pressure is P_∞, and the velocity is v_∞, so equation (10-25) becomes*

$$P + \frac{\rho v^2}{2} = P_\infty + \frac{\rho v_\infty^2}{2} = P_0 \qquad (10\text{-}26)$$

where P_0 is designated the *stagnation pressure* (i.e., the pressure at which the velocity is zero). In accordance with equation (10-24) the stagnation pressure is constant throughout the field in an irrotational flow. The velocity at the surface of the body is $v_\theta = -2v_\infty \sin \theta$, thus the surface pressure is

$$P = P_0 - 2\rho v_\infty^2 \sin^2 \theta \qquad (10\text{-}27)$$

a plot of the potential flow pressure distribution about a cylinder is shown in Figure 10.5.

10.7   CLOSURE

In this chapter we have examined potential flow. A short summary of the properties of the stream function and the velocity potential is given below.

Stream function

1. A stream function $\Psi(x, y)$ exists for each and every two-dimensional, steady, incompressible flow, whether viscous or inviscid.
2. Lines for which $\Psi(x, y) = $ constant are streamlines.
3. In cartesian coordinates,

$$v_x = \frac{\partial \Psi}{\partial y} \qquad v_y = -\frac{\partial \Psi}{\partial x} \qquad (10\text{-}28a)$$

and in general,

$$v_s = \frac{\partial \Psi}{\partial n} \qquad (10\text{-}28b)$$

where n is $90°$ counterclockwise from s.
4. The stream function identically satisfies the continuity equation.
5. For an irrotational, steady incompressible flow,

$$\nabla^2 \Psi = 0 \qquad (10\text{-}28c)$$

Velocity potential

1. The velocity potential exists if and only if the flow is irrotational. No other restrictions are required.
2. $\nabla \phi = \mathbf{v}$.
3. For irrotational, incompressible flow, $\nabla^2 \phi = 0$.
4. For steady, incompressible two-dimensional flows, lines of constant velocity potential are perpendicular to the streamlines.

---

* The stagnation pressure as given in equation (10-26) applies to incompressible flow only.

PROBLEMS

10.1 In polar coordinates, show that

$$\nabla \times \mathbf{v} = \frac{1}{r}\left[\frac{\partial(rv_\theta)}{\partial r} - \frac{\partial v_r}{\partial \theta}\right]\mathbf{e}_z$$

10.2 Determine the fluid rotation at a point in polar coordinates, using the method illustrated in Figure 10.1.

10.3 Find the stream function for a flow with a uniform free stream velocity v_∞. The free stream velocity intersects the x axis at an angle α.

10.4 In polar coordinates, the continuity equation for steady incompressible flow becomes

$$\frac{1}{r}\frac{\partial}{\partial r}(rv_r) + \frac{1}{r}\frac{\partial v_\theta}{\partial \theta} = 0$$

Derive equations (10-10), using this relation.

10.5 A vortex is a flow pattern for which the streamlines are concentric circles. Find the stream function for an irrotational vortex.

10.6 Make an analytical model of a tornado using an irrotational vortex (with velocity inversely proportional to distance from the center) outside a central core (with velocity directly proportional to distance). Assume that the core diameter is 200 ft and the static pressure at the center of the core is 38 psf below ambient pressure. Find
(a) the maximum wind velocity;
(b) the time it would take a tornado moving at 60 mph to lower the static pressure from -10 psfg to -38 psfg;
(c) the variation in total pressure across the tornado. Euler's equation may be used to relate the pressure gradient in the core to the fluid acceleration.

10.7 For the flow about a cylinder find the velocity variation along the streamline leading to the stagnation point. What is the velocity derivative $\partial v_r/\partial r$ at the stagnation point?

10.8 In problem 10.7 explain how one could obtain $\partial v_\theta/\partial \theta$ at the stagnation point, using only r and $\partial v_r/\partial r$.

10.9 At what point on the surface of the circular cylinder in a potential flow does the pressure equal the free stream pressure?

10.10 For the velocity potentials given below, find the stream function and sketch the streamlines:

(a) $\phi = v_\infty L \left[\left(\dfrac{x}{L} \right)^3 - \dfrac{3xy^2}{L^3} \right]$

(b) $\phi = v_\infty \dfrac{xy}{L}$

(c) $\phi = \dfrac{v_\infty L}{2} \ln (x^2 + y^2)$

10.11 The flow pattern associated with flow to or from a point is called source flow or sink flow. The streamlines for this flow pattern are straight lines from the source point. Determine the stream function for source flow.

10.12 Find the stream function for a source at the origin. Add this stream function to the stream function for a uniform free stream $v_x = v_\infty$ and plot the streamline $\psi = 0$.

10.13 In the above problem, how far upstream does the flow from the source reach? The mass flow rate from the source is $\dot{m}$.

10.14 Determine the pressure gradient at the stagnation point of problem 10.10a.

10.15 Calculate the total lift force on the arctic hut shown below as a function of the location of the opening. The lift force results from the difference between the inside pressure and the outside pressure. Assume potential flow and the hut is in the shape of a half-cylinder.

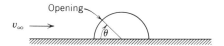

11
DIMENSIONAL ANALYSIS

An important consideration in all equations written thus far has been dimensional homogeneity. At times it has been necessary to use proper conversion factors in order that an answer be correct numerically and have the proper units. The idea of dimensional consistency can be used in another way, by a procedure known as dimensional analysis, to group the variables in a given situation into dimensionless parameters which are less numerous than the original variables. Such a procedure is very helpful in experimental work in which the very number of significant variables presents an imposing task of correlation. By combining the variables into a smaller number of dimensionless parameters, the work of experimental data reduction is considerably reduced.

This chapter will include means of evaluating dimensionless parameters both in situations in which the governing equation is known, and in those in which no equation is available. Certain dimensionless groups emerging from this analysis will be familiar, and some others will be encountered for the first time. Finally, certain aspects of similarity will be used to predict the flow behavior of equipment on the basis of experiments with scale models.

11.1 DIMENSIONS

In dimensional analysis certain dimensions must be established as fundamental, with all others expressible in terms of these. One of these fundamental dimensions is length, symbolized L. Thus area and volume may dimensionally be expressed as L^2 and L^3, respectively. A second fundamental dimension is time, symbolized t. The kinematic quantities, velocity and acceleration, may now be expressed as L/t and L/t^2, respectively.

Another fundamental dimension is mass, symbolized M. An example of a quantity whose dimensional expression involves mass is the density which would be expressed as M/L^3. Newton's second law of motion gives a relation between force and mass and allows force to be expressed dimensionally as $F = Ma = ML/t^2$. Some texts reverse this procedure and consider force fundamental, with mass expressed in terms of F, L, and t according to Newton's second law of motion. Here mass will be considered a fundamental unit.

The significant quantities in momentum transfer can all be expressed dimensionally in terms of M, L, and t; thus these comprise the fundamental dimensions we shall be concerned with presently. The dimensional analysis of energy problems in Chapter 19 will require the addition of two more fundamental dimensions, heat and temperature.

Some of the more important variables in momentum transfer and their dimensional representation in terms of M, L, and t are given in Table 11.1.

TABLE 11.1 IMPORTANT VARIABLES IN
MOMENTUM TRANSFER

Variable	Symbol	Dimension
mass	M	M
length	L	L
time	t	t
velocity	v	L/t
gravitational acceleration	g	L/t^2
force	F	ML/t^2
pressure	P	M/Lt^2
density	ρ	M/L^3
viscosity	μ	M/Lt
surface tension	σ	M/t^2
sonic velocity	a	L/t

11.2 GEOMETRIC AND KINEMATIC SIMILARITY

An application of experimental data achieved for a model to a full-sized prototype requires that certain similarities exist between the model and prototype. Two of these types of similarity are geometric and kinematic.

Geometric similarity exists between two systems if the ratio of significant dimensions is the same for the two systems. For example, if the ratio of the dimensions a/b for the model wing section in Figure 11.1 is equal in magnitude to the ratio a/b for the larger wing section they are geometrically similar. In this example there were only two significant dimensions. For more complicated geometries, geometric similarity would, of course, require all geometric ratios to be the same between a model and prototype.

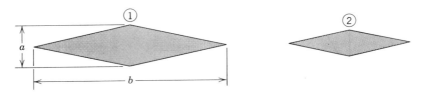

Figure 11.1   Two geometrically similar supersonic wing sections.

Kinematic similarly exists if, in geometrically similar systems (1) and (2), the velocities at the same points are related according to the relations

$$\left(\frac{v_x}{v_y}\right)_1 = \left(\frac{v_x}{v_y}\right)_2, \qquad \left(\frac{v_x}{v_z}\right)_1 = \left(\frac{v_x}{v_z}\right)_2,$$

Thus a requirement of kinematic similarity is that geometric similarity also exists.

11.3   DIMENSIONAL ANALYSIS OF THE NAVIER-STOKES EQUATION

If the differential equation describing a given flow situation is known, then dimensional homogeneity requires that each term in the equation have the same units. The ratio of one term in the equation to another must then, of necessity, be dimensionless. With knowledge of the physical meaning of the various terms in the equation we are then able to give some physical interpretation to the dimensionless parameters thus formed.

A classic example of this type of analysis involves the use of the Navier-Stokes equation in the form

$$\frac{D\mathbf{v}}{Dt} = \mathbf{g} - \frac{\nabla P}{\rho} + \nu\nabla^2\mathbf{v} \tag{9-19}$$

Each term in this expression can be expressed by a variable or combination of variables found in Table 11.1. Each term also has physical significance. The physical meaning and expression of each term are as follows:

$$\frac{D\mathbf{v}}{Dt} = \frac{\partial\mathbf{v}}{\partial t} + v_x\frac{\partial\mathbf{v}}{\partial x} + v_y\frac{\partial\mathbf{v}}{\partial y} + v_z\frac{\partial\mathbf{v}}{\partial z} \qquad \text{acceleration or inertial force, } v^2/L$$

$$\mathbf{g} \qquad \text{gravity force, } g$$

$$\frac{\nabla P}{\rho} \qquad \text{pressure force, } P/\rho L$$

$$\nu\nabla^2\mathbf{v} \qquad \text{viscous force, } \nu v/L^2$$

Each of these terms has dimensions of L/t^2, so the ratio of any two of them will produce a dimensionless group.* Dividing each of the terms on the right-side of equation (9-19) by the inertia forces, we form the following dimensionless parameters:

$$\frac{\text{gravity force}}{\text{inertia force}} = \frac{gL}{v^2}$$

$$\frac{\text{pressure force}}{\text{inertia force}} = \frac{P}{\rho v^2}$$

and

$$\frac{\text{viscous force}}{\text{inertia force}} = \frac{\nu}{Lv}$$

These dimensionless groups, as one might expect from their physical interpretation, appear often in fluid analyses, and they, or their reciprocals, are given special names as follows:

$$\frac{\text{inertia force}}{\text{gravity force}} = \frac{v^2}{gL} = \text{Fr, the } \textit{Froude number} \tag{11-1}$$

$$\frac{\text{pressure force}}{\text{inertia force}} = \frac{P}{\rho v^2} = \text{Eu, the } \textit{Euler number} \tag{11-2}$$

and

$$\frac{\text{inertia force}}{\text{viscous force}} = \frac{Lv}{\nu} = \text{Re, the } \textit{Reynolds number} \tag{11-3}$$

Note that, in addition to forming the various dimensionless groups, the dimensional analysis utilizing the governing differential equation also gives physical meaning to the parameters formed. The dimensional variables making up these parameters will vary with the particular situation. The length, velocity, and the like, to be used will be, in each case, that value which is most significant or representative. For instance, the significant length might be the diameter of a cylinder or the distance from the leading edge of a flat plate measured in the direction of flow; the applicable velocity might also be chosen in a different way for different situations. To avoid confusion it is advisable to specify clearly the reference length, reference velocity, and so on when reporting a value for any dimensionless parameter.

If, in geometrically similar systems, those parameters representing ratios of forces pertinent to the situation are equal, the systems are said to be dynamically similar. Obviously this condition would require that the pertinent dimensionless numbers be equal between two dynamically similar systems. *Dynamic similarity* is a fundamental requirement in extending experimental data from a model to its prototype.

---

* The conversion factor, g_c, must be used with the pressure term to make its units compatible with the other terms.

11.4   THE BUCKINGHAM METHOD

A more general situation in which dimensional analysis may be profitably employed is one in which there is no governing differential equation which clearly applies. In such a situation a more general procedure is required. Such a general procedure is known as the *Buckingham* method.*

The initial step in applying the Buckingham method requires the listing of the variables significant to a given problem. It is then necessary to determine the number of dimensionless parameters into which the variables may be combined. This number may be determined using the *Buckingham pi* theorem, which states:

> The number of dimensionless groups used to describe a situation involving n variables is equal to $n - r$, where r is the rank of the dimensional matrix of the variables.

Thus

$$i = n - r \tag{11-4}$$

where

i = the number of independent dimensionless groups

n = the number of variables involved

and

r = the rank of the dimensional matrix

The dimensional matrix is simply the matrix formed by tabulating the exponents of the fundamental dimensions M, L, and t, which appear in each of the variables involved.

An example of the evaluation of r and i, as well as the application of the Buckingham method, follows.

EXAMPLE 1

Determine the dimensionless groups formed from the variables involved in the flow of fluid external to a solid body. The force exerted on the body is a function of v, ρ, μ, and L (a significant dimension of the body).

A usual first step is to construct a table of the variables and their dimensions.

Variable	Symbol	Dimensions
force	F	ML/t^2
velocity	v	L/t
density	ρ	M/L^3
viscosity	μ	M/Lt
length	L	L

* E. Buckingham, *Phys. Rev.* **2**, 345 (1914).

Before determining the number of dimensionless parameters to be formed, we must know r. The dimensional matrix which applies is formed from the following tabulation:

	F	v	ρ	μ	L
M	1	0	1	1	0
L	1	1	-3	-1	1
t	-2	-1	0	-1	0

The numbers in the table represent the exponent of M, L, and t in the dimensional expression of each variable involved. For example, the dimensional expression of F is ML/t^2, hence the exponents 1, 1, and -2 are tabulated versus M, L, and t, respectively, the quantities with which they are associated. The matrix is then the array of numbers shown below:

$$\begin{pmatrix} 1 & 0 & 1 & 1 & 0 \\ 1 & 1 & -3 & -1 & 1 \\ -2 & -1 & 0 & -1 & 0 \end{pmatrix}$$

The rank, r, of a matrix is the number of rows (columns) in the largest nonzero determinant which can be formed from it. The rank is 3 in this case. Thus the number of dimensionless parameters to be formed may be found by applying equation (11-4). In this example $i = 5 - 3 = 2$.

The two dimensionless parameters will be symbolized π_1 and π_2 and may be formed in several ways. Initially, a *core group* of r variables must be chosen which will consist of those variables which will appear in each pi group and, among them, contain all of the fundamental dimensions. One way to choose a core is to exclude from it those variables whose effect one desires to isolate. In the present problem it would be desirable to have the drag force in only one dimensionless group, hence it will not be in the core. Let us arbitrarily let the viscosity be the other exclusion from the core. Our core group now consists of the remaining variables v, ρ, and L, which, we observe, include M, L, and t among them.

We now know that π_1 and π_2 both include ρ, L, and v; that one of them includes F and the other μ; and that they are both dimensionless. In order that each be dimensionless, the variables must be raised to certain exponents. Writing

$$\pi_1 = v^a \rho^b L^c F \qquad \text{and} \qquad \pi_2 = v^d \rho^e L^f \mu$$

we shall evaluate the exponents as follows. Considering each π group independently, we write

$$\pi_1 = v^a \rho^b L^c F$$

and dimensionally,

$$M^0 L^0 t^0 = 1 = \left(\frac{L}{t}\right)^a \left(\frac{M}{L^3}\right)^b (L)^c \frac{ML}{t^2}$$

Equating exponents of M, L, and t on both sides of this expression, we have, for M.

$$0 = b + 1$$

for L,

$$0 = a - 3b + c + 1$$

and for t,

$$0 = -a - 2$$

From these we find that $a = -2$, $b = -1$, and $c = -2$, giving

$$\pi_1 = \frac{F}{L^2 \rho v^2} = \frac{F/L^2}{\rho v^2} = \text{Eu}$$

Similarly for π_2 we have, in dimensional form,

$$1 = \left(\frac{L}{t}\right)^d \left(\frac{M}{L^3}\right)^e (L)^f \frac{M}{Lt}$$

and for exponents of M,

$$0 = e + 1$$

for L,

$$0 = d - 3e + f - 1$$

and for t,

$$0 = -d - 1$$

giving $d = -1$, $e = -1$. and $f = -1$. Thus for our second dimensionless group we have

$$\pi_2 = \mu/\rho v L = 1/\text{Re}$$

Dimensional analysis has enabled us to relate the original five variables in terms of only two dimensionless parameters in the form:

$$\text{Eu} = \phi(\text{Re}) \tag{11-5}$$

where $\phi(\text{Re})$ is some function of Re. The fact that the Euler number can be plotted vs. the Reynolds number has been experimentally verified, and a plot of experimental data for flow about a circular cylinder is presented in the above form in Chapter 12.

The foregoing example has illustrated the application of the Buckingham pi theorem and the subsequent steps required to correlate variables into dimensionless groups. The general rules have been expressed and must be followed in any dimensional analysis. Notice that this method does not give any physical meaning to the resulting dimensionless parameters.

11.5 MODEL THEORY

In the design and testing of large equipment involving fluid flow it is customary to build small models geometrically similar to the larger prototypes. Experimental data achieved for the models are then scaled up to the full-sized prototypes according to the requirements of geometric, kinematic, and dynamic similarity. The following example will illustrate the manner of utilizing model data to evaluate the conditions for a full-scale device.

EXAMPLE 2

Dynamic similarity may be obtained by use of a cryogenic wind tunnel in which nitrogen at low temperature and high pressures is employed as the working fluid. If nitrogen at 5 atm and 183 K is used to test the low speed aerodynamics of a prototype that has a 24.38 m wing span and is to fly at standard sea-level conditions at a speed of 60 m/s, determine:

1. The scale of the model to be tested.
2. The ratio of forces between the model and the full-scale aircraft.

Conditions of dynamic similarity should prevail. The speed of sound in nitrogen at 183 K is 275 m/s.

For dynamic similarity to exist, we know that both model and prototype must be geometrically similar and that the Reynolds number and the Mach number* must be the same. A table such as the following is helpful.

	Model	Prototype
Characteristic length	L	24.38 m
Velocity	V	60 m/s
Viscosity	μ	$1.789 \cdot 10^{-5}$ Pa $\cdot$ s
Density	ρ	1.225 kg/m^3
Speed of sound	275 m/s	340 m/s

The conditions listed for the prototype have been obtained from Appendix I. Equating Mach numbers we obtain

$$M_m = M_p$$

$$V = \frac{275}{340} 60 = 48.5 \text{ m/s}$$

Equating the Reynolds numbers of the model and the prototype we obtain

$$Re_m = Re_p$$

$$\frac{\rho 48.5 L}{\mu} = \frac{1.225 \cdot 60 \cdot 24.38}{1.789 \cdot 10^{-5}} = 1.002 \cdot 10^8$$

Using equation (7-10) we may evaluate μ for nitrogen. From Appendix K, $\epsilon/\kappa = 91.5$ K and $\sigma = 3.681$ Å for nitrogen so that $\kappa T/\epsilon = 2$ and $\Omega_\mu = 1.175$ (Appendix K). Thus

$$\mu = 2.6693 \cdot 10^{-6} \frac{\sqrt{28 \cdot 183}}{(3.681)^2 (1.175)} = 1.200 \cdot 10^{-5} \text{ Pa} \cdot \text{s}$$

The density may be approximated from the perfect gas law

$$\rho = \frac{P}{P_1} \frac{M}{M_1} \frac{T_1}{T} \rho_1$$

---

* The Mach number is a measure of the compressibility effects on fluid flow and is equal to the ratio of velocity to the speed of sound.

so that

$$\rho = 5\left(\frac{28}{28.96}\right)\left(\frac{288}{183}\right)1.225 = 7.608 \text{ kg/m}^3$$

Solving for the wing span of the model we obtain

$$L = 3.26 \text{ m } (10.7 \text{ ft})$$

The ratio of the forces on the model to the forces experienced by the prototype may be determined by noting that dynamic similarity will insure that dimensionless force coefficients for the model and the prototype will be equal. Hence

$$\left(\frac{F}{\rho V^2 A_R}\right)_{\text{model}} = \left(\frac{F}{\rho V^2 A_R}\right)_{\text{prototype}}$$

where A_R is a suitable reference area. For an aircraft this reference area is the projected wing area. The ratio of model force to prototype force is then given by

$$\frac{F_m}{F_p} = \frac{\rho_m}{\rho_p}\frac{V_m^{\ 2}}{V_p^{\ 2}}\frac{A_{R,m}}{A_{R,p}} = \frac{(\rho V^2)_m}{(\rho V^2)_p}\left(\frac{l_m}{l_p}\right)^2$$

where the ratio of reference areas can be expressed in terms of the scale ratio. Substituting numbers

$$\frac{F_m}{F_p} = \frac{7.608}{1.225}\left(\frac{48.5}{60.0}\right)^2\left(\frac{3.26}{24.38}\right)^2 = 0.0726$$

The forces on the model are seen to be 7.26% the prototype forces.

11.6   CLOSURE

The dimensional analysis of a momentum-transfer problem is simply an application of the requirement of dimensional homogeneity to a given situation. By dimensional analysis the work and time required to reduce and correlate experimental data are decreased substantially by the combination of individual variables into dimensionless π groups which are fewer in number than the original variables. The indicated relations between dimensionless parameters are then useful in expressing the performance of the systems to which they apply.

It should be kept in mind that dimensional analysis *cannot* predict which variables are important in a given situation, nor does it give any insight into the physical transfer mechanism involved. Even with these limitations, dimensional analysis techniques are a valuable aid to the engineer.

If the equation describing a given process is known, the number of dimensionless groups is automatically determined by taking ratios of the various terms in the expression to one another. This method also gives physical meaning to the groups thus obtained.

If, on the other hand, no equation applies, an empirical method, the *Buckingham* method, may be used. This is a very general approach but gives no physical meaning to the dimensionless parameters obtained from such an analysis.

The requirements of geometric, kinematic, and dynamic similarity enable one to use model data to predict the behavior of a prototype or full-size piece of equipment. *Model theory* is thus an important application of the parameters obtained in a dimensional analysis.

PROBLEMS

11.1 The power output of a hydraulic turbine depends on the diameter D of the turbine, the density ρ of water, the height H of water surface above the turbine, the gravitational acceleration g, the angular velocity ω of the turbine wheel, the discharge Q of water through the turbine, and the efficiency, η, of the turbine. By dimensional analysis, generate a set of appropriate dimensionless groups.

11.2 Through a series of tests on pipe flow, H. Darcy derived an equation for the friction loss in pipe flow as

$$h_L = f\frac{L}{D}\frac{v^2}{2g}$$

in which f is a dimensionless coefficient which depends on (a) the average velocity v of the pipe flow, (b) the pipe diameter D, (c) the fluid density ρ, (d) the fluid viscosity μ, and (e) the average pipe wall unevenness e (length). Using the Buckingham π theorem, find a dimensionless function for the coefficient f.

11.3 The pressure rise across a pump P (this term is proportional to the head developed by the pump) may be considered to be affected by the fluid density ρ, the angular velocity ω, the impeller diameter D, the volumetric rate of flow Q, and the fluid viscosity μ. Find the pertinent dimensionless groups, choosing them so the P, Q, and μ each appear in one group only. Find similar expressions, replacing the pressure rise first by the power input to the pump, then by the efficiency of the pump.

11.4 A rough method of scaling up cylindrical liquid-mixing tanks and impellers is to keep the power input per unit volume constant. If it is desired to increase the volume of a properly baffled liquid mixer by a ratio of 3, by what ratio must the tank diameter and impeller speed be changed? The mixers are geometrically similar, and both operate in the complete turbulent region. The power supplied to the mixing propeller, P, may be assumed to be a function of the propeller diameter, D, its angular velocity ω, and the liquid density ρ.

11.5 A $\frac{1}{6}$-scale model of a torpedo is tested in a water tunnel to determine drag characteristics. What model velocity corresponds to a torpedo velocity of 20 knots? If the model resistance is 10 lb, what is the prototype resistance?

11.6 The maximum pitching moment that is developed by the water on a flying boat as it lands is noted as c_{max}. The following are the variables involved in this action:

α = angle made by flight path of plane with horizontal

β = angle defining attitude of plane

M = mass of plane

L = length of hull

ρ = density of water

g = acceleration of gravity

R = radius of gyration of plane about axis of pitching

(a) According to the Buckingham pi theorem, how many independent dimensionless groups should there be which characterize this problem?
(b) What is the dimensional matrix of this problem? What is its rank?
(c) Evaluate the appropriate dimensionless parameters for this problem.

11.7 A car is traveling along a road at 22.2 m/s. Calculate the Reynolds number
(a) based on the length of the car,
(b) based on the diameter of the radio antenna.
The car length is 5.8 m and the antenna diameter is 6.4 mm.

11.8 During the development of a 300-ft ship, it is desired to test a 10% scale model in a towing tank to determine the drag characteristics of the hull. Determine how the model is to be tested if the Froude number is to be duplicated.

11.9 A 25% scale model of an undersea vehicle which has a maximum speed of 16 m/s is to be tested in a wind tunnel with a pressure of 6 atm to determine the drag characteristics of the full-scale vehicle. The model is 3 m long. Find the air speed required to test the model and find the ratio of the model drag to the full-scale drag.

11.10 Introduce the following dimensionless terms into the Navier-Stokes equation:

$$\mathbf{v}^* = \mathbf{v}/v_\infty, \text{ dimensionless velocity}$$
$$P^* = P/\rho v_\infty^2, \text{ dimensionless pressure}$$
$$t^* = tv_\infty/L, \text{ dimensionless time}$$
$$x^* = x/L, \text{ dimensionless distance}$$

The operator ∇ may then be written $\nabla = \nabla^*/L$. Show that the Navier-Stokes equation becomes

$$\frac{D\mathbf{v}^*}{Dt^*} = \frac{\mathbf{g}L}{v_\infty^{\,2}} - \nabla^* P^* + \frac{1}{\text{Re}}\nabla^{*2}\mathbf{v}^*$$

11.11 The rate at which metallic ions are electroplated from a dilute electrolytic solution onto a rotating disk electrode is usually governed by the mass diffusion rate of ions to the disk. This process is believed to be controlled by the following variables:

	Dimensions
k = mass transfer coefficient	L/t
D = diffusion coefficient	L^2/t
d = disk diameter	L
a = angular velocity	$1/t$
ρ = density	M/L^3
μ = viscosity	M/Lt

Obtain the set of dimensionless groups for these variables, where k, μ, and D are kept in separate groups. How would you accumulate and present the experimental data for this system?

11.12 The performance of a journal bearing around a rotating shaft is a function of the following variables: Q, the rate of flow lubricating oil to the bearing in volume per unit time; D, the bearing diameter; N, the shaft speed in revolutions per minute; μ, the lubricant viscosity; ρ, the lubricant density; and σ, the surface tension of the lubricating oil. Suggest appropriate parameters to be used in correlating experimental data for such a system.

11.13 The mass, M, of drops formed by a liquid discharging by gravity from a vertical tube is a function of tube diameter, D, liquid density, surface tension, and the acceleration of gravity. Determine the independent dimensionless groups that would allow the surface-tension effect to be analyzed. Neglect any effects of viscosity.

11.14 The fundamental frequency, n, of a stretched string is a function of the string length, L, its diameter, D, the mass density, ρ, and the applied tensile force, T. Suggest a set of dimensionless parameters relating these variables.

11.15 The power, P, required to run a compressor varies with compressor diameter D, angular velocity ω, volume flow rate Q, fluid density ρ, and fluid viscosity μ. Develop a relation between these variables by dimensional analysis, where fluid viscosity and angular velocity appear in only one dimensionless parameter.

11.16 An estimate is needed of the lift provided by a hydrofoil wing section when it moves through water at 60 mph. Test data are available for this purpose from experiments in a pressurized wind tunnel with an airfoil section model geometrically similar to but twice the size of the hydrofoil. If the lift F_1 is a function of the density ρ of the fluid, the velocity v of the flow, the angle of attack θ, the chord length D, and the viscosity μ, what velocity of flow in the wind tunnel would correspond to the hydrofoil velocity for which the estimate is desired? Assume the same angle of attack in both cases, that the density of the air in the pressurized tunnel was 5.0×10^{-3} slugs/ft^3, that its kinematic viscosity was 8.0×10^{-5} ft^2/s, and that the kinematic viscosity of the water is approximately 1.0×10^{-5} ft^2/s. Take the density of the water to be 1.94 slugs/ft^3.

11.17 A model of a harbor is made on the length ratio of 360 : 1. Storm waves of 2 m amplitude and 8 m/s velocity occur on the breakwater of the prototype harbor. Significant variables are the length scale, velocity and g, the acceleration of gravity. The scaling of time can be made with the aid of the length scale and velocity scaling factors.
(a) Neglecting friction, what should be the size and speed of the waves in the model?
(b) If the time between tides in the prototype is 12 hr, what should be the tidal period in the model?

11.18 A 40% scale model of an airplane is to be tested in a flow regime where unsteady flow effects are important. If the full-scale vehicle experiences the unsteady effects at a Mach number of 1 at an altitude of 40 000 ft, what pressure must the model be tested at to produce an equal Reynolds number? The model is to be tested in air at 70 F. What will the time scale of the flow about the model be relative to the full-scale vehicle?

11.19 A model ship propeller is to be tested in water at the same temperature that would be encountered by a full-scale propeller. Over the speed range considered, it is assumed that there is no dependence on the Reynolds or Euler numbers, but only on the Froude number (based on forward velocity V and propeller diameter d). In addition, it is thought that the ratio of forward to rotational speed of the propeller must be constant (the ratio V/Nd, where N is propeller rpm).
(a) With a model 0.41 m in diameter, a forward speed of 2.58 m/s and a rotational speed of 450 rpm is recorded. What are the forward and rotational speeds corresponding to a 2.45-m diameter prototype?
(b) A torque of 20 N·m is required to turn the model, and the model thrust is measured to be 245 N. What are the torque and thrust for the prototype?

12
VISCOUS FLOW

The concept of fluid viscosity was developed and viscosity defined in Chapter 7. Clearly all fluids are viscous, but in certain situations and under certain conditions a fluid may be considered ideal or inviscid, making possible an analysis by the methods of Chapter 10.

Our task in this chapter is to consider viscous fluids and the role of viscosity as it affects the flow. Of particular interest is the case of flow past solid surfaces and the interrelations between the surfaces and the flowing fluid.

12.1   REYNOLDS' EXPERIMENT

The existence of two distinct types of viscous flow is a universally accepted phenomenon. The smoke emanating from a lighted cigarette is seen to flow smoothly and uniformly for a short distance from its source and then change abruptly into a very irregular, unstable pattern. Similar behavior may be observed for water flowing slowly from a faucet.

The well-ordered type of flow occurs when adjacent fluid layers slide smoothly over one another with mixing between layers or lamina occurring only on a molecular level. It was for this type of flow that Newton's viscosity relation was derived, and in order for us to measure the viscosity, μ, this *laminar* flow must exist.

The second flow regime, in which small packets of fluid particles are transferred between layers, giving it a fluctuating nature, is called the *turbulent* flow regime.

The existence of laminar and turbulent flow, although recognized earlier, was first described quantitatively by Reynolds in 1883. His classic experiment is illustrated in Figure 12.1. Water was allowed to flow through a transparent pipe as shown at a rate controlled by a valve. A dye having the same specific gravity as water was introduced at the pipe opening and its pattern observed for

163

progressively larger flow rates of water. At low rates of flow the dye pattern was regular and formed a single line of color as shown in Figure 12.1*a*. At high flow rates, however, the dye became dispersed throughout the pipe cross section because of the very irregular fluid motion. The difference in appearance of the dye streak was, of course, due to the orderly nature of laminar flow in the first case and to the fluctuating character of turbulent flow in the latter case.

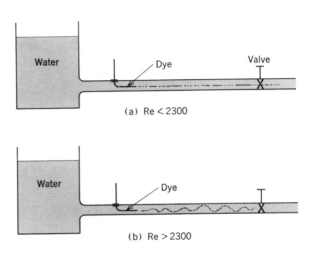

Figure 12.1   Reynolds' experiment.

The transition from laminar to turbulent flow in pipes is thus a function of the fluid velocity. Actually, Reynolds found that fluid velocity was only one variable determining the nature of pipe flow, the others being pipe diameter, fluid density, and fluid viscosity. These four variables, combined into the single dimensionless parameter

$$\mathrm{Re} \equiv \frac{D \rho v}{\mu} \qquad (12\text{-}1)$$

form the Reynolds number, symbolized Re, in honor of Osborne Reynolds and his important contributions to fluid mechanics.

For flow in circular pipes it is found that below a value for Reynolds number of 2300 the flow is *laminar*. Above this value the flow may be laminar as well, and indeed, laminar flow has been observed for Reynolds numbers as high as 40 000 in experiments wherein external disturbances were minimized. Above a Reynolds number of 2300, small disturbances will cause a transition to *turbulent* flow while below this value disturbances are damped out and laminar flow prevails. The *critical Reynolds number for pipe flow* thus is 2300.

12.2 DRAG

Reynolds' experiment clearly demonstrated the two different regimes of flow: laminar and turbulent. Another manner of illustrating these different flow regimes and their dependence upon the Reynolds number is through the consideration of drag. A particularly illustrative case is that of external flow (i.e., flow around a body as opposed to flow inside a conduit).

The drag force due to friction is caused by the shear stresses at the surface of a solid object moving through a viscous fluid. Frictional drag is evaluated by using the expression

$$\frac{F}{A} = C_f \frac{\rho v_\infty^2}{2} \qquad (12\text{-}2)$$

where F is the force; A is the area of contact between the solid body and the fluid; C_f is the coefficient of skin friction; ρ is the fluid density; and v_∞ is the free-stream fluid velocity.

The coefficient of skin friction, C_f, which is defined by equation (12-2), is dimensionless.

The total drag on an object may be due to pressure as well as frictional effects. In such a situation another coefficient, C_D, is defined as

$$\frac{F}{A_P} \equiv C_D \frac{\rho v_\infty^2}{2} \qquad (12\text{-}3)$$

where F, ρ, and v_∞ are as described above and, additionally,

$$C_D = \text{the drag coefficient}$$

and

$$A_P = \text{the projected area of the surface}$$

The value of A_P used in expressing the drag for blunt bodies is normally the maximum projected area for the body.

The quantity $\rho v_\infty^2/2$ appearing in equations (12-2) and (12-3) is frequently called the *dynamic pressure*.

Pressure drag arises from two sources.* One is induced drag, or drag due to lift. The other source is wake drag, which arises from the fact that the shear stress causes the streamlines to deviate from their inviscid flow paths, and in some cases to separate from the body altogether. This deviation in streamline pattern prevents the pressure over the rest of a body from reaching the level it would attain otherwise. As the pressure at the front of the body is now greater that that at the rear, a net rearward force develops.

In an incompressible flow, the drag coefficient depends upon the Reynolds number and the geometry of a body. A simple geometric shape that illustrates the

---

* A third source of pressure drag, wave drag, is associated with shock waves.

drag dependence upon the Reynolds number is the circular cylinder. The inviscid flow about a circular cylinder was examined in Chapter 10. The inviscid flow about a cylinder of course, produced no drag, as there existed neither frictional nor pressure drag. The variation in the drag coefficient with the Reynolds number for a smooth cylinder is shown in Figure 12.2. The flow pattern about the cylinder

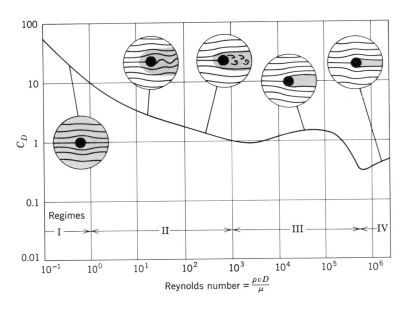

Figure 12.2 Drag coefficient for circular cylinders as a function of Reynolds number. Shaded regions indicate areas influenced by shear stress.

is illustrated for several different values of Re. The flow pattern and general shape of the curve suggest that the drag variation, and hence the effects of shear stress on the flow, may be subdivided into four regimes. The features of each regime will be examined.

REGIME 1

In this regime the entire flow is laminar and the Reynolds number small, being less than 1. Recalling the physical significance of the Reynolds number from Chapter 11 as the ratio of the inertia forces to the viscous forces, we may say that in regime 1 the viscous forces predominate. The flow pattern in this case is almost symmetric, the flow adheres to the body, and the wake is free from oscillations. In this regime of so-called *creeping flow* viscous effects predominate and extend throughout the flow field.

REGIME 2

Two illustrations of the flow pattern are shown in the second regime. As the Reynolds number is increased, small eddies form at the rear stagnation point of the cylinder. At higher values of the Reynolds number these eddies grow to the point at which they separate from the body and are swept downstream into the wake. The pattern of eddies shown in regime 2 is called a von Kármán vortex trail. This change in the character of the wake from a steady to an unsteady nature is accompanied by a change in the slope of the drag curve. The paramount features of this regime are (a) the unsteady nature of the wake, and (b) flow separation from the body.

REGIME 3

In the third regime the point of flow separation stabilizes at a point about 80° from the forward stagnation point. The wake is no longer characterized by large eddies, although it remains unsteady. The flow on the surface of the body from the stagnation point to the point of separation is laminar, and the shear stress in this interval is appreciable only in a thin layer near the body. The drag coefficient levels out at a near-constant value of approximately 1.

REGIME 4

At a Reynolds number near 5×10^5 the drag coefficient suddenly decreases to 0.3. When the flow about the body is examined, it is observed that the point of separation has moved past 90°. In addition, the pressure distribution about the cylinder (shown in Figure 12.3) up to the point of separation is fairly close to the inviscid flow pressure distribution depicted in Figure 10.5. In the figure it will be noticed that the pressure variation about the surface is a changing function of Reynolds number. The minimum point on the curves for Reynolds numbers of 10^5 and 6×10^5 are both at the point of flow separation. From this figure it is seen that separation occurs at a larger value of θ for $Re = 6 \times 10^5$ than it does for $Re = 10^5$.

The layer of flow near the surface of the cylinder is turbulent in this regime, undergoing transition from laminar flow close to the forward stagnation point. The marked decrease in drag is due to the change in the point of separation. In general, a turbulent flow resists flow separation better than a laminar flow. As the Reynolds number is large in this regime, it may be said that the inertial forces predominate over the viscous forces.

The four regimes of flow about a cylinder illustrate the decreasing realm of influence of viscous forces as the Reynolds number is increased. In regimes 3 and 4, the flow pattern over the forward part of the cylinder agrees well with the inviscid flow theory. For other geometries a similar variation in the realm of influence of viscous forces is observed and, as might be expected, agreement with inviscid-flow predictions at a given Reynolds number increases as the slenderness of the body increases. The majority of cases of engineering interest involving external flows have flow fields similar to those of regimes 3 and 4.

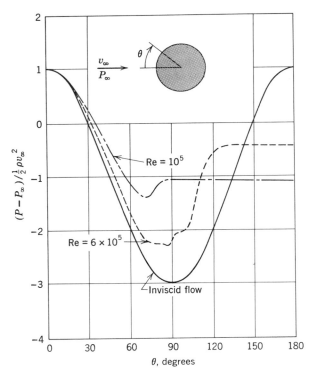

Figure 12.3 Pressure distribution on a circular cylinder
at various Reynolds numbers.

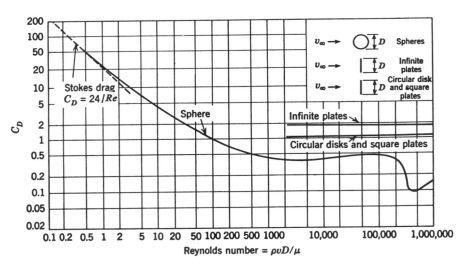

Figure 12.4 Drag coefficient versus Reynolds number for various objects.

Figure 12.4 shows the variation in the drag coefficient with the Reynolds number for a sphere, for infinite plates, and for circular disks and square plates. Note the similarity in form of the curve of C_D for the sphere to that for a cylinder in Figure 12.2. Specifically one may observe the same sharp decrease in C_D to a minimum value near a Reynolds number value of 5×10^5. This is again due to the change from laminar to turbulent flow in the boundary layer.

12.3 THE BOUNDARY-LAYER CONCEPT

The observation of a decreasing region of influence of shear stress as the Reynolds number is increased led Ludwig Prandtl to the boundary-layer concept in 1904. According to Prandtl's hypothesis, the effects of fluid friction at high Reynolds number are limited to a thin layer near the boundary of a body, hence the term *boundary layer*. Further, there is no significant pressure change across the boundary layer. This means that the pressure in the boundary layer is the same as the pressure in the inviscid flow outside the boundary layer. The significance of the Prandtl theory lies in the simplification that it allows in the analytical treatment of viscous flows. The pressure, for example, may be obtained from experiment or inviscid flow theory. Thus the only unknowns are the velocity components.

The boundary layer on a flat plate is shown in Figure 12.5. The thickness of the boundary layer, δ, is arbitrarily taken as the distance away from the surface where the velocity reaches 99% of the free-stream velocity. The thickness is exaggerated for clarity.

Figure 12.5 illustrates how the thickness of the boundary layer increases with distance x from the leading edge. At relatively small values of x, flow within the boundary layer is laminar, and this is designated as the laminar boundary-layer

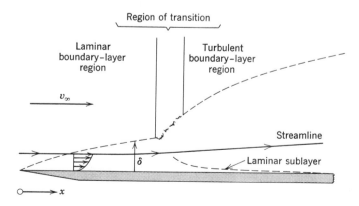

Figure 12.5 Boundary layer on a flat plate. (The thickness is exaggerated for clarity.)

region. At larger values of x the transition region is shown where fluctuations between laminar and turbulent flow occur within the boundary layer. Finally, for a certain value of x, and above, the boundary layer will always be turbulent. In the region in which the boundary layer is turbulent, there exists, as shown, a very thin film of fluid called the *laminar sublayer*, wherein flow is still laminar and large velocity gradients exist.

The criterion for the type of boundary layer present is the magnitude of Reynolds number, Re_x, known as the *local Reynolds number*, based on the distance x from the leading edge. The local Reynolds number is defined as

$$\text{Re}_x \equiv \frac{xv\rho}{\mu} \qquad (12\text{-}4)$$

For flow past a flat plate, as shown in Figure 12.5, experimental data indicate that for

(a) $\text{Re}_x < 2 \times 10^5$ the boundary layer is laminar
(b) $2 \times 10^5 < \text{Re}_x < 3 \times 10^6$ the boundary layer may be either laminar or turbulent
(c) $3 \times 10^6 < \text{Re}_x$ the boundary layer is turbulent

12.4   THE BOUNDARY-LAYER EQUATIONS

The concept of a relatively thin boundary layer at high Reynolds numbers leads to some important simplifications of the Navier–Stokes equations. For incompressible, two-dimensional flow over a flat plate, the Navier–Stokes equations are

$$\rho\left\{\frac{\partial v_x}{\partial t} + v_x\frac{\partial v_x}{\partial x} + v_y\frac{\partial v_x}{\partial y}\right\} = \frac{\partial\sigma_{xx}}{\partial x} + \frac{\partial\tau_{yx}}{\partial y} \qquad (12\text{-}5)$$

and

$$\rho\left\{\frac{\partial v_y}{\partial t} + v_x\frac{\partial v_y}{\partial x} + v_y\frac{\partial v_y}{\partial y}\right\} = \frac{\partial\tau_{xy}}{\partial x} + \frac{\partial\sigma_{yy}}{\partial y} \qquad (12\text{-}6)$$

where $\tau_{xy} = \tau_{yx} = \mu(\partial v_x/\partial y + \partial v_y/\partial x)$, $\sigma_{xx} = -P + 2\mu(\partial v_x/\partial x)$ and $\sigma_{yy} = -P + 2\mu(\partial v_y/\partial y)$. The shear stress in a thin boundary layer is closely approximated by $\mu(\partial v_x/\partial y)$. This can be seen by considering the relative magnitudes of $\partial v_x/\partial y$ and $\partial v_y/\partial x$. From Figure 12.5 we may write $v_x|_\delta/v_y|_\delta \sim \mathbb{O}(x/\delta)$, where $\mathbb{O}$ signifies the order of magnitude. Then

$$\frac{\partial v_x}{\partial y} \sim \mathbb{O}\left(\frac{v_x|_\delta}{\delta}\right) \qquad \frac{\partial v_y}{\partial x} \sim \mathbb{O}\left(\frac{v_y|_\delta}{x}\right)$$

so

$$\frac{\partial v_x/\partial y}{\partial v_y/\partial x} \sim \mathbb{O}\left(\frac{x}{\delta}\right)^2$$

which, for a relatively thin boundary layer, is a large number, and thus $\partial v_x/\partial y \gg \partial v_y/\partial x$. The normal stress at a large Reynolds number is closely approximated by the negative of the pressure as $\mu(\partial v_x/\partial x) \sim \mathbb{O}(\mu v_\infty/x) = \mathbb{O}(\rho v_\infty^2/Re_x)$; therefore $\sigma_{xx} \simeq \sigma_{yy} \simeq -P$. When these simplifications in the stresses are incorporated, the equations for flow over a flat plate become

$$\rho\left(\frac{\partial v_x}{\partial t} + v_x\frac{\partial v_x}{\partial x} + v_y\frac{\partial v_x}{\partial y}\right) = -\frac{\partial P}{\partial x} + \mu\frac{\partial^2 v_x}{\partial y^2} \tag{12-7}$$

and

$$\rho\left(\frac{\partial v_y}{\partial t} + v_x\frac{\partial v_y}{\partial x} + v_y\frac{\partial v_y}{\partial y}\right) = -\frac{\partial P}{\partial y} + \mu\frac{\partial^2 v_y}{\partial x^2} \tag{12-8}$$

Furthermore,* the terms in the second equation are much smaller than those in the first equation, and thus $\partial P/\partial y \simeq 0$; hence $\partial P/\partial x = dP/dx$, which according to Bernoulli's equation is equal to $-\rho v_\infty\, dv_\infty/dx$.

The final form of equation (12-7) becomes

$$\frac{\partial v_x}{\partial t} + v_x\frac{\partial v_x}{\partial x} + v_y\frac{\partial v_x}{\partial y} = v_\infty\frac{dv_\infty}{dx} + \nu\frac{\partial^2 v_x}{\partial y^2} \tag{12-9}$$

The above equation, and the continuity equation,

$$\frac{\partial v_x}{\partial x} + \frac{\partial v_y}{\partial y} = 0 \tag{12-10}$$

are known as the boundary-layer equations.

12.5 BLASIUS' SOLUTION FOR THE LAMINAR BOUNDARY LAYER ON A FLAT PLATE

One very important case in which an analytical solution of the equations of motion has been achieved is that for the laminar boundary layer on a flat plate in steady flow.

For flow parallel to a flat surface, $v_\infty(x) = v_\infty$, and $dP/dx = 0$ according to the Bernoulli equation. The equations to be solved are now the following:

$$v_x\frac{\partial v_x}{\partial x} + v_y\frac{\partial v_x}{\partial y} = \nu\frac{\partial^2 v_x}{\partial y^2} \tag{12-11a}$$

and

$$\frac{\partial v_x}{\partial x} + \frac{\partial v_y}{\partial y} = 0 \tag{12-11b}$$

with boundary conditions $v_x = v_y = 0$ at $y = 0$, and $v_x = v_\infty$ at $y = \infty$.

---

* The order of magnitude of each term may be considered as above. For example, $v_x(\partial v_y/\partial x) \sim \mathbb{O}(v_\infty(v_\infty/x)(\delta/x)) = \mathbb{O}(v_\infty^2\delta/x^2)$.

Blasius* obtained a solution to the set of equations (12-11) by first introducing the stream function, Ψ, as described in Chapter 10, which automatically satisfies the two-dimensional continuity equation, equation (12-11b). This set of equations may be reduced to a single ordinary differential equation by transforming the independent variables x, y to η and the dependent variables from $\Psi(x, y)$ to $f(\eta)$

where

$$\eta(x, y) = \frac{y}{2} \left(\frac{v_\infty}{\nu x} \right)^{1/2} \tag{12-12}$$

and

$$f(\eta) = \frac{\Psi(x, y)}{(\nu x v_\infty)^{1/2}} \tag{12-13}$$

The appropriate terms in equation (12-11a) may be determined from equations (12-12) and (12-13). The following expressions will result. The reader may wish to verify the mathematics involved.

$$v_x = \frac{\partial \Psi}{\partial y} = \frac{v_\infty}{2} f'(\eta) \tag{12-14}$$

$$v_y = -\frac{\partial \Psi}{\partial x} = \frac{1}{2} \left(\frac{\nu v_\infty}{x} \right)^{1/2} (\eta f' - f) \tag{12-15}$$

$$\frac{\partial v_x}{\partial x} = -\frac{v_\infty \eta}{4x} f'' \tag{12-16}$$

$$\frac{\partial v_x}{\partial y} = \frac{v_\infty}{4} \left(\frac{v_\infty}{\nu x} \right)^{1/2} f'' \tag{12-17}$$

$$\frac{\partial^2 v_x}{\partial y^2} = \frac{v_\infty}{8} \frac{v_\infty}{\nu x} f''' \tag{12-18}$$

Substitution of (12-14) through (12-18) into equation (12-11a) and cancellation gives, as a single ordinary differential equation,

$$f''' + ff'' = 0 \tag{12-19}$$

with the appropriate boundary conditions

$$f = f' = 0 \qquad \text{at } \eta = 0$$
$$f' = 2 \qquad \text{at } \eta = \infty$$

Observe that this differential equation, although ordinary, is nonlinear and that, of the end conditions on the variable $f(\eta)$, two are initial values and the third

---

* H. Blasius, Grenzshichten in Flüssigkeiten mit kleiner Reibung, *Z. Math. U. Phys. Sci.*, **1**, 1908.

is a boundary value. This equation was solved first by Blasius, using a series expansion to express the function, $f(\eta)$, at the origin and an asymptotic solution to match the boundary condition at $\eta = \infty$. Howarth* later performed essentially the same work but obtained more accurate results. Table 12.1 presents the significant numerical results of Howarth. A plot of these values is included in Figure 12.6.

TABLE 12.1   VALUES OF f', v_x/v_∞, AND f'' FOR LAMINAR FLOW PARALLEL TO A FLAT PLATE (AFTER HOWARTH)

$\eta = \dfrac{y}{2}\sqrt{\dfrac{v_\infty}{vx}}$	f'	$\dfrac{v_x}{v_\infty}$	f''
0	0	0	1.32824
0.2	0.2655	0.1328	1.3260
0.4	0.5294	0.2647	1.3096
0.6	0.7876	0.3938	1.2664
0.8	1.0336	0.5168	1.1867
1.0	1.2596	0.6298	1.9670
1.2	1.4580	0.7290	0.9124
1.4	1.6230	0.8115	0.7360
1.6	1.7522	0.8761	0.5565
1.8	1.8466	0.9233	0.3924
2.0	1.9110	0.9555	0.2570
2.2	1.9518	0.9759	0.1558
2.4	1.9756	0.9878	0.0875
2.6	1.9885	0.9943	0.0454
2.8	1.9950	0.9962	0.0217
3.0	1.9980	0.9990	0.0096
3.2	1.9992	0.9996	0.0039
3.4	1.9998	0.9999	0.0015
3.6	1.9999	1.0000	0.0005
3.8	2.0000	1.0000	0.0002
4.0	2.0000	1.0000	0.0000
5.0	2.0000	1.0000	0.0000

A simpler way of solving equation (12-19) has been suggested in Goldstein,† who presented a scheme whereby the boundary conditions on the function f are initial values.

If we define two new variables in terms of the constant, C, so that

$$\phi = f/C \qquad (12\text{-}20)$$

---

* L. Howarth, "On the Solution of the Laminar Boundary Layer Equations," *Proc. Roy. Soc. London*, **A164**, 547 (1938).

† S. Goldstein, *Modern Developments in Fluid Dynamics*, Oxford Univ. Press, London, 1938, p. 135.

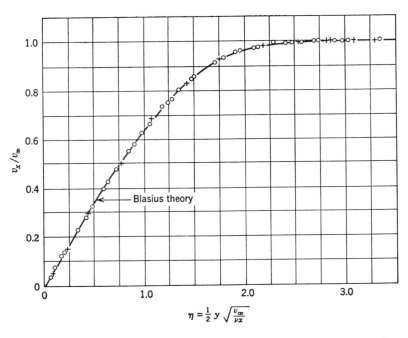

Figure 12.6 Velocity distribution in the laminar boundary layer over a flat plate. Experimental data by J. Nikuradse (monograph, Zentrale F. wiss. Berichtswesen, Berlin, 1942) for the Reynolds number range from 1.08×10^5 to 7.28×10^5.

and

$$\xi = C\eta \tag{12-21}$$

then the terms in equation (12-19) become

$$f(\eta) = C\phi(\xi) \tag{12-22}$$

$$f' = C^2 \phi' \tag{12-23}$$

$$f'' = C^3 \phi'' \tag{12-24}$$

and

$$f''' = C^4 \phi''' \tag{12-25}$$

The resulting differential equation in $\phi(\xi)$ becomes

$$\phi''' + \phi\phi'' = 0 \tag{12-26}$$

and the initial conditions on ϕ are

$$\phi = 0 \qquad \phi' = 0 \qquad \phi'' = ? \qquad \text{at } \xi = 0$$

The other boundary condition may be expressed as follows:

$$\phi'(\xi) = \frac{f'(\eta)}{C^2} = \frac{2}{C^2} \quad \text{at } \xi = \infty$$

An initial condition may be matched to this boundary condition if we let $f''(\eta = 0)$ equal some constant A; then $\phi''(\xi = 0) = A/C^3$. The constant A must have a certain value to satisfy the original boundary condition on f'. As an estimate we let $\phi''(\xi = 0) = 2$, giving $A = 2C^3$. Thus initial values of ϕ, ϕ', and ϕ'' are now specified. The estimate on $\phi''(0)$ requires that

$$\phi'(\infty) = \frac{2}{C^2} = \frac{1}{(2A)^{2/3}} \tag{12-27}$$

Thus equation (12-26) may be solved as an initial-value problem with the answer scaled according to equation (12-27) to match the boundary condition at $\eta = \infty$.

The significant results of Blasius' work are the following:

(a) The boundary thickness, δ, is obtained from Table 12.1. When $\eta = 2.5$, we have $v_x/v_\infty \cong 0.99$ thus, designating $y = \delta$ at this point, we have

$$\eta = \frac{y}{2}\sqrt{\frac{v_\infty}{\nu x}} = \frac{\delta}{2}\sqrt{\frac{v_\infty}{\nu x}} = 2.5$$

and thus

$$\delta = 5\sqrt{\frac{\nu x}{v_\infty}}$$

or

$$\frac{\delta}{x} = \frac{5}{\sqrt{\frac{v_\infty x}{\nu}}} = \frac{5}{\sqrt{\text{Re}_x}} \tag{12-28}$$

(b) The velocity gradient at the surface is given by equation (12-17)

$$\left.\frac{\partial v_x}{\partial y}\right|_{y=0} = \frac{v_\infty}{4}\left(\frac{v_\infty}{\nu x}\right)^{1/2} f''(0) = 0.332\, v_\infty \sqrt{\frac{v_\infty}{\nu x}} \tag{12-29}$$

Since the pressure does not contribute to the drag for flow over a flat plate, all the drag is viscous. The shear stress at the surface may be calculated as

$$\tau_0 = \mu \left.\frac{\partial v_x}{\partial y}\right|_{y=0}$$

Substituting equation (12-27) into this expression, we have

$$\tau_0 = \mu\, 0.332\, v_\infty \sqrt{\frac{v_\infty}{\nu x}} \tag{12-30}$$

The coefficient of skin friction may be determined by employing equation (12-2) as follows:

$$C_{fx} \equiv \frac{\tau}{\rho v_\infty^2/2} = \frac{F_d/A}{\rho v_\infty^2/2} = \frac{0.332\mu v_\infty \sqrt{\dfrac{v_\infty}{\nu x}}}{\rho v_\infty^2/2}$$

$$= 0.664 \sqrt{\frac{\nu}{x v_\infty}}$$

$$C_{fx} = \frac{0.664}{\sqrt{\mathrm{Re}_x}} \tag{12-31}$$

Equation (12-31) is a simple expression for the coefficient of skin friction at a particular value of x. For this reason the symbol C_{fx} is used, the x subscript indicating a *local coefficient*.

While it is of interest to know values of C_{fx}, it is seldom that a local value is useful; most often one wishes to calculate the total drag resulting from viscous flow over some surface of finite size. The mean coefficient of skin friction which is helpful in this regard may be determined quite simply from C_{fx} according to

$$F_d = A C_{fL} \frac{\rho v_\infty^2}{2} = \frac{\rho v_\infty^2}{2} \int_A C_{fx}\, dA$$

or the mean coefficient, designated C_{fL}, is related to C_{fx} by

$$C_{fL} = \frac{1}{A} \int_A C_{fx}\, dA$$

For the case solved by Blasius, consider a plate of uniform width W, and length L, for which

$$C_{fL} = \frac{1}{L} \int_0^L C_{fx}\, dx = \frac{1}{L} \int_0^L 0.664 \sqrt{\frac{\nu}{v_\infty}} x^{-1/2}\, dx$$

$$= 1.328 \sqrt{\frac{\nu}{L v_\infty}}$$

$$C_{fL} = \frac{1.328}{\sqrt{\mathrm{Re}_L}} \tag{12-32}$$

12.6   FLOW WITH A PRESSURE GRADIENT

In Blasius' solution for laminar flow over a flat plate, the pressure gradient was zero. A much more common flow situation involves flow with a pressure gradient. The pressure gradient plays a major role in flow separation, as can be seen with the aid of the boundary-layer equation (12-7). If we make use of the boundary conditions at the wall $v_x = v_y = 0$, at $y = 0$ equation (12-7) becomes

$$\mu \frac{\partial^2 v_x}{\partial y^2}\bigg|_{y=0} = \frac{dP}{dx} \qquad (12\text{-}33)$$

which relates the curvature of the velocity profile at the surface to the pressure gradient. Figure 12.7 illustrates the variation in v_x, $\partial v_x/\partial y$, and $\partial^2 v_x/\partial y^2$ across the boundary layer for the case of a zero pressure gradient.

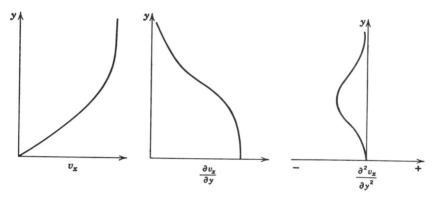

Figure 12.7   Variation in velocity and velocity derivatives across the laminar boundary layer when $dP/dx = 0$.

When $dP/dx = 0$, the second derivative of the velocity at the wall must also be zero; hence the velocity profile is linear near the wall. Further out in the boundary layer, the velocity gradient becomes smaller and gradually approaches zero. The decrease in the velocity gradient means that the second derivative of the velocity must be negative. The derivative $\partial^2 v_x/\partial y^2$ is shown as being zero at the wall, negative within the boundary layer, and approaching zero at the outer edge of the boundary layer. It is important to note that the second derivative must approach zero from the negative side as $y \to \delta$. For values of $dP/dx \neq 0$, the variation in v_x and its derivatives is shown in Figure 12.8.

A negative pressure gradient is seen to produce a velocity variation somewhat similar to that of the zero-pressure-gradient case. A positive value of dP/dx, however, requires a positive value of $\partial^2 v_x/\partial y^2$ at the wall. Since this derivative must approach zero from the negative side, at some point within the boundary layer the second derivative must equal zero. A zero second derivative, it will be recalled, is associated with an inflection point. The inflection point is shown in the velocity profile of Figure 12.8. We may now turn our attention to the subject of flow separation.

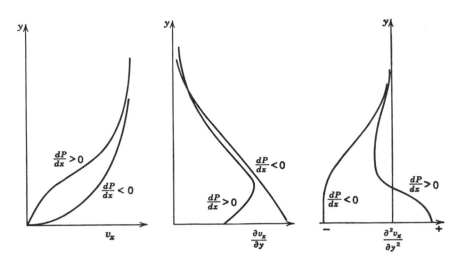

Figure 12.8   Variation in v_x and its derivatives across the boundary layer for various pressure gradients.

In order for flow separation to occur, the velocity in the layer of fluid adjacent to the wall must be zero or negative, as shown in Figure 12.9. This type of velocity profile is seen to require a point of inflection. As the only type of boundary layer flow that has an inflection point is flow with a positive pressure gradient, it may be concluded that a positive pressure gradient is necessary for flow separation. For this reason a positive pressure gradient is called an *adverse pressure gradient*. Flow can remain unseparated with an adverse pressure gradient, thus $dP/dx > 0$ is a necessary but not a sufficient condition for separation. In contrast a negative pressure gradient, in the absence of sharp corners, cannot cause flow separation. Therefore a negative pressure gradient is called a favorable pressure gradient.

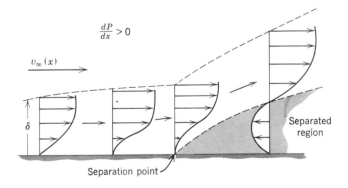

Figure 12.9   Velocity profiles in separated-flow region.

The presence of a pressure gradient also affects the magnitude of the skin friction coefficient, as can be inferred from Figure 12.8. The velocity gradient at the wall increases as the pressure gradient becomes more favorable.

12.7 VON KÁRMÁN MOMENTUM INTEGRAL ANALYSIS

The Blasius solution is obviously quite restrictive in application, applying only to the case of a laminar boundary layer over a flat surface. Any situation of practical interest more complex than this involves analytical procedures that have, to the present time, proved inferior to experiment. An approximate method providing information for systems involving other types of flow and having other geometries will now be considered.

Consider the control volume in Figure 12.10. The control volume to be analyzed is of unit depth and is bounded in the xy plane by the x axis, here drawn tangent to the surface at point 0; the y axis, the edge of the boundary layer, and a line parallel to the y axis a distance Δx away. We shall consider the case of two-dimensional, incompressible steady flow.

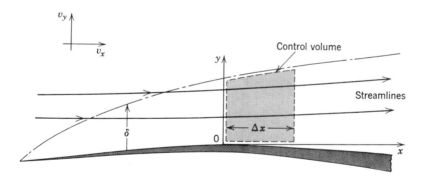

Figure 12.10   Control volume for integral analysis of the boundary layer.

A momentum analysis of the defined control volume involves the application of the x-directional scalar form of the momentum theorem,

$$\Sigma F_x = \int \int_{\text{c.s.}} v_x \rho(\mathbf{v} \cdot \mathbf{n})\, dA + \frac{\partial}{\partial t} \int \int \int_{\text{c.v.}} v_x \rho\, dV \qquad (5\text{-}5a)$$

A term-by-term analysis of the present problem yields the following:

$$\Sigma F_x = P\delta|_x - P\delta|_{x+\Delta x} + \left(P|_x + \frac{P|_{x+\Delta x} - P|_x}{2}\right)(\delta|_{x+\Delta x} - \delta|_x) - \tau_0\, \Delta x$$

where δ represents the boundary-layer thickness, and body forces are assumed negligible. The above terms represent the x-directional pressure forces on the left, right, and top sides of the control volume, and the frictional force on the bottom, respectively.

The surface integral term becomes

$$\int\int_{\text{c.s.}} v_x\rho(\mathbf{v}\cdot\mathbf{n})\,dA = \int_0^\delta \rho v_x^2\,dy\bigg|_{x+\Delta x} - \int_0^\delta \rho v_x^2\,dy\bigg|_x - v_\infty \dot{m}_{\text{top}}$$

and the accumulation term is

$$\frac{\partial}{\partial t}\int\int\int_{\text{c.v.}} v_x\rho\,dV = 0$$

since this is a steady-flow situation.

An application of the integral equation for conservation of mass will give

$$\int\int_{\text{c.s.}} \rho(\mathbf{v}\cdot\mathbf{n})\,dA + \frac{\partial}{\partial t}\int\int\int_{\text{c.v.}} \rho\,dV = 0 \tag{4-1}$$

$$\int\int_{\text{c.s.}} \rho(\mathbf{v}\cdot\mathbf{n})\,dA = \int_0^\delta \rho v_x\,dy\bigg|_{x+\Delta x} - \int_0^\delta \rho v_x\,dy\bigg|_x - \dot{m}_{\text{top}}$$

$$\frac{\partial}{\partial t}\int\int\int_{\text{c.v.}} \rho\,dV = 0$$

and the mass-flow rate into the top of the control volume, $\dot{m}_{\text{top}}$, may be evaluated as

$$\dot{m}_{\text{top}} = \int_0^\delta \rho v_x\,dy\bigg|_{x+\Delta x} - \int_0^\delta \rho v_x\,dy\bigg|_x \tag{12-34}$$

The momentum expression, including equation (12-34), now becomes

$$-(P\delta|_{x+\Delta x} - P\delta|_x) + \left(\frac{P|_{x+\Delta x} - P|_x}{2} + P|_x\right)(\delta|_{x+\Delta x} - \delta|_x) - \tau_0\,\Delta x$$

$$= \int_0^\delta \rho v_x^2\,dy\bigg|_{x+\Delta x} - \int_0^\delta \rho v_x^2\,dy\bigg|_x - v_\infty\left(\int_0^\delta \rho v_x\,dy\bigg|_{x+\Delta x} - \int_0^\delta \rho v_x\,dy\bigg|_x\right)$$

Rearranging this expression and dividing through by Δx, we get

$$-\left(\frac{P|_{x+\Delta x} - P|_x}{\Delta x}\right)\delta|_{x+\Delta x} + \left(\frac{P|_{x+\Delta x} - P|_x}{2}\right)\left(\frac{\delta|_{x+\Delta x} - \delta|_x}{\Delta x}\right) + \left(\frac{P\delta|_x - P\delta|_x}{\Delta x}\right)$$

$$= \left(\frac{\int_0^\delta \rho v_x^2\,dy|_{x+\Delta x} - \int_0^\delta \rho v_x^2\,dy|_x}{\Delta x}\right) - v_\infty\left(\frac{\int_0^\delta \rho v_x\,dy|_{x+\Delta x} - \int_0^\delta \rho v_x\,dy|_x}{\Delta x}\right) + \tau_0$$

Taking the limit as $\Delta x \to 0$ we obtain

$$-\delta\frac{dP}{dx} = \tau_0 + \frac{d}{dx}\int_0^\delta \rho v_x^2\,dy - v_\infty\frac{d}{dx}\int_0^\delta \rho v_x\,dy \tag{12-35}$$

The boundary-layer concept assumes inviscid flow outside the boundary layer, for which we may write Bernoulli's equation,

$$\frac{dP}{dx} + \rho v_\infty \frac{dv_\infty}{dx} = 0$$

which may be rearranged to the form

$$\frac{\delta}{\rho}\frac{dP}{dx} = \frac{d}{dx}(\delta v_\infty{}^2) - v_\infty\frac{d}{dx}(\delta v_\infty) \tag{12-36}$$

Notice that the left-hand sides of equations (12-35) and (12-36) are similar. We may thus relate the right-hand sides and, with proper rearrangement, get the result

$$\frac{\tau_0}{\rho} = \left(\frac{d}{dx}v_\infty\right)\int_0^\delta (v_\infty - v_x)\,dy + \frac{d}{dx}\int_0^\delta v_x(v_\infty - v_x)\,dy \tag{12-37}$$

Equation (12-37) is the von Kármán momentum integral expression, named in honor of Theodore von Kármán who first developed it.

Equation (12-37) is a general expression whose solution requires a knowledge of the velocity, v_x, as a function of distance from the surface, y. The accuracy of the final result will depend on how closely the assumed velocity profile approaches the real one.

As an example of the application of equation (12-37) let us consider the case of laminar flow over a flat plate, a situation for which an exact answer is known. In this case the free-stream velocity is constant, therefore $(d/dx)v_\infty = 0$ and equation (12-36) simplifies to

$$\frac{\tau_0}{\rho} = \frac{d}{dx}\int_0^\delta v_x(v_\infty - v_x)\,dy \tag{12-38}$$

An early solution to equation (12-38) was achieved by Pohlhausen, who assumed for the velocity profile a cubic function,

$$v_x = a + by + cy^2 + dy^3 \tag{12-39}$$

The constants a, b, c, and d may be evaluated if we know certain boundary conditions which must be satisfied in the boundary layer. These are

$$(1)\quad v_x = 0 \quad\text{at } y = 0$$

$$(2)\quad v_x = v_\infty \quad\text{at } y = \delta$$

$$(3)\quad \frac{\partial v_x}{\partial y} = 0 \quad\text{at } y = \delta$$

and

$$(4)\frac{\partial^2 v_x}{\partial y^2} = 0 \quad\text{at } y = 0$$

Boundary condition (4) results from equation (12-33), which states that the second derivative at the wall is equal to the pressure gradient. As the pressure is constant in this case, $\partial^2 v_x/\partial y^2 = 0$. Solving for a, b, c, and d from these conditions, we get

$$a = 0 \qquad b = \frac{3}{2\delta} v_\infty \qquad c = 0 \qquad d = -\frac{v_\infty}{2\delta^3}$$

which, when substituted in equation (12-39), give the form of the velocity profile

$$\frac{v_x}{v_\infty} = \frac{3}{2}\left(\frac{y}{\delta}\right) - \frac{1}{2}\left(\frac{y}{\delta}\right)^3 \tag{12-40}$$

Upon substitution, equation (12-38) becomes

$$\frac{3\nu}{2}\frac{v_\infty}{\delta} = \frac{d}{dx}\int_0^\delta v_\infty^2 \left(\frac{3}{2}\frac{y}{\delta} - \frac{1}{2}\left(\frac{y}{\delta}\right)^3\right)\left(1 - \frac{3}{2}\frac{y}{\delta} + \frac{1}{2}\left(\frac{y}{\delta}\right)^3\right) dy$$

or, after integrating,

$$\frac{3}{2}\nu\frac{v_\infty}{\delta} = \frac{39}{280}\frac{d}{dx}(v_\infty^2\delta)$$

As the free-stream velocity is constant, a simple ordinary differential equation in δ results:

$$\delta \, d\delta = \frac{140}{13}\frac{\nu \, dx}{v_\infty}$$

This, upon integration, yields

$$\frac{\delta}{x} = \frac{4.64}{\sqrt{\mathrm{Re}_x}} \tag{12-41}$$

The local skin-friction coefficient, C_{fx}, is given by

$$C_{fx} \equiv \frac{\tau_0}{\frac{1}{2}\rho v_\infty^2} = \frac{2\nu}{v_\infty^2}\frac{3}{2}\frac{v_\infty}{\delta} = \frac{0.646}{\sqrt{\mathrm{Re}_x}} \tag{12-42}$$

Integration of the local skin-friction coefficient between $x = 0$ and $x = L$ as in equation (12-31) yields

$$C_{fL} = \frac{1.292}{\sqrt{\mathrm{Re}_L}} \tag{12-43}$$

Comparing equations (12-41), (12-42), and (12-43) with exact results obtained by Blasius for the same situation, equations (12-28), (12-30), and (12.32), we observe a difference of about 7% in δ and 3% in C_f. This difference could, of course, have been smaller had the assumed velocity profile been a more accurate representation of the actual profile.

This comparison has shown the utility of the momentum integral method for the solution of the boundary layer and indicates a procedure that may be used with reasonable accuracy to obtain values for boundary-layer thickness and the coefficient of skin friction where an exact analysis is not feasible. The momentum integral method may also be used to determine the shear stress from the velocity profile.

The von Kármán momentum integral expression, equation (12-37), is usually used in a different form, particularly in cases involving a pressure gradient. Let us call the velocity outside of the boundary layer v_{x_δ} in order to emphasize that this velocity may be a function of x. When the velocity, v_{x_δ}, is constant, we may replaced v_{x_δ} by v_∞. Now let us introduce the momentum thickness, θ, and the displacement thickness, δ^*. The momentum thickness, θ, of the boundary layer is defined as follows.

$$\rho_\delta (v_{x_\delta})^2 \theta = \int_0^\infty \rho v_x (v_{x_\delta} - v_x)\, dy$$

That is,

$$\begin{bmatrix} \text{Momentum in layer} \\ \theta \text{ thick at conditions} \\ \text{outside boundary layer} \end{bmatrix} = \begin{bmatrix} \text{difference between momentum flux} \\ \text{in boundary later at free-stream} \\ \text{velocity and actual momentum flux in} \\ \text{the boundary layer} \end{bmatrix}$$

So, for incompressible flow

$$\theta = \int_0^\infty \frac{v_x}{v_{x_\delta}} \left(1 - \frac{v_x}{v_{x_\delta}} \right) dy \tag{12.44}$$

The presence of the boundary layer displaces the flow in the y direction. This displacement is measured in terms of the displacement thickness δ^*, where

$$\rho_\delta v_{x_\delta} \delta^* = \int_0^\infty (\rho_\delta v_{x_\delta} - \rho v_x)\, dy$$

$$\begin{bmatrix} \text{Mass flow at conditions} \\ \text{outside boundary layer} \\ \text{in a layer } \delta^* \text{ thick} \end{bmatrix} = \begin{bmatrix} \text{difference between mass flow} \\ \text{without boundary layer and mass} \\ \text{flow with boundary layer} \end{bmatrix}$$

So for incompressible flow,

$$\delta^* = \int_0^\infty \left(1 - \frac{v_x}{v_{x_\delta}} \right) dy \tag{12-45}$$

Substituting the above expressions into equation (12-37) and rearranging, one obtains

$$\frac{\tau_0}{\rho} = v_{x_\delta} \frac{dv_{x_\delta}}{dx} \left(2 + \frac{\delta^*}{\theta} \right) \theta + v_{x_\delta}^2 \frac{d\theta}{dx} \tag{12-46}$$

Equation (12-46) is seen to be a first-order ordinary differential equation in θ, the momentum thickness. In practice, the solution of equation (12-46) is usually accomplished by numerical integration, although a simple approximate solution valid for accelerating flows has been obtained. This solution is

$$\frac{v_{x_\delta} \theta^2}{\nu} = \frac{0.47}{v_{x_\delta}^5} \int_0^x v_{x_\delta}^5\, dx \tag{12-47}$$

12.8   CLOSURE

Viscous flow has been examined by several means in this chapter. The boundary-layer and drag coefficients have been defined and explained; the concepts of skin friction and pressure drag were presented; and two methods of boundary-layer analysis were discussed.

The concepts and ideas in this chapter are fundamental to heat and mass transfer as well as to momentum transfer. The boundary layer will be considered again in Chapters 19 and 28, and many of the results developed in this chapter will have equal significance in convective heat and mass transfer. The boundary layer is one of the most important building blocks in the entire structure of transfer processes.

PROBLEMS

12.1 If Reynolds' experiment were performed with a 38 mm-ID pipe, what flow velocity would occur at transition?

12.2 Modern subsonic aircraft have been refined to such an extent that 75% of the parasite drag (portion of total aircraft drag not directly associated with producing lift) can be attributed to friction along the external surfaces. For a typical subsonic jet the parasite drag coefficient based on wing area is 0.011. Determine the friction drag on such an aircraft
(a) at 500 mph at 35 000 ft;
(b) at 200 mph at sea level.
The wing area is 2400 ft^2.

12.3 The drag coefficient for a smooth sphere is shown below. Determine the speed at the critical Reynolds number for a 42 mm-diameter sphere in air.

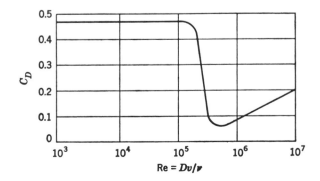

12.4 Plot a curve of drag vs. velocity for a 1.65-in.-diameter sphere in air between velocities of 50 fps and 400 fps.

12.5 Consider the flow of air at 30 m/s along a flat plate. At what distance from the leading edge will transition occur?

12.6 For the flow of air at atmospheric pressure, at 100°F, with a velocity of 88 fps, determine the magnitude of the cross-velocity component, v_y, at points 0.5, 1, 2, and 3 in. from the leading edge; plot your results. At a value of $\eta = 5.0$, $f = 8.2792$.

12.7 Is equation (12-9) valid in a region of separated flow? Explain.

12.8 Find δ^* for the Blasius solution, using the information given in problem 12.6.

12.9 Find a velocity profile for the laminar boundary layer of the form

$$\frac{v_x}{v_{x_\delta}} = c_1 + c_2 y + c_3 y^2 + c_4 y^3$$

when the pressure gradient is not zero.

12.10 Evaluate and compare with the exact solution δ, C_{fx}, and C_{fL} for the laminar boundary layer over a flat plate, using the velocity profile

$$v_x = a \sin by$$

12.11 Using equation (10-27) for pressure distribution around a circular cylinder and the velocity profile of the preceding problem, determine the boundary-layer thickness at the forward stagnation point of a circular cylinder. [Hint: as $\theta = -x/a$, at the stagnation point when x is measured along the surface of the cylinder, $v_\theta = v_\delta = 2v_\infty x/a$.]

12.12 There is fluid evaporating from a surface at which $v_x|_{y=0} = 0$, but $v_y|_{y=0} \neq 0$. Derive the von Kármán momentum relation.

12.13 For what wind velocities will a 12.7 mm-diameter cable be in the unsteady wake region of Figure 12.2?

12.14 Estimate the drag force on a 3-ft-long radio antenna with an average diameter of 0.2 in. at a speed of 60 mph.

12.15 Determine the ratio of the momentum thickness θ to the boundary-layer thickness δ for
(a) a linear velocity;
(b) the profile of problem 12.10.

12.16 Determine $\theta(x)$
  (a) on a flat plate;
  (b) for a circular cylinder, where

$$v_{x_\delta} = 2v_\infty \sin \frac{x}{a}$$

  (c) near the stagnation point of the cylinder where $x \ll a$.

12.17 The frequency of shedding vortices for a cylinder is predicted by the equation $f = 0.198 \, (v_x/D)(1 - 19.7/\text{Re})$, where f is the frequency of the vortices shed from one side of the cylinder and Re is the free-steam Reynolds number. Determine the frequency of vortex shedding of a 0.25-in.-diameter wire in a 20-mph wind. At what wind velocity will the frequency be zero?

12.18 A Ford two-door hardtop has a drag coefficient of 0.5 at road speeds, using a reference area of 2.29 m^2. Determine the horsepower required to overcome drag at a velocity of 30 m/s. Compare this figure with the case of head and tail winds of 6 m/s.

12.19 The lift coefficient is defined as $C_L \equiv \text{lift}/\frac{1}{2}\rho v_x^2 A_r$. If the lift coefficient for the auto in the previous case is 0.4, determine the lift force at a road speed of 100 mph.

12.20 What diameter circular plate would have the same drag as the auto of problem 12.18?

12.21 Estimate the normal force on a circular sign 8 ft in diameter during a hurricane wind (120 mph).

13
THE EFFECT OF TURBULENCE ON MOMENTUM TRANSFER

Turbulent flow is by far the most frequently encountered type of viscous flow, yet the analytical treatment of turbulent flow is not nearly so well developed as that of laminar flow. In this chapter turbulent flow will be examined, particularly with respect to the mechanism of momentum transfer. The role of experimental data in the formulation of an analytical approach to turbulent flow will be presented, and a comparison of laminar and turbulent flows will be made with the aid of the von Kármán integral relation.

13.1 DESCRIPTION OF TURBULENCE

In a turbulent flow the fluid and flow variables vary with time. The instantaneous velocity vector, for example, will differ from the average velocity vector in both magnitude and direction. Figure 13.1 illustrates the type of time dependence experienced by the axial component of the velocity for turbulent flow in a tube. While the velocity in Figure 13.1a is seen to be steady in its mean value, small random fluctuations in velocity occur about the mean value. Accordingly, we may express the fluid and flow variables in terms of a mean value and a fluctuating

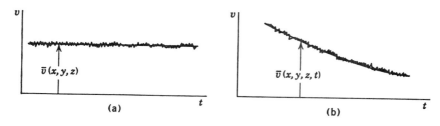

Figure 13.1 Time dependence of velocity in a turbulent flow: (a) steady mean flow; (b) unsteady mean flow.

187

value. For example, the x-directional velocity is expressed as

$$v_x = \bar{v}_x(x, y, z) + v_x'(x, y, z, t) \qquad (13\text{-}1)$$

Here $\bar{v}_x(x, y, z)$ represents the time-averaged velocity at the point (x, y, z)

$$\bar{v}_x \equiv \frac{1}{t_1} \int_0^{t_1} v_x(x, y, z, t)\, dt \qquad (13\text{-}2)$$

where t_1 is a time which is very long in comparison with the duration of any fluctuation. The mean value of $v_x'(x, y, z, t)$ is zero, as expressed by

$$\overline{v_x'} = \frac{1}{t_1} \int_0^{t_1} v_x'(x, y, z, t)\, dt = 0 \qquad (13\text{-}3)$$

Hereafter, $\bar{Q}$ will be used to designate the time average of the general property, Q, according to $\bar{Q} = 1/t_1 \int_0^{t_1} Q(x, y, z, t)\, dt$. While the mean value of the turbulent fluctuations is zero, these fluctuations contribute to the mean value of certain flow quantities. For example, the mean kinetic energy per unit volume is

$$\overline{KE} = \tfrac{1}{2}\rho \overline{[(\bar{v}_x + v_x')^2 + (\bar{v}_y + v_y')^2 + (\bar{v}_z + v_z')^2]}$$

The average of a sum is the sum of the averages; hence the kinetic energy becomes

$$\overline{KE} = \tfrac{1}{2}\rho \{ \overline{(\bar{v}_x{}^2 + 2\bar{v}_x v_x' + v_x'{}^2)} + \overline{(\bar{v}_y{}^2 + 2\bar{v}_y v_y' + v_y'{}^2)} + \overline{(\bar{v}_z{}^2 + 2\bar{v}_z v_z' + v_z'{}^2)} \}$$

or, since $\overline{\bar{v}_x v_x'} = \bar{v}_x \overline{v_x'} = 0$,

$$\overline{KE} = \tfrac{1}{2}\rho (\bar{v}_x{}^2 + \bar{v}_y{}^2 + \bar{v}_z{}^2 + \overline{v_x'{}^2} + \overline{v_y'{}^2} + \overline{v_z'{}^2}) \qquad (13\text{-}4)$$

A fraction of the total kinetic energy of a turbulent flow is seen to be associated with the magnitude of the turbulent fluctuations. It can be shown that the rms (root mean square) value of the fluctuations, $(\overline{v_x'{}^2} + \overline{v_y'{}^2} + \overline{v_z'{}^2})^{1/2}$ is a significant quantity. The level or *intensity of turbulence* is defined as

$$I \equiv \frac{\sqrt{(\overline{v_x'{}^2} + \overline{v_y'{}^2} + \overline{v_z'{}^2})/3}}{v_\infty} \qquad (13\text{-}5)$$

where v_∞ is the mean velocity of the flow. The intensity of turbulence is an extremely important parameter. Such factors as boundary-layer transition, separation, and heat- and mass-transfer rates are found to depend upon the intensity of turbulence. In model testing, simulation of turbulent flows requires not only duplication of Reynolds number but also duplication of the intensity of turbulence. Thus the measurement of turbulence is seen to be a necessity in many applications.

Of the many methods proposed for the measurement of turbulence, the hot-wire anemometer has proved the most satisfactory. This instrument utilizes a short, very thin wire placed perpendicular to the velocity component to be measured. The wire is heated to about 200°F above the stream temperature by

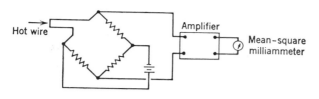

Figure 13.2 Simplified schematic of hot-wire anemometer circuit.

resistance heating. The temperature and therefore the resistance of the wire are proportional to the velocity perpendicular to it. As the velocity fluctuates it changes the wire temperature, which in turn alters the voltage drop across the wire. The wire, schematically shown in Figure 13.2 as an element in a Wheatstone bridge, may be calibrated to indicate the rms value of the velocity component normal to it.

The general discussion so far has indicated the fluctuating nature of turbulence. The random nature of turbulence lends itself to statistical analysis. We shall now turn our attention to the effect of the turbulent fluctuations on momentum transfer.

13.2 TURBULENT SHEARING STRESSES

In Chapter 7, the random molecular motion of the molecules was shown to result in a net momentum transfer between two adjacent layers of fluid. If the (molecular) random motions give rise to momentum transfer, it seems reasonable to expect that large-scale fluctuations, such as those present in a turbulent flow, will also result in a net transfer of momentum. Using an approach similar to that of section 7.3, let us consider the transfer of momentum in the turbulent flow illustrated in Figure 13.3.

The relation between the macroscopic momentum flux due to the turbulent fluctuations and the shear stress may be seen from the control-volume expression for linear momentum:

$$\Sigma \mathbf{F} = \iint_{c.s.} \mathbf{v}\rho(\mathbf{v} \cdot \mathbf{n}) \, dA + \frac{\partial}{\partial t} \iiint_{c.v.} \mathbf{v}\rho \, dV \tag{5-4}$$

The flux of x-directional momentum across the top of the control surface is

$$\iint_{\text{top}} \mathbf{v}\rho(\mathbf{v} \cdot \mathbf{n}) \, dA = \iint_{\text{top}} v_y{}' \rho(\bar{v}_x + v_x') \, dA \tag{13-6}$$

If the mean value of the momentum flux over a period of time is evaluated for the case of steady mean flow, the time derivative in equation (5-4) is zero; thus

$$\Sigma F_x = \iint \overline{v_y'\rho(\bar{v}_x + v_x')} \, dA = \iint \overline{v_y'\rho\bar{v}_x}^{\,0} \, dA + \iint \overline{\rho v_y'v_x'} \, dA \tag{13-7}$$

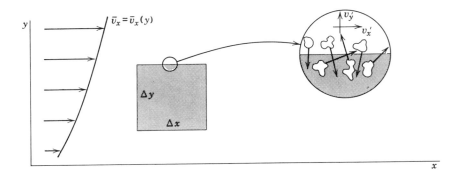

Figure 13.3 Turbulent motion at the surface of a control volume.

The presence of the turbulent fluctuations is seen to contribute a mean x directional momentum flux of $\overline{\rho v_x' v_y'}$ per unit area. Although the turbulent fluctuations are functions of position and time, their analytical description has not been achieved, even for the simplest case. The close analogy between the molecular exchange of momentum in laminar flow and the macroscopic exchange of momentum in turbulent flow suggests that the term $\overline{\rho v_x' v_y'}$ be regarded as a shear stress. Transposing this term to the left-hand side of equation (5-4) and incorporating it with the shear stress due to molecular momentum transfer, we see that the total shear stress becomes

$$\tau_{yx} = \mu \frac{d\bar{v}_x}{dy} - \overline{\rho v_x' v_y'} \tag{13-8}$$

The turbulent contribution to the shear stress is called the *Reynolds stress*. In turbulent flows it is found that the magnitude of the Reynolds stress is much greater than the molecular contribution except near the walls. Therefore $\overline{\rho v_x' v_y'}$ is a good approximation for the shear stress except in the immediate proximity of a solid boundary.

An important difference between the molecular and turbulent contributions to the shear stress is to be noted. Whereas the molecular contribution is expressed in terms of a property of the fluid and a derivative of the mean flow, the turbulent contribution is expressed solely in terms of the fluctuating properties of the flow. Further, these flow properties are not expressible in analytical terms. While Reynolds stresses exist for multidimensional flows,* the difficulties in analytically predicting even the one-dimensional case have proved insurmountable without the aid of experimental data. The reason for these difficulties may be seen by examining the number of equations and the number of unknowns involved. In the incompressible turbulent boundary layer, for example, there are two pertinent equations, momentum and continuity, and four unknowns, $\bar{v}_x$, $\bar{v}_y$, v_x', and v_y'.

---

* The existence of the Reynolds stresses may also be shown by taking the time average of the Navier–Stokes equations.

An early attempt to formulate a theory of turbulent shear stress was made by Boussinesq.* By analogy with the form of Newton's viscosity relation, Boussinesq introduced the concept relating the turbulent shear stress to the shear strain rate. The shear stress in laminar flow is $\tau_{yx} = \mu(dv_x/dy)$, thus by analogy, the Reynolds stress becomes

$$(\tau_{yx})_{\text{turb}} = A_t \frac{d\bar{v}_x}{dy}$$

where A_t is the *eddy viscosity*. Subsequent refinements have led to the introduction of the *eddy diffusivity of momentum*, $\epsilon_M \equiv A_t/\rho$, and thus

$$(\tau_{yx})_{\text{turb}} = \rho\epsilon_M \frac{d\bar{v}_x}{dy} \tag{13-9}$$

The difficulties in analytical treatment still exist, however, as the eddy diffusivity, ϵ_M, is a property of the flow and not of the fluid. By analogy with the kinematic viscosity in a laminar flow, it may be observed that the units of the eddy diffusivity are L^2/t.

13.3 THE MIXING-LENGTH HYPOTHESIS

A general similarity between the mechanism of transfer of momentum in turbulent flow and that in laminar flow permits an analog to be made for turbulent shear stress. The analog to the mean free path in molecular momentum exchange for the turbulent case, is the mixing length proposed by Prandtl† in 1925. Consider the simple turbulent flow shown in Figure 13.4.

The velocity fluctuation $v_x{}'$ is hypothesized as being due to the y-directional motion of a "lump" of fluid through a distance L. In undergoing translation the lump of fluid retains the mean velocity from its point of origin. Upon reaching a destination, a distance L from the point of origin, the lump of fluid will differ in mean velocity from that of the adjacent fluid by an amount $\bar{v}_x|_{y-L} - \bar{v}_x|_y$. If the

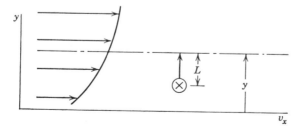

Figure 13.4   The Prandtl mixing length.

* J. Boussinesq, *Mem. Pre. par div. Sav.*, XXIII, Paris (1877).
† L. Prandtl, *ZAMM*, **5**, 136 (1925).

lump of fluid originated at $y + L$, the velocity difference would be $\bar{v}_x|_{y+L} - \bar{v}_x|_y$. The instantaneous value of $v_x'|_y$ is then $\bar{v}_x|_{y\pm L} - \bar{v}_x|_y$, the sign of L, of course, depending on the point of origin with respect to y. Further, the mixing length, although finite, is assumed to be small enough to permit the velocity difference to be written as

$$\bar{v}_x|_{\pm L} - \bar{v}_x|_y = \pm L \frac{d\bar{v}_x}{dy}$$

and thus

$$v_x' = \pm L \frac{d\bar{v}_x}{dy} \tag{13-10}$$

The concept of the mixing length is somewhat akin to that of the mean free path of a gas molecule. The important differences are its magnitude and dependence upon flow properties rather than fluid properties. With an expression for v_x' at hand, an expression for v_y' is necessary to determine the turbulent shear stress, $-\rho v_x' v_y'$.

Prandtl assumed that v_x' must be proportional to v_y'. If v_x' and v_y' were completely independent, then the time average of their product would be zero. Both the continuity equation and experimental data show that there is some degree of proportionality between v_x' and v_y'. Using the fact that $v_y' \sim v_x'$, Prandtl expressed the time average, $\overline{v_x' v_y'}$, as

$$\overline{v_x' v_y'} = -(\text{constant}) L^2 \left| \frac{d\bar{v}_x}{dy} \right| \frac{d\bar{v}_x}{dy} \tag{13-11}$$

The constant represents the unknown proportionality between v_x' and v_y' as well as their correlation in taking the time average. The minus sign and the absolute value were introduced to make the quantity $\overline{v_x' v_y'}$ agree with experimental observations. The constant in (13.11), which is unknown, may be incorporated into the mixing length, which is also unknown, giving

$$\overline{v_x' v_y'} = -L^2 \left| \frac{d\bar{v}_x}{dy} \right| \frac{d\bar{v}_x}{dy} \tag{13-12}$$

Comparison with Boussinesq's expression for the eddy diffusivity yields

$$\epsilon_M = L^2 \left| \frac{d\bar{v}_x}{dy} \right| \tag{13-13}$$

At first glance it appears that little has been gained in going from the eddy viscosity to the mixing length. There is an advantage, however, in that assumptions regarding the nature and variation of the mixing length may be made on an easier basis than assumptions concerning the eddy viscosity.

13.4 VELOCITY DISTRIBUTION FROM THE MIXING-LENGTH THEORY

One of the important contributions of the mixing-length theory is its use in correlating velocity profiles at large Reynolds numbers. Consider a turbulent flow as illustrated in Figure 13.4. In the neighborhood of the wall the mixing length is assumed to vary directly with y, and thus $L = Ky$, where K remains a dimensionless constant to be determined via experiment. The shear stress is assumed to be entirely due to turbulence and to remain constant over the region of interest. The velocity $\bar{v}_x$ is assumed to increase in the y direction, and thus $d\bar{v}_x/dy = |d\bar{v}_x/dy|$. Using these assumptions, we may write the turbulent shear stress as

$$\tau_{yx} = \rho K^2 y^2 \left(\frac{d\bar{v}}{dy}\right)^2 = \tau_0 \text{ (a constant)}$$

or

$$\frac{d\bar{v}_x}{dy} = \frac{\sqrt{\tau_0/\rho}}{Ky}$$

The quantity $\sqrt{\tau_0/\rho}$ is observed to have the units of velocity. Integration of the above equation yields

$$\bar{v}_x = \frac{\sqrt{\tau_0/\rho}}{K} \ln y + C \tag{13-14}$$

where C is a constant of integration. This constant may be evaluated by setting $\bar{v}_x = \bar{v}_{x\,max}$ at $y = h$, whereby

$$\frac{\bar{v}_{x\,max} - \bar{v}_x}{\sqrt{\tau_0/\rho}} = -\frac{1}{K}\left[\ln\frac{y}{h}\right] \tag{13-15}$$

The constant K was evaluated by Prandtl[*] and Nikuradse[†] from data on turbulent flow in tubes and found to have a value of 0.4. The agreement of experimental data for turbulent flow in smooth tubes with equation (13-15) is quite good, as can be seen from Figure 13.5.

The empirical nature of the preceding discussion cannot be overlooked. Several assumptions regarding the flow are known to be incorrect for flow in tubes, namely that the shear stress is not constant and that the geometry was treated from a two-dimensional viewpoint rather than an axisymmetric viewpoint. In view of these obvious difficulties, it is remarkable that equation (13-15) describes the velocity profile so well.

13.5 THE UNIVERSAL VELOCITY DISTRIBUTION

For turbulent flow in smooth tubes equation (13-15) may be taken as a basis for a more general development. Recalling that the term $\sqrt{\tau_0/\rho}$ has the units of velocity we may introduce a dimensionless velocity $\bar{v}_x/\sqrt{\tau_0/\rho}$. Defining

$$v^+ \equiv \frac{\bar{v}_x}{\sqrt{\tau_0/\rho}} \tag{13-16}$$

[*] L. Prandtl, *Proc. Intern. Congr. Appl. Mech.*, 2nd Congr., Zurich (1927), 62.
[†] J. Nikuradse, *VDI-Forschungsheft*, **356**, 1932.

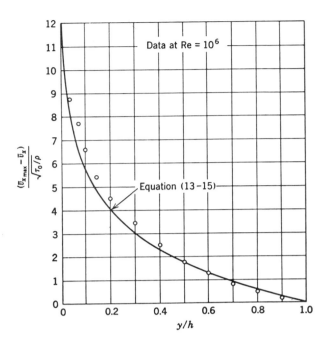

Figure 13.5 Comparison of data for flow in smooth tube with equation (13-15).

we may write equation (13-14) as

$$v^+ = \frac{1}{K}[\ln y] + C \tag{13-17}$$

The left-hand side of (13-17) is, of course, dimensionless; therefore the right-hand side of this equation must also be dimensionless. A pseudo-Reynolds number is found useful in this regard. Defining

$$y^+ \equiv \frac{\sqrt{\tau_0/\rho}}{\nu} y \tag{13-18}$$

we find that equation (13-17) becomes

$$v^+ = \frac{1}{K}\ln\frac{\nu y^+}{\sqrt{\tau_0/\rho}} + C = \frac{1}{K}(\ln y^+ + \ln \beta) \tag{13-19}$$

where the constant β is dimensionless.

Equation (13-19) indicates that for flow in smooth tubes $v^+ = f(y^+)$ or

$$v^+ \equiv \frac{\bar{v}_x}{\sqrt{\tau_0/\rho}} = f\left\{\ln\frac{y\sqrt{\tau_0/\rho}}{\nu}\right\} \tag{13-20}$$

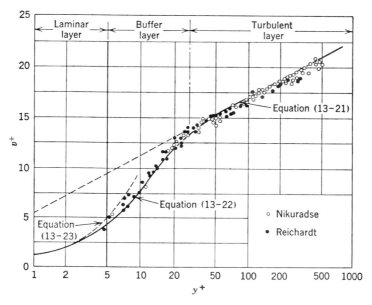

Figure 13.6 Velocity correlation for flow in circular smooth tubes at high Reynolds number (H. Reichardt, NACA TM1047, 1943).

The range of validity of equation (13-19) may be observed from a plot (see Figure 13.6) of v^+ versus $\ln y^+$, using the data of Nikuradse and Reichardt.

Three distinct regions are apparent: a turbulent core, a buffer layer, and a laminar sublayer. The velocity is correlated as follows:

for the turbulent core, $y^+ \geq 30$,

$$v^+ = 5.5 + 2.5 \ln y^+ \tag{13-21}$$

for the buffer layer, $30 \geq y^+ \geq 5$,

$$v^+ = -3.05 + 5 \ln y^+ \tag{13-22}$$

for the laminar sublayer, $5 > y^+ > 0$,

$$v^+ = y^+ \tag{13-23}$$

Equations (13-21) through (13-23) define the *universal velocity distribution*. Because of the empirical nature of these equations, there are, of course, inconsistencies. The velocity gradient, for example, at the center of the tube predicted by (13-21) is not zero. In spite of this and other inconsistencies, these equations are extremely useful for describing flow in smooth tubes.

In rough tubes, the scale of the roughness e is found to affect the flow in the turbulent core, but not in the laminar sublayer. The constant β in equation (13-19) becomes $\ln \beta = 3.4 - \ln[(e\sqrt{\tau_0/\rho})/\nu]$ for rough tubes. As the wall shear stress appears in the revised expression for $\ln \beta$, it is important to note that wall roughness affects the magnitude of the shear stress in a turbulent flow.

13.6 FURTHER EMPIRICAL RELATIONS FOR TURBULENT FLOW

Two important experimental results which are helpful in studying turbulent flows are the power-law relation for velocity profiles and a turbulent-flow shear-stress relation due to Blasius. Both of these relations are valid for flow over smooth surfaces.

For flow in smooth circular tubes it is found that over much of the cross section the velocity profile may be correlated by

$$\frac{\bar{v}_x}{\bar{v}_{x\,\text{max}}} = \left(\frac{y}{R}\right)^{1/n} \tag{13-24}$$

where R is the radius of the tube and n is a slowly varying function of Reynolds number. The exponent n is found to vary from a value of 6 at Re = 4000 to 10 at Re = 3 200 000. At Reynolds numbers of 10^5 the exponent is 7. This leads to the frequently used one-seventh-power law, $\bar{v}_x/\bar{v}_{x\,\text{max}} = (y/R)^{1/7}$. The power-law profile has also been found to represent the velocity distribution in boundary layers. For boundary layers the power law is written

$$\frac{\bar{v}_x}{\bar{v}_{x\,\text{max}}} = \left(\frac{y}{\delta}\right)^{1/n} \tag{13-25}$$

The power-law profile has two obvious difficulties: the velocity gradients at the wall and those at δ are incorrect. This expression indicates that the velocity gradient at the wall is infinite and that the velocity gradient at δ is nonzero.

In spite of these inconsistencies, the power law is extremely useful in connection with the von Kármán-integral relation, as we shall see in section 13.7.

Another useful relation is Blasius' correlation for shear stress. For pipe-flow Reynolds numbers up to 10^5 and flat-plate Reynolds numbers up to 10^7, the wall shear stress in a turbulent flow is given by

$$\tau_0 = 0.0225\rho\bar{v}_{x\,\text{max}}^2 \left(\frac{\nu}{\bar{v}_{x\,\text{max}}y_{\text{max}}}\right)^{1/4} \tag{13-26}$$

where $y_{\text{max}} = R$ in pipes and $y_{\text{max}} = \delta$ for flat surfaces.

13.7 THE TURBULENT BOUNDARY LAYER ON A FLAT PLATE

The variation in boundary-layer thickness for turbulent flow over a smooth flat plate may be obtained from the von Kármán momentum integral. The manner of approximation involved in a turbulent analysis differs from that used previously. In a laminar flow a simple polynomial was assumed to represent the velocity profile. In a turbulent flow, we have seen that the velocity profile depends upon the wall shear stress and that no single function adequately represents the velocity profile over the entire region. The procedure we shall follow in using the von Kármán integral relation in a turbulent flow is to utilize a simple profile for the integration with the Blasius correlation for the shear stress. For a zero pressure gradient the von Kármán integral relation is

$$\frac{\tau_0}{\rho} = \frac{d}{dx} \int_0^\delta v_x(v_\infty - v_x)\, dy \tag{12-38}$$

Employing the one-seventh-power law for v_x and the Blasius relation, equation (13-26) for τ_0, we see that equation (12-38) becomes

$$0.0225 v_\infty^2 \left(\frac{\nu}{v_\infty \delta}\right)^{1/4} = \frac{d}{dx} \int_0^\delta v_\infty^2 \left\{ \left(\frac{y}{\delta}\right)^{1/7} - \left(\frac{y}{\delta}\right)^{2/7} \right\} dy \tag{13-27}$$

where the free-stream velocity, v_∞, is written in place of $\bar{v}_{x\,\text{max}}$. Performing the indicated integration and differentiation, we obtain

$$0.0225 \left(\frac{\nu}{v_\infty \delta}\right)^{1/4} = \frac{7}{72} \frac{d\delta}{dx} \tag{13-28}$$

which becomes, upon integration

$$\left(\frac{\nu}{v_\infty}\right)^{1/4} x = 3.45 \delta^{5/4} + C \tag{13-29}$$

If the boundary layer is assumed to be turbulent from the leading edge, $x = 0$ (a poor assumption), the above equation may be rearranged to give

$$\frac{\delta}{x} = \frac{0.376}{\text{Re}_x^{1/5}} \tag{13-30}$$

The local skin-friction coefficient may be computed from the Blasius relation for shear stress, equation (13-26), to give

$$C_{fx} = \frac{0.0576}{Re_x^{1/5}} \tag{13-31}$$

Several things are to be noted about these expressions. First, they are limited to values of $\text{Re}_x < 10^7$, by virtue of the Blasius relation. Second, they apply only to smooth flat plates. Last, a major assumption has been made in assuming the boundary layer to be turbulent from the leading edge. The boundary layer is known to be laminar initially and to undergo transition to turbulent flow at a value of Re_x of about 2×10^5. We shall retain the assumption of a completely turbulent boundary layer for the simplicity it affords; it is recognized, however, that this assumption introduces some error in the case of a boundary layer which is not completely turbulent.

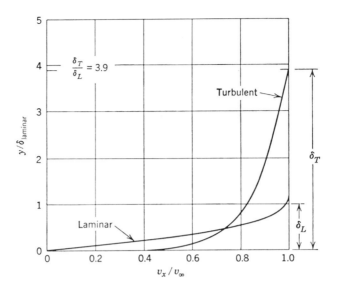

Figure 13.7 Comparison of velocity profiles in laminar and turbulent boundary layers. Reynolds number $x = 500\,000$.

A comparison of a laminar and a turbulent boundary layer can be made from Blasius' laminar-flow solution and equations (12-28), (13-30), and (13-31). At the same Reynolds number, the turbulent boundary layer is observed to be thicker, and is associated with a larger skin friction coefficient. While it would appear then that a laminar boundary layer is more desirable, the reverse is generally true. In most cases of engineering interest a turbulent boundary layer is desired because it resists separation better than a laminar boundary layer. The velocity profiles in laminar and turbulent boundary layers are compared qualitatively in Figure 13.7.

It can be seen that the turbulent boundary layer has a greater mean velocity, hence both greater momentum and energy than the laminar boundary layer. The greater momentum and energy permit the turbulent boundary layer to remain unseparated for a greater distance in the presence of an adverse pressure gradient than would be the case for a laminar boundary layer.

Consider a flat plate with transition from laminar flow to turbulent flow occurring on the plate. If transition from laminar flow to turbulent flow is assumed to occur abruptly (for computational purposes), a problem arises in how to join the laminar boundary layer to the turbulent layer at the point of transition. The prevailing procedure is to equate the momentum thicknesses, equation (12-44), at the transition point. That is, at the start of the turbulent portion of the boundary layer, the momentum thickness, θ, is equal to the momentum thickness at the end of the laminar portion of the boundary layer.

The general approach for turbulent boundary layers with a pressure gradient involves the use of the von Kármán momentum integral as given in equation (12-46). Numerical integration is required. Problems 13.12 to 13.15 develop an analytical solution for turbulent boundary layers that are not decelerting. The resulting solution, however, contains an integral that requires numerical integration for all cases except when the free-stream velocity is constant.

13.8 FACTORS AFFECTING THE TRANSITION FROM LAMINAR TO TURBULENT FLOW

The velocity profiles and momentum-transfer mechanisms have been examined for both laminar and turbulent flow regimes and found to be quite different. Laminar flow has also been seen to undergo transition to turbulent flow at certain Reynolds numbers.

So far the occurrence of transition has been expressed in terms of the Reynolds number alone, while a variety of factors other than Re actually influence transition. The Reynolds number remains, however, the principal parameter for predicting transition.

Table 13.1 indicates the influence of some of these factors on the transition Reynolds number.

TABLE 13.1 FACTORS AFFECTING THE REYNOLDS NUMBER OF
TRANSITION FROM LAMINAR TO TURBULENT FLOW

Factor	Influence
Pressure gradient	Favorable pressure gradient retards transition; unfavorable pressure gradient hastens it.
Free-stream turbulence	Free-stream turbulence decreases transition Reynolds number.
Roughness	No effect in pipes; decreases transition in external flow.
Suction	Suction greatly increases transition Re.
Wall curvatures	Convex curvature increases transition Re. Concave curvature decreases it.
Wall temperature	Cool walls increase transition Re. Hot walls decrease it.

13.9 CLOSURE

In the analytical formulation of momentum transfer in turbulent flow, it has been seen that the number of unknowns exceeds the number of equations. The result is a semi-empirical approach to the prediction of turbulent flow in which experimental data play the major role. Of necessity only a small fraction of the available experimental information has been presented, yet for the engineer such data are the key to satisfactory design. Usually there is a strong temptation to treat flows as laminar because of the simplicities involved in calculation. The prompt determination of the realm of flow, laminar or turbulent, and the use of experimental data cannot be overemphasized in this regard.

PROBLEMS

13.1 The turbulent shear stress in a two-dimensional flow is given by

$$(\tau_{yx})_{\text{turb}} = \rho \epsilon_M \frac{\partial \bar{v}_x}{\partial y} = -\rho \overline{v'_x v'_y}$$

Expanding v'_x and v'_y in a Taylor series in x and y near the wall and with the aid of the continuity equation,

$$\frac{\partial v'_x}{\partial x} + \frac{\partial v'_y}{\partial y} = 0$$

show that, near the wall, $\epsilon_M \sim y^3 +$ higher-order terms in y. How does this compare with the mixing-length theory?

13.2 Evaluate the velocity derivative, $\partial \bar{v}_x / \partial y$, for the power-law velocity profile at $y = 0$ and $y = R$.

13.3 Using the Blasius shear-stress relation (13-26) and the power-law velocity profile, determine the boundary-layer thickness on a flat plate as a function of the Reynolds number and the exponent n.

13.4 Plot the boundary-layer thickness along a flat plate for the flow of air at 30 m/s, assuming
(a) laminar flow;
(b) turbulent flow.
Indicate the probable transition point.

13.5 For a thin plate 6 in. wide and 3 ft long, estimate the friction force in air at a velocity of 40 fps, assuming
(a) turbulent flow;
(b) laminar flow.
The flow is parallel to the 6-in. dimension.

13.6 Develop an expression for the displacement thickness δ^* on a thin flat plate in turbulent flow.

13.7 For the fully developed flow of water in a smooth 0.15-m pipe at a rate of 0.006 m^3/s, determine the thickness of
(a) the laminar sublayer;
(b) the buffer layer;
(c) the turbulent core.

13.8 Using a sine profile for laminar flow and a one-seventh-power law for turbulent flow, make a dimensionless plot of the momentum and kinetic energy profiles in the boundary layer at a Reynolds number of 10^5.

13.9 Estimate the friction drag on a wing by considering the following idealization. Consider the wing to be a rectangular flat plate, 7 ft by 40 ft, with smooth surface. The wing is flying at 140 mph at 5000 ft. Determine the drag, assuming
(a) a laminar boundary layer;
(b) a turbulent boundary layer.

13.10 Compare the boundary-layer thicknesses and local skin-friction coefficients of a laminar boundary layer and a turbulent boundary layer on a smooth flat plate at a Reynolds number of 10^6. Assume both boundary layers to originate at the leading edge of the flat plate.

13.11 Use the seventh root profile and compute the drag force and boundary layer thickness on a plate 20 ft. long and 10 ft. wide (for one side) if it is immersed in a flow of water of 20 ft./s velocity. Assume turbulent flow to exist over the entire length of the plate. What would the drag be if laminar flow could be maintained over the entire surface?

13.12 Using a power law velocity profile as given in equation (13-25), determine:
(a) δ^*/δ;
(b) θ/δ;
(c) $2 + \delta^*/\theta$.

13.13 Using the Blasius correlation, equation (13-26), and the results of problem 13.12, express τ_0/ρ in terms of v_{x_δ}, θ, ν, and η.

13.14 Using the results of problems 13.12 and 13.13, show that the von Kármán momentum integral, equation (12-46), may be written as

$$\frac{4}{5}\frac{d\theta^{5/4}}{dx} + \frac{\theta^{5/4}}{v_{x_\delta}}\left(\frac{2+3n}{n}\right)\frac{dv_{x_\delta}}{dx} = 0.0225\left[\frac{n}{(n+1)(n+2)}\right]^{1/4}\left(\frac{\nu}{v_{x_\delta}}\right)^{1/4}$$

13.15 The differential equation obtained in the previous problem is a linear, first-order, ordinary differential equation in the dependent variable $\theta^{5/4}$. Obtain the general solution for $\theta^{5/4}$.

14
FLOW IN CLOSED CONDUITS

Many of the theoretical relations that have been developed in the previous chapters apply to special situations such as inviscid flow, incompressible flow, and the like. Some experimental correlations were introduced in Chapters 12 and 13 for turbulent flow in or past surfaces of simple geometry. In this chapter an application of the material that has been developed thus far will be considered with respect to a situation of considerable engineering importance, namely fluid flow, both laminar and turbulent, through closed conduits.

14.1 DIMENSIONAL ANALYSIS OF CONDUIT FLOW

As an initial approach to conduit flow we shall utilize dimensional analysis to obtain the significant parameters from the flow of an incompressible fluid in a straight, horizontal, circular pipe of constant cross section.

The significant variables and their dimensional expressions are as represented in the following table:

Variable	Symbol	Dimension
pressure drop	ΔP	M/Lt^2
velocity	v	L/t
pipe diameter	D	L
pipe length	L	L
pipe roughness	e	L
fluid viscosity	μ	M/Lt
fluid density	ρ	M/L^3

Each of the variables is familiar, with the exception of the pipe roughness, symbolized e. The roughness is included to represent the condition of the pipe surface and may be thought of as characteristic of the height of projections from the pipe wall, hence the dimension of length.

According to the Buckingham pi theorem the number of independent dimensionless groups to be formed with these variables is four. If the core group consists of the variables v, D, and ρ, then the groups to be formed are as follows:

$$\pi_1 = v^a D^b \rho^c \, \Delta P$$

$$\pi_2 = v^d D^e \rho^f L$$

$$\pi_3 = v^g D^h \rho^i e$$

and

$$\pi_4 = v^j D^k \rho^l \mu$$

Carrying out the procedure outlined in Chapter 11 to solve for the unknown exponents in each group, we see that the dimensionless parameters become

$$\pi_1 = \frac{\Delta P}{\rho v^2}$$

$$\pi_2 = \frac{L}{D}$$

$$\pi_3 = \frac{e}{D}$$

and

$$\pi_4 = \frac{v D \rho}{\mu}$$

The first π group is the Euler number. Since the pressure drop is due to fluid friction, this parameter is often written with $\Delta P / \rho$ replaced by gh_L where h_L is the "head loss," thus π_1 becomes

$$\frac{h_L}{v^2/g}$$

The third π group, the ratio of pipe roughness to diameter is the so-called relative roughness. The fourth π group is the Reynolds number, Re.

A functional expression resulting from dimensional analysis may be written as

$$\frac{h_L}{v^2/g} = \phi_1 \left(\frac{L}{D}, \frac{e}{D}, \text{Re} \right) \tag{14-1}$$

Experimental data have shown that the head loss in fully developed flow is directly proportional to the ratio L/D. This ratio may, then, be removed from the functional expression, giving

$$\frac{h_L}{v^2/g} = \frac{L}{D}\phi_2\left(\frac{e}{D}, \text{Re}\right) \tag{14-2}$$

The function ϕ_2, which varies with the relative roughness and Reynolds number is designated f, the friction factor. Expressing the head loss from equation (14-2) in terms of f, we have

$$h_L = 2f_f\frac{L}{D}\frac{v^2}{g} \tag{14-3}$$

With the factor 2 inserted in the right-hand side, equation (14-3) is the defining relation for f_f, the *Fanning friction factor*. Another friction factor in common use is the *Darcy friction factor*, f_D, defined by equation (14-4)

$$h_L = f_D\frac{L}{D}\frac{v^2}{2g} \tag{14-4}$$

Quite obviously, $f_D = 4f_f$. The student should be careful to note which friction factor he is using to properly calculate frictional head loss by either equation (14-3) or (14-4). The Fanning friction factor, f_f, will be used exclusively in this text. The student may easily verify that the Fanning friction factor is the same as the skin friction coefficient C_f.

Our task now becomes that of determining suitable relations for f_f from theory and experimental data.

14.2 FRICTION FACTORS FOR FULLY DEVELOPED LAMINAR, TURBULENT, AND TRANSITION FLOW IN CIRCULAR CONDUITS

A. LAMINAR FLOW

Some analysis has been performed already for incompressible laminar flow. Since fluid behavior can be described quite well in this regime according to Newton's viscosity relation we should expect no difficulty in obtaining a functional relationship for f_f in the case of laminar flow. Recall that, for closed conduits, the flow may be considered laminar for values of the Reynolds number less than 2300.

From Chapter 8 the Hagen-Poiseuille equation was derived for incompressible, laminar, conduit flow,

$$-\frac{dP}{dx} = 32\frac{\mu v_{\text{avg}}}{D^2} \tag{8-9}$$

Separating variables and integrating this expression along a length, L, of the passage, we get

$$-\int_{P_0}^{P} dP = 32\frac{\mu v_{\text{avg}}}{D^2}\int_0^L dx$$

and

$$\Delta P = 32\frac{\mu v_{avg}L}{D^2} \tag{14-5}$$

Recall that equation (8-9) held for the case of fully developed flow; thus v_{avg} does not vary along the length of the passage.

Forming an expression for frictional head loss from equation (14-5), we have

$$h_L = \frac{\Delta P}{\rho g} = 32\frac{\mu v_{avg}L}{g\rho D^2} \tag{14-6}$$

Combining this equation with equation (14-3), the defining relation for f_f,

$$h_L = 32\frac{\mu v_{avg}L}{g\rho D^2} = 2f_f\frac{L}{D}\frac{v^2}{g}$$

and solving for f_f, we obtain

$$f_f = 16\frac{\mu}{Dv_{avg}\rho} = \frac{16}{Re} \tag{14-7}$$

This very simple result indicates that f_f is inversely proportional to Re in the laminar flow range; the friction factor is *not* a function of pipe roughness for values of Re < 2300, but varies only with the Reynolds number.

This result has been experimentally verified and is the manifestation of viscous effects in the fluid damping out any irregularities in the flow caused by protrusions from a rough surface.

B. TURBULENT FLOW

In the case of turbulent flow in closed conduits or pipes the relation for f_f is not so simply obtained or expressed as in the laminar case. No easily derived relation such as the Hagen-Poiseuille law applies; however, some use can be made of the velocity profiles expressed in Chapter 13 for turbulent flow. All development will be based on circular conduits, thus we are primarily concerned with pipes or tubes. In turbulent flow a distinction must be made between smooth- and rough-surfaced tubes.

Smooth Tubes. The velocity profile in the turbulent core has been expressed as

$$v^+ = 5.5 + 2.5 \ln y^+ \tag{13-21}$$

where the variables v^+ and y^+ are defined according to the relations

$$v^+ \equiv \frac{\bar{v}}{\sqrt{\tau_0/\rho}} \tag{13-16}$$

and

$$y^+ \equiv \frac{\sqrt{\tau_0/\rho}}{\nu}y \tag{13-18}$$

The average velocity in the turbulent core for flow in a tube of radius R can be evaluated from equation (13-21) as follows:

$$v_{avg} = \frac{\int_0^A \bar{v} \, dA}{A}$$

$$= \frac{\sqrt{\tau_0/\rho} \int_0^R \left(2.5 \ln\left\{\frac{\sqrt{\tau_0/\rho}y}{\nu}\right\} + 5.5\right) 2\pi r \, dr}{\pi R^2}$$

Letting $y = R - r$, we obtain

$$v_{avg} = 2.5\sqrt{\tau_0/\rho} \ln\left\{\frac{\sqrt{\tau_0/\rho}R}{\nu}\right\} + 1.75\sqrt{\tau_0/\rho} \tag{14-8}$$

The functions $\sqrt{\tau_0/\rho}$ and C_f are related according to equation (12-2). Since C_f and f_f are equivalent, we may write

$$\frac{v_{avg}}{\sqrt{\tau_0/\rho}} = \frac{1}{\sqrt{f_f/2}} \tag{14-9}$$

Substitution of equation (14-9) into equation (14-8) yields

$$\frac{1}{\sqrt{f_f/2}} = 2.5 \ln\left\{\frac{R}{\nu} v_{avg}\sqrt{f_f/2}\right\} + 1.75 \tag{14-10}$$

Rearranging the argument of the logarithm into Reynolds number form, and changing to $\log_{10}$, we see that equation (14-10) reduces to

$$\frac{1}{\sqrt{f_f}} = 4.06 \log_{10}\{\text{Re }\sqrt{f_f}\} - 0.60 \tag{14-11}$$

This expression gives the relation for the friction factor as a function of Reynolds number for turbulent flow in smooth circular tubes. The preceding development was first performed by von Kármán.[*] Nikuradse,[†] from experimental data, obtained the equation

$$\frac{1}{\sqrt{f_f}} = 4.0 \log_{10}\{\text{Re}\sqrt{f_f}\} - 0.40 \tag{14-12}$$

which is very similar to equation (14-11).

---

[*] T. von Kármán, NACA TM 611, 1931.
[†] J. Nikuradse, *VDI-Forschungsheft*, **356**, 1932.

Rough Tubes. By a similar analysis to that used for smooth tubes, von Kármán developed equation (14-13) for turbulent flow in rough tubes,

$$\frac{1}{\sqrt{f_f}} = 4.06 \log_{10} \frac{D}{e} + 2.16 \tag{14-13}$$

which compares very well with the equation obtained by Nikuradse from experimental data,

$$\frac{1}{\sqrt{f_f}} = 4.0 \log_{10} \frac{D}{e} + 2.28 \tag{14-14}$$

Nikuradse's results for fully developed pipe flow indicated that the surface condition, i.e. roughness, had nothing to do with the transition from laminar to turbulent flow. Once the Reynolds number becomes large enough so that flow is fully turbulent, then either equation (14-12) or (14-14) must be used to obtain the proper value for f_f. These two equations are quite different in that equation (14-12) expresses f_f as a function of Re only and equation (14-14) gives f_f as a function only of the relative roughness. The difference is, of course, that the former equation is for smooth tubes and the latter for rough tubes. The question that naturally arises at this point is "what is 'rough'?"

It has been observed from experiment that equation (14-12) describes the variation in f_f for a range in Re, even for rough tubes. Beyond some value of Re this variation deviates from the smooth-tube equation and achieves a constant value dictated by the tube roughness as expressed by equation (14-14). The region wherein f_f varies both with Re and e/D is called the *transition region*. An empirical equation describing the variation of f_f in the transition region has been proposed by Colebrook,[*]

$$\frac{1}{\sqrt{f_f}} = 4 \log_{10} \frac{D}{e} + 2.28 - 4 \log_{10} \left(4.67 \frac{D/e}{\mathrm{Re}\sqrt{f_f}} + 1 \right) \tag{14-15}$$

Equation (14-15) is applicable to the transition region above a value of $(D/e)/(\mathrm{Re}\sqrt{f_f}) = 0.01$. Below this value the friction factor is independent of the Reynolds number, and the flow is said to be *fully turbulent*.

To summarize the development of this section, the following equations express the friction-factor variation for the surface and flow conditions specified:

For laminar flow (Re < 2300),

$$f_f = \frac{16}{\mathrm{Re}} \tag{14-7}$$

For turbulent flow (smooth pipe, Re > 3000),

$$\frac{1}{\sqrt{f_f}} = 4.0 \log_{10} \{\mathrm{Re}\sqrt{f_f}\} - 0.40 \tag{14-12}$$

[*] C. F. Colebrook, *J. Inst. Civil Engr.* (London) II, **133** (1938–39).

For turbulent flow (rough pipe, $(D/e)/(\text{Re}\sqrt{f_f}) < 0.01$),

$$\frac{1}{\sqrt{f_f}} = 4.0 \log_{10} \frac{D}{e} + 2.28 \tag{14-14}$$

And for transition flow,

$$\frac{1}{\sqrt{f_f}} = 4 \log_{10} \frac{D}{e} + 2.28 - 4 \log_{10}\left(4.67 \frac{D/e}{\text{Re}\sqrt{f_f}} + 1\right) \tag{14-15}$$

14.3 FRICTION FACTOR AND HEAD-LOSS DETERMINATION FOR PIPE FLOW

A. FRICTION FACTOR

A single friction-factor plot based upon equations (14-7), (14-12), (14-14), and (14-15) has been presented by Moody.* Figure 14.1 is a plot of the Fanning friction factor vs. the Reynolds number for a range of values of the roughness parameter e/D.

When using the friction factor plot, Figure 14.1, it is necessary to know the value of the roughness parameter which applies to a pipe of given size and material. After a pipe or tube has been in service for some time, its roughness may change considerably, making the determination of e/D quite difficult. Moody has presented a chart, reproduced in Figure 14.2, by which a value of e/D can be determined for a given size tube or pipe constructed of a particular material.

The combination of these two plots enables the frictional head loss for a length, L, of pipe having diameter D to be evaluated, using the relation

$$h_L = 2f_f \frac{L}{D} \frac{v^2}{g} \tag{14-3}$$

Recently Haaland,† has shown that over the range $10^8 \geq \text{Re} \geq 4 \cdot 10^4$, $0.05 \geq e/D \geq 0$, that the friction factor may be expressed (within ± 1.5 percent) as

$$\frac{1}{\sqrt{f_f}} = -3.6 \log_{10}\left[\frac{6.9}{\text{Re}} + \left(\frac{e}{3.7D}\right)^{10/9}\right] \tag{14-15a}$$

This expression allows explicit calculation of the friction factor.

B. HEAD LOSSES DUE TO FITTINGS

The frictional head loss calculated from equation (14-3) is only part of the total head loss which must be overcome in pipe lines and other fluid-flow circuits. Other losses may occur due to the presence of valves, elbows, and any other fittings that involve a change in the direction of flow or in the size of the flow passage. The head losses resulting from such fittings are functions of the geometry of the fitting, the Reynolds number, and the roughness. As the losses in fittings, to a first approximation, have been found to be independent of the Reynolds

* L. F. Moody, *Trans. ASME,* **66**, 671 (1944).
† S. E. Haaland, *Trans. ASME, JFE,* **105**, 89 (1983).

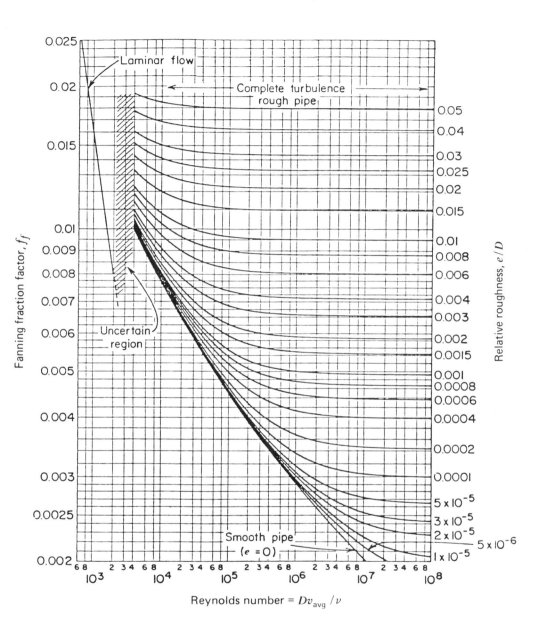

Figure 14.1 The Fanning friction factor as a function of Re and D/e.

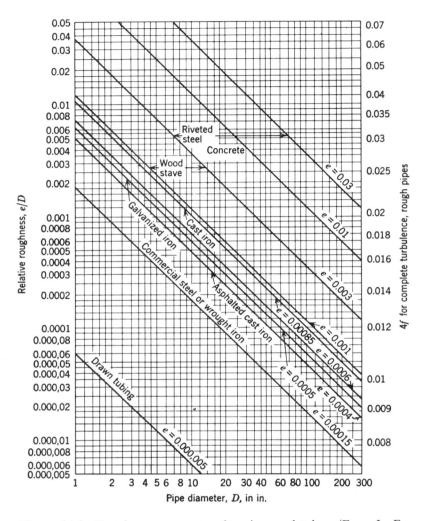

Figure 14.2 Roughness parameters for pipes and tubes. (From L. F. Moody, *Trans. ASME*, (1944).) Values of e are given in feet.

number, the head loss may be evaluated as

$$h_L = \frac{\Delta P}{\rho} = K\frac{v^2}{2g} \tag{14-16}$$

where K is a coefficient depending upon the fitting.

An equivalent method of determining the head loss in fittings is to introduce an *equivalent length*, L_{eq}, such that

$$h_L = 2f_f\frac{L_{eq}}{D}\frac{v^2}{g} \tag{14-17}$$

where L_{eq} is the length of pipe that produces a head loss equivalent to the head loss in a particular fitting. Equation (14-17) is seen to be in the same form as equation (14-3), and thus the total head loss for a piping system may be determined by adding the equivalent lengths for the fittings to the pipe length to obtain the total effective length of pipe.

Comparison of equations (14-16) and (14-17) shows that the constant K must be equal to $4f_f L_{eq}/D$. Although equation (14-17) appears to be dependent upon Reynolds number because of the appearance of the Fanning friction factor, it is not. The assumption made in both equations (14-16) and (14-17) is that the Reynolds number is large enough so that the flow is fully turbulent. The friction coefficient for a given fitting, then, is only dependent upon the roughness of the fitting. Typical values for K and L_{eq}/D are given in Table 14.1.

TABLE 14.1   FRICTION LOSS FACTORS FOR VARIOUS
PIPE FITTINGS

Fitting	K	L_{eq}/D
Globe valve, wide open	7.5	350
Angle valve, wide open	3.8	170
Gate valve, wide open	0.15	7
Gate valve, $\frac{3}{4}$ open	0.85	40
Gate valve, $\frac{1}{2}$ open	4.4	200
Gate valve, $\frac{1}{4}$ open	20	900
Standard 90° elbow	0.7	32
Short-radius 90° elbow	0.9	41
Long-radius 90° elbow	0.4	20
Standard 45° elbow	0.35	15
Tee, through side outlet	1.5	67
Tee, straight through	0.4	20
180° Bend	1.6	75

Recall that the head loss due to a sudden expansion was calculated in Chapter 6, with the result given in equation (6-13).

C. EQUIVALENT DIAMETER

Equations (14-16) and (14-17) are based upon a circular flow passage. These equations may be used to estimate the head loss in a closed conduit of any configuration if an "equivalent diameter" for a noncircular flow passage is used. An equivalent diameter is calculated according to

$$D_{eq} = 4\frac{\text{cross-sectional area of flow}}{\text{wetted perimeter}} \qquad (14\text{-}18)$$

The ratio of the cross-sectional area of flow to the wetted perimeter is called the hydraulic radius.

The reader may verify that D_{eq} corresponds to D for a circular flow passage. One type of noncircular flow passage often encountered in transfer processes is the annular area between two concentric pipes. The equivalent diameter for this configuration is determined as follows:

$$\text{Cross-sectional area} = \frac{\pi}{4}(D_0{}^2 - D_i{}^2)$$

$$\text{Wetted perimeter} = \pi(D_0 + D_i)$$

yielding

$$D_{eq} = 4\frac{\pi/4}{\pi}\frac{(D_0{}^2 - D_i{}^2)}{(D_0 + D_i)} = D_0 - D_i \qquad (14\text{-}19)$$

This value of D_{eq} may now be used to evaluate the Reynolds number, the friction factor, and the frictional head loss, using the relations and methods developed previously for circular conduits.

14.4 PIPE-FLOW ANALYSIS

Application of the equations and methods developed in the previous sections is common in engineering systems involving pipe networks. Such analyses are always straightforward but may vary as to the complexity of calculation. The following three example problems are typical, but by no means all-inclusive, of the types of problems found in engineering practice.

EXAMPLE 1

Water at 59°F flows through a straight section of a 6-in.-ID cast-iron pipe with an average velocity of 4 fps. The pipe is 120 ft long, and there is an increase in elevation of 2 ft from the inlet of the pipe to its exit.

Find the power required to produce this flow rate for the specified conditions.

The control volume in this case is the pipe and the water it encloses. Applying the energy equation to this control volume, we obtain

$$\frac{\delta Q}{dt} - \frac{\delta W_s}{dt} - \frac{\delta W_\mu}{dt} = \iint_{c.s.} \rho\left(e + \frac{P}{\rho}\right)(\mathbf{v} \cdot \mathbf{n})\, dA + \frac{\partial}{\partial t}\iiint_{c.v.} \rho e\, dV \qquad (6\text{-}10)$$

An evaluation of each term yields

$$\frac{\delta Q}{dt} = 0 \qquad \frac{\delta W_s}{dt} = \dot{W}$$

$$\iint_{c.s.} \rho\left(e + \frac{P}{\rho}\right)(\mathbf{v} \cdot \mathbf{n})\, dA = \rho A v_{avg}\left(\frac{v_2{}^2}{2} + gy_2 + \frac{P_2}{\rho} + u_2 - \frac{v_1{}^2}{2} - gy_1 - \frac{P_1}{\rho} - u_1\right)$$

$$\frac{\partial}{\partial t}\iiint_{c.v.} \rho e\, dV = 0$$

and

$$\frac{\delta W_\mu}{dt} = 0$$

The applicable form of the energy equation written on a unit mass basis is now

$$w = \frac{v_1^2 - v_2^2}{2} + g(y_1 - y_2) + \frac{P_1 - P_2}{\rho} + u_1 - u_2$$

and with the internal energy change written as gh_L, the frictional head loss, the expression for w becomes

$$w = \frac{v_1^2 - v_2^2}{2} + g(y_1 - y_2) + \frac{P_1 - P_2}{\rho} - h_L g$$

Assuming the fluid at both ends of the control volume to be at atmospheric pressure, $(P_1 - P_2)/\rho = 0$, and for a pipe of constant cross section $(v_1^2 - v_2^2)/2 = 0$, giving for w,

$$w = g(y_1 - y_2) - h_L g$$

Evaluating h_L, we have

$$\text{Re} = \frac{(\frac{1}{2})(4)}{1.22 \times 10^{-5}} = 164\,000$$

$$\frac{e}{D} = 0.0017 \quad \text{(from Figure 14.2)}$$

$$f_f = 0.006 \quad \text{(from Figure 14.1)}$$

yielding

$$h_L = \frac{2(0.006)(120 \text{ ft})(16 \text{ ft}^2/\text{s}^2)}{(0.5 \text{ ft})(32.2 \text{ ft/s}^2)} = 1.432 \text{ ft}$$

The power required to produce the specified flow conditions thus becomes

$$\frac{\delta w}{dt} = \frac{-g((-2 \text{ ft}) - 1.432 \text{ ft})}{550 \text{ ft lb}_f/\text{hp-s}} \left[\frac{62.3 \text{ lb}_m/\text{ft}^3}{32.2 \text{ lb}_m \text{ ft/s}^2 \text{ lb}_f} \left(\frac{\pi}{4}\right) \left(\frac{1}{2} \text{ ft}\right)^2 \left(4 \frac{\text{ft}}{\text{s}}\right) \right]$$

$$= 0.306 \text{ hp}$$

EXAMPLE 2

A heat exchanger is required which will be able to handle $0.0567 \text{ m}^3/\text{s}$ of water through a smooth pipe with an equivalent length of 122 m. The total pressure drop is 103 000 Pa. What size pipe is required for this application?

Once again, applying equation (6-10), we see that a term by term evaluation gives

$$\frac{\delta Q}{dt} = 0 \qquad \frac{\delta W_s}{dt} = 0 \qquad \frac{\delta W_\mu}{dt} = 0$$

$$\iint_{c.s.} \rho \left(e + \frac{P}{\rho}\right)(\mathbf{v} \cdot \mathbf{n}) \, dA = \rho A v_{\text{avg}} \left(\frac{v_2^2}{2} + gy_2 + \frac{P_2}{\rho} + u_2 - \frac{v_1^2}{2} - gy_1 - \frac{P_1}{\rho} - u_1\right)$$

$$\frac{\partial}{\partial t} \iiint_{c.v.} \rho e \, dV = 0$$

and the applicable equation for the present problem is

$$0 = \frac{P_2 - P_1}{\rho} + h_L g$$

The quantity desired, the diameter, is included in the head-loss term but cannot be solved for directly, since the friction factor also depends on D. Inserting numerical values into the above equation and solving, we obtain

$$0 = -\frac{103\,000\ \text{Pa}}{1000\ \text{kg/m}^3} + 2f_f \left(\frac{0.0567}{\pi D^2/4}\right)^2 \frac{\text{m}^2}{\text{s}^2} \cdot \frac{122}{D} \frac{\text{m}}{\text{m}} \frac{\text{g}}{\text{g}}$$

or

$$0 = -103 + 1.27 \frac{f_f}{D^5}$$

The solution to this problem must now be obtained by trial and error. A possible procedure is the following:

1. Assume a value for f_f.
2. Using this f_f, solve the above equation for D.
3. Calculate Re with this D.
4. Using e/D and the calculated Re, check the assumed value of f_f.
5. Repeat this procedure until the assumed and calculated friction factor values agree.

Carrying out these steps for the present problem, the required pipe diameter is 0.132 m (5.2 in.).

EXAMPLE 3

An existing heat exchanger has a cross section as shown in Figure 14.3 with nine 1-in.-OD tubes inside a 5-in.-ID pipe. For a 5-ft length of heat exchanger what flow rate of water can be achieved in the shell side of this unit for a pressure drop of 3 psi?

An energy-equation analysis using equation (6-10) will follow the same steps as in example 14.2, yielding, as the governing equation,

$$0 = \frac{P_2 - P_1}{\rho} + h_L g$$

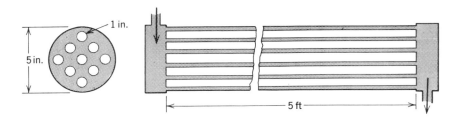

Figure 14.3   Shell-and-tube heat-exchanger configuration.

The equivalent diameter for the shell is evaluated as follows;

$$\text{Flow area} = \frac{\pi}{4}(25-9) = 4\pi \text{ in.}^2$$

$$\text{Wetted perimeter} = \pi(5+9) = 14\pi \text{ in.}$$

thus

$$D_{eq} = 4\frac{4\pi}{14\pi} = 1.142 \text{ in.}$$

Substituting the proper numerical values into the energy equation for this problem reduces it to

$$0 = -\frac{3 \text{ lb}_f/\text{in.}^2(144 \text{ in.}^2/\text{ft}^2)}{1.94 \text{ slugs/ft}^3} + 2f_f v_{avg}^2 \text{ ft}^2/\text{s}^2 \frac{5 \text{ ft}}{(1.142/12) \text{ ft}} \frac{g}{g}$$

or

$$0 = -223 + 105 f_f v_{avg}^2$$

Since f_f cannot be determined without a value of Re, which is a function of v_{avg}, a simple trial-and-error procedure such as the following might be employed:

1. Assume a value for f_f.
2. Calculate v_{avg} from the above expression.
3. Evaluate Re from this value of v_{avg}.
4. Check the assumed value of f_f using Figure 14.1.
5. If the assumed and calculated values for f_f do not agree, repeat this procedure until they do.

Employing this method, we find the velocity to be 23.6 fps, giving a flow rate for this problem of 2.06 ft^3/min (0.058 m^3/s).

Notice that in each of the last two examples in which a trial-and-error approach was used the assumption of f_f was made initially. This was not, of course, the only way to approach these problems; however, in both cases a value for f_f could be assumed within a much closer range than either D or v_{avg}.

14.5 FRICTION FACTORS FOR FLOW IN THE ENTRANCE TO A CIRCULAR CONDUIT

The development and problems in the preceding section have involved flow conditions which did not change along the axis of flow. This condition is often met, and the methods just described will be adequate to evaluate and predict the significant flow parameters.

In many real flow systems this condition is never realized. A boundary layer forms on the surface of a pipe and its thickness increases in a similar manner to that of the boundary layer on a flat plate as described in Chapter 12. The buildup of the boundary layer in pipe flow is depicted in Figure 14.4.

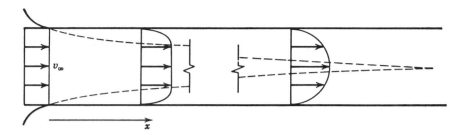

Figure 14.4   Boundary layer buildup in a pipe.

A boundary layer forms on the inside surface and occupies a larger amount of the flow area for increasing values of x, the distance downstream from the pipe entrance. At some value of x the boundary layer fills the flow area. The velocity profile will not change downstream from this point, and the flow is said to be *fully developed*. The distance downstream from the pipe entrance to where flow becomes fully developed is called the entrance length, symbolized L_e. Observe that the fluid velocity outside the boundary layer increases with x, as is required to satisfy continuity. The velocity at the center of the pipe finally reaches a value of $2v_\infty$ for fully developed laminar flow.

The entrance length required for a fully developed velocity profile to form in laminar flow has been expressed by Langhaar* according to

$$\frac{L_e}{D} = 0.0575 \text{ Re} \tag{14-20}$$

where D represents the inside diameter of the pipe. This relation, derived analytically, has been found to agree well with experiment.

There is no relation available to predict the entrance length for a fully developed turbulent velocity profile. An additional factor which affects the entrance length in turbulent flow is the nature of the entrance itself. The reader is referred to the work of Deissler† for experimentally obtained turbulent velocity profiles in the entrance region of circular pipes. A general conclusion of the results of Deissler and others is that the turbulent velocity profile becomes fully developed after a minimum distance of 50 diameters downstream from the entrance.

The reader should realize that the entrance length for the fully developed velocity profile differs considerably from that for the velocity gradient at the wall. Since the friction factor is a function of dv/dy at the pipe surface, we are also interested in this starting length.

Two conditions exist in the entrance region which cause the friction factor to be greater than in fully developed flow. The first of these is the extremely large

* H. L. Langhaar, *Trans. ASME*, **64**, A-55 (1942).
† R. G. Deissler, NACA TN 2138 (1950).

wall velocity gradient right at the entrance. This gradient decreases in the downstream direction, becoming constant before the velocity profile becomes fully developed. The other factor is the existence of a "core" of fluid outside the viscous layer whose velocity must increase as dictated by continuity. The fluid in the core is thus being accelerated, thereby producing an additional drag force whose effect is incorporated in the friction factor.

The friction factor for laminar flow in the entrance to a pipe has been studied by Langhaar.* His results indicated the friction factor to be highest in the vicinity of the entrance, then to decrease smoothly to the fully developed flow value. Figure 14.5 is a qualitative representation of this variation. Table 14.2 gives the

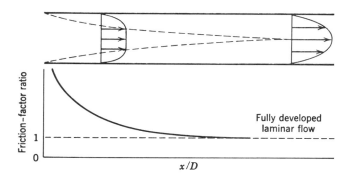

Figure 14.5 Velocity profile and friction-factor variation for laminar flow in the region near a pipe entrance.

results of Langhaar for the average friction factor between the entrance and a location a distance x from the entrance.

For turbulent flow in the entrance region the friction factor as well as the velocity profile is difficult to express. Deissler† has analyzed this situation and presented his results graphically.

Even for very high free-stream velocities there will be some portion of the entrance over which the boundary layer is laminar. The entrance configuration, as well as the Reynolds number, affects the length of pipe over which the laminar boundary layer exists before becoming turbulent. A plot similar to Figure 14.5 is presented in Figure 14.6 for turbulent-flow friction factors in the entrance region.

The foregoing description of the entrance region has been qualitative. For an accurate analytical consideration of a system involving entrance-length phenomena, Deissler's results portrayed in Figure 14.7 may be utilized.

---

* *Op cit.*
† R. G. Deissler, NACA TN 3016 (1953).

TABLE 14.2 AVERAGE FRICTION
FACTOR FOR LAMINAR FLOW IN THE
ENTRANCE TO A CIRCULAR PIPE

$\dfrac{x/D}{Re}$	$f_f\left(\dfrac{x}{D}\right)$
0.000205	0.0530
0.000830	0.0965
0.001805	0.1413
0.003575	0.2075
0.00535	0.2605
0.00838	0.340
0.01373	0.461
0.01788	0.547
0.02368	0.659
0.0341	0.845
0.0449	1.028
0.0620	1.308
0.0760	1.538

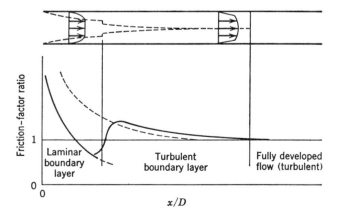

Figure 14.6 Velocity profile and friction-factor variation
in turbulent flow in the region near a pipe entrance.

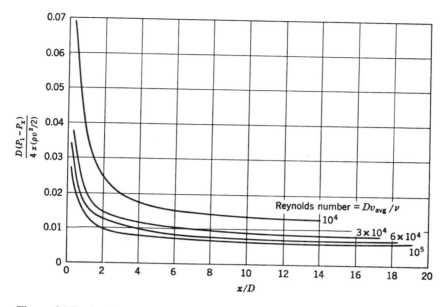

Figure 14.7 Static pressure drop due to friction and momentum change in the entrance to a smooth, horizontal, circular tube (Deissler).

It is important to realize that in many situations flow is never fully developed, thus the friction factor will be higher than that predicted from the equations for fully developed flow or the friction-factor plot.

14.6 CLOSURE

The information and techniques presented in this chapter have included applications of the theory developed in earlier chapters supported by correlations of experimental data.

The chapters to follow will be devoted to heat and mass transfer. One specific type of transfer, momentum transfer, has been considered up to this point. The student will find that he is able to apply much of the information learned in momentum transfer to counterparts in the areas of heat and mass transfer.

PROBLEMS

14.1 An oil with kinematic viscosity of 0.08×10^{-3} ft^2/s and a density of 57 lb$_m$/ft^3 flows through a horizontal tube 0.24 in. in diameter at the rate of 10 gal/hr. Determine the pressure drop in 50 ft of tube.

14.2 A lubricating line has an inside diameter of 0.1 in. and is 30 in. long. If the pressure drop is 15 psi, determine the flow rate of the oil. Use the properties given in problem 14.1.

14.3 A 280 km-long pipeline connects two pumping stations. If 0.56 m^3/s are to be pumped through a 0.62-m-diameter line, the discharge station is 250 m lower in elevation than the upstream station and the discharge pressure is to be maintained at 300 000 Pa, determine the power required to pump the oil. The oil has a kinematic viscosity of 4.5×10^{-6} m^2/s and a density of 810 kg/m^3. The pipe is constructed of commercial steel. The inlet pressure may be taken as atmospheric.

14.4 The pressure drop in a section of pipe is determined from tests with water. A pressure drop of 13 psi is obtained at a flow rate of 28.3 lb$_m$/s. If the flow is fully turbulent, what will be the pressure drop when liquid oxygen ($\rho = 70$ lb$_m$/ft^3) flows through the pipe at the rate of 35 lb$_m$/s?

14.5 Oil having a kinematic viscosity of 6.7×10^{-6} m^2/s and density of 801 kg/m^3 is pumped through a pipe of 0.71 m diameter at an average velocity of 1.1 m/s. The roughness of the pipe is equivalent to that of a "commercial steel" pipe. If pumping stations are 320 km apart, find the head loss (in meters of oil) between pumping stations and the power required.

14.6 The cold-water faucet in a house is fed from a water main through the following simplified piping system.
(a) A 160 ft length of $\frac{3}{4}$-in. ID galvanized pipe leading from the main line to the base of the faucet.
(b) Four 90° standard elbows.
(c) One wide-open angle valve (with no obstruction).
(d) The faucet. Consider the faucet to be made up of two parts: (1) a conventional globe valve and (2) a nozzle having a cross-sectional area of 0.10 in.^{2}
The pressure in the main line is 60 psig (virtually independent of flow), and the velocity there is negligible. Find the maximum rate of discharge from the faucet. As a first try, assume for the pipe $f_f = 0.007$. Neglect changes in elevation throughout the system.

14.7 Water at the rate of 118 ft^3/min flows through a smooth horizontal tube 250 ft long. The pressure drop is 4.55 psi. Determine the tube diameter.

14.8 Determine the flow rate through a 0.2-m gate valve with upstream pressure of 236 kPa when the valve is
(a) open;
(b) $\frac{1}{4}$ closed;

(c) $\frac{1}{2}$ closed;

(d) $\frac{3}{4}$ closed.

14.9 Calculate the inlet pressure to a pump 3 ft above the level of a sump. The pipe is 6 in. in diameter, 6 ft long, and made of commercial steel. The flow rate through the pump is 500 gal/min. Use the (incorrect) assumption that the flow is fully developed.

14.10 The pipe in problem 6.20 is 35 m long and made of commercial steel. Determine the flow rate.

14.11 The siphon of problem 6.17 is made of smooth rubber hose and is 23 ft long. Determine the flow rate and the pressure at point B.

14.12 A galvanized rectangular duct 8 in. square is 25 ft long and carries $600 \, \text{ft}^3/\text{min}$ of standard air. Determine the pressure drop in inches of water.

14.13 A cast-iron pipeline 2 mi long is required to carry 3 million gal of water per day. The outlet is 175 ft higher than the inlet. The costs of three sizes of pipe when in place are as follows:

10-in. diameter	$11.40 per ft
12-in. diameter	$14.70 per ft
14-in. diameter	$16.80 per ft

Power costs are estimated at $0.07 per kilowatt hour over the 20-year life of the pipeline. If the line can be bonded with 9.6% annual interest, what is the most economical pipe diameter? The pump efficiency is 80%, and the water inlet temperature is expected to be constant at 42°F.

14.14 Water flows at the rate of 500 gal/min through a 400-ft-long line. The first and last 100 ft of the line are 6-in.-diameter cast-iron pipe, while the middle 200 ft consists of two 4.24-in.-diameter cast-iron pipes. Find
(a) the average velocity in each section of pipe;
(b) the pressure drop in each length of pipe;
(c) the total pressure drop.

14.15 A 0.2-m-diameter cast-iron pipe and a 67-mm-diameter commercial steel pipe are parallel, and both run from the same pump to a reservoir. The pressure drop is 210 kPa and the lines are 150 m long. Determine the flow rate of water in each line.

14.16 Estimate the flow rate of water through 50 ft of garden hose from a 40-psig source for
(a) A $\frac{1}{2}$-in.-ID hose;
(b) A $\frac{3}{4}$-in.-ID hose.

14.17 Two water reservoirs of height $h_1 = 60$ m and $h_2 = 30$ m are connected by a pipe which is 0.35 m in diameter. The exit of the pipe is submerged at distance $h_3 = 8$ m from the reservoir surface.

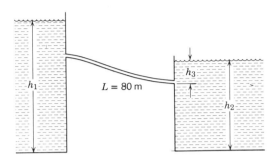

(a) Determine the flow rate through the pipe if the pipe is 80 m long and the friction factor $f_f = 0.004$. The pipe inlet is set flush with the wall.
(b) If the relative roughness $e/D = 0.004$, determine the friction factor and flow rate.

14.18 An 8-km-long, 5-m-diameter headrace tunnel at the Paute River hydroelectric project in Ecuador supplies a power station 668 m below the entrance of the tunnel. If the tunnel surface is concrete, find the pressure at the end of the tunnel if the flow rate is 90 m^3/s.

CLOSURE TO MOMENTUM TRANSFER

At this point the specific study of momentum transfer is terminated. We now direct our attention to energy transfer. The concepts, descriptions of transport mechanisms, and means of analytically describing momentum transfer will *all* be encountered in the next eight chapters. The fluxes, transport properties, and driving forces will be new, but the role that each plays in energy transfer has a direct counterpart in momentum transfer. Wherever similarities appear, they will be pointed out. The student is urged to make those associations that will strengthen his understanding of previously covered material and that will aid in comprehending that which is new. Many of the developments and equations already considered will be used without additional discussion; in such cases it may be helpful to refer occasionally to the appropriate section in earlier chapters.

15
FUNDAMENTALS OF HEAT TRANSFER

The next nine chapters deal with the transfer of energy. Gross quantities of heat added to or rejected from a system may be evaluated by applying the control-volume expression for the first law of thermodynamics as discussed in Chapter 6. The result of a first-law analysis is only a part of the required information necessary for the complete evaluation of a process or situation which involves energy transfer. The overriding consideration is, in many instances, the rate at which energy transfer takes place. Certainly in designing a plant in which heat must be exchanged with the surroundings, the size of heat-transfer equipment, the materials of which it is to be constructed, and the auxiliary equipment required for its utilization are all important considerations for the engineer. Not only must the equipment accomplish its required mission, but it must be economical to purchase and to operate.

Considerations of an engineering nature such as these require both a familiarity with the basic mechanisms of energy transfer and an ability to evaluate quantitatively these rates as well as the important associated quantities. Our immediate goal is to examine the basic mechanisms of energy transfer and to consider the fundamental equations for evaluating the rate of energy transfer.

There are three modes of energy transfer: conduction, convection, and radiation. All heat-transfer processes involve one or more of these modes. The remainder of this chapter will be devoted to an introductory description and discussion of these types of transfer.

15.1 CONDUCTION

Energy transfer by conduction is accomplished in two ways. The first mechanism is that of molecular interaction, in which the greater motion of a molecule at a higher energy level (temperature) imparts energy to adjacent molecules at lower energy levels. This type of transfer is present, to some degree, in all systems in which a temperature gradient exists and in which molecules of a solid, liquid, or gas are present.

The second mechanism of conduction heat transfer is by "free" electrons. The free-electron mechanism is significant primarily in pure-metallic solids; the concentration of free electrons varies considerably for alloys and becomes very low for nonmetallic solids. The ability of solids to conduct heat varies directly with the concentration of free electrons, thus it is not surprising that pure metals are the best heat conductors, as our experience has indicated.

Since heat conduction is primarily a molecular phenomenon, we might expect the basic equation used to describe this process to be similar to the expression used in the molecular transfer of momentum, equation (7-4). Such an equation was first stated in 1822 by Fourier in the form

$$\frac{q_x}{A} = -k\frac{dT}{dx} \qquad (15\text{-}1)$$

where q_x is the heat-transfer rate in the x direction, in watts or Btu/hr; A is the area *normal* to the direction of heat flow, in m^2 or ft^2; dT/dx is the temperature gradient in the x direction, in K/m or °F/ft; and k is the thermal conductivity, in W/(m · K) or Btu/hr ft °F. The ratio q_x/A, having the dimensions of W/m^2 or Btu/hr ft^2, is referred to as the heat flux in the x direction. A more general relation for the heat flux is equation (15-2),

$$\frac{\mathbf{q}}{A} = -k\boldsymbol{\nabla} T \qquad (15\text{-}2)$$

which expresses the heat flux as proportional to the temperature gradient. The proportionality constant is seen to be the thermal conductivity, which plays a role similar to that of the viscosity in momentum transfer. The negative sign in equation (15-2) indicates that heat flow is in the direction of a negative temperature gradient. Equation (15-2) is the vector form of the *Fourier rate equation*, often referred to as Fourier's first law of heat conduction.

The thermal conductivity, k, which is defined by equation (15-1), is assumed independent of direction in equation (15-2); thus this expression applies to an *isotropic* medium only. The thermal conductivity is a property of a conducting medium and, like the viscosity, is primarily a function of temperature, varying significantly with pressure only in the case of gases subjected to high pressures.

15.2 THERMAL CONDUCTIVITY

Since the mechanism of conduction heat transfer is one of molecular interaction, it will be illustrative to examine the motion of gas molecules from a standpoint similar to that in section 7.3.

Considering the control volume shown in Figure 15.1, in which energy transfer in the y direction is on a molecular scale only, we may utilize the first-law

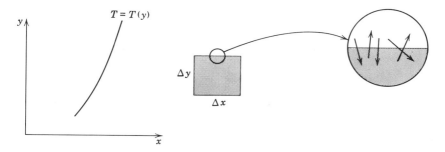

Figure 15.1 Molecular motion at the surface of a control volume.

analysis of Chapter 6 as follows. Mass transfer across the top of this control volume is considered to occur only on the molecular scale. This criterion is met for a gas in laminar flow.

Applying equation (6-10) and considering transfer only across the top face of the element considered,

$$\frac{\delta Q}{dt} - \frac{\delta W_s}{dt} - \frac{\delta W_\mu}{dt} = \iint_{c.s.} \left(e + \frac{P}{\rho}\right)\rho(\mathbf{v} \cdot \mathbf{n}) \, dA + \frac{\partial}{\partial t} \iiint_{c.v.} e\rho \, dV \quad (6\text{-}10)$$

for Z molecules crossing the plane $\Delta x \, \Delta z$ per unit time this equation reduces to

$$q_y = \sum_{n=1}^{Z} m_n c_p (T|_{y-} - T|_{y+}) \, \Delta x \, \Delta z \quad (15\text{-}3)$$

where m_n is the mass per molecule; c_p is the molecular heat capacity of the gas; Z is the frequency with which molecules will cross area $\Delta x \, \Delta z$; and $T|_{y-}$, $T|_{y+}$ are the temperatures of the gas slightly below and slightly above the plane considered, respectively. The right hand term is the summation of the energy flux associated with the molecules crossing the control surface. Noting now that $T|_{y-} = T - \partial T/\partial y|_{y_0} \, \delta$, where $y - = y_0 - \delta$, and that a similar expression may be written for $T|_{y+}$, we may rewrite equation (15-3) in the form

$$\frac{q_y}{A} = -2 \sum_{n=1}^{Z} m_n c_p \delta \frac{\partial T}{\partial y}\bigg|_{y_0} \quad (15\text{-}4)$$

where δ represents the y component of the distance between collisions. We note, as previously in Chapter 7, that $\delta = \frac{2}{3}\lambda$, where λ is the mean free path of a molecule. Using this relation and summing over Z molecules, we have

$$\frac{q_y}{A} = -\frac{4}{3}\rho c_p Z\lambda \frac{\partial T}{\partial y}\bigg|_{y_0} \quad (15\text{-}5)$$

Comparing equation (15-5) with the y component of equation (15-2),

$$\frac{q_y}{A} = -k\frac{\partial T}{\partial y}$$

it is apparent that the thermal conductivity, k, becomes

$$k = \tfrac{4}{3}\rho c_p Z \lambda$$

Utilizing further the results of the kinetic theory of gases, we may make the following substitutions:

$$Z = \frac{N\bar{C}}{4}$$

where $\bar{C}$ is the average random molecular velocity, $\bar{C} = \sqrt{8\kappa T/\pi m}$ (κ being the Boltzmann constant);

$$\lambda = \frac{1}{\sqrt{2}\,\pi N d^2}$$

where d is the molecular diameter; and

$$c_p = \frac{3}{2}\frac{\kappa}{N}$$

giving, finally,

$$k = \frac{1}{\pi^{3/2} d^2}\sqrt{\kappa^3 T/m} \tag{15-6}$$

This development, applying specifically to monatomic gases, is significant in that it shows the thermal conductivity of a gas to be independent of pressure, and to vary as the $\tfrac{1}{2}$ power of the absolute temperature. The significance of this result should not be overlooked, even though some oversimplifications were used in its development. Some relations for thermal conductivity of gases, based upon more sophisticated molecular models, may be found in Bird, Stewart, and Lightfoot.[*]

The Chapman-Enskog theory used in Chapter 7 to predict gas viscosities at low pressures has a heat-transfer counterpart. For a monatomic gas the recommended equation is

$$k = 0.0829\sqrt{(T/M)}/\sigma^2 \Omega_k \tag{15-7}$$

where k is in W/m $\cdot$ K, σ is in Angstroms, M is the molecular weight, and Ω_k is the Lennard-Jones collision integral, identical with Ω as discussed in section 7.3. Both σ and Ω_k may be evaluated from Appendices J and K.

The thermal conductivity of a liquid is not amenable to any simplified kinetic-theory development, since the molecular behavior of the liquid phase is not clearly understood and no universally accurate mathematical model presently exists. Some empirical correlations have met with reasonable success, but these are so specialized that they will not be included in this book. For a discussion of

---

[*] R. B. Bird, W. E. Stewart, and E. N. Lightfoot, *Transport Phenomena*, Wiley, New York, 1960, chap. 8.

molecular theories related to the liquid phase and some empirical correlations of thermal conductivities of liquids, the reader is referred to Reid and Sherwood.* A general observation about liquid thermal conductivities is that they vary only slightly with temperature and are relatively independent of pressure. One problem in experimentally determining values of the thermal conductivity in a liquid is making sure the liquid is free of convection currents.

In the solid phase, thermal conductivity is attributed both to molecular interaction, as in other phases, and to free electrons, which are present primarily in pure metals. The solid phase is amenable to quite precise measurements of thermal conductivity, since there is no problem with convection currents. The thermal properties of most solids of engineering interest have been evaluated, and extensive tables and charts of these properties, including thermal conductivity, are available.

The free-electron mechanism of heat conduction is directly analogous to the mechanism of electrical conduction. This realization led Wiedemann and Franz, in 1853, to relate the two conductivities in a crude way; and in 1872, Lorenz† presented the following relation, known as the Wiedemann, Franz, Lorenz equation:

$$L = \frac{k}{k_e T} = \text{constant} \qquad (15\text{-}8)$$

where k is the thermal conductivity, k_e is the electrical conductivity, T is the absolute temperature, and L is the Lorenz number.

The numerical values of the quantities in equation (15-8) are of secondary importance at this time. The significant point to note here is the simple relation between electrical and thermal conductivities and, specifically, that those materials which are good conductors of electricity are likewise good heat conductors, and vice versa.

Figure 15.2 illustrates the thermal conductivity variation with temperature of several important materials in gas, liquid, and solid phases. A more complete tabulation of thermal conductivity may be found in Appendices H and I.

The following two examples illustrate the use of the Fourier rate equation in solving simple heat-conduction problems.

EXAMPLE 1

A steel pipe having an inside diameter of 0.742 in. and a wall thickness of 0.154 in. is subjected to inside and outside surface temperature of 200° and 160°F, respectively (see Figure 15.3). Find the heat flow rate per foot of pipe length, and also the heat flux based on the inside and that based on the outside surface area.

---

* Reid and Sherwood, *The Properties of Gases and Liquids*, McGraw-Hill Book Company, New York, 1958, chap. 7.

† L. Lorenz, *Ann. Physik und Chemie* (Poggendorffs), **147**, 429 (1872).

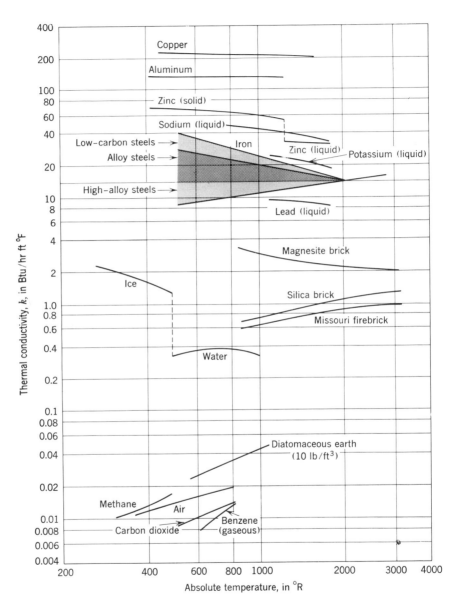

Figure 15.2 Thermal conductivity of several materials at various temperatures. (From M. Jacob and G. A. Hawkins, *Elements of Heat Transfer*, McGraw-Hill Book Company, New York, 1957, p. 23. By permission of the publishers.)

The first law of thermodynamics applied to this problem will reduce to the form $\delta Q/dt = 0$, indicating that the rate of heat transfer into the control volume is equal to the rate leaving.

Since the heat flow will be in the radial direction, the independent variable is r, and the proper form for the Fourier rate equation is

$$q_r = -kA\frac{dT}{dr}$$

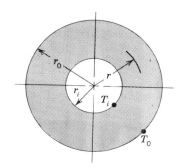

Writing $A = 2\pi rL$, we see that the equation becomes

$$q_r = -k(2\pi rL)\frac{dT}{dr}$$

Figure 15.3 Heat conduction in a radial direction with uniform surface temperatures.

where q_r is constant, which may be separated and solved as follows:

$$q_r \int_{r_i}^{r_o} \frac{dr}{r} = -2\pi kL \int_{T_i}^{T_o} dT = 2\pi kL \int_{T_o}^{T_i} dT$$

$$q_r \ln \frac{r_o}{r_i} = 2\pi kL(T_i - T_o)$$

$$q_r = \frac{2\pi kL}{\ln r_o/r_i}(T_i - T_o) \tag{15-9}$$

Substituting the given numerical values, we obtain

$$q_r = \frac{2\pi(24.8\ \text{Btu/hr ft °F})(200 - 160)\text{°F}}{\ln(1.050/0.742)}$$

$$= 17\,950\ \text{Btu/hr ft} \qquad (17\,250\ \text{W/m})$$

The inside and outside surface areas per unit length of pipe are

$$A_i = \pi\left(\frac{0.742}{12}\right)(1) = 0.194\ \text{ft}^2 \qquad (0.059\ \text{m}^2/\text{m})$$

$$A_o = \pi\left(\frac{1.050}{12}\right)(1) = 0.275\ \text{ft}^2 \qquad (0.0838\ \text{m}^2/\text{m})$$

giving

$$q_r/A_i = \frac{17\,950}{0.194} = 92\,500\ \text{Btu/hr ft}^2 \qquad (292\,000\ \text{W/m}^2)$$

and

$$q_r/A_o = \frac{17\,950}{0.275} = 65\,300\ \text{Btu/hr ft}^2 \qquad (206\,000\ \text{W/m}^2)$$

One extremely important point to be noted from the results of this example is the requirement of specifying the area upon which a heat-flux value is based. Note that for the same amount of heat flow the fluxes based upon the inside and outside surface areas differ by approximately 42%.

EXAMPLE 2

Consider a hollow cylindrical heat-transfer medium having inside and outside radii of r_i and r_o with the corresponding surface temperatures T_i and T_o. If the thermal-conductivity variation may be described as a linear function of temperature according to

$$k = k_0(1 + \beta T)$$

calculate the steady-state heat-transfer rate in the radial direction, using the above relation for the thermal conductivity, and compare the result with that using a k value calculated at the arithmetic mean temperature.

Figure 15.3 applies. The equation to be solved is now

$$q_r = -[k_o(1 + \beta T)](2\pi r L)\frac{dT}{dr}$$

which, upon separation and integration, becomes

$$q_r \int_{r_i}^{r_o} \frac{dr}{r} = -2\pi k_o L \int_{T_i}^{T_o} (1 + \beta T)\, dT$$

$$= 2\pi k_o L \int_{T_o}^{T_i} (1 + \beta T)\, dT$$

$$q_r = \frac{2\pi k_o L}{\ln r_o/r_i}\left[T + \frac{\beta T^2}{2}\right]_{T_o}^{T_i}$$

$$q_r = \frac{2\pi k_o L}{\ln r_o/r_i}\left[1 + \frac{\beta}{2}(T_i + T_o)\right](T_i - T_o) \qquad (15\text{-}10)$$

Noting that the arithmetic average value of k would be

$$k_{avg} = k_o\left[1 + \frac{\beta}{2}(T_i + T_o)\right]$$

we see that equation (15-10) could also be written as

$$q_r = \frac{2\pi k_{avg} L}{\ln r_o/r_i}(T_i - T_o)$$

Thus the two methods give identical results.

The student may find it instructive to determine what part of the problem statement of this example is responsible for this interesting result; that is, whether a different geometrical configuration or a different thermal-conductivity expression would make the results of the two types of solutions different.

15.3 CONVECTION

Heat transfer due to convection involves the energy exchange between a surface and an adjacent fluid. A distinction must be made between *forced convection*, wherein a fluid is made to flow past a solid surface by an external agent such as a fan or pump, and *free* or *natural convection* wherein warmer (or cooler) fluid next to the solid boundary causes circulation because of the density difference resulting from the temperature variation throughout a region of the fluid.

The rate equation for convective heat transfer was first expressed by Newton in 1701, and is referred to as the *Newton rate equation* or Newton's "law" of cooling. This equation is

$$q/A = h\,\Delta T \tag{15-11}$$

where q is the rate of convective heat transfer, in W or Btu/hr; A is the area normal to direction of heat flow, in m^2 or ft^2; ΔT is the temperature difference between surface and fluid, in K or °F; and h is the convective heat transfer coefficient, in $W/m^2 \cdot K$ or $Btu/hr\,ft^2\,°F$. Equation (15-11) is not a law but a definition of the coefficient h. A substantial portion of our work in the chapters to follow will involve the determination of this coefficient. It is, in general, a function of system geometry, fluid and flow properties, and the magnitude of ΔT.

Since flow properties are so important in the evaluation of the convective heat transfer coefficient, we may expect many of the concepts and methods of analysis introduced in the preceding chapters to be of continuing importance in convective heat transfer analysis; this is indeed the case.

From our previous experience we should also recall that even when a fluid is flowing in a turbulent manner past a surface, there is still a layer, sometimes extremely thin, close to the surface where flow is laminar; also, the fluid particles next to the solid boundary are at rest. Since this is always true, the mechanism of heat transfer between a solid surface and a fluid must involve conduction through the fluid layers close to the surface. This "film" of fluid often presents the controlling resistance to convective heat transfer, and the coefficient h is often referred to as the *film coefficient*.

Two types of heat transfer which differ somewhat from free or forced convection but are still treated quantitatively by equation (15-11) are the phenomena of boiling and condensation. The film coefficients associated with these two kinds of transfer are quite high. Table 15.1 represents some order-of-magnitude values of h for different convective mechanisms.

It will also be necessary to distinguish between local heat transfer coefficients, that is, those which apply at a point, and total or average values of h which apply over a given surface area. We will designate the local coefficient h_x, according to equation (15-11),

$$dq = h_x\,\Delta T\,dA$$

TABLE 15.1 APPROXIMATE VALUES OF THE CONVECTIVE
HEAT-TRANSFER COEFFICIENT

Mechanism	h, Btu/hr ft^2 °F	h, W/(m$^2 \cdot$ K)
Free convection, air	1–10	5–50
Forced convection, air	5–50	25–250
Forced convection, water	50–3000	250–15 000
Boiling water	500–5000	2500–25 000
Condensing water vapor	1000–20 000	5000–100 000

thus the average coefficient, h, is related to h_x according to the relation

$$q = \int_A h_x \, \Delta T \, dA = hA \, \Delta T \qquad (15\text{-}12)$$

The values given in Table 15.1 are average convective heat-transfer coefficients.

15.4 RADIATION

Radiant heat transfer differs from conduction and convection in that no medium is required for its propagation, indeed energy transfer by radiation is maximum when the two surfaces which are exchanging energy are separated by a perfect vacuum. The exact mechanism of radiant energy transfer is not completely understood. There is evidence to support both wave and corpuscular arguments. A remarkable fact, however, is that a relatively complex process such as radiation heat transfer can be described by a reasonably simple analytical expression.

The rate of energy emission from a perfect radiator or *black body* is given by

$$\frac{q}{A} = \sigma T^4 \qquad (15\text{-}13)$$

where q is the rate of radiant energy emission, in W or Btu/hr; A is the area of the emitting surface, in m^2 or ft^2; T is the absolute temperature, in K or °R; and σ is the Stefan-Boltzmann constant, which is equal to 5.676×10^{-8} W/m$^2 \cdot$ K^4 or 0.1714×10^{-8} Btu/hr ft^2 °R^4. The proportionally constant relating radiant-energy flux to the fourth power of the absolute temperature is named after Stefan who, from experimental observations, proposed equation (15-13) in 1879, and Boltzmann, who derived this relation theoretically in 1884. Equation (15-13) is most often referred to as the Stefan-Boltzmann law of thermal radiation.

Certain modifications will be made in equation (15-13) to account for the *net* energy transfer between two surfaces, the degree of deviation of the emitting and receiving surfaces from black-body behavior, and geometrical factors associated with radiant exchange between a surface and its surroundings. These considerations will be discussed at length in Chapter 23.

15.5 COMBINED MECHANISMS OF HEAT TRANSFER

The three modes of heat transfer have been considered separately in section 15.4. It is rare, in actual situations, for only one mechanism to be involved in the transfer of energy. It will be instructive to look at some situations in which heat transfer is accomplished by a combination of these mechanisms.

Consider the case depicted in Figure 15.4, that of steady-state conduction through a plane wall with its surfaces held at constant temperatures T_1 and T_2.

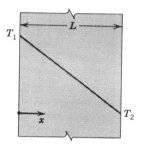

Figure 15.4   Steady-state conduction through a plane wall.

Writing the Fourier rate equation for the x direction, we have

$$\frac{q_x}{A} = -k\frac{dT}{dx} \tag{15-1}$$

Solving this equation for q_x subject to the boundary conditions $T = T_1$ at $x = 0$ and $T = T_2$ at $x = L$ we obtain

$$\frac{q_x}{A}\int_0^L dx = -k\int_{T_1}^{T_2} dT = k\int_{T_2}^{T_1} dT$$

or

$$q_x = \frac{kA}{L}(T_1 - T_2) \tag{15-14}$$

Equation (15-14) bears an obvious resemblance to the Newton rate equation,

$$q_x = hA\,\Delta T \tag{15-11}$$

We may utilize this similarity in form in a problem in which both types of energy transfer are involved.

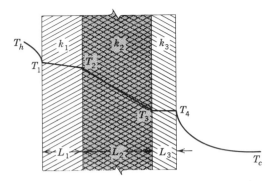

Figure 15.5 Steady-state heat transfer through a composite wall.

Consider the composite plane wall constructed of three materials in layers with dimensions as shown in Figure 15.5. We wish to express the steady-state heat-transfer rate per unit area between a hot gas at temperature T_h on one side of this wall and a cool gas at T_c on the other side. Temperature designations and dimensions are as shown in the figure. The following relations for q_x arise from the application of equations (15-11) and (15-14):

$$q_x = h_h A (T_h - T_1) = \frac{k_1 A}{L_1}(T_1 - T_2) = \frac{k_2 A}{L_2}(T_2 - T_3)$$

$$= \frac{k_3 A}{L_3}(T_3 - T_4) = h_c A(T_4 - T_c)$$

Each temperature difference is expressed in terms of q_x as follows:

$$T_h - T_1 = q_x (1/h_h A)$$
$$T_1 - T_2 = q_x (L_1/k_1 A)$$
$$T_2 - T_3 = q_x (L_2/k_2 A)$$
$$T_3 - T_4 = q_x (L_3/k_3 A)$$
$$T_4 - T_c = q_x (1/h_c A)$$

Adding these equations, we obtain

$$T_h - T_c = q_x \left(\frac{1}{h_h A} + \frac{L_1}{k_1 A} + \frac{L_2}{k_2 A} + \frac{L_3}{k_3 A} + \frac{1}{h_c A} \right)$$

and finally, solving for q_x, we have

$$q_x = \frac{T_h - T_c}{1/h_h A + L_1/k_1 A + L_2/k_2 A + L_3/k_3 A + 1/h_c A} \qquad (15\text{-}15)$$

Note that the heat-transfer rate is expressed in terms of the *overall* temperature difference. If a series electrical circuit,

is considered, we may write

$$I = \frac{\Delta V}{R_1 + R_2 + R_3 + R_4 + R_5} = \frac{\Delta V}{\sum R_i}$$

The analogous quantities in the expressions for heat flow and electrical current are apparent,

$$\Delta V \rightarrow \Delta T$$

$$I \rightarrow q_x$$

$$R_i \rightarrow 1/hA, \; L/kA$$

and each term in the denominator of equation (15-15) may be thought of as a thermal resistance due to convection or conduction. Equation (15-15) thus becomes a heat-transfer analog to Ohm's law, relating heat flow to the overall temperature difference divided by the total thermal resistance between the points of known temperature. Equation (15-15) may now be written simply as

$$q = \frac{\Delta T}{\sum R_{\text{thermal}}} \tag{15-16}$$

This relation applies to steady-state heat transfer in systems of other geometries as well. The thermal-resistance terms will change in form for cylindrical or spherical systems, but once evaluated, they can be utilized in the form indicated by equation (15-16). With specific reference to equation (15-9), it may be noted that the thermal resistance of a cylindrical conductor is

$$\frac{\ln (r_o/r_i)}{2\pi k L}$$

Another common way of expressing the heat-transfer rate for a situation involving a composite material or combination of mechanisms is with the *overall heat-transfer coefficient* defined as

$$U \equiv \frac{q_x}{A \, \Delta T} \tag{15-17}$$

where U is the overall heat-transfer coefficient having the same units as h, in $W/m^2 \cdot K$ or $Btu/hr \, ft^2 \, °F$.

The following example shows the utility of evaluating heat-transfer rates by means of an overall temperature difference.

EXAMPLE 3

Saturated steam at 0.276 MPa flows inside a steel pipe having an inside diameter of 2.09 cm and an outside diameter of 2.67 cm. The convective coefficients on the inner and outer pipe surfaces may be taken as 5680 W/m² · K and 22.7 W/m² · K, respectively. The surrounding air is at 294 K. Find the heat loss per meter of bare pipe and for a pipe having a 3.8 cm thickness of 85% magnesia insulation on its outer surface.

In the case of the bare pipe there are three thermal resistances to evaluate:

$$R_1 = R_{\text{convection inside}} = 1/h_i A_i$$
$$R_2 = R_{\text{convection outside}} = 1/h_o A_0$$
$$R_3 = R_{\text{conduction}} = \ln (r_0/r_i)/2\pi k L$$

For conditions of this problem these resistances have the values

$$R_1 = 1/[(5680 \text{ W/m}^2 \cdot \text{K})(\pi)(0.0209 \text{ m})(1 \text{ m})]$$

$$= 0.00268 \text{ K/W} \qquad\qquad \left(0.00141 \frac{\text{hr } ^\circ\text{R}}{\text{Btu}}\right)$$

$$R_2 = 1/[(22.7 \text{ W/m}^2 \cdot \text{K})(\pi)(0.0267 \text{ m})(1 \text{ m})]$$

$$= 0.525 \text{ K/W} \qquad\qquad \left(0.277 \frac{\text{hr } ^\circ\text{R}}{\text{Btu}}\right)$$

and

$$R_3 = \frac{\ln (2.67/2.09)}{2\pi(42.9 \text{ W/m} \cdot \text{K})(1 \text{ m})}$$

$$= 0.00091 \text{ K/W} \qquad\qquad \left(0.00048 \frac{\text{hr } ^\circ\text{R}}{\text{Btu}}\right)$$

The inside temperature is that of 1915 Pa saturated steam, 267°F or 404 K. The heat transfer rate per meter of pipe may now be calculated as

$$q = \frac{\Delta T}{\Sigma R} = \frac{404-294 \text{ K}}{0.528 \text{ K/W}}$$

$$= 208 \text{ W} \qquad\qquad \left(710 \frac{\text{Btu}}{\text{hr}}\right)$$

In the case of an insulated pipe, the total thermal resistance would include R_1 and R_3 evaluated above, plus additional resistances to account for the insulation. For the insulation,

$$R_4 = \frac{\ln (10.27/2.67)}{2\pi(0.0675 \text{ W/m} \cdot \text{K})(1 \text{ m})}$$

$$= 3.176 \text{K/W} \qquad\qquad \left(1.675 \frac{\text{hr } ^\circ\text{R}}{\text{Btu}}\right)$$

and for the outside surface of the insulation

$$R_5 = 1/[(22.7 \text{ W/m}^2 \cdot \text{K})(\pi)(0.1027 \text{ m})(1 \text{ m})]$$

$$= 0.1365 \text{ K/W} \qquad\qquad \left(0.0720 \frac{\text{hr } ^\circ\text{R}}{\text{Btu}}\right)$$

thus the heat loss for the insulated pipe becomes

$$q = \frac{\Delta T}{\Sigma R} = \frac{404 - 294 \text{ K}}{3.316 \text{ K/W}}$$

$$= 33.2 \text{ W} \qquad\qquad \left(113\frac{\text{Btu}}{\text{hr}}\right)$$

a reduction of approximately 85%!

It is apparent from this example that certain parts of the heat-transfer path offer a negligible resistance. If, for instance, in the case of the bare pipe, an increased rate of heat transfer were desired, the obvious approach would be to alter the outside convective resistance, which is almost 200 times the magnitude of the next-highest thermal resistance value.

Example 3 could also have been worked by using an overall heat-transfer coefficient, which would be, in general,

$$U = \frac{q_x}{A\Delta T} = \frac{\Delta T/\Sigma R}{A\Delta T} = \frac{1}{A\Sigma R}$$

or, for the specific case considered,

$$U = \frac{1}{A\{1/A_i h_i + [\ln(r_o/r_i)]/2\pi kL + 1/A_o h_o\}} \tag{15-18}$$

Equation (15-18) indicates that the overall heat-transfer coefficient, U, may have a different numerical value, depending on which area it is based upon. If, for instance, U is based upon the outside surface area of the pipe, A_o, we have

$$U_o = \frac{1}{A_o/A_i h_i + [A_o \ln(r_o/r_i)]/2\pi kL + 1/h_o}$$

thus it is necessary, when specifying an overall coefficient, to relate it to a specific area.

One other means of evaluating heat-transfer rates is by means of the *shape factor*, symbolized S. Considering the steady-state relations developed for plane and cylindrical shapes,

$$q = \frac{kA}{L}\Delta T \tag{15-14}$$

and

$$q = \frac{2\pi kL}{\ln(r_o/r_i)}\Delta T \tag{15-9}$$

if that part of each expression having to do with the geometry is separated from the remaining terms, we have, for a plane wall,

$$q = k\left(\frac{A}{L}\right)\Delta T$$

and for a cylinder,

$$q = k\left(\frac{2\pi L}{\ln(r_o/r_i)}\right)\Delta T$$

Each of the bracketed terms is the shape factor for the applicable geometry. A general relation utilizing this form is

$$q = kS \, \Delta T \tag{15-19}$$

Equation (15-19) offers some advantages when a given geometry is required because of space and configuration limitations. If this is the case, then the shape factor may be calculated and q determined for various materials displaying a range of values of k.

15.6 CLOSURE

In this chapter the basic modes of energy transfer—conduction, convection, and radiation—have been introduced, along with the simple relations expressing the rates of energy transfer associated therewith. The transport property, thermal conductivity, has been discussed and some consideration given to energy transfer in a monatomic gas at low pressure.

The rate equations for heat transfer are as follows:

Conduction: the Fourier rate equation,

$$\frac{\mathbf{q}}{A} = -k\nabla T$$

Convection: the Newton rate equation,

$$\frac{q_x}{A} = h \, \Delta T$$

Radiation: the Stefan-Boltzmann Law for energy emitted from a black surface,

$$\frac{q}{A} = \sigma T^4$$

Combined modes of heat transfer were considered, specifically with respect to the means of calculating heat-transfer rates when several transfer modes were involved. The three ways of calculating steady-state heat-transfer rates are represented by the equations

$$q_x = \frac{\Delta T}{\sum R_T} \tag{15-16}$$

where $\sum R_T$ is the total thermal resistance along the transfer path;

$$q_x = UA \, \Delta T \tag{15-17}$$

where U is the overall heat transfer coefficient; and

$$q_x = kS \, \Delta T \tag{15-19}$$

where S is the shape factor.

The equations presented will be used throughout the remaining chapters dealing with energy transfer. A primary object of the chapters to follow will be the evaluation of the heat-transfer rates for special geometries or conditions of flow, or both.

PROBLEMS

15.1 The outside walls of a house are constructed of a 4-in. layer of brick, $\frac{1}{2}$ in. of celotex, an air space $3\frac{5}{8}$ in. thick, and $\frac{1}{4}$ in. of wood panelling. If the outside surface of the brick is at 30°F and the inner surface of the panelling at 75°F, what is the heat flux if
(a) the air space is assumed to transfer heat by conduction only?
(b) the equivalent conductance of the air space is 1.8 Btu/hr ft^2 °F?
(c) the air space is filled with glass wool?

$$k_{brick} = 0.38 \text{ Btu/hr ft °F}$$

$$k_{celotex} = 0.028 \text{ Btu/hr ft °F}$$

$$k_{air} = 0.015 \text{ Btu/hr ft °F}$$

$$k_{wood} = 0.12 \text{ Btu/hr ft °F}$$

$$k_{wool} = 0.025 \text{ Btu/hr ft °F}$$

15.2 Solve problem 15.1 if, instead of the surface temperatures being known, the air temperatures outside and inside are 30°F and 75°F, and the convective heat-transfer coefficients 7 Btu/hr ft^2 °F and 2 Btu/hr ft^2 °F, respectively.

15.3 Determine the heat transfer rate, per square meter of wall area for the case of a furnace with inside air at 1340 K. The furnace wall is composed of a 0.106-m layer of fireclay brick and a 0.635-cm thickness of mild steel on its outside surface. Heat transfer coefficients on inside and outside wall surfaces are 5110 W/m^2 · K and 45 W/m^2 · K respectively; outside air is at 295 K. What will be the temperatures at each surface and at the brick-steel interface?

15.4 Given the furnace wall and other conditions as specified in problem 15.3, what thickness of celotex ($k = 0.069$ W/m · K) must be added to the furnace wall in order that the outside surface temperature of the insulation not exceed 340 K?

15.5 Plate glass, $k = 1.35$ W/m · K, initially at 850 K, is cooled by blowing air past both surfaces with an effective surface coefficient of 5 W/m^2 · K. It is necessary, in order that the glass not crack, to limit the maximum temperature gradient in the plate to 15 K/mm during the cooling process. At the start of the cooling process, what is the lowest temperature of the cooling air that can be used?

15.6 Solve problem 15.5 if all specified conditions remain the same but radiant energy exchange from the glass to the surroundings at the air temperature is also considered.

15.7 The heat loss from a boiler is to be held at a maximum of 900 Btu/hr ft^2 of wall area. What thickness of asbestos ($k = 0.10$ Btu/hr ft °F) is required if the inner and outer surfaces of the insulation are to be 1600°F and 500°F, respectively?

15.8 If, in the previous problem, a 3-in.-thick layer of kaolin brick ($k = 0.07$ Btu/hr ft °F) is added to the outside of the asbestos, what heat flux will result if the outside surface of the kaolin is 250°F? What will be the temperature at the interface between the asbestos and kaolin for this condition?

15.9 A composite wall is to be constructed of $\frac{1}{4}$-in. stainless steel ($k = 10$ Btu/hr ft °F), 3 in. of corkboard ($k = 0.025$ Btu/hr ft °F), and $\frac{1}{2}$ in. of plastic ($k = 1.5$ Btu/hr ft °F).
(a) Draw the thermal circuit for the steady-state conduction through this wall.
(b) Evaluate the individual thermal resistance of each material layer.
(c) Determine the heat flux if the steel surface is maintained at 250°F and the plastic surface held at 80°F.
(d) What are the temperatures on each surface of the corkboard under these conditions?

15.10 If, in the previous problem, the convective heat-transfer coefficients at the inner (steel) and outer surfaces are 40 Btu/hr ft^2 °F and 5 Btu/hr ft^2 °F, respectively, determine
(a) the heat flux if the gases are at 250°F and 70°F adjacent to the inner and outer surfaces;
(b) the maximum temperature reached within the plastic;
(c) which of the individual resistances is controlling.

15.11 An asbestos pad is square in cross section, measuring 5 cm on a side at its small end increasing linearly to 10 cm on a side at the large end. The pad is 15 cm high. If the small end is held at 600 K and the large end at 300 K, what heat-flow rate will be obtained if the four sides are insulated? Assume one-dimensional heat conduction. The thermal conductivity of asbestos may be taken as 0.173 W/m · K.

15.12 Solve problem 15.11 for the case of the larger cross section exposed to the higher temperature and the smaller end held at 300 K.

15.13 Solve problem 15.11 if, in addition to a varying cross-sectional area, the thermal conductivity varies according to $k = k_0(1 + \beta T)$, where $k_0 = 0.138$, $\beta = 1.95 \times 10^{-4}$, $T = $ temperature in degrees Kelvin, and k is in W/m · K. Compare this result to that using a k value evaluated at the arithmetic mean temperature.

November 7, '97   Dec 18 final

Test on that Waste on Nov 6,

#4, #5   Chapter 24 .   for Monday

chpt 24,25, ht transfer

Test on Nov 12 for 42V

Preliminary Project due Nov 17 for 42V

(will later - equations & plans)

Dec 10 final

Equations

Mass Flux $\dot{m}_{O_2}$ = $\dfrac{D_{film} \ form \ daint}{b}$

For Wed - make a plan

D_{film} = diffusion coeff.

b = film thickness

daint = driving force

f_{film} = film density

15.14 Solve problem 15.11 if the asbestos pad has a 1.905-cm steel bolt running through its center.

15.15 A 4-in.-OD pipe is to be used to transport liquid metals and will have an outside surface temperature of 1400°F under operating conditions. Insulation 6 in. thick and having a thermal conductivity expressed as

$$k = 0.08(1 - 0.0003\,T)$$

where k is in Btu/hr ft °F and T is in °F, is applied to the outside surface of the pipe.

(a) What thickness of insulation would be required for the outside insulation temperature to be no higher than 300°F?

(b) What heat-flow rate will occur under these conditions?

15.16 Water at 40°F is to flow through a $1\frac{1}{2}$-in. schedule 40 steel pipe. The outside surface of the pipe is to be insulated with a 1-in.-thick layer of 85% magnesia and a 1-in.-thick layer of packed glass wool, $k = 0.022$ Btu/hr ft °F. The surrounding air is at 100°F.

(a) Which material should be placed next to the pipe surface to produce the maximum insulating effect?

(b) What will be the heat flux on the basis of the outside pipe surface area? The convective heat transfer coefficients for the inner and outer surfaces are 100 Btu/hr ft^2 °F and 5 Btu/hr ft^2 °F, respectively.

15.17 A 1-in.-nominal-diameter steel pipe with its outside surface at 400°F is located in air at 90°F with the convective heat-transfer coefficient between the surface of the pipe and the air equal to 1.5 Btu/hr ft^2 °F. It is proposed to add insulation having a thermal conductivity of 0.06 Btu/hr ft °F to the pipe to reduce the heat loss to one-half that for the bare pipe. What thickness of insulation is necessary if the surface temperature of the steel pipe and h_o remain constant?

15.18 If, for the conditions of problem 15.17, h_o in Btu/hr ft^2 °F varies according to $h_o = 0.575/D_o^{1/4}$, where D_o is the outside diameter of the insulation in feet, determine the thickness of insulation that will reduce the heat flux to one-half that of the value for the bare pipe.

15.19 A 2-in.-thick steel plate measuring 10 in. in diameter is heated from below by a hot plate, its upper surface exposed to air at 80°F. The heat-transfer coefficient on the upper surface is 5 Btu/hr ft^2 °F and k for steel is 25 Btu/hr ft °F.

(a) How much heat must be supplied to the lower surface of the steel if its upper surface remains at 160°F? (Include radiation.)

(b) What are the relative amounts of energy dissipated from the upper surface of the steel by convection and radiation?

15.20 If, in problem 15.19, the plate is made of asbestos, $k = 0.10$ Btu/hr ft °F, what will be the temperature of the top of the asbestos if the hot plate is rated at 800 W?

15.21 A 0.20-m-thick brick wall ($k = 1.3$ W/m · K) separates the combustion zone of a furnace from its surroundings at 25 °C. For an outside wall surface temperature of 100 °C, with a convective heat transfer coefficient of 18 W/m^2 · K, what will be the inside wall surface temperature at steady-state conditions?

15.22 Solve for the inside surface temperature of the brick wall described in problem 15.21, but with the additional consideration of radiation from the outside surface to surroundings at 25°C.

15.23 The solar radiation incident on a steel plate 2 ft square is 400 Btu/hr. The plate is 1.4 in. thick and lying horizontally on an insulating surface, its upper surface being exposed to air at 90°F. If the convective heat-transfer coefficient between the top surface and the surrounding air is 4 Btu/hr ft^2 °F, what will be the steady-state temperature of the plate?

15.24 If in problem 15.23, the lower surface of the plate is exposed to air with a convective heat transfer coefficient of 3 Btu/hr ft^2 °F, what steady-state temperature will be reached
(a) assuming no radiant emission from the plate;
(b) if radiant emission away from the top surface of the plate is accounted for?

15.25 The cross-section of a storm window is shown in the sketch. How much heat will be lost through a window measuring 1.83 m by 3.66 m on a cold day when inside and outside air temperatures are, respectively, 295 K and 250 K? Convective coefficients on inside and outside surfaces of the window are 20 W/m^2 · K and 15 W/m^2 · K respectively. What temperature drop will exist across each of the glass panes? What will be the average temperature of the air between the glass panes?

Air
space
0.8 cm
wide

Window glass–0.32 cm thick

15.26 Compare the heat loss through the storm window described in the previous problem with the same conditions existing except that the window is a single pane of glass 0.32 cm thick.

16
DIFFERENTIAL EQUATIONS
OF HEAT TRANSFER

Paralleling the treatment of momentum transfer undertaken in Chapter 9, we shall now generate the fundamental equations for a differential control volume from a first-law-of-thermodynamics approach. The control-volume expression for the first law will provide our basic analytical tool. Additionally, certain differential equations already developed in previous sections will be applicable.

16.1 THE GENERAL DIFFERENTIAL EQUATION FOR ENERGY TRANSFER

Consider the control volume having dimensions Δx, Δy, and Δz as depicted in Figure 16.1. Refer to the control-volume expression for the first law of thermodynamics,

$$\frac{\delta Q}{dt} - \frac{\delta W_s}{dt} - \frac{\delta W_\mu}{dt} = \iint_{\text{c.s.}} \left(e + \frac{P}{\rho}\right)\rho(\mathbf{v} \cdot \mathbf{n})\, dA + \frac{\partial}{\partial t}\iiint_{\text{c.v.}} e\rho\, dV \qquad (6\text{-}10)$$

The individual terms are evaluated and their meaning are discussed below.

The net rate of heat added to the control volume will include all conduction effects, the net release of thermal energy within the control volume due to a chemical reaction, and the dissipation of electrical or nuclear energy. The generation effects will be included in the single term, $\dot{q}$, which is the volumetric rate of thermal energy generation having units W/m^3 or Btu/hr ft^3. Thus the first term may be expressed as

$$\frac{\delta Q}{dt} = \left[k\frac{\partial T}{\partial x}\bigg|_{x+\Delta x} - k\frac{\partial T}{\partial x}\bigg|_{x}\right]\Delta y\,\Delta z + \left[k\frac{\partial T}{\partial y}\bigg|_{y+\Delta y} - k\frac{\partial T}{\partial y}\bigg|_{y}\right]\Delta x\,\Delta z$$

$$+ \left[k\frac{\partial T}{\partial z}\bigg|_{z+\Delta z} - k\frac{\partial T}{\partial z}\bigg|_{z}\right]\Delta x\,\Delta y + \dot{q}\,\Delta x\,\Delta y\,\Delta z \qquad (16\text{-}1)$$

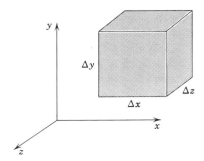

Figure 16.1 A differential control volume.

The shaft work-rate or power term will be taken as zero for our present purposes. This term is specifically related to work done by some effect within the control volume which, for the differential case, is not present. The power term is thus evaluated as

$$\frac{\delta W_s}{dt} = 0 \qquad (16\text{-}2)$$

The viscous work rate, occurring at the control surface, is formally evaluated by integrating the dot product of the viscous stress and the velocity over the control surface. As this operation is tedious, we shall express the viscous work rate as $\Lambda \, \Delta x \, \Delta y \, \Delta z$ where Λ is the viscous work rate per unit volume. The third term in equation (6-10) is thus written as

$$\frac{\delta W_\mu}{dt} = \Lambda \, \Delta x \, \Delta y \, \Delta z \qquad (16\text{-}3)$$

The surface integral includes all energy transfer across the control surface due to fluid flow. All terms associated with the surface integral have been defined previously. The surface integral is

$$\iint_{c.s.} \left(e + \frac{P}{\rho}\right)\rho(\mathbf{v} \cdot \mathbf{n}) \, dA$$

$$= \left[\rho v_x\left(\frac{v^2}{2} + gy + u + \frac{P}{\rho}\right)\bigg|_{x+\Delta x} - \rho v_x\left(\frac{v^2}{2} + gy + u + \frac{P}{\rho}\right)\bigg|_x\right] \Delta y \, \Delta z$$

$$+ \left[\rho v_y\left(\frac{v^2}{2} + gy + u + \frac{P}{\rho}\right)\bigg|_{y+\Delta y} - \rho v_y\left(\frac{v^2}{2} + gy + u + \frac{P}{\rho}\right)\bigg|_y\right] \Delta x \, \Delta z$$

$$+ \left[\rho v_z\left(\frac{v^2}{2} + gy + u + \frac{P}{\rho}\right)\bigg|_{z+\Delta z} - \rho v_z\left(\frac{v^2}{2} + gy + u + \frac{P}{\rho}\right)\bigg|_z\right] \Delta x \, \Delta y \quad (16\text{-}4)$$

The energy accumulation term, relating the variation in total energy within the control volume as a function of time, is

$$\frac{\partial}{\partial t} \iiint_{\text{c.v.}} e\rho \, dV = \frac{\partial}{\partial t}\left[\frac{v^2}{2} + gy + u\right]\rho \, \Delta x \, \Delta y \, \Delta z \qquad (16\text{-}5)$$

Equations (16-1) through (16-5) may now be combined as indicated by the general first-law expression, equation (6-10). Performing this combination and dividing through by the volume of the element, we have

$$\frac{k(\partial T/\partial x)|_{x+\Delta x} - k(\partial T/\partial x)|_x}{\Delta x} + \frac{k(\partial T/\partial y)|_{y+\Delta y} - k(\partial T/\partial y)|_y}{\Delta y}$$

$$+ \frac{k(\partial T/\partial z)|_{z+\Delta z} - k(\partial T/\partial z)|_z}{\Delta z} + \dot{q} + \Lambda$$

$$= \frac{\{\rho v_x[(v^2/2) + gy + u + (P/\rho)]|_{x+\Delta x} - \rho v_x[(v^2/2) + gy + u + (P/\rho)]|_x\}}{\Delta x}$$

$$+ \frac{\{\rho v_y[(v^2/2) + gy + u + (P/\rho)]|_{y+\Delta y} - \rho v_y[(v^2/2) + gy + u + (P/\rho)]|_y\}}{\Delta y}$$

$$+ \frac{\{\rho v_z[(v^2/2) + gy + u + (P/\rho)]|_{z+\Delta z} - \rho v_z[(v^2/2) + gy + u + (P/\rho)]|_z\}}{\Delta z}$$

$$+ \frac{\partial}{\partial t}\rho\left(\frac{v^2}{2} + gy + u\right)$$

Evaluated in the limit as Δx, Δy, and Δz approach zero, this equation becomes

$$\frac{\partial}{\partial x}\left(k\frac{\partial T}{\partial x}\right) + \frac{\partial}{\partial y}\left(k\frac{\partial T}{\partial y}\right) + \frac{\partial}{\partial z}\left(k\frac{\partial T}{\partial z}\right) + \dot{q} + \Lambda$$

$$= \frac{\partial}{\partial x}\left[\rho v_x\left(\frac{v^2}{2} + gy + u + \frac{P}{\rho}\right)\right] + \frac{\partial}{\partial y}\left[\rho v_y\left(\frac{v^2}{2} + gy + u + \frac{P}{\rho}\right)\right]$$

$$+ \frac{\partial}{\partial z}\left[\rho v_z\left(\frac{v^2}{2} + gy + u + \frac{P}{\rho}\right)\right] + \frac{\partial}{\partial t}\left[\rho\left(\frac{v^2}{2} + gy + u\right)\right] \qquad (16\text{-}6)$$

Equation (16-6) is completely general in application. Introducing the substantial derivative, we may write equation (16-6) as

$$\frac{\partial}{\partial x}\left(k\frac{\partial T}{\partial x}\right) + \frac{\partial}{\partial y}\left(k\frac{\partial T}{\partial y}\right) + \frac{\partial}{\partial z}\left(k\frac{\partial T}{\partial z}\right) + \dot{q} + \Lambda$$

$$= \boldsymbol{\nabla}\cdot(P\mathbf{v}) + \left(\frac{v^2}{2} + u + gy\right)\left(\boldsymbol{\nabla}\cdot\rho\mathbf{v} + \frac{\partial\rho}{\partial t}\right) + \frac{\rho}{2}\frac{Dv^2}{Dt} + \rho\frac{Du}{Dt} + \rho\frac{D(gy)}{Dt}$$

Utilizing the continuity equation, equation (9-2), we reduce this to

$$\frac{\partial}{\partial x}\left(k\frac{\partial T}{\partial x}\right)+\frac{\partial}{\partial y}\left(k\frac{\partial T}{\partial y}\right)+\frac{\partial}{\partial z}\left(k\frac{\partial T}{\partial z}\right)+\dot{q}+\Lambda$$

$$=\nabla\cdot P\mathbf{v}+\frac{\rho}{2}\frac{Dv^2}{Dt}+\rho\frac{Du}{Dt}+\rho\frac{D(gy)}{Dt} \qquad (16\text{-}7)$$

With the aid of equation (9-19), which is valid for incompressible flow of a fluid with constant μ, the second term on the right-hand side of equation (16-7) becomes

$$\frac{\rho}{2}\frac{Dv^2}{Dt}=-\mathbf{v}\cdot\nabla P+\mathbf{v}\cdot\rho\mathbf{g}+\mathbf{v}\cdot\mu\nabla^2\mathbf{v} \qquad (16\text{-}8)$$

Also, for incompressible flow, the first term on the right-hand side of equation (16-7) becomes

$$\nabla\cdot P\mathbf{v}=\mathbf{v}\cdot\nabla P \qquad (16\text{-}9)$$

Substituting equations (16-8) and (16-9) into equation (16-7), and writing the conduction terms as $\nabla\cdot k\nabla T$, we have

$$\nabla\cdot k\nabla T+\dot{q}+\Lambda=\rho\frac{Du}{Dt}+\rho\frac{D(gy)}{Dt}+\mathbf{v}\cdot\rho\mathbf{g}+\mathbf{v}\cdot\mu\nabla^2\mathbf{v} \qquad (16\text{-}10)$$

It will be left as an exercise for the reader to verify that equation (16-10) reduces further to the form

$$\nabla\cdot k\nabla T+\dot{q}+\Lambda=\rho c_v\frac{DT}{Dt}+\mathbf{v}\cdot\mu\nabla^2\mathbf{v} \qquad (16\text{-}11)$$

The function Λ may be expressed in terms of the viscous portion of the normal- and shear-stress terms in equations (7.13) and (7.14). For the case of incompressible flow it is written as

$$\Lambda=\mathbf{v}\cdot\mu\nabla^2\mathbf{v}+\Phi \qquad (16\text{-}12)$$

where the "dissipation function," Φ, is given by

$$\Phi=2\mu\left[\left(\frac{\partial v_x}{\partial x}\right)^2+\left(\frac{\partial v_y}{\partial y}\right)^2+\left(\frac{\partial v_z}{\partial z}\right)^2\right]$$

$$+\mu\left[\left(\frac{\partial v_x}{\partial y}+\frac{\partial v_y}{\partial x}\right)^2+\left(\frac{\partial v_y}{\partial z}+\frac{\partial v_z}{\partial y}\right)^2+\left(\frac{\partial v_z}{\partial x}+\frac{\partial v_x}{\partial z}\right)^2\right]$$

Substituting for Λ in equation (16-11), we see that the energy equation becomes

$$\nabla\cdot k\nabla T+\dot{q}+\Phi=\rho c_v\frac{DT}{Dt} \qquad (16\text{-}13)$$

From equation (16-12) Φ is seen to be a function of fluid viscosity and shear-strain rates, and is positive-definite. The effect of viscous dissipation is always to increase internal energy at the expense of potential energy or stagnation pressure. The dissipation function is negligible in all cases that we will consider; its effect becomes significant in supersonic boundary layers.

16.2   SPECIAL FORMS OF THE DIFFERENTIAL ENERGY EQUATION

The applicable forms of the energy equation for some commonly encountered situations follow. In every case the dissipation term is considered negligibly small.

I. For an incompressible fluid without energy "sources" and with constant k,

$$\rho c_v \frac{DT}{Dt} = k \nabla^2 T \tag{16-14}$$

II. For isobaric flow without energy sources and with constant k, the energy equation is

$$\rho c_v \frac{DT}{Dt} = k \nabla^2 T \tag{16-15}$$

Note that equations (16-14) and (16-15) are identical yet apply to completely different physical situations. The student may wish to satisfy himself at this point as to the reasons behind the unexpected result.

III. In a situation where there is no fluid motion all heat transfer is by conduction. If this situation exists, as it most certainly does in solids where $c_v \simeq c_p$, the energy equation becomes

$$\rho c_p \frac{\partial T}{\partial t} = \nabla \cdot k \nabla T + \dot{q} \tag{16-16}$$

Equation (16-16) applies in general to heat conduction. No assumption has been made concerning constant k. If the thermal conductivity is constant, the energy equation is

$$\frac{\partial T}{\partial t} = \alpha \nabla^2 T + \frac{\dot{q}}{\rho c_p} \tag{16-17}$$

where the ratio $k/\rho c_p$ has been symbolized by α and is designated the *thermal diffusivity*. It is easily seen that α has the units, L^2/t; in the SI system α is expressed in m^2/s, and as ft^2/hr in the English system.

If the conducting medium contains no heat sources, equation (16-17) reduces to the *Fourier field equation*,

$$\frac{\partial T}{\partial t} = \alpha \nabla^2 T \tag{16-18}$$

which is occasionally referred to as Fourier's second "law" of heat conduction.

For a system in which heat sources are present but there is no time variation, equation (16-17) reduces to the *Poisson equation,*

$$\nabla^2 T + \frac{\dot{q}}{k} = 0 \qquad (16\text{-}19)$$

The final form of the heat-conduction equation to be presented applies to a steady-state situation without heat sources. For this case the temperature distribution must satisfy the *Laplace equation,*

$$\nabla^2 T = 0 \qquad (16\text{-}20)$$

Each of equations (16-17) through (16-20) has been written in general form, thus each applies to any orthogonal coordinate system. Writing the Laplacian operator, ∇^2, in the appropriate form will accomplish the transformation to the desired coordinate system. The Fourier field equation written in rectangular coordinates is

$$\frac{\partial T}{\partial t} = \alpha \left[\frac{\partial^2 T}{\partial x^2} + \frac{\partial^2 T}{\partial y^2} + \frac{\partial^2 T}{\partial z^2} \right] \qquad (16\text{-}21)$$

in cylindrical coordinates

$$\frac{\partial T}{\partial t} = \alpha \left[\frac{\partial^2 T}{\partial r^2} + \frac{1}{r} \frac{\partial T}{\partial r} + \frac{1}{r^2} \frac{\partial^2 T}{\partial \theta^2} + \frac{\partial^2 T}{\partial z^2} \right] \qquad (16\text{-}22)$$

and in spherical coordinates

$$\frac{\partial T}{\partial t} = \alpha \left[\frac{1}{r^2} \frac{\partial}{\partial r} \left(r^2 \frac{\partial T}{\partial r} \right) + \frac{1}{r^2 \sin \theta} \frac{\partial}{\partial \theta} \left(\sin \theta \frac{\partial T}{\partial \theta} \right) + \frac{1}{r^2 \sin^2 \theta} \frac{\partial^2 T}{\partial \phi^2} \right] \qquad (16\text{-}23)$$

The reader is referred to Appendix B for an illustration of the variables in cylindrical and spherical coordinate systems.

16.3 COMMONLY ENCOUNTERED BOUNDARY CONDITIONS

In solving one of the differential equations developed thus far, the existing physical situation will dictate the appropriate initial or boundary conditions, or both, which the final solution must satisfy.

Initial conditions refer specifically to the values of T and $\mathbf{v}$ at the start of the time interval of interest. Initial conditions may be as simply specified as stating that $T|_{t=0} = T_0$ (a constant), or more complex if the temperature distribution at the start of time measurement is some function of space variables.

Boundary conditions refer to the values of T and $\mathbf{v}$ existing at specific positions on the boundaries of a system, i.e., for given values of the significant space variables. Frequently encountered boundary conditions for temperature are the case of *isothermal boundaries,* along which the temperature is constant, and *insulated boundaries,* across which no heat conduction occurs where, according to the Fourier rate equation, the temperature derivative normal to the boundary is zero. More complicated temperature functions often exist at system boundaries, and the surface temperature may also vary with time. Combinations of heat-transfer mechanisms may dictate boundary conditions as well. One situation often existing at a solid boundary is the equality between heat transfer to the surface by conduction and that leaving the surface by convection. This condition is illustrated in Figure 16.2. At the left-hand surface the boundary condition is

$$h_h(T_h - T|_{x=0}) = -k\frac{\partial T}{\partial x}\Big|_{x=0} \tag{16-24}$$

and at the right-hand surface,

$$h_c(T|_{x=L} - T_c) = -k\frac{\partial T}{\partial x}\Big|_{x=L} \tag{16-25}$$

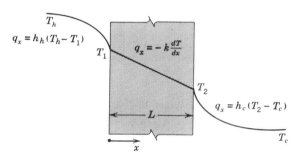

Figure 16.2 Conduction and convection at a system boundary.

It is impossible at this time to foresee all initial and boundary conditions which will be needed. The student should be aware, however, that these conditions are dictated by the physical situation. The differential equations of energy transfer are not numerous, and a specific form applying to a given situation may be found easily. It remains for the user of these equations to choose the appropriate initial and boundary conditions to make the solution meaningful.

16.4 CLOSURE

The general differential equations of energy transfer have been developed in this chapter, and some forms applying to more specific situations were presented.

Some remarks concerning initial and boundary conditions have been made as well.

In the chapters to follow, analyses of energy transfer will start with the applicable differential equation. Numerous solutions will be presented and still more assigned as student exercises. The tools for heat-transfer analysis have now been developed and examined. Our remaining task is to develop a familiarity with and facility in their use.

PROBLEMS

16.1 The Fourier field equation in cylindrical coordinates is

$$\frac{\partial T}{\partial t} = \alpha\left(\frac{\partial^2 T}{\partial r^2} + \frac{1}{r}\frac{\partial T}{\partial r} + \frac{1}{r^2}\frac{\partial^2 T}{\partial \theta^2} + \frac{\partial^2 T}{\partial z^2}\right)$$

(a) What form does this equation reduce to for the case of steady-state, radial heat transfer?
(b) Given the boundary conditions

$$T = T_i \qquad \text{at } r = r_i$$
$$T = T_0 \qquad \text{at } r = r_o$$

Solve the resulting equation from part (a) for the temperature profile.
(c) Generate an expression for the heat flow rate, q_r, using the result from part (b).

16.2 Perform the same operations as in parts a, b, and c of problem 16.1 with respect to a spherical system.

16.3 Starting with the Fourier field equation in cylindrical coordinates,
(a) Reduce this equation to the applicable form for steady-state heat transfer in the θ direction.
(b) For the conditions depicted in the figure, that is, $T = T_o$ at $\theta = 0$, $T = T_\pi$ at $\theta = \pi$, and the radial surfaces insulated, solve for the temperature profile.

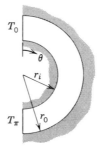

(c) Generate an expression for the heat flow rate, q_θ, using the result of part (b).

(d) What is the shape factor for this configuration?

16.4 Show that equation (16-10) reduces to the form

$$\nabla \cdot k\nabla T + \dot{q} + \Lambda = \rho c_v \frac{DT}{Dt} + \mathbf{v} \cdot \mu \nabla^2 \mathbf{v}$$

16.5 Starting with equation (16-7), show that, for a fluid with constant thermal conductivity and no energy sources, equation (16-14) and (16-15) are obtained for incompressible and isobaric conditions, respectively. (Neglect viscous dissipation.)

16.6 Solve equation (16-19) for the temperature distribution in a plane wall if the internal heat generation per unit volume varies according to $\dot{q} = \dot{q}_0 e^{-\beta x/L}$. The boundary conditions which apply are $T = T_o$ at $x = 0$ and $T = T_L$ at $x = L$.

16.7 Solve problem 16.6 for the same conditions, except that the boundary condition at $x = L$ is $dT/dx = 0$.

16.8 Solve problem 16.6 for the same conditions, except that at $x = L$, $dT/dx = \xi$ (a constant).

16.9 Use the relation $T\,ds = dh - dP/\rho$ to show that the effect of the dissipation function, Φ, is to increase the entropy, S. Is the effect of heat transfer the same as the dissipation function?

16.10 In a boundary layer where the velocity profile is given by

$$\frac{v_x}{v_\infty} = \frac{3}{2}\frac{y}{\delta} - \frac{1}{2}\left(\frac{y}{\delta}\right)^3$$

where δ is the velocity boundary layer thickness, plot the dimensionless dissipation function, $\Phi\delta^2/\mu v_\infty^2$ versus y/δ.

17
STEADY-STATE CONDUCTION

In most equipment used in transferring heat, energy flows from one fluid to another through a solid wall. Since the energy transfer through each medium is one step in the overall process, a clear understanding of the conduction mechanism of energy transfer through homogeneous solids is essential to the solutions of most heat-transfer problems.

In this chapter we shall direct our attention to steady-state heat conduction. Steady state implies that the conditions, temperature, density, and the like at all points in the conduction region are independent of time. Our analyses will parallel the approaches used for analyzing a differential fluid element in laminar flow and those which will be used in analyzing steady-state molecular diffusion. During our discussions, two types of presentations will be used: (1) the governing differential equation will be generated by means of the control-volume concept, and (2) the governing differential will be obtained by eliminating all irrelevant terms in the general differential equation for energy transfer. By using both approaches, the student will quickly recognize the significance of the various terms in the general differential equation.

17.1 ONE-DIMENSIONAL CONDUCTION

For steady-state conduction independent of any internal generation of energy, the general differential equation reduces to the Laplace equation,

$$\nabla^2 T = 0 \qquad (16\text{-}20)$$

Although this equation implies that more than one space coordinate is necessary to describe the temperature field, many problems are simpler because of the geometry of the conduction region or because of symmetries in the temperature distribution. One-dimensional cases often arise.

252

The one-dimensional, steady-state transfer of energy by conduction is the simplest process to describe, since the condition imposed upon the temperature field is an ordinary differential equation. For one-dimensional conduction, equation (16-20) reduces to

$$\frac{d}{dx}\left(x^i\frac{dT}{dx}\right)=0 \tag{17-1}$$

where $i = 0$ for rectangular coordinates, $i = 1$ for cylindrical coordinates, and $i = 2$ for spherical coordinates.

One-dimensional processes occur in flat planes, such as furnace walls; in cylindrical elements, such as steam pipes; and in spherical elements, such as nuclear-reactor pressure vessels. In this section, we shall consider steady-state conduction through simple systems in which the temperature and the energy flux are functions of a single space coordinate.

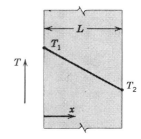

Figure 17.1 Plane wall with a one-dimensional temperature distribution.

Plane Wall. Consider the conduction of energy through a plane wall as illustrated in Figure 17.1. The one-dimensional Laplace equation is easily solved, yielding

$$T = C_1 x + C_2 \tag{17-2}$$

The two constants are obtained by applying the boundary conditions,

$$\text{at } x = 0 \qquad T = T_1$$

and

$$\text{at } x = L \qquad T = T_2$$

These constants are

$$C_2 = T_1$$

and

$$C_1 = \frac{T_2 - T_1}{L}$$

The temperature profile becomes

$$T = \frac{T_2 - T_1}{L}x + T_1$$

or

$$T = T_1 - \frac{T_1 - T_2}{L}x \tag{17-3}$$

and is linear, as illustrated in Figure 17.1.

The energy flux is evaluated, using the Fourier rate equation,

$$\frac{q_x}{A} = -k\frac{dT}{dx} \tag{15-1}$$

The temperature gradient, dT/dx, is obtained by differentiating equation (17-3) yielding

$$\frac{dT}{dx} = -\frac{T_1 - T_2}{L}$$

substituting this term into the rate equation, we obtain for a flat wall with constant thermal conductivity,

$$q_x = \frac{kA}{L}(T_1 - T_2) \tag{17-4}$$

The quantity kA/L is characteristic of a flat wall or a flat plate and is designated the *thermal conductance*. The reciprocal of the thermal conductance, L/kA, is the *thermal resistance*.

Composite Walls. The steady flow of energy through several walls in series is often encountered. A typical furnace design might include one wall for strength, an intermediate wall for insulation and the third, outer wall for appearance. This composite plane wall is illustrated in Figure 17.2.

For a solution to the system shown in this figure, the reader is referred to section 15.5.

The following example illustrates the use of the composite-wall energy-rate equation for predicting the temperature distribution in the walls.

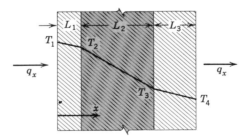

Figure 17.2   Temperature distribution for steady-state conduction of energy through a composite plane wall.

EXAMPLE 1

A furnace wall is made up of three layers, 4 in. of firebrick ($k = 0.9$ Btu/ft² hr °F/ft), followed by 9 in. of kaolin insulating brick ($k = 0.1$ Btu/ft² hr °F/ft), and finally 2 in. of masonry brick ($k = 0.4$ Btu/ft² hr °F/ft). The temperature of the inner surface is 2000°F, and the temperature of the outer surface is 200°F. What are the temperatures at the contacting wall surfaces?

The individual wall resistances per square foot of area are

$$R_1, \text{firebrick} = \frac{L_1}{Ak} = \left(\frac{4 \text{ in.}}{12 \text{ in./ft}}\right)\left(\frac{1}{1 \text{ ft}^2}\right)\left(\frac{1}{0.9 \text{ Btu/ft}^2 \text{ hr °F/ft}}\right) = 0.37 \text{ hr °F/Btu}$$

$$R_2, \text{kaolin} = \frac{L_2}{Ak_2} = \left(\frac{9}{12}\right)\left(\frac{1}{1}\right)\left(\frac{1}{0.1}\right) = 7.50 \text{ hr °F/Btu}$$

$$R_3, \text{masonry} = \frac{L_3}{Ak_3} = \left(\frac{2}{12}\right)\left(\frac{1}{1}\right)\left(\frac{1}{0.4}\right) = 0.417 \text{ hr °F/Btu}$$

The total resistance of these walls in series is equal to $0.37 + 7.5 + 0.417$ or 8.29 hr °F/Btu. The total temperature drop through the three walls is equal to $(T_1 - T_4) = 2000° - 200° = 1800°F$.

Using equation (15-16), we calculate the energy transfer rate as

$$q = \frac{T_1 - T_4}{\Sigma R_T} = \frac{1800°F}{8.29 \text{ hr °F/Btu}} = 217 \text{ Btu/hr} \qquad (63.6\text{W})$$

Since this is a steady-state situation, this energy transfer rate is constant for each part of the transfer path. The intermediate temperatures may be calculated by solving the rate equation for each wall; for example, we can determine the intermediate temperatures by the individual wall energy-rate equations,

$$T_1 - T_2 = q\frac{L_1}{Ak_1}$$

or

$$2000° - T_2 = (217 \text{ Btu/hr}) (0.37 \text{ hr °F/Btu})$$

giving

$$T_2 = 1920°F \qquad (1322 \text{ K})$$

and

$$T_3 - T_4 = q\frac{L_3}{Ak_3}$$

or

$$T_3 - 200° = (217 \text{ Btu/hr})(0.417 \text{ hr °F/Btu})$$

thus

$$T_3 = 290.5°F \qquad (417 \text{ K})$$

There are numerous situations in which a composite wall involves a combination of series and parallel energy-flow paths. An example of such a wall is illustrated in Figure 17.3, where steel is used as reinforcement for a concrete wall.

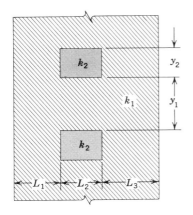

Figure 17.3 A series-parallel composite wall.

The composite wall can be divided into three sections of length L_1, L_2 and L_3, and the thermal resistance for each of these lengths may be evaluated.

The intermediate layer between planes 2 and 3 consists of two separate thermal paths in parallel; the effective thermal conductance is the sum of the conductances for the two materials. For the section of the wall of height $y_1 + y_2$ and unit depth, the resistance is

$$R_2 = \frac{1}{\dfrac{k_1 y_1}{L_2} + \dfrac{k_2 y_2}{L_2}} = L_2\left(\frac{1}{k_1 y_1 + k_2 y_2}\right)$$

The total resistance for this wall is

$$\sum R_T = R_1 + R_2 + R_3$$

or

$$\sum R_T = \frac{L_1}{k_1(y_1 + y_2)} + L_2\left(\frac{1}{k_1 y_1 + k_2 y_2}\right) + \frac{L_3}{k_1(y_1 + y_2)}$$

The electrical circuit ⎓⟍⟍⟍⊏ R_1 ⎓ R_2 ⎓ R_3 ⊐⟍⟍⟍⎓ is an analog to the composite wall. The rate of energy transferred from plane 1 to plane 4 is obtained by a modified form of equation (15-16).

$$q = \frac{T_1 - T_4}{\sum R_T} = \frac{T_1 - T_4}{\dfrac{L_1}{k_1(y_1 + y_2)} + L_2\left(\dfrac{1}{k_1 y_1 + k_2 y_2}\right) + \dfrac{L_3}{k_1(y_1 + y_2)}} \qquad (17\text{-}5)$$

It is important to recognize that this equation is only an approximation. Actually, there is a significant temperature distribution in the y direction close to the material which has the higher thermal conductivity.

In our discussions of composite walls, no allowance was made for a temperature drop at the contact face between two different solids. This assumption is not always valid, since there will often be vapor spaces caused by rough surfaces, or even oxide films on the surfaces of metals. These additional contact resistances must be accounted for in a precise energy-transfer equation.

Long, Hollow Cylinder. Radial energy flow by conduction through a long, hollow cylinder is another example of one-dimensional conduction. The radial heat flow for this configuration was evaluated in example 1 of Chapter 15 as

$$\frac{q_r}{L} = \frac{2\pi k}{\ln(r_o/r_i)}(T_i - T_o) \tag{17-6}$$

where r_i is the inside radius; r_o is the outside radius; T_i is the temperature on the inside surface; and T_o is the temperature on the outside surface. The resistance concept may again be used; the thermal resistance of the hollow cylinder is

$$R = \frac{\ln(r_o/r_i)}{2\pi k L} \tag{17-7}$$

The radial temperature distribution in a long, hollow cylinder may be evaluated by using equation (17-1) in cylindrical form,

$$\frac{d}{dr}\left(r\frac{dT}{dr}\right) = 0 \tag{17-8}$$

Solving this equation subject to the boundary conditions,

$$\text{at } r = r_i \qquad T = T_i$$

and

$$\text{at } r = r \qquad T = T_0$$

we see that temperature profile is

$$T(r) = T_i - \frac{T_i - T_o}{\ln(r_o/r_i)} \ln\frac{r}{r_i} \tag{17-9}$$

Thus the temperature in a long, hollow cylinder is a logarithmic function of radius r, while for the plane wall the temperature distribution is linear.

The following example illustrates the analysis of radial energy conduction through a long, hollow cylinder.

EXAMPLE 2

A long steam pipe of outside radius r_2 is covered with thermal insulation having an outside radius of r_3. The temperature of the outer surface of the pipe, T_2, and the temperature of the surrounding air, T_∞, are fixed. The energy loss per unit area of outside surface of the insulation is described by the Newton rate equation,

$$\frac{q_r}{A} = h(T_3 - T_\infty) \tag{15-11}$$

Can the energy loss increase with an increase in the thickness of insulation? If possible, under what conditions will this situation arise? Figure 17.4 may be used to illustrate this composite cylinder.

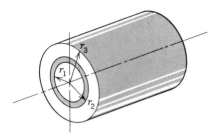

Figure 17.4 A series composite hollow cylinder.

In example 3 of Chapter 15, the thermal resistance of a hollow cylindrical element was shown to be

$$R = \frac{\ln (r_o/r_i)}{2\pi k L} \tag{17-10}$$

In the present example, the total difference in temperature is $T_2 - T_\infty$ and the two resistances, due to the insulation and the surrounding air film, are

$$R_2 = \frac{\ln (r_3/r_2)}{2\pi k_2 L}$$

for the insulation, and

$$R_3 = \frac{1}{hA} = \frac{1}{h 2\pi r_3 L}$$

for the air film.

Substituting these terms into the radial heat flow equation and rearranging, we obtain

$$q_r = \frac{2\pi L (T_2 - T_\infty)}{[\ln (r_3/r_2)]/k + 1/h r_3} \tag{17-11}$$

The dual effect of increasing the resistance to energy transfer by conduction and simultaneously increasing the surface area as r_3 is increased suggests that, for a pipe of given size, a particular outer radius exists for which the heat loss is maximum. Since the ratio r_3/r_2 increases logarithmically, and the term $1/r_3$ decreases as r_3 increases, the relative importance of each resistance term will change as the insulation thickness is varied. In this example, L, T_2, T_∞, k_2, h, and r_2 are considered constant. Differentiating equation (17-11) with respect to r_3, we obtain

$$\frac{dq_r}{dr_3} = -\frac{2\pi L (T_2 - T_\infty)\left(\dfrac{1}{k_2 r_3} - \dfrac{1}{h r_3^2}\right)}{\left[\dfrac{1}{k_2}\ln\left(\dfrac{r_3}{r_2}\right) + \dfrac{1}{h r_3}\right]^2} \tag{17-12}$$

The radius of insulation associated with the maximum energy transfer, the *critical radius*, found by setting $dq_r/dr_3 = 0$; equation (17-12) reduces to

$$(r_3)_{\text{critical}} = \frac{k_2}{h} \tag{17-13}$$

In the case of 85% magnesia insulation ($k = 0.0692$ W/m·K) and a typical value for the heat transfer coefficient in natural convection ($h = 34$ W/m²·K), the critical radius is calculated as

$$r_{crit} = \frac{k}{h} = \frac{0.0692 \text{ W/m·K}}{34 \text{ W/m}^2\text{·K}} = 0.0020 \text{ m} \qquad (0.0067 \text{ ft})$$

$$= 0.20 \text{ cm} \qquad (0.0787 \text{ in.})$$

These very small numbers indicate that the critical radius will be exceeded in any practical problem. The question then is whether the critical radius given by equation (17-13) represents a maximum or a minimum condition for q. The evaluation of the second derivative, d^2q_r/dr_3^2, when $r_3 = k/h$ yields a negative result, thus r_{crit} is a maximum condition. It now follows that q_r will be decreased for any value of r_3 greater than 0.0020 m.

Hollow Sphere. Radial heat flow through a hollow sphere is another example of one-dimensional conduction. For constant thermal conductivity, the modified Fourier rate equation,

$$q_r = -k\frac{dT}{dr}A$$

applies, where A = area of a sphere = $4\pi r^2$, giving

$$q_r = -4\pi k r^2 \frac{dT}{dr} \tag{17-14}$$

This relation, when integrated between the boundary conditions,

$$\text{at } T = T_i \qquad r = r_i$$

and

$$\text{at } T = T_o \qquad r = r_o$$

yields

$$q = \frac{4\pi k(T_i - T_o)}{\dfrac{1}{r_i} - \dfrac{1}{r_o}} \tag{17-15}$$

The hyperbolic temperature distribution,

$$T = T_i - \left(\frac{T_i - T_o}{1/r_i - 1/r_o}\right)\left(\frac{1}{r_i} - \frac{1}{r}\right) \tag{17-16}$$

is obtained by using the same procedure which was followed to obtain equation (17-9). The student may find it instructive to perform the necessary mathematical steps necessary to derive equations (17-15) and (17-16).

Variable Thermal Conductivity. If the thermal conductivity of the medium through which the energy is transferred varies significantly, the preceding equations in this section do not apply. Since Laplace's equation involves the assumption of constant thermal conductivity, a new differential equation must be determined from the general equation for heat transfer. For steady-state conduction in the x direction without internal generation of energy, the equation which applies is

$$\frac{d}{dx}\left(k\frac{dT}{dx}\right) = 0 \tag{17-17}$$

where k may be a function of T.

In many cases the thermal conductivity may be a linear function of temperature over a considerable range. The equation of such a straight-line function may be expressed by

$$k = k_o(1 + \beta T)$$

where k_o and β are constants for a particular material. In general, for materials satisfying this relation, β is negative for good conductors and positive for good insulators. Other relations for varying k have been experimentally determined for specific materials. The evaluation of the rate of energy transfer when the material has a varying thermal conductivity was illustrated in Example 2 in Chapter 15.

17.2 ONE-DIMENSIONAL CONDUCTION WITH INTERNAL GENERATION OF ENERGY

In certain systems, such as electric resistance heaters and nuclear fuel rods, heat is generated within the conducting medium. As one might expect, the generation of energy within the conducting medium produces temperature profiles different than those for simple conduction.

In this section, we shall consider two simple example cases: steady-state conduction in a circular cylinder with uniform or homogeneous energy generation, and steady-state conduction in a plane wall with variable energy generation. Carslaw and Jaeger* and Jakob† have written excellent treatises dealing with more complicated problems.

Cylindrical Solid with Homogeneous Energy Generation. Consider a cylindrical solid with internal energy generation as shown in Figure 17.5. The cylinder will be considered long enough so that only radial conduction occurs. The density

---

* H. S. Carslaw and J. C. Jaeger, *Conduction of Heat in Solids*, 2nd Edition, Oxford Univ. Press, New York, 1959.
† M. Jakob, *Heat Transfer*, Vol. I, Wiley, New York, 1949.

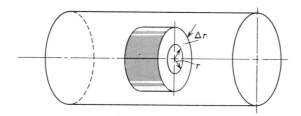

Figure 17.5 Annular element in a long, circular cylinder with internal heat generation.

ρ, the heat capacity c_p, and the thermal conductivity of the material will be considered constant. The energy balance for the element shown is

$$\left\{\begin{array}{c}\text{Rate of energy}\\\text{conduction into}\\\text{the element}\end{array}\right\} + \left\{\begin{array}{c}\text{rate of energy}\\\text{generation within}\\\text{the element}\end{array}\right\} - \left\{\begin{array}{c}\text{rate of energy}\\\text{conduction out}\\\text{of the element}\end{array}\right\}$$

$$= \left\{\begin{array}{c}\text{rate of accumula-}\\\text{tion of energy}\\\text{within the element}\end{array}\right\} \qquad (17\text{-}18)$$

Applying the Fourier rate equation and letting $\dot{q}$ represent the rate of energy generated per unit volume, we may express equation (17-18) by the algebraic expression

$$-k(2\pi rL)\frac{\partial T}{\partial r}\Big|_r + \dot{q}(2\pi rL\,\Delta r) - \left[-k(2\pi rL)\frac{\partial T}{\partial r}\Big|_{r+\Delta r}\right] = \rho c_p \frac{\partial T}{\partial t}(2\pi rL\,\Delta r)$$

Dividing each term by $2\pi rL\,\Delta r$, we obtain

$$\dot{q} + \frac{k[r(\partial T/\partial r)|_{r+\Delta r} - r(\partial T/\partial r)|_r]}{r\,\Delta r} = \rho c_p \frac{\partial T}{\partial t}$$

In the limit as Δr approaches zero, the following differential equation is generated:

$$\dot{q} + \frac{k}{r}\frac{\partial}{\partial r}\left(r\frac{\partial T}{\partial r}\right) = \rho c_p \frac{\partial T}{\partial t} \qquad (17\text{-}19)$$

For steady-state conditions, the accumulation term is zero; when we eliminate this term from the above expression, the differential equation for a solid cylinder with homogeneous energy generation becomes

$$\dot{q} + \frac{k}{r}\frac{d}{dr}\left(r\frac{dT}{dr}\right) = 0 \qquad (17\text{-}20)$$

The variables in this equation may be separated and integrated to yield

$$rk\frac{dT}{dr} + \dot{q}\frac{r^2}{2} = C_1$$

or

$$k\frac{dT}{dr} + \dot{q}\frac{r}{2} = \frac{C_1}{r}$$

Because of the symmetry of the solid cylinder, a boundary condition which must be satisfied stipulates that the temperature gradient must be finite at the center of the cylinder, where $r = 0$. This can only be true if $C_1 = 0$. Accordingly, the above relation reduces to

$$k\frac{dT}{dr} + \dot{q}\frac{r}{2} = 0 \tag{17-21}$$

A second integration will now yield

$$T = -\frac{\dot{q}r^2}{4k} + C_2 \tag{17-22}$$

If the temperature T is known at any radial value, such as a surface, the second constant, C_2, may be evaluated. This, of course, provides the completed expression for the temperature profile. The energy flux in the radial direction may be obtained from

$$\frac{q_r}{A} = -k\frac{dT}{dr}$$

by substituting equation (17-21), yielding

$$\frac{q_r}{A} = \dot{q}\frac{r}{2}$$

or

$$q_r = (2\pi rL)\dot{q}\frac{r}{2} = \pi r^2 L\dot{q} \tag{17-23}$$

Plane Wall with Variable Energy Generation. The second case associated with energy generation involves a temperature-dependent, energy-generating process. This situation develops when an electric current is passed through a conducting medium possessing an electrical resistivity which varies with temperature. In our discussion, we shall assume that the energy-generation term varies linearly with temperature, and that the conducting medium is a flat plate with temperature T_L at both surfaces. The internal energy generation is described by

$$\dot{q} = \dot{q}_L[1 + \beta(T - T_L)] \tag{17-24}$$

where $\dot{q}_L$ is the generation rate at the surface and β is a constant.

With this model for the generation function, and since both surfaces have the same temperature, the temperature distribution within the flat plate is symmetric about the midplane. The plane wall and its coordinate system are illustrated in Figure 17.6. The symmetry of the temperature distribution requires a zero temperature gradient at $x = 0$. With steady-state conditions, the differential equation may be obtained by eliminating the irrelevant terms in the general differential equation for heat transfer. Equation (16–19) for the case of steady-state conduction in the x direction in a stationary solid with constant thermal conductivity becomes

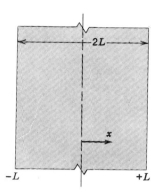

Figure 17.6 Flat plate with temperature-dependent energy generation.

$$\frac{d^2T}{dx^2}+\frac{\dot{q}_L}{k}[1+\beta(T-T_L)]=0$$

The boundary conditions are

$$\text{at } x = 0 \qquad \frac{dT}{dx}=0$$

and

$$\text{at } x = \pm L \qquad T = T_L$$

These relations may be expressed in terms of a new variable, $\theta = T - T_L$, by

$$\frac{d^2\theta}{dx^2}+\frac{\dot{q}_L}{k}(1+\beta\theta)=0$$

or

$$\frac{d^2\theta}{dx^2}+C+s\theta=0$$

where $C = \dot{q}_L/k$ and $s = \beta\dot{q}_L/k$. The boundary conditions are

$$\text{at } x = 0 \qquad \frac{d\theta}{dx}=0$$

and

$$\text{at } x = \pm L \qquad \theta = 0$$

The integration of this differential equation is simplified by a second change in variables; inserting ϕ for $C + s\theta$ into the differential equation and the boundary conditions, we obtain

$$\frac{d^2\phi}{dx^2} + s\phi = 0$$

for

$$x = 0 \qquad \frac{d\phi}{dx} = 0$$

and

$$x = \pm L \qquad \phi = C$$

The solution is

$$\phi = C + s\theta = A \cos (x \sqrt{s}) + B \sin (x \sqrt{s})$$

or

$$\theta = A_1 \cos (x \sqrt{s}) + A_2 \sin (x \sqrt{s}) - \frac{C}{s}$$

The temperature distribution becomes

$$T - T_L = \frac{1}{\beta} \left[\frac{\cos (x \sqrt{s})}{\cos (L \sqrt{s})} - 1 \right] \qquad (17\text{-}25)$$

where $s = \beta \dot{q}_L / k$ is obtained by applying the two boundary conditions.

The cylindrical and spherical examples of one-dimensional temperature-dependent generation are more complex; solutions to these may be found in the technical literature.

17.3 HEAT TRANSFER FROM EXTENDED SURFACES

A very useful application of one-dimensional heat-conduction analysis is that of describing the effect of extended surfaces. It is possible to increase the energy transfer between a surface and an adjacent fluid by increasing the amount of surface area in contact with the fluid. This increase in area is accomplished by adding extended surfaces which may be in the forms of fins or spines of various cross sections.

The one-dimensional analysis of extended surfaces may be formulated in general terms by considering the situation depicted in Figure 17.7.

The shaded area represents a portion of the extended surface which has variable cross-sectional area, $A(x)$, and surface area, $S(x)$, which are functions of x alone. For steady-state conditions the first law of thermodynamics, equation (6-10), reduces to the simple expression

$$\frac{\delta Q}{dt} = 0$$

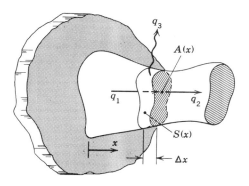

Figure 17.7 An extended surface of general configuration.

thus, in terms of the heat flow rates designated in the figure, we may write

$$q_1 = q_2 + q_3 \tag{17-26}$$

The quantities q_1 and q_2 are conduction terms, while q_3 is a convective heat-flow rate. Evaluating each of these in the appropriate way and substituting into equation (17-26), we obtain

$$kA\frac{dT}{dx}\bigg|_{x+\Delta x} - kA\frac{dT}{dx}\bigg|_x - hS(T - T_\infty) = 0 \tag{17-27}$$

where T_∞ is the fluid temperature. Expressing the surface area, $S(x)$, in terms of the width, Δx, times the perimeter, $P(x)$, and dividing through by Δx, we obtain

$$\frac{kA(dT/dx)|_{x+\Delta x} - kA(dT/dx)|_x}{\Delta x} - hP(T - T_\infty) = 0$$

Evaluating this equation in the limit as $\Delta x \to 0$, we obtain the very general differential equation

$$\frac{d}{dx}\left(kA\frac{dT}{dx}\right) - hP(T - T_\infty) = 0 \tag{17-28}$$

One should note, at this point, that the temperature gradient, dT/dx, and the surface temperature, T, are expressed such that T is a function of x only. This treatment assumes the temperature to be "lumped" in the transverse direction. This is physically realistic when the cross section is thin or when the material thermal conductivity is large. Both of these conditions apply in the case of fins. More will be said about the "lumped parameter" approach in Chapter 18. This approximation in the present case leads to equation (17-28), an ordinary differential equation. If we did not make this simplifying analysis we would have a distributed parameter problem that would require solving a partial differential equation.

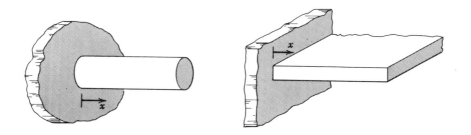

Figure 17.8   Two examples of extended surfaces with constant cross section.

A wide range of possible forms exist when equation (17-28) is applied to specific geometries. Three possible applications and the resulting equations are described in the following paragraphs.

(1) Fins or Spines of Uniform Cross Section. For either of the cases shown in Figure 17.8 the following are true: $A(x) = A$, and $P(x) = P$, both constants. If, additionally, both k and h are taken to be constant, equation (17-28) reduces to

$$\frac{d^2 T}{dx^2} - \frac{hP}{kA}(T - T_\infty) = 0 \qquad (17\text{-}29)$$

(2) Straight Surfaces with Linearly Varying Cross Section. Two configurations for which A and P are not constant are shown in Figure 17.9. If the area and perimeter both vary in a linear manner from the primary surface, $x = 0$, to some lesser value at the end, $x = L$, both A and P may be expressed as

$$A = A_0 - (A_0 - A_L)\frac{x}{L} \qquad (17\text{-}30)$$

and

$$P = P_0 - (P_0 - P_L)\frac{x}{l} \qquad (17\text{-}31)$$

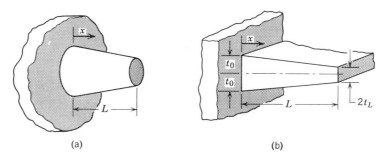

(a)   (b)

Figure 17.9   Two examples of straight extended surfaces with variable cross section.

In the case of the rectangular fin shown in Figure 17.9b, the appropriate values of A and P are

$$A_0 = 2t_0 W \qquad\qquad A_L = 2t_L W$$
$$P_0 = 2[2t_0 + W] \qquad P_L = 2[2t_L + W]$$

where t_0 and t_L represent the semithickness of the fin evaluated at $x = 0$ and $x = L$, respectively, and W is the total depth of the fin.

For constant h and k, equation (17-28) applied to extended surfaces with cross-sectional area varying linearly becomes

$$\left[A_0 - (A_0 - A_L)\frac{x}{L}\right]\frac{d^2T}{dx^2} - \frac{A_0 - A_L}{L}\frac{dT}{dx} - \frac{h}{k}\left[P_0 - (P_0 - P_L)\frac{x}{L}\right](T - T_\infty) = 0 \tag{17-32}$$

(3) Curved Surfaces of Uniform Thickness. A common type of extended surface is that of the circular fin of constant thickness as depicted in Figure 17.10.

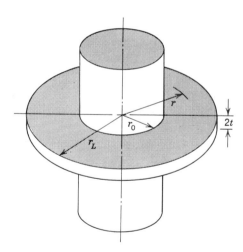

Figure 17.10 A curved fin of constant thickness.

The appropriate expressions for A and P, in this case, are

and

$$\left.\begin{array}{l} A = 4\pi rt \\[12pt] P = 4\pi r \end{array}\right\} \quad r_0 \le r \le r_L$$

When these expressions are substituted into equation (17-28), the applicable differential equation, considering k and h constant, is

$$\frac{d^2T}{dr^2}+\frac{1}{r}\frac{dT}{dr}-\frac{h}{kt}(T-T_\infty)=0 \tag{17-33}$$

In each of the cases considered the thermal conductivity and convective heat-transfer coefficient were assumed constant. When the variable nature of these quantities is considered, the resulting differential equations become still more complex than those developed thus far.

Solutions for the temperature profile in the case of the straight fin of constant cross section will now be considered; equation (17-29) applies.

The general solution to equation (17-29) may be written

$$\theta = c_1 e^{mx} + c_2 e^{-mx} \tag{17-34}$$

or

$$\theta = A\cosh mx + B\sinh mx \tag{17-35}$$

where $m^2 = hP/kA$ and $\theta = T - T_\infty$. The evaluation of the constants of integration requires that two boundary conditions be known. The three sets of boundary conditions which we shall consider are as follows:

(a)	$T = T_0$	at $x = 0$
	$T = T_L$	at $x = L$
(b)	$T = T_0$	at $x = 0$
	$\dfrac{dT}{dx}=0$	at $x = L$

and

(c)	$T = T_0$	at $x = 0$
	$-k\dfrac{dT}{dx}=h(T-T_\infty)$	at $x = L$

The first boundary condition of each set is the same and stipulates that the temperature at the base of the extended surface is equal to that of the primary surface. The second boundary condition relates the situation at a distance L from the base. In set (a) the condition is that of a known temperature at $x = L$. In set (b) the temperature gradient is zero at $x = L$. In set (c) the requirement is that heat flow to the end of an extended surface by conduction be equal to that leaving this position by convection.

The temperature profile, associated with the first set of boundary conditions, is

$$\frac{\theta}{\theta_0}=\frac{T-T_\infty}{T_0-T_\infty}=\left(\frac{\theta_L}{\theta_0}-e^{-mL}\right)\left(\frac{e^{mx}-e^{-mx}}{e^{mL}-e^{-mL}}\right)+e^{-mx} \tag{17-36}$$

A special case of this solution applies when L becomes very large, that is, $L\to\infty$, for which equation (17-36) reduces to

$$\frac{\theta}{\theta_0} = \frac{T - T_\infty}{T_0 - T_\infty} = e^{-mx} \tag{17-37}$$

The constants, c_1 and c_2, obtained by applying set (b), yield, for the temperature profile,

$$\frac{\theta}{\theta_0} = \frac{T - T_\infty}{T_0 - T_\infty} = \frac{e^{mx}}{1 + e^{2mL}} + \frac{e^{-mx}}{1 + e^{-2mL}} \tag{17-38}$$

An equivalent expression to equation (17-38) but in more compact form is

$$\frac{\theta}{\theta_0} = \frac{T - T_\infty}{T_0 - T_\infty} = \frac{\cosh\,[m(L - x)]}{\cosh\,mL} \tag{17-39}$$

Note that, in either equation (17-38) or (17-39), as $L \to \infty$ the temperature profile approaches that expressed in equation (17-37).

The application of set (c) of the boundary conditions yields, for the temperature profile,

$$\frac{\theta}{\theta_0} = \frac{T - T_\infty}{T_0 - T_\infty} = \frac{\cosh\,[m(L - x)] + (h/mk)\sinh\,[m(L - x)]}{\cosh\,mL + (h/mk)\sinh\,mL} \tag{17-40}$$

It may be noted that this expression reduces to equation (17-39) if $d\theta/dx = 0$ at $x = L$ and to equation (17-37) if $T = T_\infty$ at $L = \infty$.

The expressions for $T(x)$ which have been obtained are particularly useful in evaluating the total heat transfer from an extended surface. This total heat transfer may be determined by either of two approaches. The first is to integrate the convective heat-transfer expression over the surface according to

$$q = \int_S h[T(x) - T_\infty]\,dS = \int_S h\theta\,dS \tag{17-41}$$

The second method involves evaluating the energy conducted into the extended surface at the base as expressed by

$$q = -kA\frac{dT}{dx}\bigg|_{x=0} \tag{17-42}$$

The latter of these two expressions is easier to evaluate; accordingly we will use this equation in the following development.

Using equation (17-36), we find that the heat transfer rate, when set (a) of the boundary conditions applies, is

$$q = kAm\theta_0\left[1 - 2\frac{\theta_L/\theta_0 - e^{-mL}}{e^{mL} - e^{-mL}}\right] \tag{17-43}$$

If the length L is very long, this expression becomes

$$q = kAm\theta_0 = kAm(T_0 - T_\infty) \tag{17-44}$$

Substituting equation (17-39) [obtained by using set (b) of the boundary conditions] into equation (17-42) we obtain

$$q = kAm\theta_0 \tanh mL \qquad (17\text{-}45)$$

Equation (17-40), utilized in equation (17-42), yields for q the expression

$$q = kAm\theta_0 \frac{\sinh mL + (h/mk) \cosh mL}{\cosh mL + (h/mk) \sinh mL} \qquad (17\text{-}46)$$

The equations for the temperature profile and total heat transfer for extended surfaces of more involved configuration have not been considered. Certain of these cases will be left as exercises for the reader.

A question that is logically asked at this point is, "What benefit is accrued by the addition of extended surfaces?" A term which aids in answering this question is the *fin efficiency*, symbolized η_f, defined as the ratio of the actual heat transfer from an extended surface to the maximum possible heat transfer from the surface. The maximum heat transfer would occur if the temperature of the extended surface were equal to the base temperature, T_0, at all points.

Figure 17.11 is a plot of η_f as a function of a significant parameter for both straight and circular fins of constant thickness (when fin thickness is small, $t \ll r_L - r_0$.)

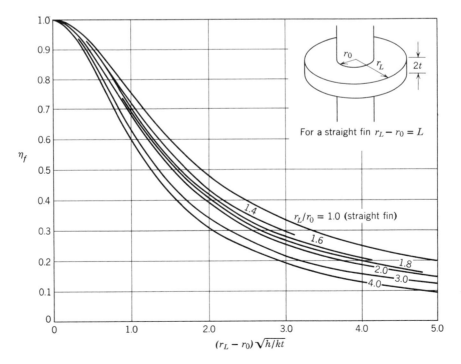

Figure 17.11　Fin efficiency for straight and circular fins of constant thickness.

The total heat transfer from a finned surface is

$$q_{total} = q_{primary\ surface} + q_{fin}$$

$$= A_0 h (T_0 - T_\infty) + A_f h (T - T_\infty) \tag{17-47}$$

The second term in equation (17-47) is the actual heat transfer from the fin surface in terms of the variable surface temperature. This may be written in terms of the fin efficiency, yielding

$$q_{total} = A_0 h (T_0 - T_\infty) + A_f h \eta_f (T_0 - T_\infty)$$

or

$$q_{total} = h (A_0 + A_f \eta_f)(T_0 - T_\infty) \tag{17-48}$$

In this expression A_0 represents the exposed area of the primary surface, A_f is the total fin surface area, and the heat transfer coefficient, h, is assumed constant.

The application of equation (17-48) as well as an idea of the effectiveness of fins is illustrated in Example 3.

EXAMPLE 3

Water and air are separated by a mild-steel plane wall. It is proposed to increase the heat-transfer rate between these fluids by adding straight rectangular fins of 0.05-in. thickness and 1-in. length, spaced 0.5 in. apart. The airside and waterside heat-transfer coefficients may be assumed constant at 2 and 45 Btu/hr ft^2 °F, respectively. What per cent increase in heat transfer can be realized if fins are placed on (a) the waterside? (b) the airside? (c) both sides?

For a 1-ft by 1-ft section the areas of the primary surface and fins are

$$A_0 = 1\ \text{ft}^2 - (24\ \text{fins})(1\ \text{ft})\left(\frac{0.05}{12}\frac{\text{ft}}{\text{fin}}\right)$$

$$= 0.9\ \text{ft}^2$$

$$A_f = (24\ \text{fins})(1\ \text{ft})(2 \times \tfrac{1}{12}\ \text{ft}) + 0.1\ \text{ft}^2$$

$$= 4.1\ \text{ft}^2$$

Employing Figure 17.11, we find that the fin effectiveness for the airside is $\eta_f|_{air} = 0.9$, and that for the waterside $\eta_f|_{water} = 0.4$. Evaluating the heat-transfer rate with fins on the airside, we have

$$q = h_a\ \Delta T_a [A_0 + \eta_{fa} A_f]$$

$$= 2\ \Delta T_a [0.9 + 0.9(4.1)] = 9.18\ \Delta T_{air}$$

and on the waterside,

$$q = h_w\ \Delta T_w [A_0 + \eta_{fw} A_f]$$

$$= 45\ \Delta T_w [0.9 + 0.4(4.1)] = 114\ \Delta T_{water}$$

The ΔT_{air} and ΔT_{water} represent the temperature differences between the particular fluid and the steel surface at temperature T_0.

The reciprocals of the coefficients are the thermal resistances of the finned surfaces in air and water.

Without fins the total rate of heat transfer in terms of the overall temperature difference, $\Delta T = T_w - T_{air}$, and neglecting the conductive resistance of the steel wall, is

$$q = \frac{\Delta T_{total}}{\frac{1}{2} + \frac{1}{45}} = 1.915 \, \Delta T_{total}$$

With fins on the airside the heat transfer is

$$q = \frac{\Delta T_{total}}{1/9.18 + 1/45} = 7.96 \, \Delta T_{total}$$

an increase of 316%!

When fins are added to the waterside alone,

$$q = \frac{\Delta T_{total}}{\frac{1}{2} + \frac{1}{114}} = 1.966 \, \Delta T_{total}$$

an increase of 3.0%.

With fins added to both sides the total heat-flow rate is

$$q = \frac{\Delta T_{total}}{1/9.18 + 1/114} = 8.91 \, \Delta T_{total}$$

an increase of 365%.

It is apparent that, when fins are added to the airside, a greater increase occurs than for fins added to the waterside alone.

17.4　TWO- AND THREE-DIMENSIONAL SYSTEMS

In sections 17.2 and 17.3 we discussed systems in which the temperature and the energy transfer were functions of a single space variable. Although many problems fall into this category, there are many other systems involving complicated geometry or temperature boundary conditions, or both, for which two or even three spatial coordinates are necessary to describe the temperature field.

In this section we shall review some of the methods for analyzing heat transfer by conduction in two- and three-dimensional systems. The problems will mainly involve two-dimensional systems, since they are less cumbersome to solve yet illustrate the techniques of analysis.

Analytical Solution. An analytical solution to any transfer problem must satisfy the differential equation describing the process as well as the prescribed boundary conditions. Many mathematical techniques have been used to obtain solutions for particular energy conduction situations in which a partial differential

equation describes the temperature field. Carslaw and Jaeger* and Boelter et al.†
have written excellent treatises which deal with the mathematical solutions for
many of the more complex conduction problems. Since most of this material is too
specialized for an introductory course, a solution will be obtained to one of the
first cases analyzed by Fourier‡ in the classical treatise which established the
theory of energy transfer by conduction. This solution of a two-dimensional
conduction medium employs the mathematical method of separation of variables.

Consider a thin, infinitely long rectangular plate which is free of heat sources,
as illustrated in Figure 17.12. For a thin plate $\partial T/\partial z$ is negligible, and the

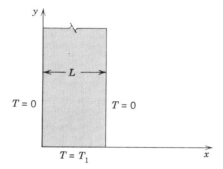

Figure 17.12 Model for two-dimensional conduction analysis.

temperature is a function of x and y only. The solution will be obtained for the
case in which the two edges of the plate are maintained at zero temperature and
the bottom is maintained at T_1 as shown. The steady-state temperature distribution
in the plate of constant thermal conductivity must satisfy the differential
equation,

$$\frac{\partial^2 T}{\partial x^2} + \frac{\partial^2 T}{\partial y^2} = 0 \qquad (17\text{-}49)$$

and the boundary conditions,

$$T = 0 \qquad \text{at } x = 0 \text{ for all values of } y$$

$$T = 0 \qquad \text{at } x = L \text{ for all values of } y$$

$$T = T_1 \qquad \text{at } y = 0 \text{ for } 0 \le x \le L$$

* H. S. Carslaw and J. C. Jaeger, *Conduction of Heat in Solids*, Second Edition, Oxford Univ. Press, New York, 1959.
† L. M. K. Boelter, V. H. Cherry, H. A. Johnson, and R. C. Martinelli, *Heat Transfer Notes*, McGraw-Hill Book Company, New York, 1965.
‡ J. B. J. Fourier, *Theorie Analytique de la Chaleur*, Gauthier-Villars, Paris, 1822.

and

$$T = 0 \quad \text{at } y = \infty \text{ for } 0 \le x \le L$$

Equation (17-49) is a linear, homogeneous partial differential equation. This type of equation usually can be integrated by assuming that the temperature distribution, $T(x, y)$, is of the form

$$T(x, y) = X(x) Y(y) \tag{17-50}$$

where $X(x)$ is a function of x only and $Y(y)$ is a function of y only. Substituting this equation into equation (17-49), we obtain an expression in which the variables are separated,

$$-\frac{1}{X} \frac{d^2 X}{dx^2} = \frac{1}{Y} \frac{d^2 Y}{dy^2} \tag{17-51}$$

Since the left-hand side of equation (17-51) is independent of y and the equivalent right-hand side is independent of x, it follows that both must be independent of x and y, and hence must be equal to a constant. If we designate this constant λ^2, two ordinary differential equations result:

$$\frac{d^2 X}{dx^2} + \lambda^2 X = 0 \tag{17-52}$$

and

$$\frac{d^2 Y}{dy^2} - \lambda^2 Y = 0 \tag{17-53}$$

These differential equations may be integrated, yielding

$$X = A \cos \lambda x + B \sin \lambda x$$

and

$$Y = C e^{\lambda y} + D e^{-\lambda y}$$

According to equation (17-50) the temperature distribution is defined by the relation

$$T(x, y) = XY = (A \cos \lambda x + B \sin \lambda x)(C e^{\lambda y} + D e^{-\lambda y}) \tag{17-54}$$

where A, B, C, and D are constants to be evaluated from the four boundary conditions. The condition that $T = 0$ at $x = 0$ requires that $A = 0$. Similarly, $\sin \lambda x$ must be zero at $x = L$; accordingly, λL must be an integral multiple of π or $\lambda = n\pi/L$. Equation (17-54) is now reduced to

$$T(x, y) = B \sin \left(\frac{n\pi x}{L}\right)(C e^{n\pi y/L} + D e^{-n\pi y/L}) \tag{17-55}$$

The requirement that $T = 0$ at $y = \infty$ stipulates that C must be zero. A combination of B and D into the single constant E reduces equation (17-55) to

$$T(x, y) = E e^{-n\pi y/L} \sin \left(\frac{n\pi x}{L}\right)$$

This expression satisfies the differential equation for any integer n greater than or equal to zero. The general solution is obtained by summing all possible solutions, giving

$$T = \sum_{n=1}^{\infty} E_n e^{-n\pi y/L} \sin\left(\frac{n\pi x}{L}\right) \tag{17-56}$$

The last boundary condition, $T = T_1$ at $y = 0$, is used to evaluate E_n according to the expression

$$T_1 = \sum_{n=1}^{\infty} E_n \sin\left(\frac{n\pi x}{L}\right) \qquad \text{for } 0 \le x \le L$$

The constants E_n are the Fourier coefficients for such an expansion and are given by

$$E_n = \frac{4T_1}{n\pi} \qquad \text{for } n = 1, 3, 5, \ldots$$

and

$$E_n = 0 \qquad \text{for } n = 2, 4, 6, \ldots$$

The solution to this two-dimensional conduction problem is

$$T = \frac{4T_1}{\pi} \sum_{n=0}^{\infty} \frac{e^{[-(2n+1)\pi y]/L}}{2n+1} \sin\frac{(2n+1)\pi x}{L} \tag{17-57}$$

The isotherms and energy flow lines are plotted in Figure 17.13. The isotherms are shown in the figure as solid lines, and the dotted lines, which are orthogonal to the

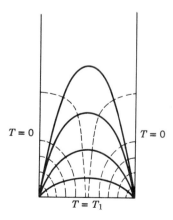

$T = 0$ $T = 0$

$T = T_1$

Figure 17.13 Isotherms and energy flow lines for the rectangular plate in figure 17.12.

isotherms, are energy-flow lines. Note the similarity to the lines of constant velocity potential and stream function as discussed in momentum transfer.

The separation of variables method can be extended to three-dimensional cases by assuming T to be equal to the product $X(x)Y(y)Z(z)$ and substituting this expression for T into the applicable differential equation. When the variables are separated, three second-order ordinary differential equations are obtained which may be integrated subject to the given boundary conditions.

Analytical solutions are useful when they can be obtained. There are, however, practical problems with complicated geometry and boundary conditions which cannot be solved analytically. As an alternative approach, one must turn to graphical or numerical methods.

Graphical Solutions by Flux Plotting. An approximate solution to the two-dimensional Laplace equation for a homogeneous medium with constant thermal conductivity,

$$\frac{\partial^2 T}{\partial x^2}+\frac{\partial^2 T}{\partial y^2}=0$$

can be obtained graphically by plotting the potential field. This method has proved particularly applicable for systems which have isothermal boundaries.

The basic principles of the graphical method will be initially illustrated by considering the simple case of an infinitely long flat plate as illustrated in Figure 17.14. With constant surface temperatures, T_1 and T_2, the isotherms must lie

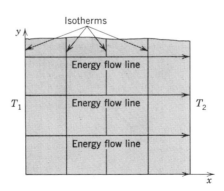

Figure 17.14 Isotherms and constant energy-flow lines in an infinitely long, flat plate.

parallel to the surfaces as shown. Since constant-energy flow lines are orthogonal to the isotherms, they will be parallel horizontal lines.

In constructing the flux plot for the more complex situations, the general procedure is to divide the body into a network of curvilinear squares consistent with the boundary conditions, by trial and error. The flux plot is complete when the following requirements are satisfied:

1. Isotherms and energy flow lines intersect each other at right angles while forming a network of curvilinear squares.
2. Diagonals of the curvilinear squares bisect each other at 90° and bisect any boundary corner.
3. Isotherms are parallel to constant-temperature boundaries.
4. Energy flow lines are perpendicular to constant-temperature boundaries.
5. Energy flow lines, leading to a corner of a constant-temperature boundary, bisect the angle between the surfaces of the boundary formed at the corner.

Flux plotting is approximate, the accuracy depending upon the patience of the sketcher and his ability to satisfy the above requirements. Bewley* has made the following helpful suggestions which may reduce the amount of trial and error in constructing the flux plot:

1. Note conditions of symmetry. Lines of symmetry are flow lines and divide the potential field into compartments.
2. Mark all known isotherms.
3. At each corner of an isothermal boundary, draw a short line bisecting the angle. All such lines are the beginning of constant-energy flow lines.
4. Tentatively extend these lines to other isotherms, realizing that these lines are orthogonal to the isotherms.
5. Start isotherms in a region where flow lines are uniformly spaced if such a region exists.
6. Begin with a crude network and first find the approximate location of the isothermal and constant-energy flow lines.
7. In this first attempt, it will usually be difficult to make the flow lines orthogonal to the isotherms and yet have a network of curvilinear squares. To satisfy these requirements, individual or simultaneous adjustments may have to be made in the location of the lines.

To evaluate the energy transfer rate, it is necessary that the isotherms and constant-energy flow lines form a network of curvilinear squares. With this spacing, a constant amount of energy will flow between any two adjacent energy flow lines. A portion of a flux plot in a system of unit thickness is shown in Figure 17.15. With N energy flow tubes and Δq energy flow per flow tube, the total energy transfer rate may be evaluated by

$$q_{total} = N \, \Delta q \qquad (17\text{-}58)$$

For the flow tube shown in Figure 17.15, the temperature gradient is $\Delta T/\Delta n$ and the area normal to the direction of energy flow is Δm. By the Fourier rate

---

*L. V. Bewley, *Two Dimensional Fields in Electrical Engineering*, The Macmillan Company, New York, 1948.

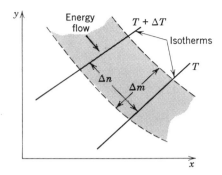

Figure 17.15 Temperature-energy flux plot element.

equation, the energy flow per flow tube is

$$\Delta q = -k\,\Delta m \frac{\Delta T}{\Delta n}$$

or for the curvilinear squares where $\Delta m = \Delta n$,

$$\Delta q = -k\,\Delta T \qquad (17\text{-}59)$$

This is true *regardless of the size of the squares*. Therefore, for a curvilinear-square network, the overall temperature difference, $T_h - T_c$, divided by the number of squares in a flow tube, M, will give the temperature gradient per flow tube, as expressed by the relation

$$\frac{\Delta T_{\text{overall}}}{M} = \frac{(T_h - T_c)}{M} \qquad (17\text{-}60)$$

and the rate of energy flow per flow tube is

$$\Delta q = k\frac{(T_h - T_c)}{M} \qquad (17\text{-}61)$$

For N tubes, the energy transfer rate is

$$q = N\,\Delta q = k\frac{N}{M}(T_h - T_c)$$

or

$$q = Sk(T_h - T_c) \qquad (17\text{-}62)$$

where the *shape factor, S*, is equal to N/M for the two-dimensional system.

The following example illustrates the use of a flux plot for a more complex case.

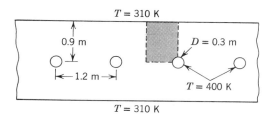

Figure 17.16 Isothermal heating element embedded in a plane wall.

EXAMPLE 4

Determine the energy transferred from heating elements imbedded in a large heating wall with isothermal surfaces as illustrated in Figure 17.16. The thermal conductivity of the wall is 0.7 W/m · K.

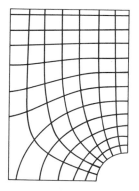

Figure 17.17 Shaded area of Figure 17.16.

The shaded area represents conditions of symmetry within the wall. The flux plot for this area is shown in Figure 17.17. In this plot, the number of flow tubes, N, equals 7, and the number of squares in a flow tube, M, equals 11.5. The shape factor, S, is equal to

$$S = \frac{N}{M} = \frac{7}{11.5} = 0.608 \text{ m}^2/\text{m}$$

and the energy transferred by steady-state conduction through the shaded area per unit depth is

$$q = Sk(T_h - T_c)$$

$$q = (0.608 \text{ m}^2/\text{m})(0.7 \text{ W/m} \cdot \text{K})(400 \text{ K} - 310 \text{ K})$$

or

$$q \cong 38.3 \text{ W per meter of depth}$$
$$(39.8 \text{ Btu/hr per foot of depth})$$

It is important to note that in the flux network one can obtain fractions of flow tubes and of squares in the flow tubes. Referring back to Figure 17.16, we can see that each heating element transfers energy through four equivalent areas. Accordingly, each heating element transfers 4(38.3) or 153.2 W.

The shape factors, S, for several simple geometries have been obtained via conformal mapping and are presented in Table 17.1.

TABLE 17.1 CONDUCTION SHAPE FACTORS

Shape	Shape factor, S $q/L = kS(T_i - T_o)$
  Concentric circular cylinders	$\dfrac{2\pi}{\ln(r_o/r_i)}$
  Eccentric circular cylinders	$\dfrac{2\pi}{\cosh^{-1}\left(\dfrac{1 + \rho^2 - \epsilon^2}{2\rho}\right)}$ $\rho \equiv r_i/r_o,\ \epsilon \equiv e/r_o$
  Circular cylinder in a hexagonal cylinder	$\dfrac{2\pi}{\ln(r_o/r_i) - 0.10669}$
  Circular cylinder in a square cylinder	$\dfrac{2\pi}{\ln(r_o/r_i) - 0.27079}$
  Infinite cylinder buried in semi-infinite medium	$\dfrac{2\pi}{\cosh^{-1}(\rho/r)}$

Analogical Solution. The similarity between any two or more transport phenomena permits the analysis of each process by analogous mathematical methods. The two-dimensional Laplace equation is used to describe potential fields for several phenomena. The potential distribution in an electrostatic field is described by

$$\frac{\partial^2 E}{\partial x^2} + \frac{\partial^2 E}{\partial y^2} = 0$$

the distribution in a temperature field is described by

$$\frac{\partial^2 T}{\partial x^2} + \frac{\partial^2 T}{\partial y^2} = 0$$

and the concentration distribution is described in Chapter 25 by

$$\frac{\partial^2 c_A}{\partial x^2} + \frac{\partial^2 c_A}{\partial y^2} = 0$$

There are many conduction problems for which solutions have not been obtained by mathematical techniques. Solutions to some of these problems have been determined in an analogous system and then reexpressed in terms of the thermal problem. The analog field plotter is an instrument which is easily used to obtain field distribution described by the two-dimensional Laplace equation.

The analog field plotter uses a thin sheet of electrically conducting paper cut to the scaled shape of the energy-conducting medium. An electrical current flow pattern can be established in the paper by means of suitably attached and energized electrodes. Reconsider Example 4. Figure 17.18 is a simplified schematic representation of the circuit diagram of the equipment. Boundary conditions corresponding to isotherms are obtained in the electrical field by attaching copper wires to the paper or by painting the areas with highly conductive silver paint and

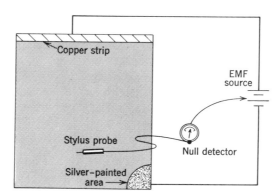

Figure 17.18 Arrangement of analog field plotter for solving Example 4.

then attaching an emf source as illustrated in Figure 17.18. The plain edges of the conducting paper correspond to insulated surfaces in the temperature field.

Lines of constant voltage are found by moving the stylus across the paper, making small perforations in the paper whenever the null detector indicates that the stylus is at a specified voltage. The voltage level of the particular potential being evaluated is established by choosing a slider position on the voltage-dividing potentiometer of the null detector. Selecting equal increments of voltage makes adjacent perforation lines analogous to isotherms separated by the same temperature difference.

Since the constant-energy flow lines are orthogonal to the potential lines, they can be sketched freehand; the result will be a network of curvilinear squares as obtained in flux plotting. The constant-flow lines can also be traced by simply reversing the conducting and insulating portions of the boundary. The evaluation of the energy transferred by steady-state conduction involves the same equations used in flux plotting.

NUMERICAL SOLUTIONS

Each of the solution techniques discussed thus far for multidimensional conduction has considerable utility when conditions permit its use. Analytical solutions require relatively simple functions and geometries; flux plotting requires equipotential boundaries. When the situation of interest becomes sufficiently complex or when boundary conditions preclude the use of simple solution techniques, one must turn to numerical solutions.

With the presence of digital computers to accomplish the large number of manipulations inherent in numerical solutions rapidly and accurately this approach is now very common. In this section we shall introduce the concepts of numerical problem formulation and solution. A more complete and detailed discussion of numerical solutions to heat conduction problems may be found in Carnahan et al.* and in Welty.†

Shown in Figure 17.19 is a two-dimensional representation of an element within a conducting medium. The element or "node" i, j is centered in the figure along with its adjacent nodes. The designation, i, j, implies a general location in a two-dimensional system where i is a general index in the x direction and j is the y index. Adjacent node indices are shown in Figure 17.19. The grid is set up with constant node width, Δx, and constant height, Δy. It may be convenient to make the grid "square," that is, $\Delta x = \Delta y$, but for now we will allow these dimensions to be different.

A direct application of equation (6-10) to node i, j yields

$$\frac{\delta Q}{dt} = \frac{\partial}{\partial t} \iiint_{\text{c.v.}} e\rho\, dV \tag{17-63}$$

---

* B. Carnahan, H. A. Luther, and J. O. Wilkes, *Applied Numerical Methods*, Wiley, New York, 1969.
† J. R. Welty, *Engineering Heat Transfer*, Wiley, New York, 1974.

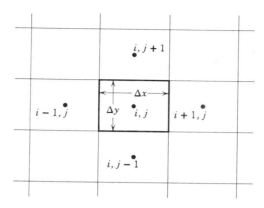

Figure 17.19 Two-dimensional volume element in a conducting medium.

The heat input term, $\delta Q/dt$, may be evaluated allowing for conduction into node i, j from the adjacent nodes and by energy generation within the medium. Evaluating $\delta Q/dt$ in this manner we obtain

$$\frac{\delta Q}{dt} = k\frac{\Delta y}{\Delta x}(T_{i-1,j} - T_{i,j}) + k\frac{\Delta y}{\Delta x}(T_{i+1,j} - T_{i,j})$$

$$+ k\frac{\Delta x}{\Delta y}(T_{i,j-1} - T_{i,j}) + k\frac{\Delta x}{\Delta y}(T_{i,j+1} - T_{i,j}) + \dot{q}\,\Delta x\,\Delta y \quad (17\text{-}64)$$

The first two terms in this expression relate conduction in the x direction, the third and fourth express y-directional conduction, and the last is the generation term. All of these terms are positive; heat transfer is assumed positive.

The rate of energy increase within node i, j may be written simply as

$$\frac{\partial}{\partial t}\iiint_{\text{c.v.}} e\rho\,dV = \left[\frac{\rho cT|_{t+\Delta t} - \rho cT|_t}{\Delta t}\right]\Delta x\,\Delta y \quad (17\text{-}65)$$

Equation (17-63) indicates that the expressions given by equations (17-64) and (17-65) may be equated. Setting these expressions equal to each other and simplifying we have

$$k\frac{\Delta y}{\Delta x}[T_{i-1,j} + T_{i+1,j} - 2T_{i,j}] + k\frac{\Delta x}{\Delta y}[T_{i,j-1} + T_{i,j+1} - 2T_{i,j}]$$

$$+ \dot{q}\,\Delta x\,\Delta y = \left[\frac{\rho cT_{i,j}|_{t+\Delta t} - \rho cT_{i,j}|_t}{\Delta t}\right]\Delta x\,\Delta y \quad (17\text{-}66)$$

This expression will be considered in more complete form in the next chapter. For the present we will not consider time-variant terms, moreover we will consider the nodes to be square, that is, $\Delta x = \Delta y$. With these simplifications equation (17-66) becomes

$$T_{i-1,j} + T_{i+1,j} + T_{i,j-1} + T_{i,j+1} - 4T_{i,j} + \dot{q}\frac{\Delta x^2}{k} = 0 \qquad (17\text{-}67)$$

In the absence of internal generation equation (17-67) may be solved for $T_{i,j}$ to yield

$$T_{i,j} = \frac{T_{i-1,j} + T_{i+1,j} + T_{i,j-1} + T_{i,j+1}}{4} \qquad (17\text{-}68)$$

or, the temperature of node i, j is the arithmetic mean of the temperatures of its adjacent nodes. A simple example showing the use of equation (17-68) in solving a two-dimensional heat conduction problem follows.

EXAMPLE 5

A hollow square duct of the configuration shown has its surfaces maintained at 200 K and 100 K respectively. Determine the steady-state heat transfer rate between the hot and

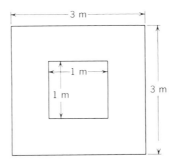

cold surface of this duct; the wall material has a thermal conductivity of 1.21 W/m · K. We may take advantage of the eightfold symmetry of this figure to lay out the simple square grid shown below.

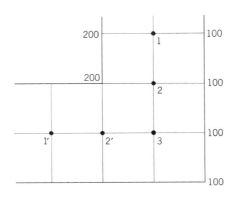

The grid chosen is square with $\Delta x = \Delta y = \frac{1}{2}$ m. Three interior node points are thus identified; their temperatures may be determined by proper application of equation (17-68). Writing the proper expressions for T_1, T_2, and T_3 using equation (17-68) as a guide we have

$$T_1 = \frac{200 + 100 + 2T_2}{4}$$

$$T_2 = \frac{200 + 100 + T_1 + T_3}{4}$$

$$T_3 = \frac{100 + 100 + 2T_2}{4}$$

This set of three equations and three unknowns may be solved quite easily to yield the following: $T_1 = 145.83$ K, $T_2 = 141.67$ K, $T_3 = 120.83$ K.

The temperatures just obtained may now be used to find heat transfer. Implicit in the procedure of laying out a grid of the sort we have specified is the assumption that heat flows in the x and y directions between nodes. On this basis heat transfer occurs from the hot surface to the interior only to nodes 1 and 2; heat transfer occurs to the cooler surface from nodes 1, 2 and 3. We should also recall that the section of duct that has been analyzed is one-eighth of the total thus, of the heat transfer to and from node 1, only one-half should be properly considered as a part of the element analyzed.

We now solve for the heat transfer rate from the hotter surface, and write

$$q = \frac{k(200 - T_1)}{2} + k(200 - T_2)$$

$$= k\left[\left(\frac{200 - 145.83}{2}\right) + (200 - 141.67)\right]$$

$$= 85.415\,k \quad (q \text{ in W/m}, k \text{ in W/m} \cdot \text{K})$$

A similar accounting for the heat flow from nodes 1, 2, and 3 to the cooler surface is written

$$q = \frac{k(T_1 - 100)}{2} + k(T_2 - 100) + k(T_3 - 100)$$

$$= k\left[\left(\frac{145.83 - 100}{2}\right) + (141.67 - 100) + (120.83 - 100)\right]$$

$$= 85.415\,k \quad (q \text{ in W/m}, k \text{ in W/m} \cdot \text{K})$$

Observe that these two different means of solving for q yield identical results. This is obviously a requirement of the analysis and serves as a check on the formulation and numerical work.

The example may now be concluded. The total heat transfer per meter of duct is calculated as

$$q = 8\,(85.415 \text{ K})(1.21 \text{ W/m} \cdot \text{K})$$

$$= 826.8 \text{ W/m}$$

Example 5 has illustrated, in simple fashion, the numerical approach to solving two dimensional steady-state conduction problems. It is apparent that any added complexity in the form of more involved geometry; other types of boundary conditions such as convection, radiation, specified heat flux, etc.; or simply a greater number of interior nodes; will render a problem too complex for hand calculation. Techniques for formulating such problems and some solution techniques are described by Welty.* A technique known as "relaxation" has been used in past years, however digital computers now comprise the most convenient and fastest means for solving such problems.

In this section, we have considered four techniques for solving two- and three-dimensional steady-state conduction problems. Each of these approaches has certain requirements which limit their use. The analytical solution is recommended for problems of simple geometrical shapes and simple boundary conditions. Systems with complex geometry but with isothermal boundaries may be easily treated by the graphical solutions involving flux plotting. Numerical techniques may be used to solve complex problems involving nonuniform boundary conditions and variable physical properties. The experimental analogical approach may also be used for complex problems. With an analog field plotter, solutions can be rapidly obtained for many varying conditions.

17.5 CLOSURE

In this chapter, we have considered solutions to steady-state conduction problems. The defining differential equations were frequently established by generating the equation through the use of the control-volume expression for the conservation of energy as well as by using the general differential equation for energy transfer. It is hoped that this approach will provide the student with an insight into the various terms contained in the general differential equation and thus enable one to decide, for each situation, which terms are relevant.

One-dimensional systems with and without internal generation of energy were considered. Analytical, graphical flux plotting, analogical and numerical solutions were discussed as techniques used in solving two- and three-dimensional conduction problems.

PROBLEMS

17.1 A composite wall is to be constructed of $\frac{1}{4}$ in. of stainless steel ($k = 10$ Btu/hr ft °F), 3 in. of corkboard ($k = 0.025$ Btu/hr ft °F), and $\frac{1}{2}$ in. of plastic ($k = 1.5$ Btu/hr ft °F). Determine the thermal resistance of this wall if it is bolted together by $\frac{1}{2}$-in.-diameter bolts on 6-in. centers made of
(a) stainless steel;
(b) aluminum ($k = 120$ Btu/hr ft °F).

* J. R. Welty, op. cit.

17.2 It is desired to transport liquid metal through a pipe imbedded in a wall at a point where the temperature is 650 K. A 1.2-m thick wall constructed of a material having a thermal conductivity varying with temperature according to $k = 0.073$ $(1 + 0.0054$ $T)$, where T is in K and k is in W/m · K, has its inside surface maintained at 925 K. The outside surface is exposed to air at 300 K with a convective heat-transfer coefficient of 23 W/m^2 · K. How far from the hot surface should the pipe be located? What is the heat flux for the wall?

17.3 A 2-in. schedule-40 steel pipe carries saturated steam at 60 psi through a laboratory which is 60 ft long. The pipe is insulated with 1.5 in. of 85% magnesia which costs $0.75 per foot. How long must the steam line be in service to justify the insulation cost if the heating cost for the steam is $0.68 per 10^5 Btu? The outside-surface convective heat-transfer coefficient may be taken as 5 Btu/hr ft^2 °F.

17.4 A furnace wall is to be designed to transmit a maximum heat flux of 200 Btu/hr ft^2 of wall area. The inside and outside wall temperatures are to be 2000°F and 300°F, respectively. Determine the most economical arrangement of bricks measuring 9 by $4\frac{1}{2}$ by 3 in. if they are made from two materials, one with a k of 0.44 Btu/hr ft °F and a maximum usable temperature of 1500°F and the other with a k of 0.94 Btu/hr ft °F and a maximum usable temperature of 2200°F. Bricks made of each material cost the same amount and may be laid in any manner.

17.5 Determine the percent increase in heat flux if, in addition to the conditions specified in problem 17.4, there are two $\frac{3}{4}$-in.-diameter steel bolts extending through the wall per square foot of wall area (k for steel = 22 Btu/hr ft °F).

17.6 A 2.5-cm thick sheet of plastic ($k = 2.42$ W/m · K) is to be bonded to a 5-cm thick aluminum plate. The glue which will accomplish the bonding is to be held at a temperature of 325 K to achieve the best adherence, and the heat to accomplish this bonding is to be provided by a radiant source. The convective heat-transfer coefficient on the outside surfaces of both the plastic and aluminum is 12 W/m^2 · K, and the surrounding air is at 295 K. What is the required heat flux if it is applied to the surface of (a) the plastic? (b) the aluminum?

17.7 Saturated steam at 40 psia flows at 5 fps through a schedule-40, $1\frac{1}{2}$-in. steel pipe. The convective heat-transfer coefficient for condensing steam on the inside surface may be taken as 1500 Btu/hr ft^2 °F. The surrounding air is at 80°F, and the outside surface coefficient is 3 Btu/hr ft^2 °F. Determine the following:
(a) The heat loss per 10 ft of bare pipe.

(b) The heat loss per 10 ft of pipe insulated with 2 in. of 85% magnesia.
(c) The mass of steam condensed in 10 ft of bare pipe.

17.8 The steady-state expression for heat conduction through a plane wall is $q = (kA/L)\,\Delta T$ as given by equation (17-4). For steady-state heat conduction through a hollow cylinder, an expression similar to equation (17-4) is

$$q = \frac{k\bar{A}}{r_o - r_i}\,\Delta T$$

where $\bar{A}$ is the "log-mean" area defined as

$$\bar{A} = 2\pi\frac{r_o - r_i}{\ln{(r_o/r_i)}}$$

(a) Show that $\bar{A}$ as defined above satisfies the equations for steady-state radial heat transfer in a hollow cylindrical element.
(b) If the arithmetic mean area, $\pi(r_o + r_i)$, is used rather than the logarithmic mean, calculate the resulting percent error for values of r_o/r_i of 1.5, 3, and 5.

17.9 Evaluate the appropriate "mean" area for steady-state heat conduction in a hollow sphere which satisfies an equation of the form

$$q = \frac{k\bar{A}}{r_o - r_i}\,\Delta T$$

Repeat part (b) of problem 17.8 for the spherical case.

17.10 A furnace wall consisting of 0.25 m of fire clay brick, 0.20 m of kaolin, and a 0.10-m outer layer of masonry brick is exposed to furnace gas at 1370 K with air at 300 K adjacent to the outside wall. The inside and outside convective heat transfer coefficients are 115 and 23 W/m$^2 \cdot$ K, respectively. Determine the heat loss per square foot of wall and the temperature of the outside wall surface under these conditions.

17.11 Given the conditions of problem 17.10, except that the outside temperature of the masonry brick cannot exceed 325 K, by how much must the thickness of kaolin be adjusted to satisfy this requirement?

17.12 A 10-kW heater using Nichrome wire is to be designed. The surface of the Nichrome is to be limited to a maximum temperature of 1650 K. Other design criteria for the heater are:

minimum convective heat-transfer coefficient: 850 W/m$^2 \cdot$ K
minimum temperature of the surrounding medium (air): 370 K

The resistivity of Nichrome is 110 $\mu\Omega$-cm and the power to the heater is available at 12 volts.
(a) What size wire is required if the heater is to be in one piece 0.6 m long?

(b) What length of 14-gage wire is necessary to satisfy these design criteria?

(c) How will the answers to parts (a) and (b) change if $h = 1150$ W/m$^2 \cdot$ K?

17.13 A heater composed of Nichrome wire wound back and forth and closely spaced is covered on both sides with a $\frac{1}{8}$-in. thickness of asbestos ($k = 0.15$ Btu/hr ft °F) and then with a $\frac{1}{8}$-in. thickness of stainless steel ($k = 10$ Btu/hr ft °F). If the center temperature of this sandwich construction is considered constant at 1000°F and the outside convective heat-transfer coefficient is 3 Btu/hr ft^2 °F, how much energy must be supplied in W/ft^2 to the heater? What will be the outside temperature of the stainless steel?

17.14 Liquid nitrogen at 77 K is stored in an insulated spherical container that is vented to the atmosphere. The container is made of a thin-walled material with an outside diameter of 0.5 m; 25 mm of insulation ($k = 0.002$ W/m $\cdot$ K) covers its outside surface. The latent heat of nirogen is 200 kJ/kg; its density, in the liquid phase, is 804 kg/m^3. For surroundings at 25°C and with a convective coefficient of 18 W/m$^2 \cdot$ K at the outside surface of the insulation, what will be the rate of liquid nitrogen boil-off?

17.15 What additional thickness of insulation will be necessary to reduce the boil-off rate of liquid nitrogen to one-half of the rate corresponding to the conditions of problem 17.14? All values and dimensions specified in problem 17.14 apply.

17.16 A 1-in.-OD steel tube has its outside wall surface maintained at 250°F. It is proposed to increase the rate of heat transfer by adding fins of $\frac{3}{32}$-in. thickness and $\frac{3}{4}$ in. long to the outside tube surface. Compare the increase in heat transfer achieved by adding 12 longitudinal straight fins or circular fins with the same total surface area as the 12 longitudinal fins. The surrounding air is at 80°F, and the convective heat-transfer coefficient is 6 Btu/hr ft^2 °F.

17.17 Solve the previous problem if the convective heat-transfer coefficient is increased to 60 Btu/hr ft^2 °F by forcing air past the tube surface.

17.18 A cylindrical rod 3 cm in diameter is partially inserted into a furnace with one end exposed to the surrounding air, which is at 300 K. The temperatures measured at two points 7.6 cm apart are 399 K and 365 K, respectively. If the convective heat-transfer coefficient is 17 W/m$^2 \cdot$ K, determine the thermal conductivity of the rod material.

17.19 Heat is to be transferred from water to air through an aluminum wall. It is proposed to add rectangular fins 0.05 in. thick and $\frac{3}{4}$ in. long spaced 0.08 in. apart to the aluminum surface to aid in transferring heat. The heat-transfer coefficients on the air and water sides are 3 Btu/hr ft^2 °F and 25 Btu/hr ft^2 °F, respectively. Evaluate the percent increase in heat transfer if these fins are added to (a) the airside; (b) the waterside; (c) both sides. What conclusons may be reached regarding this result?

17.20 An iron bar used for a chimney support is exposed to hot gases at 625 K with the associated convective heat-transfer coefficient of 740 W/m$^2 \cdot$ K. The bar is attached to two opposing chimney walls, which are at 480 K. The bar is 1.9 cm in diameter and 45 cm long. Determine the maximum temperature in the bar.

17.21 A copper rod $\frac{1}{4}$ in. in diameter and 3 ft. long runs between two bus bars, which are at 60°F. The surrounding air is at 60°F, and the convective heat transfer coefficient is 6 Btu/hr ft^2 °F. Assuming the electrical resistivity of copper to be constant at 1.72×10^{-6} ohm-cm, determine the maximum current the copper may carry if its temperature is to remain below 150°F.

17.22 A 13 cm by 13 cm steel angle with the dimensions shown is attached to a wall with a surface temperature of 600 K. The surrounding air is at 300 K, and the convective heat transfer coefficient between the angle surface and the air is 45 W/m$^2 \cdot$ K.

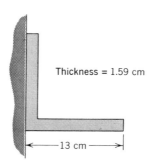

(a) Plot the temperature profile in the angle, assuming a negligible temperature drop through the side of the angle attached to the wall.
(b) Determine the heat loss from the sides of the angle projecting out from the wall.

17.23 A steel I-beam with a cross-sectional area as shown has its lower and upper surfaces maintained at 700 K and 370 K, respectively.
(a) Assuming a negligible temperature change through both flanges, develop an expression for the temperature variation in the web as a function of the distance from the upper flange.

(b) Plot the temperature profile in the web if the convective heat-transfer coefficient between the steel surface and the surrounding air is 57 W/m² · K. The air temperature is 300 K.

(c) What is the net heat transfer at the upper and lower ends of the web?

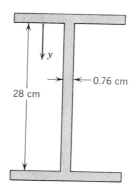

17.24 Repeat problem 17.23 for the case of an aluminum beam.

17.25 A 2-in.-OD stainless-steel tube has 16 longitudinal fins spaced around its outside surface as shown. The fins are $\frac{1}{16}$ in. thick and extend 1 in. from the outside surface of the tube.

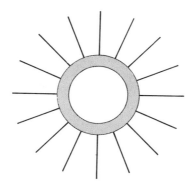

(a) If the outside surface of the tube wall is at 250°F, the surrounding air is at 80°F, and the convective heat-transfer coefficient is 8 Btu/hr ft² °F, determine the percent increase in heat transfer for the finned pipe over that for the unfinned pipe.

(b) Determine the same information as in part (a) for values of h of 2, 5,

15, 50, and 100 Btu/hr ft^2 °F. Plot the percent increase in q versus h. What conclusions can be reached concerning this plot?

17.26 Repeat problem 17.25 for the case of an aluminum pipe-and-fin arrangement.

17.27 The temperatures at the inner and outer surfaces of a plane wall of thickness L are held at the constant values T_0 and T_L, respectively, where $T_0 > T_L$. The wall material has a thermal conductivity which varies linearly according to $k = k_0(1 + \beta T)$, k_0 and β being constants. At what position will the actual temperature profile differ most from that which would exist in the case of constant thermal conductivity?

17.28 Solve problem 17.27 for the case of a hollow cylinder with boundary conditions $T = T_0$ at $r = R_0$ and $T = T_L$ at $r = R_0 + L$.

17.29 A copper bus bar measuring 5 cm by 10 cm by 2.5 m long is in a room in which the air is maintained at 300 K. The bus bar is supported by two plastic pedestals to which it is attached by an adhesive. The pedestals are square in cross section, measuring 8 cm on a side. The pedestals are mounted on a wall whose temperature is 300 K. If 1 kW of energy is

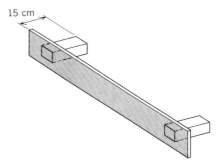

dissipated in the copper bar, what will be its equilibrium temperature? The convective heat-transfer coefficient for all surfaces may be taken as 23 W/m^2 · K. The thermal conductivity of the plastic is 2.6 W/m · K. Neglect thermal radiation.

17.30 Solve the previous problem if each plastic pedestal has a 1.9 cm steel bolt running through the center.

17.31 Copper wire having a diameter of $\frac{3}{16}$ in. is insulated with a 4-in. layer of material having a thermal conductivity of 0.14 Btu/hr ft °F. The outer surface of the insulation is maintained at 70°F. How much current may pass through the wire if the insulation temperature is limited to a maximum of 120°F? The resistivity of copper is 1.72×10^{-6} ohm-cm.

17.32 What would be the result for problem 17.31 if the fluid surrounding the insulated wire were maintained at 70°F with a convective heat-transfer coefficient between the insulation and the fluid of 4 Btu/hr ft^2 °F? What would be the surface temperature of the insulation under these conditions?

17.33 Work problem 17.31 for the case of aluminum rather than copper. The resistivity of aluminum is 2.83×10^{-6} ohm-cm.

17.34 Find the rate of heat transfer from a 3-in.-OD pipe placed eccentrically inside a 6-in.-ID cylinder with the axis of the smaller pipe displaced 1 in. from the axis of the large cylinder. The space between the cylindrical surfaces is filled with rock wool ($k = 0.023$ Btu/hr ft °F). The surface temperatures at the inside and outside surfaces are 400°F and 100°F, respectively.

17.35 Two 8 cm-diameter pipes have their outside surfaces at 590 K and 370 K, respectively. They are aligned with their axes parallel, and their centers are both 16 cm below the surface of a concrete slab ($k = 18$ W/m · K) whose surface is at 300 K. Determine the heat transfer between the pipes per 30 m of length.

17.36 A tunnel measuring 3 ft in width by 6 ft high is dug in permafrost ($k = 0.06$ Btu/hr ft °F) with the top of the tunnel 2 ft below the surface. Determine the heat loss to the surface if the tunnel walls are at 40°F and the surface of the permafrost is at −60°F. Compare this result with that obtained for a buried cylinder with a diameter of 5 ft having its axis at a depth of 4 ft.

17.37 Determine the heat flow per foot for the configuration shown, using the numerical procedure for a grid size of $1\frac{1}{2}$ ft. The material has a thermal

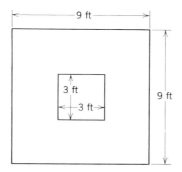

conductivity of 0.15 Btu/hr ft °F. The inside and outside temperatures are at the uniform values of 200°F and 0°F, respectively.

17.38 Repeat the previous problem, using a grid size of 1 ft.

17.39 A 5-in. standard steel angle is attached to a wall with a surface temperature of 600°F. The angle supports a 4.375-in. by 4.375-in. section of building

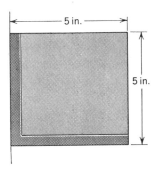

brick whose mean thermal conductivity may be taken as 0.38 Btu/hr ft °F. The convective heat-transfer coefficient between all surfaces and the surrounding air is 8 Btu/hr ft^2 °F. The air temperature is 80°F. Using numerical methods, determine
(a) the total heat loss to the surrounding air;
(b) the location and value of the minimum temperature in the brick.

17.40 A Calrod heating element ($k = 117$ Btu/hr ft °F) with a diameter of 0.496 in. has a heated length of 12 in. This heating element is encased in the center of an aluminum block 1 ft long and 3 in. square. The temperature at the aluminum-Calrod interface is 600°F, and the outside aluminum surface is at 200°F. Determine the heat loss if end effects are neglected.

17.41 Solve problem 17.40 if all conditions remain the same except that, because of an error in construction, the heater is cast with its center displaced $\frac{1}{2}$ in. off the center of the aluminum.

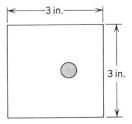

17.42 Saturated steam at 400°F is transported through the 1-ft pipe shown in the figure, which may be assumed to be at the steam temperature. The pipe is centered in the 2-ft-square duct, whose surface is at 100°F. If the space between the pipe and duct is filled with powdered 85% magnesia insulation, how much steam will condense in a 50-ft length of pipe?

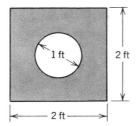

17.43 A 32.4 cm-OD pipe, 145 cm long, is buried with its centerline 1.2 m below the surface of the ground. The ground surface is at 280 K and the mean thermal conductivity of the soil is 0.66 W/m · K. If the pipe surface is at 370 K, what is the heat loss per day from the pipe?

18
UNSTEADY-STATE CONDUCTION

Transient processes, in which the temperature at a given point varies with time, will be considered in this chapter. Since the transfer of energy is directly related to the temperature gradient, these processes involve an *unsteady-state* flux of energy.

Transient conduction processes are commonly encountered in engineering design. These design problems generally fall into two categories: the process which ultimately reaches steady-state conditions, and the process which is operated a relatively short time in a continually changing temperature environment. Examples of this second category would include metal stock or ingots undergoing heat treatment and missile components during re-entry into the earth's atmosphere.

In this chapter, we shall consider problems and their solutions which deal with unsteady-state heat transfer within systems both with and without internal energy sources.

18.1 ANALYTICAL SOLUTIONS

The solution of an unsteady-state conduction problem is, in general, more difficult than that for a steady-state problem because of the dependence of temperature on both time and position. The solution is approached by establishing the defining differential equation and the boundary conditions. In addition, the initial temperature distribution in the conducting medium must be known. By finding the solution to the partial differential equation which satisfies the initial and boundary conditions, the variation in the temperature distribution with time is established, and the flux of energy at a specific time can then be evaluated.

In heating or cooling a conducting medium, the rate of energy transfer is dependent upon both the internal and surface resistances, the limiting cases being represented either by negligible internal resistance or by negligible surface resistance. Both of these cases will be considered, as well as the more general case in which both resistances are important.

LUMPED PARAMETER ANALYSIS—SYSTEMS WITH NEGLIGIBLE INTERNAL RESISTANCE

Equation (16-17) will be the starting point for transient conduction analysis; it is repeated below for reference.

$$\frac{\partial T}{\partial t} = \alpha \nabla^2 T + \frac{\dot{q}}{\rho c_p} \qquad (16\text{-}17)$$

Recall that, in the derivation of this expression, thermal properties were taken to be independent of position and time, however the rate of internal generation, $\dot{q}$, can vary in both.

It is frequently the case that temperature within a medium varies significantly in fewer than all three space variables. A circular cylinder, heated at one end with a fixed boundary condition, will show a temperature variation in the axial and radial directions as well as time. If the cylinder has a length which is large compared to its diameter or, if it is composed of a material with high thermal conductivity, temperature will vary with axial position and time only. If a metallic specimen, initially with uniform temperature, is suddenly exposed to surroundings at a different temperature, it may be that size, shape, and thermal conductivity may combine in such a way that the temperature within the material varies with time only, that is, is not a significant function of position. These conditions are characteristic of a "lumped" system, where the temperature of a body varies only with time; this case is the easiest of all to analyze. Because of this we will consider, as our first transient conduction case, that of a completely lumped-parameter system.

Shown in Figure 18.1 we have a spherical metallic specimen, initially at uniform temperature, T_0, after it has been immersed in a hot oil at temperature, T_∞, for a period of time t. It is presumed that the temperature of the metallic sphere

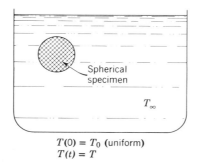

$$T(0) = T_0 \text{ (uniform)}$$
$$T(t) = T$$

Figure 18.1

is uniform at any given time. A first-law analysis using equation (6-10), applied to a spherical control volume coinciding with the specimen in question will reduce to

$$\frac{\delta Q}{dt} = \frac{\partial}{\partial t} \int\int\int_{c.v.} e\rho \, dV \qquad (18\text{-}1)$$

The rate of heat addition to the control volume, $\delta Q/dt$, is due to convection from the oil and is written as

$$\frac{\delta Q}{dt} = hA(T_\infty - T) \qquad (18\text{-}2)$$

The rate of energy increase within the specimen, $\partial/\partial t \iiint_{c.v.} e\rho \, dV$, with constant properties, may be expressed as

$$\frac{\partial}{\partial t} \int\int\int_{c.v.} e\rho \, dV = \rho V c_p \frac{dT}{dt} \qquad (18\text{-}3)$$

Equating these expressions as indicated by equation (18-1) we have, with slight rearrangement,

$$\frac{dT}{dt} = \frac{hA(T_\infty - T)}{\rho V c_p} \qquad (18\text{-}4)$$

We may now obtain a solution for the temperature variation with time by solving equation (18-4) subject to the initial condition, $T(0) = T_0$, and obtain

$$\frac{T - T_\infty}{T_0 - T_\infty} = e^{-hAt/\rho c_p V} \qquad (18\text{-}5)$$

The exponent is observed to be dimensionless. A rearrangement of terms in the exponent may be accomplished as follows

$$\frac{hAt}{\rho c_p V} = \left(\frac{hV}{kA}\right)\left(\frac{A^2 k}{\rho V^2 c_p} t\right) = \left(\frac{hV/A}{k}\right)\left[\frac{\alpha t}{(V/A)^2}\right] \qquad (18\text{-}6)$$

Each of the bracketed terms in equation (18-6) is dimensionless. The ratio, V/A, having units of length, is also seen to be a part of each of these new parametric forms. The first of the new nondimensional parameters formed is the *Biot modulus*, abbreviated Bi,

$$\text{Bi} = \frac{hV/A}{k} \qquad (18\text{-}7)$$

By analogy with the concepts of thermal resistance, discussed at length earlier, the Biot modulus is seen to be the ratio of $(V/A)/k$, the conductive (internal) resistance to heat transfer, to $1/h$, the convective (external) resistance to heat transfer. The magnitude of Bi thus has some physical significance in relating where the greater resistance to heat transfer occurs. A large value of Bi indicates that the conductive resistance controls, that is, there is more capacity for heat to leave the surface by convection than to reach it by conduction. A small value for Bi represents the case where internal resistance is negligibly small and there is more capacity to transfer heat by conduction than there is by convection. In this latter case the controlling heat transfer phenomenon is convection, and temperature gradients within the medium are quite small. An extremely small internal temperature gradient is the basic assumption in a lumped-parameter analysis.

A natural conclusion to the foregoing discussion is that the magnitude of the Biot modulus is a reasonable measure of the likely accuracy of a lumped-parameter analysis. A commonly used rule of thumb is that the error inherent in a lumped-parameter analysis will be less than 5% for a value of Bi less than 0.1. The evaluation of the Biot modulus should thus be the first thing done when analyzing an unsteady-state conduction situation.

The other bracketed term in equation (18-6) is the *Fourier modulus,* abbreviated Fo, where

$$\text{Fo} = \frac{\alpha t}{(V/A)^2} \tag{18-8}$$

The Fourier modulus is frequently used as a nondimensional time parameter.

The lumped-parameter solution for transient conduction may now be written as

$$\frac{T - T_\infty}{T_0 - T_\infty} = e^{-\text{BiFo}} \tag{18-9}$$

Equation (18-9) is portrayed graphically in Figure 18.2. The use of equation (18-9) is illustrated in the following example.

EXAMPLE 1

A long copper wire, $\frac{1}{4}$-in. in diameter, was held in an air stream with a temperature $T_\infty = 100°F$. After 30 s, the average temperature of the wire increased from 50°F to 80°F. Make an estimate of the average unit surface conductance, h.

In order to determine whether or not equation (18-5) may be used the Biot modulus must be evaluated. The value of Bi is calculated as

$$\text{Bi} = \frac{hV/A}{k} = \frac{h\left(\dfrac{\pi D^2 L}{4\pi DL}\right)}{223\ \text{Btu/hr ft °F}} = \frac{h(1/4 \times 1/12\ \text{ft})}{4(223\ \text{Btu/hr ft °F})} = 2.34 \times 10^{-5} h$$

Setting $\text{Bi} = 0.1$ which is the limiting value of Bi for a lumped-parameter analysis to be valid, and solving for h we have

$$h = 0.1/2.34 \times 10^{-5} = 4460\ \text{Btu/hr ft}^2\ \text{°F}$$

Thus, a lumped-parameter analysis will be sufficiently accurate so long as $h <$ 4460 Btu/hr ft °F. Proceeding to solve equation (18-5) for h we obtain

$$h = \frac{\rho c_p V}{tA} \ln \frac{T_0 - T_\infty}{T - T_\infty}$$

$$= \frac{(555\ \text{lb}_\text{m}/\text{ft}^3)(0.092\ \text{Btu/lb}_\text{m}\text{°F})}{30/3600\ \text{hr}} \left(\frac{\pi D^2 L}{4\pi DL}\ \text{ft}\right) \ln \frac{50 - 100}{80 - 100}$$

$$= 29.2\ \text{Btu/hr ft}^2\ \text{°F} \qquad (166\ \text{W/m}^2 \cdot \text{K})$$

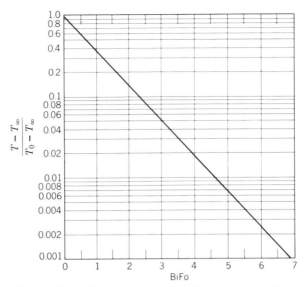

Figure 18.2 Time-temperature history of a body at initial temperature, T_0, exposed to an environment at T_∞; lumped-parameter case.

Heating a Body Under Conditions of Negligible Surface Resistance. A second class of time-dependent energy-transfer processes is encountered when the surface resistance is small relative to the overall resistance, that is, Bi is $\gg 0.1$. For this process, the temperature of the surface, T_s, is constant for all time, $t > 0$, and its value is essentially equal to the ambient temperature, T_∞.

To illustrate the analytical method of solving this class of transient heat-conduction problems, consider a large flat plate of uniform thickness L. The initial temperature distribution through the plate will be assumed to be an arbitrary function of x. The solution for the temperature history must satisfy the Fourier field equation,

$$\frac{\partial T}{\partial t} = \alpha \nabla^2 T \tag{16-18}$$

or for one-directional energy flow,

$$\frac{\partial T}{\partial t} = \alpha \frac{\partial^2 T}{\partial x^2} \tag{18-10}$$

and the initial and boundary conditions

$$T = T_0(x) \qquad \text{at } t = 0 \qquad \text{for } 0 \le x \le L$$

$$T = T_s \qquad \text{at } x = 0 \qquad \text{for } t > 0$$

and

$$T = T_s \qquad \text{at } x = L \qquad \text{for } t > 0$$

For convenience, let $Y = (T - T_s)/(T_0 - T_s)$, where T_0 is an arbitrarily chosen reference temperature; the partial differential equation may be rewritten in terms of the new temperature variable as

$$\frac{\partial Y}{\partial t} = \alpha \frac{\partial^2 Y}{\partial x^2} \tag{18-11}$$

and the initial and boundary conditions become

$$Y = Y_0(x) \qquad \text{at } t = 0 \qquad \text{for } 0 \leq x \leq L$$

$$Y = 0 \qquad \text{at } x = 0 \qquad \text{for } t > 0$$

and

$$Y = 0 \qquad \text{at } x = L \qquad \text{for } t > 0$$

Solving equation (18-11) by the method of separation of variables leads to product solutions of the form

$$Y = (C_1 \cos \lambda x + C_2 \sin \lambda x) e^{-\alpha \lambda^2 t}$$

The constants C_1 and C_2 and the parameter λ are obtained by applying the initial and boundary conditions. The complete solution is

$$Y = \frac{2}{L} \sum_{n=1}^{\infty} \sin\left(\frac{n\pi}{L}x\right) e^{-(n\pi/2)^2 \text{Fo}} \int_0^L Y_0(x) \sin\frac{n\pi}{L} x \, dx \tag{18-12}$$

where $\text{Fo} = \alpha t/(L/2)^2$. Equation (18-12) points out the necessity for knowing the initial temperature distribution in the conducting medium, $Y_0(x)$, before the complete temperature history may be evaluated. Consider the special case in which the conducting body has a uniform initial temperature, $Y_0(x) = Y_0$. With this temperature distribution, equation (18-12) reduces to

$$\frac{T - T_s}{T_0 - T_s} = \frac{4}{\pi} \sum_{n=1}^{\infty} \frac{1}{n} \sin\left(\frac{n\pi}{L}x\right) e^{-(n\pi/2)^2 \text{Fo}} \qquad n = 1, 3, 5, \ldots \tag{18-13}$$

The temperature history at the center of the infinite plane, as well as the central temperature history in other solids, is illustrated in Figure 18.3. The central temperature history for the plane wall, infinite cylinder, and sphere is presented in Appendix F, in "Heissler charts." These charts cover a much greater range in the Fourier modulus than Figure 18.3.

The heat rate, q, at any plane in the conducting medium may be evaluated by

$$q_x = -kA\frac{\partial T}{\partial x} \tag{18-14}$$

In the case of the infinite flat plate with an initial uniform temperature distribution of T_0, the heat rate at any time t is

$$q_x = 4\left(\frac{kA}{L}\right)(T_s - T_0) \sum_{n=1}^{\infty} \cos\left(\frac{n\pi}{L}x\right) e^{-(n\pi/2)^2 \text{Fo}} \qquad n = 1, 3, 5, \ldots \tag{18-15}$$

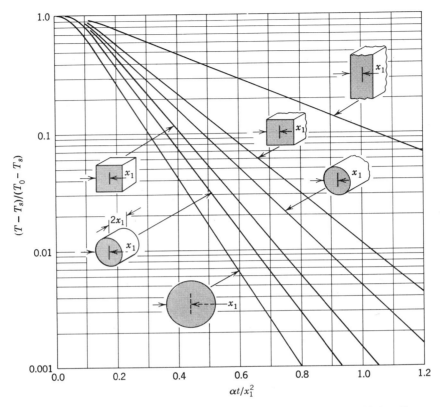

Figure 18.3 Central temperature history of various solids with initial uniform temperature, T_0, and constant surface temperature, T_s. (From P. J. Schneider, *Conduction Heat Transfer*, Addison-Wesley Publishing Co., Inc., Reading, Mass., 1955, p. 249. By permission of the publishers.)

In the following example, the use of the central temperature-history figure will be illustrated.

EXAMPLE 2

A concrete cylinder, 0.1 m in length and 0.1 m in diameter, is initially at room temperature, 292 K. It is suspended in a steam environment where water vapor at 373 condenses on all surfaces with an effective film coefficient, h, of 8500 W/m^2 · K. Determine the time required for the center of this stubby cylinder to reach 310 K. If the cylinder were sufficiently long so that it could be considered infinite, how long would it take?

For the first case, the finite cylinder, the Biot number is evaluated as

$$\text{Bi} = \frac{h(V/A)}{k} = \frac{h\left(\dfrac{\pi D^2 L}{4}\right)}{k\left(\pi DL + \dfrac{\pi D^2}{2}\right)} = \frac{h(DL/4)}{k(L + D/2)}$$

$$= \frac{(8500 \text{ W/m}^2 \cdot \text{K})(0.1 \text{ m})(0.1 \text{ m})/4}{1.21 \text{ W/m} \cdot \text{K}(0.1 + 0.1/2) \text{ m}}$$

$$= 117$$

For this large value, Figure 18.3 may be used. The second line from the bottom in this figure applies to a cylinder with height equal to diameter, as in this case. The ordinate is

$$\frac{T - T_s}{T_0 - T_s} = \frac{310 - 373}{292 - 373} = 0.778$$

and the corresponding abscissa value is approximately 0.11. The time required may now be determined as

$$\frac{\alpha t}{x_1{}^2} = 0.11$$

Thus,

$$t = 0.11 \frac{(0.05\text{m})^2}{5.95 \times 10^{-7} \text{ m}^2/\text{s}} = 462 \text{ s}$$

$$= 7.7 \text{ min}$$

In the case of an infinitely long cylinder, the fourth line from the bottom applies. The Biot number in this case is

$$\text{Bi} = \frac{h(V/A)}{k} = \frac{h\left(\dfrac{\pi D^2 L}{4}\right)}{k(\pi DL)} = \frac{h\dfrac{D}{4}}{k}$$

$$= \frac{(8500 \text{ W/m}^2 \cdot \text{K})(0.1 \text{ m})/4}{1.21 \text{ W/m} \cdot \text{K}} = 176$$

which is even larger than the finite cylinder case. Figure 18.3 will again be used. The ordinate value of 0.778 yields, for the abscissa, a value of approximately 0.13. The required time, in this case, is

$$t = \frac{0.13(0.05 \text{ m})^2}{5.95 \times 10^{-7} \text{ m}^2/\text{s}} = 546 \text{ s}$$

$$= 9.1 \text{ min}$$

Heating a Body with Finite Surface and Internal Resistances. The most general cases of transient heat-conduction processes involve significant values of internal and surface resistances. The solution for the temperature history without internal generation must satisfy the Fourier field equation, which may be expressed for one-dimensional heat flow by

$$\frac{\partial T}{\partial t} = \alpha \frac{\partial^2 T}{\partial x^2} \tag{18-7}$$

A case of considerable practical interest is one in which a body having a uniform temperature is placed in a new fluid environment with its surfaces suddenly and simultaneously exposed to the fluid at temperature, T_∞. In this case, the temperature history must satisfy the initial, symmetry and convective boundary conditions

$$T = T_0 \qquad\qquad \text{at } t = 0$$

$$\frac{\partial T}{\partial x} = 0 \qquad\qquad \text{at the centerline of the body}$$

and

$$-\frac{\partial T}{\partial x} = \frac{h}{k}(T - T_\infty) \qquad\qquad \text{at the surface}$$

One method of solution for this class of problems involves separation of variables, which results in product solutions as previously encountered when only the internal resistance was involved.

Solutions to this case of time-dependent energy-transfer processes have been obtained for many geometries. Excellent treatises discussing these solutions have been written by Carslaw and Jaeger[*] and by Ingersoll, Zobel, and Ingersoll.[†] If we reconsider the infinite flat plate of thickness, $2x_1$, when inserted into a medium at constant temperature, T_∞, but now include a constant surface conductance, h, the following solution is obtained

$$\frac{T - T_\infty}{T_0 - T_\infty} = 2 \sum_{n=1}^{\infty} \frac{\sin \delta_n \cos (\delta_n x / x_1)}{\delta_n + \sin \delta_n \cos \delta_n} e^{-\delta_n^2 \text{Fo}} \tag{18-16}$$

where δ_n is defined by the relation

$$\delta_n \tan \delta_n = \frac{hx_1}{k} \tag{18-17}$$

The temperature history for this relatively simple geometrical shape is a function of three dimensionless quantities: $\alpha t / x_1^2$, hx_1/k, and the relative distance, x/x_1. The evaluation of the temperature profile from this analytical equation is very time-consuming.

[*] H. S. Carslaw and J. C. Jaeger, *Conduction of Heat in Solids*, Oxford University Press, 1947.
[†] L. R. Ingersoll, O. J. Zobel, and A. C. Ingersoll, *Heat Conduction (With Engineering and Geological Applications)*, McGraw-Hill Book Company, New York, 1948.

Heat Transfer to a Semi-Infinite Wall. An analytical solution to the one-dimensional heat-conduction equation for the case of the semi-infinite wall has some utility in engineering computations. Consider the situation illustrated in Figure 18.4. A large plane wall initially at a constant temperature T_0 is subjected

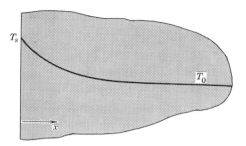

Figure 18.4   Temperature distribution in a semi-infinite wall at time t.

to a surface temperature T_s, where $T_s > T_0$. The differential equation to be solved is

$$\frac{\partial T}{\partial t} = \alpha \frac{\partial^2 T}{\partial x^2} \tag{18-10}$$

and the initial and boundary conditions are

$$T = T_0 \quad \text{at } t = 0 \qquad \text{for all } x$$
$$T = T_s \quad \text{at } x = 0 \qquad \text{for all } t$$

and

$$T \to T_0 \quad \text{as } x \to \infty \qquad \text{for all } t$$

The solution to this problem may be accomplished in a variety of ways, among which are the Laplace transformation and the Fourier transformation. We shall use an alternative procedure, which is less involved mathematically. The variables in equation (18-10) may be expressed in dimensionless form by analogy with the previous case. Thus we may write

$$\frac{T - T_0}{T_s - T_0} = f\left(\frac{x}{x_1}, \frac{\alpha t}{x_1^2}\right)$$

however, in this problem there is no finite characteristic dimension, x_1, and thus $(T - T_0)/(T_s - T_0) = f(\alpha t / x^2)$, or with equal validity, $(T - T_0)/(T_s - T_0) = f(x/\sqrt{\alpha t})$. If $\eta = x/2\sqrt{\alpha t}$ is selected as the independent variable and the dependent variable $Y = (T - T_0)/(T_s - T_0)$ is used, substitution into equation (18-10) yields the ordinary differential equation

$$d^2 Y/d\eta^2 + 2\eta \, dY/d\eta = 0 \tag{18-18}$$

with the transformed boundary and initial conditions

$$Y \to 0 \qquad \text{as } \eta \to \infty$$

and

$$Y = 1 \qquad \text{at } \eta = 0$$

The first condition above is the same as the initial condition $T = T_0$ at $t = 0$, and the boundary condition $T \to T_0$ as $x \to \infty$. Equation (18-18) may be integrated once to yield

$$\ln \frac{dY}{d\eta} = c_1 - \eta^2$$

or

$$\frac{dY}{d\eta} = c_2 e^{-\eta^2}$$

and integrated once more to yield

$$Y = c_3 + c_2 \int e^{-\eta^2} \, d\eta \tag{18-19}$$

The integral is related to a frequently encountered form called the *error function*, designated "erf," where

$$\operatorname{erf} \phi \equiv \frac{2}{\sqrt{\pi}} \int_0^\phi e^{-\eta^2} \, d\eta$$

and erf $(0) = 0$, erf $(\infty) = 1$. A short table of erf ϕ is given in Appendix L. Applying the boundary conditions to equation (18-19), we obtain

$$Y = 1 - \operatorname{erf} \left(\frac{x}{2\sqrt{\alpha t}} \right)$$

or

$$\frac{T - T_0}{T_s - T_0} = 1 - \operatorname{erf} \left(\frac{x}{2\sqrt{\alpha t}} \right)$$

or

$$\frac{T_s - T}{T_s - T_0} = \operatorname{erf} \left(\frac{x}{2\sqrt{\alpha t}} \right) \tag{18-20}$$

This equation is extremely simple to use and quite valuable.

Consider a finite wall of thickness L subject to the surface temperature T_s. Until the temperature change at $x = L$ exceeds some nominal amount, say

$(T - T_0)/(T_s - T_0)$ equal to 0.5%, the solution for the finite and infinite walls will be the same. The value of $L/(2\sqrt{\alpha t})$ corresponding to a 0.5% change in $(T - T_0)/(T_s - T_0)$ is $L/(2\sqrt{\alpha t}) \simeq 2$, so for $L/(2\sqrt{\alpha t}) > 2$, equation (18-20) may be used for finite geometry with little or no error. For the case of finite surface resistance, the solution to equation (18-10) for a semi-infinite wall is

$$\frac{T_\infty - T}{T_\infty - T_0} = \text{erf}\,\frac{x}{2\sqrt{\alpha t}} + \exp\left(\frac{hx}{k} + \frac{h^2 \alpha t}{k^2}\right)\left[1 - \text{erf}\left(\frac{h\sqrt{\alpha t}}{k} + \frac{x}{2\sqrt{\alpha t}}\right)\right] \quad (18\text{-}21)$$

This equation may be used to determine the temperature distribution in finite bodies for small times in the same manner as equation (18-20). The surface temperature is particularly easy to obtain from the above equation, if we let $x = 0$, and the heat transfer rate may be determined from

$$\frac{q}{A} = h(T_s - T_\infty)$$

18.2 TEMPERATURE-TIME CHARTS FOR SIMPLE GEOMETRIC SHAPES

For unsteady-state energy transfer in several simple shapes with certain restrictive boundary conditions, the equations describing temperature profiles have been evaluated and have been presented in a wide variety of charts to facilitate their use. Two forms of these charts are available in Appendix F.

Solutions are presented in Appendix F for the flat plate, sphere, and long cylinder in terms of four dimensionless ratios:

$$Y, \text{unaccomplished temperature change} = \frac{T_\infty - T}{T_\infty - T_0}$$

$$X, \text{relative time} = \frac{\alpha t}{x_1^2}$$

$$n, \text{relative position} = \frac{x}{x_1}$$

and

$$m, \text{relative resistance} = \frac{k}{hx_1}$$

where x_1 is the radius or semithickness of the conducting medium. These charts may be used to evaluate temperature profiles for cases involving transport of energy into or out of the conducting medium if the following conditions are met:

(a) Fourier's field equation describes the process; i.e., constant thermal diffusivity and no internal heat source.

(b) The conducting medium has a uniform initial temperature, T_0.

(c) The temperature of the boundary or the adjacent fluid is changed to a new value, T_∞, for $t \geq 0$.

For flat plates where the transport takes place from only one of the faces, the relative time, position, and resistance are evaluated as if the thickness were twice the true value.

Although the charts were drawn for one-dimensional transport, they may be combined to yield solutions for two- and three-dimensional problems. The following is a summary of these combined solutions.

    1. For transport in a rectangular bar with insulated ends

$$Y_{\text{bar}} = Y_a Y_b \qquad (18\text{-}22)$$

where Y_a is evaluated with width $x_1 = a$, and Y_b is evaluated with thickness $x_1 = b$.

    2. For transport in a rectangular parallelepiped,

$$Y_{\text{parallelepiped}} = Y_a Y_b Y_c \qquad (18\text{-}23)$$

where Y_a is evaluated with width $x_1 = a$, Y_b is evaluated with thickness $x_1 = b$, and Y_c is evaluated with depth $x_1 = c$.

    3. For transport in a cylinder, including both ends,

$$Y_{\substack{\text{cylinder} \\ \text{plus ends}}} = Y_{\text{cylinder}} Y_a \qquad (18\text{-}24)$$

where Y_a is evaluated by using the flat-plate chart, and thickness $x_1 = a$.

The use of temperature-time charts is demonstrated in the following examples.

EXAMPLE 3

A flat wall of fire-clay brick, 0.5 m thick and originally at 200 K, has one of its faces suddenly exposed to a hot gas at 1200 K. If the heat-transfer coefficient on the hot side is 7.38 W/m² · K and the other face of the wall is insulated so that no heat passes out of that face, determine (a) the time necessary to raise the center of the wall to 600 K; (b) the temperature of the insulated wall face at the time evaluated in (a).

From the table of physical properties given in Appendix H, the following values are listed:

$$k = 1.125 \text{ W/m} \cdot \text{K}$$

$$c_p = 919 \text{ J/kg} \cdot \text{K}$$

$$\rho = 2310 \text{ kg/m}^3$$

and

$$\alpha = 5.30 \times 10^{-7} \text{ m}^2/\text{s}$$

The insulated face limits the energy transfer into the conducting medium to only one direction. This is equivalent to heat transfer from a 1-m thick wall, where x is then measured from the line of symmetry, the insulated face. The relative position, x/x_1, is $1/2$.

The relative resistance, k/hx_1 is $1.125/[(7.38)(0.5)]$ or 0.305. The unaccomplished change, $Y = (T_\infty - T)/(T_\infty - T_0)$, is equal to $(1200 - 600)/(1200 - 200)$, or 0.6. From Figure F.7, in Appendix F, the abscissa, $\alpha t/x_1^2$, is 0.35 under these conditions. The time required to raise the centerline to 600°F is

$$t = \frac{0.35x_1^2}{\alpha} = \frac{0.35(0.5)^2}{5.30 \times 10^{-7}} = 1.651 \times 10^5 \text{ s} \qquad \text{or} \qquad 45.9 \text{ hr}$$

The relative resistance and the relative time for (b) will be the same as in part (a). The relative position, x/x_1, will be 0. Using these values and Figure F.1, Appendix F, we find the unaccomplished change, Y, to be 0.74. Using this value, the desired temperature can be evaluated by

$$\frac{T_s - T}{T_s - T_0} = \frac{1200 - T}{1200 - 200} = 0.74$$

or

$$T = 460 \text{ K} \qquad (368°F)$$

EXAMPLE 4

A billet of steel 1 ft in diameter by 2 ft in length, initially at 700°F, is immersed in an oil bath maintained at 100°F. If the surface conductance is 6 Btu/hr ft^2 °F, determine the temperature at the center of the billet after 1 hr.

From Appendix H, the following average values may be used for the pertinent physical properties in this problem:

$$k = 22 \text{ Btu/hr ft °F}$$

$$c_p = 0.11 \text{ Btu/lb °F}$$

$$\rho = 490 \text{ lb/ft}^3$$

and

$$\alpha = 0.408 \text{ ft}^2/\text{hr}$$

The unaccomplished change will be determined by equation (18-21). To evaluate Y_a, the following dimensionless ratios are used:

$$X = \frac{\alpha t}{x_1^2} = \frac{(0.408)(1)}{(1)^2} = 0.408$$

$$n = \frac{x}{x_1} = 0$$

and

$$m = \frac{k}{hx_1} = \frac{22}{(6)(1)} = 3.67$$

By Figure F.1, we find Y_a to be 0.9. To evaluate Y_{cylinder}, the following dimensionless ratios are used:

$$X = \frac{\alpha t}{x_1^2} = \frac{(0.408)(1)}{(0.5)^2} = 1.63$$

$$n = \frac{x}{x_1} = 0$$

and

$$m = \frac{k}{hx_1} = \frac{22}{(6)(0.5)} = 7.33$$

From Figure F.2, we find $Y_{cylinder}$ to be 0.7. For energy transfer through the cylindrical walls and ends,

$$Y = Y_{cylinder} Y_a,$$

or

$$Y = (0.7)(0.9) = 0.63$$

thus

$$\frac{T_\infty - T}{T_\infty - T_0} = 0.63$$

or

$$\frac{100 - T}{100 - 700} = 0.63$$

and

$$T = 478\,°F \quad (521\text{ K})$$

As illustrated in Example 3, the temperature at any given plane within the conducting medium can be evaluated for any specific time. Once the entire profile is known, the instantaneous heat rate, $(q/A)|_t$ at any plane can be evaluated by finding the slope of the temperature profile at the given plane; substituting this value into the Fourier rate equation

$$\frac{q_x}{A} = -k\frac{\partial T}{\partial x} \tag{15-1}$$

18.3 NUMERICAL METHODS FOR TRANSIENT CONDUCTION ANALYSIS

In many time-dependent or unsteady-state conduction processes actual initial and/or boundary conditions do not correspond to those mentioned earlier with regard to analytical solutions. An initial temperature distribution may be nonuniform in nature; ambient temperature, surface conductance, or system geometry may be variable or quite irregular. For such complex cases, numerical techniques may be used to achieve solutions. More recently, with sophisticated computing codes available, numerical solutions are being obtained for heat transfer problems of all types, and this trend will doubtlessly continue. It is likely that many users of this book will be involved in code development for such analysis.

Some numerical work was introduced in Chapter 17, dealing with two-dimensional, steady-state conduction. In this section we will consider variation in time as well as position.

To begin our discussion, the reader is referred to equation (17-66) and the development leading up to it. For the case of no internal generation of energy, equation (17-66) reduces to

$$k\frac{\Delta y}{\Delta x}(T_{i-1,j} + T_{i+1,j} - 2T_{i,j}) + k\frac{\Delta x}{\Delta y}(T_{i,j-1} + T_{i,j+1} - 2T_{i,j})$$

$$= \left(\frac{\rho c_p T_{i,j}|_{t+\Delta t} - \rho c_p T_{i,j}|_t}{\Delta t}\right) \Delta x \, \Delta y \qquad (18\text{-}25)$$

This expression applies to two dimensions; however, it can be extended easily to three dimensions.

The time-dependent term on the right of equation (18-25) is written such that the temperature at node i, j is presumed known at time t; this equation can then be solved to find $T_{i,j}$ at the end of time interval Δt. Since $T_{i,j|t+\Delta t}$ appears only once in this equation, it can be evaluated quite easily. This means of evaluating $T_{i,j}$ at the end of a time increment is designated an "explicit" technique. A more thorough discussion of explicit solutions is given by Welty.*

Equation (18-25) may be solved to evaluate the temperature at node i, j for all values of i, j that comprise the region of interest. For large numbers of nodes it is clear that a great number of calculations are needed and that much information must be stored for use in subsequent computation. Digital computers obviously provide the only feasible way to accomplish solutions.

We will next consider the one-dimensional form of equation (18-25). For a space increment Δx, the simplified expression becomes

$$\frac{k}{\Delta x}(T_{i-1}|_t + T_{i+1}|_t - 2T_i|_t) = \left(\frac{\rho c_p T_i|_{t+\Delta t} - \rho c_p T_i|_t}{\Delta t}\right) \Delta x \qquad (18\text{-}26)$$

where the j notation has been dropped. The absence of variation in the y direction allows several terms to be deleted. We next consider properties to be constant and represent the ratio $k/\rho c_p$ as α. Solving for $T_{i|t+\Delta t}$, we obtain

$$T_i|_{t+\Delta t} = \frac{\alpha \, \Delta t}{(\Delta x)^2}(T_{i+1}|_t + T_{i-1}|_t) + \left(1 - \frac{2\alpha \, \Delta t}{(\Delta x)^2}\right)T_1|_t \qquad (18\text{-}27)$$

The ratio, $\alpha \, \Delta t/(\Delta x)^2$, a form resembling the Fourier modulus, is seen to arise naturally in this development. This grouping relates the time step, Δt, to the space increment, Δx. The magnitude of this grouping will, quite obviously, have an effect

---

* J. R. Welty, *Engineering Heat Transfer*, SI Edition, Wiley, New York, 1978.

on the solution. It has been determined that equation (18-27) is numerically "stable" when

$$\frac{\alpha \, \Delta t}{(\Delta x)^2} \le \frac{1}{2} \qquad (18\text{-}28)$$

For a discussion of numerical stability the reader is referred to Carnahan et al.*

The choice of a time step involves a trade-off between solution accuracy—a smaller time step will produce greater accuracy—and computation time—a solution will be achieved more rapidly for larger values of Δt. When computing is done by machine, a small time step will likely be used without major difficulty.

An examination of equation (18-27) indicates considerable simplification to be achieved if the equality in equation (18-28) is used. For the case with $\alpha \, \Delta t/(\Delta x^2) = 1/2$, equation (18-27) becomes

$$T_{i|t+\Delta t} = \frac{T_{i+1|t} + T_{i-1|t}}{2} \qquad (18\text{-}29)$$

Equations (18-28) and (18-29) can be used in a graphical solution technique known as the Schmidt plot. Equation (18-29), the solution algorithm, indicates the temperature at node i, after a time increment Δt has elapsed, to be the arithmetic mean of the temperatures at adjacent nodes, Δx units away, at the start of the time interval. The following example illustrates the use of a Schmidt plot to solve a transient conduction problem.

EXAMPLE 5

A thick asbestos slab is initially at a uniform temperature of 100°F. If its surface temperature is raised to—and held at—1500°F, beyond what distance from the heated surface will the temperature remain below 300°F after an hour of elapsed time?

A temperature-vs.-thickness graph is drawn as illustrated in Figure 18.5. Equal spatial intervals of convenient size are drawn in the conducting medium. After one time interval, plane (1) is at temperature $T_1^{\,1}$ as found by drawing a straight line between 1500°F on temperature reference plane (0) and 100°F on temperature reference plane (2). This time interval is designated by ①. During the second time interval, ② the temperature on reference plane (2) is raised to $T_2^{\,2}$, this is again evaluated by drawing a straight line between the temperatures on the two adjacent temperature reference planes. During time interval ③, the temperatures at the two reference planes (1) and (3) are elevated. The graphical evaluation of the temperature profile can continue over $n \, \Delta t$ time intervals. With each construction, the temperature profile across the conducting medium is established for that specific time. The temperature is 300°F at a distance of 4.3 Δx from the heated surface after the 8th time interval. Since the total time involved in this transient heating problem is 1 hr, $\Delta t = \frac{1}{8}$ hr. The thermal diffusivity of asbestos, α, is equal

---

* B. Carnahan, H. A. Lather, and J. O. Wilkes, *Applied Numerical Methods*, Wiley, New York, 1969.

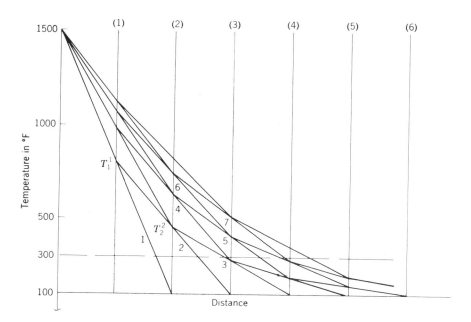

Figure 18.5 Schmidt plot for Example 5.

to $0.01 \text{ ft}^2/\text{hr}$. Substituting these values into equation (18-27), we can determine the length of each spatial interval as

$$\Delta t = \frac{(\Delta x)^2}{2\alpha}$$

$$(\Delta x)^2 = 2\alpha \, \Delta t = (2)(0.01)(\tfrac{1}{8}) = 0.0025 \text{ ft}^2$$

$$\Delta x = 0.05 \text{ ft}$$

The limiting depth of asbestos is at least $4.3 \, \Delta L = 4.3(0.05) = 0.215 \text{ ft} = 2.58 \text{ in}$. As a check on the accuracy of this result, we may compare with the analytical solution given by equation (18-20). Doing so, we obtain

$$\frac{T_s - T}{T_s - T_0} = \frac{1500 - 300}{1500 - 100} = 0.8571$$

$$= \text{erf}\left[\frac{x}{2\sqrt{(0.01 \text{ ft}^2/\text{hr})(1 \text{ hr})}}\right] = \text{erf}(5x)$$

From Appendix L, erf 0.8571 corresponds to an argument of 1.058. The value of x is thus determined as $1.058/5$ or 0.212 ft or 2.54 in.

The Schmidt plot may also be used when the surface temperature is not constant because of convective energy transfer at the surface of the conducting

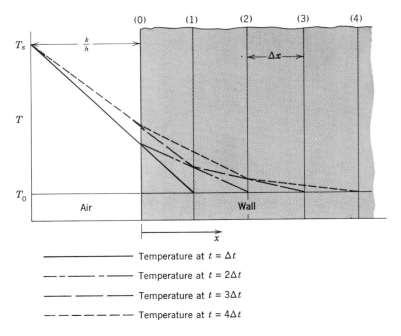

Figure 18.6 Schmidt plot with surface resistance.

medium. The flux of energy at the surface may be described in terms of the rate equation for conduction and convection:

$$(q_x/A)_{\text{surface}} = -k\frac{\partial T}{\partial x}\bigg|_{\text{surface}} = h(T_s - T_0) \tag{18-29}$$

The temperature gradient at the reference plane (0) is

$$-\frac{\partial T}{\partial x}\bigg|_0 = \frac{(T_s - T_0)}{k/h} = \frac{T_s - T_0}{\Delta x^*} \tag{18-30}$$

where Δx^* is the fictitious wall thickness which must be added to the actual wall to account for the resistance at the surface to convective heat transfer. This fictitious length is illustrated in Figure 18.6. If the surface conductance h changes with time, the fictitious distance k/h can be altered after each time interval Δt.

18.4 AN INTEGRAL METHOD FOR ONE-DIMENSIONAL UNSTEADY CONDUCTION

The von Kármán momentum integral approach to the hydrodynamic boundary layer has a counterpart in conduction. Figure 18.7 shows a portion of a

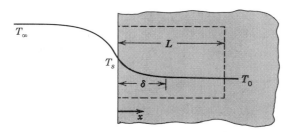

Figure 18.7 A portion of a semi-infinite wall used in integral analysis.

semi-infinite wall, originally at uniform temperature T_0, exposed to a fluid at temperature T_∞, with the surface of the wall at any time at temperature T_s.

At any time t, heat transfer from the fluid to the wall affects the temperature profile within the wall. The "penetration distance," designated δ, is the distance from the surface wherein this effect is manifested. At distance δ the temperature gradient, $\partial T/\partial x$, is taken as zero.

Applying the first law of thermodynamics, equation (6-10), to a control volume extending from $x = 0$ to $x = L$, where $L > \delta$, we have

$$\frac{\delta Q}{dt} - \frac{\delta W_s}{dt} - \frac{\delta W_\mu}{dt} = \iint_{\text{c.s.}} \left(e + \frac{P}{\rho}\right)\rho(\mathbf{v} \cdot \mathbf{n})\, dA + \frac{\partial}{\partial t}\iiint_{\text{c.v.}} e\rho\, dV \qquad (6\text{-}10)$$

with

$$\frac{\delta W_s}{dt} = \frac{\delta W_\mu}{dt} = \iint_{\text{c.s.}} \left(e + \frac{P}{\rho}\right)\rho(\mathbf{v} \cdot \mathbf{n})\, dA = 0$$

The applicable form of the first law is now

$$\frac{\delta Q}{dt} = \frac{\partial}{\partial t}\iiint_{\text{c.v.}} e\rho\, dV$$

Considering all variables to be functions of x alone, we may express the heat flux as

$$\frac{q_x}{A} = \frac{d}{dt}\int_0^L \rho u\, dx = \frac{d}{dt}\int_0^L \rho c_p T\, dx \qquad (18\text{-}32)$$

The interval from 0 to L will now be divided into two increments, giving

$$\frac{q_x}{A} = \frac{d}{dt}\left[\int_0^\delta \rho c_p T\, dx + \int_\delta^L \rho c_p T_0\, dx\right]$$

and, since T_0 is constant, this becomes

$$\frac{q_x}{A} = \frac{d}{dt}\left[\int_0^\delta \rho c_p T\, dx + \rho c_p T_0(L - \delta)\right]$$

The integral equation to be solved is now

$$\frac{q_x}{A} = \frac{d}{dt} \int_0^{\delta} \rho c_p T \, dx - \rho c_p T_0 \frac{d\delta}{dt} \qquad (18\text{-}33)$$

If a temperature profile of the form $T = T(x, \delta)$ is assumed, equation (18-33) will produce a differential equation in $\delta(t)$, which may be solved, and one may use this result to express the temperature profile as $T(x, t)$.

The solution of equation (18-33) is subject to three different boundary conditions at the wall, $x = 0$, in the sections to follow.

Case 1. Constant Wall Temperature

The wall, initially at uniform temperature T_0, has its surface maintained at temperature T_s for $t > 0$. The temperature profile at two different times is illustrated in Figure 18.8. Assuming the temperature profile to be parabolic of the form

$$T = A + Bx + Cx^2$$

and requiring that the following boundary conditions

$$T = T_s \qquad \text{at } x = 0$$

$$T = T_0 \qquad \text{at } x = \delta$$

and

$$\frac{\partial T}{\partial x} = 0 \qquad \text{at } x = \delta$$

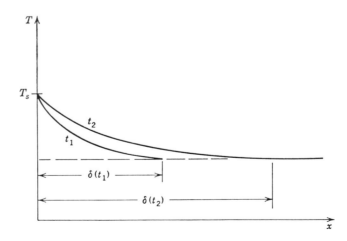

Figure 18.8 Temperature profiles at two times after the surface temperature is raised to T_s.

be satisfied, we see that the expression for $T(x)$ becomes

$$\frac{T-T_0}{T_s-T_0} = \left(1-\frac{x}{\delta}\right)^2$$
(18-34)

The heat flux at the wall may now be evaluated as

$$\frac{q_x}{A} = -k\frac{\partial T}{\partial x}\bigg|_{x=0} = 2\frac{k}{\delta}(T_s - T_0)$$
(18-35)

which may be substituted into the integral expression along with equation (18-33), yielding

$$2\frac{k}{\delta}(T_s - T_0) = \frac{d}{dt}\int_0^\delta \rho c_p\left[T_0+(T_s-T_0)\left(1-\frac{x}{\delta}\right)^2\right]dx - \rho c_p T_0\frac{d\delta}{dt}$$

and, after dividing through by ρc_p, both quantities being considered constant, we have

$$2\frac{\alpha}{\delta}(T_s - T_0) = \frac{d}{dt}\int_0^\delta \left[T_0+(T_s-T_0)\left(1-\frac{x}{\delta}\right)^2\right]dx - T_0\frac{d\delta}{dt}$$
(18-36)

After integration, equation (18-36) becomes

$$\frac{2\alpha}{\delta}(T_s - T_0) = \frac{d}{dt}\left[(T_s-T_0)\frac{\delta}{3}\right]$$

and cancelling $(T_s - T_0)$, we obtain

$$6\alpha = \delta\frac{d\delta}{dt}$$
(18-37)

and thus the penetration depth becomes

$$\delta = \sqrt{12\alpha t}$$
(18-38)

The corresponding temperature profile may be obtained from equation (18-34) as

$$\frac{T-T_0}{T_s-T_0} = \left[1-\frac{x}{\sqrt{3}(2\sqrt{\alpha t})}\right]^2$$
(18-39)

which compares reasonably well with the exact result,

$$\frac{T-T_0}{T_s-T_0} = 1-\text{erf}\,\frac{x}{2\sqrt{\alpha t}}$$
(18-40)

Figure 18.9 shows a comparison of these two results.

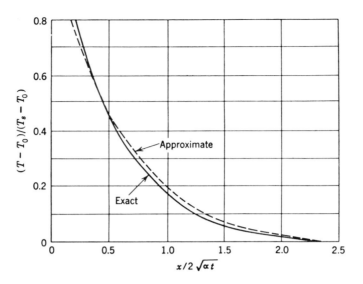

Figure 18.9 A comparison of exact and approximate results for one-dimensional conduction with a constant wall temperature.

Case 2. A Specified Heat Flux at the Wall

In this case the appropriate boundary conditions are

$$T = T_0 \qquad \text{at } x = \delta$$

$$\frac{\partial T}{\partial x} = 0 \qquad \text{at } x = \delta$$

and

$$\frac{\partial T}{\partial x} = -\frac{F(t)}{k} \qquad \text{at } x = 0$$

where the heat flux at the wall is expressed as the general function $F(t)$.

If the parabolic temperature profile is used, the above boundary conditions yield

$$T - T_0 = \frac{[F(t)](\delta - x)^2}{2k\delta} \tag{18-41}$$

which, when substituted into equation (18-33), yields

$$\frac{d}{dt}\left(\frac{F(t)\delta^2}{6k}\right) = \frac{\alpha F(t)}{k} \tag{18-42}$$

and

$$\delta(t) = \sqrt{6\alpha}\left[\frac{1}{F(t)}\int_0^t F(t)\,dt\right]^{1/2} \tag{18-43}$$

For a constant heat flux of magnitude q_0/A the resulting expression for T_s is

$$T_s - T_0 = \frac{q_0}{Ak} \sqrt{\frac{3}{2}\alpha t} \qquad (18\text{-}44)$$

which differs by approximately 8% from the exact expression

$$T_s - T_0 = \frac{1.13 q_0}{Ak} \sqrt{\alpha t} \qquad (18\text{-}45)$$

Case 3. Convection at the Surface

The wall temperature is variable in this case; however, it may be easily determined. If the temperature variation within the medium is expressed generally as

$$\frac{T - T_0}{T_s - T_0} = \phi\left(\frac{x}{\delta}\right) \qquad (18\text{-}46)$$

we note that the temperature gradient at the surface becomes

$$\left.\frac{\partial T}{\partial x}\right|_{x=0} = -\frac{T_s - T_0}{\delta} N \qquad (18\text{-}47)$$

where N is a constant depending upon the form of $\phi(x/\delta)$.

At the surface we may write

$$\left.\frac{q}{A}\right|_{x=0} = -k \left.\frac{\partial T}{\partial x}\right|_{x=0} = h(T_\infty - T_s)$$

which becomes, upon substituting equation (18-47),

$$T_s - T_0 = \frac{h\delta}{Nk}(T_\infty - T_0) \qquad (18\text{-}48)$$

or

$$T_s = \frac{T_0 + (h\delta/Nk)T_\infty}{1 + h\delta/Nk} \qquad (18\text{-}49)$$

We may now write

$$\frac{T_s - T_0}{T_\infty - T_0} = \frac{h\delta/Nk}{1 + h\delta/Nk} \qquad (18\text{-}50)$$

and

$$\frac{T_\infty - T_s}{T_\infty - T_0} = \frac{1}{1 + h\delta/Nk} \qquad (18\text{-}51)$$

The appropriate substitutions into the integral equation and subsequent solution follow the same procedures as in cases (a) and (b); the details of this solution are left as a student exercise.

The student should recognize the marked utility of the integral solution for solving one-dimensional unsteady-state conduction problems. Temperature profile expressions more complex than a parabolic form may be assumed; however, additional boundary conditions are needed in such cases to evaluate the constants. The similarity between the penetration depth and the boundary-layer thickness from the integral analysis of Chapter 12 should also be noted.

18.5 CLOSURE

In this chapter some of the techniques for solving transient or unsteady-state heat-conduction problems have been presented and discussed. Situations considered included cases of negligible internal resistance, negligible surface resistance, and those for which both resistances were significant.

For flat slabs, cylinders, and spheres, with a uniform initial temperature, whose surfaces are suddenly exposed to surroundings at a different temperature, charts are available for evaluating the temperature at any position and time. A graphical technique, the Schmidt plot, for one-dimensional transient systems was introduced. An integral method for solving one-dimensional transient conduction problems was also presented.

PROBLEMS

18.1 A type-304 stainless-steel billet, 6 in. in diameter, is passing through a 20-ft-long heat-treating furnace. The initial billet temperature is 200°F, and it must be raised to a minimum temperature of 1500°F before working. The heat-transfer coefficient between the furnace gases and the billet surface is 15 Btu/hr ft^2 °F, and the furnace gases are at 2300°F. At what minimum velocity must the billet travel through the furnace to satisfy these conditions?

18.2 A household iron has a stainless-steel sole plate which weighs 3 lb and has a surface area of 0.5 ft^2. The iron is rated at 500 W. If the surroundings are at a temperature of 80°F, and the convective heat-transfer coefficient between the sole plate and surroundings is 3 Btu/hr ft^2 °F, how long will it take for the iron to reach 240°F after it is plugged in?

18.3 A copper bus bar is initially at 400°F. The bar measures 0.2 ft by 0.5 ft and is 10 ft long. If the edges are suddenly all reduced to 100°F, how long will it take for the center to reach a temperature of 250°F?

18.4 In the curing of rubber tires, the "vulcanization" process requires that a tire carcass, originally at 295 K, be heated so that its central layer reach a minimum temperature of 410 K. This heating is accomplished by introducing steam at 435 K to both sides. Determine the time required, after introducing steam, for a 3-cm-thick tire carcass to reach the specified central temperature condition. Properties of rubber that may be used are the following: $k = 0.151$ W/m · K, $c_p = 200$ J/kg · K, $\rho = 1201$ kg/m³, $\alpha = 6.19 \times 10^{-8}$ m²/s.

18.5 If a rectangular block of rubber (see problem 18.4 for properties) is set out in air at 297 K to cool after being heated to a uniform temperature of 420 K, how long will it take for the rubber surface to reach 320 K? The dimensions of the block are 0.6 m high by 0.3 m long by 0.45 m wide. The block sits on one of the 0.3 m by 0.45 m bases; the adjacent surface may be considered an insulator. The effective heat transfer coefficient at all exposed surface is 6.0 W/m² · K. What will the maximum temperature within the rubber block be at this time?

18.6 Rework the previous problem for the case when air is blown by the surfaces of the rubber block with an effective surface coefficient of 230 W/m² · K resulting.

18.7 Consider a hot dog to have the following dimensions and properties: diameter = 20 mm, $c_p = 3.35$ kJ/kg · K, $\rho = 880$ kg/m³, and $k = 0.5$ W/m · K. For the hot dog initially at 5°C, exposed to boiling water at 100°C, with a surface coefficient of 90 W/m² · K, what will be the cooking time if the required condition is for the center temperature to reach 80°C?

18.8 Buckshot, 0.2 in. in diameter, is quenched in 90°F oil from an initial temperature of 400°F. The buckshot is made of lead and takes 15 s to fall from the oil surface to the bottom of the quenching bath. If the convective heat-transfer coefficient between the lead and oil is 40 Btu/hr ft² °F, what will be the temperature of the shot as it reaches the bottom of the bath?

18.9 Cast-iron cannonballs used in the War of 1812 were occasionally heated for some extended time so that, when fired at houses or ships, they would set them afire. If one of these so-called "hot shot" were at a uniform temperature of 2000°F, how long after being exposed to air at 0°F with an outside convective heat-transfer coefficient of 16 Btu/hr ft² °F, would be required for the surface temperature to drop to 600°F? What would be the center temperature at this time? The ball diameter is 6 in. The following properties of cast iron may be used:

$$k = 23 \text{ Btu/hr ft °F}$$

$$c_p = 0.10 \text{ Btu/lb}_m \text{ °F}$$

$$\rho = 460 \text{ lb}_m/\text{ft}^3$$

18.10 A rocket-engine nozzle is coated with a ceramic material having the following properties: $k = 1.73$ Btu/hr ft °F, $\alpha = 0.35$ ft^2/hr. The convective heat-transfer coefficient between the nozzle and the gases, which are at 3000°F, is 200 Btu/hr ft^2 °F. How long after startup will it take for the temperature at the ceramic surface to reach 2700°F? What will be the temperature at a point $\frac{1}{2}$ in. from the surface at this time? The nozzle is initially at 0°F.

18.11 For an asbestos cylinder with both height and diameter of 13 cm initially at a uniform temperature of 295 K placed in a medium at 810 K with an associated convective heat-transfer coefficient of 22.8 W/m^2 · K, determine the time required for the center of the cylinder to reach 530 K if end effects are neglected.

18.12 Given the cylinder in problem 18.11, construct a plot of the time for the midpoint temperature to reach 530 K as a function of H/D, where H and D are the height and diameter of the cylinder, respectively.

18.13 A copper cylinder with a diameter of 3 in. is initially at a uniform temperature of 70°F. How long after being placed in a medium at 1000°F with an associated convective heat-transfer coefficient of 4 Btu/hr ft^2 °F will the temperature at the center of the cylinder reach 500°F, if the height of the cylinder is (a) 3 in.? (b) 6 in.? (c) 12 in.? (d) 24 in.? (e) 5 ft?

18.14 A cylinder 2 ft high with a diameter of 3 in. is initially at the uniform temperature of 70°F. How long after the cylinder is placed in a medium at 1000°F, with associated convective heat-transfer coefficient of 4 Btu/hr ft^2 °F, will the center temperature reach 500°F if the cylinder is made from
 (a) copper, $k = 212$ Btu/hr ft °F?
 (b) aluminum, $k = 130$ Btu/hr ft °F?
 (c) zinc, $k = 60$ Btu/hr ft °F?
 (d) mild steel, $k = 25$ Btu/hr ft °F?
 (e) stainless steel, $k = 10.5$ Btu/hr ft °F?
 (f) asbestos, $k = 0.087$ Btu/hr ft °F?

18.15 One estimate of the original temperature of the earth is 7000°F. Using this value and the following properties for the earth's crust, Lord Kelvin obtained an estimate of 9.8×10^7 yr for the earth's age:

$$\alpha = 0.0456 \text{ ft}^2/\text{hr}$$

$$T_s = 0°F$$

$$\left.\frac{\partial T}{\partial y}\right|_{y=0} = 0.02°F/\text{ft, (measured)}$$

Comment on Lord Kelvin's result by considering the exact expression for unsteady-state conduction in one dimension:

$$\frac{T-T_s}{T_0-T_s} = \operatorname{erf}\frac{x}{2\sqrt{\alpha t}}$$

18.16 Determine an expression for the depth below the surface of a semi-infinite solid at which the rate of cooling is maximum. Substitute the information given in problem 18.15 to estimate how far below the earth's surface this cooling rate is maximum.

18.17 If the temperature profile through the ground is linear, increasing from 35°F at the surface by 0.5°F per foot of depth, how long will it take for a pipe buried 10 ft below the surface to reach 32°F if the outside air temperature is suddenly dropped to 0°F? The thermal diffusivity of soil may be taken as 0.02 ft^2/hr, its thermal conductivity is 0.8 Btu/hr ft °F, and the convective heat-transfer coefficient between the soil and the surrounding air is 1.5 Btu/hr ft^2 °F.

18.18 Soil, having a thermal diffusivity of 5.16×10^{-7} m^2/s, has its surface temperature suddenly raised and maintained at 1100 K from its initial uniform value of 280 K. Determine the temperature at a depth of 0.25 m after a period of 5 hr has elapsed at this surface condition.

18.19 A brick wall ($\alpha = 0.016$ ft^2/hr) with a thickness of $1\frac{1}{2}$ ft is initially at a uniform temperature of 80°F. How long after the wall surfaces are raised to 300°F and 600°F, respectively, will it take for the temperature at the center of the wall to reach 300°F?

18.20 A masonry brick wall 0.45 m thick has a temperature distribution at time, $t = 0$, which may be approximated by the expression $T(\text{K}) = 520 + 330\sin \pi(x/L)$ where L is the wall width and x is the distance from either surface. How long after both surfaces of this wall are exposed to air at 280 K will the center temperature of the wall be 360 K? The convective coefficient at both surfaces of the wall may be taken as 14 W/m^2 · K. What will the surface temperature be at this time?

18.21 Water, initially at 40°F, is contained within a thin-walled cylindrical vessel having a diameter of 18 in. Plot the temperature of the water vs. time up to 1 hr if the water and container are immersed in an oil bath at a constant temperature of 300°F. Assume that the water is well stirred and that the convective heat-transfer coefficient between the oil and cylindrical surface is 40 Btu/hr ft^2 °F. The cylinder is immersed to a depth of 2 ft.

18.22 The convective heat-transfer coefficient between a large brick wall and air at 100°F is expressed as $h = 0.44(T - T_\infty)^{1/3}$ Btu/hr ft^2 °F. If the wall is initially at a uniform temperature of 1000°F, estimate the temperature of the surface after 1 hr, 6 hr, 24 hr. Use the Schmidt graphical method, and assume the wall to be a semi-infinite medium. How does this answer compare with the exact result?

18.23 If the heat flux into a solid is given as $F(t)$, show that the penetration depth δ for a semi-infinite solid is of the form

$$\delta = (\text{constant})\sqrt{\alpha}\left[\frac{\int_0^t F(t)\,dt}{F(t)}\right]^{1/2}$$

18.24 Air at 65°F is blown against a pane of glass $\frac{1}{8}$ in. thick. If the glass is initially at 30°F and has frost on the outside, estimate the length of time required for the frost to begin to melt.

18.25 A brick wall at 90°F is subject to air at 60°F; the film coefficient is 5.0 Btu/hr ft^2 °F. Considering the wall to be semi-infinite, determine the surface temperature after 10 hr. How far into the wall has the temperature change penetrated? How much heat has the wall lost?

18.26 How long will a 1-ft-thick concrete wall subject to a surface temperature of 1500°F on one side maintain the other side below 130°F? The wall is initially at 70°F.

18.27 A stainless steel bar is initially at a temperature of 25°C. Its upper surface is suddenly exposed to an airstream at 200°C, with a corresponding convective coefficient of 22 W/m^2 · K. If the bar is considered semiinfinite, how long will it take for the temperature at a disance of 50 mm from the surface to reach 100°C?

18.28 If the stainless steel bar in problem 18.27 is very long, with square cross section measuring 100 mm on a side, and its sides and bottom insulated, how long after exposure to the 200°C airstream will the center temperature reach 100°C? Solve this problem numerically.

18.29 A thick plate made of stainless steel is initially at a uniform temperature of 300°C. The surface is suddenly exposed to a coolant at 20°C with a convective surface coefficient of 110 W/m^2 · K. Evaluate the temperature after 3 min of elapsed time at
(a) the surface;
(b) a depth of 50 mm.
Work this problem both analytically and numerically.

19
CONVECTIVE
HEAT TRANSFER

Heat transfer by convection is associated with energy exchange between a surface and an adjacent fluid. There are very few energy-transfer situations of practical importance in which fluid motion is not in some way involved. This effect has been eliminated as much as possible in the preceding chapters, but will now be considered in some depth.

The rate equation for convection has been expressed previously as

$$\frac{q}{A} = h \, \Delta T \qquad (15\text{-}11)$$

where the heat flux, q/A, occurs by virtue of a temperature difference. This simple equation is the defining relation for h, the convective heat-transfer coefficient. The determination of the coefficient h is, however, not at all a simple undertaking. It is related to the mechanism of fluid flow, the properties of the fluid, and the geometry of the specific system of interest.

In light of the intimate involvement between the convective heat-transfer coefficient and fluid motion, we may expect many of the considerations from momentum transfer to be of interest. In the analyses to follow, much use will be made of the developments and concepts of Chapters 4 through 14.

19.1 FUNDAMENTAL CONSIDERATIONS IN CONVECTIVE HEAT TRANSFER

As mentioned in Chapter 12, the fluid particles immediately adjacent to a solid boundary are stationary, and a thin layer of fluid close to the surface will be in laminar flow regardless of the nature of the free stream. Thus molecular energy exchange or conduction effects will always be present, and play a major role in any convection process. If fluid flow is laminar, then all energy transfer between a surface and contacting fluid or between adjacent fluid layers is by molecular

325

means. If, on the other hand, flow is turbulent, then there is bulk mixing of fluid particles between regions at different temperatures, and the heat transfer rate is increased. The distinction between laminar and turbulent flow will thus be a major consideration in any convective situation.

There are two main classifications of convective heat transfer. These have to do with the driving force causing fluid to flow. *Natural* or *free convection* designates the type of process wherein fluid motion results from the heat transfer. When a fluid is heated or cooled, the associated density change and buoyant effect produce a natural circulation in which the affected fluid moves of its own accord past the solid surface, the fluid which replaces it is similarly affected by the energy transfer, and the process is repeated. *Forced convection* is the classification used to describe those convection situations in which fluid circulation is produced by an external agency such as a fan or a pump.

The hydrodynamic boundary layer, analyzed in Chapter 12, plays a major role in convective heat transfer, as one would expect. Additionally, we shall define and analyze the *thermal boundary layer*, which will also be vital to the analysis of a convective energy-transfer process.

There are four methods of evaluating the convective heat-transfer coefficient which will be discussed in this book. These are as follows:

(a) dimensional analysis, which to be useful requires experimental results;
(b) exact analysis of the boundary layer;
(c) approximate integral analysis of the boundary layer; and
(d) analogy between energy and momentum transfer.

19.2 SIGNIFICANT PARAMETERS IN CONVECTIVE HEAT TRANSFER

Certain parameters will be found useful in the correlation of convection data and in the functional relations for the convective heat-transfer coefficients. Some parameters of this type have been encountered earlier; these include the Reynolds and the Euler numbers. Several of the new parameters to be encountered in energy transfer will arise in such a manner that their physical meaning is obscure. For this reason, we shall devote a short section to the physical interpretation of two such terms.

The molecular diffusivities of momentum and energy have been defined previously as

$$\text{momentum diffusivity:} \quad \nu \equiv \frac{\mu}{\rho}$$

and

$$\text{thermal diffusivity:} \quad \alpha \equiv \frac{k}{\rho c_p}$$

That these two are designated similarly would indicate that they must also play similar roles in their specific transfer modes. This is indeed the case, as we shall see several times in the developments to follow. For the moment we should note that both have the same dimensions, those of L^2/t; thus their ratio must be dimensionless. This ratio, that of the molecular diffusivity of momentum to the molecular diffusivity of heat, is designated the *Prandtl number*,

$$\Pr \equiv \frac{\nu}{\alpha} = \frac{\mu c_p}{k} \tag{19-1}$$

The Prandtl number is observed to be a combination of fluid properties; thus Pr itself may be thought of as a property. The Prandtl number is primarily a function of temperature and is tabulated in Appendix I, at various temperatures for each fluid listed.

The temperature profile for a fluid flowing past a surface is depicted in Figure 19.1. In the figure the surface is at a higher temperature than the fluid. The

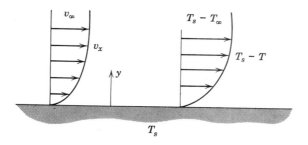

Figure 19.1   Temperature and velocity profiles for a fluid flowing past a heated plate.

temperature profile which exists is due to the energy exchange resulting from this temperature difference. For such a case the heat-transfer rate between the surface and the fluid may be written as

$$q_y = hA(T_s - T_\infty) \tag{19-2}$$

and, since heat transfer at the surface is by conduction,

$$q_y = -kA\frac{\partial}{\partial y}(T - T_s)\big|_{y=0} \tag{19-3}$$

These two terms must be equal; thus

$$h(T_s - T_\infty) = -k\frac{\partial}{\partial y}(T - T_s)\big|_{y=0}$$

which may be rearranged to give

$$\frac{h}{k} = \frac{\partial(T_s - T)/\partial y|_{y=0}}{T_s - T_\infty} \tag{19-4}$$

Equation (19-4) may be made dimensionless if a length parameter is introduced. Multiplying both sides by a representative length, L, we have

$$\frac{hL}{k} = \frac{\partial(T_s - T)/\partial y|_{y=0}}{(T_s - T_\infty)/L} \tag{19-5}$$

The right-hand side of equation (19-5) is now the ratio of the temperature gradient at the surface to an overall or reference temperature gradient. The left-hand side of this equation is written in a manner similar to that for the Biot modulus encountered in Chapter 18. It may be considered a ratio of conductive thermal resistance to the convective thermal resistance of the fluid. This ratio is referred to as the *Nusselt number*,

$$\text{Nu} \equiv \frac{hL}{k} \tag{19-6}$$

where the thermal conductivity is that of the fluid as opposed to that of the solid, which was the case in the evaluation of the Biot modulus.

These two parameters, Pr and Nu, will be encountered many times in the work to follow.

19.3 DIMENSIONAL ANALYSIS OF CONVECTIVE ENERGY TRANSFER

Forced Convection. The specific forced-convection situation, which we shall now consider, is that of fluid flowing in a closed conduit at some average velocity, v, with a temperature difference existing between the fluid and the tube wall.

The important variables, their symbols, and dimensional representations are listed below. It is necessary to include two more dimensions—Q, heat, and T, temperature—to the fundamental group considered in Chapter 11; thus all

Variable	Symbol	Dimensions
Tube diameter	D	L
Fluid density	ρ	M/L^3
Fluid viscosity	μ	M/Lt
Fluid heat capacity	c_p	Q/MT
Fluid thermal conductivity	k	Q/tLT
Velocity	v	L/t
Heat-transfer coefficient	h	Q/tL^2T

variables must be expressed dimensionally as some combination of M, L, t, Q, and T. The above variables include terms descriptive of the system geometry, thermal and flow properties of the fluid, and the quantity of primary interest, h.

Utilizing the Buckingham method of grouping the variables as presented in Chapter 11, the required number of dimensionless groups is found to be 3. Note that the rank of the dimensional matrix is 4, one less than the total number of fundamental dimensions.

Choosing D, k, μ, and v as the four variables comprising the core, we find that the three π groups to be formed are

$$\pi_1 = D^a k^b \mu^c v^d \rho$$

$$\pi_2 = D^e k^f \mu^g v^h c_p$$

and

$$\pi_3 = D^i k^j \mu^k v^l h$$

Writing π_1 in dimensional form,

$$1 = (L)^a \left(\frac{Q}{LtT}\right)^b \left(\frac{M}{Lt}\right)^c \left(\frac{L}{t}\right)^d \frac{M}{L^3}$$

and equating the exponents of the fundamental dimensions on both sides of this equation, we have for

$$L: \quad 0 = a - b - c + d - 3$$

$$Q: \quad 0 = b$$

$$t: \quad 0 = -b - c - d$$

$$T: \quad 0 = -b$$

and

$$M: \quad 0 = c + 1$$

Solving these equations for the four unknowns yields

$$a = 1 \qquad c = -1$$
$$b = 0 \qquad d = 1$$

and π_1 becomes

$$\pi_1 = \frac{Dv\rho}{\mu}$$

which is the Reynolds number. Solving for π_2 and π_3 in the same way will give

$$\pi_2 = \frac{\mu c_p}{k} = \text{Pr} \qquad \text{and} \qquad \pi_3 = \frac{hD}{k} = \text{Nu}$$

The result of a dimensional analysis of forced-convection heat-transfer in a circular conduit indicates that a possible relation correlating the important variables is of the form

$$\text{Nu} = f_1(\text{Re}, \text{Pr}) \tag{19-7}$$

If, in the preceding case, the core group had been chosen to include ρ, μ, c_p, and v, the analysis would have yielded the groups $Dv\rho/\mu$, $\mu c_p/k$, and $h/\rho v c_p$. The first two of these we recognize as Re and Pr. The third is the *Stanton number.*

$$\text{St} \equiv \frac{h}{\rho v c_p} \tag{19-8}$$

This parameter could also have been formed by taking the ratio Nu/(Re Pr). An alternative correlating relation for forced convection in a closed conduit is thus

$$\text{St} = f_2(\text{Re, Pr}) \tag{19-9}$$

Natural Convection. In the case of natural-convection heat transfer from a vertical plane wall to an adjacent fluid, the variables will differ significantly from those used in the preceding case. The velocity no longer belongs in the group of variables, since it is a result of other effects associated with the energy transfer. New variables to be included in the analysis are those accounting for fluid circulation. They may be found by considering the relation for buoyant force in terms of the density difference due to the energy exchange.

The coefficient of thermal expansion, β, is given by

$$\rho = \rho_0(1 - \beta \, \Delta T) \tag{19-10}$$

where ρ_0 is the bulk fluid density, ρ is the fluid density inside the heated layer, and ΔT is the temperature difference between the heated fluid and the bulk value. The buoyant force per unit volume, F_{buoyant}, is

$$F_{\text{buoyant}} = (\rho_0 - \rho)g$$

which becomes, upon substituting equation (19-10),

$$F_{\text{buoyant}} = \beta g \rho_0 \, \Delta T \tag{19-11}$$

Equation (19-11) suggests the inclusion of the variables β, g, and ΔT into the list of those important to the natural convection situation.

The list of variables for the problem under consideration is given below.

Variable	Symbol	Dimensions
Significant length	L	L
Fluid density	ρ	M/L^3
Fluid viscosity	μ	M/Lt
Fluid heat capacity	c_p	Q/MT
Fluid thermal conductivity	k	Q/LtT
Fluid coefficient of thermal expansion	β	$1/T$
Gravitational acceleration	g	L/t^2
Temperature difference	ΔT	T
Heat-transfer coefficient	h	Q/L^2tT

The Buckingham Pi theorem indicates that the number of independent dimensionless parameters applicable to this problem is $9 - 5 = 4$. Choosing $L, \mu, k, g,$ and β as the core group, we see that the π groups to be formed are

$$\pi_1 = L^a \mu^b k^c \beta^d g^e c_p$$

$$\pi_2 = L^f \mu^g k^h \beta^i g^j \rho$$

$$\pi_3 = L^k \mu^l k^m \beta^n g^o \, \Delta T$$

and

$$\pi_4 = L^p \mu^q k^r \beta^s g^t h$$

Solving for the exponents in the usual way, we obtain

$$\pi_1 = \frac{\mu c_p}{k} = \mathrm{Pr} \qquad\qquad \pi_3 = \beta \, \Delta T$$

$$\pi_2 = \frac{L^3 g \rho^2}{\mu^2} \qquad \text{and} \qquad \pi_4 = \frac{hL}{k} = \mathrm{Nu}$$

The product of π_2 and π_3, which must be dimensionless, is $(\beta g \rho^2 L^3 \, \Delta T)/\mu^2$. This parameter, used in correlating natural-convection data, is the *Grashof number*,

$$\mathrm{Gr} \equiv \frac{\beta g \rho^2 L^3 \, \Delta T}{\mu^2} \tag{19-12}$$

From the preceding brief dimensional-analyses considerations, we have obtained the following possible forms for correlating convection data:

(a) *Forced convection* $\qquad\qquad \mathrm{Nu} = f_1(\mathrm{Re}, \mathrm{Pr})$ $\qquad\qquad\qquad$ (19-7)

or

$$\mathrm{St} = f_2(\mathrm{Re}, \mathrm{Pr}) \tag{19-9}$$

(b) *Natural convection* $\qquad\qquad \mathrm{Nu} = f_3(\mathrm{Gr}, \mathrm{Pr})$ $\qquad\qquad\qquad$ (19-13)

The similarity between the correlations of equations (19-7) and (19-13) is apparent. In equation (19-13) Gr has replaced Re in the correlation indicated by equation (19-7). It should be noted that the Stanton number can be used only in correlating forced-convection data. This becomes obvious when we observe the velocity, v, contained in the expression for St.

19.4 EXACT ANALYSIS OF THE LAMINAR BOUNDARY LAYER

An exact solution for a special case of the hydrodynamic boundary layer was discussed in section 12.5. Blasius' solution for the laminar boundary layer on a flat plate may be extended to include the convective heat-transfer problem for the same geometry and laminar flow.

The boundary-layer equations considered previously include the two-dimensional, incompressible continuity equation

$$\frac{\partial v_x}{\partial x}+\frac{\partial v_y}{\partial y}=0 \tag{12-10}$$

and the equation of motion in the x direction,

$$\frac{\partial v_x}{\partial t}+v_x\frac{\partial v_x}{\partial x}+v_y\frac{\partial v_x}{\partial y}=v_\infty\frac{dv_\infty}{dx}+\nu\frac{\partial^2 v_x}{\partial y^2} \tag{12-9}$$

Recall that the y-directional equation of motion gave the result of constant pressure through the boundary layer. The proper form of the energy equation will thus be equation (16-14), for isobaric flow, written in two-dimensional form as

$$\frac{\partial T}{\partial t}+v_x\frac{\partial T}{\partial x}+v_y\frac{\partial T}{\partial y}=\alpha\left(\frac{\partial^2 T}{\partial x^2}+\frac{\partial^2 T}{\partial y^2}\right) \tag{19-14}$$

With respect to the thermal boundary layer depicted in Figure 19.2, $\partial^2 T/\partial x^2$ is much smaller in magnitude than $\partial^2 T/\partial y^2$.

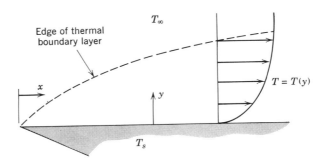

Figure 19.2 The thermal boundary layer for laminar flow past a flat surface.

In steady, incompressible, two-dimensional, isobaric flow the energy equation which applies is now

$$v_x\frac{\partial T}{\partial x}+v_y\frac{\partial T}{\partial y}=\alpha\frac{\partial^2 T}{\partial y^2} \tag{19-15}$$

From Chapter 12 the applicable equation of motion with uniform free-stream velocity is

$$v_x\frac{\partial v_x}{\partial x}+v_y\frac{\partial v_x}{\partial y}=\nu\frac{\partial^2 v_x}{\partial y^2} \tag{12-11a}$$

and the continuity equation

$$\frac{\partial v_x}{\partial x} + \frac{\partial v_y}{\partial y} = 0 \tag{12-11b}$$

The latter two of the above equations were originally solved by Blasius to give the results discussed in Chapter 12. The solution was based upon the boundary conditions

$$\frac{v_x}{v_\infty} = \frac{v_y}{v_\infty} = 0 \qquad \text{at } y = 0$$

and

$$\frac{v_x}{v_\infty} = 1 \qquad \text{at } y = \infty$$

The similarity in form between equations (19-15) and (12-11a) is obvious. This situation suggests the possibility of applying the Blasius solution to the energy equation. In order that this be possible the following conditions must be satisfied:

1. The coefficients of the second-order terms must be equal. This requires that $\nu = \alpha$ or that $Pr = 1$.

2. The boundary conditions for temperature must be compatible with those for the velocity. This may be accomplished by changing the dependent variable from T to $(T - T_s)/(T_\infty - T_s)$. The boundary conditions now are

$$\frac{v_x}{v_\infty} = \frac{v_y}{v_\infty} = \frac{T - T_s}{T_\infty - T_s} = 0 \qquad \text{at } y = 0$$

$$\frac{v_x}{v_\infty} = \frac{T - T_s}{T_\infty - T_s} = 1 \qquad \text{at } y = \infty$$

Imposing these conditions upon the set of equations (19-15) and (12-11a), we may now write the results obtained by Blasius for the energy-transfer case. Using the nomenclature of Chapter 12,

$$f' = 2\frac{v_x}{v_\infty} = 2\frac{T - T_s}{T_\infty - T_s} \tag{19-16}$$

$$\eta = \frac{y}{2}\sqrt{\frac{v_\infty}{\nu x}} = \frac{y}{2x}\sqrt{\frac{xv_\infty}{\nu}} = \frac{y}{2x}\sqrt{Re_x} \tag{19-17}$$

and applying the Blasius result, we obtain

$$\frac{df'}{d\eta}\bigg|_{y=0} = f''(0) = \frac{d[2(v_x/v_\infty)]}{d[(y/2x)\sqrt{Re_x}]}\bigg|_{y=0}$$

$$= \frac{d\{2[(T - T_s)/(T_\infty - T_s)]\}}{d[(y/2x)\sqrt{Re_x}]}\bigg|_{y=0} = 1.328 \tag{19-18}$$

It should be noted that according to equation (19-16), the dimensionless velocity profile in the laminar boundary layer is identical with the dimensionless temperature profile. This is a consequence of having $Pr = 1$. A logical consequence of this situation is that the hydrodynamic and thermal boundary layers are of equal thickness. It is significant that the Prandtl numbers for most gases are sufficiently close to unity that the hydrodynamic and thermal boundary layers are of similar extent.

We may now obtain the temperature gradient at the surface:

$$\left.\frac{\partial T}{\partial y}\right|_{y=0} = (T_\infty - T_s)\left[\frac{0.332}{x}\,Re_x^{1/2}\right] \tag{19-19}$$

Application of the Newton and Fourier rate equations now yields

$$\frac{q_y}{A} = h_x(T_s - T_\infty) = -k\left.\frac{\partial T}{\partial y}\right|_{y=0}$$

from which

$$h_x = -\frac{k}{T_s - T_\infty}\left.\frac{\partial T}{\partial y}\right|_{y=0} = \frac{0.332k}{x}\,Re_x^{1/2} \tag{19-20}$$

or

$$\frac{h_x x}{k} = Nu_x = 0.332\,Re_x^{1/2} \tag{19-21}$$

Pohlhausen* considered the same problem with the additional effect of a Prandtl number other than unity. He was able to show the relation between the thermal and hydrodynamic boundary layers in laminar flow to be approximately given by

$$\frac{\delta}{\delta_t} = Pr^{1/3} \tag{19-22}$$

The additional factor of $Pr^{1/3}$ multiplied by η allows the solution to the thermal boundary layer to be extended to Pr values other than unity. A plot of the dimensionless temperature versus $\eta\,Pr^{1/3}$ is shown in Figure 19.3. The temperature variation given in this form leads to an expression for the convective heat-transfer coefficient similar to equation (19-20). At $y = 0$ the gradient is

$$\left.\frac{\partial T}{\partial y}\right|_{y=0} = (T_\infty - T_s)\left[\frac{0.332}{x}\,Re_x^{1/2}\,Pr^{1/3}\right] \tag{19-23}$$

which, when used with the Fourier and Newton rate equations, yields

$$h_x = 0.332\frac{k}{x}\,Re_x^{1/2}\,Pr^{1/3} \tag{19-24}$$

---

* E. Pohlhausen, *ZAMM*, **1**, 115 (1921).

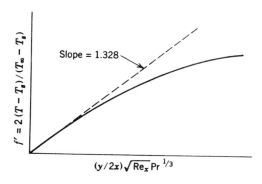

Figure 19.3 Temperature variation for laminar flow over a flate plate.

or

$$\frac{h_x x}{k} = \mathrm{Nu}_x = 0.332 \, \mathrm{Re}_x^{1/2} \, \mathrm{Pr}^{1/3} \qquad (19\text{-}25)$$

The inclusion of the factor $\mathrm{Pr}^{1/3}$ in these equations extends the range of application of equations (19-20) and (19-21) to situations in which the Prandtl number differs considerably from 1.

The mean heat-transfer coefficient applying over a plate of width w and length L may be obtained by integration. For a plate of these dimensions

$$q_y = hA(T_s - T_\infty) = \int_A h_x(T_s - T_\infty)\, dA$$

$$h(wL)(T_s - T_\infty) = 0.332kw \, \mathrm{Pr}^{1/3}(T_s - T_\infty) \int_0^L \frac{\mathrm{Re}_x^{1/2}}{x}\, dx$$

$$hL = 0.332k \, \mathrm{Pr}^{1/3}\left(\frac{v_\infty \rho}{\mu}\right)^{1/2} \int_0^L x^{-1/2}\, dx$$

$$= 0.664k \, \mathrm{Pr}^{1/3}\left(\frac{v_\infty \rho}{\mu}\right)^{1/2} L^{1/2}$$

$$= 0.664k \, \mathrm{Pr}^{1/3} \, \mathrm{Re}_L^{1/2}$$

The mean Nusselt number becomes

$$\mathrm{Nu}_L = \frac{hL}{k} = 0.664 \, \mathrm{Pr}^{1/3} \, \mathrm{Re}_L^{1/2} \qquad (19\text{-}26)$$

and it is seen that

$$\mathrm{Nu}_L = 2 \, \mathrm{Nu}_x \qquad \text{at } x = L \qquad (19\text{-}27)$$

In applying the results of the foregoing analysis it is customary to evaluate all fluid properties at the *film temperature*, which is defined as

$$T_f = \frac{T_s + T_\infty}{2} \qquad (19\text{-}28)$$

the arithmetic mean between the wall and bulk fluid temperatures.

19.5 APPROXIMATE INTEGRAL ANALYSIS OF THE THERMAL BOUNDARY LAYER

The application of the Blasius solution to the thermal boundary layer in section 19.4 was convenient although very limited in scope. For flow other than laminar or for a configuration other than a flat surface, another method must be utilized to estimate the convective heat-transfer coefficient. An approximate method for analysis of the thermal boundary layer employs the integral analysis as used by von Kármán for the hydrodynamic boundary layer. This approach was discussed in Chapter 12.

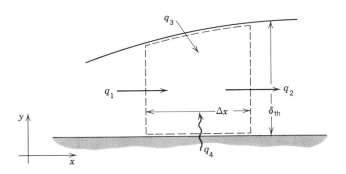

Figure 19.4 Control volume for integral energy analysis.

Consider the control volume designated by the dashed lines in Figure 19.4, applying to flow parallel to a flat surface with no pressure gradient, having width Δx, a height equal to the thickness of the thermal boundary layer, δ_t, and a unit depth. An application of the first law of thermodynamics in integral form,

$$\frac{\delta Q}{dt} - \frac{\delta W_s}{dt} - \frac{\delta W_\mu}{dt} = \iint_{\text{c.s.}} (e + P/\rho)\rho(\mathbf{v} \cdot \mathbf{n})\, dA + \frac{\partial}{\partial t} \iiint_{\text{c.v.}} e\rho\, dV \quad (6\text{-}10)$$

yields the following under steady-state conditions:

$$\frac{\delta Q}{dt} = -k\,\Delta x \frac{\partial T}{\partial y}\Big|_{y=0}$$

$$\frac{\delta W_s}{dt} = \frac{\delta W_\mu}{dt} = 0$$

$$\iint_{c.s.} (e + P/\rho)\rho(\mathbf{v} \cdot \mathbf{n})\,dA = \int_0^{\delta_t} \left(\frac{v_x^2}{2} + gy + u + \frac{P}{\rho}\right)\rho v_x\,dy\Big|_{x+\Delta x}$$

$$- \int_0^{\delta_t} \left(\frac{v_x^2}{2} + gy + u + \frac{P}{\rho}\right)\rho v_x\,dy\Big|_x$$

$$- \frac{d}{dx} \int_0^{\delta_t} \left[\rho v_x\left(\frac{v_x^2}{2} + gy + u + \frac{P}{\rho}\right)\Big|_{\delta_t}\right]dy\,\Delta x$$

and

$$\frac{\partial}{\partial t}\iiint_{c.v.} e\rho\,dV = 0$$

In the absence of significant gravitational effects the convective-energy-flux terms become

$$\frac{v_x^2}{2} + u + \frac{P}{\rho} = h_0 \approx c_p T_0$$

where h_0 is the stagnation enthalpy and c_p is the constant-pressure heat capacity. The stagnation temperature will now be written merely as T (without subscript) to avoid confusion. The complete energy expression is now

$$-k\,\Delta x\frac{\partial T}{\partial y}\Big|_{y=0} = \int_0^{\delta_t} \rho v_x c_p T\,dy\Big|_{x+\Delta x} - \int_0^{\delta_t} \rho v_x c_p T\,dy\Big|_x - \rho c_p\,\Delta x\frac{d}{dx}\int_0^{\delta_t} v_x T_\infty\,dy$$

$$(19\text{-}29)$$

Equation (19-29) can also be written as $q_4 = q_2 - q_1 - q_3$ where these quantities are shown in Figure 19.4. In equation (19-29) T_∞ represents the free-stream stagnation temperature. If flow is incompressible, and an average value of c_p is used, the product ρc_p may be taken outside the integral terms in this equation. Dividing both sides of equation (19-29) by Δx and evaluating the result in the limit as Δx approaches zero, we obtain

$$\frac{k}{\rho c_p}\frac{\partial T}{\partial y}\Big|_{y=0} = \frac{d}{dx}\int_0^{\delta_t} v_x(T_\infty - T)\,dy \qquad (19\text{-}30)$$

Equation (19-30) is analogous to the momentum integral relation, equation (12-37), with the momentum terms replaced by their appropriate energy counterparts. This equation may be solved if both a velocity and a temperature profile are known. Thus for the energy equation both the variation in v_x and in T with y must be assumed. This contrasts slightly with the momentum integral solution in which the velocity profile alone was assumed.

An assumed temperature profile must satisfy the boundary conditions

(1) $T - T_s = 0$ at $y = 0$

(2) $T - T_s = T_\infty - T_s$ at $y = \delta_t$

(3) $\dfrac{\partial}{\partial y}(T - T_s) = 0$ at $y = \delta_t$

(4) $\dfrac{\partial^2}{\partial y^2}(T - T_s) = 0$ at $y = 0$ [see equation (19-15)]

If a power-series expression for the temperature variation is assumed in the form

$$T - T_s = a + by + cy^2 + dy^3$$

the application of the boundary conditions will result in the expression for $T - T_s$,

$$\frac{T - T_s}{T_\infty - T_s} = \frac{3}{2}\left(\frac{y}{\delta_t}\right) - \frac{1}{2}\left(\frac{y}{\delta_t}\right)^3 \tag{19-31}$$

If the velocity profile is assumed in the same form, then the resulting expression, as obtained in Chapter 12, is

$$\frac{v}{v_\infty} = \frac{3}{2}\frac{y}{\delta} - \frac{1}{2}\left(\frac{y}{\delta}\right)^3 \tag{12-40}$$

Substituting equations (19-31) and (12-40) into the integral expression and solving, we obtain the result

$$\mathrm{Nu}_x = 0.36\,\mathrm{Re}_x^{1/2}\,\mathrm{Pr}^{1/3} \tag{19-32}$$

which is approximately 8% larger than the exact result expressed in equation (19-25).

This result, although inexact, is sufficiently close to the known value to indicate that the integral method may be used with confidence in situations in which an exact solution is not known. It is interesting to note that equation (19-32) again involves the parameters predicted from dimensional analysis.

19.6 ENERGY- AND MOMENTUM-TRANSFER ANALOGIES

Many times in our consideration of heat transfer thus far we have noted the similarities to momentum transfer both in the transfer mechanism itself and in the manner of its quantitative description. This section will deal with these analogies and use them to develop relations to describe energy transfer.

Osborne Reynolds first noted the similarities in mechanism between energy and momentum transfer in 1874.* In 1883 he presented† the results of his work on frictional resistance to fluid flow in conduits, thus making possible the quantitative analogy between the two transport phenomena.

As we have noted in the previous sections, for flow past a solid surface with a Prandtl number of unity, the dimensionless velocity and temperature gradients are related as follows:

$$\frac{d}{dy}\frac{v_x}{v_\infty}\bigg|_{y=0} = \frac{d}{dy}\left(\frac{T-T_s}{T_\infty-T_s}\right)\bigg|_{y=0} \tag{19-33}$$

For $\mathrm{Pr} = \mu c_p/k = 1$, we have $\mu c_p = k$ and we may write equation (19-33) as

$$\mu c_p \frac{d}{dy}\left(\frac{v_x}{v_\infty}\right)\bigg|_{y=0} = k\frac{d}{dy}\left(\frac{T-T_s}{T_\infty-T_s}\right)\bigg|_{y=0}$$

which may be transformed to the form

$$\frac{\mu c_p}{v_\infty}\frac{dv_x}{dy}\bigg|_{y=0} = -\frac{k}{T_s-T_\infty}\frac{d}{dy}(T-T_s)\big|_{y=0} \tag{19-34}$$

Recalling a previous relation for the convective heat-transfer coefficient

$$\frac{h}{k} = \frac{d}{dy}\left[\frac{(T_s-T)}{(T_s-T_\infty)}\right]\bigg|_{y=0} \tag{19-4}$$

it is seen that the entire right-hand side of equation (19-34) may be replaced by h, giving

$$h = \frac{\mu c_p}{v_\infty}\frac{dv_x}{dy}\bigg|_{y=0} \tag{19-35}$$

Introducing next the coefficient of skin friction

$$C_f \equiv \frac{\tau_0}{\rho v_\infty^2/2} = \frac{2\mu}{\rho v_\infty^2}\frac{dv_x}{dy}\bigg|_{y=0}$$

we may write equation (19-35) as

$$h = \frac{C_f}{2}(\rho v_\infty c_p)$$

which, in dimensionless form, becomes

$$\frac{h}{\rho v_\infty c_p} \equiv \mathrm{St} = \frac{C_f}{2} \tag{19-36}$$

Equation (19-36) is the *Reynolds analogy* and is an excellent example of the similar nature of energy and momentum transfer. For those situations satisfying

* O. Reynolds, *Proc. Manchester Lit. Phil. Soc.*, **14**:7 (1874).
† O. Reynolds, *Trans. Roy. Soc.* (London), **174A**, 935 (1883).

the basis for the development of equation (19-36), a knowledge of the coefficient of frictional drag will enable the convective heat-transfer coefficient to be readily evaluated.

The restrictions on the use of the Reynolds analogy should be kept in mind; they are (1) $Pr = 1$, and (2) no form drag. The former of these was the starting point in the preceding development and obviously must be satisfied. The latter is sensible when one considers that, in relating two transfer mechanisms, the manner of expressing them quantitatively must remain consistent. Obviously the description of drag in terms of the coefficient of skin friction requires that the drag be wholly viscous in nature. Thus equation (19-36) is applicable only for those situations in which form drag is not present. Some possible areas of application would be flow parallel to plane surfaces or flow in conduits. The coefficient of skin friction for conduit flow has already been shown to be equivalent to the Fanning friction factor, which may be evaluated by using Figure 14.1.

The restriction that $Pr = 1$ makes the Reynolds analogy of limited use. Colburn* has suggested a simple variation of the Reynolds analogy form that allows its application to situations where the Prandtl number is other than unity. The Colburn analogy expression is

$$St \, Pr^{2/3} = \frac{C_f}{2} \tag{19-37}$$

which obviously reduces to the Reynolds analogy when $Pr = 1$.

Colburn applied this expression to a wide range of data for flow and geometries of different types and found it to be quite accurate for conditions where (1) no form drag exists, and (2) $0.5 < Pr < 50$. The Prandtl number range is extended to include gases, water, and several other liquids of interest. The Colburn analogy is particularly helpful for evaluating heat transfer in internal forced flows. It can be easily shown that the exact expression for a laminar boundary layer on a flat plate reduces to equation (19-37).

The Colburn analogy is often written as

$$j_H = \frac{C_f}{2} \tag{19-38}$$

where

$$j_H = St \, Pr^{2/3} \tag{19-39}$$

is designated the Colburn j factor for heat transfer. A mass-transfer j factor, j_D, will be discussed in Chapter 28.

Note that for $Pr = 1$, the Colburn and Reynolds analogies are the same. Equation (19-37) is thus an extension of the Reynolds analogy for fluids having Prandtl numbers other than unity, within the range 0.5 to 50 as specified above. High and low Prandtl number fluids falling outside this range would be heavy oils at one extreme and liquid metals at the other.

* A. P. Colburn, *Trans. A.I.Ch.E.*, **29**, 174 (1933).

19.7   TURBULENT FLOW CONSIDERATIONS

The effect of turbulent flow on energy transfer is directly analogous to the similar effects on momentum transfer as discussed in Chapter 13. Consider the temperature profile variation in Figure 19.5 to exist in turbulent flow. The

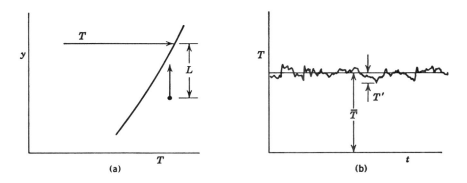

Figure 19.5   Turbulent-flow temperature variation.

distance moved by a fluid "packet" in the y direction, which is normal to the direction of bulk flow, is denoted by L, the Prandtl mixing length. The packet of fluid moving through the distance L retains the mean temperature from its point of origin, and upon reaching its destination, the packet will differ in temperature from that of the adjacent fluid by an amount $T|_{y \pm L} - T|_y$. The mixing length is assumed small enough to permit the temperature difference to be written as

$$T|_{y \pm L} - T|_y = \pm L \frac{dT}{dy}\bigg|_y \qquad (19\text{-}40)$$

We now define the quantity T' as the fluctuating temperature, synonymous with the fluctuating velocity component, v_x', described in Chapter 13. The instantaneous temperature is the sum of the mean and fluctuating values, as indicated in Figure 19.5(b), or, in equation form,

$$T = \bar{T} + T' \qquad (19\text{-}41)$$

Any significant amount of energy transfer in the y direction, for bulk flow occurring in the x direction, is accomplished because of the fluctuating temperature, T'; thus it is apparent from equations (19-40) and (19-41) that

$$T' = \pm L \frac{d\bar{T}}{dy} \qquad (19\text{-}42)$$

The energy flux in the y direction may now be written as

$$\frac{q_y}{A}\bigg|_y = \rho c_p T v_y'$$ (19-43)

where v_y' may be either positive or negative. Substituting for T its equivalent, according to equation (19-41),

$$\frac{q_y}{A}\bigg|_y = \rho c_p v_y'(\bar{T} + T')$$

and taking the time average, we obtain, for the y-directional energy flux due to turbulent effects,

$$\frac{q_y}{A}\bigg|_{turb} = \rho c_p \overline{(v_y' T')}$$ (19-44)

or, with T' in terms of the mixing length,

$$\frac{q_y}{A}\bigg|_{turb} = \rho c_p \overline{v_y' L \frac{d\bar{T}}{dy}}$$ (19-45)

The total energy flux due to both microscopic and turbulent contributions may be written as

$$\frac{q_y}{A} = -\rho c_p [\alpha + \overline{|v_y' L|}] \frac{d\bar{T}}{dy}$$ (19-46)

Since α is the molecular diffusivity of heat, the quantity $\overline{|v_y' L|}$ is the *eddy diffusivity of heat*, designated ϵ_H. This quantity is exactly analogous to the eddy diffusivity of momentum, ϵ_M, as defined in equation (13-9). In a region of turbulent flow, $\epsilon_H \gg \alpha$ for all fluids except liquid metals.

Since the Prandtl number is the ratio of the molecular diffusivities of momentum and heat, an analogous term, the *turbulent Prandtl number*, can be formed by the ratio ϵ_M/ϵ_H. Utilizing equations (19-46) and (13-13), we have

$$\text{Pr}_{turb} = \frac{\epsilon_M}{\epsilon_H} = \frac{L^2 |dv_x/dy|}{|Lv_y'|} = \frac{L^2 |dv_x/dy|}{L^2 |dv_x/dy|} = 1$$ (19-47)

Thus in a region of fully turbulent flow the effective Prandtl number is unity, and the Reynolds analogy applies in the absence of form drag.

In terms of the eddy diffusivity of heat, the heat flux can be expressed as

$$\frac{q_y}{A}\bigg|_{turb} = -\rho c_p \epsilon_H \frac{d\bar{T}}{dy}$$ (19-48)

The total heat flux, including both molecular and turbulent contributions, thus becomes

$$\frac{q_y}{A} = -\rho c_p (\alpha + \epsilon_H) \frac{d\bar{T}}{dy}$$ (19-49)

Equation (19-49) applies both to the region wherein flow is laminar, for which $\alpha \gg \epsilon_H$, and to that for which flow is turbulent and $\epsilon_H \gg \alpha$. It is in this latter region that the Reynolds analogy applies. Prandtl* achieved a solution which includes the influences of both the laminar sublayer and the turbulent core. In his analysis solutions were obtained in each region and then joined at $y = \xi$, the hypothetical distance from the wall which is assumed to be the boundary separating the two regions.

Within the laminar sublayer the momentum and heat flux equations reduce to

$$\tau = \rho \nu \frac{dv_x}{dy} \quad \text{(a constant)}$$

and

$$\frac{q_y}{A} = -\rho c_p \alpha \frac{dT}{dy}$$

Separating variables and integrating between $y = 0$ and $y = \xi$, we have, for the momentum expression

$$\int_0^{v_x|_\xi} dv_x = \frac{\tau}{\rho \nu} \int_0^\xi dy$$

and for the heat flux,

$$\int_{T_s}^{T_\xi} dT = -\frac{q_y}{A \rho c_p \alpha} \int_0^\xi dy$$

Solving for the velocity and temperature profiles in the laminar sublayer yields

$$v_x|_\xi = \frac{\tau \xi}{\rho \nu} \tag{19-50}$$

and

$$T_s - T_\xi = \frac{q_y \xi}{A \rho c_p \alpha} \tag{19-51}$$

Eliminating the distance ξ between these two expressions gives

$$\frac{\rho \nu v_x|_\xi}{\tau} = \frac{\rho A c_p \alpha}{q_y}(T_s - T_\xi) \tag{19-52}$$

Directing our attention now to the turbulent core where the Reynolds analogy applies, we may write equation (19-36)

$$\frac{h}{\rho c_p (v_\infty - v_x|_\xi)} = \frac{C_f}{2} \tag{19-36}$$

* L. Prandtl, *Zeit. Physik.*, **11**, 1072 (1910).

and, expressing h and C_f in terms of their defining relations, we obtain

$$\frac{q_y/A}{\rho c_p (v_\infty - v_x|_\xi)(T_\xi - T_\infty)} = \frac{\tau}{\rho (v_\infty - v_x|_\xi)^2}$$

Simplifying and rearranging this expression, we have

$$\frac{\rho(v_\infty - v_x|_\xi)}{\tau} = \rho A c_p \frac{(T_\xi - T_\infty)}{q_y} \tag{19-53}$$

which is a modified form of the Reynolds analogy applying from $y = \xi$ to $y = y_{max}$.
Eliminating T_ξ between equations (19-52) and (19-53), we have

$$\frac{\rho}{\tau}\left[v_\infty + v_x|_\xi \left(\frac{\nu}{\alpha} - 1\right) \right] = \frac{\rho A c_p}{q_y}(T_s - T_\infty) \tag{19-54}$$

Introducing the coefficient of skin friction

$$C_f = \frac{\tau}{\rho v_\infty^2/2}$$

and the convective heat-transfer coefficient,

$$h = \frac{q_y}{A(T_s - T_\infty)}$$

we may reduce equation (19-54) to

$$\frac{v_\infty + v_x|_\xi(\nu/\alpha - 1)}{v_\infty^2 C_f/2} = \frac{\rho c_p}{h}$$

Inverting both sides of this expression and making it dimensionless, we obtain

$$\frac{h}{\rho c_p v_\infty} \equiv St = \frac{C_f/2}{1 + (v_x|_\xi/v_\infty)[(\nu/\alpha) - 1]} \tag{19-55}$$

This equation involves the ratio ν/α, which has been defined previously as the Prandtl number. For a value of $Pr = 1$, equation (19-55) reduces to the Reynolds analogy. For $Pr \neq 1$ the Stanton number is a function of C_f, Pr, and the ratio $v_x|_\xi/v_\infty$. It would be convenient to eliminate the velocity ratio; this may be accomplished by recalling some results from Chapter 13.

At the edge of the laminar sublayer,

$$v^+ = y^+ = 5$$

and by definition $v^+ = v_x/(\sqrt{\tau/\rho})$. Thus for the case at hand

$$v^+ = v_x|_\xi/(\sqrt{\tau/\rho}) = 5$$

Again introducing the coefficient of skin friction in the form

$$C_f = \frac{\tau}{\rho v_\infty^2/2}$$

we may write

$$\sqrt{\frac{\tau}{\rho}} = v_\infty \sqrt{\frac{C_f}{2}}$$

which, when combined with the previous expression given for the velocity ratio, gives

$$\frac{v_x|_\xi}{v_\infty} = 5\sqrt{\frac{C_f}{2}} \qquad (19\text{-}56)$$

Substitution of equation (19-56) into (19-55) gives

$$St = \frac{C_f/2}{1 + 5\sqrt{C_f/2}(\text{Pr} - 1)} \qquad (19\text{-}57)$$

which is known as the *Prandtl analogy*. This equation is written entirely in terms of measurable quantities.

von Kármán[*] extended Prandtl's work to include the effect of the transition or buffer layer in addition to the laminar sublayer and turbulent core. His result, the von Kármán analogy, is expressed as

$$St = \frac{C_f/2}{1 + 5\sqrt{C_f/2}\{\text{Pr} - 1 + \ln[1 + \frac{5}{6}(\text{Pr} - 1)]\}} \qquad (19\text{-}58)$$

Note that, just as for the Prandtl analogy, equation (19-58) reduces to the Reynolds analogy for a Prandtl number of unity.

The application of the Prandtl and von Kármán analogies is, quite logically, restricted to those cases in which there is negligible form drag. These equations yield the most accurate results for Prandtl numbers greater than unity.

An illustration of the use of the four relations developed in this section is given in the example below.

EXAMPLE 1

Water at 50°F enters a heat-exchanger tube having an inside diameter of 1 in. and a length of 10 ft. The water flows at 20 gal/min. For a constant wall temperature of 210°F estimate the exit temperature of the water using (a) the Reynolds analogy, (b) the Colburn analogy, (c) the Prandtl analogy, and (d) the von Kármán analogy. Entrance effects are to be neglected, and the properties of water may be evaluated at the arithmetic-mean bulk temperature.

[*] T. von Kármán, *Trans. ASME,* **61**, 705 (1939).

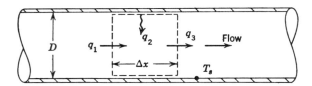

Figure 19.6 Analog analysis of water flowing in a circular tube.

Considering a portion of the heat-exchanger tube shown in Figure 19.6, we see that an application of the first law of thermodynamics to the control volume indicated will yield the result that

$$
\left\{\begin{array}{c}\text{Rate of heat}\\\text{transfer into c.v.}\\\text{by fluid flow}\end{array}\right\}+\left\{\begin{array}{c}\text{rate of heat}\\\text{transfer into c.v.}\\\text{by convection}\end{array}\right\}=\left\{\begin{array}{c}\text{rate of heat}\\\text{transfer out of c.v.}\\\text{by fluid flow}\end{array}\right\}
$$

If these heat-transfer rates are designated as q_1, q_2, and q_3, they may be evaluated as follows:

$$
q_1 = \rho\frac{\pi D^2}{4}v_x c_p T\big|_x
$$

$$
q_2 = h\pi D\,\Delta x(T_s - T)
$$

and

$$
q_3 = \rho\frac{\pi D^2}{4}v_x c_p T\big|_{x+\Delta x}
$$

The substitution of these quantities into the energy balance expression gives

$$
\rho\frac{\pi D^2}{4}v_x c_p[T\big|_{x+\Delta x} - T\big|_x] - h\pi D\,\Delta x(T_s - T) = 0
$$

which may be simplified and rearranged into the form

$$
\frac{D}{4}\frac{T\big|_{x+\Delta x} - T\big|_x}{\Delta x} + \frac{h}{\rho v_x c_p}(T - T_s) = 0 \tag{19-59}
$$

Evaluated in the limit as $\Delta x \to 0$, equation (19-59) reduces to

$$
\frac{dT}{dx} + \frac{h}{\rho v_x c_p}\frac{4}{D}(T - T_s) = 0 \tag{19-60}
$$

Separating the variables, we have

$$
\frac{dT}{T - T_s} + \frac{h}{\rho v_x c_p}\frac{4}{D}dx = 0
$$

and integrating between the limits indicated, we obtain

$$\int_{T_0}^{T_L} \frac{dT}{T - T_s} + \frac{h}{\rho v_x c_p} \frac{4}{D} \int_0^L dx = 0$$

$$\ln \frac{T_L - T_s}{T_0 - T_s} + \frac{h}{\rho v_x c_p} \frac{4L}{D} = 0 \tag{19-61}$$

Equation (19-61) may now be solved for the exit temperature T_L. Observe that the coefficient of the right-hand term, $h/\rho v_x c_p$, is the Stanton number. This parameter has been achieved quite naturally from our analysis.

The coefficient of skin friction may be evaluated with the aid of Figure 14.1. The velocity is calculated as

$$v_x = 20 \text{ gal/min } (\text{ft}^3/7.48 \text{ gal})[144/(\pi/4)(1^2)]\text{ft}^2(\text{min}/60 \text{ s}) = 8.17 \text{ fps}$$

Initially, we will assume the mean bulk temperature to be 90°F. The film temperature will then be 150°F, at which $\nu = 0.474 \times 10^{-5} \text{ ft}^2/\text{s}$. The Reynolds number is, thus,

$$\text{Re} = \frac{Dv_x}{\nu} = \frac{(1/12 \text{ ft})(8.17 \text{ ft/s})}{0.474 \times 10^{-5} \text{ ft}^2/\text{s}} = 144\,000$$

At this value of Re, the friction factor, f_f, assuming smooth tubing, is 0.0042. For each of the four analogies, the Stanton number is evaluated as follows:

(a) *Reynolds analogy,*

$$\text{St} = \frac{C_f}{2} = 0.0021$$

(b) *Colburn analogy,*

$$\text{St} = \frac{C_f}{2}\text{Pr}^{-2/3} = 0.0021(2.72)^{-2/3} = 0.00108$$

(c) *Prandtl analogy,*

$$\text{St} = \frac{C_f/2}{1 + 5\sqrt{C_f/2}(\text{Pr} - 1)}$$

$$= \frac{0.0021}{1 + 5\sqrt{0.0021}(1.72)} = 0.00151$$

(d) *Von Kármán analogy,*

$$\text{St} = \frac{C_f/2}{1 + 5\sqrt{C_f/2}\{\text{Pr} - 1 + \ln[1 + \frac{5}{6}(\text{Pr} - 1)]\}}$$

$$= \frac{0.0021}{1 + 5\sqrt{0.0021}\{2.72 - 1 + \ln[1 + \frac{5}{6}(2.72 - 1)]\}}$$

$$= 0.00131$$

Substituting these results into equation (19-61), we obtain, for T_L, the following results:

(a) $T_L = 152°F$
(b) $T_L = 115°F$
(c) $T_L = 132°F$
(d) $T_L = 125°F$

Some fine tuning of these results may be necessary to adjust the physical property values for the calculated film temperatures. In none of these cases is the assumed film temperature different than the calculated one by more than 6°F, so the results are not going to change much.

The Reynolds analogy value is much different from the other results obtained. This is not surprising, since the Prandtl number was considerably above a value of one. The last three analogies yielded quite consistent results. The Colburn analogy is the simplest to use and is preferable from that standpoint.

19.8 CLOSURE

The fundamental concepts of convection heat transfer have been introduced in this chapter. New parameters pertinent to convection are the Prandtl, Nusselt, Stanton, and Grashof numbers.

Four methods of analyzing a convection heat-transfer process have been discussed. These are as follows:

(1) dimensional analysis coupled with experiment;
(2) exact analysis of the boundary layer;
(3) integral analysis of the boundary layer;
and (4) analogy between momentum and energy transfer.

Several empirical equations for the prediction of convective heat-transfer coefficients will be given in the chapters to follow.

PROBLEMS

19.1 Using dimensional analysis, demonstrate that the parameters

$$\frac{T - T_\infty}{T_0 - T_\infty} \quad \frac{x}{L} \quad \frac{\alpha t}{L^2} \quad \text{and} \quad \frac{hL}{k}$$

are possible combinations of the appropriate variables in describing unsteady-state conduction in a plane wall.

19.2 Dimensional analysis has shown the following parameters to be significant for forced convection:

$$\frac{x v_\infty \rho}{\mu} \quad \frac{\mu c_p}{k} \quad \frac{hx}{k} \quad \frac{h}{\rho c_p v_\infty}$$

Evaluate each of these parameters at 340 K, for air, water, benzene, mercury, and glycerin. The distance x may be taken as 0.3 m, $v_\infty = 15$ m/s, and $h = 34$ W/m$^2 \cdot$ K.

19.3 Plot the parameters $xv_\infty\rho/\mu$, $\mu c_p/k$, hx/k, and $h/\rho c_p v_\infty$ versus temperature for air, water, and glycerin, using the values for x, h, and v from problem 19.2.

19.4 Nitrogen at 100°F and 1 atm flows at a velocity of 10 fps. A flat plate 6 in. wide, at a temperature of 200°F, is aligned parallel to the direction of flow. At a position 4 ft from the leading edge, determine the following:
(a) δ; (b) δ_t; (c) C_{fx}; (d) C_{fL}; (e) h_x; (f) h; (g) total drag force; (h) total heat transfer.

19.5 The fuel plates in a nuclear reactor are 4 ft long and stacked with a $\frac{1}{2}$-in. gap between them. The heat flux along the plate surfaces varies sinusoidally according to the equation

$$\frac{q}{A} = \alpha + \beta \sin\frac{\pi x}{L}$$

where $\alpha = 250$ Btu/hr ft^2, $\beta = 1500$ Btu/hr ft^2, x is the distance from the leading edge of the plates, and L is the total plate length. If air at 120°F, 80 psi, flowing at a mass velocity of 6000 lb$_m$/hr ft^2, is used to cool the plates, prepare plots showing
(a) the heat flux versus x;
(b) the mean air temperature versus x.

19.6 Given the information in problem 19.5, determine the total heat transferred for a stack of plates with a combined surface area of 640 ft^2, each plate being 4 ft wide.

19.7 Shown in the figure is the case of a fluid flowing parallel to a flat plate, where, for a distance X from the leading edge, the plate and fluid are at the same temperature. For values of $x > X$ the plate is maintained at a constant temperature, T_s, where $T_s > T_\infty$. Assuming a cubic profile for both the

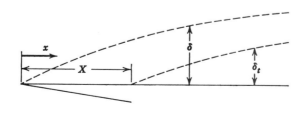

hydrodynamic and the thermal boundary layers, show that the ratio of the thickness, ξ, is expressed as

$$\xi = \frac{\delta_t}{\delta} \cong \frac{1}{Pr^{1/3}}\left[1-\left(\frac{X}{x}\right)^{3/4}\right]^{1/3}$$

Also show that the local Nusselt number can be expressed as

$$Nu_x \cong 0.33\left(\frac{Pr}{1-(X/x)^{3/4}}\right)^{1/3} Re_x^{1/2}$$

19.8 Show that, for the case of natural convection adjacent to a plane vertical wall, the appropriate integral equations for the hydrodynamic and thermal boundary layers are

$$\alpha\frac{\partial T}{\partial y}\bigg|_{y=0} = \frac{d}{dx}\int_0^{\delta_t} v_x(T_\infty - T)\,dy$$

and

$$-\nu\frac{\partial v_x}{\partial y}\bigg|_{y=0} + \beta g\int_0^{\delta_t}(T-T_\infty)\,dy = \frac{d}{dx}\int_0^\delta v_x^2\,dy$$

19.9 Using the integral relations from problem 19.8, and assuming the velocity and temperature profiles of the form

$$\frac{v}{v_x} = \left(\frac{y}{\delta}\right)\left(1-\frac{y}{\delta}\right)^2$$

and

$$\frac{T-T_\infty}{T_s-T_\infty} = \left(1-\frac{y}{\delta}\right)^2$$

where δ is the thickness of both the hydrodynamic and thermal boundary layers, show that the solutions in terms of δ and v_x from each integral equation reduce to

$$\frac{2\alpha}{\delta} = \frac{d}{dx}\left(\frac{\delta v_x}{30}\right)$$

and

$$-\frac{\nu v_x}{\delta} + \beta g\,\Delta T\frac{\delta}{3} = \frac{d}{dx}\left(\frac{\delta v_x^2}{105}\right)$$

Next, assuming that both δ and v_x vary with x according to

$$\delta = Ax^a \qquad \text{and} \qquad v_x = Bx^b$$

show that the resulting expression for δ becomes

$$\delta/x = 3.94 Pr^{-1/2}(Pr+0.953)^{1/4}\, Gr_x^{-1/4}$$

and that the local Nusselt number is

$$Nu_x = 0.508\, Pr^{1/2}(Pr+0.953)^{-1/4}\, Gr_x^{1/4}$$

19.10 Using the relations from problem 19.9, determine, for the case of air at 310 K adjacent to a vertical wall with its surface at 420 K,
(a) the thickness of the boundary layer at $x = 15$ cm, 30 cm, 1.5 m,
(b) the magnitude of h_x at 15 cm, 30 cm, 1.5 m.

19.11 Determine the total heat transfer from the vertical wall described in problem 19.10 to the surrounding air per meter of width if the wall is 2.5 m high.

19.12 Simplified relations for natural convection in air are of the form

$$h = \alpha(\Delta T/L)^\beta$$

where α, β are constants; L is a significant length, in ft; ΔT is $T_s - T_\infty$, in °F; and h is the convective heat-transfer coefficient, Btu/hr ft^2 °F. Determine the values for α and β for the plane vertical wall, using the equation from problem 19.9.

19.13 Using the appropriate integral formulas for flow parallel to a flat surface with a constant free-stream velocity, develop expressions for the local Nusselt number in terms of Re_x and Pr for velocity and temperature profiles of the form

$$v = a + by, \qquad T - T_s = \alpha + \beta y$$

19.14 Repeat problem 19.13 for velocity and temperature profiles of the form
$$v = a + by + cy^2 \qquad T - T_s = \alpha + \beta y + \gamma y^2$$

19.15 Repeat problem 19.13 for velocity and temperature profiles of the form
$$v = a \sin by \qquad T - T_s = \alpha \sin \beta y$$

19.16 For the case of a turbulent boundary layer on a flat plate, the velocity profile has been shown to follow closely the form

$$\frac{v}{v_\infty} = \left(\frac{y}{\delta}\right)^{1/7}$$

Assuming a temperature profile of the same form, that is,

$$\frac{T - T_s}{T_\infty - T_s} = \left(\frac{y}{\delta_t}\right)^{1/7}$$

and assuming that $\delta = \delta_t$, use the integral relation for the boundary layer to solve for h_x and Nu_x. The temperature gradient at the surface may be considered similar to the velocity gradient at $y = 0$ given by equation (13-26).

19.17 Water, at 60°F, enters a 1-in. ID tube which is used to cool a nuclear reactor. The water flow rate is 30 gal/min. Determine the total heat transfer and exiting water temperature for a 15-ft-long tube if the tube surface temperature is a constant value of 300°F. Compare the answer obtained, using the Reynolds and Colburn analogies.

19.18 Water at 60°F enters a 1-in. ID tube which is used to cool a nuclear reactor. The water flow rate is 30 gal/min. Determine the total heat transfer, exiting water temperature, and the wall temperature at the exit of a 15-ft long tube if the tube wall condition is one of uniform heat flux of 500 Btu/hr ft^2.

19.19 Solve problem 19.18 for the case of a wall heat flux varying according to

$$\frac{q}{A} = \alpha + \beta \sin \frac{\pi x}{L}$$

where $\alpha = 250$ Btu/hr ft^2, $\beta = 1500$ Btu/hr ft^2, x is the distance from the entrance, and L is the tube length.

19.20 Work problem 19.17 for the case in which the flowing fluid is air at 15 fps.

19.21 Work problem 19.18 for the case in which the flowing fluid is air at 15 fps.

19.22 Work problem 19.17 for the case in which the flowing fluid is sodium entering the pipe at 200°F.

19.23 Work problem 19.18 for the case in which the flowing fluid is sodium entering the pipe at 200°F.

19.24 Use the results of problem 19.7 along with those of Chapter 12 to determine δ, C_{fx}, δ_t, and h_x at a distance of 40 cm from the leading edge of a flat plate. Air with a free stream velocity of 5 m/s and $T_\infty = 300$ K flows parallel to the plate surface. The first 20 cm of the plate is unheated; the surface temperature is maintained at 400 K beyond that point.

19.25 Glycerin flows parallel to a flat plate measuring 2 ft by 2 ft with a velocity of 10 fps. Determine values for the mean convective heat-transfer coefficient and the associated drag force imposed on the plate for glycerin temperatures of 30, 50, and 180°F. What heat flux will result, in each case, if the plate temperature is 50°F above that of the glycerin?

19.26 Given the information in problem 19.25, construct a plot of local heat-transfer coefficient vs. position along the plate for glycerin temperatures of 30, 50, and 80°F.

20
CONVECTIVE HEAT-TRANSFER CORRELATIONS

Convective heat transfer was treated from an analytical point of view in Chapter 19. While the analytic approach is very meaningful, it may not offer a practical solution to every problem. There are many situations for which no mathematical models have as yet been successfully applied. Even in those cases for which an analytical solution is possible, it is necessary to verify the results by experiment. In this chapter we shall present some of the most useful correlations of experimental heat-transfer data available. Most correlations are in the forms indicated by dimensional analysis.

The sections to follow include discussion and correlations for natural convection, forced convection for internal flow, and forced convection for external flow, respectively. In each case, those analytical relations which are available are presented along with the most satisfactory empirical correlations for a particular geometry and flow condition.

20.1 NATURAL CONVECTION

The mechanism of energy transfer by natural convection involves the motion of a fluid past a solid boundary which is the result of the density differences resulting from the energy exchange. Because of this, it is quite natural that the heat-transfer coefficients and their correlating equations will vary with the geometry of a given system.

Vertical plates. The natural convection system most amenable to analytical treatment is that of a fluid adjacent to a vertical wall.

Standard nomenclature for a two-dimensional consideration of natural convection adjacent to a vertical plane surface is indicated in Figure 20.1. The x direction is commonly taken along the wall, with y measured normal to the plane surface.

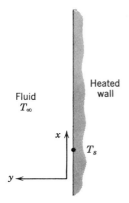

Schmidt and Beckmann* measured the temperature and velocity of air at different locations near a vertical plate and found a significant variation in both quantities along the direction parallel to the plate. The variations of velocity and temperature for a 12.5-cm-high vertical plate are shown in Figures 20.2 and 20.3 for the conditions $T_s = 65°C$, $T_\infty = 15°C$.

Figure 20.1 Coordinate system for the analysis of natural convection adjacent to a heated vertical wall.

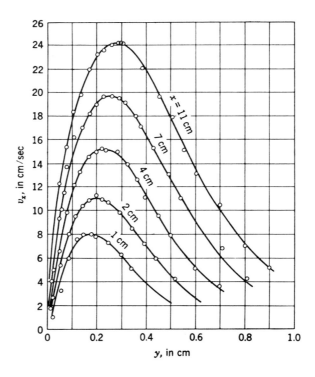

Figure 20.2 Velocity distribution in the vicinity of a vertical heated plate in air.

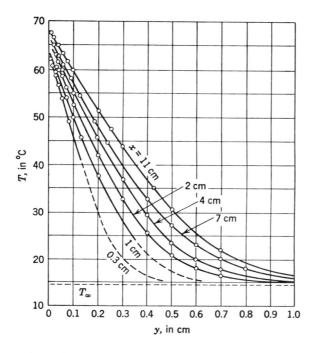

Figure 20.3 Temperature distribution in the vicinity of a vertical heated plate in air.

The two limiting cases for vertical plane walls are those with constant surface temperature and with constant wall heat flux. The former of these cases have been solved by Ostrach* and the latter by Sparrow and Gregg.†

Ostrach, employing a similarity transformation with governing equations of mass conservation, motion, and energy in a free-convection boundary layer, obtained an expression for local Nusselt number of the form

$$\mathrm{Nu}_x = f(\mathrm{Pr})\left(\frac{\mathrm{Gr}_x}{4}\right)^{1/4} \qquad (20\text{-}1)$$

The coefficient, $f(\mathrm{Pr})$, varies with Prandtl number, with values given in Table 20.1.

TABLE 20.1 VALUES OF THE COEFFICIENT $f(\mathrm{Pr})$ FOR USE IN EQUATION (20-1)

Pr	0.01	0.072	1	2	10	100	1000
$f(\mathrm{Pr})$	0.081	0.505	0.567	0.716	1.169	2.191	3.966

* S. Ostrach, NACA Report 1111, 1953.
† E. M. Sparrow and J. L. Gregg, *Trans. A.S.M.E.*, **78**, 435 (1956).

We usually find the mean Nusselt number, Nu_L, to be of more value than Nu_x. Using an integration procedure, as discussed earlier, an expression for Nu_L may be determined using equation (20-1). The mean heat transfer coefficient for a vertical surface of height, L, is related to the local value according to

$$h_L = \frac{1}{L} \int_0^L h_x \, dx$$

Inserting equation (20-1) appropriately, we proceed to

$$h_L = \frac{k}{L} f(\text{Pr}) \left[\frac{\beta g_{\Delta T}}{4\nu^2} \right]^{1/4} \int_0^L x^{-1/4} \, dx$$

$$= \left(\frac{4}{3} \right)\left(\frac{k}{L} \right) f(\text{Pr}) \left[\frac{\beta g L^3 \, \Delta T}{4\nu^2} \right]^{1/4}$$

and, in dimensionless form, we have

$$Nu_L = \frac{4}{3} f(\text{Pr})\left(\frac{Gr_L}{4} \right)^{1/4} \tag{20-2}$$

Sparrow and Gregg's* results for the constant wall heat flux case compare within 5% to those of Ostrach for like values of Pr. Equations (20-1) and (20-2), along with coefficients from Table 20.1, may thus be used, with reasonable accuracy, to evaluate any vertical plane surface regardless of wall conditions, provided boundary layer flow is laminar.

Fluid properties, being temperature dependent, will have some effect on calculated results. It is important therefore that properties involved in equations (20-1) and (20-2) be evaluated at the film temperature.

$$T_f = \frac{T_w + T_\infty}{2}$$

As with forced convection, turbulent flow will also occur in free convection boundary layers. When turbulence is present, an analytical approach is quite difficult, and we must rely heavily on correlations of experimental data.

Transition from laminar to turbulent flow in natural convection boundary layers adjacent to vertical plane surfaces has been determined to occur at, or near,

$$Gr_t \, Pr = Ra_t \cong 10^9 \tag{20-3}$$

where the subscript, t, indicates transition. The product, $Gr \, Pr$, is often referred to as Ra, the **Rayleigh** number.

Churchill and Chu[†] have correlated a large amount of experimental data for natural convection adjacent to vertical planes over 13 orders of magnitude

---

* E. M. Sparrow and J. L. Gregg, *Trans. A.S.M.E.*, **78**, 435 (1956).

† S. W. Churchill and H. S. Chu, *Int. J. Heat & Mass Tr.*, **18**, 1323 (1975).

of Ra. They propose a single equation for Nu_L that applies to all fluids. This powerful equation is

$$Nu_L = \left\{ 0.825 + \frac{0.387 \, Ra_L^{1/6}}{[1 + (0.492/Pr)^{9/16}]^{8/27}} \right\}^2 \qquad (20\text{-}4)$$

Churchill and Chu show this expression to provide accurate results for both laminar and turbulent flows. Some improvement was found for the laminar range ($Ra_L < 10^9$) by using the following equation:

$$Nu_L = 0.68 + \frac{0.670 \, Ra_L^{1/4}}{[1 + (0.492/Pr)^{9/16}]^{4/9}} \qquad (20\text{-}5)$$

Vertical Cylinders. For the case of cylinders with their axes vertical, the expressions presented for plane surfaces can be used provided the curvature effect is not too great. The criterion for this is expressed in equation (20-6); specifically, a vertical cylinder can be evaluated using correlations for vertical plane walls when

$$\frac{D}{L} \geq \frac{35}{Gr_L^{1/4}} \qquad (20\text{-}6)$$

Physically, this represents the limit where boundary layer thickness is small relative to cylinder diameter, D.

Horizontal Plates. The correlations suggested by McAdams[*] are well accepted for this geometry. A distinction is made regarding whether the fluid is hot or cool, relative to the adjacent fluid, and whether the surface faces up or down. It is clear that the induced buoyancy will be much different for a hot surface facing up than down. McAdams' correlations are, for a hot surface facing up or cold surface facing down,

$$10^5 < Ra_L < 2 \times 10^7 \qquad Nu_L = 0.54 \, Ra_L^{1/4} \qquad (20\text{-}7)$$

$$2 \times 10^7 < Ra_L < 3 \times 10^{10} \qquad Nu_L = 0.14 \, Ra_L^{1/3} \qquad (20\text{-}8)$$

and for a hot surface facing down or cold surface facing up,

$$3 \times 10^5 < Ra_L < 10^{10} \qquad Nu_L = 0.27 \, Ra_L^{1/4} \qquad (20\text{-}9)$$

In each of these correlating equations the film temperature, T_f, should be used for fluid property evaluation. The length scale, L, is the ratio of plate surface area to perimeter.

For plane surfaces inclined at an angle, θ, with the vertical, equations (20-4) and (20-5) may be used, with modification, for values of θ up to 60°. Churchill and Chu[†] suggest replacing g by $g \cos \theta$ in equation (20-5) when boundary layer

[*] W. H. McAdams, *Heat Transmission*, Third Edition, Chapter 7, McGraw-Hill Book Company, New York, 1957.
[†] S. W. Churchill and H. H. S. Chu, *Int. J. Heat & Mass Tr.*, **18**, 1323 (1975).

flow is laminar. With turbulent flow equation (20-4) may be used without modification.

Horizontal Cylinders. With cylinders of sufficient length that end effects are insignificant, two correlations are recommended. Churchill and Chu* suggest the following correlation,

$$Nu_D = \left\{ 0.60 + \frac{0.387 \, Ra_D}{[1 + (0.559/Pr)^{9/16}]^{8/27}} \right\}^2 \tag{20-10}$$

over the Rayleigh number range $10^{-5} < Ra_D < 10^{12}$.

A simpler equation has been suggested by Morgan,† in terms of variable coefficients

$$Nu_D = C \, Ra_D^{\,n} \tag{20-11}$$

where values of c and n are specified as functions of Ra_D in Table 20.2.

TABLE 20.2 VALUES OF CONSTANTS C AND n IN EQUATION (20-11)

	C	n
$10^{-10} < Ra_D < 10^{-2}$	0.675	0.058
$10^{-2} < Ra_D < 10^{2}$	1.02	0.148
$10^{2} < Ra_D < 10^{4}$	0.850	0.188
$10^{4} < Ra_D < 10^{7}$	0.480	0.250
$10^{7} < Ra_D < 10^{12}$	0.125	0.333

The film temperature should be used in evaluating fluid properties in the above equations.

Spheres. The correlation suggested by Yuge‡ is recommended for the case with $Pr \approx 1$, and $1 < Ra_D < 10^5$.

$$Nu_D = 2 + 0.43 \, Ra_D^{\,1/4} \tag{20-12}$$

We may notice that, for the sphere, as Ra approaches zero, heat transfer from the surface to the surrounding medium is by conduction. This problem may be solved to yield a limiting value for Nu_D equal to 2. This result is obviously compatible with equation (20-12).

* S. W. Churchill and H. H. S. Chu, *Int. J. Heat & Mass Tr.*, **18**, 1049 (1975).

† V. T. Morgan, *Advances in Heat Transfer*, Vol. II, T. F. Irvine and J. P. Hartnett, Eds., Academic Press, New York, 1975, pp. 199–264.

‡ T. Yuge, *J. Heat Transfer*, **82**, 214 (1960).

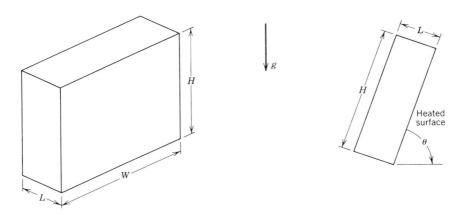

Figure 20.4   The rectangular enclosure.

Rectangular Enclosures. Shown in Figure 20.4 is the configuration and nomenclature pertinent to rectangular enclosures. These cases have become much more important in recent years due to their application in solar collectors. Clearly, heat transfer will be affected by the angle of tilt, θ; by the aspect ratio, H/L; and by the usual dimensionless parameters, Pr and Ra_L.

In each of the correlations to follow the temperature of the hotter of the two large surfaces is designated T_1, and the cooler surface is at temperature T_2. Fluid properties are evaluated at the film temperature, $T_f = (T_1 + T_2)/2$. Convective heat flux is expressed as

$$\frac{q}{A} = h(T_1 - T_2) \tag{20-13}$$

Case 1.   Horizontal Enclosures, $\theta = 0$

With the bottom surface heated a critical Rayleigh member has been determined by several investigators to exist. For cases where

$$\mathrm{Ra}_L = \frac{\beta g L^3 (T_1 - T_2)}{\alpha \nu} > 1700$$

conditions within an enclosure are thermally unstable and natural convection will occur. A correlation for this case has been proposed by Globe and Dropkin[*] in the form

$$\mathrm{Nu}_L = 0.069 \, \mathrm{Ra}_L^{1/3} \, \mathrm{Pr}^{0.074} \tag{20-14}$$

for the range $3 \times 10^5 < \mathrm{Ra}_L < 7 \times 10^9$.

When $\theta = 180°$, that is, the upper surface is heated, or when $\mathrm{Ra}_L < 1700$, heat transfer is by conduction; thus, $\mathrm{Nu}_L = 1$.

[*] S. Globe and D. Dropkin, *J. Heat Transfer*, **81C**, 24 (1959).

Case 2. Vertical Enclosures, $\theta = 90°$

For aspect ratios less than 10, Catton* suggests the use of the following correlations

$$\mathrm{Nu}_L = 0.18\left(\frac{\mathrm{Pr}}{0.2 + \mathrm{Pr}}\,\mathrm{Ra}_L\right)^{0.29} \qquad (20\text{-}15)$$

when $1 < H/L < 2$, $10^{-3} < \mathrm{Pr} < 10^5$, $10^3 < \mathrm{Ra}_L\mathrm{Pr}/(0.2 + \mathrm{Pr})$

and

$$\mathrm{Nu}_L = 0.22\left(\frac{\mathrm{Pr}}{0.22 + \mathrm{Pr}}\,\mathrm{Ra}_L\right)^{0.28}\left(\frac{H}{L}\right)^{-1/4} \qquad (20\text{-}16)$$

when $2 < H/L < 10$, $\mathrm{Pr} < 10^5$, $\mathrm{Ra}_L < 10^{10}$.

For higher values of H/L, the correlations of MacGregor and Emery† are recommended. These are

$$\mathrm{Nu}_L = 0.42\,\mathrm{Ra}_L^{1/4}\,\mathrm{Pr}^{0.012}(H/L)^{-0.3} \qquad (20\text{-}17)$$

for $10 < H/L < 40$, $1 < \mathrm{Pr} < 2 \times 10^4$, $10^4 < \mathrm{Ra}_L < 10^7$,

and

$$\mathrm{Nu}_L = 0.046\,\mathrm{Ra}_L^{1/3} \qquad (20\text{-}18)$$

for $10 < H/L < 40$, $1 < \mathrm{Pr} < 20$, $10^6 < \mathrm{Ra}_L < 10^9$.

Case 3. Tilted Vertical Enclosures, $0 < \theta < 90°$

Numerous recent publications have dealt with this configuration. Correlations for this case, when the aspect ratio is large $(H/L > 12)$, are the following:

$$\mathrm{Nu}_L = 1 + 1.44\left[1 - \frac{1708}{\mathrm{Ra}_L\cos\theta}\right]\left[1 - \frac{1708(\sin 1.8\theta)^{1.6}}{\mathrm{Ra}_L\cos\theta}\right]$$
$$+ \left[\left(\frac{\mathrm{Ra}_L\cos\theta}{5830}\right)^{1/3} - 1\right] \qquad (20\text{-}19)$$

when $H/L \geq 12$, $0 < \theta < 70°$. In applying this relationship any bracketed term with a negative value should be set equal to zero. Equation (20-19) was suggested by Hollands, *et al.*‡ With enclosures nearing the vertical, Ayyaswamy and Catton§ suggest the relationship

$$\mathrm{Nu}_L = \mathrm{Nu}_{LV}(\sin\theta)^{1/4} \qquad (20\text{-}20)$$

* I. Catton, *Proc. 6th Int. Heat Tr. Conference, Toronto, Canada,* **6**, 13 (1978).
† P. K. MacGregor and A. P. Emery, *J. Heat Transfer,* **91**, 391 (1969).
‡ K. G. T. Hollands, S. E. Unny, G. D. Raithby, and L. Konicek, *J. Heat Transfer,* **98**, 189 (1976).
§ P. S. Ayyaswamy and I. Catton, *J. Heat Transfer,* **95**, 543 (1973).

for all aspect ratios, and $70° < \theta < 90°$. The value $\theta = 70°$ is termed the "critical" tilt angle for vertical enclosures with $H/L > 12$. For smaller aspect ratios the critical angle of tilt is also smaller. A recommended review article on the subject of inclined rectangular cavities is that of Buchberg, Catton, and Edwards.*

EXAMPLE 1

Determine the surface temperature of a cylindrical tank measuring 2.5 ft in diameter and 4 ft high. The tank contains a transformer immersed in an oil bath that produces a uniform surface condition. Assume all heat loss to be by natural convection to the surrounding air at 70°F. The total electric energy dissipated by the transformer is 1.5 kW.
Surface areas that apply are:

$$A_{\text{top}} = A_{\text{bottom}} = \frac{\pi}{4}(2.5 \text{ ft})^2 = 4.91 \text{ ft}^2$$

$$A_{\text{side}} = \pi(2.5 \text{ ft})(4 \text{ ft}) = 31.4 \text{ ft}^2$$

Total heat transfer is the sum of the contributions from the three surfaces. This may be written as

$$q_{\text{total}} = [h_t(4.91) + h_b(4.91) + h_s(31.4)](T - 70)$$

with subscripts t, b, and s referring to top, bottom, and side.
Our expression for q_{total} will next be rewritten in the form

$$q_{\text{total}} = \left[\text{Nu}_t \frac{k}{2.5}(4.91) + \text{Nu}_b \frac{k}{2.5}(4.91) + \text{Nu}_s \frac{k}{4}(31.4) \right](T - 70)$$

or

$$q_{\text{total}} = [1.96 \, \text{Nu}_t + 1.96 \, \text{Nu}_b + 7.85 \, \text{Nu}_s]k \, (T - 70)$$

A complication exists in solving this equation, since the unknown is surface temperature. Our procedure must involve trial and error, where we will initially assume a surface temperature for property evaluation and then solve for T. This new value for T_{surface} will then be used and the procedure continued until the resulting temperature agrees with the one used in evaluating fluid properties.

To begin the problem, we assume that $T_{\text{surface}} = 230°F$. Properties will thus be evaluated at $T_f = 150°F$. For air at 150°F we have $\nu = 0.209 \times 10^{-3} \text{ ft}^2/\text{s}$, $k = 0.0167$ Btu/hr ft °F, $\alpha = 1.06 \text{ ft}^2/\text{hr}$, Pr $= 0.71$, and $\beta g/D^2 = 1.22 \times 10^6/\text{ft}^3$ °F.
For the vertical surface, we determine Gr to be

$$\text{Gr} = \frac{\beta g}{\nu^2} L^3 \, \Delta T$$

$$= (1.22 \times 10^6)(4)^3(160) = 1.25 \times 10^{10}$$

*H. Buchberg, I. Catton and D. K. Edwards, *J. Heat Transfer*, **98**, 182 (1976).

According to equation (20-6), we may neglect the effect of curvature if

$$\frac{D}{L} \geq \frac{35}{(1.25 \times 10^{10})^{1/4}} = 0.105$$

In the present case $D/L = 2.5/4 = 0.625$; thus the vertical surface will be treated using the equations for plane walls.

We must now solve for Nu_t, Nu_b, and Nu_s. Equations (20-8), (20-9), and (20-4) will be used. Equation (20-8) is the proper choice for the top, since it is clear that $\text{Ra}_D > 2 \times 10^7$.

The three values for Nu are determined as follows:

Nu_t: $L = \dfrac{A}{P} = \dfrac{\pi D^2/4}{\pi D} = \dfrac{D}{4} = 0.675 \text{ ft}$

$$\text{Nu}_t = 0.14[(1.22 \times 10^6)(0.625)^3(160)(0.71)]^{1/3}$$

$$= 45.3$$

Nu_b: $\text{Nu}_b = 0.27[(1.22 \times 10^6)(0.625)^3(160)(0.71)]^{1/4}$

$$= 20.6$$

Nu_s: $\text{Nu}_L = 0.825 + \left\{\dfrac{0.387[(1.22 \times 10^6)(4)^3(160)(0.71)]^{1/6}}{[1 + (0.492/0.71)^{9/16}]^{8/27}}\right\}^2$

$$= 243$$

The solution for T is now

$$T = 70 + \frac{(1.5 \text{ kW})(3413 \text{ Btu/kW hr})/0.0167 \text{ Btu/hr ft °F}}{1.96(45.3) + 1.96(20.6) + 7.85(243)}$$

$$= 70 + 151 = 221°F$$

With this value for T_{surface} the film temperature is approximately 146°F. Properties of air at this temperature are:

$$\nu = 0.207 \times 10^{-3} \text{ ft}^2/\text{s}, \ k = 0.0166 \text{ Btu/hr ft °F},$$

$$\alpha = 1.049 \text{ ft}^2/\text{hr}, \ \text{Pr} = 0.71, \ \beta g/\nu^2 = 1.26 \times 10^6/\text{ft °F}$$

New values of Nu become:

$$\text{Nu}_t = 0.14[(1.26 \times 10^6)(0.625)^3(151)(0.71)]^{1/3} = 44.9$$

$$\text{Nu}_D = 0.27[(1.26 \times 10^6)(0.625)^3(151)(0.71)]^{1/4} = 28.0$$

$$\text{Nu}_s = \left\{0.825 + \frac{0.387[(1.26 \times 10^6)(4)^3(151)(0.71)]^{1/6}}{\left[1 + \left(\dfrac{0.492}{0.71}\right)^{9/16}\right]^{8/27}}\right\}^2$$

$$= 241$$

The revised value for T then is

$$T = 70 + \frac{1.5(3413)/0.0166}{1.96(44.9) + 1.96(28) + 7.85(241)}$$

$$= 70 + 152 = \underline{222°F}$$

This value is obviously close enough; our solution is complete.

20.2 FORCED CONVECTION FOR INTERNAL FLOW

Undoubtedly the most important convective heat-transfer process from an industrial point of view is that of heating or cooling a fluid which is flowing inside a closed conduit. The momentum transfer associated with this type of flow was studied in Chapter 14. Many of the concepts and terminology of that chapter will be used in this section without further discussion.

Energy transfer associated with forced convection inside closed conduits will be considered separately for laminar and turbulent flow. The reader will recall that the critical Reynolds number for conduit flow is approximately 2300.

Laminar flow. The first analytical solution for laminar flow forced convection inside tubes was formulated by Graetz* in 1885. The assumptions basic to the Graetz solution are as follows:

(1) The velocity profile is parabolic and fully developed before any energy exchange between the tube wall and the fluid occurs.
(2) All properties of the fluid are constant.
(3) The surface temperature of the tube is constant at a value T_s during the energy transfer.

Considering the system as depicted in Figure 20.5, we may write the velocity profile as

$$v_x = v_{max} \left[1 - \left(\frac{r}{R} \right)^2 \right] \tag{8-7}$$

or, recalling that $v_{max} = 2v_{avg}$, we may write

$$v_x = 2v_{avg} \left[1 - \left(\frac{r}{R} \right)^2 \right] \tag{20-21}$$

The applicable form of the energy equation written in cylindrical coordinates, assuming radial symmetry, and neglecting $\partial^2 T / \partial x^2$ (axial conduction) in

* L. Graetz, *Ann. Phys. u. Chem.*, **25**, 337 (1885).

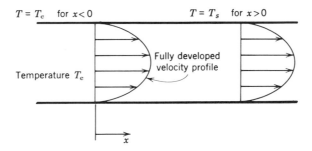

Figure 20.5 Boundary and flow conditions for the Graetz solution.

comparison to the radial variation in temperature is

$$v_x \frac{\partial T}{\partial x} = \alpha \left[\frac{1}{r} \frac{\partial}{\partial r} \left(r \frac{\partial T}{\partial r} \right) \right] \tag{20-22}$$

Substituting equation (20-21) for v_x into equation (20-22) gives

$$2v_{avg} \left[1 - \left(\frac{r}{R} \right)^2 \right] \frac{\partial T}{\partial x} = \alpha \left[\frac{1}{r} \frac{\partial}{\partial r} \left(r \frac{\partial T}{\partial r} \right) \right] \tag{20-23}$$

which is the equation to be solved subject to the boundary conditions

$$T = T_e \qquad \text{at } x = 0 \quad \text{for} \quad 0 \le r \le R$$

$$T = T_s \qquad \text{at } x > 0, \qquad r = R$$

and

$$\frac{\partial T}{\partial r} = 0 \qquad \text{at } x > 0, \qquad r = 0$$

The solution to equation (20-23) takes the form

$$\frac{T - T_e}{T_s - T_e} = \sum_{n=0}^{\infty} c_n f\left(\frac{r}{R} \right) \exp\left[-\beta_n^2 \frac{\alpha}{Rv_{avg}} \frac{x}{R} \right] \tag{20-24}$$

The terms c_n, $f(r/R)$, and β_n are all coefficients to be evaluated by using appropriate boundary conditions.

The argument of the exponential exclusive of β_n, that is, $(\alpha/Rv_{avg})(x/R)$, may be rewritten as

$$\frac{4}{(2Rv_{avg}/\alpha)(2R/x)} = \frac{4}{(Dv_{avg}\rho/\mu)(c_p\mu/k)(D/x)}$$

or, in terms of dimensionless parameters already introduced, this becomes

$$\frac{4}{\text{Re Pr } D/x} = \frac{4x/D}{\text{Pe}}$$

The product of Re and Pr is often referred to as the *Peclet number*, Pe. Another parameter encountered in laminar forced convection is the *Graetz number*, Gz, defined as

$$\text{Gz} \equiv \frac{\pi}{4}\frac{D}{x}\text{Pe}$$

Detailed solutions of equation (20-24) are found in the literature, and Knudsen and Katz* summarize these quite well. Figure 20.6 presents the results

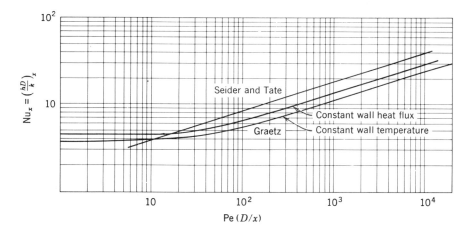

Figure 20.6 Variation in the local Nusselt number for laminar flow in tubes.

of the Graetz solution graphically for two different boundary conditions at the wall, these being (1) a constant wall temperature, and (2) uniform heat input at the wall.

Experimental data for laminar flow in tubes have been correlated by Seider and Tate† by the equation

$$\text{Nu}_D = 1.86\left(\text{Pe}\frac{D}{L}\right)^{1/3}\left(\frac{\mu_b}{\mu_w}\right)^{0.14} \tag{20-25}$$

The Seider–Tate relation is also shown in Figure 20.7 along with the two Graetz

* J. G. Knudsen and D. L. Katz, *Fluid Dynamics and Heat Transfer*, McGraw-Hill Book Company, New York, 1958, p. 370.
† F. N. Seider and G. E. Tate, *Ind. Eng. Chem.*, **28**, 1429 (1936).

results. These results cannot be compared directly since the Graetz results yield local values of h_x and the Seider–Tate equation gives mean values of the heat transfer coefficient. The last part of equation (20-25), the ratio of the fluid viscosity at the arithmetic mean bulk temperature to that at the temperature of the wall, takes into account the significant effect that variable fluid viscosity has on the heat transfer rate. All properties other than μ_w are evaluated at the bulk fluid temperature.

Turbulent flow. When considering energy exchange between a conduit surface and a fluid in turbulent flow we must resort to correlations of experimental data as suggested by dimensional analysis. The three most-used equations of this nature and the restrictions on their use are as follows.

Dittus and Boelter[*] proposed the following equation of the type suggested earlier by dimensional analysis, equation (19-7):

$$Nu_D = 0.023 \, Re_D^{0.8} \, Pr^n \tag{20-26}$$

where

(1) $n = 0.4$ if the fluid is being heated, $n = 0.3$ is the fluid is being cooled;
(2) all fluid properties are evaluated at the arithmetic-mean bulk temperature;
(3) the value of Re_D should be $> 10^4$;
(4) Pr is in the range $0.7 < Pr < 100$; and
(5) $L/D > 60$.

Colburn[†] proposed an equation using the Stanton number, St, in place of Nu_D as related in equation (19-9). His equation is

$$St = 0.023 \, Re_D^{-0.2} \, Pr^{-2/3} \tag{20-27}$$

where

(1) Re_D and Pr are evaluated at the *film* temperature, and St is evaluated at the bulk temperature;
(2) Re_D, Pr, and L/D should have values within the following limits:

$$Re_D > 10^4 \qquad 0.7 < Pr < 160 \qquad \text{and} \qquad L/D > 60$$

To account for high Prandtl number fluids, such as oils, Seider and Tate[‡] proposed the equation

$$St = 0.023 \, Re_D^{-0.2} \, Pr^{-2/3} \left(\frac{\mu_b}{\mu_w}\right)^{0.14} \tag{20-28}$$

[*] F. W. Dittus and L. M. K. Boelter, University of California, *Publ. Eng.*, **2**, 443 (1930).
[†] A. P. Colburn, *Trans. A.I.Ch.E.*, **29**, 174 (1933).
[‡] E. N. Seider and G. E. Tate, *Ind. Eng. Chem.*, **28**, 1429 (1936).

where

(1) All fluid properties except μ_w are evaluated at bulk temperature;
(2) $Re_D > 10^4$;
(3) $0.7 < Pr < 17\,000$;

and

(4) $L/D > 60$.

Of the three equations presented, the first two are most often used for those fluids whose Prandtl numbers are within the specified range. The Dittus–Boelter equation is simpler to use than the Colburn equation because of the fluid property evaluation at the bulk temperature.

The following examples illustrate the use of some of the expressions presented in this section.

EXAMPLE 2

Hydraulic fluid (MIL-M-5606) with a fully developed velocity profile flows through a 2-ft-long copper tube having a 1-in.-ID at an average velocity of 10 fpm. The oil enters at 70°F. Steam condenses on the outside surface of the tube; the associated heat-transfer coefficient is 2000 Btu/hr ft² °F. Find the rate of heat transfer to the oil.

In order to use a film temperature or an average bulk temperature for fluid property evaluation, one must know the exit temperature. Equation (19-61) will be useful in this determination:

$$\ln \frac{T_L - T_s}{T_0 - T_s} + 4 \frac{L}{D} \frac{h}{\rho v c_p} = 0 \tag{19-61}$$

The heat-transfer rate may be approximated as follows if the thermal resistance of the copper tube tube is neglected:

$$q = \frac{A_{\text{surf}} \Delta T}{1/h_i + 1/h_o} = \rho A v c_p (T_L - T_0)$$

To determine the type of flow we must evaluate Re; a bulk oil temperature of 100°F is assumed,

$$Re_D = \frac{Dv}{\nu} = \frac{(\tfrac{1}{12}\,\text{ft})(\tfrac{10}{60}\,\text{fps})}{10.7 \times 10^{-5}\,\text{ft}^2/\text{s}} = 130$$

and the flow is in the laminar range. The film coefficient, h_i, may be determined from equation (20-25) as follows:

$$h_i = \frac{k}{D} Nu_D = \frac{k}{D} 1.86 \left(Pe \frac{D}{L} \right)^{0.33} \left(\frac{\mu_b}{\mu_w} \right)^{0.14}$$

As a first assumption we shall assume $T_{\text{bulk}} = 100°F$ and $T_{\text{wall}} = 210°F$. Substituting the appropriate values from Appendix I at these temperatures, we have

$$h_i = \left(\frac{0.0690 \text{ Btu/hr ft °F}}{\frac{1}{12} \quad \text{ft}}\right)(1.86)\left[(130)(136)\frac{\frac{1}{12}}{2}\right]^{0.33}\left[\frac{556}{250}\right]^{0.14}$$

$$= 15.56 \text{ Btu/hr ft}^2 \text{ °F} \qquad (88.4 \text{ W/m}^2 \cdot \text{K})$$

Substituting this value of h into equation (19-61),

$$\ln\frac{T_s - T_L}{T_s - T_0} + \left(\frac{4 \times 2}{\frac{1}{12}}\right)\frac{15.56 \text{ Btu/hr ft}^2 \text{ °F}}{(52) \times \frac{1}{6} \times 0.467 \times 3600} = 0$$

we obtain

$$\frac{T_s - T_L}{T_s - T_0} = e^{-0.102} = 0.903$$

or

$$T_L = 210 - 0.903(140) = 83.6\text{°F} \qquad (302 \text{ K})$$

The average bulk temperature is thus

$$\frac{70 + 83.6}{2} = 76.8\text{°F} \qquad (298 \text{ K})$$

The calculation of h_i, using this temperature, yields

$$h_i = 12.38 \text{ Btu/hr ft}^2 \text{ °F}$$

$$T_L = 81\text{°F} \qquad T_b = 75.5\text{°F}$$

This agreement is sufficient; with this value for h_i the heat flow rate is

$$q = (52.4 \text{ lb}_\text{m}/\text{ft}^3)\left(\frac{\pi}{4} \times \frac{1}{144} \text{ ft}^2\right)(10 \times 60 \text{ ft/hr})(0.443 \text{ Btu/lb}_\text{m} \text{ °F})(11\text{°F})$$

$$= 836 \text{ Btu/hr} \qquad (245 \text{ W})$$

EXAMPLE 3

Air at 1 atmosphere and a temperature of 60°F enters a $\frac{1}{2}$-in.-ID tube with a velocity of 80 fps. The wall is maintained at a constant temperature of 210°F by condensing steam. Find the convective heat-transfer coefficient for this situation if the tube is 5 ft long.

As in example 2 it will be necessary to calculate the exit fluid temperature by

$$\ln\frac{T_L - T_s}{T_0 - T_s} + 4\frac{L}{D}\frac{h}{\rho v c_p} = 0 \qquad (19\text{-}61)$$

Another expression to be satisfied is

$$q = hA_s(T_s - T) = \rho A v c_p(T_L - T_0)$$

Evaluating the Reynolds number at the tube entrance, we have

$$\text{Re} = \frac{Dv}{\nu} = \frac{[(\frac{1}{2})/12] \text{ ft}(80 \text{ fps})}{0.159 \times 10^{-3} \text{ ft}^2/\text{s}} = 21\,000$$

The flow is turbulent and Re is sufficiently high that equation (20-26), (20-27), or (20-28) may be used.

Employing equation (20-27) and assuming an exit temperature of 190°F, giving a mean bulk temperature of 125°F, and a film temperature of 167°F, we obtain

$$St = \frac{h}{\rho v c_p} = 0.023 \, Re^{-0.2} \, Pr^{-2/3}$$

$$= 0.023 \left[\frac{(\frac{1}{24} \, ft)(80 \, fps)(0.0764 \, lb_m/ft^3)}{1.45 \times 10^{-5} \, lb_m/ft \, s} \right]^{-0.2} (0.694)^{-2/3}$$

$$= 0.023(0.1416)(1.276) = 0.00416$$

Substitution into equation (19-61) yields

$$\frac{T_L - T_s}{T_0 - T_s} = \exp\left[-4\left(\frac{5}{\frac{1}{24}}\right)(0.00416)\right]$$

$$= \exp[-1.99] = 0.136$$

and

$$T_L = 210 - 0.136(150) = 189.6°F \qquad (361 \, K)$$

This agreement is excellent. The heat transfer coefficient is

$$h = \rho v c_p(St)$$

$$= (0.0764 \, lb_m/ft^3)(80 \, fps)(0.240 \, Btu/lb_m \, °F)(0.00416)(3600 \, s/hr)$$

$$= 22.0 \, Btu/hr \, ft^2 \, °F \qquad (125 \, W/m^2 \cdot K)$$

For flow in short passages the correlations presented thus far must be modified to account for variable velocity and temperature profiles along the axis of flow. Deissler* has analyzed this region extensively for the case of turbulent flow. The following equations may be used to modify the heat-transfer coefficients in passages for which $L/D < 60$:

for $2 < L/D < 20$,

$$\frac{h_L}{h_\infty} = 1 + (D/L)^{0.7} \qquad (20-29)$$

and for $20 < L/D < 60$,

$$\frac{h_L}{h_\infty} = 1 + 6D/L \qquad (20-30)$$

Both of these expressions are approximations relating the appropriate coefficient, h_L, in terms of h_∞, where h_∞ is the value calculated for $L/D > 60$.

---

* R. G. Deissler, *Trans. A.S.M.E.*, **77**, 1221 (1955).

20.3 FORCED CONVECTION FOR EXTERNAL FLOW

Numerous situations exist in practice in which one is interested in analyzing or describing heat transfer associated with the flow of a fluid past the exterior surface of a solid. The sphere and cylinder are the shapes of greatest engineering interest, with heat transfer between these surfaces and a fluid in crossflow frequently encountered.

The reader will recall the nature of momentum-transfer phenomena discussed in Chapter 12 relative to external flow. The analysis of such flow and of heat transfer in these situations is complicated when the phenomenon of boundary-layer separation is encountered. Separation will occur in those cases in which an adverse pressure gradient exists; such a condition will exist for most situations of engineering interest.

Flow Parallel to Plane Surfaces. This condition is amenable to analysis and has already been discussed in Chapter 19. The significant results are repeated here for completeness.

We recall that, in this case, the boundary layer flow regimes are laminar for $\mathrm{Re}_x < 2 \times 10^5$ and turbulent for $3 \times 10^6 < \mathrm{Re}_x$. For the laminar range

$$\mathrm{Nu}_x = 0.332\,\mathrm{Re}_x^{1/2}\,\mathrm{Pr}^{1/3} \tag{19-25}$$

and

$$\mathrm{Nu}_L = 0.664\,\mathrm{Re}_L^{1/2}\,\mathrm{Pr}^{1/3} \tag{19-26}$$

With turbulent flow in the boundary layer, an application of the Colburn analogy

$$\mathrm{St}_x\mathrm{Pr}^{2/3} = \frac{C_{fx}}{2} \tag{19-37}$$

along with equation (13-31) yields

$$\mathrm{Nu}_x = 0.0288\,\mathrm{Re}_x^{4/5}\,\mathrm{Pr}^{1/3} \tag{20-31}$$

A mean Nusselt number can be calculated using this expression for Nu_x. The resulting expression is

$$\mathrm{Nu}_L = 0.360\,\mathrm{Re}_L^{4/5}\,\mathrm{Pr}^{1/3} \tag{20-32}$$

Fluid properties should be evaluated at the film temperature when using these equations.

Cylinders in crossflow. Eckert and Soehngen[*] evaluated local Nusselt numbers at various positions on a cylindrical surface past which flowed an air stream with a range in Reynolds numbers from 20 to 600. Their results are shown in

---

[*] E. R. G. Eckert and E. Soehngen, *Trans. A.S.M.E.*, **74**, 343 (1952).

Figure 20.7. A much higher Reynolds number range was investigated by Giedt*
whose results are shown in Figure 20.8.

Figures 20.7 and 20.8 show a smooth variation in the Nusselt number near
the stagnation point. At low Reynolds numbers the film coefficient decreases
almost continuously from the stagnation point, the only departure being a slight
rise in the separated-wake region of the cylinder. At higher Reynolds numbers, as
illustrated in Figure 20.7, the film coefficient reaches a second maximum, which is
greater than the stagnation-point value. The second peak in the Nusselt number at
high Reynolds numbers is due to the fact that the boundary layer undergoes
transition from laminar to turbulent flow. In the bottom curves of Figure 20.7 the
laminar boundary layer separates from the cylinder near 80° from the stagnation
point, and no large change in the Nusselt number occurs. The effect of higher
Reynolds number is twofold. First, the separation point moves past 90° as the
boundary layer becomes turbulent, thus less of the cylinder is engulfed in the
wake. A second effect is that the Nusselt number reaches a value which is higher
than the stagnation-point value. The increase is due to the greater conductance of
the turbulent boundary layer.

It is quite apparent from the figures that the convective heat-transfer
coefficient varies in an irregular, complex manner in external flow about a
cylinder. It is likely, in practice, that an average h for the entire cylinder is desired.
McAdams† has plotted the data of 13 separate investigations for the flow of air
normal to single cylinders and found excellent agreement when plotted as Nu_D
versus Re_D. His plot is reproduced in Figure 20.9.

A widely used correlation for these data is of the form

$$Nu_D = B \, Re^n \, Pr^{1/3} \qquad (20\text{-}33)$$

where the constants B and n are functions of the Reynolds number. Values for
these constants are given in Table 20.3. The film temperature is appropriate for
physical property evaluation.

TABLE 20.3 VALUES OF B AND n FOR
USE IN EQUATION (20-33)

Re_D	B	n
0.4–4	0.989	0.330
4–40	0.911	0.385
40–4000	0.683	0.466
4000–40,000	0.193	0.618
40,000–400,000	0.027	0.805

---

* W. H. Giedt, *Trans. A.S.M.E.*, **71**, 378 (1949).

† W. H. McAdams, *Heat Transmission*, Third Edition, McGraw-Hill Book Company, New York,
1949.

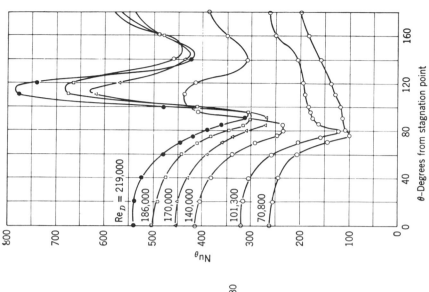

Figure 20.8 Local Nusselt numbers for crossflow about a circular cylinder at high Reynolds numbers. (From W. H. Giedt, *Trans. A.S.M.E.*, **71**, 378 (1949). By permission of the publishers.)

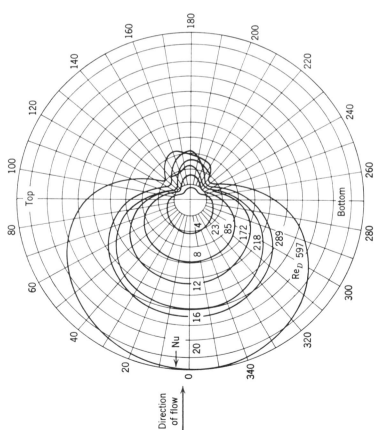

Figure 20.7 Local numbers for crossflow about a circular cylinder at low Reynolds numbers. (From E. R. G. Eckert and E. Soehngen, *Trans. A.S.M.E.*, **74**, 346 (1952). By permission of the publishers.)

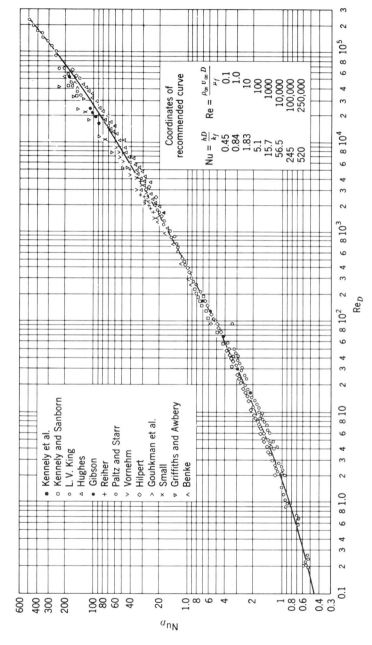

Figure 20.9 Nu versus Re for flow normal to single cylinders. (From W. H. McAdams, *Heat Transmission*, Third Edition, McGraw-Hill Book Company, New York, 1954, p. 259.)

Churchill and Bernstein* have recommended a single correlating equation covering conditions for which $Re_D\,Pr > 0.2$. This correlation is expressed in equation (20-34).

$$Nu_D = 0.3 + \frac{0.62\,Re_D^{1/2}\,Pr^{1/3}}{[1 + (0.4/Pr)^{2/3}]^{1/4}}\left[1 + \left(\frac{Re_D}{28\,200}\right)^{5/8}\right]^{4/5} \qquad (20\text{-}34)$$

Single spheres. Local convective heat-transfer coefficients at various positions relative to the forward stagnation point for flow past a sphere are plotted in Figure 20.10, following the work of Cary.†

McAdams‡ has plotted the data of several investigators relating Nu_D versus Re_D for air flowing past spheres. His plot is duplicated in Figure 20.11.

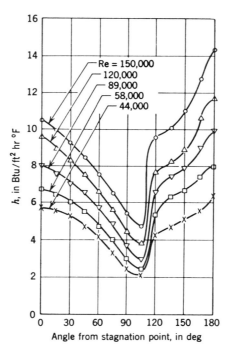

Figure 20.10 Local heat transfer coefficients for flow past a sphere. (From J. R. Cary, *Trans. A.S.M.E.*, 75, 485 (1953). By permission of the publishers.)

* S. W. Churchill and M. Bernstein, *J. Heat Transfer*, **99**, 300 (1977).
† J. R. Cary, *Trans. A.S.M.E.*, **75**, 483 (1953).
‡ W. H. McAdams, *op. cit.*

A recent correlation proposed by Whitaker[*] is recommended for the following conditions: $0.71 < \text{Pr} < 380$, $3.5 < \text{Re}_D < 7.6 \times 10^4$, $1.0 < \mu_\infty / \mu_s < 3.2$. All properties are evaluated at T_∞ except for μ_s, which is the value at the surface temperature. Whitaker's correlation is

$$\text{Nu}_D = 2 + (0.4 \, \text{Re}_D{}^{192} + 0.06 \, \text{Re}_D{}^{2/3}) \, \text{Pr}^{0.4} (\mu_\infty / \mu_s)^{1/4} \qquad (20\text{-}35)$$

An important case is that of falling liquid drops, modeled as spheres. The correlation of Ranz and Marshall,[†] for this case, is

$$\text{Nu}_D = 2 + 0.6 \, \text{Re}_D{}^{1/2} \, \text{Pr}^{1/3} \qquad (20\text{-}36)$$

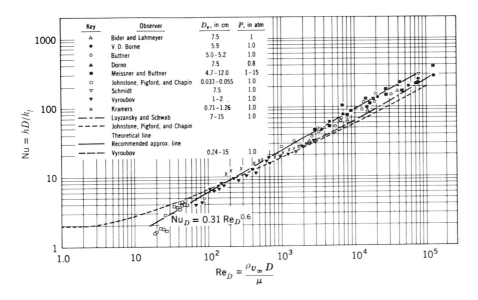

Figure 20.11 Nu_D versus Re_D for air-flow past single spheres. (From McAdams, *Heat Transmission*, Third Edition, McGraw-Hill Book Company, New York, 1954, p. 266. By permission of the publishers.)

Tube banks in crossflow. When a number of tubes are placed together in a bank or bundle, as might be encountered in a heat exchanger, the effective heat-transfer coefficient is affected by the tube arrangement and spacing, in addition to those factors already considered for flow past single cylinders. Several investigators have made significant contributions to the analysis of these configurations.

Since fluid flow through and past tube bundles involves an irregular flow path, some investigators have chosen significant lengths other than D, the tube

[*] S. Whitaker, *A.I.Ch.E. J.*, **18**, 361 (1972).
[†] W. Ranz and W. Marshall, *Chem. Engr. Progr.*, **48**, 141 (1952).

diameter, to use in calculating Reynolds numbers. One such term is the equivalent diameter of a tube bundle, D_{eq}, defined as

$$D_{eq} = \frac{4(S_L S_T - \pi D^2/4)}{\pi D} \qquad (20\text{-}37)$$

where S_L is the center-to-center distance between tubes *along* the direction of flow, S_T is the center-to-center distance between tubes *normal* to the flow direction, and D is the OD of a tube.

Bergelin, Colburn, and Hull[*] studied the flow of liquids past tube bundles in the region of laminar flow with $1 < \text{Re} < 1000$. Their results plotted as $\text{St Pr}^{2/3} (\mu_w/\mu_b)^{0.14}$ versus Re for various configurations are presented in Figure 20.12. In that figure all fluid properties except μ_w are evaluated at the average bulk temperature.

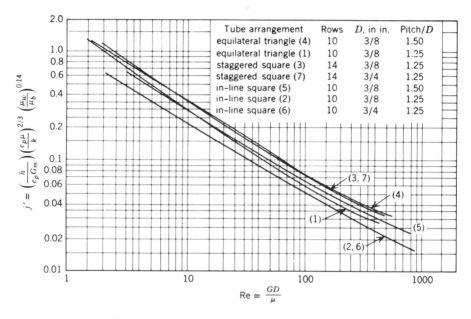

Tube arrangement	Rows	D, in in.	Pitch/D
equilateral triangle (4)	10	3/8	1.50
equilateral triangle (1)	10	3/8	1.25
staggered square (3)	14	3/8	1.25
staggered square (7)	14	3/4	1.25
in-line square (5)	10	3/8	1.50
in-line square (2)	10	3/8	1.25
in-line square (6)	10	3/4	1.25

Figure 20.12 Convective heat-transfer exchange between liquids in laminar flow and tube bundles. (From O. P. Bergelin, A. P. Colburn, and H. L. Hull, Univ. of Delaware, Engr. Dept., Station Bulletin 10.2, 1950, p. 8. By permission of the publishers.)

For liquids in transition flow across tube bundles, Bergelin, Brown, and Doberstein[†] extended the work just mentioned for five of the tube arrangements

---

[*] O. P. Bergelin, A. P. Colburn, and H. L. Hull, Univ. Delaware, Eng. Expt. Sta. Bulletin No. 2 (1950).
[†] O. P. Bergelin, G. A. Brown, and S. C. Doberstein, *Trans. A.S.M.E.*, **74**, 953 (1952).

to include values of Re up to 10^4. Their results are presented both for energy transfer and friction factor versus Re in Figure 20.13.

In addition to the greater Reynolds number range, Figure 20.13 involves Re calculated by using the tube diameter, D, as opposed to Figure 20.12, in which D_{eq}, defined by equation (20-37), was used.

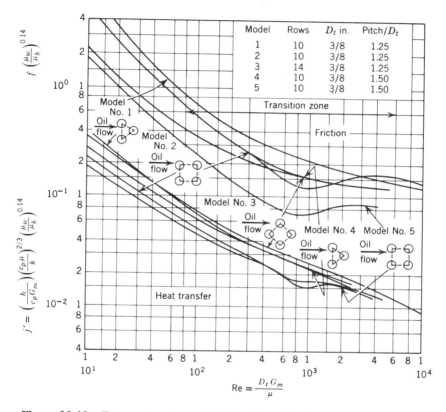

Figure 20.13 Energy transfer and frictional loss for liquids in laminar and transition flow past tube bundles. (From O. P. Bergelin, G. A. Brown, and S. C. Doberstein, *Trans. A.S.M.E.*, **74**, 1958 (1952). By permission of the publishers.)

20.4 CLOSURE

Many of the more useful experimentally developed correlations for predicting convective heat-transfer coefficients have been presented in this chapter. Those graphs and equations presented are a small part of the information of this type available in the literature. The information included should, in any case, allow most of the more common convection heat-transfer coefficients to be predicted with some confidence.

The convection phenomena considered have included the following:

Natural convection past vertical and horizontal surfaces, plus some useful simplified expressions for air;

Forced convection for internal flow, including laminar and turbulent flow correlations; and
Forced convection for external flow with cylinders, spheres, and tube bundles being the types of surfaces so considered.

The reader is reminded to observe any special considerations relative to the equations and plots in this chapter. Such considerations include whether to evaluate fluid properties at the bulk or film temperature, what significant length is used in a given correlation, and what is the allowable Prandtl and Reynolds number range for a given set of data.

PROBLEMS

20.1 A 750-W immersion heater in the form of a cylinder with $\frac{3}{4}$-in. diameter and 6 in. in length is placed in 95°F stagnant water. Calculate the surface temperature of the heater if it is oriented with its axis
(a) vertical;
(b) horizontal.

20.2 Repeat problem 20.1 if the stagnant liquid is
(a) bismuth at 700°F;
(b) hydraulic fluid at 0°F.

20.3 An immersion heater, rated at 1000 W, is in the form of a rectangular solid with dimensions 16 cm by 10 cm by 1 cm. Determine the surface temperature of the heater if it is oriented in 295 K water with:
(a) the 16-cm dimension vertical,
(b) the 10-cm dimension vertical.

20.4 A 2-in. copper cylinder, 6 in. in length, at a uniform temperature of 200°F, is plunged vertically into a large tank of water at 50°F.
(a) How long will it take for the outside surface of the cylinder to reach 100°F?
(b) How long will it take for the center of the cylinder to reach 100°F?
(c) What is the surface temperature when the center temperature is 100°F? Heat transfer from ends of the cylinder may be neglected.

20.5 Rubber balls are molded into spheres and cured at 360 K. Following this operation they are allowed to cool in room air. What will be the elapsed time for the surface temperature of a solid rubber ball to reach 320 K when the surrounding air temperature is 295 K? Consider balls with diameters of 7.5 cm, 5 cm, and 1.5 cm. Properties of rubber that may be used are the following: ($k = 0.24$ W/m · K, $\rho = 1120$ kg/m³, $c_p = 1020$ J/kg · K).

20.6 Determine the required time for the rubber balls described in the previous problem to reach the condition such that the center temperature is 320 K. What will be the surface temperature when the center temperature reaches 320 K?

20.7 A 1-in. 16-BWG copper tube has its outside surface maintained at 240°F. If this tube is located in still air at 60°F, what heat flux will be achieved if the tube is oriented
(a) horizontally?
(b) vertically?
The tube length is 10 ft.

20.8 Solve problem 20.7 if the medium surrounding the tube is stagnant water at 60°F.

20.9 A valve on a hot-water line is opened just enough to allow a flow of 0.06 fps. The water is maintained at 180°F, and the inside wall of the $\frac{1}{2}$-in. schedule-40 water line is at 80°F. What is the total heat loss through 5 ft of water line under these conditions? What is the exit water temperature?

20.10 When the valve on the water line in problem 20.9 is opened wide, the water velocity is 35 fps. What is the heat loss per 5 ft of water line in this case if the water and pipe temperatures are the same as specified in problem 20.6?

20.11 A 0.6 m diameter spherical tank contains liquid oxygen at 78 K. This tank is covered with 5 cm of glass wool. Determine the rate of heat gain if the tank is surrounded by air at 278 K. The tank is constructed of stainless steel 0.32 cm thick.

20.12 A "swimming-pool" nuclear reactor, consisting of 30 rectangular plates measuring 1 ft in width and 3 ft high, spaced $2\frac{1}{2}$ in. apart, is immersed in water at 80°F. If 200°F is the maximum allowable plate temperature, what is the maximum power level at which the reactor may operate?

20.13 A solar energy collector measuring 20×20 ft is installed on a roof in a horizontal position. The incident solar energy flux is 200 Btu/hr ft², and the collector surface temperature is 150°F. What fraction of incident solar energy is lost by convection to the stagnant surrounding air at a temperature of 50°F? What effect on the convective losses would result if the collector were crisscrossed with ridges spaced 1 ft apart?

20.14 Given the conditions for problem 20.13, determine the fraction of incident solar energy lost by convection to the surrounding air at 283 K flowing parallel to the collector surface at a velocity of 6.1 m/s.

20.15 Steam at 400 psi, 800°F flows through an 8-in. schedule-140 steel pipe at a rate of 10 000 lb$_m$/hr. Estimate the value of h which applies at the inside pipe surface.

20.16 If the steam line described in problem 20.15 is bare and surrounded by still air at 70°F, what total heat transfer would be predicted from a 20-ft length of bare pipe? Consider the bare pipe to be a black surface and the surroundings black at 70°F.

20.17 Solve problem 20.16 if the bare pipe is located so that 295 K air flows normal to the pipe axis at a velocity of 6.5 m/s.

20.18 Solve problem 20.16 if 3 in. of insulation having a thermal conductivity of 0.060 Btu/hr ft °F is applied to the outside of the pipe. Neglect radiation from the insulation. What will be the outside surface temperature of the insulation?

20.19 What thickness of insulation having a thermal conductivity as given in problem 20.18 must be added to the steam pipe of problem 20.15 in order that the outside temperature of the insulation not exceed 250°F?

20.20 If insulation having a thermal conductivity of 0.060 Btu/hr ft °F is added to the outside of the steam pipe described in problem 20.15, what thickness will be necessary if radiation losses from the outside surface of the insulation account for no more than 15% of the total? The surroundings may be considered black at 70°F. What is the temperature at the outside surface of the insulation under these conditions?

20.21 Oil at 300 K is heated by steam condensing at 372 K on the outside of steel pipes with ID = 2.09 cm, OD = 2.67 cm. The oil flow rate is 1.47×10^5 kg/s; six tubes, each 2.5 m long, are used. The properties of oil to be used are as follows:

T, K	ρ, kg/m^3	c_p, J/kg · K	k, W/m · K	μ, Pa · s
300	910	1.84×10^3	0.133	0.0414
310	897	1.92×10^3	0.131	0.0228
340	870	2.00×10^3	0.130	7.89×10^{-3}
370	865	2.13×10^3	0.128	3.72×10^{-3}

Determine the total heat transfer to the oil and its temperature at the heater exit.

20.22 Solve problem 20.21 if, instead of six tubes, three tubes each 5 m in length are used to heat the oil.

20.23 Air at 60°F and atmospheric pressure flows inside a 1-in. 16-BWG copper tube whose surface is maintained at 240°F by condensing steam. Find the temperature of the air after passing through 20 ft of tubing if its entering velocity is 40 fps.

20.24 A 1-in. 16-BWG copper tube, 10 ft long, has its outside surface maintained at 240°F. Air at 60°F and atmospheric pressure is forced past this tube with a velocity of 40 fps. Determine the heat flux from the tube to the air if the flow of air is
(a) parallel to the tube;
(b) normal to the tube axis.

20.25 Solve problem 20.24 if the medium flowing past the tube in forced convection is water at 60°F.

20.26 Solve problem 20.24 if the medium flowing past the tube in forced convection is MIL-M-5606 hydraulic fluid.

20.27 An industrial heater is composed of a tube bundle consisting of horizontal $\frac{3}{8}$-in. OD tubes in a staggered array with tubes arranged in equilateral triangle fashion having a pitch-to-diameter ratio of 1.5. If water at 160°F flows at 20 ft/s past the tubes with constant surface temperature of 212°F, what will be the effective heat-transfer coefficient?

20.28 For the heater consisting of the tube bank described in problem 20.27, evaluate the heat transferred to the water if the tube array consists of 6 rows of tubes in the flow direction with 8 tubes per row. The tubes are 5 ft long.

20.29 Cast-iron cannonballs used in the War of 1812 were occasionally heated for some extended time so that, when fired at houses and ships, they would set them afire. If one of these so-called "hot shot" with a 15 cm diameter were at a uniform temperature of 1300 K, what heat flux value would exist if it were suddenly placed in still air at 270 K? The following properties of cast iron may be used:

$$k = 39.8 \text{ W/m} \cdot \text{K}$$

$$c_p = 4.8 \text{ J/kg} \cdot \text{K}$$

$$\rho = 7370 \text{ kg/m}^3$$

20.30 Given the information in problem 20.27, construct a plot of convective heat-transfer coefficient vs. temperature for values of T_{surface} between 420 K and 1300 K. How long would it take for the surface temperature of a cannonball to reach 600 K? What would be its center temperature at this time?

20.31 Solve problem 20.27 with all conditions as given except that the "hot shot" is traveling through the air at 270 K with a velocity of 150 m/s.

20.32 Air at 25 psia is to be heated from 60°F to 100°F in a smooth, $\frac{3}{4}$-in.-ID tube whose surface is held at a constant temperature of 120°F. What is the length of tube required for an air velocity of 25 fps? At 15 fps?

20.33 Copper wire with a diameter of 0.5 cm is covered with a 0.65-cm layer of insulating material having a thermal conductivity of 0.242 W/m · K. The air adjacent to the insulation is at 290 K. If the wire carries a current of 400 amps, determine:
(a) the convective heat transfer coefficient between the insulation surface and the surrounding air;
(b) the temperatures at the insulation-copper interface and at the outside surface of the insulation.
The resistivity of copper is 1.72×10^{-6} ohm-cm.

20.34 Work problem 20.31 for an aluminum conductor of the same size (resistivity of aluminum $= 2.83 \times 10^{-6}$ ohm-cm).

20.35 What would be the results of problem 20.31 if a fan provided an air flow normal to the conductor axis at a velocity of 9 m/s?

20.36 Air is transported through a rectangular duct measuring 2 ft by 4 ft. The air enters at 120°F and flows with a mass velocity of 6 lb_m/s ft^2. If the duct walls are at a temperature of 80°F, how much heat is lost by the air per foot of duct length? What is the corresponding temperature decrease of the air per foot?

20.37 A 0.5-in. diameter jet of water at 70 fps is directed against a 200°F steel plate. If the water is at 50°F, how long will it take the surface of the plate to reach 70°F?

20.38 Compare the value of the Nusselt number at the stagnation point of a circular cylinder in Figure 20.8 with the values calculated by using equation (20-50).

21
BOILING AND
CONDENSATION

Energy-transfer processes associated with the phenomena of boiling and condensation may achieve relatively high heat-transfer rates, while the accompanying temperature differences may be quite small. These phenomena, associated with the change in phase between a liquid and a vapor, are more involved and thus more difficult to describe than the convective heat-transfer processes discussed in the preceding chapters. This is due to the additional considerations of surface tension, latent heat of vaporization, surface characteristics, and other properties of two-phase systems which were not involved in earlier considerations. The processes of boiling and condensation deal with opposite effects relative to the change in phase between a liquid and its vapor. These phenomena will be considered separately in the sections to follow.

21.1 BOILING

Boiling heat transfer is associated with a change in phase from liquid to vapor. Extremely high heat fluxes may be achieved in conjunction with boiling phenomena, making the application particularly valuable where a small amount of space is available to accomplish a relatively large energy transfer. One such application is the cooling of nuclear reactors. The advent of this application has spurred the interest in boiling, and concentrated research in this area in recent years has shed much light on the mechanism and behavior of the boiling phenomenon.

There are two basic types of boiling: *pool boiling* and *flow boiling*. Pool boiling is that which occurs on a heated surface submerged in a liquid pool which is not agitated. Flow boiling occurs in a flowing stream, and the boiling surface may itself be a portion of the flow passage. The flow of liquid and vapor associated with flow boiling is an important type of two-phase flow.

Regimes of boiling. An electrically heated horizontal wire submerged in a pool of water at its saturation temperature is a convenient system to illustrate the regimes of boiling heat transfer. A plot of the heat flux associated with such a system as the ordinate vs. the temperature difference between the heated surface and the saturated liquid is depicted in Figure 21.1. There are six different regimes of boiling associated with the behavior exhibited in this figure.

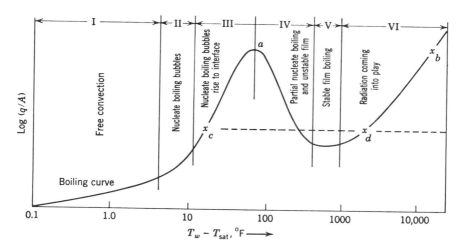

Figure 21.1 Pool boiling on a horizontal wire at atmospheric pressure.

In regime I the wire surface temperature is a very few degrees higher than the surrounding saturated liquid. Natural convection currents circulate the superheated liquid, and evaporation occurs at the free liquid surface as the superheated liquid reaches that position.

An increase in wire temperature is accompanied by the formation of vapor bubbles on the wire surface. These bubbles form at certain surface sites, where vapor bubble nuclei are present, break off, and condense before reaching the free liquid surface. This is the process occurring in regime II.

At still higher wire surface temperature, as in regime III, larger and more numerous bubbles form, break away from the wire surface, rise, and reach the free surface. Regimes II and III are associated with *nucleate boiling.*

Beyond the peak of this curve the transition boiling regime is entered. This is region IV on the curve. In this regime a vapor film forms around the wire, and portions of this film break off and rise, briefly exposing a portion of the wire surface. This film collapse and reformation and this unstable nature of the film is characteristic of the transition regime. When present, the vapor film provides a considerable resistance to heat transfer, thus the heat flux decreases.

When the surface temperature reaches a value of approximately 400°F above the saturated liquid the vapor film around the wire becomes stable. This is region V, the *stable film boiling* regime.

For surface temperatures of 1000°F or greater above that of the saturated liquid, radiant energy transfer comes into play, and the heat flux curve rises once more. This is designated region VI in Figure 21.1.

The curve in Figure 21.1 can be achieved if the energy source is a condensing vapor. If, however, electrical heating is used, then regime IV will probably not be obtained because of wire burnout. As the energy flux is increased ΔT increases through regions I, II, and III. When the peak value of q/A is exceeded slightly, the required amount of energy cannot be transferred by boiling. The result is an increase in ΔT accompanied by a further decrease in the possible q/A. This condition continues until point c is reached. Since ΔT at point c is extremely high, the wire will long since have reached its melting point. Point a on the curve is often referred to as the "burnout point" for these reasons.

Since the mechanism of energy removal is intimately associated with buoyant forces, the magnitude of the body-force intensity will affect both the mechanism and the magnitude of boiling heat transfer. Other than normal gravitational effects are encountered in space vehicles.

Note the somewhat anomalous behavior exhibited by the heat flux associated with boiling. One normally considers a flux to be proportional to the driving force, thus the heat flux might be expected to increase continuously as the temperature difference between the heated surface and the saturated liquid increases. This, of course, is not the case; the very high heat fluxes associated with moderate temperature differences in the nucleate boiling regime are much higher than the heat fluxes resulting from much higher temperature differences in the film boiling regime. The reason for this is the presence of the vapor film, which covers and insulates the heating surface in the latter case.

Correlations of boiling heat-transfer data. Since the fluid behavior in a boiling situation is very difficult to describe, there is no adequate analytical solution available for boiling heat transfer. Various correlations of experimental data have been achieved for the different boiling regimes; the most useful of these follow.

In the natural convection regime, *regime I* of Figure 21.1, the correlations presented in Chapter 20 for natural convection may be used.

Regime II, the regime of partial nucleate boiling and partial natural convection, is a combination of regime I and III, and the results for each of these two regimes may be superposed to describe a process in regime II.

The nucleate boiling regime, *regime III*, is of great engineering importance because of the very high heat fluxes possible with moderate temperature differences. That data for this regime are correlated by equations of the form

$$\mathrm{Nu}_b = \phi(\mathrm{Re_b}, \mathrm{Pr}_L) \tag{21-1}$$

The parameter Nu_b in equation (21-1) is a Nusselt number defined as

$$Nu_b \equiv \frac{(q/A)D_b}{(T_s - T_{sat})k_L} \qquad (21-2)$$

where q/A is the total heat flux, D_b is the maximum bubble diameter as it leaves the surface, $T_s - T_{sat}$ is the *excess temperature* or the difference between the surface and saturated-liquid temperatures, and k_L is the thermal conductivity of the liquid. The quantity, Pr_L, is the Prandtl number for the liquid. The bubble Reynolds number, Re_b, is defined as

$$Re_b \equiv \frac{D_b G_b}{\mu_L} \qquad (21-3)$$

where G_b is the average mass velocity of the vapor leaving the surface, and μ_L is the liquid viscosity.

The mass velocity, G_b, may be determined from

$$G_b = \frac{q/A}{\rho_v h_{fg}} \rho_L \qquad (21-4)$$

where h_{fg} is the latent heat of vaporization.

Rohsenow* has used equation (21-1) to correlate Addoms'† pool boiling data for a 0.024-in.-diameter platinum wire immersed in water. This correlation is shown in Figure 21.2, and is expressed in equation form as

$$\frac{q}{A} = \mu_L h_{fg} \left[\frac{g(\rho_L - \rho_v)}{\sigma} \right]^{1/2} \left[\frac{c_{pL}(T_s - T_{sat})}{C_{sf} h_{fg} Pr_L^{1.7}} \right]^3 \qquad (21-5)$$

where c_{pL} is the heat capacity for the liquid, and the other terms have their usual meanings.

The coefficients C_{sf} in equation (21-5) varies with the surface-fluid combination. The curve drawn in Figure 21.2 is for $C_{sf} = 0.013$. A table of C_{sf} for various combinations of fluid and surface is presented by Rohsenow and Choi‡ and duplicated here as Table 21.1.

From earlier discussion it is clear that the burnout point has considerable importance. The "critical heat flux" is the value of q/A represented by point a in Figure 21.1. An analysis of conditions at burnout modified by experimental results is expressed in equation (21-6) as

$$q/A|_{critical} = 0.18 h_{fg} \rho_v \left[\frac{\sigma g(\rho_L - \rho_v)}{\rho_v^2} \right]^{1/4} \qquad (21.6)$$

---

*W. M. Rohsenow, *A.S.M.E. Trans.*, **74**, 969 (1952).
†J. N. Addoms, D.Sc. Thesis, Chemical Engineering Department, Massachusetts Institute of Technology, June 1948.
‡W. M. Rohsenow and H. Y. Choi, *Heat, Mass, and Momentum Transfer*, Prentice-Hall, Inc., Englewood Cliffs, N.J., 1961.

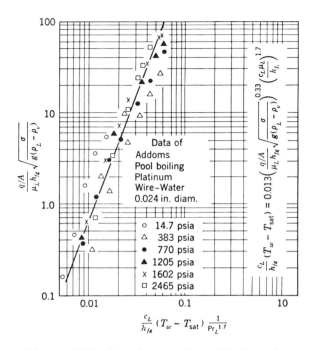

Figure 21.2 Correlation of pool-boiling data. (From W. M. Rohsenow and H. Choi, *Heat, Mass, and Momentum Transfer*, Prentice-Hall, Inc., Englewood Cliffs, N.J., 1961, p. 224. By permission of the publishers.)

TABLE 21.1   VALUES OF C_{sf} FOR EQUATION (21-6)

Surface/fluid combination	C_{sf}
water/nickel	0.006
water/platinum	0.013
water/copper	0.013
water/brass	0.006
CCl_4/copper	0.013
benzene/chromium	0.010
n-pentane/chromium	0.015
ethyl alcohol/chromium	0.0027
isopropyl alcohol/copper	0.0025
35% K_2CO_3/copper	0.0054
50% K_2CO_3/copper	0.0027
n-butyl alcohol/copper	0.0030

The interested reader is referred to the work of Zuber* for a discussion of this subject.

Regime IV, that of unstable film boiling, is not of great engineering interest, and no satisfactory correlation has been found for this region as yet.

The stable-film-boiling region, *regime V*, requires quite high surface temperatures, thus few experimental data have been reported for this region.

Stable film boiling on the surface of horizontal tubes and vertical plates has been studied both analytically and experimentally by Bromley.†‡ Considering conduction alone through the film on a horizontal tube, Bromley obtained the expression

$$h = 0.62 \left[\frac{k_v^3 \rho_v (\rho_L - \rho_v) g (h_{fg} + 0.4\, c_{pv} \Delta T)}{D_o \mu_v (T_s - T_{sat})} \right]^{1/4} \tag{21-7}$$

where all terms are self-explanatory except D_o, which is the outside diameter of the tube.

A modification in equation (21-7) was proposed by Berenson§ to provide a similar correlation for stable film boiling on a horizontal surface. In Berenson's correlation the tube diameter, D_o, is replaced by the term $[\sigma/g(\rho_L - \rho_v)]^{1/2}$, and the recommended expression is

$$h = 0.425 \left[\frac{k_{vf}^3 \rho_{vf} (\rho_L - \rho_v) g (h_{fg} + 0.4 c_{pv} \Delta T)}{\mu_{vf} (T_s - T_{sat}) \sqrt{\sigma/g(\rho_L - \rho_v)}} \right]^{1/4} \tag{21-8}$$

where k_{vf}, ρ_{vf}, and μ_{vf} are to be evaluated at the film temperature as indicated.

Hsu and Westwater‖ considered film boiling for the case of a vertical tube. Their test results were correlated by the equation

$$h \left[\frac{\mu_v^2}{g \rho_v (\rho_L - \rho_v) k_v^3} \right]^{1/3} = 0.0020\, \text{Re}^{0.6} \tag{21-9}$$

where

$$\text{Re} = \frac{4 \dot{m}}{\pi D_o \mu_v} \tag{21-10}$$

$\dot{m}$ being the flow rate of vapor in lb_m/hr at the upper end of the tube, and the other terms being identical to those in equation (21-7). Hsu¶ states that heat-transfer rates for film boiling are higher for vertical tubes than for horizontal tubes when all other conditions remain the same.

In *regime VI*, the correlations for film boiling still apply; however, the superimposed contribution of radiation is appreciable, becoming dominant at

* N. Zuber, *Trans. A.S.M.E.*, **80**, 711 (1958).

† L. A. Bromley, *Chem. Engr. Prog.*, **46**, 5, 221 (May 1950).

‡ L. A. Bromley et al., *Ind. Engr. Chem.*, **45**, 2639 (1953).

§ P. Berenson, A.I.Ch.E. Paper No. 18, Heat Transfer Conference, Buffalo, N.Y., August 14–17, 960.

‖ Y. Y. Hsu and J. W. Westwater, *A.I.Ch.E.J.*, **4**, 59 (1958).

¶ S. T. Hsu, *Engineering Heat Transfer*, Van Nostrand, Princeton, N.J., 1963.

extremely high values of ΔT. Without any appreciable flow of liquid the two contributions may be combined, as indicated by equation (21-11) below.

The contribution of radiation to the total heat-transfer coefficient may be expressed as

$$h = h_c\left(\frac{h_c}{h}\right)^{1/3} + h_r \tag{21-11}$$

where h is the total heat transfer coefficient, h_c is the coefficient for the boiling phenomenon, and h_r is an effective radiant heat transfer coefficient considering exchange between two parallel planes with the liquid between assigned a value for unity for its emissivity. This term is discussed in Chapter 23.

When there is appreciable flow of either the liquid or the vapor the foregoing correlations are unsatisfactory. The description of *flow boiling* or *two-phase flow* will not be discussed here. The interested reader is referred to the recent literature for pertinent discussion of these phenomena. It is evident that for vertical surfaces or large-diameter horizontal tubes the density difference between liquid and vapor will produce significant local velocities. Any correlation which neglects flow contributions should, therefore, be used with caution.

21.2 CONDENSATION

Condensation occurs when a vapor contacts a surface which is at a temperature below the saturation temperature of the vapor. When the liquid condensate forms on the surface it will flow under the influence of gravity.

Normally the liquid wets the surface, spreads out, and forms a film. Such a process is called *film condensation.* If the surface is not wetted by the liquid, then droplets form and run down the surface, coalescing as they contact other condensate droplets. This process is designated *dropwise condensation.* After a condensate film has been developed in filmwise condensation, additional condensation will occur at the liquid-vapor interface, and the associated energy transfer must occur by conduction through the condensate film. Dropwise condensation, on the other hand, always has some surface present as the condensate drop forms and runs off. Dropwise condensation is, therefore, associated with the higher heat-transfer rates of the two types of condensation phenomena. Dropwise condensation is very difficult to achieve or maintain commercially; therefore, all equipment is designed on the basis of filmwise condensation.

Film condensation: the Nusselt model. In 1916 Nusselt* achieved an analytical result for the problem of filmwise condensation of a pure vapor on a vertical wall. The meanings of the various terms in this analysis will be made clear by

---

* W. Nusselt, *Zeitschr. d. Ver. deutsch. Ing.,* **60**, 514 (1916).

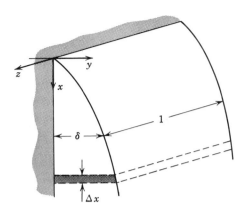

Figure 21.3 Filmwise condensation on a
vertical plane wall.

referring to Figure 21.3. In this figure the film thickness, δ, is seen to be zero at the
top of the vertical wall, $x = 0$, and to increase with increasing values of x.

The initial assumption made by Nusselt was that of wholly laminar flow in the
condensate film. Under these conditions the velocity profile may be easily
obtained from equation (8-12),

$$v_x = \frac{\rho g L^2 \sin \theta}{\mu}\left[\frac{y}{L} - \frac{1}{2}\left(\frac{y}{L}\right)^2\right] \tag{8-12}$$

For the present application $\sin \theta = 1$, and $L = \delta$. It is also necessary to modify the
density for the present case. In the derivation of equation (8-12) the density of the
gas or vapor at the liquid surface was neglected. This may be true in many cases of
a condensation process; however, the process may occur at a sufficiently high
pressure that the vapor density, ρ_v, is significant in comparison to that of the liquid,
ρ_L. To account for this possibility the density function to be used in the present
case is $\rho_L - \rho_v$ instead of simply ρ_L. The resulting expression for the velocity profile
in the condensate film at a particular distance x from the top of the wall becomes

$$v_x = \frac{(\rho_L - \rho_v)g\delta^2}{\mu}\left[\frac{y}{\delta} - \frac{1}{2}\left(\frac{y}{\delta}\right)^2\right] \tag{21-12}$$

The flow rate per unit width, Γ, at any value $x > 0$ is

$$\Gamma = \int_0^\delta v_x \, dy$$

$$= \frac{(\rho_L - \rho_v)g\delta^3}{3\mu} \tag{21-13}$$

A differential change, $d\Gamma$, in the flow rate is evaluated from this expression to be

$$d\Gamma = \frac{(\rho_L - \rho_v)g\delta^2 \, d\delta}{\mu} \tag{21-14}$$

This result has been obtained from momentum considerations alone. We shall now, as Nusselt did originally, look at the related energy transfer.

Since the flow of condensate is assumed to be laminar, it is not unreasonable to consider energy transfer through the film from the temperature at the vapor-liquid interface, T_{sat}, to the wall-liquid boundary at temperature T_w to be purely by conduction. On this basis the temperature profile is linear and the heat flux to the wall is

$$\frac{q_y}{A} = k\frac{(T_{sat} - T_w)}{\delta} \tag{21-15}$$

This same amount of energy must be transferred from the vapor as it condenses and then cools to the average liquid temperature. Relating these two effects, we may write

$$\frac{q_y}{A} = k\frac{(T_{sat} - T_w)}{\delta} = \rho_L \left[h_{fg} + \frac{1}{\rho_L \Gamma} \int_0^\delta \rho_L v_x c_{pL}(T_{sat} - T) \, dy \right]\frac{d\Gamma}{dx}$$

which, if a linear temperature variation in y is utilized, becomes

$$\frac{q_y}{A} = \frac{k(T_{sat} - T_w)}{\delta} = \rho_L[h_{fg} + \tfrac{3}{8}c_{pL}(T_{sat} - T_w)]\frac{d\Gamma}{dx} \tag{21-16}$$

Solving equation (21-16) for $d\Gamma$, we have

$$d\Gamma = \frac{k(T_{sat} - T_w) \, dx}{\rho_L \delta[h_{fg} + \tfrac{3}{8}c_{pL}(T_{sat} - T_w)]} \tag{21-17}$$

which may now be equated to the result in equation (21-14), giving

$$\frac{(\rho_L - \rho_v)g}{\mu}\delta^2 \, d\delta = \frac{k(T_{sat} - T_w)}{\rho_L \delta[h_{fg} + \tfrac{3}{8}c_{pL}(T_{sat} - T_w)]} dx$$

Simplifying this result and solving for δ, we obtain

$$\delta = \left[\frac{4k\mu(T_{sat} - T_w)x}{\rho_L g(\rho_L - \rho_v)[h_{fg} + \tfrac{3}{8}c_{pL}(T_{sat} - T_w)]} \right]^{1/4} \tag{21-18}$$

We may now solve for the heat transfer coefficient, h, from the expression

$$h = \frac{q_y/A}{T_{sat} - T_w} = \frac{k}{\delta}$$

The substitution of equation (21-18) into this expression yields

$$h_x = \left\{ \frac{\rho_L g k^3(\rho_L - \rho_v)[h_{fg} + \tfrac{3}{8}c_{pL}(T_{sat} - T_w)]}{4\mu(T_{sat} - T_w)x} \right\}^{1/4} \tag{21-19}$$

The average heat-transfer coefficient for a surface of length L is determined from

$$h = \frac{1}{L} \int_0^L h_x \, dx$$

which, when equation (21-19) is substituted, becomes

$$h = 0.943 \left\{ \frac{\rho_L g k^3 (\rho_L - \rho_v)[h_{fg} + \frac{3}{8} c_{pL}(T_{\text{sat}} - T_w)]}{L \mu (T_{\text{sat}} - T_w)} \right\}^{1/4} \qquad (21\text{-}20)$$

The latent heat term, h_{fg}, in equation (21-20) and those preceding it, should be evaluated at the saturation temperature. Liquid properties should all be taken at the film temperature.

An expression similar to equation (21-20) may be achieved for a surface inclined at an angle θ from the horizontal if $\sin \theta$ is introduced into the bracketed term. This extension obviously has a limit and should not be used when θ is small, that is, when the surface is horizontal. For such a condition the analysis is quite simple; example 1 illustrates such a case.

Rohsenow* performed a modified integral analysis of this same problem, obtaining a result which differs only in that the term $[h_{fg} + \frac{3}{8} c_{pL}(T_{\text{sat}} - T_w)]$ is replaced by $[h_{fg} + 0.68 c_{pL}(T_{\text{sat}} - T_w)]$. Rohsenow's results agree well with experimental data achieved for values of $\text{Pr} > 0.5$ and $c_{pL}(T_{\text{sat}} - T_w)/h_{fg} < 1.0$.

EXAMPLE 1

A square pan with its bottom surface maintained at 350 K is exposed to water vapor at 1 atm pressure and 373 K. The pan has a lip all around so the condensate that forms cannot flow away. How deep will the condensate film be after 10 min have elapsed at this condition?

We will employ a "pseudo-steady-state" approach to solve this problem. An energy balance at the vapor-liquid interface will indicate that the heat flux and rate of mass condensed, $\dot{m}_{\text{cond}}$, are related as

$$\left. \frac{q}{A} \right|_{\text{in}} = \frac{\dot{m}_{\text{cond}} h_{fg}}{A}$$

The condensation rate, $\dot{m}_{\text{cond}}$, may be expressed as follows:

$$\dot{m}_{\text{cond}} = \rho \dot{V}_{\text{cond}} = \rho A \frac{d\delta}{dt}$$

where $d\delta/dt$ is the rate at which the condensate film thickness, δ, grows. The heat flux at the interface may now be expressed as

$$\left. \frac{q}{A} \right|_{\text{in}} = \rho g_{fg} \frac{d\delta}{dt}$$

* W. M. Rohsenow, *A.S.M.E. Trans.*, **78**, 1645 (1956).

This heat flux is now equated to that which must be conducted through the film to the cool pan surface. The heat flux expression that applies is

$$\frac{q}{A}\bigg|_{out} = \frac{k_L}{\delta}(T_{sat} - T_s)$$

This is a steady-state expression; that is, we are assuming δ to be constant. If δ is not rapidly varying, this "pseudo-steady-state" approximation will give satisfactory results. Now, equating the two heat fluxes, we have

$$\rho h_{fg}\frac{d\delta}{dt} = \frac{k_L}{\delta}(T_{sat} - T_s)$$

and, progressing, the condensate film thickness is seen to vary with time according to

$$\delta\frac{d\delta}{dt} = \frac{k_L}{\rho h_{fg}}(T_{sat} - T_s)$$

$$\int_0^\delta \delta \, d\delta = \frac{k_L}{\rho h_{fg}}(T_{sat} - T_s)\int_0^t dt$$

$$\delta = \left[\frac{k_L}{2\rho h_{fg}}(T_{sat} - T_s)\right]^{1/2} t^{1/2}$$

A quantitative answer to our example problem now yields the result

$$\delta = \left[\frac{(0.674 \text{ W/m} \cdot \text{K})(23 \text{ K})(600 \text{ s})}{2(966 \text{ kg/m}^3)(2250 \text{ kJ/kg})}\right]^{1/2}$$

$$= 1.46 \text{ mm}$$

Film condensation: turbulent-flow analysis. It is logical to expect the flow of the condensate film to become turbulent for relatively long surfaces or for high condensation rates. The criterion for turbulent flow is, as we should expect, a Reynolds number for the condensate film. In terms of an equivalent diameter the applicable Reynolds number is

$$\text{Re} = \frac{4A}{P}\frac{\rho_L v}{\mu_f} \tag{21-21}$$

where A is the condensate flow area, P is the wetted perimeter, and v is the velocity of condensate. The critical value of Re in this case is approximately 2000.

The first attempt to analyze the case of turbulent flow of a condensate film was that of Colburn,[*] who used the same j factor determined for internal pipe flow. On the basis partly of analysis and partly of experiment, Colburn formulated the plot shown in Figure 21.4. The data points shown are those of Kirkbride.[†] The correlating equations for the two regions shown are for $4\Gamma_c/\mu_f < 2000$,

$$h_{avg} = 1.51\left(\frac{k^3\rho^2 g}{\mu^2}\right)_f^{1/3}\left(\frac{4\Gamma_c}{\mu_f}\right)^{-1/3} \tag{21-22}$$

[*] A. P. Colburn, *Ind. Eng. Chem.*, **26**, 432 (1934).
[†] C. G. Kirkbride, *Ind. Eng. Chem.*, **26**, 4 (1930).

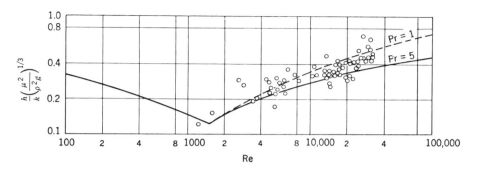

Figure 21.4 Film condensation including the regions of both laminar and turbulent flow.

and for $4\Gamma_c/\mu_f > 2000$,

$$h_{avg} = 0.045 \frac{(k^3\rho^2 g/\mu^2)_f^{1/3}(4\Gamma_c/\mu_f)\, \text{Pr}^{1/3}}{[(4\Gamma_c/\mu_f)^{4/5} - 364] + 576\, \text{Pr}^{1/3}} \qquad (21\text{-}23)$$

In these expressions Γ_c is the mass flow rate per unit width of surface; that is, $\Gamma_c = \rho_L v_{avg}\delta$, δ being film thickness and v_{avg} the average velocity. The term $4\Gamma_c/\mu_f$ is thus a Reynolds number for a condensate film on a plane vertical wall. McAdams[*] recommends a simpler expression for the turbulent range, $\text{Re}_\delta > 2000$, as

$$h = 0.0077\left[\frac{\rho_L g(\rho_L - \rho_v)k_L^3}{\mu_L^2}\right]^{1/3} \text{Re}_\delta^{0.4} \qquad (21\text{-}24)$$

Film condensation: analysis of the horizontal cylinder. An analysis by Nusselt[†] produced the following expression for the mean heat-transfer coefficient for a horizontal cylinder:

$$h_{avg} = 0.725\left\{\frac{\rho_L g(\rho_L - \rho_v)k^3[h_{fg} + \tfrac{3}{8}c_{pL}(T_{sat} - T_w)]}{\mu D(T_{sat} - T_w)}\right\}^{1/4} \qquad (21\text{-}25)$$

The similarity between equation (21-25) for a horizontal tube and equation (21-20) for a vertical tube is marked. Combining these expressions and cancelling similar terms, we obtain the result that

$$\frac{h_{vert}}{h_{horiz}} = \frac{0.943}{0.725}\left(\frac{D}{L}\right)^{1/4} = 1.3\left(\frac{D}{L}\right)^{1/4} \qquad (21\text{-}26)$$

[*] W. H. McAdams, *Heat Transmission*, Third edition, McGraw-Hill Book Company, New York, 1954.
[†] W. Nusselt, *Zeitschr. d. Ver. deutsch. Ing.*, **60**, 569 (1916).

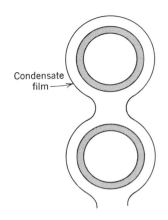

Figure 21.5 Condensation on a horizontal tube bank.

For the case of equal heat-transfer coefficients the relation between D and L is

$$\frac{L}{D} = 2.86 \qquad (21\text{-}27)$$

or, equal amounts of energy can be transferred from the same tube in either the vertical or the horizontal position if the ratio L/D is 2.86. For L/D values greater than 2.86 the horizontal position has the greater heat-transfer capability.

Film condensation: banks of horizontal tubes. For a bank of horizontal tubes there is, naturally, a different value of h for each tube, since the condensate film from one tube will drop on that tube next below it in the line. This process is depicted in Figure 21.5.

Nusselt also considered this situation analytically and achieved, for a vertical bank of n tubes in line, the expression

$$h_{\text{avg}} = 0.725 \left\{ \frac{\rho_L g (\rho_L - \rho_v) k^3 [h_{fg} + \tfrac{3}{8} c_{pL}(T_{\text{sat}} - T_w)]}{n \, D \mu (T_{\text{sat}} - T_w)} \right\}^{1/4} \qquad (21\text{-}28)$$

This equation yields a mean heat-transfer coefficient averaged over all n tubes.

Observing that experimental data exceeded those values predicted from equation (21-28), Chen[*] modified this expression to include the effect of condensation on the liquid layer between the tubes. His resulting equation is

$$h_{\text{avg}} = 0.725 \left[1 + 0.02 \frac{c_{pL}(T_{\text{sat}} - T_w)}{h_{fg}} (n-1) \right]$$
$$\times \left\{ \frac{\rho_L g (\rho_L - \rho_v) k^3 [h_{fg} + \tfrac{3}{8} c_{pL}(T_{\text{sat}} - T_w)]}{n \, D \mu (T_{\text{sat}} - T_w)} \right\}^{1/4} \qquad (21\text{-}29)$$

which is valid for values of $c_{pL}(T_{\text{sat}} - T_w)(n-1)/h_{fg} > 2$. Chen's equation agrees reasonably well with experimental data for condensation on vertical banks of horizontal tubes.

Drop condensation. Dropwise condensation, as mentioned earlier, is associated with higher heat-transfer coefficients than the filmwise condensation phenomenon. For dropwise condensation to occur, the surface must not be "wetted" by the condensate. Normally this requires that metal surfaces be specially treated.

---

[*] M. M. Chen, *A.S.M.E.* (*Trans.*) Series C, **83**, 48 (1961).

Dropwise condensation is an attractive phenomenon for applications where extremely large heat-transfer rates are desired. At present it is difficult to maintain this condition for several reasons. Because of its uncertain nature and the conservative approach of a design based on lower heat-transfer coefficients, filmwise condensation is the typoe predominantly used in design.

21.3 CLOSURE

The phenomena of boiling and condensation have been examined in this chapter. Each condition has a prominent place in engineering practice and both are difficult to describe analytically. Several empirical correlations for these phenomena for various surfaces oriented in different ways have been presented.

Boiling is normally described as nucleate type, film type, or a combination of the two. Very high heat-transfer rates are possible in the nucleate-boiling regime with relatively low temperature differences between the primary surface and the saturation temperature of the liquid. Film boiling is associated with a higher temperature difference yet a lower rate of heat transfer. This anomalous behavior is peculiar to the boiling phenomenon.

Condensation is categorized as either filmwise or dropwise. Dropwise condensation is associated with much higher heat-transfer coefficients than filmwise; however, it is difficult both to achieve and to maintain. Thus filmwise condensation is of primary interest. Analytical solutions have been presented, along with empirical results, for filmwise condensation on vertical and horizontal plates and cylinders and for banks of horizontal cylinders.

PROBLEMS

The surface tension of water, a needed quantity in several of the following problems, is related to temperature according to the expression $\sigma = 0.1232[1 - 0.00146\,T]$ where σ is in N/m and T is in K. In the English system with σ given in lb_f/ft and T in °R the surface tension may be calculated from $\sigma = (8.44 \times 10^{-3})[1 - 0.00082\,T]$.

21.1 An electrically heated square plate measuring 20 cm on a side is immersed vertically in water at atmospheric pressure. As the electrical energy supplied to the plate is increased, its surface temperature rises above that of the adjacent saturated water. At low power levels the heat-transfer mechanism is natural convection, then becoming a nucleate boiling phenomenon at higher ΔT's. At what value of ΔT are the heat fluxes due to boiling and natural convection the same? Plot $q/A|_{convection}$, $q/A|_{boiling}$, and $q/A|_{total}$ versus ΔT values from 250 K to 300 K.

21.2 Plot values of the heat-transfer coefficient for the case of pool boiling of water on horizontal metal surfaces at 1 atmosphere total pressure and surface temperatures varying from 390 K to 450 K Consider the following metals: (a) nickel; (b) copper; (c) platinum; (d) brass.

21.3 A cylindrical copper heating element 2 ft long and $\frac{1}{2}$ in. in diameter is immersed in water. The system pressure is maintained at 1 atmosphere, and the tube surface is held at 280°F. Determine the nucleate-boiling heat-transfer coefficient and the rate of heat dissipation for this system.

21.4 If the cylinder described in problem 21.3 were initially heated to 500°F, how long would it take for the center of the cylinder to cool to 240°F if it were constructed of
(a) copper?
(b) brass?
(c) nickel?

21.5 Four immersion heaters in the shape of cylinders 15 cm long and 2 cm in diameter are immersed in a water bath at 1 atmosphere total pressure. Each heater is rated at 500 W. If the heaters operate at rated capacity, estimate the temperature of the heater surface. What is the convective heat-transfer coefficient in this case?

21.6 A horizontal circular cylinder 1 in. in diameter has its outside surface at a temperature of 1200°F. This tube is immersed in saturated water at a pressure of 40 psi. Estimate the heat flux due to film boiling which may be achieved with this configuration. At 40 psi the temperature of saturated water is 267°F.

21.7 Estimate the heat-transfer rate per foot of length from a 0.02-in. diameter nichrome wire immersed in water at 240°F. The wire temperature is 2200°F.

21.8 Two thousand watts of electrical energy are to be dissipated through copper plates measuring 5 cm by 10 cm by 0.6 cm thick immersed in water at 390 K. How many plates would you recommend? Substantiate all of the design criteria used.

21.9 A steel plate is removed from a heat-treating operation at 600 K and is immediately immersed into a water bath at 373 K.
(a) Construct a plot of heat flux vs. plate temperature for this system.
(b) Construct a plot of convective heat-transfer coefficient vs. plate temperature.
(c) For a mild-steel plate 3 cm thick and 30 cm square plot the plate temperature vs. time.

21.10 Water, flowing in a pipe, is to receive heat at a rate of 3×10^6 Btu/hr ft^2 of pipe surface. The pipe has an inside diameter of $\frac{3}{4}$ in. and is 4 ft long. If the water is to be at 212°F throughout its residence in the pipe, what rate of water flow would you suggest for safe operation? Support your results with all design criteria used.

21.11 Saturated steam at atmospheric pressure is enclosed within a vertical $\frac{1}{2}$-in. diameter pipe whose surface is at 160°F. Construct a plot of the amount of pipe cross section filled by condensate vs. distance from the top of the pipe. What happens as the ratio of the area occupied by the condensate approaches the cross-sectional area of the pipe?

21.12 Saturated steam at atmospheric pressure condenses on the outside surface of a 1-m-long tube with 150-mm diameter. The surface temperature is maintained at 91°C. Evaluate the condensation rate if the pipe is oriented
(a) vertically;
(b) horizontally.

21.13 Water flowing at a rate of 4000 kg/hr through a 16.5-mm ID tube enters at 20°C. The tube outside diameter is 19 mm. Saturated atmospheric steam condenses on the outside of the tube. For a horizontal brass tube 2 m long, evaluate
(a) the convective coefficient on the water side;
(b) the convective coefficient on the condensate side;
(c) the exit water temperature;
(d) the condensation rate.

21.14 Saturated steam at atmospheric pressure flows at a rate of 0.042 kg/s/m between two vertical surfaces maintained at 340 K which are separated by a distance of 1 cm. How tall may this configuration be if the steam velocity is not to exceed 15 m/s?

21.15 A circular pan has its bottom surface maintained at 200°F and is situated in saturated steam at 212°F. Construct a plot of condensate depth in the pan vs. time up to 1 hr for this situation. The sides of the pan may be considered non-conducting.

21.16 Saturated steam at 365 K condenses on a 2 cm tube whose surface is maintained at 340 K. Determine the rate of condensation and the heat transfer coefficient for the case of a 1.5 m-long tube oriented (a) vertically; (b) horizontally.

21.17 If eight tubes of the size designated in problem 21.14 are oriented horizontally in a vertical bank, what heat-transfer rate will occur?

21.18 Determine the heat-transfer coefficient for a horizontal $\frac{5}{8}$-in.-OD tube with its surface maintained at 100°F surrounded by steam at 200°F.

21.19 If eight tubes of the size designated in problem 21.16 are arranged in a vertical bank and the flow is assumed laminar, determine: (a) the average heat-transfer coefficient for the bank; (b) the heat-transfer coefficient for the first, third, and eighth tubes.

21.20 Given the conditions of problem 21.16, what height of vertical wall will cause the film at the bottom of the tube to be turbulent?

21.21 A vertical flat surface 2 ft high is maintained at 60°F. If saturated ammonia at 85°F is adjacent to the surface, what heat-transfer coefficient will apply to the condensation process? What total heat transfer will occur?

21.22 A square pan measuring 40 cm on a side and having a 2 cm-high lip on all sides has its surface maintained at 350 K. If this pan is situated in saturated steam at 372 K, how long will it be before condensate spills over the lip if the pan is
(a) horizontal?
(b) inclined at 10° to the horizontal?
(c) inclined at 30° to the horizontal?

21.23 A square pan with sides measuring 1 ft and a perpendicular lip extending 1 in. above the base is oriented with its base at an angle of 20° from the horizontal. The pan surface is kept at 180°F and it is situated in an atmosphere of 210°F steam. How long will it be before condensate spills over the lip of the pan?

22
HEAT-TRANSFER EQUIPMENT

A device whose primary purpose is the transfer of energy between two fluids is called a *heat exchanger*. Heat exchangers are usually classified into three categories:

(1) regenerators;

(2) open-type exchangers;

and

(3) closed-type exchangers or recuperators.

Regenerators are exchangers in which the hot and cold fluids flow alternately through the same space with as little physical mixing between the two streams as possible. The amount of energy transfer is dependent upon the fluid and flow properties of the fluid stream as well as the geometry and thermal properties of the surface. The required analytical tools for handling this type of heat exchanger have been developed in the preceding chapters.

Open-type heat exchangers are, as implied in their designation, devices wherein physical mixing of the two fluid streams actually occurs. Hot and cold fluids enter open-type heat exchangers and leave as a single stream. The nature of the exit stream is predicted by continuity and the first law of thermodynamics; no rate equations are necessary for the analysis of this type of exchanger.

The third type of heat exchanger, the recuperator, is the one of primary importance and the one to which we shall direct most of our attention. In the recuperator the hot and cold fluid streams do not come into direct contact with each other but are separated by a tube wall or a surface which may be flat or curved in some manner. Energy exchange is thus accomplished from one fluid to a surface by convection, through the wall or plate by conduction, and then by convection from the surface to the second fluid. Each of these energy-transfer processes has been considered separately in the preceding chapters. We shall, in the sections to follow, investigate the conditions under which these three energy-transfer processes act in series with one another, resulting in a continuous change in the temperature of at least one of the fluid streams involved.

We shall be concerned with a thermal analysis of these exchangers. A complete design of such equipment involves an analysis of pressure drop, using techniques from Chapter 14, as well as material and structural considerations which are not within the scope of this text.

22.1   TYPES OF HEAT EXCHANGERS

In addition to being considered a closed-type exchanger, a recuperator is classified according to its configuration and the number of passes made by each fluid stream as it traverses the heat exchanger.

A *single-pass* heat exchanger is one in which each fluid flows through the exchanger only once. An additional descriptive term identifies the relative directions of the two streams, the terms used being *parallel flow* or *cocurrent flow* if the fluids flow in the same direction, *countercurrent flow* or simply *counterflow* if the fluids flow in opposite directions, and *crossflow* if the two fluids flow at right angles to one another. A common single-pass configuration is the double-pipe arrangement shown in Figure 22.1. A crossflow arrangement is shown in Figure 22.2.

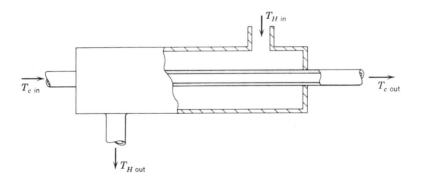

Figure 22.1   A double-pipe heat exchanger.

Variations on the crossflow configuration occur when one or the other, or both fluids are mixed. The arrangement shown in Figure 22.2 is one in which neither fluid is mixed. If the baffles or corrugations were not present, the fluid streams would be unseparated or mixed. In a condition such as that depicted in the figure, the fluid leaving at one end of the sandwich arrangement will have a nonuniform temperature variation from one side to the other, since each section contacts an adjacent fluid stream at a different temperature. It is normally desirable to have one or both fluids unmixed.

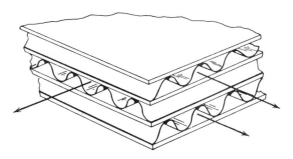

Figure 22.2   A crossflow heat exchanger.

In order to accomplish as much transfer of energy in as little space as possible, it is desirable to utilize multiple passes of one or both fluids. A popular configuration is the *shell-and-tube* arrangement shown in Figure 22.3. In this figure the *tube-side fluid* makes two passes while the *shell-side fluid* makes one pass. Good mixing of the shell-side fluid is accomplished with the baffles shown. Without these baffles the fluid becomes stagnant in certain parts of the shell, the flow is

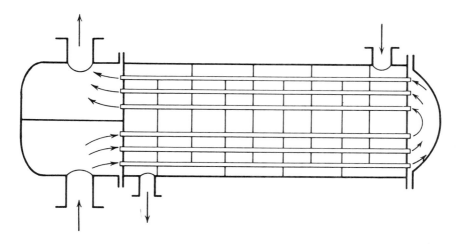

Figure 22.3   Shell-and-tube heat exchanger.

partially channelled past these stagnant or "dead" regions, and less-than-optimum performance is achieved. Variations on the number of tube-and-shell passes are encountered in numerous applications; seldom are more than two shell-side passes used.

A number of more recent heat-transfer applications require more compact configurations than that afforded by the shell-and-tube arrangement. The subject of "compact heat exchangers" has been investigated and reported both carefully and quite thoroughly by Kays and London.* Typical compact arrangements are shown in Figure 22.4.

The analysis of shell-and-tube, compact, or any multiple-pass heat exchanger is quite involved. Since each is a composite of several single-pass arrangements, we shall initially focus our attention on the single-pass heat exchanger.

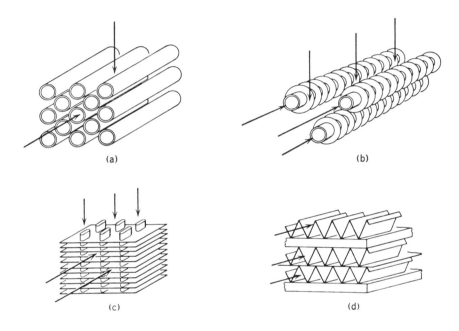

(a)

(b)

(c)

(d)

Figure 22.4 Compact heat-exchanger configurations.

22.2 SINGLE-PASS HEAT-EXCHANGER ANALYSIS: THE LOG-MEAN TEMPERATURE DIFFERENCE

It is useful, when considering parallel or counterflow single-pass heat exchangers, to draw a simple sketch depicting the general temperature variation experienced by each fluid stream. There are four such profiles in this category, all of which are shown and labeled in Figure 22.5. Each of these may be found in a double-pipe arrangement.

---

* W. M. Kays and A. L. London, *Compact Heat Exchangers*, Second Edition, McGraw-Hill Book Company, 1964.

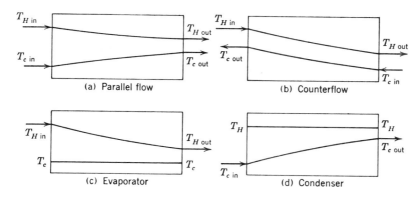

Figure 22.5 Temperature profiles for single-pass, double-pipe heat exchangers.

In (c) and (d) of Figure 22.5 one of the two fluids remains at constant temperature while exchanging heat with the other fluid whose temperature is changing. This situation occurs when energy transfer results in a change of phase rather than of temperature as in the cases of evaporation and condensation shown. The direction of flow of the fluid undergoing a change in phase was not depicted in the figure, since it is of no consequence to the analysis. If the situation occurs where the complete phase change such as condensation occurs within the exchanger along with some subcooling, then the diagram will appear as in Figure 22.6. In such a case the direction of flow of the condensate stream is important. For purposes of analysis this process may be considered the superposition of a condenser and a counterflow exchanger, as depicted in the diagram.

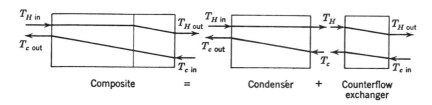

Figure 22.6 Temperature profile in a condenser with subcooling.

Also quite noticeable from parts (a) and (b) of Figure 22.5 is the significant difference in temperature profile exhibited by the parallel and counterflow arrangements. It is apparent that the exit temperatures of the hot and cold fluids in the parallel-flow case approach the same value. It is a simple exercise to show that this temperature is the one resulting if the two fluids are mixed in an open-type heat exchanger.

In the counterflow arrangement it is possible for the hot fluid to leave the exchanger at a temperature below that at which the cold fluid leaves. This situation obviously corresponds to a case of greater total energy transfer per unit area of heat exchanger surface than would be obtained if the same fluids entered a parallel-flow configuration. The obvious conclusion to this discussion is that the counterflow configuration is the most desirable of the single-pass arrangements. It is thus the single pass counterflow arrangement to which we shall direct our attention.

The detailed analysis of a single-pass counterflow heat exchanger which follows is referred to the diagram and nomenclature of Figure 22.7.

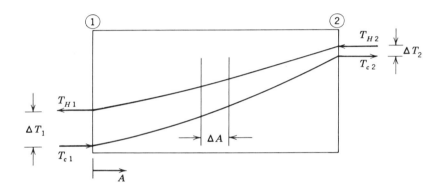

Figure 22.7 Diagram of temperature vs. contact area for single-pass counterflow analysis.

The abscissa of this figure is area. For a double-pipe arrangement the heat-transfer area varies linearly with distance from one end of the exchanger; in the case shown the zero reference is the end of the exchanger at which the hot fluid enters.

With reference to a general increment of area, ΔA, between the ends of this unit, a first-law-of-thermodynamics analysis of the two fluid streams will yield

$$\Delta q = (\dot{m}c_p)_c \Delta T_c$$

and

$$\Delta q = (\dot{m}c_p)_H \Delta T_H$$

As the incremental area approaches differential size, we may write

$$dq = (\dot{m}c_p)_c \, dT_c = C_c \, dT_c \qquad (22\text{-}1)$$

and

$$dq = (\dot{m}c_p)_H \, dT_H = C_H \, dT_H \qquad (22\text{-}2)$$

where the capacity coefficient, C, is introduced in place of the more cumbersome product, $\dot{m}c_p$.

Writing equation (15-17) for the energy transfer between the two fluids at this location, we have

$$dq = U\,dA(T_H - T_c) \tag{22-3}$$

which utilizes the overall heat-transfer coefficient, U, introduced in Chapter 15. Designating $T_H - T_c$ as ΔT, we have

$$d(\Delta T) = dT_H - dT_c \tag{22-4}$$

and substituting for dT_H and dT_c from equations (22-1) and (22-2), we obtain

$$d(\Delta T) = dq\left(\frac{1}{C_H} - \frac{1}{C_c}\right) = \frac{dq}{C_H}\left(1 - \frac{C_H}{C_c}\right) \tag{22-5}$$

We should also note that dq is the same in each of these expressions, thus equations (22-1) and (22-2) may be equated and integrated from one end of the exchanger to the other, yielding, for the ratio C_H/C_c,

$$\frac{C_H}{C_c} = \frac{T_{c2} - T_{c1}}{T_{H2} - T_{H1}} \tag{22-6}$$

which may be substituted into equation (22-5) and rearranged as follows:

$$d(\Delta T) = \frac{dq}{C_H}\left(1 - \frac{T_{c2} - T_{c1}}{T_{H2} - T_{H1}}\right) = \frac{dq}{C_H}\left(\frac{T_{H2} - T_{H1} - T_{c2} + T_{c1}}{T_{H2} - T_{H1}}\right)$$

$$= \frac{dq}{C_H}\left(\frac{\Delta T_2 - \Delta T_1}{T_{H2} - T_{H1}}\right) \tag{22-7}$$

Combining equations (22-3) and (22-7), and noting that $C_H(T_{H2} - T_{H1}) = q$, we have, for constant U,

$$\int_{\Delta T_1}^{\Delta T_2} \frac{d(\Delta T)}{\Delta T} = \frac{U}{q}(\Delta T_2 - \Delta T_1)\int_0^A dA \tag{22-8}$$

which, upon integration, becomes

$$\ln\frac{\Delta T_2}{\Delta T_1} = \frac{UA}{q}(\Delta T_2 - \Delta T_1)$$

This result is normally written as

$$q = UA\frac{\Delta T_2 - \Delta T_1}{\ln\dfrac{\Delta T_2}{\Delta T_1}} \tag{22-9}$$

The driving force, on the right-hand side of equation (22-9), is seen to be a particular sort of mean temperature difference between the two fluid streams. This ratio, $(\Delta T_2 - \Delta T_1)/\ln(\Delta T_2/\Delta T_1)$ is designated ΔT_{lm}, the *logarithmic-mean temperature difference*, and the expression for q is written simply as

$$q = UA \, \Delta T_{lm} \tag{22-10}$$

Even though equation (22-10) was developed for the specific case of counterflow, it is equally valid for any of the single-pass operations depicted in Figure 22.5.

It was mentioned earlier, but bears repeating, that equation (22-10) is based upon a constant value of the overall heat-transfer coefficient, U. This coefficient will not, in general, remain constant; however, calculations based upon a value of U taken midway between the ends of the exchanger are usually accurate enough. If there is considerable variation in U from one end of the exchanger to the other, then a step-by-step numerical integration is necessary, equations (22-1), (22-2), and (22-3) being evaluated repeatedly over a number of small-area increments.

It is also possible that the temperature differences in equation (22-9), evaluated at either end of a counterflow exchanger, are equal. In such a case the log-mean temperature difference is indeterminate, that is,

$$\frac{\Delta T_2 - \Delta T_1}{\ln \left(\Delta T_2 / \Delta T_1 \right)} = \frac{0}{0}, \qquad \text{if } \Delta T_1 = \Delta T_2$$

In such a case L'Hôpital's rule may be applied as follows:

$$\lim_{\Delta T_2 \to \Delta T_1} \frac{\Delta T_2 - \Delta T_1}{\ln \left(\Delta T_2 / \Delta T_1 \right)} = \lim_{\Delta T_2 / \Delta T_1 \to 1} \left[\frac{\Delta T_1 \{ (\Delta T_2 / \Delta T_1) - 1 \}}{\ln \left(\Delta T_2 / \Delta T_1 \right)} \right]$$

when the ratio $\Delta T_2 / \Delta T_1$ is designated by the symbol F, we may write

$$= \lim_{F \to 1} \Delta T \left(\frac{F - 1}{\ln F} \right)$$

Differentiating numerator and denominator with respect to F yields the result that

$$\lim_{\Delta T_2 \to \Delta T_1} \frac{\Delta T_2 - \Delta T_1}{\ln \left(\Delta T_2 / \Delta T_1 \right)} = \Delta T$$

or that equation (22-10) may be used in the simple form

$$q = UA \, \Delta T \tag{22-11}$$

From the foregoing simple analysis it should be apparent that equation (22-11) may be used and achieve reasonable accuracy so long as ΔT_1 and ΔT_2 are not vastly different. It turns out that a simple arithmetic mean is within 1% of the logarithmic-mean temperature difference for values of $(\Delta T_2 / \Delta T_1) < 1.5$.

EXAMPLE I

Light lubricating oil ($c_p = 2090$ J/kg · K) is cooled by allowing it to exchange energy with water in a small heat exchanger. The oil enters and leaves the heat exchanger at 375 K and 350 K, respectively, and flows at a rate of 0.5 kg/s. Water at 280 K is available in

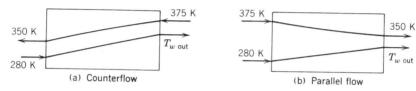

Figure 22.8 Single-pass temperature profiles for counterflow and parallel flow.

sufficient quantity to allow 0.201 kg/s to be used for cooling purposes. Determine the required heat transfer area for (a) counterflow, and (b) parallel flow operation (see Figure 22.8). The overall heat transfer coefficient may be taken as 250 W/m² · K.

The outlet water temperature is determined by applying equations (22-1) and (22-2),

$$q = (0.5 \text{ kg/s})(2090 \text{ J/kg} \cdot \text{K})(25 \text{ K}) = 26\,125 \text{ W}$$

$$= (0.201 \text{ kg/s})(4177 \text{ J/kg} \cdot \text{K})(T_{w\,out} - 280 \text{ K})$$

from which we obtain

$$T_{w\,out} = 280 + \frac{(0.201)(2090)(25)}{(0.5)(4177)} = 311.1 \text{ K} \qquad (100°F)$$

This result applies to both parallel flow and counterflow. For the counterflow configuration, ΔT_{lm} is calculated as

$$\Delta T_{lm} = \frac{70 - 63.9}{\ln \dfrac{70}{63.9}} = 66.9 \text{ K} \qquad (120.4°F)$$

and applying equation (22-10), we see that the area required to accomplish this energy transfer is

$$A = \frac{26\,125 \text{ W}}{(250 \text{ W/m}^2 \cdot \text{K})(66.9 \text{ K})} = 1.562 \text{ m}^2 \qquad (16.81 \text{ ft}^2)$$

Performing similar calculations for the parallel-flow situation, we obtain

$$\Delta T_{lm} = \frac{95 - 38.9}{\ln \dfrac{95}{38.9}} = 62.8 \text{ K} \qquad (113°F)$$

$$A = \frac{26\,125 \text{ W}}{(250 \text{ W/m}^2 \cdot \text{K})(62.8 \text{ K})} = 1.66 \text{ m}^2 \qquad (17.9 \text{ ft}^2)$$

The area required to transfer 26 125 W is seen to be lower for the counterflow arrangement by approximately 7%.

22.3 CROSSFLOW AND SHELL-AND-TUBE HEAT-EXCHANGER ANALYSIS

More complicated flow arrangements than the ones considered in the previous sections are much more difficult to treat analytically. Correction factors to be

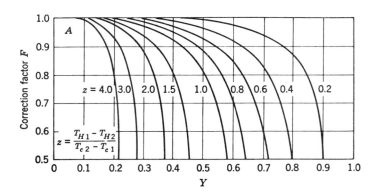

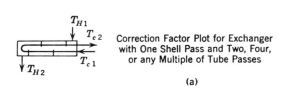

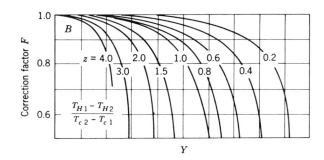

Correction Factor Plot for Exchanger
with One Shell Pass and Two, Four,
or any Multiple of Tube Passes

(a)

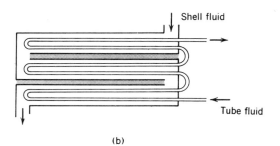

(b)

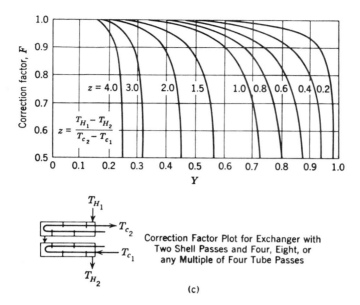

Correction Factor Plot for Exchanger with
Two Shell Passes and Four, Eight, or
any Multiple of Four Tube Passes

(c)

Figure 22.9 Correction factors for three shell-and-tube heat exchanger configurations. (a) One shell pass and two or a multiple of two tube passes. (b) One shell pass and three or a multiple of three tube passes. (c) Two shell pass and two or a multiple of two tube passes. (From R. A. Bowman, A. C. Mueller, and W. M. Nagle, *Trans. A.S.M.E.*, **62**, 284, 285 (1940). By permission of the publishers.) Correction factors, F, based on counterflow LMTD.

used with equation (22-10) have been presented in chart form by Bowman, Mueller, and Nagle* and by the Tubular Exchanger Manufacturers Association.† Figures 22.9 and 22.10 present correction factors for six types of heat-exchanger configurations. The first three are for different shell-and-tube configurations, the latter three are for different crossflow conditions.

The parameters in Figures 22.9 and 22.10 are evaluated as follows:

$$Y = \frac{T_{t\,\text{out}} - T_{t\,\text{in}}}{T_{s\,\text{in}} - T_{t\,\text{in}}} \qquad (22\text{-}12)$$

$$Z = \frac{(\dot{m}c_p)_{\text{tube}}}{(\dot{m}c_p)_{\text{shell}}} = \frac{C_t}{C_s} = \frac{T_{s\,\text{in}} - T_{s\,\text{out}}}{T_{t\,\text{out}} - T_{t\,\text{in}}} \qquad (22\text{-}13)$$

* R. A. Bowman, A. C. Mueller, and W. M. Nagle, *A.S.M.E.* (*Trans.*), **62**, 283 (1940).
† Tubular Exchanger Manufacturers Association, Standards, TEMA, Third Edition, New York, 1952.

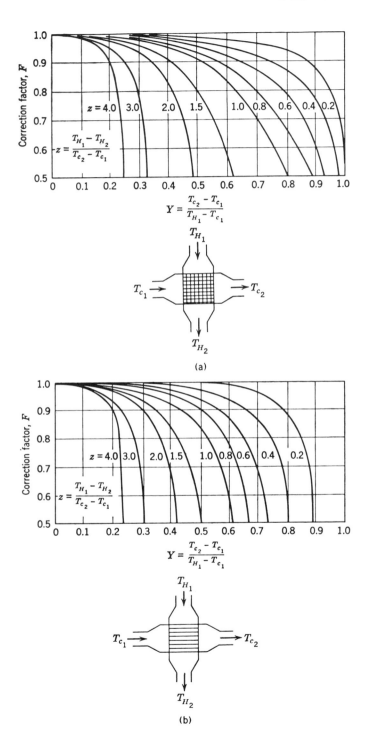

(a)

(b)

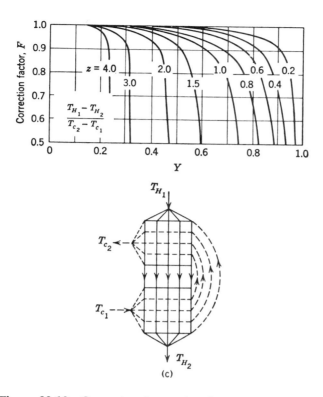

Figure 22.10 Correction factors for three crossflow heat-exchanger configurations. (a) Crossflow, single-pass, both fluids unmixed. (b) Crossflow, single-pass, one fluid unmixed. (c) Crossflow, tube passes mixed; fluid flows over first and second passes in series. (From R. A. Bowman, A. C. Mueller, and W. M. Nagle, *Trans. A.S.M.E.*, **62**, 288, 289 (1940). By permission of the publishers.)

where the subscripts s and t refer to the shell-side and tube-side fluids, respectively. The quantity read on the ordinate of each plot, for given values of Y and Z, is F, the correction factor to be applied to equation (22-10), and thus these more complicated configurations may be treated in much the same way as the single-pass double-pipe case. The reader is cautioned to apply equation (22-10), using the factor F as in equation (22-14)

$$q = UA(F \, \Delta T_{lm}) \qquad (22\text{-}14)$$

with the logarithmic mean temperature difference calculated on the basis of *counterflow.*

The manner of using Figures 22.9 and 22.10 may be illustrated by referring to the following example:

EXAMPLE 2

In the oil-water energy transfer described in Example 1, compare the result obtained with the result that would be obtained if the heat exchanger were

(a) Crossflow, water-mixed

(b) Shell-and-tube with four tube-side passes, oil being the tube-side fluid.

For part (a) Figure 22.10b must be used. The parameters needed to use this figure are

$$Y = \frac{T_{t \, out} - T_{t \, in}}{T_{s \, in} - T_{t \, in}} = \frac{25}{95} = 0.263$$

and

$$Z = \frac{T_{s \, in} - T_{s \, out}}{T_{t \, out} - T_{t \, in}} = \frac{31.1}{25} = 1.244$$

and from the figure we read $F = 0.96$. The required area for part (a) is thus equal to $(1.562)/(0.96) = 1.63 \text{ m}^2$.

The values of Y and Z determined above are the same in part (b), yielding a value of F equal to 0.97. The area for part (b) becomes $(1.562)/(0.97) = 1.61 \text{ m}^2$.

22.4 THE NUMBER-OF-TRANSFER-UNITS (NTU) METHOD OF HEAT-EXCHANGER ANALYSIS AND DESIGN

Earlier mention was made of the work of Kays and London* with particular reference to compact heat exchangers. The book "Compact Heat Exchangers," by Kays and London, also presents charts useful for heat exchanger design on a different basis than discussed thus far.

Nusselt,† in 1930, proposed the method of analysis based upon the heat exchanger effectiveness $\mathscr{E}$. This term is defined as the ratio of the actual heat transfer in a heat exchanger to the maximum possible heat transfer that would take place if infinite surface area were available. By referring to a temperature profile diagram for counterflow operation as in Figure 22.11 it is seen that, in general, one fluid undergoes a greater total temperature change than the other. It is apparent that the fluid experiencing the larger change in temperature is the one having the smaller capacity coefficient, which we designate C_{min}. If $C_c = C_{min}$, as in Figure 22.11a, and if there is infinite area available for energy transfer, the exit temperature of the cold fluid will equal the inlet temperature of the hot fluid. According to the definition of the effectiveness, we may write

$$\mathscr{E} = \frac{C_H(T_{H \, in} - T_{H \, out})}{C_c(T_{c \, out} - T_{c \, in})_{max}} = \frac{C_{max}(T_{H \, in} - T_{H \, out})}{C_{min}(T_{H \, in} - T_{c \, in})} \tag{22-15}$$

If the hot fluid is the minimum fluid, as in Figure 22.11b, the expression for $\mathscr{E}$

---

* W. M. Kays and A. L. London, *op. cit.*

† W. Nusselt, *Tech. Mechanik and Thermodynamik,* **12** (1930).

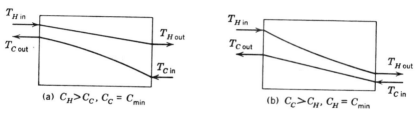

Figure 22.11 Temperature profiles for counterflow heat exchangers.

becomes

$$\mathscr{E} = \frac{C_c(T_{c\,out} - T_{c\,in})}{C_H(T_{H\,in} - T_{H\,out})_{max}} = \frac{C_{max}(T_{c\,out} - T_{c\,in})}{C_{min}(T_{H\,in} - T_{c\,in})} \tag{22-16}$$

Notice that the denominators in both equations (22-15) and (22-16) are the same and that, in each case, the numerator represents the actual heat transfer. It is thus possible to write a fifth expression for q as

$$q = \mathscr{E}C_{min}(T_{H\,in} - T_{c\,in}) \tag{22-17}$$

which, along with the integrated forms of equations (22-1) and (22-2), as well as equations (22-10) and (22-14), expresses q, the rate of heat transfer, in all of its useful forms as far as heat-exchanger analysis and design are concerned. Equation (22-17) is conspicuous among these others, since the temperature difference appearing is that between the inlet streams alone. This is a definite advantage when a given heat exchanger is to be used under conditions other than those for which it was designed. The exit temperatures of the two streams are then needed quantities, and equation (22-17) is obviously the easiest means of attaining this knowledge if one can determine the value of $\mathscr{E}$.

   To determine $\mathscr{E}$ for a single-pass case we initially write equation (22-17) in the form

$$\mathscr{E} = \frac{C_H(T_{H\,in} - T_{H\,out})}{C_{min}(T_{H\,in} - T_{c\,in})} = \frac{C_c(T_{c\,out} - T_{c\,in})}{C_{min}(T_{H\,in} - T_{c\,in})} \tag{22-18}$$

The appropriate form for equation (22-18) depends on which of the two fluids has the smaller value of C. We shall consider the cold fluid to be the minimum fluid and consider the case of counterflow. For these conditions equation (22-10) may be written as follows (numerical subscripts correspond to the situation shown in Figure 22.7):

$$q = C_c(T_{c2} - T_{c1}) = UA\frac{(T_{H1} - T_{c1}) - (T_{H2} - T_{c2})}{\ln\left[(T_{H1} - T_{c1})/(T_{H2} - T_{c2})\right]} \tag{22-19}$$

The entering temperature of the hot fluid, T_{H2}, may be written in terms of $\mathscr{E}$ by use of equation (22-18), yielding

$$T_{H2} = T_{c1} + \frac{1}{\mathscr{E}}(T_{c2} - T_{c1}) \tag{22-20}$$

and also

$$T_{H2} - T_{c2} = T_{c1} - T_{c2} + \frac{1}{\mathscr{E}}(T_{c2} - T_{c1})$$

$$= \left(\frac{1}{\mathscr{E}} - 1\right)(T_{c2} - T_{c1}) \tag{22-21}$$

From the integrated forms of equations (22-1) and (22-2) we have

$$\frac{C_c}{C_H} = \frac{T_{H2} - T_{H1}}{T_{c2} - T_{c1}}$$

which may be rearranged to the form

$$T_{H1} = T_{H2} - \frac{C_{min}}{C_{max}}(T_{c2} - T_{c1})$$

or

$$T_{H1} - T_{c1} = T_{H2} - T_{c1} - \frac{C_{min}}{C_{max}}(T_{c2} - T_{c1}) \tag{22-22}$$

Combining this expression with equation (22-20), we obtain

$$T_{H1} - T_{c1} = \frac{1}{\mathscr{E}}(T_{c2} - T_{c1}) - \frac{C_{min}}{C_{max}}(T_{c2} - T_{c1})$$

$$= \left(\frac{1}{\mathscr{E}} - \frac{C_{min}}{C_{max}}\right)(T_{c2} - T_{c1}) \tag{22-23}$$

Now substituting equations (22-21) and (22-23) into equation (22-19) and rearranging, we have

$$\ln \frac{1/\mathscr{E} - C_{min}/C_{max}}{1/\mathscr{E} - 1} = \frac{UA}{C_{min}}\left(1 - \frac{C_{min}}{C_{max}}\right)$$

Taking the antilog of both sides of this expression and solving for $\mathscr{E}$, we have, finally,

$$\mathscr{E} = \frac{1 - \exp\left[-\dfrac{UA}{C_{min}}\left(1 - \dfrac{C_{min}}{C_{max}}\right)\right]}{1 - C_{min}/C_{max}\,\exp\left[\dfrac{UA}{C_{min}}\left(1 - \dfrac{C_{min}}{C_{max}}\right)\right]} \tag{22-24}$$

The ratio UA/C_{min} is designated the *number of transfer units*, abbreviated NTU. Equation (22-24) was derived on the basis that $C_c = C_{min}$; if we had initially considered the hot fluid to be minimum, the same result would have been achieved. Thus equation (22-25),

$$\mathscr{E} = \frac{1 - \exp\left[-\text{NTU}\left(1 - \dfrac{C_{min}}{C_{max}}\right)\right]}{1 - (C_{min}/C_{max})\exp\left[-\text{NTU}\left(1 - \dfrac{C_{min}}{C_{max}}\right)\right]} \tag{22-25}$$

is valid for counterflow operation in general. For parallel flow an analogous development to the preceding will yield

$$\mathscr{E} = \frac{1 - \exp\left[-NTU\left(1 + \dfrac{C_{min}}{C_{max}}\right)\right]}{1 + C_{min}/C_{max}} \tag{22-26}$$

Kays and London* have put equations (22-25) and (22-26) into chart form, along with comparable expressions for the effectiveness of several shell-and-tube and crossflow arrangements. Figures 22.12 and 22.13 are charts for $\mathscr{E}$ as functions of NTU for various values of the parameter C_{min}/C_{max}.

With the aid of these figures equation (22-17) may be used both as an original design equation and as a means of evaluating existing equipment when it operates at other than design conditions.

The utility of the NTU approach is illustrated in the following example.

EXAMPLE 3

Repeat the calculations for Examples 22.1 and 22.2 to determine the required heat-transfer area for the specified conditions if the configurations are

(a) counterflow;
(b) parallel flow;
(c) crossflow, water-mixed;

and

(d) shell-and-tube with four tube-side passes.

It is first necessary to determine the capacity coefficients for the oil and water,

$$C_{oil} = (\dot{m}c_p)_{oil} = (0.5 \text{ kg/s})(2090 \text{ J/kg} \cdot \text{K}) = 1045 \text{ J/s} \cdot \text{K}$$

and

$$C_{water} = (\dot{m}c_p)_w = (0.201 \text{ kg/s})(4177 \text{ J/kg} \cdot \text{K}) = 841.2 \text{ J/s} \cdot \text{K}$$

thus the water is the minimum fluid. From equation (22-16) the effectiveness is evaluated as

$$\mathscr{E} = \frac{26\,125 \text{ W}}{(841.2 \text{ J/kg} \cdot \text{s})(95 \text{ K})} = 0.327$$

By using the appropriate chart in Figures 22.12 and 22.13 the appropriate NTU values and, in turn, the required area may be evaluated for each heat-exchanger configuration:

(a) *counterflow*

$$NTU = 0.47$$

$$A = \frac{(0.47)(841.2)}{250} = 1.581 \text{ m}^2$$

---

* W. M. Kays and A. L. London, *op. cit.*

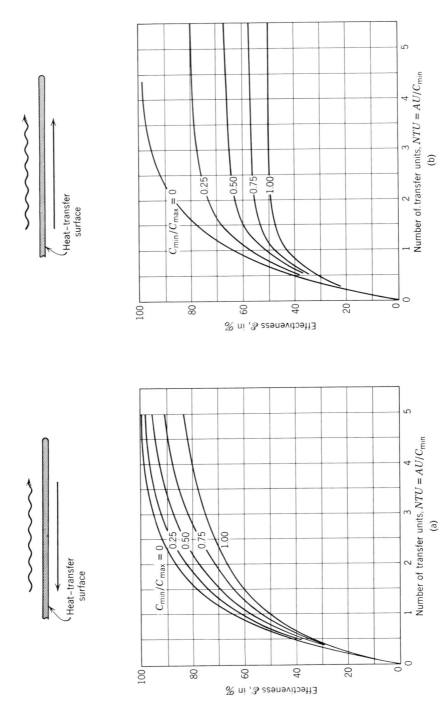

Figure 22.12 Heat-exchanger effectiveness for three shell-and-tube configurations. (a) Counterflow. (b) Parallel flow.

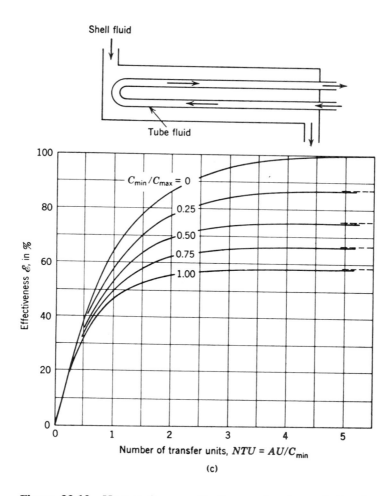

Shell fluid

Tube fluid

$C_{min}/C_{max} = 0$

0.25

0.50

0.75

1.00

Effectiveness $\mathscr{E}$, in %

Number of transfer units, $NTU = AU/C_{min}$

(c)

Figure 22.12 Heat-exchanger effectiveness for three shell-and-tube configurations. (c) One shell pass and two or a multiple of two tube passes.

(b) *parallel flow*

$$NTU = 0.50$$

$$A = \frac{(0.50)(841.2)}{250} = 1.682 \text{ m}^2$$

(c) *crossflow, water-mixed*

$$NTU = 0.48$$

$$A = \frac{(0.48)(841.2)}{250} = 1.615 \text{ m}^2$$

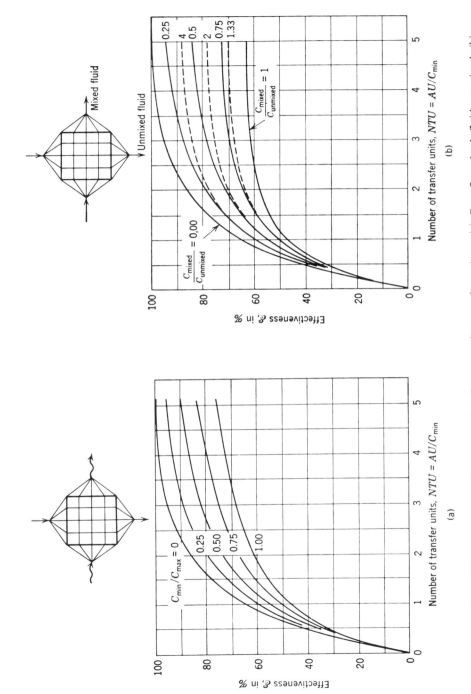

Figure 22.13 Heat-exchanger effectiveness for three crossflow configurations. (a) Crossflow, one fluid mixed. (b) Crossflow, both fluids unmixed.

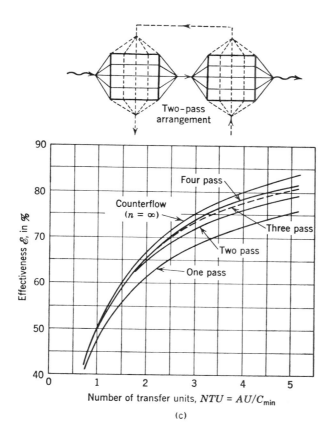

Figure 22.13 Heat-exchanger effectiveness for three crossflow configurations. (c) Crossflow, multiple pass.

and (d) *shell-and-tube, four tube-side passes*

$$NTU = 0.49$$

$$A = \frac{(0.49)(841.2)}{250} = 1.649 \text{ m}^2$$

These results are comparable to those obtained earlier, with some possible inaccuracies involved in reading the chart.

The NTU method offers no distinct advantage over the procedure introduced earlier, using the log-mean temperature difference, when performing calculations of the type involved in the preceding examples. In Example 4, however, the NTU approach is clearly superior.

EXAMPLE 4

In the energy exchange between water and lubricating oil as considered in the preceding examples, a crossflow exchanger with the shell-side fluid (water) mixed is constructed with a heat-transfer area of 16.5 ft^2. A new pump is attached to the water supply line enabling the water flow rate to be increased to $2200 \text{ lb}_m/\text{hr}$. What will be the exit temperatures of the oil and water streams at this new operating condition?

If one were to use the ΔT_{lm} method, a trial-and-error procedure would have to be used, since ΔT_{lm}, Y, and F are all dependent on one or both exit stream temperatures. Using the NTU method, it is first necessary to calculate the capacity coefficients:

$$C_{oil} = (4000 \text{ lb}_m/\text{hr})(0.5 \text{ Btu/lb}_m \text{ °F}) = 2000 \text{ Btu/hr °F}$$

and

$$C_w = (2200 \text{ lb}_m/\text{hr})(1 \text{ Btu/lb}_m \text{ °F}) = 2200 \text{ Btu/hr °F}$$

Oil is now the "minimum" fluid. Using $C_{oil} = C_{min}$, we obtain

$$\text{NTU} = \frac{UA}{C_{min}} = \frac{(45 \text{ Btu/hr ft}^2 \text{ °F})(16.5 \text{ ft}^2)}{2000 \text{ Btu/hr °F}} = 0.371$$

and from Figure 22.13 the effectiveness is

$$\mathscr{E} = 0.3$$

Using equation (22-17), we may evaluate the total heat-transfer rate,

$$q = (0.3)(2000 \text{ Btu/hr °F})(150°\text{F}) = 90\,000 \text{ Btu/hr}$$

an increase of over 12%. This value may now be used in equations (22-1) and (22-2) to yield the required answers,

$$T_{oil\,out} = 200°\text{F} - \frac{90\,000 \text{ Btu/hr}}{2000 \text{ Btu/hr °F}} = 155°\text{F}$$

and

$$T_{water\,out} = 50°\text{F} + \frac{90\,000 \text{ Btu/hr}}{2200 \text{ Btu/hr °F}} = 90.9°\text{F}$$

22.5 ADDITIONAL CONSIDERATIONS IN HEAT-EXCHANGER DESIGN

After a heat exchanger has been in service for some time, its performance may change as a result of the buildup of scale on a heat-transfer surface or of the deterioration of the surface by a corrosive fluid. When the nature of the surface is altered in some way as to affect the heat-transfer capability the surface is said to be "fouled."

When a fouling resistance exists, the thermal resistance is increased and a heat exchanger will transfer less energy than the design value. It is extremely difficult to predict the rate of scale buildup or the effect such buildup will have upon heat transfer. Some evaluation can be done after a heat exchanger has been in service for some time by comparing its performance with that when the surfaces were clean. The thermal resistance of the scale is determined by

$$R_{sc} = \frac{1}{U_f} - \frac{1}{U_0} \tag{22-27}$$

where U_0 is the overall heat transfer coefficient of the clean exchanger, U_f is the overall heat transfer coefficient of the fouled exchanger, and R_{sc} is the thermal resistance of the scale.

Fouling resistances that have been obtained from experiments may be used to roughly predict the overall heat-transfer coefficient by incorporation into an expression similar to equation (15-19). The following equation includes the fouling resistances, R_i on the inside tube surface and R_o on the outside tube surface

$$U_0 = \frac{1}{A_0/A_i h_i + R_i + [A_0 \ln (r_o/r_i)]/2\pi k + R_o + 1/h_o} \tag{22-28}$$

Fouling resistances to be used in equation (22-28) have been compiled by the Tubular Exchanger Manufacturers Association.* Some useful values are given in Table 22.1.

TABLE 22.1   HEAT-EXCHANGER FOULING
RESISTANCES

Fluid	Fouling resistances, hr ft² °F/Btu
Distilled water	0.0005
Sea water, below 125°F	0.0005
above 125°F	0.001
Boiler feed water, treated	0.001
City or well water, below 125°F	0.001
above 125°F	0.002
Refrigerating liquids	0.001
Refrigerating vapors	0.002
Liquid gasoline, organic vapors	0.0005
Fuel oil	0.005
Quenching oil	0.004
Steam, non-oil-bearing	0.0005
Industrial air	0.002

* Tubular Exchanger Manufacturers Association, TEMA Standards, Third Edition, New York (1952).

It is often useful to have "ball-park" figures on heat-exchanger size, flow rates, and the like. The most difficult quantity to estimate quickly is the overall heat-transfer coefficient, U. Mueller* has prepared the very useful table of approximate U values which is reproduced here as Table 22.2.

TABLE 22.2 APPROXIMATE VALUES FOR OVERALL HEAT-TRANSFER COEFFICIENTS

Fluid combination	U, Btu/hr ft^2 °F
Water to compressed air	10–30
Water to water, jacket water coolers	150–275
Water to brine	100–200
Water to gasoline	60–90
Water to gas oil or distillate	35–60
Water to organic solvents, alcohol	50–150
Water to condensing alcohol	45–120
Water to lubricating oil	20–60
Water to condensing oil vapors	40–100
Water to condensing or boiling Freon-12	50–150
Water to condensing ammonia	150–250
Steam to water, instantaneous heater	400–600
storage-tank heater	175–300
Steam to oil, heavy fuel	10–30
light fuel	30–60
light petroleum distillate	50–200
Steam to aqueous solutions	100–600
Steam to gases	5–50
Light organics to light organics	40–75
Medium organics to medium organics	20–60
Heavy organics to heavy organics	10–40
Heavy organics to light organics	10–60
Crude oil to gas oil	30–55

22.6 CLOSURE

The basic equations and procedures for heat-exchanger design were presented and developed in this chapter. All heat-exchanger design and analysis involve one or more of the following equations:

$$dq = C_c \, dT_c \tag{22-1}$$

---

* A. C. Mueller, Purdue Univ. Engr. Expt. Sta. Engr. Bulletin Res. Ser. 121 (1954).

$$dq = C_H \, dT_H \tag{22-2}$$

$$dq = U \, dA \, (T_H - T_c) \tag{22-3}$$

$$q = UA \, \Delta T_{lm} \tag{22-10}$$

and

$$q = \mathscr{E} C_{min} (T_{Hin} - T_{cin}) \tag{22-17}$$

Charts were presented by which single-pass techniques could be extended to include the design and analysis of crossflow and shell-and-tube configurations.

The two methods for heat-exchanger design utilize either equation (22-10) or (22-17). Either is reasonably rapid and straightforward for designing an exchanger. Equation (22-17) is a simpler and more direct approach when analyzing an exchanger which operates at other than design conditions.

PROBLEMS

22.1 A single tube-pass heat exchanger is to be designed to heat water by condensing steam in the shell. The water is to pass through the smooth horizontal tubes in turbulent flow, and the steam is to be condensed dropwise in the shell. The water flow rate, the initial and final water temperatures, the condensation temperature of the steam, and the available tube-side pressure drop (neglecting entrance and exit losses) are all specified. In order to determine the optimum exchanger design, it is desired to know how the total required area of the exchanger varies with the tube diameter selected. Assuming that the water flow remains turbulent, and that the thermal resistance of the tube wall and the steam-condensate film is negligible, determine the effect of tube diameter on the total area required in the exchanger.

22.2 One hundred thousand pounds per hour of water are to pass through a heat exchanger which is to raise the water temperature from 140°F to 200°F. Combustion products having a specific heat of 0.24 Btu/lb$_m$ °F are available at 800°F. The overall heat-transfer coefficient is 12 Btu/hr ft^2 °F. If 100 000 lb$_m$/hr of the combustion products are available, determine
(a) the exit temperature of the flue gas;
(b) the required heat-transfer area for a counterflow exchanger.

22.3 A shell-and-tube heat exchanger having one shell pass and eight tube passes is to heat kerosene from 80 to 130°F. The kerosene enters at a rate of 2500 lb$_m$/hr. Water, entering at 200°F and a rate of 900 lb$_m$/hr, is to flow on the shell side. The overall heat-transfer coefficient is 260 Btu/hr ft^2 °F. Determine the required heat-transfer area.

22.4 An oil having a specific heat of 1880 J/kg · K enters a single-pass counterflow heat exchanger at a rate of 2 kg/s and a temperature of 400 K. It is to be cooled to 350 K. Water is available to cool the oil at a rate of 2 kg/s and a temperature of 280 K. Determine the surface area required if the overall heat-transfer coefficient is 230 W/m^2 · K.

22.5 If the overall heat-transfer coefficient, initial fluid temperatures, and total heat-transfer area determined in problem 22.4 remain the same, find the exit oil temperature if the configuration is changed to
(a) crossflow, both fluids unmixed;
(b) shell-and-tube with two tube passes and one shell pass.

22.6 A condenser unit is of a shell-and-tube configuration with steam condensing at 85°C in the shell. The coefficient on the condensate side is 10,600 W/m^2 · K. Water at 20°C enters the tubes, which make 2 passes through the single shell unit. The water leaves the unit at a temperature of 38°C. An overall heat transfer coefficient of 4600 W/m^2 · K may be assumed to apply. The heat transfer rate is 2×10^6 kW. What must be the required length of tubes for this case?

22.7 Air at 103 kPa and 290 K flows in a long rectangular duct with dimensions 10 cm by 20 cm. A 2.5-m length of this duct is maintained at 395 K, and the average exit air temperature from this section is 300 K. Calculate the air flow rate and the total heat transfer.

22.8 Water enters a counterflow, double-pipe heat exchanger at a rate of 150 lb$_m$/min and is heated from 60°F to 140°F by an oil with a specific heat of 0.45 Btu/lb$_m$ °F. The oil enters at 240°F and leaves at 80°F. The overall heat-transfer coefficient is 50 Btu/hr ft^2 °F.
(a) What heat transfer area is required?
(b) What area is required if all conditions remain the same except that a shell-and-tube heat exchanger is used with the water making one shell pass and the oil making two tube passes?
(c) What exit water temperature would result if, for the exchanger of part (a), the water flow rate were decreased to 120 lb$_m$/min?

22.9 Compressed air is used in a heat-pump system to heat water which is subsequently used to warm a house. The house demand is 95 000 Btu/hr. Air enters the exchanger at 200°F and leaves at 120°F, water enters and leaves the exchanger at 90°F and 125°F, respectively. Choose from the following alternative units the one which is most compact.
(a) A counterflow surface with $U = 30$ Btu/hr ft^2 °F and a surface-to-volume ratio of 130 ft^2/ft^3.
(b) A crossflow configuration with the water unmixed and air mixed having $U = 40$ Btu/hr ft^2 °F and a surface-to-volume ratio of 100 ft^2/ft^3.

(c) A crossflow unit with both fluids unmixed with $U = 50$ Btu/hr ft² °F and surface-to-volume ratio of 90 ft²/ft³.

22.10 A shell-and-tube heat exchanger with 2 shell passes and 4 tube passes is used to exchange energy between two pressurized water streams. One stream flowing at 5000 lb$_m$/hr is heated from 75°F to 220°F. The hot stream flows at 2400 lb$_m$/hr and enters at 400°F. If the overall heat-transfer coefficient is 300 W/m² · K, determine the required surface area.

22.11 For the heat exchanger described in problem 22.10 it is observed, after a long period of operation, that the cold stream leaves at 184°F instead of at the design value of 220°F. This is for the same flow rates and entering temperatures of both streams. Evaluate the fouling factor that exists at the new conditions.

22.12 Saturated steam at 373 K is to be condensed in a shell-and-tube heat exchanger (it is to enter as steam at 373 K and leave as condensate at approximately 373 K). If the NTU rating for the condenser is given by the manufacturer as being 1.25 in this service for a circulating water flow of 0.07 kg/s, and circulating water is available at 280 K, what will be the

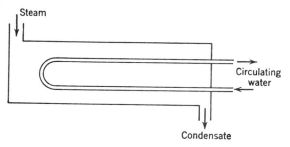

Steam

Circulating water

Condensate

approximate maximum flow rate of steam in kg/s that can be condensed? What will be the leaving temperature of the circulating water under these conditions? Take the heat of vaporization to be 2256 kJ/kg and c_p to be 4.18 kJ/kg · K.

22.13 A water-to-oil heat exchanger has entering and exiting temperatures of 255 K and 340 K, respectively, for the water and 305 K and 350 K, respectively, for the oil. What is the effectiveness of this heat exchanger?

22.14 Water at 50°F is available for cooling at a rate of 400 lb$_m$/hr. It enters a double-pipe heat exchanger with a total area of 18 ft². Oil, with $c_p = 0.45$ Btu/lb$_m$ °F, enters the exchanger at 250°F. The exiting water temperature is limited at 212°F, and the oil must leave the exchanger at no more than 160°F. Given the value of $U = 60$ Btu/hr ft² °F, find the maximum flow of oil which may be cooled with this unit.

23
RADIATION HEAT TRANSFER

The mechanism of radiation heat transfer has no analogy in either momentum or mass transfer. Radiation heat transfer is extremely important in many phases of engineering design such as boilers, home heating, and spacecraft. In this chapter we will concern ourselves first with understanding the nature of thermal radiation. Next we will discuss properties of surfaces and consider how system geometry influences radiant heat transfer. Finally we'll illustrate some techniques for solving relatively simple problems where surfaces and some gases participate in radiant energy exchange.

23.1 NATURE OF RADIATION

The transfer of energy by radiation has several unique characteristics when contrasted with conduction or convection. First, matter is not required for radiant heat transfer; indeed the presence of a medium will impede radiation transfer between surfaces. Cloud cover is observed to reduce maximum daytime temperatures and to increase minimum evening temperatures, both of which are dependent upon radiant energy transfer between earth and space. A second unique aspect of radiation is that both the amount of radiation *and* the quality of the radiation depend upon temperature. In conduction and convection the amount of heat transfer was found to depend upon the temperature difference; in radiation, the amount of heat transfer depends upon both the temperature difference between two bodies and the temperature level. In addition, radiation from a hot object will be different in quality than radiation from a body at a lower temperature. The color of incandescent objects is observed to change as the temperature is changed. The changing optical properties of radiation with temperature are of paramount importance in determining the radiant energy exchange between bodies.

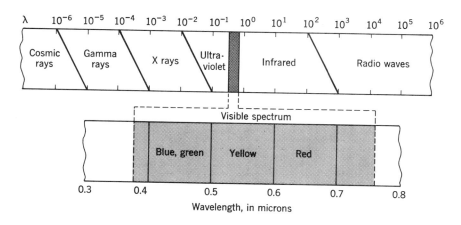

Figure 23.1 The electromagnetic spectrum.

Radiation travels at the speed of light, having both wave properties and particle-like properties. The electromagnetic spectrum shown in Figure 23.1 illustrates the tremendous range of frequency and wavelength over which radiation occurs.

The unit of wavelength which we shall use in discussing radiation is the micron, symbolized μ. One micron is 10^{-6} meter or $3.94(10)^{-5}$ in. The frequency, ν, of radiation is related to the wavelength λ, by $\lambda\nu = c$, where c is the speed of light. Short-wavelength radiation such as gamma rays and x-rays is associated with very high energies. To produce radiation of this type we must disturb the nucleus or the inner-shell electrons of an atom. Gamma rays and x-rays also have great penetrating ability; surfaces which are opaque to visible radiation are easily traversed by gamma and x-rays. Very-long-wavelength radiation, such as radio waves, also may pass through solids; however, the energy associated with these waves is much less than that for short wavelength radiation. In the range from $\lambda = 0.38$ to 0.76 microns, radiation is sensed by the optical nerve of the eye and is what we call light. Radiation in the visible range is observed to have little penetrating power except in some liquids, plastics, and glasses. The radiation between wavelengths of 0.1 and 100 microns is termed *thermal* radiation. The thermal band of the spectrum includes a portion of the ultraviolet and all of the infrared regions.

23.2 THERMAL RADIATION

In this chapter we shall deal exclusively with thermal radiation when discussing radiation heat transfer. Thermal radiation is observed to exhibit primarily the same optical properties as one of its subgroups, visible light, and we shall utilize

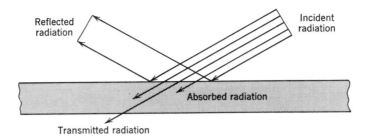

Figure 23.2   Fate of radiation incident upon a surface.

the wave-like properties of thermal radiation more than the particle-like properties.

Thermal radiation incident upon a surface as shown in Figure 23.2 may be either absorbed, reflected, or transmitted.

If ρ, α, and τ are the fractions of the incident radiation which are reflected, absorbed and transmitted, respectively, then

$$\rho + \alpha + \tau = 1 \tag{23-1}$$

where ρ is called the *reflectivity*, α is called the *absorptivity*, and τ is called the *transmissivity*.

There are two types of reflection that can occur, specular reflection and diffuse reflection. In *specular reflection*, the angle of incidence of the radiation is equal to the angle of reflection. The reflection shown in Figure 23.2 is specular reflection. Most bodies do not reflect in a specular manner, they reflect radiation in all directions. *Diffuse reflection* is sometimes likened to a situation in which the incident thermal radiation is absorbed and then re-emitted from the surface, still retaining its initial wavelength.

Absorption of thermal radiation in solids takes place in a very short distance, on the order of one micron in electrical conductors and about 0.05 in. in electrical nonconductors, the difference being caused by the different population of energy states in electrical conductors which can absorb energy at thermal radiation frequencies.

For most solids, the transmissivity is zero, and thus they may be called *opaque* to thermal radiation. Equation (23-1) becomes, for an opaque body, $\rho + \alpha = 1$.

The ideally absorbing body, for which $\alpha = 1$, is called a *black body*. A black body neither reflects nor transmits any thermal radiation. Since we see reflected light (radiation), a so-called black body will appear black, no light being reflected from it. A small hole in a large cavity closely approaches a black body, regardless of the nature of the interior surface. Radiation incident to the hole has very little opportunity to be reflected back out of the hole. Black bodies may also be made of bright objects, as can be shown by looking at a stack of razor blades, sharp edge forward.

The *total emissive power*, E, of a surface is defined as the total rate of thermal energy emitted via radiation from a surface in all directions and at all wavelengths per unit surface area. The total emissive power is also referred to elsewhere as the emittance or the total hemispheric intensity. Closely related to the total emissive power is the emissivity. The *emissivity*, ϵ, is defined as the ratio of the total emissive power of a surface to the total emissive power of an ideally radiating surface at the same temperature. The ideal radiating surface is also called a black body, so we may write

$$\epsilon = \frac{E}{E_b} \tag{23-2}$$

where E_b is the total emissive power of a black body. As the total emissive power includes radiant-energy contributions from all wavelengths, the *monochromatic emissive power*, E_λ, may also be defined. The radiant energy E_λ contained between wavelengths λ and $\lambda + d\lambda$ is the monochromatic emissive power; thus

$$dE = E_\lambda \, d\lambda, \quad \text{or} \quad E = \int_0^\infty E_\lambda \, d\lambda$$

The monochromatic emissivity, ϵ_λ, is simply $\epsilon_\lambda = E_\lambda/E_{\lambda,b}$, where $E_{\lambda,b}$ is the monochromatic emissive power of a black body at wavelength λ at the same temperature. A monochromatic absorptivity, α_λ, may be defined in the same manner as the monochromatic emissivity. The *monochromatic absorptivity* is defined as the ratio of the incident radiation of wavelength λ which is absorbed by a surface to the incident radiation absorbed by a black surface.

A relation between the absorptivity and the emissivity is given by Kirchhoff's law. Kirchhoff's law states that, for a system in thermodynamic equilibrium, the following equality holds for each surface:

$$\epsilon_\lambda = \alpha_\lambda \tag{23-3}$$

Thermodynamic equilibrium requires that all surfaces be at the same temperature so that there is no net heat transfer. The utility of Kirchhoff's law lies in its use for situations in which the departure from equilibrium is small. In such situations the emissivity and the absorptivity may be assumed to be equal. For radiation between bodies at greatly different temperatures, such as between the earth and the sun, Kirchhoff's law does not apply. A frequent error in using Kirchhoff's law arises from confusing thermal equilibrium with steady-state conditions. Steady state means that time derivatives are zero, while equilibrium refers to the equality of temperatures.

23.3 THE INTENSITY OF RADIATION

In order to characterize the quantity of radiation that travels from a surface along a specified path, the concept of a single ray is not adequate. The amount of energy traveling in a given direction is determined from I, the *intensity* of radiation. With reference to Figure 23.3, we are interested in knowing the rate at which radiant energy is emitted from a representative portion, dA, of the surface shown in a prescribed direction. Our perspective will be that of an observer at point P looking at dA. Standard spherical coordinates will be used, these being r, the radial coordinate; θ, the zenith angle shown in Figure 23.3; and ϕ, the azimuthal angle, which will be discussed shortly. If a unit area of surface, DA, emits a total energy dq, then the *intensity of radiation* is given by

$$I \equiv \frac{d^2 q}{dA \, d\Omega \cos \theta} \tag{23-4}$$

where $d\Omega$ is a differential solid angle, that is, a portion of space. Note that with the eye located at point P, in Figure 23.3, the apparent size of the emitting area is $dA \cos \theta$. It is important to remember that the intensity of radiation is independent of direction for a diffusely radiating surface. Rearranging equations (23-4), we see that the relation between the total emissive power, $E = dq/dA$, and the intensity, I, is

$$\frac{dq}{dA} = E = \int I \cos \theta \, d\Omega = I \int \cos \theta \, d\Omega \tag{23-5}$$

The relation is seen to be purely geometric for a diffusely radiating $(I \neq I(\theta))$ surface. Consider an imaginary hemisphere of radius r covering the plane surface on which dA is located. The solid angle $d\Omega$ intersects the hatched area on the

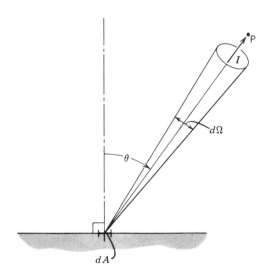

Figure 23.3   The intensity of radiation.

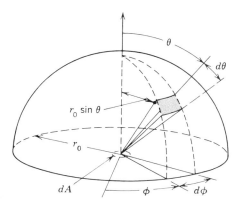

Figure 23.4   Integration of intensity over solid angles.

hemisphere as shown in Figure 23.4. A solid angle is defined by $\Omega = A/r^2$ or $d\Omega = dA/r^2$, and thus

$$d\Omega = \frac{(r \sin \theta \, d\phi)(r \, d\theta)}{r^2} = \sin \theta \, d\theta \, d\phi$$

The total emissive power per unit area becomes

$$E = I \int \cos \theta \, d\Omega = I \int_0^{2\pi} \int_0^{\pi/2} \cos \theta \sin \theta \, d\theta \, d\phi$$

or simply

$$E = \pi I \qquad\qquad (23\text{-}6)$$

If the surface does not radiate diffusely, then

$$E = \int_0^{2\pi} \int_0^{\pi/2} I \cos \theta \sin \theta \, d\theta \, d\phi$$

The relation between the intensity of radiation, I, and the total emissive power is an important step in determining the total emissive power.

   Radiation intensity is fundamental in formulating a quantitative description of radiant heat transfer but its definition, as already discussed, is cumbersome. Equation (23-6) relates intensity to emissive power which, potentially, is much easier to describe. We will now consider the means of such a description.

23.4 PLANCK'S LAW OF RADIATION

Planck* introduced the quantum concept in 1900 and with it the idea that radiation is emitted not in a continuous energy state but in discrete amounts or quanta. The intensity of radiation emitted by a black body, derived by Planck, is

$$I_{b,\lambda} = \frac{2c^2 h \lambda^{-5}}{\exp\left(\dfrac{ch}{\kappa \lambda T}\right) - 1}$$

where $I_{b,\lambda}$ is the intensity of radiation from a black body between wavelengths λ and $\lambda + d\lambda$, c is the speed of light, h is Planck's constant, κ is the Boltzmann constant, and T is the temperature. The total emissive power between wavelengths λ and $\lambda + d\lambda$ is then

$$E_{b,\lambda} = \frac{2\pi c^2 h \lambda^{-5}}{\exp\left(\dfrac{ch}{\kappa \lambda T}\right) - 1} \tag{23-7}$$

Figure 23.5 illustrates the spectral energy distribution of energy of a black body as given by equation (23-7).

In Figure 23.5 the area under the curve of $E_{b,\lambda}$ versus λ (the total emitted energy) is seen to increase rapidly with temperature. The peak energy is also observed to occur at shorter and shorter wavelengths as the temperature is increased. For a black body at 10 000°F (the approximate temperature of the sun) most of the emitted energy is in the visible region. Equation (23-7) expresses, functionally, $E_{b,\lambda}$ as a function of wavelength and temperature. Dividing both sides of this equation by T^5, we get

$$\frac{E_{b\lambda}}{T^5} = \frac{2\pi^2 h (\lambda T)^{-5}}{\exp\left(\dfrac{ch}{\lambda T}\right) - 1} \tag{23-8}$$

where the quantity $E_{b\lambda}/T^5$ is expressed as a function of the λT product, which can be treated as a single independent variable. This functional relationship is plotted in Figure 23.5b, and discrete values of $E_{b\lambda}/\sigma T^5$ are given in Table 23.1. The constant, σ, will be discussed in the next section.

The peak energy is observed to be emitted at $\lambda T = 5215.6$ microns °R $(2897.6\mu$ K), as can be determined by maximizing equation (23-8). The relation, $\lambda_{max} T = 5215.6\mu$ °R, is called Wien's displacement law. Wien obtained this result in 1893, seven years prior to Planck's development.

We are often interested in knowing how much emission occurs in a specific portion of the total wavelength spectrum. This is conveniently expressed as a fraction of the total emissive power. The fraction between wavelengths λ_1 and

* M. Planck, *Verh. d. deut. physik. Gesell.*, **2**, 237 (1900).

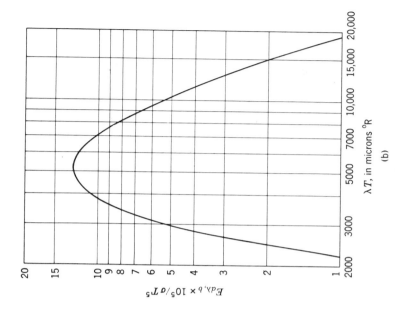

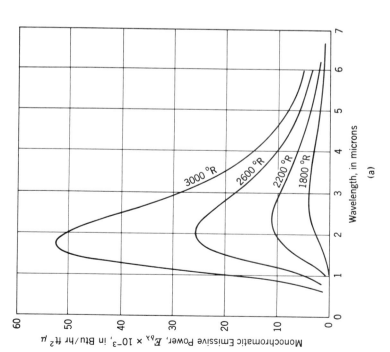

Figure 23.5 Spectral energy distribution for a black body.

TABLE 23.1   PLANCK RADIATION FUNCTIONS

$\lambda T [\mu^\circ R]$	$\dfrac{E_{b\lambda} \times 10^5}{\sigma T^5}$	$F_{0-\lambda T}$	$\lambda T [\mu^\circ R]$	$\dfrac{E_{b\lambda} \times 10^5}{\sigma T^5}$	$F_{0-\lambda T}$
1000.0	0.000039	0.0000	10400.0	5.142725	0.7183
1200.0	0.001191	0.0000	10600.0	4.921745	0.7284
1400.0	0.012008	0.0000	10800.0	4.710716	0.7380
1600.0	0.062118	0.0000	11000.0	4.509291	0.7472
1800.0	0.208018	0.0003	11200.0	4.317109	0.7561
2000.0	0.517405	0.0010	11400.0	4.133804	0.7645
2200.0	1.041926	0.0025	11600.0	3.959010	0.7726
2400.0	1.797651	0.0053	11800.0	3.792363	0.7803
2600.0	2.761875	0.0098	12000.0	3.633505	0.7878
2800.0	3.882650	0.0164	12200.0	3.482084	0.7949
3000.0	5.093279	0.0254	12400.0	3.337758	0.8017
3200.0	6.325614	0.0368	12600.0	3.200195	0.8082
3400.0	7.519353	0.0507	12800.0	3.069073	0.8145
3600.0	8.626936	0.0668	13000.0	2.944084	0.8205
3800.0	9.614973	0.0851	13200.0	2.824930	0.8263
4000.0	10.463377	0.1052	13400.0	2.711325	0.8318
4200.0	11.163315	0.1269	13600.0	2.602997	0.8371
4400.0	11.714711	0.1498	13800.0	2.499685	0.8422
4600.0	12.123821	0.1736	14000.0	2.401139	0.8471
4800.0	12.401105	0.1982	14200.0	2.307123	0.8518
5000.0	12.559492	0.2232	14400.0	2.217411	0.8564
5200.0	12.613057	0.2483	14600.0	2.131788	0.8607
5400.0	12.576066	0.2735	14800.0	2.050049	0.8649
5600.0	12.462308	0.2986	15000.0	1.972000	0.8689
5800.0	12.284687	0.3234	16000.0	1.630989	0.8869
6000.0	12.054971	0.3477	17000.0	1.358304	0.9018
6200.0	11.783688	0.3715	18000.0	1.138794	0.9142
6400.0	11.480102	0.3948	19000.0	0.960883	0.9247
6600.0	11.152254	0.4174	20000.0	0.815714	0.9335
6800.0	10.807041	0.4394	21000.0	0.696480	0.9411
7000.0	10.450309	0.4607	22000.0	0.597925	0.9475
7200.0	10.086964	0.4812	23000.0	0.515964	0.9531
7400.0	9.721078	0.5010	24000.0	0.447405	0.9579
7600.0	9.355994	0.5201	25000.0	0.389739	0.9621
7800.0	8.994419	0.5384	26000.0	0.340978	0.9657
8000.0	8.638524	0.5561	27000.0	0.299540	0.9689
8200.0	8.290014	0.5730	28000.0	0.264157	0.9717
8400.0	7.950202	0.5892	29000.0	0.233807	0.9742
8600.0	7.620072	0.6048	30000.0	0.207663	0.9764
8800.0	7.300336	0.6197	40000.0	0.074178	0.9891
9000.0	6.991475	0.6340	50000.0	0.032617	0.9941
9200.0	6.693786	0.6477	60000.0	0.016479	0.9965
9400.0	6.407408	0.6608	70000.0	0.009192	0.9977
9600.0	6.132361	0.6733	80000.0	0.005521	0.9984
9800.0	5.868560	0.6853	90000.0	0.003512	0.9989
10000.0	5.615844	0.6968	100000.0	0.002339	0.9991
10200.0	5.373989	0.7078			

λ_2 is designated $F_{\lambda_1-\lambda_2}$ and may be expressed as

$$F_{\lambda_1-\lambda_2} = \frac{\int_{\lambda_1}^{\lambda_2} E_{b\lambda}\,d\lambda}{\int_0^{\infty} E_{b\lambda}\,d\lambda} = \frac{\int_{\lambda_1}^{\lambda_2} E_{b\lambda}\,d\lambda}{\sigma T^4} \tag{23-9}$$

Equation (23-9) is conveniently broken into two integrals as follows.

$$F_{\lambda_1-\lambda_2} = \frac{1}{\sigma T^4}\left(\int_0^{\lambda_2} E_{b\lambda}\,d\lambda - \int_0^{\lambda_1} E_{b\lambda}\,d\lambda\right)$$

$$= F_{0-\lambda_2} - F_{0-\lambda_1} \tag{23-10}$$

So, at a given temperature, the fraction of emission between any two wavelengths can be determined by subtraction.

This process can be simplified if the temperature is eliminated as a separate variable. This may be accomplished by using the fraction $E_{b\lambda}/\sigma T^5$, as discussed. Equation (23-10) may be modified in this manner to yield

$$F_{\lambda_1 T-\lambda_2 T} = \int_0^{\lambda_2 T} \frac{E_{b\lambda}}{\sigma T^5}\,d(\lambda T) - \int_0^{\lambda_1 T} \frac{E_{b\lambda}}{\sigma T^5}\,d(\lambda T)$$

$$= F_{0-\lambda_2 T} - F_{0-\lambda_1 T} \tag{23-11}$$

Values of $F_{0-\lambda T}$ are given as functions of the product, λT, in Table 23.1.

23.5 STEFAN-BOLTZMANN LAW

Planck's law of radiation may be integrated over wavelengths from zero to infinity to determine the total emissive power. The result is

$$E_b = \int_0^{\infty} E_{b,\lambda}\,d\lambda = \frac{2\pi^5\kappa^4 T^4}{15c^2 h^3} = \sigma T^4 \tag{23-12}$$

where σ is called the Stefan-Boltzmann constant and has the value $\sigma = 5.676 \times 10^{-8}\,\text{W/m}^2 \cdot \text{K}^4$ ($0.1714 \times 10^{-8}\,\text{Btu/hr ft}^2\,{}^{\circ}\text{R}^4$). This constant is observed to be a combination of other physical constants. The Stefan-Boltzmann relation, $E_b = \sigma T^4$, was obtained prior to Planck's law, via experiment by Stefan in 1879 and via a thermodynamic derivation by Boltzmann in 1884. The exact value of the Stefan-Boltzmann constant, σ, and its relation to other physical constants were obtained after the presentation of Planck's law in 1900.

23.6 EMISSIVITY AND ABSORPTIVITY OF SOLID SURFACES

Whereas thermal conductivity, specific heat, density, and viscosity are the important physical properties of matter in heat conduction and convection, emissivity and absorptivity are the controlling properties in heat exchange by radiation.

From preceding sections it is seen that, for black-body radiation, $E_b = \sigma T^4$. For actual surfaces, $E = \epsilon E_b$, following the definition of emissivity. The emissivity of the surface, so defined, is a gross factor, since radiant energy is being sent out from a body not only in all directions but also other various wavelengths. For actual surfaces, the emissivity may vary with wavelength as well as the direction of emission. Consequently, we have to differentiate the monochromatic emissivity ϵ_λ and the directional emissivity ϵ_θ from the total emissivity ϵ.

Monochromatic emissivity. By definition, the monochromatic emissivity of an actual surface is the ratio of its monochromatic emissive power to that of a black surface at the same temperature. Figure 23.6 represents a typical distribution of the intensity of radiation of two such surfaces at the same temperature over

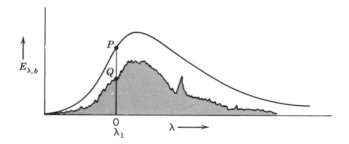

Figure 23.6   Emissivity at various wavelengths.

various wavelengths. The monochromatic emissivity at a certain wavelength, λ_1, is seen to be the ratio of two ordinates such as $\overline{OQ}$ and $\overline{OP}$. That is,

$$\epsilon_{\lambda_1} = \frac{\overline{OQ}}{\overline{OP}}$$

which is equal to the monochromatic absorptivity α_{λ_1} from radiation of a body at the same temperature. This is the direct consequence of Kirchhoff's law. The total emissivity of the surface is given by the ratio of the cross-hatched area shown in Figure 23.6 to that under the curve for the black-body radiation.

Directional emissivity. The cosine variation discussed previously, equation (23-5), is strictly applicable to radiation from a black surface but only fulfilled approximately by materials present in nature. This is due to the fact that the emissivity (averaged over all wavelengths) of actual surfaces is not a constant in all directions. The variation of emissivity of materials with the direction of emission can be conveniently represented by polar diagrams.

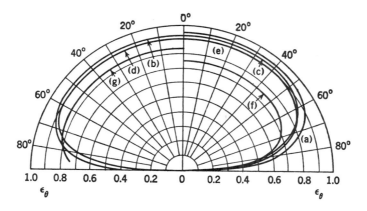

Figure 23.7 Emissivity variation with direction for nonconductors. (a) Wet ice. (b) Wood. (c) Glass. (d) Paper. (e) Clay. (f) Copper oxide. (g) Aluminum oxide.

If the cosine law is fulfilled, the distribution curves should take the form of semicircles. Most nonconductors have much smaller emissivities for emission angles in the neighborhood of 90° (see Figure 23.7).

Deviation from the cosine law is even greater for many conductors (see Figure 23.8). The emissivity stays fairly constant in the neighborhood of the normal direction of emission; as the emission angle is increased, it first increases and then decreases as the former approaches 90°.

The average total emissivity may be determined by using the following expression:

$$\epsilon = \int_0^{\pi/2} \epsilon_\theta \sin 2\theta \, d\theta$$

The emissivity, ϵ, is, in general, different from the normal emissivity, ϵ_n (emissivity in the normal direction). It has been found that for most bright metallic surfaces,

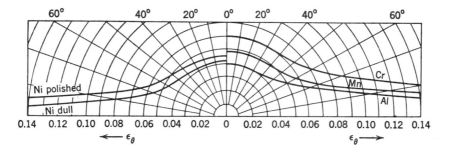

Figure 23.8 Emissivity variation with direction for conductors.

TABLE 23.2   THE RATIO ϵ/ϵ_n FOR BRIGHT
METALLIC SURFACES

Aluminum, bright rolled (338°F)	$\dfrac{0.049}{0.039} = 1.25$
Nickel, bright matte (212°F)	$\dfrac{0.046}{0.041} = 1.12$
Nickel, polished (212°F)	$\dfrac{0.053}{0.045} = 1.18$
Manganin, bright rolled (245°F)	$\dfrac{0.057}{0.048} = 1.19$
Chromium, polished (302°F)	$\dfrac{0.071}{0.058} = 1.22$
Iron, bright etched (302°F)	$\dfrac{0.158}{0.128} = 1.23$
Bismuth, bright (176°F)	$\dfrac{0.340}{0.336} = 1.08$

the total emissivity is approximately 20% higher than ϵ_n. Table 23.2 lists the ratio of ϵ/ϵ_n for a few representative bright metallic surfaces. For nonmetallic or other surfaces, the ratio ϵ/ϵ_n is slightly less than unity. Because of the inconsistency which can often be found among various sources, the normal emissivity values can be used, without appreciable error, for total emissivity (see Table 23.3).

A few generalizations may be made concerning the emissivity of surfaces:

(a) In general, emissivity depends on surface conditions.

(b) The emissivity of highly polished metallic surfaces is very low.

(c) The emissivity of all metallic surfaces increases with temperature.

(d) The formation of a thick oxide layer and roughening of the surface increase the emissivity appreciably.

(e) The ratio ϵ/ϵ_n is always greater than unity for bright metallic surfaces. The value 1.2 can be taken as a good average.

TABLE 23.3   THE RATIO ϵ/ϵ_n FOR NONMETALLIC
AND OTHER SURFACES

Copper oxide (300°F)	0.96
Fire clay (183°F)	0.99
Paper (200°F)	0.97
Plywood (158°F)	0.97
Glass (200°F)	0.93
Ice (32°F)	0.95

(f) The emissivities of nonmetallic surfaces are much higher than for metallic surfaces and show a decrease as temperature increases.

(g) The emissivities of colored oxides of heavy metals like Zn, Fe, and Cr are much larger than emissivities of white oxides of light metals like Ca, Mg, and Al.

Absorptivity. The absorptivity of a surface depends on the factors affecting the emissivity and, in addition, on the quality of the incident radiation. It may be remarked once again that Kirchhoff's law holds strictly true under thermal equilibrium. That is, if a body at temperature T_1 is receiving radiation from a black body also at temperature T_1, then $\alpha = \epsilon$. For most materials, in the usual range of temperature encountered in practice (from room temperature up to about 2000°F) the simple relationship $\alpha = \epsilon$ holds with good accuracy. However, if the incident radiation is that from a very-high-temperature source, say solar radiation (~10 000°F), the emissivity and absorptivity of ordinary surfaces may differ widely. White metal oxides usually exhibit an emissivity (and absorptivity) value of about 0.95 at ordinary temperature, but their absorptivity drops sharply to 0.15 if these oxides are exposed to solar radiation. Contrary to the above, freshly polished metallic surfaces have an emissivity value (and absorptivity under equilibrium conditions) of about 0.05. When exposed to solar radiation, their absorptivity increases to 0.2 or even 0.4.

Under these latter circumstances a double-subscript notation, $\alpha_{1,2}$, may be employed, the first subscript referring to the temperature of the receiving surface and the second subscript to the temperature of the incident radiation.

Gray surfaces. Like emissivity, the monochromatic absorptivity, α_λ, of a surface may vary with wavelength. If α_λ is a constant and thus independent of λ,

TABLE 23.4 NORMAL TOTAL EMISSIVITY OF VARIOUS SURFACES
(COMPILED BY H. C. HOTTEL)[†]

Surface	T, °F[‡]	Emissivity
A. Metals and their oxides		
Aluminum:		
Highly polished plate, 98.3% pure	440–1070	0.039–0.057
Commercial sheet	212	0.09
Oxidized at 1110°F	390–1110	0.11–0.19
Heavily oxidized	200–940	0.20–0.31
Brass:		
Polished	100–600	0.10
Oxidized by heating at 1110°F	390–1110	0.61–0.59

† By permission from W. H. McAdams (ed.), *Heat Transmission*, Third Edition, McGraw-Hill Book Company, 1954. Table of normal total emissivity compiled by H. C. Hottel.
‡ When temperatures and emissivities appear in pairs separated by dashes, they correspond and linear interpolation is permissible.

TABLE 23.4 (CONT'D.)

Surface	T, °F	Emissivity
Chromium (see nickel alloys for Ni-Cr steels):		
Polished	100–2000	0.08–0.36
Copper		
Polished	212	0.052
Plate heated at 1110°F	390–1110	0.57
Cuprous oxide	1470–2010	0.66–0.54
Molten copper	1970–2330	0.16–0.13
Gold:		
Pure, highly polished	440–1160	0.018–0.035
Iron and steel (not including stainless):		
Metallic surfaces (or very thin oxide layer)		
Iron, polished	800–1880	0.14–0.38
Cast iron, polished	392	0.21
Wrought iron, highly polished	100–480	0.28
Oxidized surfaces		
Iron plate, completely rusted	67	0.69
Steel plate, rough	100–700	0.94–0.97
Molten surfaces		
Cast iron	2370–2550	0.29
Mild steel	2910–3270	0.28
Lead:		
Pure (99.96%), unoxidized	260–440	0.057–0.075
Gray oxidized	75	0.28
Nickel alloys:		
Chromnickel	125–1894	0.64–0.76
Copper-nickel, polished	212	0.059
Nichrome wire, bright	120–1830	0.65–0.79
Nichrome wire, oxidized	120–930	0.95–0.98
Platinum:		
Pure, polished plate	440–1160	0.054–0.104
Strip	1700–2960	0.12–0.17
Filament	80–2240	0.036–0.192
Wire	440–2510	0.073–0.182
Silver:		
Polished, pure	440–1160	0.020–0.032
Polished	100–700	0.022–0.031
Stainless steels:		
Polished	212	0.074
Type 310 (25 Cr; 20 Ni)		
Brown, splotched, oxidized from furnace service	420–980	0.90–0.97
Tin:		
Bright tinned iron	76	0.043 and 0.064
Bright	122	0.06
Commercial tin-plated sheet iron	212	0.07, 0.08

TABLE 23.4　(CONT'D.)

Surface	T, °F	Emissivity
Tungsten:		
Filament, aged	80–6000	0.032–0.35
Filament	6000	0.39
Polished coat	212	0.066
Zinc:		
Commercial 99.1% pure, polished	440–620	0.045–0.053
Oxidized by heating at 750°F	750	0.11

B. Refractories, building materials, paints, and miscellaneous

Asbestos:		
Board	74	0.96
Paper	100–700	0.93–0.94
Brick		
Red, rough, but no gross irregularities	70	0.93
Brick, glazed	2012	0.75
Building	1832	0.45
Fireclay	1832	0.75
Carbon:		
Filament	1900–2560	0.526
Lampblack-waterglass coating	209–440	0.96–0.95
Thin layer of same on iron plate	69	0.927
Glass:		
Smooth	72	0.94
Pyrex, lead, and soda	500–1000	0.95–0.85
Gypsum, 0.02 in. thick on smooth or blackened plate	70	0.903
Magnesite refractory brick	1832	0.38
Marble, light gray, polished	72	0.93
Oak, planed	70	0.90
Paints, lacquers, varnishes:		
Snow-white enamel varnish on rough iron plate	73	0.906
Black shiny lacquer, sprayed on iron	76	0.875
Black shiny shellac on tinned iron sheet	70	0.821
Black matte shellac	170–295	0.91
Black or white lacquer	100–200	0.80–0.95
Flat black lacquer	100–200	0.96–0.98
Oil paints, 16 different, all colors	212	0.92–0.96
Al paint, after heating to 620°F	300–600	0.35
Plaster, rough lime	50–190	0.91
Roofing paper	69	0.91
Rubber:		
Hard, glossy plate	74	0.94
Soft, gray, rough (reclaimed)	76	0.86
Water	32–212	0.95–0.963

the surface is called *gray*. For a gray surface, the total average absorptivity will be independent of the spectral-energy distribution of the incident radiation. Consequently, the emissivity, ϵ, may be used in place of α, even though the temperature of the incident radiation and the receiver are not the same. Good approximations of a gray surface are slate, tar board, and dark linoleum. Table 23.4 lists emissivities, at various temperatures, for several materials.

23.7 RADIANT HEAT TRANSFER BETWEEN BLACK BODIES

The exchange of energy between black bodies is dependent upon the temperature difference and the geometry with the geometry, in particular, playing a dominant role. Consider the two surfaces illustrated in Figure 23.9. The radiant energy emitted from a black surface at dA_1 and received at dA_2 is

$$dq_{1\to2} = I_{b_1} \cos \theta_1 \, d\Omega_{1-2} \, dA_1$$

where $d\Omega_{1-2}$ is the solid angle subtended by dA_2 as seen from dA_1. Thus

$$d\Omega_{1-2} = \cos \theta_2 \frac{dA_2}{r^2}$$

and as $I_{b_1} = E_{b_1}/\pi$, the heat transfer from 1 to 2 is

$$dq_{1\to2} = E_{b_1} \, dA_1 \left\{ \frac{\cos \theta_1 \cos \theta_2 \, dA_2}{\pi r^2} \right\}$$

The bracketed term is seen to depend solely upon geometry. In exactly the same

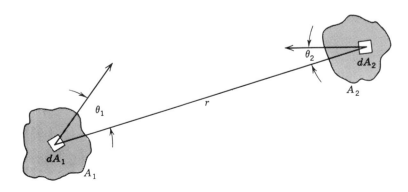

Figure 23.9   Radiant heat transfer between two surfaces.

manner the energy emitted by dA_2 and captured by dA_1, may be determined. This is

$$dq_{2 \to 1} = E_{b_2} \, dA_2 \left\{ \frac{\cos \theta_2 \cos \theta_1 \, dA_1}{\pi r^2} \right\}$$

The net heat transfer between surfaces dA_1 and dA_2 is then simply

$$dq_{1-2 \, net} = dq_{1 \rightleftharpoons 2} = dq_{1 \to 2} - dq_{2 \to 1}$$

or

$$dq_{1 \rightleftharpoons 2} = (E_{b_1} - E_{b_2}) \frac{\cos \theta_1 \cos \theta_2 \, dA_1 \, dA_2}{\pi r^2}$$

Integrating over surfaces 1 and 2, we obtain

$$q_{1 \rightleftharpoons 2} = (E_{b_1} - E_{b_2}) \int_{A_1} \int_{A_2} \frac{\cos \theta_1 \cos \theta_2 \, dA_2 \, dA_1}{\pi r^2}$$

the insertion of A_1 / A_1 yields

$$q_{1 \rightleftharpoons 2} = (E_{b_1} - E_{b_2}) A_1 \left[\frac{1}{A_1} \int_{A_1} \int_{A_2} \frac{\cos \theta_1 \cos \theta_2 \, dA_2 \, dA_1}{\pi r^2} \right] \qquad (23\text{-}13)$$

The bracketed term in the above equation is called the *view factor* F_{1-2}. If we had used A_2 as a reference, then the view factor would be F_{21}. Clearly, the net heat transfer is not affected by these operations, and thus $A_1 F_{12} = A_2 F_{21}$. This simple but extremely important expression is called the **reciprocity** relationship.

A physical interpretation of the view factor may be obtained from the following argument. As the total energy leaving surface A_1 is $E_b A_1$, the amount of heat which surface A_2 receives is $E_{b_1} A_1 F_{12}$. The amount of heat lost by surface A_2 is $E_{b_2} A_2$, while the amount that reaches A_1 is $E_{b_2} A_2 F_{21}$. The net rate of heat transfer between A_1 and A_2 is the difference or $E_{b_1} A_1 F_{12} - E_{b_2} A_2 F_{21}$. This may be arranged to yield $(E_{b_1} - E_{b_2}) A_1 F_{12}$. Thus the view factor F_{12} can be interpreted as the fraction of black-body energy leaving A_1 which reaches A_2. Clearly the view factor cannot exceed unity.

Before some specific view factors are examined, there are several generalizations worthy of note concerning view factors.

1. The *reciprocity relation*, $A_1 F_{12} = A_2 F_{21}$, is always valid.
2. The view factor is independent of temperature. It is purely geometric.
3. For an enclosure, $F_{11} + F_{12} + F_{13} + \cdots = 1$.

In many cases the view factor may be determined without integration. An example of such a case follows.

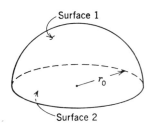

Figure 23.10 Hemisphere and plane.

EXAMPLE 1

Consider the view factor between a hemisphere and a plane as shown in Figure 23.10. Determine the view factors F_{11}, F_{12}, and F_{21}.

The view factor F_{21} is unity, as surface 2 sees only surface 1. For surface 1 we may write $F_{11} + F_{12} = 1$ and $A_1 F_{12} = A_2 F_{21}$. As $F_{21} = 1$, $A_2 = \pi r_0^2$, and $A_1 = 2\pi r_0^2$, the above relations give

$$F_{12} = F_{21}\frac{A_2}{A_1} = (1)\left(\frac{\pi r_0^2}{2\pi r_0^2}\right) = \frac{1}{2}$$

and

$$F_{11} = 1 - F_{12} = \frac{1}{2}$$

The view factor F_{12} can, in general, be determined by integration. As

$$F_{12} \equiv \frac{1}{A_1}\int_{A_1}\int_{A_2}\frac{\cos\theta_1\cos\theta_2\,dA_2\,dA_1}{\pi r^2} \qquad (23\text{-}14)$$

this integration process becomes quite tedious, and the view factor for a complex geometry is seen to require numerical methods. In order to illustrate the analytical evaluation of view factors, consider the view factor between the differential area dA_1 and the parallel plane A_2 shown in Figure 23.11. The view factor $F_{dA_1 A_2}$ is given by

$$F_{dA_1 A_2} = \frac{1}{dA_1}\int_{A_1}\int_{A_2}\frac{\cos\theta_1\cos\theta_2\,dA_2\,dA_1}{\pi r^2}$$

and as $A_2 \gg dA_1$ the view of dA_2 from dA_1 is independent of the position on dA_1, hence

$$F_{dA_1 A_2} = \frac{1}{\pi}\int_{A_2}\frac{\cos\theta_1\cos\theta_2}{r^2}\,dA_2$$

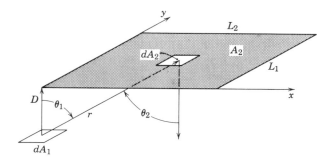

Figure 23.11 Differential area and parallel finite area.

Also, it may be noted that $\theta_1 = \theta_2$ and $\cos \theta = D/r$, where $r^2 = D^2 + x^2 + y^2$. The resulting integral becomes

$$F_{dA_1 A_2} = \frac{1}{\pi} \int_0^{L_1} \int_0^{L_2} \frac{D^2 \, dx \, dy}{(D^2 + x^2 + y^2)^2}$$

or

$$F_{dA_1 A_2} = \frac{1}{2\pi} \left\{ \frac{L_1}{\sqrt{D^2 + L_1^2}} \tan^{-1} \frac{L_2}{\sqrt{D^2 + L_1^2}} + \frac{L_2}{\sqrt{D^2 + L_2^2}} \tan^{-1} \frac{L_1}{\sqrt{D^2 + L_2^2}} \right\}$$

(23-15)

The view factor given by equation (23-15) is shown graphically in Figure 23.12.

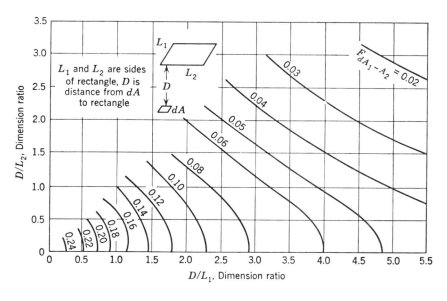

Figure 23.12 View factor for a surface element and a rectangular surface parallel to it. (From H. C. Hottel, "Radiant Heat Transmission," *Mech. Engrg.*, 52 (1930). By permission of the publishers.)

Figures 23.13, 23.14, and 23.15 also illustrate some view factors for simple geometries.

EXAMPLE 2

Determine the view factor from a 1-ft square to a parallel rectangular plane 10 ft by 12 ft centered 8 ft above the 1-ft square.

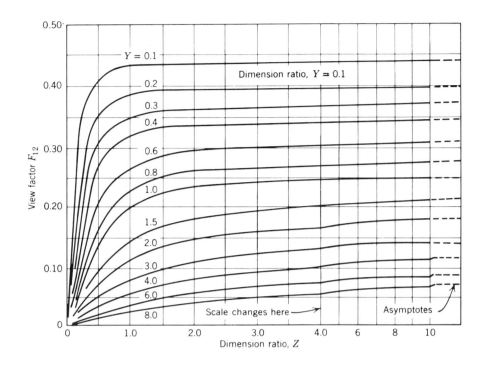

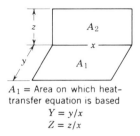

A_1 = Area on which heat-transfer equation is based

$$Y = y/x$$
$$Z = z/x$$

Figure 23.13   View factor for adjacent rectangles in perpendicular planes. (From H. C. Hottel, "Radiant Heat Transmission," *Mech. Engrg.*, 52 (1930). By permission of the publishers.)

     The smaller area may be considered a differential area, and Figure 23.12 may be used. The 10-ft by 12-ft area may be divided into four 5-ft by 6-ft rectangles directly over the smaller area. Thus the total view factor is the sum of the view factors to each subdivided rectangle. Using $D = 8$, $L_1 = 6$, $L_2 = 5$, we find that the view factor from Figure 23.12 is 0.09. The total view factor is the sum of the view factors or 0.36.

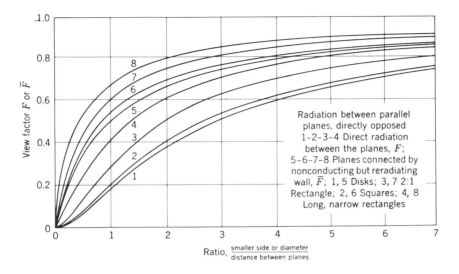

Figure 23.14 View factors for equal and parallel squares, rectangles and disks. The curves labeled 5, 6, 7, and 8 allow for continuous variation in the sidewall temperatures from top to bottom. (From H. C. Hottel, "Radiant Heat Transmission," *Mech. Engrg.*, 52 (1930). By permission of the publishers.)

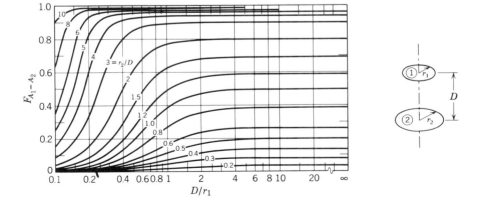

Figure 23.15 View factors for parallel opposed circular disks of unequal size.

VIEW FACTOR ALGEBRA

View factors between combinations of differential- and finite-size areas have been expressed in equation form thus far. Some generalizations can be made that will be useful in evaluating radiant energy exchange in cases that, at first glance, seem quite difficult.

In an enclosure all energy leaving one surface, designated i, will be incident on the other surfaces that it can "see." If there are n surfaces in total, with j designating any surface that receives energy from i, we may write

$$\sum_{i=1}^{n} F_{ij} = 1 \tag{23-16}$$

A general form of the reciprocity relationship may be written as

$$A_i F_{ij} = A_j F_{ji} \tag{23-17}$$

these two expressions form the basis of a technique designated *view factor algebra*.

A simplified notation will be introduced, using the symbol G_{ij}, defined as

$$G_{ij} \equiv A_i F_{ij}$$

Equations (23-16) and (23-17) may now be written as

$$\sum G_{ij} = A_i \tag{23-18}$$

$$G_{ij} = G_{ji} \tag{23-19}$$

The quantity G_{ij} is designated the *geometric flux*. Relations involving geometric fluxes are dictated by energy conservation principles.

Some special symbolism will now be explained. If surface 1 "sees" two surfaces, designated 2 and 3, we may write

$$G_{1-(2+3)} = G_{1-2} + G_{1-3} \tag{23-20}$$

This relation says simply that the energy leaving surface 1 and striking both surfaces 2 and 3 is the total of that striking each separately. Equation (23-20) can be reduced further to

$$A_1 F_{1-(2+3)} = A_1 F_{12} + A_1 F_{13}$$

or

$$F_{1-(2+3)} = F_{12} + F_{13}$$

A second expression, involving four surfaces, is reduced to

$$G_{(1+2)-(3+4)} = G_{1-(3+4)} + G_{2-(3+4)}$$

which decomposes further to the form

$$G_{(1+2)-(3+4)} = G_{1-3} + G_{1-4} + G_{2-3} + G_{2-4}$$

Examples of how view factor algebra can be used follow.

EXAMPLE 3

Determine the view factors, F_{1-2}, for the finite areas shown.

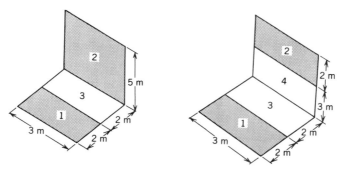

Inspection indicates that, in case (a), view factors F_{2-3} and $F_{2-(1+3)}$ can be read directly from Figure 23.13. The desired view factor, f_{1-2}, can be obtained using view factor algebra in the following steps.

$$G_{2-(1+3)} = G_{2-1} + G_{2-3}$$

Thus,

$$G_{2-1} = G_{2-(1+3)} - G_{2-3}$$

Finally, by reciprocity, we may solve for F_{1-2} according to

$$G_{1-2} = G_{2-1} = G_{2-(1+3)} - G_{2-3}$$

$$A_1 F_{1-2} = A_2 F_{2-(1+3)} - A_2 F_{2-3}$$

$$F_{1-2} = \frac{A_2}{A_1}[F_{2-(1+3)} - F_{2-3}]$$

From Figure 23.13 we read

$$F_{2-(1+3)} = 0.15 \qquad F_{2-3} = 0.10$$

Thus, for configuration (a), we obtain

$$F_{1-2} = \frac{5}{2}(0.15 - 0.10) = 0.125$$

Now, for case (b), the solution steps are

$$G_{1-2} = G_{1-(2+4)} - G_{1-4}$$

which may be written as

$$F_{1-2} = F_{1-(2+4)} - F_{1-4}$$

The result from part (a) can now be utilized to write

$$F_{1-(2+4)} = \frac{A_2 + A_4}{A_1}[F_{(2+4)-(1+3)} - F_{(2+4)-3}]$$

$$F_{1-4} = \frac{A_4}{A_1}[F_{4-(1+3)} - F_{4-3}]$$

Each of the view factors on the right side of these two expressions may be evaluated from Figure 23.13; the appropriate values are

$$F_{(2+4)-(1+3)} = 0.15 \qquad F_{4-(1+3)} = 0.22$$

$$F_{(2+4)-3} = 0.10 \qquad F_{4-3} = 0.165$$

Making these substitutions, we have

$$F_{1-(2+4)} = \tfrac{5}{2}(0.15 - 0.10) = 0.125$$

$$F_{1-4} = \tfrac{3}{2}(0.22 - 0.165) = 0.0825$$

The solution to case (b) now becomes

$$F_{1-2} = 0.125 - 0.0825 = 0.0425$$

23.8   RADIANT EXCHANGE IN BLACK ENCLOSURES

As pointed out earlier, a surface which views n other surfaces may be described according to

$$F_{11} + F_{12} + \cdots + F_{1i} + \cdots + F_{1n} = 1$$

or

$$\sum_{i=1}^{n} F_{1i} = 1 \tag{23-21}$$

Obviously, the inclusion of A_1 with equation (23-12) yields

$$\sum_{i=1}^{n} A_1 F_{1i} = A_1 \tag{23-22}$$

Between any two black surfaces the radiant heat exchange rate is given by

$$q_{12} = A_1 F_{12}(E_{b_1} - E_{b_2}) = A_2 F_{21}(E_{b_1} - E_{b_2}) \tag{23-23}$$

For surface 1 and any other surface, designated i, in a black enclosure the radiant exchange is given as

$$q_{1i} = A_1 F_{1i}(E_{b_1} - E_{b_i}) \tag{23-24}$$

For an enclosure where surface 1 views n other surfaces, we may write for the net heat transfer with 1,

$$q_{1-\text{others}} = \sum_{i=1}^{n} q_{1i} = \sum_{i=1}^{n} A_1 F_{1i}(E_{b_1} - E_{b_i}) \tag{23-25}$$

Equation (23-25) can be thought of as an analog to Ohm's law where the quantity of transfer, q; the potential driving force, $E_{b_1} - E_{b_i}$; and the thermal resistance, $1/A_1 F_{1i}$; have electrical counterparts I, ΔV, and R respectively.

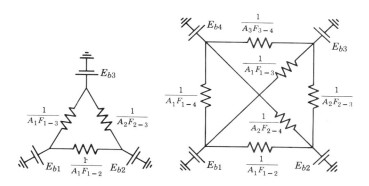

Figure 23.16

Figure 23.16 depicts the analogous electrical circuits for enclosures with 3 and 4 surfaces, respectively.

The solution to a three-surface problem, that is, to find q_{12}, q_{13}, q_{23}, although somewhat tedious, can be accomplished in reasonable time. When analyzing enclosures with four and more surfaces an analytical solution becomes impractical. In such situations one would resort to numerical methods or actually set up the electrical analog and measure currents at different locations in the corresponding circuit. For detailed examples of such solutions the interested reader may refer to Welty.*

23.9 RADIANT EXCHANGE WITH RERADIATING SURFACES PRESENT

The circuit diagrams shown in Figure 23.16 show a path to ground at each of the junctions. The thermal analog is a surface which has some external influence whereby its temperature is maintained at a certain level by the addition or rejection of energy. Such a surface is in contact with its surroundings and will conduct heat by virtue of an imposed temperature difference across it.

In radiation applications we encounter surfaces which effectively are insulated from the surroundings. Such a surface will reemit all radiant energy that is absorbed—usually in a diffuse fashion. These surfaces thus act as reflectors and their temperatures "float" at some value which is required for the system to be in equilibrium. Figure 23.17 shows a physical situation and the corresponding electric analog for a three-surface enclosure with one being a nonabsorbing reradiating surface.

* J. R. Welty, *Engineering Heat Transfer*, Wiley, New York, 1974.

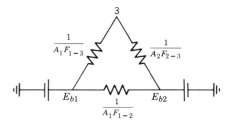

Figure 23.17

Evaluating the net heat transfer between the two black surfaces, q_{1-2}, we have

$$q_{12} = \frac{E_{b_1} - E_{b_2}}{R_{equiv}}$$

$$= \left[A_1 F_{12} + \frac{1}{1/A_1 F_{13} + 1/A_2 F_{23}} \right] (E_{b_1} - E_{b_2})$$

$$= A_1 \left[F_{12} + \frac{1}{1/F_{13} + A_1/A_2 F_{23}} \right] (E_{b_1} - E_{b_2})$$

$$= A_1 \bar{F}_{12} (E_{b_1} - E_{b_2}) \qquad (23\text{-}26)$$

The resulting expression, equation (23-26), contains a new term, $\bar{F}_{12}$, the *reradiating view factor*. This new factor, $\bar{F}_{12}$, is seen equivalent to the square-bracketed term in the previous expression which includes direct exchange between surfaces 1 and 2, F_{12}, plus terms which account for the energy which is exchanged between these surfaces via the intervening reradiating surface. It is apparent that $\bar{F}_{12}$ will always be greater than F_{12}. Figure 23.14 allows reradiating view factors to be read directly for some simple geometries. In other situations where curves such as in this figure are not available the electrical analog may be used with the simple modification that no path to ground exists at the reradiating surface.

23.10   RADIANT HEAT TRANSFER
BETWEEN GRAY SURFACES

In the case of surfaces that are not black, determination of heat transfer becomes more involved. For gray bodies, that is, surfaces for which the absorptivity and emissivity are independent of wavelength, considerable simplifications can be made. The net heat transfer from the surface shown in Figure 23.18 is determined by the difference between the radiation leaving the surface and the radiation incident upon the surface. The *radiosity*, J, is defined as the rate at which

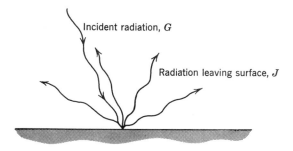

Figure 23.18   Heat transfer at a surface.

radiation leaves a given surface per unit area. The *irradiation, G,* is defined as the rate at which radiation is incident on a surface per unit area. For a gray body, the radiosity, irradiation and the total emissive power are related by

$$J = \rho G + \epsilon E_b \tag{23-27}$$

where ρ is the reflectivity and ϵ is the emissivity. The net heat transfer from a surface is

$$\frac{q_{net}}{A} = J - G = \epsilon E_b + \rho G - G = \epsilon E_b - (1 - \rho)G \tag{23-28}$$

In most cases it is useful to eliminate G from equation (23-28). This yields

$$\frac{q_{net}}{A} = \epsilon E_b - (1 - \rho)\frac{(J - \epsilon E_b)}{\rho}$$

As $\alpha + \rho = 1$ for an opaque surface,

$$\frac{q_{net}}{A} = \frac{\epsilon E_b}{\rho} - \frac{\alpha J}{\rho} \tag{23-29}$$

When the emissivity and absorptivity can be considered equal, an important simplification may be made in equation (23-29). Setting $\alpha = \epsilon$, we obtain

$$q_{net} = \frac{A\epsilon}{\rho}(E_b - J) \tag{23-30}$$

which suggests an analogy with Ohm's law, $V = IR$, where the net heat leaving a surface can be thought of in terms of a current, the difference $E_b - J$ may be likened to a potential difference, and the quotient $\rho/\epsilon A$ may be termed a resistance. Figure 23.19 illustrates this analogy.

Now the net exchange of heat via radiation between two surfaces will depend upon their radiosities and their relative "views" of each other. From equation (23-17) we may write

$$q_{1 \rightleftharpoons 2} = A_1 F_{12}(J_1 - J_2) = A_2 F_{21}(J_1 - J_2)$$

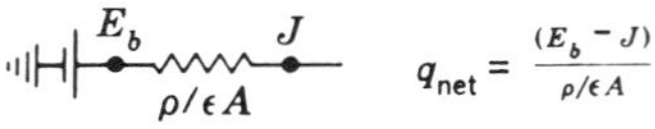

$$\text{I}|\text{H}\underset{\rho/\epsilon A}{\overset{E_b \quad\quad J}{\bullet\!\!-\!\!\text{WWW}\!\!-\!\!\bullet}} \qquad q_{net} = \frac{(E_b - J)}{\rho/\epsilon A}$$

Figure 23.19 Electrical analogy for radiation from a surface.

We may now write the net heat exchange in terms of the different "resistances" offered by each part of the heat transfer path as follows.

Rate of heat leaving surface 1:

$$q = \frac{A_1 \epsilon_1}{\rho_1}(E_{b_1} - J_1)$$

Rate of heat exchange between surfaces 1 and 2: $q = A_1 F_{12}(J_1 - J_2)$

Rate of heat received at surface 2:

$$q = \frac{A_2 \epsilon_2}{\rho_2}(J_2 - E_{b_2})$$

If surfaces 1 and 2 view each other and no others then each of the q's in the above equations is equivalent. In such a case an additional expression for q can be written in terms of the overall driving force, $E_{b_1} - E_{b_2}$. Such an expression is

$$q = \frac{E_{b_1} - E_{b_2}}{\rho_1/A_1\epsilon_1 + 1/A_1 F_{12} + \rho_2/A_2\epsilon_2}$$

where the terms in the denominator are the equivalent resistances due to the characteristics of surface 1, geometry, and the characteristics of surface 2, respectively. The electrical analog to this equation is portrayed in Figure 23.20.

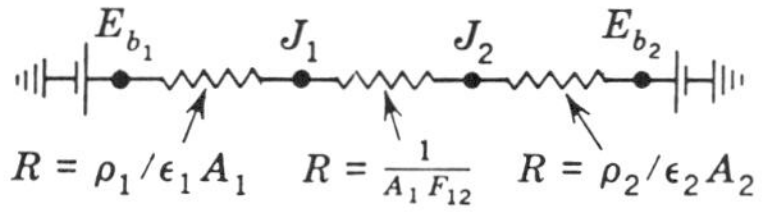

$$\text{I}|\text{H}\overset{E_{b_1} \quad\quad J_1 \quad\quad J_2 \quad\quad E_{b_2}}{\bullet\!\!-\!\!\text{www}\!\!-\!\!\bullet\!\!-\!\!\text{www}\!\!-\!\!\bullet\!\!-\!\!\text{www}\!\!-\!\!\bullet}\text{H}|\text{I}$$
$$R = \rho_1/\epsilon_1 A_1 \quad R = \frac{1}{A_1 F_{12}} \quad R = \rho_2/\epsilon_2 A_2$$

Figure 23.20 Equivalent network for gray-body relations between two surfaces.

The assumptions required to use the electrical analog approach to solve radiation problems are the following:
1. Each surface must be gray,
2. Each surface must be isothermal,
3. Kirchhoff's law must apply, that is $\alpha = \epsilon$,
4. There is no heat absorbing medium between the participating surfaces.

Examples 4 and 5, which follow, illustrate features of gray-body problem solutions.

EXAMPLE 4

Two parallel gray surfaces maintained at temperatures T_1 and T_2 view each other. Each surface is sufficiently large that they may be considered infinite. Generate an expression for the net heat transfer between these surfaces.

A simple series electrical circuit is useful in solving this problem. The circuit and important quantities are shown here.

$$R_1 = \rho_1/A_1\epsilon_1 \qquad R_2 = 1/A_1F_{12} \qquad R_3 = \rho_2/A_2\epsilon_2 \qquad E_{b2} = \sigma T_2^{\,4}$$

$$E_{b1} = \sigma T_1^{\,4} \qquad\qquad J_1 \qquad\qquad J_2$$

Utilizing Ohm's law, we obtain the expression

$$q_{12} = \frac{E_{b1} - E_{b2}}{\Sigma R} = \frac{\sigma(T_1^{\,4} - T_2^{\,4})}{\dfrac{\rho_1}{A_1\epsilon_1} + \dfrac{1}{A_1F_{12}} + \dfrac{\rho_2}{A_2\epsilon_2}}$$

Now, noting that for infinite parallel planes $A_1 = A_2 = A$ and $F_{12} = F_{21} = 1$ and writing $\rho_1 = 1 - \epsilon_1$ and $\rho_2 = 1 - \epsilon_2$, we obtain the result

$$q_{12} = \frac{A\sigma(T_1^{\,4} - T_2^{\,4})}{\dfrac{1 - \epsilon_1}{\epsilon_1} + 1 + \dfrac{1 - \epsilon_2}{\epsilon_2}}$$

$$= \frac{A\sigma(T_1^{\,4} - T_2^{\,4})}{\dfrac{1}{\epsilon_1} + \dfrac{1}{\epsilon_2'} - 1}$$

EXAMPLE 5

Two 7-ft square and parallel plates are situated 7 ft apart. Plate A_1 is maintained at a temperature of 1540°F, and plate A_2 is maintained at 540°F. Determine the net rate of heat transfer from the high-temperature plate under the following conditions:

(a) The plates are black and the surroundings are at 0°R and black.

(b) The plates are black, and the region enclosing the plates consists of reradiating walls.

(c) The plates have emissivities of 0.4 and 0.8, respectively, and the surroundings are at 0°R and black.

The equivalent circuits for parts (a), (b), and (c) are shown in Figure 23.21.

The determination of the heat flux is seen to require evaluation of the quantities F_{12}, F_{1R}, and $\bar{F}_{12}$. These are as follows:

$$F_{12} \text{ from Figure 23.14 is } 0.20$$

$$\bar{F}_{12} \text{ from Figure 23.14 is } 0.54$$

and

$$F_{1R} = 1 - F_{12} = 0.80$$

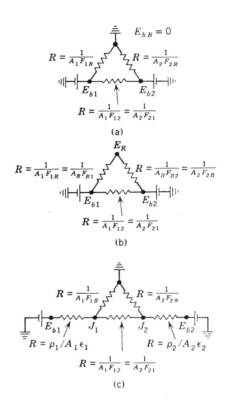

(a)

(b)

(c)

Figure 23.21   Equivalent circuits for example 3.

Part (a). The net rate at which heat leaves plate 1 is

$$q_{1\,\text{net}} = q_{1\rightleftharpoons 2} + q_{1\rightleftharpoons R}$$

or

$$q_{1\,\text{net}} = A_1 F_{12}(E_{b_1} - E_{b_2}) + A_1 F_{1R} E_{b_1}$$

so

$$q_{1\,\text{net}} = (49 \times 0.20 \times 25\,710) + (49 \times 0.80 + 27\,424)\ \text{Btu/hr}$$

$$= 252\,000 + 1\,075\,000 = 1\,327\,000\ \text{Btu/hr}$$

Part (b). When the reradiating walls are introduced, the heat flux becomes

$$q_{1\,\text{net}} = q_{1\rightleftharpoons 2} = (E_{b_1} - E_{b_2})\left(A_1 F_{12} + \frac{1}{(1/A_1 F_{1R}) + (1/A_2 F_{2R})} \right)$$

or as $A_1 = A_2$ and $F_{2R} = F_{1R}$,

$$q_{1\,net} = A_1(E_{b_1} - E_{b_2})\left(F_{12} + \frac{F_{1R}}{2}\right)$$

Now for surface 1, we have $F_{12} + F_{1R} = 1$, and thus $(F_{1R}/2) = (1 - F_{12})/2$, yielding

$$q_{1\,net} = A_1(E_{b_1} - E_{b_2})\left(\frac{F_{12} + 1}{2}\right) = 755\,000 \text{ Btu/hr}$$

An alternate solution to part (b) can be obtained using the expression $q_{1\,net} = A_1\bar{F}_{12}(E_{b_1} - E_{b_2})$. The value of $\bar{F}_{12}$ from Figure 23.15 yields $q_{1\,net} = 680\,000$ Btu/hr. Values of $\bar{F}_{12}$ obtained from Figure 23.15 allow for a continuous distribution in temperature along the reradiating surface from T_1 to T_2. The analysis using the circuit analog assumes the reradiating wall to be at a constant temperature. In reality the temperature does vary thus the circuit analog produces some error, in this case approximately 10%.

Part (c). The analysis of the circuit shown in Figure 23.21(c) yields $q_{1\,net} = 436\,000$ Btu/hr.

23.11 RADIATION FROM GASES

So far, the interaction of radiation with gases has been neglected. Gases emit and absorb radiation in discrete energy bands dictated by the allowed energy states within the molecule. As the energy associated with, say, the vibrational or rotational motion of a molecule may have only certain values, it follows that the amount of energy emitted or absorbed by a molecule will have a frequency, $\nu = \Delta E/h$, corresponding to the difference in energy ΔE between allowed states. Thus while the energy emitted by a solid will comprise a continuous spectrum, the radiation emitted and absorbed by a gas will be restricted to bands. Figure 23.22 illustrates the emission bands of carbon dioxide and water vapor relative to black-body radiation at 1500°F.

The emission of radiation for these gases is seen to occur in the infrared region of the spectrum.

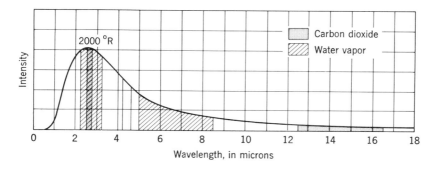

Figure 23.22 Emission bands of CO_2 and H_2O.

For nonluminous gases, the inert gases and diatomic gases of symmetrical composition such as O_2, N_2, and H_2 may be considered transparent to thermal radiation. Important types of media which absorb and emit radiation are polyatomic gases such as CO_2 and H_2O and unsymmetrical molecules such as CO. These gases are also associated with the products of combustion of hydrocarbons. The determination of the absorption and emission of radiation is very difficult, as it involves the temperature, composition, density, and geometry of the gas. There are several simplifications which allow estimation of radiation in gases to be made in a straightforward manner. These idealizations are as follows:

1. The gas is in thermodynamic equilibrium. The state of the gas may therefore be characterized locally by a single temperature.

2. The gas may be considered gray. This simplification allows the absorption and emission of radiation to be characterized by one parameter as $\alpha = \epsilon$ for a gray body.

In the range of temperatures associated with the products of hydrocarbon combustion, the gray gas emissivities of H_2O and CO_2 may be obtained from the results of Hottel. A hemispherical mass of gas at 1 atmosphere pressure was used by Hottel to evaluate the emissivity. While the graphs apply strictly only to a hemispherical gas mass of radius L, other shapes can be treated by consideration of a mean beam length L as given in Table 23.5. For geometries not covered in the

TABLE 23.5   MEAN BEAM LENGTH, L, FOR VARIOUS GEOMETRIES†

Shape	L
Sphere	$\frac{2}{3}$ × diameter
Infinite cylinder	1 × diameter
Space between infinite parallel planes	1.8 × distance between planes
Cube	$\frac{2}{3}$ × side
Space outside infinite bank of tubes with centers on equilateral triangles; tube diameter equals clearance	2.8 × clearance
Same as (5) except tube diameter equals one-half clearance	3.8 × clearance

† From H. C. Hottel, "Radiation," Chap. IV in W. H. McAdams (ed.), *Heat Transmission*, Third Edition, McGraw-Hill Book Company, New York, 1964. By permission of the publishers.

table, the mean beam length may be approximated by the relation $L = 3.4(\text{volume})/(\text{surface area})$.

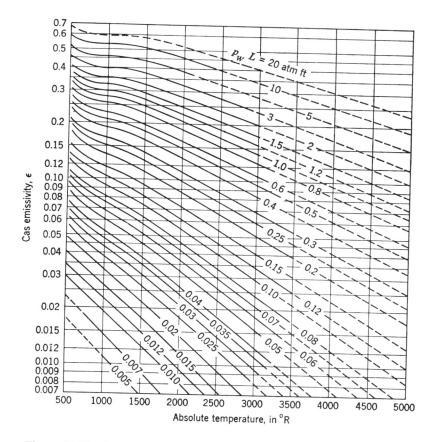

Figure 23.23 Emissivity of water vapor at one atmosphere total pressure and near-zero partial pressure.

Figure 23.23 gives the emissivity of a hemispherical mass of water vapor at 1 atmosphere total pressure and near-zero partial pressure as a function of temperature and the product p_wL, where p_w is the partial pressure of the water vapor. For pressures other than atmospheric Figure 23.24 gives the correction factor, C_w, which is the ratio of the emissivity at total pressure P to the emissivity at a total pressure of 1 atmosphere. Figures 23.25 and 23.26 give the corresponding data for CO_2.

From Figure 23.22 it may be seen that the emission bands of CO_2 and H_2O overlap. When both carbon dioxide and water vapor are present, the total emissivity may be determined from the relation

$$\epsilon_{total} = \epsilon_{H_2O} + \epsilon_{CO_2} - \Delta\epsilon$$

where $\Delta\epsilon$ is given in Figure 23.27.

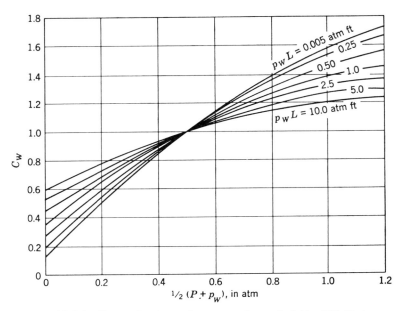

Figure 23.24 Correction factor for converting emissivity of H_2O at one atmosphere total pressure to emissivity at P atmospheres total pressure.

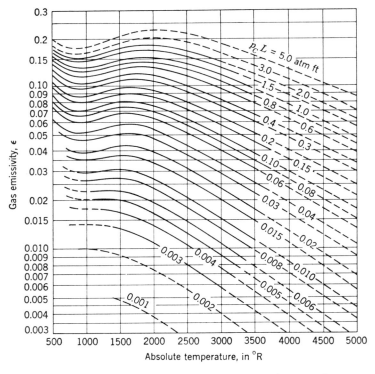

Figure 23.25 Emissivity of CO_2 at one atmosphere total pressure and near-zero partial pressure.

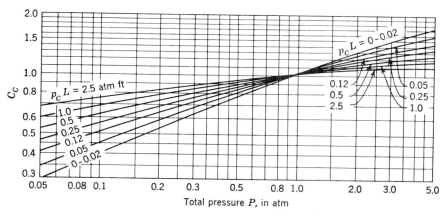

Figure 23.26 Correction factor for converting emissivity of CO_2 at one atmosphere total pressure to emissivity at P atmospheres total pressure.

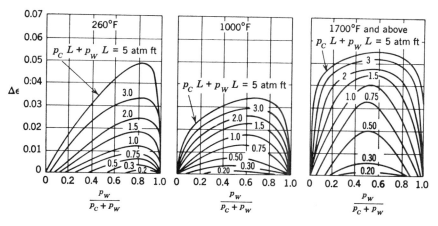

Figure 23.27 Correction to gas emissivity due to spectral overlap of H_2O and CO_2.

The results presented here for the gray gas are gross simplifications. For a more complete treatment, textbooks by Sparrow and Cess,[*] by Wiebelt,[†] and by Vincenti and Kruger[‡] present the fundamentals of nongray-gas radiation, as well as extensive bibliographies.

[*] E. M. Sparrow and R. D. Cess, *Radiation Heat Transfer*, Brooks/Cole, Belmont, Calif., 1966.

[†] J. A. Wiebelt, *Engineering Radiation Heat Transfer*, Holt, Rinehart and Winston, Inc., New York, 1966.

[‡] W. G. Vincenti and C. H. Kruger, Jr., *Introduction to Physical Gas Dynamics*, Wiley, New York, 1965.

23.12 THE RADIATION HEAT-TRANSFER COEFFICIENT

Frequently in engineering analysis, convection and radiation occur simultaneously rather than as isolated phenomena. An important approximation in such cases is the linearization of the radiation contribution so that

$$h_{total} = h_{convection} + h_{radiation} \qquad (23\text{-}32)$$

where

$$h_r \equiv \frac{q_r/A_1}{(T - T_R)} = \mathscr{F}_{1\text{-}2}\left[\frac{\sigma(T_1^{\ 4} - T_2^{\ 4})}{T - T_R}\right] \qquad (23\text{-}33)$$

Here T_R is a reference temperature, and T_1 and T_2 are the respective surface temperatures. In effect, equation (23-33) represents a straight-line approximation to the radiant heat transfer as illustrated in Figure 23.33. The factor, $\mathscr{F}$, accounts for geometry and surface condition of the radiating and absorbing surface.

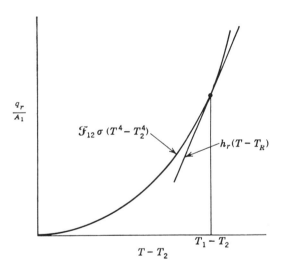

Figure 23.28 Tangent approximation for h_r.

By constructing a tangent to the relation curve at $T = T_1$, the following relations are obtained for h_r and T_R:

$$h_r = 4\sigma T_1^{\ 3}\mathscr{F}_{1\text{-}2} \qquad (23\text{-}34)$$

and

$$T_R = T_1 - \frac{T_1^{\ 4} - T_2^{\ 4}}{4T_1^{\ 3}} \qquad (23\text{-}35)$$

23.13 CLOSURE

While the subject of radiation does not have any analogy in momentum or mass transfer, this mode of heat transfer is extremely important in engineering practice. The role of geometry in radiation is seen to be a dominant one, and considerable effort and care must be exercised in determining the view factors. The use of the gray-body idealization results in a simplified approach to radiant energy exchange of great utility.

The student should, however, keep sight of the fact that both Kirchhoff's law and the gray-body concept are simplifications of the general problem of radiant energy exchange.

PROBLEMS

23.1 The sun is approximately 93 million miles distant from the earth, and its diameter is 860 000 miles. On a clear day solar irradiation at the earth's surface has been measured at 360 Btu/hr ft^2 and an additional 90 Btu/hr ft^2 are absorbed by the earth's atmosphere. With this information, estimate the sun's surface temperature.

23.2 A satellite may be considered spherical with its surface properties roughly those of aluminum. Its orbit may be considered circular at a height of 500 miles above the earth. Taking the satellite diameter as 50 in., estimate the temperature of the satellite skin. The earth may be considered to be at a uniform temperature of 50°F, and the emissivity of the earth may be taken as 0.95. Solar irradiation may be taken as 450 Btu/hr ft^2 of satellite disc area.

23.3 An opaque gray surface with $\epsilon = 0.3$ is irradiated with 1000 W/cm. For an effective convective heat-transfer coefficient of 12 W/m$^2 \cdot$ K applying, and air at 20°C adjacent to the plate, what will be the net heat flux to or from a 30°C surface?

23.4 A sheet-metal box in the shape of a 0.70 m cube has a surface emissivity of 0.7. The box encloses electronic equipment which dissipates 1200 W of energy. If the surroundings are taken to be black at 280 K, and the top and sides of the box are considered to radiate uniformly, at what temperature will the box surface be?

23.5 A tungsten filament, radiating as a gray body, is heated to a temperature of 4000°R. At what wavelength is the emissive power maximum? What portion of the total emission lies within the visible-light range, 0.3 to 0.75 microns?

23.6 Determine the fraction of total energy emitted by a black body which lies in the wavelength band between 0.8 μ and 5.0 μ, for surface temperatures of 500 K, 2000 K, 3000 K, and 4500 K.

23.7 The sun's temperature is approximately 5800 K and the visible light range is taken to be between 0.4 μ and 0.7 μ. What fraction of solar emission is visible? What fraction of solar emission lies in the ultraviolet range? The infrared range? At what wavelength is solar emissive power a maximum?

23.8 The circular base of the cylindrical enclosure shown may be considered a reradiating surface. The cylindrical walls have an effective emissivity of 0.80 and are maintained at 540°F. The top of the enclosure is open to the surroundings, which are maintained at 40°F. What is the net rate of radiant transfer to the surroundings?

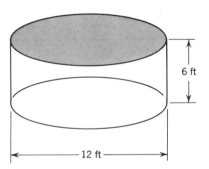

23.9 The hemispherical cavity shown in the figure has an inside surface temperature of 700 K. A plate of refractory material is placed over the cavity with a circular hole of 5 cm diameter in the center. How much energy will be lost through the hole if the cavity is
(a) black;
(b) gray with an emissivity of 0.7?
What will be the temperature of the refractory under each condition?

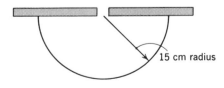

23.10 A room measuring 12 ft by 20 ft by 8 ft high has its floor and ceiling temperatures maintained at 85°F and 65°F, respectively. Assuming the walls to be reradiating and all surfaces to have an emissivity of 0.8, determine the net energy exchange between the floor and ceiling.

23.11 Two parallel rectangles have emissivities of 0.6 and 0.9, respectively. These rectangles are 1.2 m wide and 2.4 m high and are 0.6 m apart. The plate having $\epsilon = 0.6$ is maintained at 1000 K and the other is at 420 K. The surroundings may be considered to absorb all energy which escapes the two-plate system. Determine
(a) the total energy lost from the hot plate;
(b) the radiant energy interchange between the two plates.

23.12 If a third rectangular plate with both surfaces having an emissivity of 0.8 is placed between the two plates described in problem 23.10, how will the answer to part a of problem 23.10 be affected? Draw the thermal circuit for this case.

23.13 Two disks are oriented on parallel planes separated by a distance of 10 in., as shown in the accompanying figure. The disk to the right is 4 in. in diameter and is at a temperature of 500°F. The disk to the left has an inner ring cut out such that it is annular in shape with inner and outer diameters of 2.5 in. and 4 in., respectively. The disk surface temperature is 210°F. Find the heat exchange between these disks if:
(a) they are black;
(b) they are gray $\epsilon_1 = 0.6$, $\epsilon_2 = 0.3$.

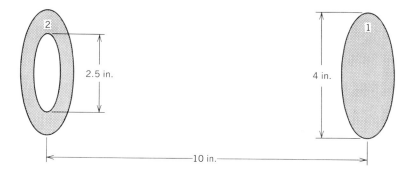

23.14 Evaluate the net heat transfer between the disks described in problem 23.13 if they are bases of a cylinder with the side wall considered a nonconducting, reradiating surface. How much energy will be lost through the hole?

23.15 Evaluate the heat transfer leaving disk 1 for the geometry shown in problem 23.13. In this case the two disks comprise the bases of a cylinder with side wall at constant temperature of 350°F. Evaluate for the case of
(a) the side wall is black,
(b) the side wall is gray with $\epsilon = 0.2$.
Determine the rate of heat loss through the hole in each case.

23.16 A circular duct 2 ft long with a diameter of 3 in. has a thermocouple in its center with a surface area of 0.3 in.^{2} The duct walls are at 200°F, and the thermocouple indicates 310°F. Assuming the convective heat-transfer coefficient between the thermocouple and gas in the duct to be 30 Btu/hr ft^2 °F, estimate the actual temperature of the gas. The emissivity of the duct walls may be taken as 0.8 and that of the thermocouple as 0.6.

23.17 A heating element in the shape of a cylinder is maintained at 2000°F and placed at the center of a half-cylindrical reflector as shown. The rod diameter is 2 in. and that of the reflector is 18 in. The emissivity of the heater surface is 0.8, and the entire assembly is placed in a room maintained at 70°F. What is the radiant energy loss from the heater per foot of length? How does this compare to the loss from the heater without the reflector present?

23.18 A 12 ft long, 3-in.-OD iron pipe $\epsilon = 0.7$, passes horizontally through a $12 \times 14 \times 9$ ft room whose walls are maintained at 70°F and have an emissivity of 0.8. The pipe surface is at a temperature of 205°F. Compare the radiant energy loss from the pipe with that due to convection to the surrounding air at 70°F.

23.19 A 7.5 cm-diameter hole is drilled in a 10 cm-thick iron plate. If the plate temperature is 700 K and the surroundings are at 310 K, determine the energy loss through the hole. The hole sides may be considered to be black.

23.20 If the 7.5 cm-diameter hole in problem 23.15 were drilled to a depth of 5 cm, what heat loss would result?

23.21 A small ($\frac{1}{4}$ in. diameter $\times$ 1 in. long) metal test specimen is suspended by very fine wires in a large evacuated tube. The metal is maintained at a temperature of 2500°F, at which temperature it has an emissivity of approximately 0.2. The water-cooled walls and ends of the tube are maintained at 50°F. In the upper end is a small ($\frac{1}{4}$-in.-diameter) silica glass

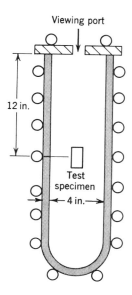

Viewing port

12 in.

Test
specimen

← 4 in. →

viewing port. The inside surfaces of the steel tube are newly galvanized.
Room temperature is 70°F. Estimate:
(a) the view factor from the specimen to the window;
(b) the total net heat-transfer rate by radiation from the test specimen;
(c) the energy radiated through the viewing port.

23.22 A duct with square cross section measuring 20 cm by 20 cm has water
vapor at 1 atmosphere and 600 K flowing through it. One wall of the duct is
held at 420 K and has an emissivity of 0.8. The other three walls may be
considered refractory surfaces. Determine the rate of radiant energy
transfer to the cold wall from the water vapor.

23.23 A gas at mixture at 1000 K and a pressure of 5 atmospheres is introduced
into an evacuated spherical cavity with a diameter of 3 m. The cavity walls
are black and initially at a temperature of 600 K. What initial rate of heat
transfer will occur between the gas and spherical walls if the gas contains
15% CO_2 with the remainder of the gas being nonradiating?

23.24 A gas consisting of 20% CO_2 and 80% oxygen and nitrogen leaves a lime
kiln at 2000°F and enters a square duct measuring 6 in. by 6 in. in cross
section. The specific heat of the gas is 0.28 Btu/lb$_m$ °F, and it is to be cooled
to 1000°F in the duct, whose inside surface is maintained at 800°F, and
whose walls have an emissivity of 0.9. The mass velocity of the kiln gas is
0.4 lb$_m$/ft^2 s and the convective heat-transfer coefficient between the gas
and duct walls is 1.5 Btu/hr ft^2 °F.
(a) Determine the required length of duct to cool the gas to 1000°F.

(b) Determine the ratio of radiant energy transfer to that by convection.

(c) At what temperature would the gas leave the duct if the length of the duct were twice the value determined in part (a)?

(Courtesy of the American Institute of Chemical Engineers.)

HINT:

As the response of the gas to emission and absorption of radiant energy differs, an approximation for the radiant energy exchange between the enclosure and gas contained within an arbitrary control volume is given by $A_w F_{w-g} \sigma \epsilon_w (\epsilon_g T_g^4 - \alpha_g T_w^4)$.

CLOSURE TO ENERGY TRANSFER

Comments similar to those made at the end of Chapter 14 are appropriate here. We shall now direct our attention to mass transfer with its associated fluxes, transport properties, driving forces, and other particular terminology. Almost all of the material in the next eight chapters has a counterpart in earlier sections. The student is again reminded that such similarities should ease the understanding of mass transfer and strengthen that of momentum and energy transfer.

24
FUNDAMENTALS
OF MASS TRANSFER

The previous chapters dealing with the transport phenomena of momentum and heat transfer have dealt with one-component phases which possessed a natural tendency to reach equilibrium conditions. When a system contains two or more components whose concentrations vary from point to point, there is a natural tendency for mass to be transferred, minimizing the concentration differences within the system. The transport of one constituent from a region of higher concentration to that of a lower concentration is called *mass transfer.*

Many of our day-by-day experiences involve mass transfer. A lump of sugar added to a cup of black coffee eventually dissolves and then diffuses uniformly throughout the coffee. Water evaporates from ponds to increase the humidity of the passing air stream. Perfume presents a pleasant fragrance which is imparted throughout the surrounding atmosphere.

Mass transfer plays an important role in many industrial processes: the removal of pollutants from plant discharge streams by absorption, the stripping of gases from wastewater, neutron diffusion within nuclear reactors, the diffusion of adsorbed substances within the pores of activated carbon, the rate of catalyzed chemical and biological reactions, and air conditioning are typical examples.

If we consider the lump of sugar added to the cup of black coffee, experience teaches us that the length of time required to distribute the sugar will depend upon whether the liquid is quiescent or whether it is mechanically agitated by a spoon. The mechanism of mass transfer, as we have also observed in heat transfer, depends upon the dynamics of the system in which it occurs. Mass can be transferred by random molecular motion in quiescent fluids, or it can be transferred from a surface into a moving fluid, aided by the dynamic characteristics of the flow. These two distinct modes of transport, molecular mass transfer and convective mass transfer, are analogous to conduction heat transfer and convective heat transfer. Each of these modes of mass transfer will be described and analyzed. As in the case of heat transfer, we should immediately realize that the two mechanisms often act simultaneously. However, in the confluence of the two

471

modes of mass transfer, one mechanism can dominate quantitatively so that approximate solutions involving only the dominant mode need be used.

24.1 MOLECULAR MASS TRANSFER

As early as 1815 Parrot observed qualitatively that whenever a gas mixture contains two or more molecular species, whose relative concentrations vary from point to point, an apparently natural process results which tends to diminish any inequalities of composition. This macroscopic transport of mass, independent of any convection within the system, is defined as *molecular diffusion.*

In the specific case of gaseous mixtures a logical explanation of this transport phenomenon can be deduced from the kinetic theory of gases. At temperatures above absolute zero, individual molecules are in a state of continual yet random motion. Within dilute gas mixtures each solute molecule behaves independently of the other solute molecules, since it seldom encounters them. Collisions between the solute and the solvent molecules are continually occurring. As a result of the collisions the solute molecules move along a zigzag path, sometimes toward a region of higher concentration, sometimes toward a lower concentration.

Let us consider a hypothetical section passing normal to the concentration gradient within an isothermal, isobaric gaseous mixture containing solute and solvent molecules. The two thin, equal elements of volume above and below the section will contain the same number of molecules, as stipulated by Avogadro's law. Although it is not possible to state which way any particular molecule will travel in a given interval of time, a definite number of the molecules in the lower element of the volume will cross the hypothetical section from below, and the same number of molecules will leave the upper element and cross the section from above. With the existence of a concentration gradient, there are more solute molecules in one of the elements of volume than in the other; accordingly, an overall net transfer from a region of higher concentration to one of lower concentration will result. The net flow of each molecular species occurs in the direction of a negative concentration gradient.

As pointed out in Chapters 7 and 15, the molecular transport of momentum and the transport of energy by conduction are also due to random molecular motion. Accordingly, one should expect that the three transport phenomena will depend upon many of the same characteristic properties, such as mean-free path, and that the theoretical analyses of all three phenomena will have much in common.

THE FICK RATE EQUATION

The laws of mass transfer show the relation between the flux of the diffusing substance and the concentration gradient responsible for this mass transfer. Unfortunately, the quantitative description of molecular diffusion is considerably more complex than the analogous descriptions for the molecular transfer of momentum and energy which occur in a one-component phase. Since *mass transfer*, or *diffusion* as it is also called, occurs only in mixtures, its evaluation must involve an examination of the effect of each component. For example, we will often desire to know the diffusion rate of a specific component relative to the velocity of the mixture in which it is moving. Since each component may possess a different mobility, the mixture velocity must be evaluated by averaging the velocities of all of the components present.

In order to establish a common basis for future discussions, let us first consider definitions and relations which are often used to explain the role of components within a mixture.

Concentrations. In a multicomponent mixture the concentration of a molecular species can be expressed in many ways. Figure 24.1 shows an elemental volume dV which contains a mixture of components, including species A. Since each molecule of each species has a mass, a *mass concentration* for each species, as well as for the mixture, can be defined. For species A, *mass* concentration, ρ_A, is defined as the mass of A per unit volume of the mixture. The total mass concentration or *density*, ρ, is the total mass of the mixture contained in the unit volume; that is

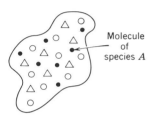

Figure 24.1 Elemental volume containing a multicomponent mixture.

$$\rho = \sum_{i=1}^{n} \rho_i \qquad (24\text{-}1)$$

where n is the number of species in the mixture. The *mass fraction*, ω_A, is the mass concentration of species A divided by the total mass density

$$\omega_A = \frac{\rho_A}{\displaystyle\sum_i^n \rho_i} = \frac{\rho_A}{\rho} \qquad (24\text{-}2)$$

The sum of the mass fractions, by definition, must be 1;

$$\sum_{i=1}^{n} \omega_i = 1 \qquad (24\text{-}3)$$

The *molar concentration* of species A, c_A, is defined as the number of moles of A present per unit volume of the mixture. By definition, one mole of any species contains a mass equivalent to its molecular weight; the mass concentration and molar concentration terms are related by the following relation

$$c_A = \frac{\rho_A}{M_A} \qquad (24\text{-}4)$$

where M_A is the molecular weight of species A. When dealing with a gas phase, concentrations are often expressed in terms of partial pressures. Under conditions in which the ideal gas law, $p_A V = n_A RT$, applies, the molar concentration is

$$c_A = \frac{n_A}{V} = \frac{p_A}{RT} \qquad (24\text{-}5)$$

where p_A is the partial pressure of the species A in the mixture, n_A is the number of moles of species A, V is the gas volume, T is the absolute temperature, and R is the gas constant. The total molar concentration, c, is the total moles of the mixture contained in the unit volume; that is,

$$c = \sum_{i=1}^{n} c_i \qquad (24\text{-}6)$$

or for a gaseous mixture that obeys the ideal gas law, $c = n_{\text{total}}/V = P/RT$, where P is the total pressure. The mole fraction for liquid or solid mixtures, x_A, and for gaseous mixtures, y_A, are the molar concentrations of species A divided by the total molar density

$$x_A = \frac{c_A}{c} \quad \text{(liquids and solids)}$$

$$y_A = \frac{c_A}{c} \quad \text{(gases)} \qquad (24\text{-}7)$$

For a gaseous mixture that obeys the ideal gas law, the mole fraction, y_A, can be written in terms of pressures,

$$y_A = \frac{c_A}{c} = \frac{p_A/RT}{P/RT} = \frac{p_A}{P} \qquad (24\text{-}8)$$

Equation (24-8) is an algebraic representation of Dalton's law for gas mixtures. The sum of the mole fractions, by definition must be 1;

$$\sum_{i=1}^{n} x_i = 1$$

$$\sum_{i=1}^{n} y_i = 1 \qquad (24\text{-}9)$$

A summary of the various concentration terms and of the interrelations for a binary system containing species A and B is given in Table 24.1.

EXAMPLE 1

The composition of air is often given in terms of only the two principal species in the gas mixture

$$\text{oxygen,} \quad O_2, \quad y_{O_2} = 0.21$$

$$\text{nitrogen,} \quad N_2, \quad y_{N_2} = 0.79$$

TABLE 24.1 CONCENTRATIONS IN A BINARY MIXTURE OF A AND B.

Mass concentrations

ρ = total mass density of the mixture
ρ_A = mass density of species A
ρ_B = mass density of species B
ω_A = mass fraction of species $A = \rho_A/\rho$
ω_B = mass fraction of species $B = \rho_B/\rho$
$\rho = \rho_A + \rho_B$
$1 = \omega_A + \omega_B$

Molar concentrations

Liquid or solid mixture	*Gas mixture*
c = molar density of mixture = n/V	$c = n/V = P/RT$
c_A = molar density of species $A = n_A/V$	$c_A = n_A/V = p_A/RT$
c_B = molar density of species $B = n_B/V$	$c_B = n_B/V = p_B/RT$
x_A = mole fraction of species A	$y_A = c_A/c = n_A/n = p_A/P$
$\quad = c_A/c = n_A/n$	
x_B = mole fraction of species B	$y_B = c_B/c = n_B/n = p_B/P$
$\quad = c_B/c = n_B/n$	
$c = c_A + c_B$	$c = c_A + c_B = \dfrac{p_A}{RT} + \dfrac{p_B}{RT} = \dfrac{P}{RT}$
$1 = x_A + x_B$	$1 = y_A + y_B$

Interrelations

$$\rho_A = c_A M_A$$

$$x_A \quad \text{or} \quad y_A = \frac{\omega_A/M_A}{\omega_A/M_A + \omega_B/M_B} \qquad (24\text{-}10)$$

$$\omega_A = \frac{x_A M_A}{x_A M_A + x_B M_B} \quad \text{or} \quad \frac{y_A M_A}{y_A M_A + y_B M_B} \qquad (24\text{-}11)$$

Determine the mass fraction of both oxygen and nitrogen and the mean molecular weight of the air when it is maintained at 25°C (298 K) and 1 atm (1.013×10^5 Pa). The molecular weight of oxygen is 0.032 kg/mol and of nitrogen is 0.028 kg/mol.

As a basis for our calculations, consider 1 mole of the gas mixture;

$$\text{oxygen present} = (1 \text{ mol})(0.21) = 0.21 \text{ mol}$$

$$= (0.21 \text{ mol})\frac{(0.032 \text{ kg})}{\text{mol}} = 0.00672 \text{ kg}$$

nitrogen present $= (1 \text{ mol})(0.79) = 0.79 \text{ mol}$

$$= (0.79 \text{ mol})\frac{(0.028 \text{ kg})}{\text{mol}} = 0.0221 \text{ kg}$$

total mass present $= 0.00672 + 0.0221 = 0.0288 \text{ kg}$

$$\omega_{O_2} = \frac{0.00672 \text{ kg}}{0.0288 \text{ kg}} = 0.23$$

$$\omega_{N_2} = \frac{0.0221 \text{ kg}}{0.0288 \text{ kg}} = 0.77$$

Since 1 mole of the gas mixture has a mass of 0.0288 kg, the mean molecular weight of the air must be 0.0288. When one takes into account the other constituents that are present in air, the mean molecular weight of air is often rounded off to 0.029 kg/mol.

This problem could also be solved using the ideal gas law, $PV = nRT$. At ideal conditions, 0°C or 273 K and 1 atm or 1.013×10^5 Pa pressure, the gas constant is evaluated to be

$$R = \frac{PV}{RT} = \frac{(1.013 \times 10^5 \text{ Pa})(22.4 \text{ m}^3)}{(1 \text{ kg mol})(273 \text{ K})} = 8.314 \frac{\text{Pa} \cdot \text{m}^3}{\text{mol} \cdot \text{K}} \tag{24-12}$$

The volume of the gas mixture, at 298 K, is

$$V = \frac{nRT}{P} = \frac{(1 \text{ mol})\left(8.314\dfrac{\text{Pa} \cdot \text{m}^3}{\text{mol} \cdot \text{K}}\right)(298 \text{ K})}{1.013 \times 10^5 \text{ Pa}}$$

$$= 0.0245 \text{ m}^3$$

The concentrations are

$$c_{O_2} = \frac{0.21 \text{ mol}}{0.0245 \text{ m}^3} = 8.57 \frac{\text{mol O}_2}{\text{m}^3}$$

$$c_{N_2} = \frac{0.79 \text{ mol}}{0.0245 \text{ m}^3} = 32.3 \frac{\text{mol N}_2}{\text{m}^3}$$

$$c = \sum_{i=1}^{n} c_i = 8.57 + 32.3 = 40.9 \text{ mol/m}^3$$

The total density, ρ, is

$$\rho = \frac{0.0288 \text{ kg}}{0.0245 \text{ m}^3} = 1.180 \text{ kg/m}^3$$

and the mean molecular weight of the mixture is

$$M = \frac{\rho}{c} = \frac{1.180 \text{ kg/m}^3}{40.9 \text{ mol/m}^3} = 0.0288 \text{ kg/mol}$$

Velocities. In a multicomponent system the various species will normally move at different velocities; accordingly, an evaluation of a velocity for the gas mixture requires the averaging of the velocities of each species present.

The *mass-average velocity* for a multicomponent mixture is defined in terms of the mass densities and velocities of all components by

$$\mathbf{v} = \frac{\sum\limits_{i=1}^{n} \rho_i \mathbf{v}_i}{\sum\limits_{i=1}^{n} \rho_i} = \frac{\sum\limits_{i=1}^{n} \rho_i \mathbf{v}_i}{\rho} \tag{24-13}$$

where $\mathbf{v}_i$ denotes the absolute velocity of species i relative to stationary coordinate axes. This is the velocity which would be measured by a pitot tube and is the velocity which was previously encountered in the equations of momentum transfer. The *molar-average velocity* for a multicomponent mixture is defined in terms of the molar concentrations of all components by

$$\mathbf{V} = \frac{\sum\limits_{i=1}^{n} c_i \mathbf{v}_i}{\sum\limits_{i=1}^{n} c_i} = \frac{\sum\limits_{i=1}^{n} c_i \mathbf{v}_i}{c} \tag{24-14}$$

The velocity of a particular species relative to the mass-average or molar-average velocity is termed a *diffusion velocity*. We can define two different diffusion velocities

$\mathbf{v}_i - \mathbf{v}$, the diffusion velocity of species i relative to the mass-average velocity

and

$\mathbf{v}_i - \mathbf{V}$, the diffusion velocity of species i relative to the molar-average velocity.

According to Fick's law, a species can have a velocity relative to the mass or molar-average velocity only if gradients in the concentration exist.

Fluxes. The mass (or molar) flux of a given species is a vector quantity denoting the amount of the particular species, in either mass or molar units, that passes per given increment of time through a unit area normal to the vector. The flux may be defined with reference to coordinates which are fixed in space, coordinates which are moving with the mass-average velocity, or coordinates which are moving with the molar-average velocity.

The basic relation for molecular diffusion defines the molar flux relative to the molar-average velocity, $\mathbf{J}_A$. An empirical relation for this molar flux, first postulated by Fick[*] and, accordingly, often referred to as Fick's first law, defines the diffusion of component A in an isothermal, isobaric system. For diffusion in

---

[*] A. Fick, *Ann. Physik.*, **94**, 59 (1855).

only the z direction, the Fick rate equation is

$$J_{A,z} = -D_{AB}\frac{dc_A}{dz} \tag{24-15}$$

where $J_{A,z}$ is the molar flux in the z direction relative to the molar average velocity, dc_A/dz is the concentration gradient in the z direction, and D_{AB}, the proportionality factor, is the *mass diffusivity* or *diffusion coefficient* for component A diffusing through component B.

A more general flux relation which is not restricted to isothermal, isobaric systems was proposed by de Groot* who chose to write

$$\text{Flux} = -\begin{pmatrix}\text{overall}\\\text{density}\end{pmatrix}\begin{pmatrix}\text{diffusion}\\\text{coefficient}\end{pmatrix}\begin{pmatrix}\text{concentration}\\\text{gradient}\end{pmatrix}$$

or

$$J_{A,z} = -cD_{AB}\frac{dy_A}{dz} \tag{24-16}$$

Since the total concentration c is constant under isothermal, isobaric conditions, equation (24-15) is a special form of the more general relation (24-16). An equivalent expression for $j_{A,z}$, the mass flux in the z direction relative to the mass average velocity, is

$$j_{A,z} = -\rho D_{AB}\frac{d\omega_A}{dz} \tag{24-17}$$

where $d\omega_A/dz$ is the concentration gradient in terms of the mass fraction. When the density is constant, this relation simplifies to

$$j_{A,z} = -D_{AB}\frac{d\rho_A}{dz}$$

For a binary system with a constant average velocity in the z direction, the molar flux in the z direction relative to the molar average velocity may also be expressed by

$$J_{A,z} = c_A(v_{A,z} - V_z) \tag{24-18}$$

Equating expressions (24-16) and (24-18), we obtain

$$J_{A,z} = c_A(v_{A,z} - V_z) = -cD_{AB}\frac{dy_A}{dz}$$

which, upon rearrangement, yields

$$c_A v_{A,z} = -cD_{AB}\frac{dy_A}{dz} + c_A V_z$$

---

*S. R. de Groot, *Thermodynamics of Irreversible Processes*, North-Holland, Amsterdam, 1951.

For this binary system, V_z can be evaluated by equation (24-14) as

$$V_z = \frac{1}{c}(c_A v_{A,z} + c_B v_{B,z})$$

or

$$c_A V_z = y_A(c_A v_{A,z} + c_B v_{B,z})$$

Substituting this expression into our relation, we obtain

$$c_A v_{A,z} = -c D_{AB}\frac{dy_A}{dz} + y_A(c_A v_{A,z} + c_B v_{B,z}) \qquad (24\text{-}19)$$

Since the component velocities, $v_{A,z}$ and $v_{B,z}$, are velocities relative to the fixed z axis, the quantities $c_A v_{A,z}$ and $c_B v_{B,z}$ are fluxes of components A and B relative to a fixed z coordinate; accordingly, we symbolize this new type of flux which is relative to a set of stationary axes by

$$\mathbf{N}_A = c_A \mathbf{v}_A$$

and

$$\mathbf{N}_B = c_B \mathbf{v}_B$$

Substituting these symbols into equation (24-19), we obtain a relation for the flux of component A relative to the z axis,

$$N_{A,z} = -c D_{AB}\frac{dy_A}{dz} + y_A(N_{A,z} + N_{B,z}) \qquad (24\text{-}20)$$

This relation may be generalized and written in vector form as

$$\mathbf{N}_A = -c D_{AB}\nabla y_A + y_A(\mathbf{N}_A + \mathbf{N}_B) \qquad (24\text{-}21)$$

It is important to note that the molar flux, $\mathbf{N}_A$, is a resultant of the two vector quantities:

$-c D_{AB}\nabla y_A$ the molar flux, $\mathbf{J}_A$, resulting from the concentration gradient. This term is referred to as the *concentration gradient contribution*;

and

$y_A(\mathbf{N}_A + \mathbf{N}_B) = c_A \mathbf{V}$ the molar flux resulting as component A is carried in the bulk flow of the fluid. This flux term is designated the *bulk motion contribution*.

Either or both quantities can be a significant part of the total molar flux, $\mathbf{N}_A$. Whenever equation (24-21) is applied to describe molar diffusion, the vector nature of the individual fluxes, $\mathbf{N}_A$ and $\mathbf{N}_B$, must be considered and then, in turn, the direction of each of two vector quantities must be evaluated.

If species A were diffusing in a multicomponent mixture, the expression equivalent to equation (24-21) would be

$$\mathbf{N}_A = -cD_{AM}\nabla y_A + y_A \sum_{i=1}^{n} \mathbf{N}_i$$

where D_{AM} is the diffusion coefficient of A in the mixture.

The mass flux, $\mathbf{n}_A$, relative to a fixed spatial coordinate system, is defined for a binary system in terms of mass density and mass fraction by

$$\mathbf{n}_A = -\rho D_{AB}\nabla \omega_A + \omega_A(\mathbf{n}_A + \mathbf{n}_B) \tag{24-22}$$

where

$$\mathbf{n}_A = \rho_A \mathbf{v}_A$$

and

$$\mathbf{n}_B = \rho_B \mathbf{v}_B$$

Under isothermal, isobaric conditions, this relation simplifies to

$$\mathbf{n}_A = -D_{AB}\nabla \rho_A + \omega_A(\mathbf{n}_A + \mathbf{n}_B)$$

As previously noted, the flux is a resultant of two vector quantities:

$-D_{AB}\nabla\rho_A$,   the mass flux, $\mathbf{j}_A$, resulting from a concentration gradient; the *concentration gradient contribution*.

$w_A(\mathbf{n}_A + \mathbf{n}_B) = \rho_A \mathbf{v}$,   the mass flux resulting as component A is carried in the bulk flow of the fluid; the *bulk motion contribution*.

If a balloon, filled with a color dye, is dropped into a large lake, the dye will diffuse radially as a concentration gradient contribution. When a stick is dropped into a moving stream, it will float downstream by the bulk motion contribution. If the dye-filled balloon were dropped into the moving stream, the dye would diffuse radially while being carried downstream; thus both contributions participate simultaneously in the mass transfer.

The four equations defining the fluxes, $\mathbf{J}_A$, $\mathbf{j}_A$, $\mathbf{N}_A$ and $\mathbf{n}_A$ are equivalent statements of the Fick rate equation. The diffusion coefficient, D_{AB}, is identical in all equations. Any one of these equations is adequate to describe molecular diffusion; however, certain fluxes are easier to use for specific cases. The mass fluxes, $\mathbf{n}_A$ and $\mathbf{j}_A$, are used when the Navier-Stokes equations are also required to describe the process. Since chemical reactions are described in terms of moles of the participating reactants, the molar fluxes, $\mathbf{J}_A$ and $\mathbf{N}_A$, are used to describe mass-transfer operations in which chemical reactions are involved. The fluxes relative to coordinates fixes in space, $\mathbf{n}_A$ and $\mathbf{N}_A$, are often used to describe engineering operations within process equipment. The fluxes $\mathbf{J}_A$ and $\mathbf{j}_A$ are used to describe the mass transfer in diffusion cells used for measuring the diffusion coefficient. Table 24.2 summarizes the equivalent forms of the Fick rate equation.

RELATED TYPES OF MOLECULAR MASS TRANSFER

According to the second law of thermodynamics, systems not in equilibrium will tend to move toward equilibrium with time. A generalized driving force in

TABLE 24.2   EQUIVALENT FORMS OF THE MASS FLUX EQUATION FOR BINARY SYSTEM A AND B

Flux	Gradient	Fick rate equation	Restrictions
$\mathbf{n}_A$	$\nabla \omega_A$	$\mathbf{n}_A = -\rho D_{AB} \nabla \omega_A + \omega_A(\mathbf{n}_A + \mathbf{n}_B)$	
	$\nabla \rho_A$	$\mathbf{n}_A = -D_{AB} \nabla \rho_A + \omega_A(\mathbf{n}_A + \mathbf{n}_B)$	Constant ρ
$\mathbf{N}_A$	∇y_A	$\mathbf{N}_A = -c D_{AB} \nabla y_A + y_A(\mathbf{N}_A + \mathbf{N}_B)$	
	∇c_A	$\mathbf{N}_A = -D_{AB} \nabla c_A + y_A(\mathbf{N}_A + \mathbf{N}_B)$	Constant c
$\mathbf{j}_A$	$\nabla \omega_A$	$\mathbf{j}_A = -\rho D_{AB} \nabla \omega_A$	
	$\nabla \rho_A$	$\mathbf{j}_A = -D_{AB} \nabla \rho_A$	Constant ρ
$\mathbf{J}_A$	∇y_A	$\mathbf{J}_A = -c D_{AB} \nabla y_A$	
	∇c_A	$\mathbf{J}_A = -D_{AB} \nabla c_A$	Constant c

chemical thermodynamic terms is $-d\mu_c/dz$ where μ_c is the *chemical potential*. The molar diffusion velocity of component A is defined in terms of the chemical potential by

$$v_{A,z} - V_z = u_A \frac{d\mu_c}{dz} = -\frac{D_{AB}}{RT} \frac{d\mu_c}{dz} \tag{24-23}$$

where μ_c is the "mobility" of component A, or the resultant velocity of the molecule while under the influence of a unit driving force. Equation (24-23) is known as the Nernst-Einstein relation. The molar flux of A becomes

$$J_{A,z} = c_A(v_{A,z} - V_z) = -c_A \frac{D_{AB}}{RT} \frac{d\mu_c}{dz} \tag{24-24}$$

Equation (24-24) may be used to define all molecular mass-transfer phenomena. As an example, consider the conditions specified for equation (24-15); the chemical potential of a component in a homogeneous ideal solution at constant temperature and pressure is defined by

$$\mu_c = \mu^0 + RT \ln c_A \tag{24-25}$$

where μ^0 is a constant, the chemical potential of the standard state. When we substitute this relation into equation (24-24), the Fick rate equation for a homogeneous phase is obtained

$$J_{A,z} = -D_{AB} \frac{dc_A}{dz} \tag{24-15}$$

There are a number of other physical conditions, in addition to differences in concentration, which will produce a chemical potential gradient: temperature differences, pressure differences, and differences in the forces created by external fields, such as gravity, magnetic, and electrical fields. We can, for example, obtain mass transfer by applying a temperature gradient to a multicomponent system; this transport phenomenon, the *Soret effect* or *thermal diffusion*, although normally small relative to other diffusion effects, is used successfully in the separation of isotopes. Components in a liquid mixture can be separated with a centrifuge by *pressure diffusion*. There are many well-known examples of mass fluxes being induced in a mixture subjected to an external force field: separation by sedimentation under the influence of gravity, electrolytic precipitation due to an electrostatic force field, and magnetic separation of mineral mixtures through the action of a magnetic force field. Although these mass-transfer phenomena are important, they are very specific processes.

The molecular mass transfer, resulting from concentration differences and described by Fick's law, results from the random molecular motion over small mean free paths, independent of any containment walls. The diffusion of fast neutrons and molecules in extremely small pores or at very low gas density cannot be described by this relationship.

Neutrons, produced in a nuclear fission process, initially possess high kinetic energies and are termed *fast neutrons* because of their high velocities; that is, up to 15 million meters per second. At these high velocities, neutrons pass through the electronic shells of other atoms or molecules with little hindrance. To be deflected, the fast neutrons must collide with a nucleus, which is a very small target compared to the volume of most atoms and molecules. The mean free path of fast neutrons is approximately one million times greater than the free paths of gases at ordinary pressures. After the fast neutrons are slowed down through elastic-scattering collisions between the neutrons and the nuclei of the reactor's moderator, these slower moving neutrons, *thermal neutrons*, migrate from positions of higher concentration to positions of lower concentration, and their migration is described by Fick's law of diffusion.

If the density of the gas is low, or if the pores through which the gas is traveling are quite small, the molecules will collide with the walls more frequently than with each other. This is known as Knudsen flow or *Knudsen diffusion*. Upon hitting the wall, the molecules are momentarily adsorbed and then given off in random directions. The gas flux is reduced by the wall collisions. By use of the kinetic theory, one may develop relations for the Knudsen diffusion in gases. The proportionality constant which relates the Knudsen flux to the concentration gradient is independent of pressure; the proportionality constant for molecular diffusion in gases, as described by Fick's law, will soon be shown to be inversely proportional to pressure.

The mass transfer resulting from concentration differences and described by Fick's law is the principal process encountered by engineers and the mass-transfer phenomenon we shall treat in later chapters. It is important, however, for us to realize that these other related molecular transport phenomena exist and that they may play important roles in future mass-transfer operations.

24.2 THE DIFFUSION COEFFICIENT

Fick's law proportionality, D_{AB}, is known as the diffusion coefficient. Its fundamental dimensions, which may be obtained from equation (24-15),

$$D_{AB} = \frac{-J_{A,z}}{dc_A/dz} = \left(\frac{M}{L^2 t}\right)\left(\frac{1}{M/L^3 \cdot 1/L}\right) = \frac{L^2}{t}$$

are identical to the fundamental dimensions of the other transport properties: kinematic viscosity, ν, and thermal diffusivity, α, or its equivalent ratio, $k/\rho c_p$. The mass diffusivity has been reported in cm^2/s; the SI units are m^2/s which is a factor 10^{-4} smaller. In the English system ft^2/hr are commonly used. Conversion between these systems involves the simple relations:

$$\frac{D_{AB}(cm^2/s)}{D_{AB}(m^2/s)} = 10^4$$

$$\frac{D_{AB}(ft^2/hr)}{D_{AB}(cm^2/s)} = 3.87 \tag{24-26}$$

The diffusion coefficient depends upon the pressure, temperature, and composition of the system. Experimental values for the diffusivities of gases, liquids and solids are tabulated in Appendix Tables J.1, J.2, and J.3, respectively. As one might expect from consideration of the mobility of the molecules, the diffusion coefficients are generally higher for gases (in the range of 5×10^{-6} to 1×10^{-5} m^2/s), than for liquids (in the range of 10^{-10} to 10^{-9} m^2/s) which are higher than the values reported for solids (in the range of 10^{-14} to 10^{-10} m^2/s).

In the absence of experimental data, semitheoretical expressions have been developed which give approximations, sometimes as valid as experimental values due to the difficulties encountered in their measurement.

GAS MASS DIFFUSIVITY

Theoretical expressions for the diffusion coefficient in low-density gaseous mixtures as a function of the system's molecular properties were derived by Jeans,* Chapman,† and Sutherland,‡ using the kinetic theory of gases. Using the reasoning of these earlier scientists to explain molecular transport phenomena, we can examine the motion of the gas molecules as we did in sections 7.3 and 15.2, and then derive an expression relating the diffusion coefficient to the properties of the gaseous systems.

---

* Sir James Jeans, *Dynamical Theory of Gases*, Cambridge Univ. Press, London, 1921.
† S. Chapman and T. G. Cowling, *Mathematical Theory of Non-Uniform Gases*, Cambridge Univ. Press, London, 1959.
‡ W. Sutherland, *Phil. Mag.*, **36**, 507 (1893); **38**, 1 (1894).

Consider the control volume shown in Figure 24.2. If we specify a static gas or one in laminar flow in the x direction, we can consider the mass transfer of species A in the y direction to occur only on the molecular scale. Applying equation (4-1),

$$\iint_{\text{c.s.}} \rho(\mathbf{v} \cdot \mathbf{n})\, dA + \frac{\partial}{\partial t} \iiint_{\text{c.v.}} \rho\, dV = 0 \qquad (4\text{-}1)$$

to a steady flow of mass across the top face of the element, we obtain

$$\iint_{\text{c.s.}} \rho(\mathbf{v} \cdot \mathbf{n})\, dA = 0$$

This equation simply states that the upward mass flux must equal the downward mass flux.

As a first approximation, let us consider a system containing molecules of equal size and mass, possessing equal average velocities. Only a gas mixture made up of isotopes of the same element would approximately resemble this system. Reexamining our derived equation on a microscopic basis, we may conclude that the number of molecules crossing the top face from below must equal the number of molecules crossing from above. Since a concentration of species A exists as shown in Figure 24.2, more molecules of species A will be transported across the control surface from above than from below. This produces a net flux of A molecules in the y direction.

As done previously in Chapters 7 and 15, we shall use the following equations derived from the kinetic theory of low density gases:

$$\bar{C} = \sqrt{\frac{8\kappa T}{\pi m}}$$

$$\lambda = \frac{1}{\sqrt{2}\pi d^2 N}$$

$$Z = \tfrac{1}{4} N \bar{C}$$

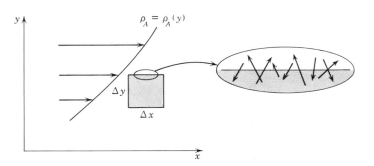

Figure 24.2 Molecular motion at the surface of a control volume.

where $\bar{C}$ is the average random moleculer velocity, $\bar{C}/4$ is the velocity of an individual molecule as it passes through area $\Delta x \, \Delta z$, λ is the mean free path, κ is the Boltzmann constant, m is the mass of a molecule, d is the diameter of the spherical molecules, N is the molecular concentration, and Z is the frequency with which a molecule will arrive at area $\Delta x \, \Delta z$. The equation of continuity written in terms of the moving molecules is

$$\sum_n m_n \frac{\bar{C}}{4} \Delta x \, \Delta z|_{y-} - \sum_n m_n \frac{\bar{C}}{4} \Delta x \, \Delta z|_{y+} = 0$$

or when summed over the N molecules in a unit volume,

$$\rho \frac{\bar{C}}{4} \Delta x \, \Delta z|_{y-} - \rho \frac{\bar{C}}{4} \Delta x \, \Delta z|_{y+} = 0$$

If we are to count only the molecules of A crossing this surface, the equation must account for a net mass flux in the y direction,

$$j_{A,y} = \frac{\bar{C}}{4} \rho_A|_{y-} - \frac{\bar{C}}{4} \rho_A|_{y+}$$

$$j_{A,y} = (\rho_A|_{y-} - \rho_A|_{y+}) \frac{\bar{C}}{4} \tag{24-27}$$

Again, as in Chapters 7 and 15, we shall assume the concentration profile to be essentially linear for a distance of several mean free paths. Then

$$\rho_A|_{y-} = \rho_A - \frac{\partial \rho_A}{\partial y} \delta$$

and

$$\rho_A|_{y+} = \rho_A + \frac{\partial \rho_A}{\partial y} \delta$$

where

$$y_- = y - \delta$$

and

$$y_+ = y + \delta$$

The substitution of these relations for $\rho_A|_{y-}$ and $\rho_A|_{y+}$ into equation (24-27) yields

$$j_{A,y} = -2 \frac{\bar{C}}{4} \delta \frac{\partial \rho_A}{\partial y} \tag{24-28}$$

where δ represents the y component of the distance between collisions. It is related to the mean free path of a molecule, λ, by the relation

$$\delta = \tfrac{2}{3} \lambda$$

Our equation now becomes

$$j_{A,y} = -\tfrac{1}{3}\bar{C}\lambda\frac{\partial\rho_A}{\partial y}$$ (24-29)

If we compare equation (24-29) with equation (24-17),

$$j_{A,y} = -D_{AB}\frac{\partial\rho_A}{\partial y}$$ (24-17)

it is apparent that the diffusion coefficient for a mixture of similar molecules, that is, A and isotope A^*, is

$$D_{AA^*} = \tfrac{1}{3}\bar{C}\lambda$$ (24-30)

This diffusion coefficient is often referred to as the *self-diffusion coefficient* and is used to explain the diffusion of tracer molecules.

The substitution of kinetic results for $\bar{C}$ and λ into equation (24-30) gives

$$D_{AA^*} = \frac{2}{3\pi^{3/2}\,d^2 N}\left(\frac{\kappa T}{m}\right)^{1/2}$$ (24-31)

For an ideal gas, N can be replaced, using the relation

$$N\kappa T = cRT = P$$

giving

$$D_{AA^*} = \frac{2}{3\pi^{3/2}\,d^2 P}\left(\frac{\kappa^3 T^3}{m}\right)^{1/2}$$ (24-32)

Equation (24-32) shows that the diffusion coefficient can be expressed entirely in terms of properties of the gas. Unlike the other two molecular transport coefficients, viscosity and thermal conductivity, the diffusion coefficient is dependent on pressure as well as on a higher order of the absolute temperature. The significance of this result should not be overlooked even though an oversimplified model was used in the development.

Theoretical expressions for the diffusion coefficient in low-density gas mixtures as a function of the system's molecular properties were derived by Jeans,* Chapman,† and Sutherland‡ using the kinetic theory of gases. Modern versions of the kinetic theory have attempted to account for the forces of attraction and repulsion between the molecules. Hirschfelder, Bird, and Spotz,§ using the Lennard-Jones potential to evaluate the influence of the intermolecular forces,

---

* Sir James Jeans, *Dynamical Theory of Gases*, Cambridge University Press, London (1921).
† S. Chapman and T. G. Cowling, *Mathematical Theory of Non-Uniform Gases*, Cambridge University Press, London (1959).
‡ W. Sutherland, *Phil. Mag.*, **36**, 507 (1893); **38**, 1 (1894).
§ J. O. Hirschfelder, R. B. Bird, and E. L. Spotz, *Chem. Revs.*, **44**, 205–231 (1949).

presented an equation for the diffusion coefficient for gas pairs of nonpolar,* nonreacting molecules

$$D_{AB} = \frac{0.001858 T^{3/2} \left[\frac{1}{M_A} + \frac{1}{M_B}\right]^{1/2}}{P \sigma_{AB}^2 \Omega_D} \tag{24-33}$$

where D_{AB} is the mass diffusivity of A through B, in cm^2/s; T is the absolute temperature, in K; M_A, M_B are the molecular weights of A and B, respectively; P is the absolute pressure, in atmospheres; σ_{AB} is the "collision diameter," a Lennard-Jones parameter, in Angstroms; and Ω_D is the "collision integral" for molecular diffusion, a dimensionless function of the temperature and of the intermolecular potential-field for one molecule of A and one molecule of B. Table K.1 in the appendix lists Ω_D as a function of $\kappa T/\varepsilon_{AB}$; κ is the Boltzmann constant, which is 1.38×10^{-16} ergs/K; and ε_{AB} is the energy of molecular interaction for the binary system A and B, a Lennard-Jones parameter, in ergs; see equation (24-40). Unlike the other two molecular transport coefficients, viscosity and thermal conductivity, the diffusion coefficient is dependent on pressure as well as on a higher order of the absolute temperature. When the transport process in a single component phase was examined, we did not find any composition dependency in equation (24-32) or in the similar equations for viscosity and thermal conductivity.

The Lennard-Jones parameters, σ and ϵ_A, are usually obtained from viscosity data. Unfortunately, this information is available for only a very few pure gases. Appendix Table K.2 tabulates these values. In the absence of experimental data, the values for pure components may be estimated from the following empirical relations:

$$\sigma = 1.18 \, V_b^{1/3} \tag{24-34}$$

$$\sigma = 0.841 \, V_c^{1/3} \tag{24-35}$$

$$\sigma = 2.44 \left(\frac{T_c}{P_c}\right)^{1/3} \tag{24-36}$$

$$\epsilon_A/\kappa = 0.77 \, T_c \tag{24-37}$$

and

$$\epsilon_A/\kappa = 1.15 \, T_b \tag{24-38}$$

where V_b is the molecular volume at the normal boiling point, in $(cm)^3/g$ mole (this is evaluated by using Table 24.3); V_c is the critical molecular volume, in $(cm)^3/g$ mole; T_c is the critical temperature, in K; T_b is the normal boiling temperature, in K; and P_c is the critical pressure, in atmospheres.

---

*For an introductory discussion of polar and nonpolar structure, see R. A. Alberty and F. Daniels, *Physical Chemistry*, Fifth Edition, John Wiley & Sons, New York (1980) or P. W. Atkins, *Physical Chemistry*, Second Edition, W. H. Freeman, San Francisco (1982).

For a binary system composed of nonpolar molecular pairs, the Lennard-Jones parameters of the pure component may be combined empirically by the following relations:

$$\sigma_{AB} = \frac{\sigma_A + \sigma_B}{2} \qquad (24\text{-}39)$$

and

$$\epsilon_{AB} = \sqrt{\epsilon_A \epsilon_B} \qquad (24\text{-}40)$$

These relations must be modified for polar-polar and polar-nonpolar molecular pairs; the proposed modifications are discussed by Hirschfelder, Curtiss, and Bird.*

The Hirschfelder equation (24-33) is often used to extrapolate experimental data. For moderate ranges of pressure, up to 25 atmospheres, the diffusion coefficient varies inversely as the pressure. Higher pressures apparently require dense gas corrections; unfortunately, no satisfactory correlation is available for high pressures. Equation (24-33) also states that the diffusion coefficient varies with the temperature as $T^{3/2}/\Omega_D$ varies. Simplifying equation (24-33), we can predict the diffusion coefficient at any temperature and at any pressure below 25 atmospheres from a known experimental value by

$$D_{AB_{T_2,P_2}} = D_{AB_{T_1,P_1}} \left(\frac{P_1}{P_2}\right) \left(\frac{T_2}{T_1}\right)^{3/2} \frac{\Omega_D|_{T_1}}{\Omega_D|_{T_2}} \qquad (24\text{-}41)$$

In Appendix Table J.1, experimental values of the product $D_{AB}P$ are listed for several gas pairs at a particular temperature. Using equation (24-41), we may extend these values to other temperatures.

EXAMPLE 2

Evaluate the diffusion coefficient of carbon dioxide in air at 20°C and atmospheric pressure. Compare this value with the experimental value reported in Appendix Table J.1.

From Table K.2 of the appendix values of σ and ϵ/κ are obtained:

	σ, in Å	ϵ_A/κ, in K
Carbon dioxide	3.996	190
Air	3.617	97

The various parameters for equation (24-33) may be evaluated as follows:

$$\sigma_{AB} = \frac{\sigma_A + \sigma_B}{2} = \frac{3.996 + 3.617}{2} = 3.806 \text{ Å}$$

---

* J. O. Hirschfelder, C. F. Curtiss, and R. B. Bird, *Molecular Theory of Gases and Liquids*, John Wiley & Sons, Inc., New York, 1954.

$$\epsilon_{AB}/\kappa = \sqrt{(\epsilon_A/\kappa)(\epsilon_B/\kappa)} = \sqrt{(190)(97)} = 136$$

$$T = 20 + 273 = 293 \text{ K}$$

$$P = 1 \text{ atm}$$

$$\frac{\epsilon_{AB}}{\kappa T} = \frac{136}{293} = 0.463$$

$$\frac{\kappa T}{\epsilon_{AB}} = 2.16$$

$$\Omega_D \text{ (Table K.1)} = 1.047$$

$$M_{CO_2} = 44$$

and

$$M_{\text{Air}} = 29$$

Substituting these values into the equation, we obtain

$$D_{AB} = \frac{0.001858 T^{3/2}(1/M_A + 1/M_B)^{1/2}}{P\sigma_{AB}{}^2 \Omega_D}$$

$$= \frac{(0.001858)(293)^{3/2}(1/44 + 1/29)^{1/2}}{(1)(3.806)^2(1.047)} = 0.147 \text{ cm}^2/\text{s}$$

From Table J.1 for CO_2 in air at 273 K, 1 atmosphere, we have

$$D_{AB} = 0.136 \text{ cm}^2/\text{s}$$

Equation (24-41) will be used to correct for the differences in temperature,

$$\frac{D_{AB,T_1}}{D_{AB,T_2}} = \left(\frac{T_1}{T_2}\right)^{3/2}\left(\frac{\Omega_D|_{T_2}}{\Omega_D|_{T_1}}\right)$$

Values for Ω_D may be evaluated as follows

$$\text{at } T_2 = 273 \qquad \epsilon_{AB}/\kappa T = \frac{136}{273} = 0.498 \qquad \Omega_D|_{T_2} = 1.074$$

$$\text{at } T_1 = 293 \qquad \Omega_D|_{T_1} = 1.047 \qquad \text{(previous calculations)}$$

The corrected value for the diffusion coefficient at 20°C is

$$D_{AB,T_1} = \left(\frac{293}{273}\right)^{3/2}\frac{(1.074)}{(1.047)}(0.136) = 0.155 \text{ cm}^2/\text{s} \quad (1.55 \times 10^{-5} \text{ m}^2/\text{s})$$

We readily see that the temperature dependency of the "collision integral" is very small. Accordingly, most scaling of diffusivities relative to temperature only include the ratio $(T_1/T_2)^{3/2}$.

Equation (24-33) was developed for dilute gases consisting of nonpolar, spherical monatomic molecules. However, this equation gives good results for most nonpolar, binary gas systems over a wide range of temperatures.* Other empirical equations have been proposed† for estimating the diffusion coefficient for nonpolar, binary gas systems at low pressures. The empirical correlation recommended by Fuller, Schettler, and Giddings permits the evaluation of the diffusivity when reliable Lennard-Jones parameters, σ_i and ε_i, are unavailable. The Fuller correlation is

$$D_{AB} = \frac{10^{-3}T^{1.75}\left(\dfrac{1}{M_A}+\dfrac{1}{M_B}\right)^{1/2}}{P[(\Sigma v)_A^{1/3}+(\Sigma v)_B^{1/3}]^2} \tag{24-42}$$

where D_{AB} is in cm^2/s, T is in K, and P is in atmospheres. To determine the v terms, the authors recommend the addition of the atomic and structural diffusion-volume increments v reported in Table 24-3.

EXAMPLE 3

Reevaluate the diffusion coefficient of carbon dioxide in air at 20°C and atmospheric pressure using the Fuller, Schettler, and Giddings equation and compare the new value with the one reported in example 2.

$$\begin{aligned}
D_{AB} &= \frac{10^{-3}T^{1.75}\left(\dfrac{1}{M_A}+\dfrac{1}{M_B}\right)^{1/2}}{P[(\Sigma v)_A^{1/3}+(\Sigma v)_B^{1/3}]^2} \\[2mm]
&= \frac{10^{-3}(293)^{1.75}\left(\dfrac{1}{44}+\dfrac{1}{29}\right)^{1/2}}{(1)[26.9)^{1/3}+(20.1)^{1/3}]^2} \\[2mm]
&= 0.152 \text{ cm}^2/\text{s}
\end{aligned}$$

This value compares very favorably to the value evaluated with Hirschfelder equation, 0.155 cm^2/s, and its determination was easily accomplished.

Brokaw‡ has suggested a method for estimating diffusion coefficient for binary gas mixtures containing polar compounds. The Hirschfelder equation (24-33) is still used; however, the collision integral is evaluated by

$$\Omega_D = \Omega_{D_0} + \frac{0.196\delta_{AB}^2}{T^*} \tag{24-43}$$

where

$$\delta_{AB} = (\delta_A\delta_B)^{1/2}$$

$$\delta = \frac{1.94\times10^3 \mu_p^2}{V_b T_b} \tag{24-44}$$

* R. C. Reid, J. M. Prausnitz, and T. K. Sherwood, *The Properties of Gases and Liquids*, Third Edition, McGraw-Hill Book Company, New York, 1977, Chapter 11.
† J. H. Arnold, *J. Am. Chem. Soc.*, **52**, 3937 (1930). E. R. Gilliland, *Ind. Eng. Chem.*, **26**, 681 (1934). J. C. Slattery and R. B. Bird, *A.I.Ch.E. J.*, **4**, 137 (1958). D. F. Othmer and H. T. Chen, *Ind. Eng. Chem. Process Des. Dev.*, **1**, 249 (1962). R. G. Bailey, *Chem. Engr.*, **82**(6), 86, (1975). E. N. Fuller, P. D. Schettler, and J. C. Giddings, *Ind. Eng. Chem.*, **58**(5), 18 (1966).
‡ R. S. Brokaw, *Ind. Engr. Chem. Process Des. Dev.*, **8**, 240 (1969).

TABLE 24.3 ATOMIC DIFFUSION VOLUMES FOR USE IN ESTIMAT-
ING D_{AB} BY METHOD OF FULLER, SCHETTLER, AND GIDDINGS

Atomic and Structure Diffusion-Volume Increments, v			
C	16.5	Cl	19.5
H	1.98	S	17.0
O	5.48	Aromatic ring	−20.2
N	5.69	Heterocyclic ring	−20.2

Diffusion Volumes for Simple Molecules, v					
H_2	7.07	Ar	16.1	H_2O	12.7
D_2	6.70	Kr	22.8	$CClF_2$	114.8
He	2.88	CO	18.9	SF_6	69.7
N_2	17.9	CO_2	26.9	Cl_2	37.7
O_2	16.6	N_2O	35.9	Br_2	67.2
Air	20.1	NH_3	14.9	SO_2	41.1

μ_p = dipole moment, debyes

V_b = liquid molar volume of the specific compound
at its boiling point, cm^3/g mol

T_b = normal boiling point, K

and

$$T^* = \kappa T / \varepsilon_{AB}$$

where

$$\frac{\varepsilon_{AB}}{\kappa} = \left(\frac{\varepsilon_A}{\kappa} \frac{\varepsilon_B}{\kappa} \right)^{1/2}$$

$$\varepsilon/\kappa = 1.18(1 + 1.3\,\delta^2)T_b \qquad (24\text{-}45)$$

δ is evaluated with (24-44). And

$$\Omega_{Do} = \frac{A}{(T^*)^B} + \frac{C}{\exp{(DT^*)}} + \frac{E}{\exp{(FT^*)}} + \frac{G}{\exp{(HT^*)}} \qquad (24\text{-}46)$$

with

$$A = 1.060\,36 \qquad E = 1.035\,87$$
$$B = 0.156\,10 \qquad F = 1.529\,96$$
$$C = 0.193\,00 \qquad G = 1.764\,74$$
$$D = 0.476\,35 \qquad H = 3.894\,11$$

The collision diameter, σ_{AB}, is evaluated with

$$\sigma_{AB} = (\sigma_A \sigma_B)^{1/2} \qquad (24\text{-}47)$$

with each component's characteristic length evaluated by

$$\sigma = \left(\frac{1.585 V_b}{1 + 1.3 \, \delta^2}\right)^{1/3} \qquad (24\text{-}48)$$

Reid, Prausnitz, and Sherwood* noted that the Brokaw equation is fairly reliable, permitting the evaluation of the diffusion coefficients for gases involving polar compounds with errors less than 15%.

Mass transfer in gas mixtures of several components can be described by theoretical equations involving the diffusion coefficients for the various binary pairs involved in the mixture. Hirschfelder, Curtiss, and Bird† present an expression in its most general form. Wilke‡ has simplified the theory and has shown that a close approximation to the correct form is given by the relation

$$D_{1\text{-mixture}} = \frac{1}{y_2'/D_{1\text{-}2} + y_3'/D_{1\text{-}3} + \ldots + y_n'/D_{1\text{-}n}} \qquad (24\text{-}49)$$

where $D_{1\text{-mixture}}$ is the mass diffusivity for component 1 in the gas mixture; $D_{1\text{-}n}$ is the mass diffusivity for the binary pair, component 1 diffusing through component n; and y_n' is the mole fraction of component n in the gas mixture evaluated on a component-1-free basis, that is,

$$y_2' = \frac{y_2}{y_2 + y_3 + \ldots + y_n}$$

EXAMPLE 4

Determine the diffusivity of carbon monoxide through a gas mixtture in which the mole fractions of each component are

$$y_{O_2} = 0.20$$

$$y_{N_2} = 0.70$$

$$y_{CO} = 0.10$$

The gas mixture is at 298 K and 2 atmosphere total pressure.

From Appendix Table J.1, we find

$$D_{CO\text{-}O_2} = 0.185 \times 10^{-4} \text{ m}^2/\text{s at 273 K, 1 atm}$$

$$D_{CO\text{-}N_2} = 0.192 \times 10^{-4} \text{ m}^2/\text{s at 288 K, 1 atm}$$

* R. C. Reid, J. M. Prausnitz, and T. K. Sherwood, *The Properties of Gases and Liquids*, Third Edition, McGraw-Hill Book Company, New York, 1977, Chapter 11.
† J. A. Hirschfelder, C. F. Curtiss, and R. B. Bird, *Molecular Theory of Gases and Liquids*, Wiley, New York, p. 718.
‡ C. R. Wilke, *Chem. Engr. Prog.*, **46**, 95–104 (1950).

The two binary diffusion coefficients may be corrected for temperature and pressure differences by using equation (24-41),

$$\frac{D_{AB \text{ condition 1}}}{D_{AB \text{ condition 2}}} = \left(\frac{T_1}{T_2}\right)^{3/2}\left(\frac{P_2}{P_1}\right)$$

For 298 K and 2 atm, we have

$$D_{CO-O_2} = \left(\frac{298}{273}\right)^{3/2}\left(\frac{1}{2}\right)(0.185 \times 10^{-4} \text{ m}^2/\text{s}) = 0.105 \times 10^{-4} \text{ m}^2/\text{s}$$

and

$$D_{CO-N_2} = \left(\frac{298}{288}\right)^{3/2}\left(\frac{1}{2}\right)(0.192 \times 10^{-4} \text{ m}^2/\text{s}) = 0.101 \times 10^{-4} \text{ m}^2/\text{s}$$

The compositions of oxygen and nitrogen on a CO-free basis are

$$y_{O_2}' = \frac{0.20}{1-0.10} = 0.22$$

$$y_{N_2}' = \frac{0.70}{1-0.10} = 0.78$$

Upon substituting these values into equation (24-42), we obtain

$$D_{CO-O_2,N_2} = \frac{1}{\dfrac{0.22}{0.105 \times 10^{-4}} + \dfrac{0.78}{0.101 \times 10^{-4}}}$$

$$= 0.102 \times 10^{-4} \text{ m}^2/\text{s} \qquad (0.395 \text{ ft}^2/\text{hr})$$

LIQUID MASS DIFFUSIVITY

In contrast to the case for gases, where we have available an advanced kinetic theory for explaining molecular motion, theories of the structure of liquids and their transport characteristics are still inadequate to permit a rigorous treatment. Inspection of published experimental values for liquid diffusion coefficients in Appendix J.2 reveals that they are several orders of magnitude smaller than gas diffusion coefficients and that they depend on concentration due to the changes in viscosity with concentration and changes in the degree of ideality of the solution.

Certain molecules diffuse as molecules, while others which are designated as electrolytes ionize in solutions and diffuse as ions. For example, sodium chloride, NaCl, diffuses in water as the ions Na^+ and Cl^-. Though each ion has a different mobility, the electrical neutrality of the solution indicates the ions must diffuse at the same rate; accordingly, it is possible to speak of a diffusion coefficient for molecular electrolytes such as NaCl. However, if several ions are present, the diffusion rates of the individual cations and anions must be considered, and molecular diffusion coefficients have no meaning. Needless to say, separate correlations for predicting the relation between the liquid mass diffusivities and the properties of the liquid solution will be required for electrolytes and nonelec-trolytes.

Two theories, the Eyring "hole" theory and the hydrodynamical theory, have been postulated as possible explanations for diffusion of nonelectrolyte solutes in low-concentration solutions. In the Eyring concept the ideal liquid is treated as a quasi-crystalline lattice model interspersed with holes. The transport phenome-non is then described by a unimolecular rate process involving the jumping of solute molecules into the holes within the lattice model. These jumps are empirically related to Eyring's theory of reaction rate.* The hydrodynamical theory states that the liquid diffusion coefficient is related to the solute molecule's mobility; that is, to the net velocity of the molecule while under the influence of a unit driving force. The laws of hydrodynamics provide relations between the force and the velocity. An equation that has been developed from the hydrodynamical theory is the Stokes-Einstein equation,

$$D_{AB} = \frac{\kappa T}{6\pi r \mu_B} \tag{24-50}$$

where D_{AB} is the diffusivity of A in dilute solution in B; κ is the Boltzmann constant; T is the absolute temperature; r is the solute particle radius; and μ_B is the solvent viscosity. This equation has been fairly successful in describing the diffusion of colloidal particles or large round molecules through a solvent which behaves as a continuum relative to the diffusing species.

The results of the two theories can be rearranged into the general form

$$\frac{D_{AB}\mu_B}{\kappa T} = f(V) \tag{24-51}$$

in which $f(V)$ is a function of the molecular volume of the diffusing solute. Empirical correlations, using the general form of equation (24-51), have been developed which attempt to predict the liquid diffusion coefficient in terms of the solute and solvent properties. Wilke and Chang† have proposed the following correlation for nonelectrolytes in an infinitely dilute solution:

$$\frac{D_{AB}\mu_B}{T} = \frac{7.4 \times 10^{-8}(\Phi_B M_B)^{1/2}}{V_A^{0.6}} \tag{24-52}$$

where D_{AB} is the mass diffusivity of A diffusing through liquid solvent B, in cm^2/s; μ_B is the viscosity of the solution, in centipoises; T is absolute temperature, in K; M_B is the molecular weight of the solvent; V_A is the molal volume of solute at normal boiling point, in cm^3/g mol; and Φ_B is the "association" parameter for solvent B.

---

* S. Glasstone, K. J. Laidler, and H. Eyring, *Theory of Rate Processes*, McGraw-Hill Book Company, New York, 1941, Chap. IX.
† C. R. Wilke and P. Chang, *A.I.Ch.E. J.*, **1**, 264 (1955).

TABLE 24.4   MOLECULAR VOLUMES AT NORMAL BOILING POINT FOR SOME COMMONLY ENCOUNTERED COMPOUNDS

Compound	Molecular volume, cm^3/g mole	Compound	Molecular volume, in cm^3/g mole
Hydrogen, H_2	14.3	Nitric oxide, NO	23.6
Oxygen, O_2	25.6	Nitrous oxide, N_2O	36.4
Nitrogen, N_2	31.2	Ammonia, NH_3	25.8
Air	29.9	Water, H_2O	18.9
Carbon monoxide, CO	30.7	Hydrogen sulfide, H_2S	32.9
Carbon dioxide, CO_2	34.0	Bromine, Br_2	53.2
Carbonyl sulfide, COS	51.5	Chlorine, Cl_2	48.4
Sulfur dioxide, SO_2	44.8	Iodine, I_2	71.5

Molecular volumes at normal boiling points, V_A, for some commonly encountered compounds, are tabulated in Table 24.4. For other compounds, the atomic volumes of each element present are added together as per the molecular formulae. Table 24.5 lists the contributions for each of the constituent atoms. When certain ring structures are involved, corrections must be made to account

TABLE 24.5   ATOMIC VOLUMES FOR COMPLEX MOLECULAR VOLUMES FOR SIMPLE SUBSTANCES†

Element	Atomic volume, in cm^3/g mole	Element	Atomic volume, in cm^3/g mole
Bromine	27.0	Oxygen, except as noted below	7.4
Carbon	14.8		
Chlorine	21.6	Oxygen, in methyl esters	9.1
Hydrogen	3.7		
Iodine	37.0	Oxygen, in methyl ethers	9.9
Nitrogen, double bond	15.6		
Nitrogen, in primary amines	10.5	Oxygen, in higher ethers and other esters	11.0
Nitrogen, in secondary amines	12.0	Oxygen, in acids	12.0
		Sulfur	25.6

† G. Le Bas, *The Molecular Volumes of Liquid Chemical Compounds*, Longmans, Green & Company, Ltd., London, 1915.

for the specific ring configuration; the following corrections are recommended:

for three-membered ring, as ethylene oxide	deduct 6
for four-membered ring, as cyclobutane	deduct 8.5
for five-membered ring, as furan	deduct 11.5
for pyridine	deduct 15
for benzene ring	deduct 15
for naphthalene ring	deduct 30
for anthracene ring	deduct 47.5

Recommended values of the association parameter, Φ_B, are given below for a few common solvents.

Solvent	Φ_B
water	2.26*
methanol	1.9
ethanol	1.5
benzene, ether, heptane, and other unassociated solvents	1.0

EXAMPLE 5

Estimate the liquid diffusion coefficient of ethanol, C_2H_5OH, in a dilute solution of water at 10°C. The molecular volume of ethanol may be evaluated by using values from Table 24.5 as follows:

$$V_{C_2H_5OH} = 2V_C + 6V_H + V_O$$

$$V_{C_2H_5OH} = 2(14.8) + 6(3.7) + 7.4 = 59.2 \text{ cm}^3/\text{mol}$$

At 10°C, the viscosity of a solution containing 0.05 mole of alcohol/liter of water is 1.45 centipoises; the remaining parameters to be used are

$$T = 283 \text{ K}$$

$$\Phi_B \text{ for water} = 2.26$$

and

$$M_B \text{ for water} = 18$$

Substituting these values into equation (24-52), we obtain

$$D_{C_2H_5OH-H_2O} = \left(\frac{7.4 \times 10^{-8}(2.26 \times 18)^{1/2}}{(59.2)^{0.6}}\right)\left(\frac{283}{1.45}\right)$$

$$= 7.96 \times 10^{-6} \text{ cm}^2/\text{s} \quad (7.96 \times 10^{-10} \text{ m}^2/\text{s})$$

---

* The correction of Φ_B is recommended by R. C. Reid, J. M. Praunsnitz, and T. K. Sherwood, *The Properties of Gases and Liquids*, Third Edition, McGraw-Hill Book Company, New York, 1977, p. 578.

This value is in good agreement with the experimental value of $8.3 \times 10^{-10}\, \text{m}^2/\text{s}$ reported in Appendix J.

Hayduk and Laudie* have proposed a much simpler equation for evaluating infinite dilution diffusion coefficients of nonelectrolytes *in water*;

$$D_{AB} = 13.26 \times 10^{-5}\, \mu_B^{-1.14}\, V_A^{-0.589} \qquad (24\text{-}53)$$

where D_{AB} is the mass diffusivity of A through liquid B, in cm^2/s; μ_B is the viscosity of water, in centipoises; and V_A is the molal volume of the solute at normal boiling point, in $\text{cm}^3/\text{g}\,\text{mol}$. This relation is much simpler to use and gives similar results to the Wilke-Chang equation. If we substitute the values used in example 4 into the Hayduk and Laudie relationship, we would obtain a diffusion coefficient for ethanol in a dilute water solution of $7.85 \times 10^{-6}\, \text{cm}^2/\text{s}$; this value is essentially the same value obtained using the Wilke-Chang equation.

Scheibel† has proposed that the Wilke-Chang relation be modified to eliminate the association factor, Φ_B, yielding

$$\frac{D_{AB}\mu_B}{T} = \frac{K}{V_A^{1/3}} \qquad (24\text{-}54)$$

where K is determined by

$$K = (8.2 \times 10^{-8})\left[1 + \left(\frac{3V_B}{V_A}\right)^{2/3}\right]$$

except

1. For benzene as a solvent, if $V_A < 2V_B$, use $K = 18.9 \times 10^{-8}$.
2. For other organic solvents, if $V_A < 2.5V_B$, use $K = 17.5 \times 10^{-8}$.

Reid, Prausnitz, and Sherwood‡ recommend this equation for solutes diffusing into organic solvents; however, they noted that this equation might evaluate values that had errors up to 20%.

The properties of electrically conducting solutions have been studied intensively for more than 75 years. Even so, the known relations between electrical conductance and the liquid diffusion coefficient are valid only for dilute solutions of salts in water. The diffusion coefficient of a univalent salt in dilute solution is given by the Nernst equation,

$$D_{AB} = \frac{2RT}{(1/\lambda_+^0 + 1/\lambda_-^0)\mathscr{F}} \qquad (24\text{-}55)$$

where D_{AB} is the diffusion coefficient based on the molecular concentration of A, in cm^2/s; R is the gas constant, 8.316 joules/(K)(g mole); T is absolute temperature, in K; λ_+^0, λ_-^0 are the limiting (zero concentration) ionic conductances in

* W. Hayduk and H. Laudie, *A.I.Ch.E. J.*, **20**, 611, (1974).
† E. G. Scheibel, *Ind. Eng. Chem.*, **46**, 2007, (1954).
‡ R. C. Reid, J. M. Prausnitz, and T. K. Sherwood, *The Properties of Gases and Liquids*, Third Edition, McGraw-Hill Book Company, New York, 1977, Chapter 11.

(amp/cm^2) (volt/cm) (g equivalent/cm^3); and $\mathscr{F}$ is Faraday's constant, 96 500 coulombs/g equivalent. This equation has been extended to polyvalent ions by replacing the numerical constant 2 by $(1/n^+ + 1/n^-)$, where n^+ and n^- are the valences of the cation and anion, respectively.

SOLID MASS DIFFUSIVITY

In any discussion of molecular movement in the solid state the explanation of mass transfer is automatically divided into two major fields of interest, the diffusion of gases or liquids into the pores of the solid and the interdiffusion of solid constituents by atomic movement. The first class of diffusion plays a major role in catalysis and is important to the chemical engineer. Metallurgists are the chief investigators of the diffusion of atoms within solids.

Pore diffusion may occur by one or more of three mechanisms: Fick diffusion, Knudsen diffusion, and surface diffusion. If the pores are large and the gas is relatively dense, the mass transfer will be by Fick diffusion. Within the catalyst, the diffusion paths are tortuous, irregularly shaped channels; accordingly, the flux is less than it would be in a uniform pore of the same length and mean radius. The mass flux is described in terms of an "effective" diffusion coefficient by

$$\mathbf{J}_A = -cD_{A,\text{eff}} \nabla y_A \qquad (24\text{-}56)$$

The magnitude of the coefficient depends on the variables influencing the diffusing phase as temperature and pressure, and on the catalyst properties, such as fractional void space, θ, a length angle factor, L', and a shape factor, S',

$$D_{A,\text{eff}} = \frac{D_{AB}\theta}{2L'S'} = \frac{D_{AB}\theta}{\tau} \qquad (24\text{-}57)$$

where τ is the tortuosity, a factor that describes the relationship between the actual path length relative to the nominal length of the porous media. Experimental values of θ, τ, and $D_{A,\text{eff}}$ are reported by Satterfield.*

Knudsen diffusion occurs when the size of the pores is of the order of the mean free path of the diffusing molecule. A relationship which describes Knudsen diffusion and its diffusion coefficient has been developed using the kinetic theory of gases.

$$N_A = \frac{D_{K,\text{eff}}}{x_0}(c_{A_1} - c_{A_2}) = \frac{D_{K,\text{eff}}}{RTx_0}(p_{A_1} - p_{A_2}) \qquad (24\text{-}58)$$

and

$$D_{k,\text{eff}} = \frac{19\,400\,\theta^2}{\tau S'\rho}\sqrt{\frac{T}{M}} \qquad (24\text{-}59)$$

* C. S. Scatterfield, *Heterogenous Catalysis in Practice*, McGraw-Hill Book Company, New York, 1980.

where x_0 is the length of the diffusion path, ρ is the density of the catalyst particle, and M is the molecular weight of the diffusing gas.

Surface diffusion takes places when molecules which have been adsorbed are transported along the surface as a result of two-dimensional surface concentration gradients.* Normally, surface diffusion plays a minor role in diffusion within a catalyst solid unless there is a large amount of adsorption.

The introduction of a foreign atom can frequently alter important properties of a metal. The hardening of steel is an industrial process based upon the diffusion of carbon and other elements through iron. Semiconductors have been produced by the diffusion of "doping" atoms into crystals. Mechanisms have been postulated for diffusion in solids including vacancy diffusion, interstitial diffusion, interstitialcy diffusion, and diffusion by interchange of adjoining molecules.

All crystals, in thermal equilibrium at temperatures above absolute zero, have some unoccupied lattice sites or vacancies. An atom can jump from a lattice position into a neighboring vacancy as shown in Figure 24.3a. The diffusing atom

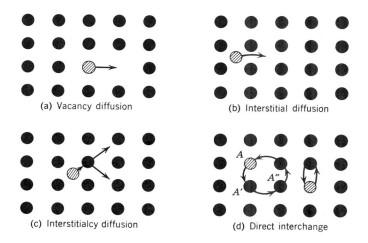

(a) Vacancy diffusion

(b) Interstitial diffusion

(c) Interstitialcy diffusion

(d) Direct interchange

Figure 24.3   Diffusion within a crystal.

continues to move through the crystal by a series of exchanges with vacancies which appear adjacent to it from time to time. This normally requires a distortion of the lattice. This mechanism has been mathematically described by assuming a unimolecular rate process and applying Eyring's "activated-state" concept, as discussed in the "hole" theory for liquid diffusion. The resulting equation is complex and will not be presented. It relates the diffusivity in terms of the

*V. G. Levich, *Physicochemical Hydrodynamics*, Prentice-Hall, Inc., Englewood Cliffs, N.J., 1962.

geometric relation between the lattice positions, the length of the jump path, the vibration frequency of the atom in the direction of the jump, and the energy of activation associated with the jump. The *vacancy diffusion* is the dominant diffusion mechanism in face-centered cubic metals and alloys; the mechanism has been used to explain diffusion in body-centered cubic material, as well as in ionic compounds and oxides.

An atom moves in *interstitial diffusion* by jumping from one interstitial site to a neighboring one as illustrated in Figure 24.3b. This normally involves a dilation or distortion of the lattice; accordingly, the mathematical treatment involving Eyring's unimolecular rate theory is also used to define the interstitial diffusivity. This diffusion coefficient is defined in terms of the same variables as in the case of vacancy diffusion.

When the diffusing interstitial atom has a size comparable to the lattice atoms, a process other than direct interstitial diffusion takes place. By pushing a neighboring lattice atom into an adjacent interstitial site, the interstitial atom can move onto a normal lattice site. This mechanism, as shown in Figure 24.3c, is called *interstitialcy diffusion.*

A fourth mechanism, which is proposed to explain self-diffusion in metals and alloys, involves the *direct interchange* of more atoms by a ring rotation. The three atoms rotate as if on a merry-go-round, and after two shifts, atom A is moved to a new position A'. It then rotates with the next two atoms, finally arriving at A''. The ring mechanism is not known to occur in any metal or alloy, but it has been suggested as a mechanism which could explain some apparent anomalies in the diffusion coefficients for metals with body-centered lattices.

Excellent references are available for a more detailed discussion on the mass transfer of atoms through crystals.*

Table J.3 in the appendix list a few values of binary diffusivities in solids. A technical journal, initially published under the title of *Diffusion Data* and currently listed under the title *Diffusion and Defect Data*, is an excellent reference for diffusivities in solids.

24.3 CONVECTIVE MASS TRANSFER

Mass transfer due to convection involves transfer between a moving fluid and a surface or between two relatively immiscible moving fluids. This mode of transfer depends both on the transport properties and on the dynamic characteristics of the flowing fluid.

* *Atom Movement*, American Society for Metals, 1951. R. M. Barrer, *Diffusion In and Through Solids*, Cambridge University Press, London, 1941. P. G. Shewmon, *Diffusion of Solids*, McGraw-Hill Book Company, New York, 1963. L. A. Girifalco, *Atomic Migration in Crystals*, Blaisdell, New York, 1964. B. L. Sharma, *Diffusion in Semiconductors*, Trans. Tech. Publications, D-3392, Clausthal-Zellerfeld, Germany, 1970. S. Mrowec, *Defects and Diffusion in Solids, An Introduction*, Elsevier Scientific Publishing Co., New York, 1980.

As in the case of convective heat transfer, a distinction must be made between two types of flow. When an external pump or other similar device causes the fluid motion, the process is called *forced convection*. If the fluid motion is due to a density difference, which may have resulted from a concentration or a temperature difference, the process is called *free* or *natural convection*.

The rate equation for convective mass transfer, generalized in a manner analogous to Newton's "law" of cooling, equation (15-11) is

$$N_A = k_c \, \Delta c_A \tag{24-60}$$

where N_A is the molar mass transfer species A measured relative to fixed spatial coordinates; Δc_A is the concentration difference between the boundary surface concentration and the average concentration of the fluid stream of the diffusing species A; and k_c is the convective mass-transfer coefficient.

As in the case of molecular mass transfer, convective mass transfer occurs in the direction of a decreasing concentration. Equation (24-60) defines the coefficient k_c in terms of the mass flux and the concentration difference from the beginning to the end of the diffusion path. The coefficient, therefore, includes the characteristics of the laminar and turbulent flow regions of the fluid in whatever proportions may be involved. Chapters 28 and 29 will consider the methods of determining this coefficient. It is, in general, a function of system geometry, fluid and flow properties, and the concentration difference Δc_A.

From our experiences in dealing with a fluid flowing past a surface, we can recall that there is always a layer, sometimes extremely thin, close to the surface where the flow is laminar, and that fluid particles next to the solid boundary are at rest. Since this is always true, the mechanism of mass transfer between a surface and a fluid must involve molecular mass transfer through the stagnant and laminar flowing fluid layers. The controlling resistance to convective mass transfer is often the result of this "film" of fluid and the coefficient, k_c, is accordingly referred to as the *film coefficient*.

It is important for the student to recognize the close similarity between the convective mass-transfer coefficient and the convective heat-transfer coefficient. This immediately suggests that the techniques developed for evaluating the convective heat-transfer coefficient may be reapplied for convective mass transfer. A complete discussion of convective mass-transfer coefficients and their evaluation will be given in Chapter 28.

24.4 CLOSURE

In this chapter the two modes of mass transport, molecular and convective mass transfer, have been introduced. Since diffusion of mass involves a multicomponent mixture, fundamental relations were presented for concentrations and velocities of the individual species as well as for the mixture. The molecular transport property, D_{AB}, the diffusion coefficient or mass diffusivity in gas, liquid, and solid systems, has been discussed and correlating equations presented.

The rate equations for the mass transfer of species A in a binary mixture are as follows:

molecular mass transfer:

$$\mathbf{J}_A = -cD_{AB}\nabla y_A$$ molar flux relative to the molar average velocity

$$\mathbf{j}_A = -\rho D_{AB}\nabla \omega_A$$ mass flux relative to the mass average velocity

$$\mathbf{N}_A = -cD_{AB}\nabla y_A + y_A(\mathbf{N}_A + \mathbf{N}_B)$$ molar flux relative to fixed spatial coordinates

$$\mathbf{n}_A = -\rho D_{AB}\nabla \omega_A + \omega_A(\mathbf{n}_A + \mathbf{n}_B)$$ mass flux relative to fixed spatial coordinates

convective mass transfer:

$$N_A = k_c \Delta c_A$$

PROBLEMS

24.1 A mixture of noble gases, containing helium, argon, and xenon, is at a total pressure of 1.7×10^5 Pa and 300 K.
(a) If the mixture has equal mole fractions of each of the gases, determine (1) the composition in terms of mass fractions; (2) the molar concentration; and (3) the mass density.
(b) If the mxiture has equal mass fractions of each of the gases, determine (1) the corresponding mole fractions; and (2) the average molecular weight of the gas mixture.

24.2 Liquified natural gas, LNG, is to be shipped from the Kenai Peninsula in Alaska by an ocean carrier to a location at Newport, Oregon. The molar composition of the commercial LNG is

methane, CH_4,	93.5 mole %
ethane, C_2H_6,	4.6%
propane, C_3H_8,	1.2%
carbon dioxide, CO_2,	0.7%

determine
(a) weight fraction of ethane;
(b) average molecular weight of the LNG mixture;
(c) density of the gas mixture when heated to 207 K and 1.4×10^5 Pa;

(d) partial pressure of methane when the total pressure on the system is 1.4×10^5 Pa;

(e) mass fraction of carbon dioxide in parts per million.

24.3 Air is contained in a 1500-ft^3 container at 250°F and 1.5 atmosphere pressure. Determine the following properties of the dry gas mixture:
(a) mole fraction of oxygen;
(b) volume fraction of oxygen;
(c) weight of the mixture;
(d) mass density of nitrogen;
(e) mass density of oxygen;
(f) mass density of the mixture;
(g) molar density of the mixture;
(h) average molecular weight of the mixture;
(i) partial pressure of oxygen.

24.4 Starting with Fick's equation for the diffusion of A through a binary mixture of A and B

$$N_A = -cD_{AB}\nabla y_A + y_A(N_A + N_B)$$

derive the following relations, stating the assumptions made in the derivations:
(a) $n_A = -D_{AB}\nabla\rho_A + w_A(n_A + n_B)$;
(b) $J_A = -D_{AB}\nabla c_A$;
(c) $j_A = -\rho D_{AB}\nabla w_A$ (It is important to realize that j_A does not equal $J_A M_A$; why?)

24.5 Starting with the Fick equation for the diffusion of A through a binary gas mixture of A and B, prove
(a) $N_A + N_B = cV$;
(b) $n_A + n_B = \rho v$;
(c) $j_A + j_B = 0$.

24.6 A gas mixture at a total pressure of 1.5×10^5 Pa and 295 K contains 20% H_2, 40% O_2, and 40% H_2O by volume. The absolute velocities of each species are -10 m/s, -2 m/s, and 12 m/s, respectively, all in the direction of the z-axis.
(a) Determine the mass average velocity, v, and the molar average velocity, V, for the mixture.
(b) Determine the four fluxes: $j_{O_2,z}$, $n_{O_2,z}$, $J_{O_2,z}$, $N_{O_2,z}$.

24.7 Air, stored in a 1000-ft^3 container at 150°F and 1.0 atm pressure, is saturated with water vapor. Determine the following properties of the gas mixture:
(a) mole fraction of oxygen;

(b) weight of the gas mixture;
(c) mass density of nitrogen;
(d) mass density of the mixture;
(e) molar density of the mixture;
(f) average molecular weight of the mixture;
(g) partial pressure of oxygen.

24.8 Prove that the value of D_{AB} is constant for the eight forms of the Fick rate equation given in Table 24.2. This is the same as showing the equivalency of each form of the Fick rate equation.

24.9 For binary mixture of A and B, using only the definitions of concentrations, velocities, and fluxes, show that

(a)
$$x_A = \frac{w_A/M_A}{w_A/M_A + w_B/M_B}$$

(b)
$$dx_A = \frac{dw_A}{M_A M_B (w_A/M_A + w_B/M_B)^2}$$

(c)
$$dw_A = \frac{M_A M_B\, dx_A}{(x_A M_A + x_B M_B)^2}$$

24.10 Stefan and Maxwell explained the diffusion of A through B in terms of the driving force, dc_A, the resistances that must be overcome in the molecular mass transfer, and a proportionality constant, β. The following equation expresses mathematically the resistances for an isothermal, isobaric binary gaseous system:

$$-dc_A = \beta \frac{\rho_A}{M_A} \frac{\rho_B}{M_B} (v_{A,z} - v_{B,z})\, dz$$

Wilke* extended this theory to a multicomponent gas mixture. The appropriate form of the Maxwell-type equation was assumed to be

$$\frac{-dc_A}{dz} = \beta_{AB} \frac{\rho_A}{M_A} \frac{\rho_B}{M_B} (v_{A,z} - v_{B,z}) + \beta_{AC} \frac{\rho_A}{M_A} \frac{\rho_C}{M_C} (v_{A,z} - v_{C,z})$$

$$+ \beta_{AD} \frac{\rho_A}{M_A} \frac{\rho_D}{M_D} (v_{A,z} - v_{D,z}) + \ldots$$

Using this relation, verify equation (24-49).

* C. Wilke, *Chem. Eng. Prog.*, **46**, 95 (1950).

24.11 A gas mixture flowing through a conduit has the following molar composition:

CO	3%
CO_2	7%
O_2	11%
N_2	79%

A pitot tube, connected to a manometer filled with water, is used to measure the average velocity of the stream. If the velocities of the individual components are 1000 ft/min for CO, 600 ft/min for CO_2, 900 ft/min for O_2, and 1000 ft/min for N_2, what is the manometer reading in inches? The gas mixture is at 70°F and 1 atm pressure.

24.12 Determine the value of the following gas diffusivities:
(a) carbon dioxide/air at 310 K and 1.5×10^5 Pa;
(b) ethanol/air at 325 K and 2.0×10^5 Pa;
(c) carbon monoxide/air at 298 K and 1.0×10^5 Pa;
(d) carbon tetrachloride/air at 298 K and 1.013×10^5 Pa;
(e) sulfur dioxide/air at 300 K and 1.5×10^5 Pa.

24.13 An absorption tower has been proposed to remove selectively ammonia from an exhaust gas stream. Estimate the diffusivity of ammonia in air at 1.013×10^5 Pa and 373 K using the Brokaw equation. The dipole moment for ammonia is 1.46 debye. Compare the evaluated value with experimental value reported in appendix Table J.1.

24.14 Determine the diffusivity of carbon dioxide through a gas mixture having the following composition:

O_2	7%
CO	10%
CO_2	15%
N_2	68%

The gas mixture will be at 273 K and 1.5×10^5 Pa.

24.15 Determine the diffusivity of methane in air using (a) the Hirschfelder equation, and (b) the Wilke equation for a gas mixture. The air is at 100°C and 1.5 atmosphere pressure.

24.16 Moth balls are frequently made of naphthalene. Upon sublimation, the naphthalene diffuses into the surrounding air. Estimate the diffusivity of naphthalene in air at 1 atm and 70°F using the Hirschfelder, Bird, and Spotz equation. Correct the value reported for this system in Appendix

J to the specified 70°F and 1 atm and compare with the previous evaluated value. The critical properties of naphthalene are

$$V_c = 3.1847 \text{ ml/g}$$

$$T_c = 469°C$$

$$P_c = 29,792 \text{ torr}$$

24.17 An absorption tower has been proposed to remove selectively two pollutants, hydrogen sulfide and sulfur dioxide, from an exhaust gas stream containing

H_2S	2%
SO_2	4%
air	94%

Estimate the diffusivity of
(a) hydrogen sulfide in the gas mixture
(b) sulfur dioxide in the gas mixture
if the gas mixture is at 373 K and 1.013×10^5 Pa.

24.18 Evaluate the diffusion coefficient of hydrogen chloride in air at 298 K and 1.013×10^5 Pa. Hydrogen chloride has a dipole moment of 1.03 debye.

24.19 Determine the diffusivity of ammonia in water at 15°C using
(a) the Wilke-Chang equation;
(b) the Hayduk and Laudie equation.
Compare the results with the experimental value reported in Appendix J.2.

24.20 Water supplies are often treated with chlorine as one of the processing steps in treating wastewater. Estimate the liquid diffusion coefficient of chlorine in an infinitely dilute solution of water at 289 K using
(a) the Wilke–Chang equation;
(b) the Hayduk and Laudie equation.
Compare the results with the experimental value reported in Appendix J.2.

24.21 The liquid diffusivity of carbon tetrachloride in methanol was reported by Reid and Sherwood* to be 1.7×10^{-9} m²/s at 288 K. Use this experimental value to evaluate the association parameter, Φ_B, of methanol. The viscosity of methanol at 288 K is 0.62 centipoises.

24.22 Estimate the diffusivity
(a) of water in *n*-butanol at 15°C;
(b) of *n*-butanol in water at 15°C; compare this value with experimental value reported in Appendix J.2.

* R. C. Reid and T. K. Sherwood, *The Properties of Gases and Liquids*, McGraw-Hill Book Company, New York, 1958, Chapter 8.

24.23 Estimate the liquid diffusivity of the following solutes that are transferred through dilute solutions:
(a) oxygen in ethanol at 20°C;
(b) oxygen in water at 20°C;
(c) carbon tetrachloride in benzene at 25°C;
(d) methanol in water at 15°C;
(e) water in methanol at 15°C.
When possible, compare the estimated value with the experimental values reported in Appendix Table J.2.

24.24 Larson* measured the diffusivity of chloroform in air at 25°C and 1 atm and reported its value as 0.093 cm^2/s. Evaluate the diffusion coefficient by the Hirschfelder equation and compare it with the experimental value.

24.25 Evaluate the diffusion coefficient of hydrogen sulfide in sulfur dioxide at 298 K and 1.5×10^5 Pa using the Brokaw equation for binary gas mixtures containing polar compounds. The dipole moments for hydrogen sulfide is 1.10 debye and for sulfur dioxide is 1.6 debye.

24.26 Use the Wilke equation for evaluating the diffusion coefficient in a gas mixture to determine the diffusivity of nitrogen through a gas mixture; the mixture is at 100 °C and 1.5 atm pressure and has the following composition:

O_2	6 mole%
CO	11 mole%
CO_2	16 mole%
N_2	67 mole%

* E. M. Larson, MS thesis, Oregon State University, 1964.

25
DIFFERENTIAL EQUATIONS
OF MASS TRANSFER

In Chapter 9 the general differential equations for momentum transfer were derived by use of a differential control volume concept. By an analogous treatment, the general differential equations for heat transfer were generated in Chapter 16. Once again, we shall use this approach to develop the differential equations for mass transfer. By making a mass balance over a differential control volume, we shall establish the equation of continuity for a given species.

Additional differential equations will be obtained when we insert, into the continuity equation, mass flux relationships developed in the previous chapter.

25.1 THE DIFFERENTIAL EQUATION
FOR MASS TRANSFER

Consider the control volume, $\Delta x \, \Delta y \, \Delta z$, through which a mixture including component A is flowing, as shown in Figure 25.1. The control volume expression for the conservation of mass is

$$\iint_{c.s.} \rho(\mathbf{v} \cdot \mathbf{n}) \, dA + \frac{\partial}{\partial t} \iiint_{c.v.} \rho \, dV = 0 \qquad (4\text{-}1)$$

which may be stated in words as

$$\left\{ \begin{array}{c} \text{Net rate of mass} \\ \text{efflux from} \\ \text{control volume} \end{array} \right\} + \left\{ \begin{array}{c} \text{net rate of accumulation} \\ \text{of mass within control} \\ \text{volume} \end{array} \right\} = 0$$

If we consider the conservation of a given species A, this relation should also include a term which accounts for the production or disappearance of A by chemical reaction within the volume. The general relation for a mass balance of

508

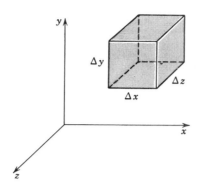

Figure 25.1 A differential control volume.

species A for our control volume may be stated as

$$\left\{\begin{array}{l}\text{Net rate of mass}\\\text{efflux of } A \text{ from}\\\text{control volume}\end{array}\right\} + \left\{\begin{array}{l}\text{net rate of accum-}\\\text{ulation of } A \text{ within}\\\text{control volume}\end{array}\right\} - \left\{\begin{array}{l}\text{rate of chemical}\\\text{production of } A\\\text{within the control}\\\text{volume}\end{array}\right\} = 0 \tag{25-1}$$

The individual terms will be evaluated for constituent A, and a discussion of their meanings will be given below.

The net rate of mass efflux from the control volume may be evaluated by considering the mass transferred across control surfaces. For example, the mass of A transferred across the area $\Delta y\,\Delta z$ at x will be $\rho_A v_{A,x}\,\Delta y\,\Delta z|_x$, or in terms of the flux vector, $\mathbf{n}_A = \rho_A \mathbf{v}_A$, it would be $n_{A,x}\,\Delta y\,\Delta z|_x$. The net rate of mass efflux of constituent A will be

$$\text{in the } x \text{ direction: } n_{A,x}\,\Delta y\,\Delta z|_{x+\Delta x} - n_{A,x}\,\Delta y\,\Delta z|_x$$

$$\text{in the } y \text{ direction: } n_{A,y}\,\Delta x\,\Delta z|_{y+\Delta y} - n_{A,y}\,\Delta x\,\Delta z|_y$$

and

$$\text{in the } z \text{ direction: } n_{A,z}\,\Delta x\,\Delta y|_{z+\Delta z} - n_{A,z}\,\Delta x\,\Delta y|_z$$

The rate of accumulation of A in the control volume is

$$\frac{\partial \rho_A}{\partial t}\,\Delta x\,\Delta y\,\Delta z$$

If A is produced within the control volume by a chemical reaction at a rate r_A, where r_A has the units (mass of A produced)/(volume)(time), the rate of production of A is

$$r_A\,\Delta x\,\Delta y\,\Delta z$$

This production term is analogous to the energy generation term which appeared in the differential equation for energy transfer, as discussed in Chapter 16.

Substituting each term in equation (25-1), we obtain

$$n_{A,x} \, \Delta y \, \Delta z|_{x+\Delta x} - n_{A,x} \, \Delta y \, \Delta z|_x + n_{A,y} \, \Delta x \, \Delta z|_{y+\Delta y}$$

$$- n_{A,y} \, \Delta x \, \Delta z|_y + n_{A,z} \, \Delta x \, \Delta y|_{z+\Delta z} - n_{A,z} \, \Delta x \, \Delta y|_z$$

$$+ \frac{\partial \rho_A}{\partial t} \Delta x \, \Delta y \, \Delta z - r_A \, \Delta x \, \Delta y \, \Delta z = 0 \qquad (25\text{-}2)$$

Dividing through by the volume, $\Delta x \, \Delta y \, \Delta z$, and cancelling terms, we have

$$\frac{n_{A,x}|_{x+\Delta x} - n_{A,x}|_x}{\Delta x} + \frac{n_{A,y}|_{y+\Delta y} - n_{A,y}|_y}{\Delta y} + \frac{n_{A,z}|_{z+\Delta z} - n_{A,z}|_z}{\Delta z} + \frac{\partial \rho_A}{\partial t} - r_A = 0 \qquad (25\text{-}3)$$

Evaluated in the limit as Δx, Δy, and Δz approach zero, this yields

$$\frac{\partial}{\partial x} n_{A,x} + \frac{\partial}{\partial y} n_{A,y} + \frac{\partial}{\partial z} n_{A,z} + \frac{\partial \rho_A}{\partial t} - r_A = 0 \qquad (25\text{-}4)$$

Equation (25-4) is the *equation of continuity for component A*. Since $n_{A,x}$, $n_{A,y}$, and $n_{A,z}$ are the rectangular components of the mass flux vector, $\mathbf{n}_A$, equation (25-4) may be written

$$\nabla \cdot \mathbf{n}_A + \frac{\partial \rho_A}{\partial t} - r_A = 0 \qquad (25\text{-}5)$$

A similar equation of continuity may be developed for a second constituent B in the same manner. The differential equations are

$$\frac{\partial}{\partial x} n_{B,x} + \frac{\partial}{\partial y} n_{B,y} + \frac{\partial}{\partial z} n_{B,z} + \frac{\partial \rho_B}{\partial t} - r_B = 0 \qquad (25\text{-}6)$$

and

$$\nabla \cdot \mathbf{n}_B + \frac{\partial \rho_B}{\partial t} - r_B = 0 \qquad (25\text{-}7)$$

where r_B is the rate at which B will be produced within the control volume by a chemical reaction. Adding equations (25-5) and (25-7), we obtain

$$\nabla \cdot (\mathbf{n}_A + \mathbf{n}_B) + \frac{\partial(\rho_A + \rho_B)}{\partial t} - (r_A + r_B) = 0 \qquad (25\text{-}8)$$

For a binary mixture of A and B, we have

$$\mathbf{n}_A + \mathbf{n}_B = \rho_A \mathbf{v}_A + \rho_B \mathbf{v}_B = \rho \mathbf{v}$$

$$\rho_A + \rho_B = \rho$$

and

$$r_A = -r_B$$

by the law of conservation of mass. Substituting these relations into (25-8), we obtain

$$\nabla \cdot \rho \mathbf{v} + \frac{\partial \rho}{\partial t} = 0 \qquad (25\text{-}9)$$

This is the *equation of continuity for the mixture*. Equation (25-9) is identical to the equation of continuity (9-2) for a homogeneous fluid.

The equation of continuity for the mixture and for a given species can be written in terms of the substantial derivative. As shown in Chapter 9, the continuity equation for the mixture can be rearranged and written

$$\frac{D\rho}{Dt} + \rho \nabla \cdot \mathbf{v} = 0 \qquad (9\text{-}5)$$

Through similar mathematical manipulations, the equation of continuity for species A in terms of the substantial derivative may be derived. This equation is

$$\frac{\rho D \omega_A}{Dt} + \nabla \cdot \mathbf{j}_A - r_A = 0 \qquad (25\text{-}10)$$

We could follow the same development in terms of molar units. If R_A represents the rate of molar production of A per unit volume and R_B represents the rate of molar production of B per unit volume, the molar-equivalent equations are

for component A,

$$\nabla \cdot \mathbf{N}_A + \frac{\partial c_A}{\partial t} - R_A = 0 \qquad (25\text{-}11)$$

for component B,

$$\nabla \cdot \mathbf{N}_B + \frac{\partial c_B}{\partial t} - R_B = 0 \qquad (25\text{-}12)$$

and for the mixture,

$$\nabla \cdot (\mathbf{N}_A + \mathbf{N}_B) + \frac{\partial (c_A + c_B)}{\partial t} - (R_A + R_B) = 0 \qquad (25\text{-}13)$$

For the binary mixture of A and B, we have

$$\mathbf{N}_A + \mathbf{N}_B = c_A \mathbf{v}_A + c_B \mathbf{v}_B = c \mathbf{V}$$

and

$$c_A + c_B = c$$

However, only when the stoichiometry of the reaction is

$$A \rightleftharpoons B$$

which stipulates that one molecule of B is produced for each mole of A disappearing, can we stipulate that $R_A = -R_B$. In general, the equation of continuity for the mixture in molar units is

$$\mathbf{\nabla} \cdot c\mathbf{V} + \frac{\partial c}{\partial t} - (R_A + R_B) = 0 \tag{25-14}$$

25.2 SPECIAL FORMS OF THE DIFFERENTIAL MASS-TRANSFER EQUATION

Special forms of the equation of continuity applicable to commonly encountered situations follow. In order to use the equations for evaluating the concentration profiles, we replace the fluxes, $\mathbf{n}_A$ and $\mathbf{N}_A$, by the appropriate expressions developed in Chapter 24. These expressions are

$$\mathbf{N}_A = -cD_{AB}\mathbf{\nabla}y_A + y_A(\mathbf{N}_A + \mathbf{N}_B) \tag{24-21}$$

or its equivalent,

$$\mathbf{N}_A = -cD_{AB}\mathbf{\nabla}y_A + c_A\mathbf{V}$$

and

$$\mathbf{n}_A = -\rho D_{AB}\mathbf{\nabla}\omega_A + \omega_A(\mathbf{n}_A + \mathbf{n}_B) \tag{24-22}$$

or its equivalent,

$$\mathbf{n}_A = -\rho D_{AB}\mathbf{\nabla}\omega_A + \rho_A\mathbf{v}$$

Substituting equation (24-22) into equation (25-5), we obtain

$$-\mathbf{\nabla} \cdot \rho D_{AB}\mathbf{\nabla}\omega_A + \mathbf{\nabla} \cdot \rho_A\mathbf{v} + \frac{\partial \rho_A}{\partial t} - r_A = 0 \tag{25-15}$$

and substituting equation (24-21) into equation (25-11), we obtain

$$-\mathbf{\nabla} \cdot cD_{AB}\mathbf{\nabla}y_A + \mathbf{\nabla} \cdot c_A\mathbf{V} + \frac{\partial c_A}{\partial t} - R_A = 0 \tag{25-16}$$

Either equation (25-15) or equation (25-16) may be used to describe concentration profiles within a diffusing system. Both equations are completely general; however, they are relatively unwieldy. These equations can be simplified by making restrictive assumptions. Important forms of the equation of continuity, with their qualifying assumptions, include:

(i) If the density, ρ, and the diffusion coefficient, D_{AB}, can be assumed constant, equation (25-15) becomes

$$-D_{AB}\nabla^2\rho_A + \rho_A \overset{0}{\cancel{\mathbf{\nabla} \cdot \mathbf{v}}} + \mathbf{v} \cdot \mathbf{\nabla}\rho_A + \frac{\partial \rho_A}{\partial t} - r_A = 0$$

Dividing each term by the molecular weight of A and rearranging, we obtain

$$\mathbf{v} \cdot \nabla c_A + \frac{\partial c_A}{\partial t} = D_{AB} \nabla^2 c_A + R_A \qquad (25\text{-}17)$$

(ii) If there is no production term, $R_A = 0$, and if the density and diffusion coefficient are assumed constant, equation (25-17) reduces to

$$\frac{\partial c_A}{\partial t} + \mathbf{v} \cdot \nabla c_A = D_{AB} \nabla^2 c_A \qquad (25\text{-}18)$$

We recognize that $(\partial c_A/\partial t) + \mathbf{v} \cdot \nabla c_A$ is the substantial derivative of c_A; rewriting the left-hand side of equation (25-18), we obtain

$$\frac{Dc_A}{Dt} = D_{AB} \nabla^2 c_A \qquad (25\text{-}19)$$

which is analogous to equation (16-14) from heat transfer,

$$\frac{DT}{Dt} = \frac{k}{\rho c_P} \nabla^2 T \qquad (16\text{-}14)$$

or

$$\frac{DT}{Dt} = \alpha \nabla^2 T$$

where α is the thermal diffusivity. The similarity between these two equations is the basis for the analogies drawn between heat and mass transfer.

(iii) In a situation in which there is no fluid motion, $\mathbf{v} = 0$, no production term, $R_A = 0$, and no variation in the diffusivity or density, equation (25-18) reduces to

$$\frac{\partial c_A}{\partial t} = D_{AB} \nabla^2 c_A \qquad (25\text{-}20)$$

Equation (25-20) is commonly referred to as *Fick's second "law" of diffusion*. The assumption of no fluid motion restricts its applicability to diffusion in solids, or stationary liquids, and for binary systems of gases or liquids, where $\mathbf{N}_A$ is equal in magnitude, but acting in the opposite direction to $\mathbf{N}_B$; that is, the case of equimolar counterdiffusion. Equation (25-20) is analogous to Fourier's second "law" of heat conduction,

$$\frac{\partial T}{\partial t} = \alpha \nabla^2 T \qquad (16\text{-}18)$$

(iv) Equations (25-17), (25-18), and (25-20) may be simplified further when the process to be defined is a steady-state process; that is, $\partial c_A/\partial t = 0$. For constant density and a constant diffusion coefficient the equation becomes

$$\mathbf{v} \cdot \nabla c_A = D_{AB} \nabla^2 c_A + R_A \qquad (25\text{-}21)$$

For constant density, constant diffusivity and no chemical production, $R_A = 0$, we obtain

$$\mathbf{v} \cdot \nabla c_A = D_{AB} \nabla^2 c_A \tag{25-22}$$

If additionally, $\mathbf{v} = 0$, the equation reduces to

$$\nabla^2 c_A = 0 \tag{25-23}$$

Equation (25-23) is the *Laplace equation* in terms of molar concentration.

Each of the equations (25-15) through (25-23) has been written in vector form, thus each applies to any orthogonal coordinate system. By writing the Laplacian operator, ∇^2, in the appropriate form, the transformation of the equation to the desired coordinate system is accomplished. Fick's second "law" of diffusion written in rectangular coordinates is

$$\frac{\partial c_A}{\partial t} = D_{AB} \left[\frac{\partial^2 c_A}{\partial x^2} + \frac{\partial^2 c_A}{\partial y^2} + \frac{\partial^2 c_A}{\partial z^2} \right] \tag{25-24}$$

in cylindrical coordinates is

$$\frac{\partial c_A}{\partial t} = D_{AB} \left[\frac{\partial^2 c_A}{\partial r^2} + \frac{1}{r} \frac{\partial c_A}{\partial r} + \frac{1}{r^2} \frac{\partial^2 c_A}{\partial \theta^2} + \frac{\partial^2 c_A}{\partial z^2} \right] \tag{25-25}$$

and in spherical coordinates is

$$\frac{\partial c_A}{\partial t} = D_{AB} \left[\frac{1}{r^2} \frac{\partial}{\partial r} \left(r^2 \frac{\partial c_A}{\partial r} \right) + \frac{1}{r^2 \sin \theta} \frac{\partial}{\partial \theta} \left(\sin \theta \frac{\partial c_A}{\partial \theta} \right) + \frac{1}{r^2 \sin \theta} \frac{\partial^2 c_A}{\partial \phi^2} \right] \tag{25-26}$$

The general differential equation for mass transfer of component A, or the equation of continuity of A, written in rectangular coordinates is

$$\frac{\partial c_A}{\partial t} + \left[\frac{\partial N_{A,x}}{\partial x} + \frac{\partial N_{A,y}}{\partial y} + \frac{\partial N_{A,z}}{\partial z} \right] = R_A \tag{25-27}$$

in cylindrical coordinates is

$$\frac{\partial c_A}{\partial t} + \left[\frac{1}{r} \frac{\partial}{\partial r} (r N_{A,r}) + \frac{1}{r} \frac{\partial N_{A,\theta}}{\partial \theta} + \frac{\partial N_{A,z}}{\partial z} \right] = R_A \tag{25-28}$$

and in spherical coordinates is

$$\frac{\partial c_A}{\partial t} + \left[\frac{1}{r^2} \frac{\partial}{\partial r} (r^2 N_{A,r}) + \frac{1}{r \sin \theta} \frac{\partial}{\partial \theta} (N_{A,\theta} \sin \theta) + \frac{1}{r \sin \theta} \frac{\partial N_{A,\phi}}{\partial \phi} \right] = R_A \tag{25-29}$$

25.3 COMMONLY ENCOUNTERED BOUNDARY CONDITIONS

A mass transfer process may be described by solving one of the differential equations of mass transfer, using the appropriate initial or boundary conditions, or both, to determine the constants of integration. The initial and boundary conditions used in mass transfer are very similar to those used in section 16.3 for energy transfer. You may wish to refer to that section for discussion of initial and boundary conditions.

The initial condition in mass transfer processes is the concentration of the diffusing species at the start of the time interval of interest expressed in either molar or mass concentration units. The concentration may be simply equal to a constant,

$$\text{at } t = 0, \qquad c_A = c_{A_0} \text{ in molar units}$$

$$\text{at } t = 0, \qquad \rho_A = \rho_{A_0} \text{ in mass units}$$

or be more complex if the concentration distribution is a function of space variables at the start of the time measurement.

The boundary conditions most generally encountered include:

1. The concentration of a surface may be specified. This concentration may be in terms of molar concentration, $c_A = c_{A_1}$; of mole fractions, $y_A = y_{A_1}$ in gases or $x_A = x_{A_1}$ in liquids and solids; of mass concentration, $\rho_A = \rho_{A_1}$; or of mass fraction, $\omega_A = \omega_{A_1}$. When the system is a gas, the concentration is related to the partial pressure by Dalton's law; accordingly, the concentration may be in terms of partial pressure, $p_A = p_{A_1} = y_{A_1}P$. For the specific case of diffusion from a liquid into a gas phase, the boundary condition at the liquid surface is defined for an ideal liquid solution by Raoult's law to be $p_{A_1} = x_A P_A$, where x_A is the mole fraction in the liquid, and P_A is the vapor pressure of species A evaluated at the temperature of the liquid.

2. The mass flux at a surface may be specified; for example, $j_A = j_{A_1}$ or $N_A = N_{A_1}$. Cases of engineering interest include where the flux is zero due to an impermeable surface and where the mass flux is specified as, for example,

$$j_{A,z \text{ surface}} = -\rho D_{AB} \frac{d\omega_A}{dz}\Big|_{z=0}$$

3. The rate of a chemical reaction may be specified; for example, if component A disappears at the boundary by a first-order chemical reaction, we may write $N_{A_1} = k_1 c_{A_1}$, where k_1 is the rate constant for a first-order reaction. When the diffusing species disappears at the boundary by an instantaneous reaction, the concentration of that species is often assumed to be zero.

4. When a fluid is flowing over the phase for which the differential equation of mass is written, the species may be lost from the phase of interest by convective mass transfer. The boundary mass flux is then defined by

$$N_{A_1} = k_c(c_{A_1} - c_{A\infty}) \tag{25-30}$$

where $c_{A\infty}$ is the concentration in the fluid stream, c_{A_1} is the concentration in the fluid adjacent to the surface, and k_c is the convection mass-transfer coefficient defined in section 24.3.

We will apply these initial and boundary conditions in solving molecular-diffusion problems in Chapters 26 and 27. Examples on reducing the general differential equation for mass transfer to only the relevant terms follow.

EXAMPLE 1

In a cylindrical nuclear fuel rod which contains fissionable material, the rate of production of neutrons is proportional to the neutron concentration. Use one of the general differential equations for mass transfer to write the differential equation which describes the mass transfer process. List any obvious boundary conditions.

For component A, equation (25-11) stipulates

$$\nabla \cdot \mathbf{N}_A + \frac{\partial c_A}{\partial t} - R_A = 0$$

Since the rate of production is proportional to the neutron concentration, $R_A = kc_A$. For diffusion in solids, where the bulk motion contribution is zero,

$$\mathbf{N}_A = -D_{AB}\nabla c_A$$

Upon substituting these relations into equation (25-11), we obtain

$$\frac{\partial c_A}{\partial t} = D_{AB}\left[\frac{\partial^2 c_A}{\partial r^2} + \frac{1}{r}\frac{\partial c_A}{\partial r} + \frac{1}{r^2}\frac{\partial^2 c_A}{\partial \theta^2} + \frac{\partial^2 c_A}{\partial z^2}\right] + kc_A$$

If the cylinder is relatively long compared to the radius, $\partial^2 c_A/\partial z^2 = 0$ and if the concentration does not vary with the angle θ, $\partial^2 c_A/\partial \theta^2 = 0$; the equation reduces to

$$\frac{\partial c_A}{\partial t} = D_{AB}\left[\frac{\partial^2 c_A}{\partial r^2} + \frac{1}{r}\frac{\partial c_A}{\partial r}\right] + kc_A$$

The only obvious boundary condition is

$$\left.\frac{\partial c_A}{\partial r}\right|_{r=0} = 0$$

which requires the concentration of the diffusing species to be finite at the center of the rod.

EXAMPLE 2

In a hot combustion chamber, oxygen diffuses through an air film to a carbon surface where it reacts according to the following equation:

$$3C + 2O_2 \rightarrow 2CO + CO_2$$

(a) With the assumption that the carbon surface is flat, reduce the general differential equation for mass transfer to write the specific differential equation that describes this steady-state process.

For oxygen, equation (25-11) stipulates

$$\nabla \cdot \mathbf{N}_{O_2} + \frac{\partial c_{O_2}}{\partial t} - R_{O_2} = 0$$

Since this is a steady-state process, $\partial c_{O_2}/\partial t = 0$. The reaction occurs at one of the boundaries and not uniformly along the diffusion path; accordingly, $R_{O_2} = 0$. With diffusion in only the z direction, $\nabla \cdot \mathbf{N}_{O_2}$ reduces to $dN_{O_2,z}/dz$ and the differential equation becomes

$$\frac{dN_{O_2,z}}{dz} = 0$$

(b) Write Fick's law in terms of only oxygen.

Fick's law for molecular diffusion stipulates

$$N_{O_2,z} = -cD_{O_2\text{-mixture}}\frac{dy_{O_2}}{dz} + y_{O_2}(N_{O_2,z} + N_{CO,z} + N_{CO_2,z} + N_{N_2,z})$$

According to the reaction, two moles of oxygen enter the film and two moles of carbon monoxide leave; accordingly, $N_{O_2,z} = -N_{CO,z}$. The reaction also predicts that as the two moles of oxygen enter, one mole of carbon dioxide leaves; this stipulates $\frac{1}{2}N_{O_2,z} = -N_{CO_2,z}$. The net flux of nitrogen is zero. With these stipulations, the bulk contribution term becomes

$$y_{O_2}(N_{O_2,z} + N_{CO,z} + N_{CO_2,z} + N_{N_2,z}) = y_{O_2}(N_{O_2,z} - N_{O_2,z} - \tfrac{1}{2}N_{O_2,z} + 0)$$

$$= -\frac{y_{O_2}}{2}N_{O_2,z}$$

and Fick's law reduces to

$$N_{O_2,z} = -cD_{O_2\text{-mixture}}\frac{dy_{O_2}}{dz} - \frac{y_{O_2}}{2}N_{O_2,z}$$

or

$$N_{O_2,z} = -\frac{cD_{O_2\text{-mixture}}}{1 + \dfrac{y_{O_2}}{2}}\frac{dy_{O_2}}{dz}$$

This equation can either be substituted into the differential equation that is defined in part (a) or integrated directly in order to obtain the flux of O_2 in the z direction.

25.4 CLOSURE

The general differential equation of mass transfer has been developed in this chapter for a diffusing component in a mixture. Special forms of the general equation, applying to specific situations, were presented. The boundary conditions most generally encountered in mass-transfer processes have been listed.

PROBLEMS

25.1 Derive equation (25-11) for component A in terms of molar units, starting with the control-volume expression for the conservation of mass.

25.2 Transform equation (25-27) from rectangular coordinates

$$\frac{\partial c_A}{\partial t} + \frac{\partial N_{A,x}}{\partial x} + \frac{\partial N_{A,y}}{\partial y} + \frac{\partial N_{A,z}}{\partial z} = R_A$$

into an equivalent equation written in cylindrical coordinates.

25.3 Show that equation (25-5) may be written in the form

$$\frac{\partial \rho_A}{\partial t} + (\nabla \cdot \rho_A \mathbf{v}) - D_{AB}\nabla^2 \rho_A = r_A$$

25.4 The following sketch illustrates the gas-phase diffusion in the neighborhood of a catalytic surface. Hot gases of heavy hydrocarbons diffuse to the catalytic surface where they are cracked (i.e. decomposed)

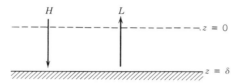

by the reaction

$$H \to 3L$$

the light products diffuse back into the gas stream.

(a) Reduce the general differential equation for mass transfer to write the specific differential equation that will describe this steady-state transfer process if the catalyst is considered a flat surface.

(b) Repeat (a) for a spherical, catalytic surface.

(c) Determine the Fick's law relationship in terms of only compound H.

25.5 An Arnold diffusion cell is a simple device that is used to measure gas diffusion coefficients. A pool of pure liquid A is maintained at the bottom of a small diameter tube. A gas, which is insoluble in the liquid, flows across the mouth of the tube carrying away the vapors of A, which diffuse through the gas space above the liquid pool. Under isothermal, isobaric conditions, the evaporation of A is a steady-state process.

Reduce the general differential equation for mass

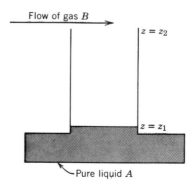

transfer to write the specific differential equation that will describe this mass-transfer process.

What would be the form of the Fick's law relationship for species A that would be substituted into the differential equation? Give two boundary conditions that might be used to solve the resulting differential equation.

25.6 A hemisphere drop of water, lying on a flat surface, evaporates by molecular diffusion through an "effective film" of air 0.5 cm thick surrounding the drop. Reduce the general differential equation for mass transfer to write the specific differential equation that will describe this mass-transfer process.

What would be the form of the Fick's law relationship written in terms of only the diffusing water vapor? Give two boundary conditions that might be used to solve the resulting differential equation.

25.7 A large deep lake, which initially had a uniform oxygen concentration of 1 kg/m^3, has its surface concentration suddenly raised and maintained at a 9 kg/m^3 concentration level.

Reduce the general differential equation for mass transfer to write the specific differential equation for

(a) the transfer of oxygen into the lake without the presence of a chemical reaction;

(b) the transfer of oxygen into the lake that occurs with the simultaneous disappearance of oxygen by a first-order biological reaction.

25.8 The moisture in the hot, humid air surrounding a cold-water pipeline continually diffuses to the cold surface where it condenses.

Reduce the general differential equation for mass transfer to write the specific differential equation that will describe this steady-state mass-transfer process. Give two boundary conditions that might be used to solve the resulting differential equation.

25.9 Use the general differential equation for mass transfer to write the specific differential equation that will describe the diffusion of a microorganism that was initially placed in a stagnant fluid or gel; the microorganism has a cell division as it diffuses that follows the first-order reaction

$$A \to 2A$$

Give two boundary conditions that could be used to solve the differential equation.

25.10 Water vapor diffuses through a 0.1-in. layer of air to a beaker of sulfuric acid where it is instantaneously absorbed. The concentration of the water vapor at the outer edge of the air layer is $0.15 \text{ lb}_m \text{ H}_2\text{O/lb}_m$ dry air. Use the general differential equation for mass transfer to write the differential equation for this steady-state, unidirectional mass-transfer process. List the two boundary conditions in terms of mole fraction of water vapor. What would be the form of Fick's law for this process, written in terms of the diffusing water vapor?

25.11 An oak lawn-mower roller, having an initial moisture content of 55 wt%, is placed in a drying kiln where its surface moisture content is maintained at 20 wt%. Reduce the general differential equation that will describe the drying of the interior of the oak roller.

25.12 The following sketch illustrates the gas-phase diffusion in the neighborhood of a catalytic surface:

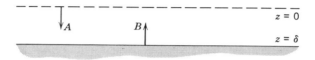

Component A diffuses through a stagnant film to the catalytic surface, where it is instantaneously converted into species B by the reaction

$$A \rightarrow B$$

When species B diffuses into the stagnant film, it begins to decompose by the first-order reaction

$$B \rightarrow A$$

The rate of formation of component A is equal to

$$R_A = k_1 y_B \frac{\text{moles } A \text{ produced}}{(\text{time})(\text{volume})}$$

where y_B is the concentration of B expressed in mole fraction.

Use the general differential equation for mass transfer to write the differential equation that describes this diffusion process. List the boundary conditions that could be used in solving the differential equation.

25.13 A liquid flows over a thin, flat sheet of a slightly soluble solid. Over the region in which diffusion is occurring, the liquid velocity may be assumed to be parallel to the plate and to be given by $v = ay$, where y is the distance from the plate and a is a constant. Show that the equation governing the mass transfer, with certain simplifying assumptions, is

$$D\left(\frac{\partial^2 c_A}{\partial x^2} + \frac{\partial^2 c_A}{\partial y^2}\right) = ay\frac{\partial c_A}{\partial x}$$

List the simplifying assumptions.

25.14 The diffusivity of isopropyl alcohol in water was measured in a Loschmidt diffusion cell. The vertical cylinder, 20 cm long, was divided into halves by a removable disk. The lower half was initially filled with water and the upper half with the alcohol. When the disk is removed, there is one-dimensional interdiffusion of the two liquids. Write the differential equation for the concentration distribution of isopropyl alcohol and state the necessary boundary conditions that would be required to solve the differential equation.

25.15 An early mass-transfer study of oxygen transport in human tissue won a Nobel prize for August Krough. By considering a tissue cylinder surrounding each blood vessel, he proposed the diffusion of oxygen away from the blood vessel into the annular tissue was accompanied by a zero-order reaction; that is, $R_A = -m$, where m is a constant. This reaction was necessary to explain the metabolic consumption of the oxygen to produce carbon dioxide.

Use the general differential for mass transfer to write the specific differential equation that will describe the diffusion of oxygen in the human tissue. What would be the form of Fick's relationship written in terms of only the diffusing oxygen?

25.16 In a hot combustion chamber, oxygen diffuses through air to a flat carbon surface where it reacts to make CO and/or CO_2.

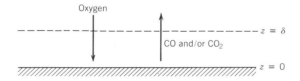

Reduce the general differential equation for mass transfer to write a specific differential equation for oxygen that will describe the steady-state mass transfer process. The concentration of oxygen at $z = \delta$ is 21 mole percent. The reaction at the surface may be assumed to occur instantaneously. No reaction occurs in the gas film.

What would be the form of Fick's law if

(a) only carbon monoxide is produced on the carbon surface;

(b) only carbon dioxide is produced on the carbon surface;

(c) the following instantaneous reaction occurs on the surface

$$4C + 3O_2 \rightarrow 2CO + 2CO_2$$

25.17 Derive the equation of continuity for component A, using an infinitely long cylindrical control volume and cylindrical coordinates.

25.18 A fluidized coal reactor has been proposed for a new power plant. If the coal can be assumed to be spherical, reduce the general differential equation for mass transfer to obtain a specific differential equation for describing the steady-state diffusion of oxygen to the surface of the coal particle.

Determine the Fick's law relationship for the flux of oxygen from the surrounding air environment if

(a) only carbon monoxide, CO, is produced at the surface of the carbon particle;

(b) only carbon dioxide, CO_2, is produced at the surface of the carbon particle.

If the reaction at the surface of the carbon particle is instantaneous, give two boundary conditions that might be used in solving the differential equation.

25.19 Use the general differential equation for mass transfer to write the specific differential equation for describing the steady-state diffusion of oxygen to the spherical particle of coal that is described in problem 25.18, when carbon monoxide is produced at the surface and as it diffuses away from the particle it reacts with oxygen to form carbon dioxide.

25.20 A charcoal briquet is formed with an initial moisture content of ρ_{A0}. The briquet is approximately spherical with a radius of R. It is placed into a forced-air dryer, which produces a lower surface moisture content of $\rho_{A,s}$. Reduce the general differential equation for mass transfer to obtain a differential equation that will describe the drying of the interior of the briquet.

26
STEADY-STATE
MOLECULAR DIFFUSION

In this chapter we shall direct our attention to describing the steady-state transfer of mass from a differential point of view. To do this, the differential equation and the boundary conditions which describe the physical situation must be established. The approach will parallel those previously used in Chapter 8 for the analysis of a differential fluid element in laminar flow, and in Chapter 17 for analyzing steady-state heat conduction. During our discussions, two types of presentation will be used: (1) defining the control volume for a specific situation, we shall generate the governing differential equation, and (2) using the general differential equations for mass transfer, we shall eliminate the terms that are irrelevant and, in so doing, obtain the governing differential equation. By using both types of approach, the student should become more familiar with the various terms in the general differential equation of mass transfer.

Reconsidering the general differential equation for mass transfer

$$\nabla \cdot \mathbf{N}_A + \frac{\partial c_A}{\partial t} - R_A = 0 \qquad (25\text{-}11)$$

we recognize three terms may be involved in mass transfer: R_A, the rate of chemical production of species A within the phase through which the mass is being transferred; $\partial c_A/\partial t$, the accumulation of A within the phase; and $\nabla \cdot \mathbf{N}_A$, the net rate of mass efflux of species A. These terms may or may not be involved in a particular mass-transfer process; for example, in the steady-state transfer of mass, the concentration at a given point does not vary with time so $\partial c_A/\partial t$ is zero. To gain confidence in treating mass-transfer processes, we will initially treat the simplest case, steady-state diffusion which is free of chemical production. We will then obtain solutions for increasingly complex mass-transfer operations.

26.1 ONE-DIMENSIONAL MASS TRANSFER, INDEPENDENT OF CHEMICAL REACTION

In this section, steady-state molecular mass transfer through simple systems in which the concentration and the mass flux are functions of a single space coordinate will be considered. Although all four fluxes, $\mathbf{N}_A$, $\mathbf{n}_A$, $\mathbf{J}_A$, and $\mathbf{j}_A$, may be used to describe mass-transfer operations, only the molar flux relative to a set of axes fixed in space, $\mathbf{N}_A$, will be used in the following discussions. In a binary system, the z component of this flux is expressed by equation (24-20),

$$N_{A,z} = -cD_{AB}\frac{dy_A}{dz} + y_A(N_{A,z} + N_{B,z})\qquad(24\text{-}20)$$

DIFFUSION THROUGH A STAGNANT GAS FILM

The diffusion coefficient or mass diffusivity for a gas may be experimentally measured in an Arnold diffusion cell. This cell is illustrated schematically in Figure 26.1. The narrow tube, which is partially filled with pure liquid A, is maintained at a constant temperature and pressure. Gas B, which flows across the open end of the tube, has a negligible solubility in liquid A and is also chemically inert to A. Component A vaporizes and diffuses into the gas phase; the rate of vaporization may be physically measured and may also be mathematically expressed in terms of the molar mass flux.

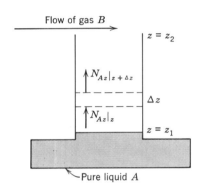

Figure 26.1 Arnold diffusion cell.

Consider the control volume $S\,\Delta z$, where S is the uniform cross-sectional area of the tube. A mass balance over this control volume for a steady-state operation yields

$$SN_{A,z}|_{z+\Delta z} - SN_{A,z}|_z = 0$$

Dividing through by the volume, $S\,\Delta z$, and evaluating in the limit as Δz approaches zero, we obtain the differential equation,

$$\frac{d}{dz}N_{A,z} = 0\qquad(26\text{-}1)$$

This relation stipulates a constant molar flux of A throughout the gas phase from z_1 to z_2.

Equation (26-1) could have been obtained from the general differential equation for mass transfer,

$$\nabla \cdot \mathbf{N}_A + \frac{\partial c_A}{\partial t} - R_A = 0\qquad(25\text{-}11)$$

or

$$\left[\frac{\partial}{\partial x}N_{A,x} + \frac{\partial}{\partial y}N_{A,y} + \frac{\partial}{\partial z}N_{A,z}\right] + \frac{\partial c_A}{\partial t} - R_A = 0$$

For a steady-state process, $\partial c_A / \partial t = 0$, and when there is no chemical production of A, we have $R_A = 0$. For diffusion in the z direction only, we are concerned with the z component of the mass flux vector, $\mathbf{N}_A$. Thus, for our physical situation, equation (25-11) reduces to equation (26-1).

A similar differential equation could also be generated for component B,

$$\frac{d}{dz} N_{B,z} = 0 \qquad (26\text{-}2)$$

and, accordingly, the molar flux of B is also constant over the entire diffusion path from z_1 to z_2. Considering only the plane at z_1 and the restriction that gas B is insoluble in liquid A, we realize $N_{B,z}$ at plane z_1 is zero, and conclude that $N_{B,z}$, the net flux of B, is zero throughout the diffusion path; accordingly, component B is a *stagnant* gas.

The constant molar flux of A was described in Chapter 24 by the equation

$$N_{A,z} = -cD_{AB}\frac{dy_A}{dz} + y_A(N_{A,z} + N_{B,z}) \qquad (24\text{-}20)$$

this equation reduces, when $N_{B,z} = 0$, to

$$N_{A,z} = -\frac{cD_{AB}}{1 - y_A}\frac{dy_A}{dz} \qquad (26\text{-}3)$$

This equation may be integrated between the two boundary conditions:

$$\text{at } z = z_1 \qquad y_A = y_{A_1}$$

and

$$\text{at } z = z_2 \qquad y_A = y_{A_2}$$

Assuming the diffusion coefficient to be independent of concentration, and realizing from equation (26-1) that $N_{A,z}$ is constant along the diffusion path, we obtain, by integrating,

$$N_{A,z} \int_{z_1}^{z_2} dz = cD_{AB} \int_{y_{A_1}}^{y_{A_2}} -\frac{dy_A}{1 - y_A} \qquad (26\text{-}4)$$

Solving for $N_{A,z}$, we obtain

$$N_{A,z} = \frac{cD_{AB}}{(z_2 - z_1)} \ln \frac{(1 - y_{A_2})}{(1 - y_{A_1})} \qquad (26\text{-}5)$$

The log-mean average concentration of component B is defined as

$$y_{B,lm} = \frac{y_{B_2} - y_{B_1}}{\ln (y_{B_2}/y_{B_1})}$$

or, in the case of a binary mixture, this equation may be expressed in terms of component A as follows:

$$y_{B,lm} = \frac{(1 - y_{A_2}) - (1 - y_{A_1})}{\ln [(1 - y_{A_2})/(1 - y_{A_1})]} = \frac{y_{A_1} - y_{A_2}}{\ln [(1 - y_{A_2})/(1 - y_{A_1})]} \qquad (26\text{-}6)$$

Inserting equation (26-6) into equation (26-5), we obtain

$$N_{A,z} = \frac{cD_{AB}}{z_2 - z_1} \frac{(y_{A_1} - y_{A_2})}{y_{B,lm}} \qquad (26\text{-}7)$$

Equation (26-7) may also be written in terms of pressures. For an ideal gas,

$$c = \frac{n}{V} = \frac{P}{RT}$$

and

$$y_A = \frac{p_A}{P}$$

The equation equivalent to equation (26-7) is

$$N_{A,z} = \frac{D_{AB}P}{RT(z_2 - z_1)} \frac{(p_{A_1} - p_{A_2})}{p_{B,lm}} \qquad (26\text{-}8)$$

Equations (26-7) and (26-8) are commonly referred to as equations for *steady-state diffusion of one gas through a second stagnant gas*. Many mass-transfer operations involve the diffusion of one gas component through another nondiffusing component; *absorption* and *humidification* are typical operations defined by these two equations.

Equation (26-8) has also been used to describe the convective mass-transfer coefficients by the "*film concept*" or *film theory*. In Figure 26.2, the flow of gas over

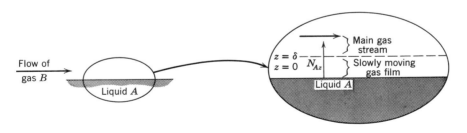

Figure 26.2 Film model for mass transfer of component A into a moving gas stream.

a liquid surface is illustrated. The "film concept" is based upon a model in which the entire resistance to diffusion from the liquid surface to the main gas stream is assumed to occur in a stagnant or laminar film of constant thickness δ. In other words, for this model, δ, is a fictitious length which represents the thickness of a fluid layer offering the same resistance to molecular diffusion as is encountered in the combined process of molecular diffusion and diffusion due to the mixing by the moving fluid. If this model is accurate, the convective mass-transfer coefficient may be expressed in terms of the gas diffusion coefficient. If $z_2 - z_1$ is set equal to δ, equation (26-8) becomes

$$N_{A,z} = \frac{D_{AB}P}{RTp_{B,lm}\delta}(p_{A_1} - p_{A_2})$$

and from equation (25-30), we have

$$N_{A,z} = k_c(c_{A_1} - c_{A_2})$$

and

$$N_{A,z} = \frac{k_c}{RT}(p_{A_1} - p_{A_2})$$

Comparison reveals that the film coefficient is expressed as

$$k_c = \frac{D_{AB}P}{p_{B,lm}\delta} \tag{26-9}$$

when the diffusing component is transported through a nondiffusing gas. Although this model is physically unrealistic, the "film concept" has had educational value in supplying a simple picture of a complicated process. The film concept has proved frequently misleading in suggesting that the convective mass-transfer coefficient is always directly proportional to the mass diffusivity. Other models for the convective coefficient will be discussed in this chapter and in Chapter 28. At that time we will find that k_c is a function of the diffusion coefficient raised to an exponent varying from 0.5 to 1.0.

Frequently, in order to complete the description of the physical operation in which mass is being transported, it is necessary to express the concentration profile. Recalling equation (26-1),

$$\frac{d}{dz}N_{A,z} = 0 \tag{26-1}$$

and equation (26-3),

$$N_{A,z} = -\frac{cD_{AB}}{1-y_A}\frac{dy_A}{dz} \tag{26-3}$$

we can obtain the differential equation which describes the variation in concentration along the diffusing path; this equation is

$$\frac{d}{dz}\left(-\frac{cD_{AB}}{1-y_A}\frac{dy_A}{dz}\right) = 0 \tag{26-10}$$

Since c and D_{AB} are constant under isothermal and isobaric conditions, the equation reduces to

$$\frac{d}{dz}\left(\frac{1}{1-y_A}\frac{dy_A}{dz}\right) = 0 \tag{26-11}$$

This second-order equation may be integrated twice with respect to z to yield

$$-\ln(1-y_A) = c_1 z + c_2 \qquad (26\text{-}12)$$

The two constants of integration are evaluated, using the boundary conditions:

$$\text{at } z = z_1 \qquad y_A = y_{A_1}$$

and

$$\text{at } z = z_2 \qquad y_A = y_{A_2}$$

Substituting the resulting constants into equation (26-12), we obtain the following expression for the concentration profile of component A

$$\left(\frac{1-y_A}{1-y_{A_1}}\right) = \left(\frac{1-y_{A_2}}{1-y_{A_1}}\right)^{(z-z_1)/(z_2-z_1)} \qquad (26\text{-}13)$$

or, since $y_A + y_B = 1$,

$$\left(\frac{y_B}{y_{B_1}}\right) = \left(\frac{y_{B_2}}{y_{B_1}}\right)^{(z-z_1)/(z_2-z_1)} \qquad (26\text{-}14)$$

Equations (26-13) and (26-14) describe logarithmic concentration profiles for both species. The average concentration of one of the species along the diffusion path may be evaluated, as an example for species B, by

$$\bar{y}_B = \frac{\displaystyle\int_{z_1}^{z_2} y_B \, dz}{\displaystyle\int_{z_1}^{z_2} dz} \qquad (26\text{-}15)$$

Upon substitution of equation (26-14) into equation (26-15), we obtain

$$\bar{y}_B = y_{B_1} \frac{\displaystyle\int_{z_1}^{z_2} \left(\frac{y_{B_2}}{y_{B_1}}\right)^{(z-z_1)/(z_2-z_1)} dz}{z_2 - z_1}$$

$$= \frac{(y_{B_2} - y_{B_1})(z_2 - z_1)}{\ln(y_{B_2}/y_{B_1})(z_2 - z_1)} = \frac{y_{B_2} - y_{B_1}}{\ln(y_{B_2}/y_{B_1})}$$

$$= y_{B,lm} \qquad (26\text{-}6)$$

The following example problem illustrates the application of the foregoing analysis to a mass-transfer situation.

EXAMPLE 1

Through the accidental opening of a valve, water has been spilled on the floor of an industrial plant in a remote, difficult-to-reach area. It is desired to estimate the time required to evaporate the water into the surrounding quiescent air. The water layer is

0.04 in. thick and may be assumed to remain at a constant temperature of 75°F. The air is also at 75°F and at 1 atmosphere pressure, with an absolute humidity of 0.002 lb of water per lb of dry air. The evaporation is assumed to take place by molecular diffusion through a gas film 0.20 in. thick.

Basis. 1 ft² of surface area

$$\text{Volume of water evaporated} = (1 \text{ ft}^2)\frac{0.04 \text{ in.}}{12 \text{ in./ft}} = 0.0033 \text{ ft}^3$$

$$\text{Weight of water evaporated} = (0.0033 \text{ ft}^3)\left(7.48\frac{\text{gal}}{\text{ft}^3}\right)\left(8.34\frac{\text{lb}_m}{\text{gal}}\right)$$

$$= 0.206 \text{ lb}_m$$

$$\text{Moles of water evaporated} = \frac{0.206 \text{ lb}_m}{18 \text{ lb}_m/\text{lb mole}} = 0.0114 \text{ lb mole}$$

The moles of water evaporated per square foot per unit time may be expressed by

$$N_{A,z} = \frac{cD_{AB}}{(z_2 - z_1)}\frac{(y_{A_1} - y_{A_2})}{y_{B,lm}} \tag{26-7}$$

The total molar concentration in the gas phase, c, can be evaluated by the ideal gas law at the stated temperature and pressure

$$PV = nRT$$

or

$$c = \frac{n}{V} = \frac{P}{RT}$$

To evaluate the gas constant, R, we shall use the standard conditions, where 1 lb mole occupies 359 ft³ at 1 atmosphere pressure and 492°R,

$$R = \frac{PV}{nT} = \frac{(1 \text{ atm})(359 \text{ ft}^3)}{(1 \text{ lb mole})(492°\text{R})} = 0.73\frac{\text{atm ft}^3}{\text{lb mole }°\text{R}}$$

For the conditions of 1 atm and 75°F or 535°R,

$$c = \frac{P}{RT} = \frac{1 \text{ atm}}{(0.73 \text{ atm ft}^3/\text{lb mole}°\text{R})(535°\text{R})} = 0.00256\frac{\text{lb mole}}{\text{ft}^3}$$

From Appendix J, the gas diffusion coefficient of water vapor in air at 298 K and 1 atmosphere is found to be 0.260 cm²/s. This value can be scaled to our temperature of 75°F or 535°R by equation (24-41),

$$D_{AB}|_{T_2} = D_{AB}|_{T_1}\left(\frac{T_2}{T_1}\right)^{3/2} \tag{24-41}$$

Since $T_1 = 298$ K or 537°R, we have

$$D_{AB \text{ at } 75°\text{F}} = 0.260\left(\frac{535}{537}\right)^{3/2} = 0.259 \text{ cm}^2/\text{s}$$

Equation (24-26) is used to convert the units,

$$D_{AB} = \left(0.259\frac{\text{cm}^2}{\text{s}}\right)\left(3.87\frac{\text{ft}^2/\text{hr}}{\text{cm}^2/\text{s}}\right) = 1.00 \text{ ft}^2/\text{hr}$$

The concentration values can be evaluated from a water–air humidity chart. At 75°F, the saturated humidity is 0.0189 lb H$_2$O/lb dry air or

$$\left(0.0189\frac{\text{lb}_\text{m} \text{ water}}{\text{lb}_\text{m} \text{ dry air}}\right)\left(\frac{\text{lb mole water}}{18 \text{ lb}_\text{m} \text{ water}}\right)\left(\frac{29 \text{ lb}_\text{m} \text{ air}}{\text{lb mole air}}\right) = 0.0304\frac{\text{lb mole water}}{\text{lb mole air}}$$

and, converted to a mole fraction basis,

$$y_{A_1} = \frac{0.0304}{1.0304} = 0.0295$$

The humidity in the air stream is

$$\left(0.002\frac{\text{lb}_\text{m} \text{ water}}{\text{lb}_\text{m} \text{ dry air}}\right)\left(\frac{\text{lb mole water}}{18 \text{ lb}_\text{m} \text{ water}}\right)\left(\frac{29 \text{ lb}_\text{m} \text{ air}}{\text{lb mole air}}\right) = 0.00322\frac{\text{lb mole water}}{\text{lb mole air}}$$

or

$$y_{A_2} = \frac{0.00322}{1.0032} = 0.0032$$

accordingly,

$$y_{A_1} - y_{A_2} = 0.0263$$

For a binary system,

$$y_{B_1} = 1 - y_{A_1} = 0.9705$$

and

$$y_{B_2} = 1 - y_{A_2} = 0.9968$$

By equation (26-6),

$$y_{B,lm} = \frac{y_{B_2} - y_{B_1}}{\ln (y_{B_2}/y_{B_1})} = \frac{0.9968 - 0.9705}{\ln (0.9968/0.9705)} = 0.983$$

The diffusion path, $z_2 - z_1$, equals 0.20 in./(12 in./ft) = 0.0167 ft. Substituting these values into equation (26-7), we obtain

$$N_{A,z} = \frac{(0.00256 \text{ lb mole/ft}^3)(1.00 \text{ ft}^2/\text{hr})}{0.0167 \text{ ft}}\left(\frac{0.0263}{0.983}\right)$$

$$= 0.00410 \text{ lb mole/ft}^2 \text{ hr}$$

Since we have 0.0114 lb mole to be evaporated per square foot, the time required for the evaporation is

$$t = \frac{0.0114 \text{ lb mole/ft}^2}{0.00410 \text{ lb mole/ft}^2 \text{ hr}} = 2.78 \text{ hr} \quad (10\,010 \text{ s})$$

PSEUDO-STEADY-STATE DIFFUSION THROUGH A STAGNANT GAS FILM

In many mass-transfer operations, one of the boundaries may move with time. If the length of the diffusion path changes a small amount over a long period

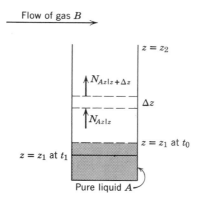

Figure 26.3 Arnold diffusion cell with moving liquid surface.

of time, a pseudo-steady-state diffusion model may be used. When this condition exists, equation (26-7) describes the mass flux in the stagnant gas film. Reconsider Figure 26.1, with a moving liquid surface as illustrated in Figure 26.3. Two surface levels are shown, one at time t_0, and the second at time t_1. If the difference in the level of liquid A over the time interval considered is only a small fraction of the total diffusion path, and $t_1 - t_0$ is a relatively long period of time, at any instant in that period the molar flux in the gas phase may be evaluated by

$$N_{A,z} = \frac{cD_{AB}(y_{A_1} - y_{A_2})}{z y_{B,lm}} \tag{26-7}$$

where $z_2 - z_1$ equals z, the length of the diffusion path at time t.

The molar flux $N_{A,z}$ is related to the amount of A leaving the liquid by

$$N_{A,z} = \frac{\rho_{A,L}}{M_A} \frac{dz}{dt} \tag{26-16}$$

where $\rho_{A,L}/M_A$ is the molar density of A in the liquid phase. Under pseudo-steady-state conditions, equations (26-7) and (26-16) may be combined to give

$$\frac{\rho_{A,L}}{M_A} \frac{dz}{dt} = \frac{cD_{AB}(y_{A_1} - y_{A_2})}{z \, y_{B,lm}} \tag{26-17}$$

Equation (26-17) may be integrated from $t = 0$ to $t = t$ and from $z = z_{t_0}$ to $z = z_t$ as follows

$$\int_{t=0}^{t} dt = \frac{\rho_{A,L} y_{B,lm}/M_A}{cD_{AB}(y_{A_1} - y_{A_2})} \int_{z_{t_0}}^{z_t} z \, dz$$

This yields

$$t = \frac{\rho_{A,L} y_{B,lm}/M_A}{cD_{AB}(y_{A_1} - y_{A_2})} \left(\frac{z_t^2 - z_{t_0}^2}{2} \right) \tag{26-18}$$

Rearranging this expression, we obtain the equation commonly used to evaluate the gas diffusion coefficient from Arnold cell experimental data. This equation is

$$D_{AB} = \frac{\rho_{A,L} y_{B,lm}/M_A}{c(y_{A_1} - y_{A_2})t}\left(\frac{z_t^2 - z_{t_0}^2}{2}\right) \tag{26-19}$$

A diffusion coefficient is evaluated from experimental data in the following example.

EXAMPLE 2

E. M. Larson,* using an Arnold cell, measured the diffusivity of chloroform in air at 25°C and one atmosphere pressure. The liquid density of chloroform at 25°C is 1.485 g/cm³, and its vapor pressure at 25°C is 200 mm Hg. At time $t = 0$, the liquid chloroform surface was 7.40 cm from the top of the tube, and after 10 hours the liquid surface had dropped 0.44 cm. If the concentration of chloroform is zero at the top of the tube, what would be the gas diffusion coefficient of chloroform in air?

At 25°C, the vapor pressure of chloroform = 200 mm Hg. By Dalton's law,

$$y_{A_1} = \frac{P_{A_1}}{P} = \frac{200 \text{ mm Hg}}{760 \text{ mm Hg}} = 0.263$$

$$y_{B_1} = 1 - 0.263 = 0.737$$

Pure B is flowing across the top of the tube; accordingly,

$$y_{B_2} = 1 \qquad y_{A_2} = 0$$

and

$$y_{A_1} - y_{A_2} = 0.263$$

By equation (26-6),

$$y_{B,lm} = \frac{y_{B_2} - y_{B_1}}{\ln(y_{B_2}/y_{B_1})} = \frac{1.0 - 0.737}{\ln(1.0/0.737)} = 0.862$$

The molecular weight of chloroform is 119.39 g/mol. The molar density of chloroform is evaluated by

$$\rho_A/M_A = (1.485 \text{ g/cm}^3)(\text{mol}/119.39 \text{ g}) = 0.0124 \text{ mol/cm}^3$$

At the stipulated temperature and pressure, the ideal gas law

$$PV = nRT$$

may be used to express the concentration as

$$c = \frac{n}{V} = \frac{P}{RT}$$

---

* E. M. Larson, Diffusion Coefficients of Chlorinated Hydrocarbons in Air, M.S. thesis in Chemical Engineering, Oregon State University, 1964.

To evaluate the gas constant, R, we shall use the standard conditions, where 1 g mole occupies 22 400 cm^3 at 1 atmosphere pressure and 273 K,

$$R = \frac{PV}{nT} = \frac{(1\ \text{atm})(22\ 400\ \text{cm}^3)}{(1\ \text{mol})(273\ \text{K})} = 82.06 \frac{\text{atm cm}^3}{\text{mol K}}$$

At the cell conditions, the molar concentration of the gas phase is

$$c = \frac{P}{RT} = \frac{1\ \text{atm}}{(82.06\ \text{atm cm}^3/\text{mol K})(298\ \text{K})} = 4.09 \times 10^{-5}\ \text{mol/cm}^3$$

The two liquid levels are

$$z_{t_0} = 7.40\ \text{cm}$$

and

$$z_t = 7.84\ \text{cm}$$

and the time is

$$(10)(3600) = 36\ 000\ \text{s}$$

Substituting these values into equations (26-19), we obtain

$$D_{AB} = \left(\frac{(0.0124\ \text{mol/cm}^3)(0.862)}{(4.09 \times 10^{-5}\ \text{mol/cm}^3)(0.263)} \right) \left(\frac{(7.84\ \text{cm})^2 - (7.40\ \text{cm})^2}{(2)(36\ 000\ \text{s})} \right)$$

$$= 0.093\ \text{cm}^2/\text{s} \quad (9.3 \times 10^{-6}\ \text{m}^2/\text{s})$$

EQUIMOLAR COUNTERDIFFUSION

A physical situation which is encountered in the *distillation* of two constituents whose molar latent heats of vaporization are essentially equal, stipulates that the flux of one gaseous component is equal to but acting in the opposite direction from the other gaseous component; that is, $N_{A,z} = -N_{B,z}$. Equation (25-11),

$$\nabla \cdot \mathbf{N}_A + \frac{\partial c_A}{\partial t} - R_A = 0 \tag{25-11}$$

for the case of steady-state mass transfer without chemical reaction, may be reduced to

$$\nabla \cdot \mathbf{N}_A = 0$$

For the transfer in the z direction, this equation reduces to

$$\frac{d}{dz} N_{A,z} = 0$$

This relation stipulates that $N_{A,z}$ is constant along the path of transfer. The molar flux, $N_{A,z}$, for a binary system at constant temperature and pressure is described by

$$N_{A,z} = -D_{AB} \frac{dc_A}{dz} + y_A (N_{A,z} + N_{B,z}) \tag{24-20}$$

The substitution of the restriction, $N_{A,z} = -N_{B,z}$, into the above equation gives an equation describing the flux of A when *equimolar counterdiffusion* conditions exist

$$N_{A,z} = -D_{AB}\frac{dc_A}{dz} \tag{26-20}$$

Equation (26-20) may be integrated, using the boundary conditions

$$\text{at } z = z_1 \qquad c_A = c_{A_1}$$

and

$$\text{at } z = z_2 \qquad c_A = c_{A_2}$$

giving

$$N_{A,z}\int_{z_1}^{z_2} dz = -D_{AB}\int_{c_{A_1}}^{c_{A_2}} dc_A$$

from which we obtain

$$N_{A,z} = \frac{D_{AB}}{(z_2 - z_1)}(c_{A_1} - c_{A_2}) \tag{26-21}$$

When the ideal gas law is obeyed, the molar concentration of A is related to the partial pressure of A by

$$c_A = \frac{n_A}{V} = \frac{p_A}{RT}$$

Substituting this expression for c_A into equation (26-21), we obtain

$$N_{A,z} = \frac{D_{AB}}{RT(z_2 - z_1)}(p_{A_1} - p_{A_2}) \tag{26-22}$$

Equations (26-21) and (26-22) are commonly referred to as the *equations for steady-state equimolar counterdiffusion.*

The concentration profile for equimolar counterdiffusion processes may be obtained by substituting equation (26-20) into the differential equation which describes transfer in the z direction

$$\frac{d}{dz}N_{A,z} = 0$$

or

$$\frac{d^2c_A}{dz^2} = 0$$

This second-order equation may be integrated twice with respect to z to yield

$$c_A = C_1 z + C_2$$

The two constants of integration are evaluated, using the boundary conditions

$$\text{at } z = z_1 \qquad c_A = c_{A_1}$$
$$\text{at } z = z_2 \qquad c_A = c_{A_2}$$

to obtain the linear concentration profile

$$\frac{c_A - c_{A_1}}{c_{A_1} - c_{A_2}} = \frac{z - z_1}{z_1 - z_2} \tag{26-23}$$

It is interesting to note that when we consider the "film concept" for mass transfer with equimolar counterdiffusion, the definition of the convective mass-transfer coefficient is different from that for diffusion in a stagnant gas film. In the case of equimolar counterdiffusion

$$k^0 = \frac{D_{AB}}{\delta} \tag{26-24}$$

The superscript on the mass transfer coefficient is used to designate that there is no net molar transfer into the film due to the equimolar counterdiffusion. Comparing equation (26-24) with equation (26-9), we realize that these two defining equations yield the same results only when the concentration of A is very small and $p_{B,lm}$ is essentially equal to P.

An example illustrating the application of equimolar counterdiffusion analysis follows.

EXAMPLE 3

A simple distillation tower consists of a very large, vertical tube supplied from below with a binary vapor of benzene and toluene. The vapors leaving the top of the tube are condensed, and part of the product is returned to flow as a thin liquid film down the inner wall of the tube. At one plane in the tube, the vapor contains 85.3 mole percent benzene and the adjacent liquid film contains 70 mole percent benzene. The temperature at this point is 86.8°C. The diffusional resistance to mass transfer between the vapor-liquid interface and the bulk conditions of the vapor stream is assumed to be equivalent to the diffusional resistance of a stagnant layer of gas 0.1 in. thick. Since the tube is large, this relatively thin layer appears as a unidirectional film, unaffected by the curvature of the tube. The molar latent heats of vaporization of benzene and toluene are essentially equal, thus $N_{\text{toluene}_z} = -N_{\text{benzene}_z}$. It is desired to calculate the rate of interchange of benzene and toluene between vapor and liquid if the tower is operated at atmospheric pressure.

Equation (26-22) is applicable to the existing physical situation:

$$N_{A,z} = \frac{D_{AB}}{RT(z_2 - z_1)}(p_{A_1} - p_{A_2}) \tag{26-22}$$

The diffusion coefficient for benzene diffusing through toluene may be evaluated by the Hirschfelder equation. At 86.8°C (359.8 K) and 1 atm (1.013×10^5 Pa) pressure, the value for the coefficient is

$$D_{\text{benzene-toluene}} = 5.06 \times 10^{-6} \text{ m}^2/\text{s}$$

The partial pressure of toluene, immediately above the liquid surface, may be evaluated by Raoult's law, $p_{\text{toluene}}|_{z_1} = x_{\text{toluene}}|_{z_1} P_{\text{toluene}}$; the vapor pressure of toluene at 359.8 K is 4.914×10^4 Pa and the mole fraction of toluene in the liquid is equal to $1 - 0.70$ or 0.30. Accordingly, we obtain

$$p_{\text{toluene}}|_{z_1} = (0.30)(4.914 \times 10^4 \text{ Pa}) = 1.474 \times 10^4 \text{ Pa}$$

By Dalton's law, the concentration in the vapor stream is

$$p_{\text{toluene}}|_{z_2} = y_{\text{toluene}}|_{z_2} P$$
$$= (1 - y_{\text{benzene}}|_{z_2}) P$$
$$= (0.147)(1.013 \times 10^5) = 1.489 \times 10^4 \text{ Pa}$$

Then, $p_{A_1} - p_{A_2} = (1.474 - 1.489) \times 10^4 = -1.50 \times 10^2$ Pa. The length of the diffusion path, $z_2 - z_1$, is 0.1 in. or 2.54×10^{-3} m. The gas constant in SI units was evaluated in example 1, Chapter 24, to be 8.314 Pa $\cdot$ m^3/mol $\cdot$ K.

The rate of toluene transfer may be calculated by substituting the above values into equation (26-22)

$$N_{\text{toluene}}|_z = \frac{D_{\text{toluene-benzene}}}{RT(z_2 - z_1)}(p_{\text{toluene}}|_1 - p_{\text{toluene}}|_2)$$

$$= \frac{(5.06 \times 10^{-6} \text{ m}^2/\text{s})(-1.50 \times 10^2 \text{ Pa})}{(8.314 \text{ Pa} \cdot \text{m}^3/\text{mol} \cdot \text{K})(359.8 \text{ K})(2.54 \times 10^{-3} \text{ m})}$$

$$= -9.99 \times 10^{-5} \text{ mol/m}^2 \cdot \text{s} \qquad (-7.36 \times 10^{-5} \text{ lb mole (hr)(ft}^2))$$

The negative sign indicates that toluene moves from z_2 to z_1 or, in terms of the distillation tower, from the gas stream to the gas-liquid interface.

26.2 ONE-DIMENSIONAL SYSTEMS ASSOCIATED WITH CHEMICAL REACTION

Many diffusional operations involve the simultaneous diffusion of a molecular species and the disappearance or appearance of the species through a chemical reaction either within or at the boundary of the phase of interest. We distinguish between the two types of chemical reactions, defining the reaction which occurs uniformly throughout a given phase as a *homogeneous reaction* and the reaction which takes place in a restricted region within or at a boundary of the phase as a *heterogeneous reaction*.

The rate of appearance of species A by a homogeneous reaction appears in the general differential equation of mass transfer as the source term, R_A,

$$\nabla \cdot \mathbf{N}_A + \frac{\partial c_A}{\partial t} - R_A = 0 \tag{25-11}$$

The rate of appearance of A by a heterogeneous reaction does not appear in the general differential equation, since the reaction does not occur within the control volume; instead it enters the analysis as a boundary condition.

In this section, we shall consider two simple cases involving both types of chemical reactions. For a treatment of more complicated problems the student is referred to the two excellent treatises by Crank* and Jost.†

SIMULTANEOUS DIFFUSION AND HETEROGENEOUS, FIRST-ORDER CHEMICAL REACTIONS: DIFFUSION WITH VARYING AREA

Many industrial processes involve the diffusion of a reactant to an interface, where a chemical reaction occurs. Since both diffusion and reaction steps are involved in the overall process, the relative rates of each step, like individual members of a track relay team, are important. When the reaction rate is relatively rapid compared to the rate of diffusion, the process is said to be *diffusion-controlled*. In contrast, the process is said to be *reaction-controlled* when the rate of mass transfer is limited by the reaction step.

In many power plants, pulverized coal particles are fluidized by air into a hot combustion chamber, where oxygen from the air reacts with the coal to produce carbon monoxide or carbon dioxide. This process, which is used to produce energy via the heat of combustion, is an example of a simultaneous diffusion and heterogeneous reaction process which is diffusion-controlled.

Let us consider the diffusion of oxygen to the surface of a spherical particle of coal, where it reacts to form carbon monoxide by the reaction

$$2C + O_2 \rightarrow 2CO$$

Under steady-state conditions, two moles of carbon monoxide will diffuse back through the gas film surrounding the coal particle for each mole of oxygen diffusing into the coal surface. The physical situation is illustrated in Figure 26.4.

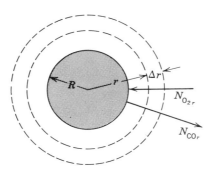

Figure 26.4 Diffusion through a spherical film.

* J. Crank, *The Mathematics of Diffusion*, Oxford Univ. Press, London, 1957.

† W. Jost, *Diffusion in Solids, Liquids and Gases*, Academic Press Inc., New York, 1952.

As an initial model, let us assume that no reaction between oxygen and carbon monoxide occurs within the gas film. A mass balance for oxygen over the spherical element yields

$$N_{O_2,r} 4\pi r^2|_r - N_{O_2,r} 4\pi r^2|_{r+\Delta r} = 0 \tag{26-25}$$

Dividing through by the volume, $4\pi r^2 \Delta r$, and evaluating in the limit as Δr approaches zero, we obtain the differential equation

$$\frac{1}{r^2} \frac{d}{dr}(r^2 N_{O_2,r}) = 0 \tag{26-26}$$

This equation specifies that $r^2 N_{O_2,r}$ is constant over the diffusion path

$$r^2 N_{O_2,r}|_{r_1} = r^2 N_{O_2,r}|_{r_2} = r^2 N_{O_2,r}|_r$$

By comparing equations (26-26) and (26-1), we observe that the relationship which remains constant over the diffusion path is entirely different for diffusion through the varying area than for diffusion through a constant area. *Mass balances must be made for each geometrical shape since each shape will have its own characteristic differential equation.*

We may also consider the general differential equation for mass transfer in spherical coordinates, equation (25-29), and eliminate the irrelevant terms in order to obtain the same equation. This differential equation states

$$\frac{\partial c_{O_2}}{\partial t} + \left[\frac{1}{r^2} \frac{\partial}{\partial r}(r^2 N_{O_2,r}) + \frac{1}{r \sin \theta} \frac{\partial}{\partial \theta}(N_{O_2,\theta} \sin \theta) + \frac{1}{r \sin \theta} \frac{\partial N_{O_2,\phi}}{\partial \phi} \right] = R_{O_2} \tag{25-29}$$

For steady-state conditions, $\partial c_{O_2}/\partial t = 0$, and when there is no net production of O_2 by a chemical reaction *within* the control volume, $R_{O_2} = 0$. The mass flux is in the r direction only; accordingly, $N_{O_2,\theta}$ and $N_{O_2,\phi}$ are zero. For these specified conditions, equation (25-29) reduces to the previously established differential equation

$$\frac{1}{r^2} \frac{d}{dr}(r^2 N_{O_2,r}) = 0 \tag{26-26}$$

or

$$\frac{d}{dr}(r^2 N_{O_2,r}) = 0$$

A similar mass balance for carbon monoxide on the same spherical element yields

$$\frac{d}{dr}(r^2 N_{CO,r}) = 0 \tag{26-27}$$

The stoichiometry of the chemical reaction states that two moles of carbon monoxide will be produced and diffuse back through the gas film for each mole of oxygen diffusing to the surface. Accordingly, the flux of carbon monoxide is related to the flux of oxygen by

$$-N_{CO,r} = 2N_{O_2,r} \tag{26-28}$$

The negative sign is required to indicate the diffusion is in the opposite direction. In the binary system, the flux of oxygen for the r direction is evaluated, using Fick's equation

$$N_{O_2,r} = -cD_{O_2-air}\frac{dy_{O_2}}{dr} + y_{O_2}(N_{O_2,r} + N_{CO,r} + N_{N_2,r})$$

or for our specific situation, since N_2 is nondiffusing,

$$N_{O_2,r} = -cD_{O_2-air}\frac{dy_{O_2}}{dr} + y_{O_2}(N_{O_2,r} - 2N_{O_2,r})$$

This equation simplifies to

$$N_{O_2,r} = \frac{-cD_{O_2-air}}{1 + y_{O_2}}\frac{dy_{O_2}}{dr} \tag{26-29}$$

By substituting equation (26-29) into equation (26-26), the mass flux and the concentration may be determined with the boundary conditions

$$\text{at } r = R \qquad y_{O_2} = y_{O_2}|_R$$

and

$$\text{at } r = \infty \qquad y_{O_2} = 0.21$$

The total mass transfer is easily calculated by defining the rate of mass transfer, W_{O_2}, in terms of the flux at a given radius

$$W_{O_2} = 4\pi r^2 N_{O_2,r}|_R \tag{26-30}$$

From equation (26-26), we recognize that W_{O_2} is a constant along the diffusion path. Substituting equation (26-29) into equation (26-30), we obtain

$$W_{O_2} = 4\pi r^2 N_{O_2,r}|_R = 4\pi r^2 N_{O_2,r}|_r$$

$$W_{O_2} = -4\pi r^2 c\frac{D_{O_2-air}}{1 + y_{O_2}}\frac{dy_{O_2}}{dr}$$

or

$$W_{O_2}\frac{dr}{r^2} = -4\pi cD_{O_2-air}\frac{dy_{O_2}}{1 + y_{O_2}} \tag{26-31}$$

This equation may be integrated over the defined diffusion path to give

$$W_{O_2} = +4\pi cD_{O_2-air}R \ln\left(\frac{1 + y_{O_2}|_R}{1 + 0.21}\right) \tag{26-32}$$

Knowing the rate at which oxygen is transferred to the coal surface, we can determine the rate of combustion of coal and, in turn, the rate of energy released from the combustion reaction.

Other models may be proposed for this process. For example, if carbon dioxide were produced at the surface by the reaction

$$C + O_2 \rightarrow CO_2$$

the flux of carbon dioxide would be equal but in the opposite direction to the flux of oxygen, $N_{O_2,r} = -N_{CO_2,r}$. Substituting this relation into Fick's equation,

$$N_{O_2,r} = -cD_{O_2-air}\frac{dy_{O_2}}{dr} + y_{O_2}(N_{O_2,r} + N_{CO_2,r} + N_{N_2,r})$$

we obtain a new expression when $N_{N_2,r}$ is zero,

$$N_{O_2,r} = -cD_{O_2-air}\frac{dy_{O_2}}{dr} \tag{26-33}$$

The total mass transfer of oxygen, W_{O_2}, is determined in a manner analogous to that used to obtain equation (26-32). This is a suggested exercise for the student. Another model could be proposed in which the diffusing carbon monoxide is oxidized to carbon dioxide. In this case, a homogeneous reaction occurs in the gas phase; consequently, an R_A term must be included in the differential equation describing the process.

It is important to note that in heterogeneous reactions, information on the rate of the chemical reaction can provide an important boundary condition,

$$\text{at } r = R \qquad N_{A,r}|_R = -k_s c_A|_R \tag{26-34}$$

where k_s is the reaction rate constant related to the surface and the negative sign indicates that species A is disappearing on the surface. If the chemical reaction is considered to be instantaneous relative to the simultaneous diffusion step, the concentration of the diffusing component at the reaction surface, $c_A|_R$ is assumed to be zero; for example, equation (26-32) becomes

$$W_{O_2} = 4c\pi D_{O_2-air} R \ln\left(\frac{1}{1+0.21}\right) \tag{26-35}$$

This is a negative quantity since oxygen is being transferred in a negative r direction. If the reaction is not instantaneous at the surface, the concentration at the surface may be obtained from equation (26-34)

$$y_A|_R = \frac{-N_A|_R}{k_s c}$$

Substituting this value into equation (26-32), we obtain a transcendental equation for $N_{A,r}$

$$W_{O_2} = 4\pi R^2 N_{O_2,r}|_R = 4\pi c D_{O_2-air} R \ln\left[\frac{1 - \dfrac{N_{O_2,r}|_R}{ck_s}}{1+0.21}\right]$$

When k_s is large, the logarithm of $1 - (N_{O_2,r}|_R/ck_s)$ may be expanded in a Taylor series and the resulting equation may be simplified to give

$$W_{O_2} = \frac{4\pi c D_{O_2-air} R}{1 + \dfrac{D_{O_2-air}}{\cdot k_s R}} \ln\left[\frac{1}{1+0.21}\right] \tag{26-36}$$

This equation accounts for the combined surface reaction and diffusion process.

In the following example, we will consider a pseudo-steady state combustion of a pulverized coal particle.

EXAMPLE 4

A fluidized coal reactor has been proposed for a new power plant. If operated at 1145 K, the process will be limited by the diffusion of oxygen counterflow to the carbon monoxide, CO, formed at the particle surface. Assume that the coal is pure carbon with a density of 1.28×10^3 kg/m^3 and that the particle is spherical with an initial diameter of 1.50×10^{-4} m. Air (21% O_2 and 79% N_2) exists several diameters away from the sphere. Under the conditions of the combustion process, the diffusivity of oxygen in the gas mixture may be evaluated for the temperature as 1.3×10^{-4} m^2/s. If a steady-state process is assumed, calculate the time necessary to reduce the diameter of the carbon to 5.0×10^{-5} m.

Equation (26-35) describes the instantaneous mass transfer of oxygen to the surface of the coal particle

$$W_{O_2} = 4\pi c D_{O_2\text{-air}} R \ln\left(\frac{1}{1.21}\right) \qquad (26\text{-}35)$$

The surface reaction, $2C + O_2 \rightarrow 2CO$, stipulates that 2 atoms of carbon will disappear per each mole of oxygen reaching the surface; accordingly,

$$W_C = -2W_{O_2,r} = 2[4\pi c D_{O_2\text{-air}} R \ln (1.21)]$$

$$= 8\pi c D_{O_2\text{-air}} R \ln (1.21)$$

A total carbon balance, {input} = {output} + {accumulation}, written in terms of moles of carbon/time, stipulates

$$0 = 8\pi c D_{O_2\text{-air}} R \ln (1.21) + \frac{\rho_C}{M_C}\frac{dV}{dt}$$

or

$$0 = 8\pi c D_{O_2\text{-air}} R \ln (1.21) + \frac{\rho_C}{M_C} 4\pi R^2 \frac{dR}{dt}$$

This simplifies to

$$dt = \frac{-\rho_C R\, dR}{2M_C c D_{O_2\text{-air}} \ln (1.21)} \qquad (26\text{-}37)$$

This equation can be integrated between the limits

$$\text{at } t = 0 \qquad R = R_{\text{initial}}$$

$$\text{at } t = t_{\text{final}} \qquad R = R_{\text{final}}$$

to give

$$t_{\text{final}} = \frac{\rho_C(R_{\text{initial}}^2 - R_{\text{final}}^2)}{4M_C c D_{O_2\text{-air}} \ln (1.21)}$$

The gas molar composition, c, can be replaced by P/RT from the ideal gas law; upon substitution of the known values, we find

$$(1.28 \times 10^6 \text{ g/m}^3)(8.314 \text{ Pa} \cdot \text{m}^3/\text{mol} \cdot \text{K})(1145 \text{ K})$$

$$t_{\text{final}} = \frac{\times [(0.75 \times 10^{-4} \text{ m})^2 - (2.5 \times 10^{-5} \text{ m})^2]}{4(12 \text{ g/mol})(1.013 \times 10^5 \text{ Pa})(1.3 \times 10^{-4} \text{ m}^2/\text{s})[\ln(1.21)]}$$

$$= 0.50 \text{ s}$$

DIFFUSION WITH A HOMOGENEOUS, FIRST-ORDER CHEMICAL REACTION

In the unit operation of absorption, one of the constituents of a gas mixture is preferentially dissolved in a contacting liquid. Depending upon the chemical nature of the involved molecules, the absorption may or may not involve chemical reactions. When there is a production or disappearance of the diffusing component, equation (25-11) may be used to analyze the mass transfer within the liquid phase. The following analysis illustrates mass transfer which is accompanied by a homogeneous chemical reaction.

Consider a layer of the absorbing medium as illustrated in Figure 26.5. At the liquid surface the composition of A is c_{A_0}. The thickness of the film, δ, is defined so

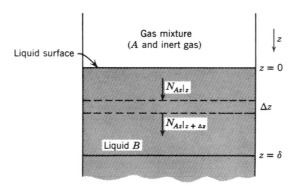

Figure 26.5 Absorption with homogeneous chemical reaction.

that beyond this film the concentration of A is always zero; that is, $c_{A_\delta} = 0$. If there is very little fluid motion within the film, and if the concentration of A in the film is assumed small, the molar flux within the film is described by

$$N_{A,z} = -D_{AB}\frac{dc_A}{dz} \tag{26-38}$$

For one-directional steady-state mass transfer, the general differential equation of mass transfer reduces to

$$\frac{d}{dz}N_{A,z} - R_A = 0 \tag{26-39}$$

The disappearance of component A by a first-order reaction is defined by

$$-R_A = k_1 c_A \tag{26-40}$$

where k_1 is the chemical reaction rate constant. Substitution of equations (26-38) and (26-40) into equation (26-39) gives a second-order differential equation which describes simultaneous mass transfer accompanied by a first-order chemical reaction

$$-\frac{d}{dz}\left(D_{AB}\frac{dc_A}{dz}\right) + k_1 c_A = 0 \tag{26-41}$$

or with a constant diffusion coefficient, this reduces to

$$-D_{AB}\frac{d^2 c_A}{dz^2} + k_1 c_A = 0 \tag{26-42}$$

The general solution to equation (26-42) is

$$c_A = c_1 \cosh\sqrt{k_1/D_{AB}}\,z + c_2 \sinh\sqrt{k_1/D_{AB}}\,z \tag{26-43}$$

The boundary conditions,

$$\text{at } z = 0 \qquad c_A = c_{A_0}$$

and

$$\text{at } z = \delta \qquad c_A = 0$$

permit the evaluation of the two constants of integration. The constant c_1 is equal to c_{A_0}, and c_2 is equal to $-(c_{A_0})/(\tanh\sqrt{k_1/D_{AB}}\,\delta)$, where δ is the thickness of the liquid film. Substituting these constants into equation (26-43), we obtain an equation for the concentration profile,

$$c_A = c_{A_0} \cosh\sqrt{k_1/D_{AB}}\,z - \frac{c_{A_0}\sinh\sqrt{k_1/D_{AB}}\,z}{\tanh\sqrt{k_1/D_{AB}}\,\delta} \tag{26-44}$$

The molar mass flux at the liquid surface can be determined by differentiating equation (26-44) and evaluating the derivative, $(dc_A/dz)|_{z=0}$. The derivative of c_A with respect to z is

$$\frac{dc_A}{dz} = +c_{A_0}\sqrt{k_1/D_{AB}}\sinh\sqrt{k_1/D_{AB}}\,z - \frac{c_{A_0}\sqrt{k_1/D_{AB}}\cosh\sqrt{k_1/D_{AB}}\,z}{\tanh\sqrt{k_1/D_{AB}}\,\delta}$$

which, when z equals zero, becomes

$$\left.\frac{dc_A}{dz}\right|_{z=0} = 0 - \frac{c_{A_0}\sqrt{k_1/D_{AB}}}{\tanh\sqrt{k_1/D_{AB}}\,\delta} = -\frac{c_{A_0}\sqrt{k_1/D_{AB}}}{\tanh\sqrt{k_1/D_{AB}}\,\delta} \tag{26-45}$$

Substituting equation (26-45) into equation (26-38) and multiplying by δ/δ, we obtain

$$N_{A,z}|_{z=0} = \frac{D_{AB}c_{A_0}}{\delta}\left[\frac{\sqrt{k_1/D_{AB}}\,\delta}{\tanh\sqrt{k_1/D_{AB}}\,\delta}\right] \tag{26-46}$$

It is interesting to consider the simpler mass-transfer operation involving the absorption of A into liquid B without an accompanying chemical reaction. The molar flux of A is easily determined by integrating equation (26-38) between the two boundary conditions, giving

$$N_{A,z} = \frac{D_{AB}c_{A_0}}{\delta} \tag{26-47}$$

It is apparent by comparing the two equations that the term $[(\sqrt{k_1/D_{AB}}\,\delta)/(\tanh\sqrt{k_1/D_{AB}}\,\delta)]$ shows the influence of the chemical reaction. This term is a dimensionless quantity, often called the *Hatta number*.[*]

As the rate of the chemical reaction increases, the reaction rate constant, k_1, increases and the hyperbolic tangent term, $\tanh\sqrt{k_1/D_{AB}}\,\delta$, approaches the value of 1.0. Accordingly, equation (26-46) reduces to

$$N_{A,z}|_{z=0} = \sqrt{D_{AB}k_1}\,(c_{A_0} - 0)$$

A comparison of this equation with equation (25-30)

$$N_{A,z} = k_c(c_{A_1} - c_{A_2}) \tag{25-30}$$

reveals that the film coefficient, k_c, is proportional to the diffusion coefficient raised to the $\frac{1}{2}$ power. With a relatively rapid chemical reaction, component A will disappear after penetrating only a short distance into the absorbing medium; thus a second model for convective mass transfer has been proposed, the *penetration theory model*, in which k_c is considered a function of D_{AB} raised to the $\frac{1}{2}$ power. In our earlier discussion of another model for convective mass transfer, the film theory model, the mass-transfer coefficient was a function of the diffusion coefficient raised to the first power. We shall reconsider the penetration model in section 26.4 and also in Chapter 28, when we discuss convective mass-transfer coefficients.

---

[*] S. Hatta, *Technol. Repts. Tohoku Imp. Univ.*, **10**, 119 (1932).

26.3 TWO- AND THREE-DIMENSIONAL SYSTEMS

In sections 26.1 and 26.2 we have discussed problems in which the concentration and the mass transfer were functions of a single space variable. Although many problems fall into this category, there are systems involving irregular boundaries or nonuniform concentrations along the boundary for which the one-dimensional treatment may not apply. In such cases, the concentration profile may be a function of two or even three spatial coordinates.

In this section we shall review some of the methods for analyzing molecular mass transfer in two- and three-dimensional systems. Since the transfer of heat by conduction is analogous to molecular mass transfer, we shall find the analytical, graphical, analogical, and numerical techniques described in Chapter 17 to be directly applicable.

ANALYTICAL SOLUTION

An analytical solution to any transfer problem must satisfy the general differential equation describing the transfer as well as the boundary conditions specified by the physical situation. A complete treatment of the analytical solutions for two- and three-dimensional systems requires a prior knowledge of partial differential equation and complex variable theory. Since most of this material is too advanced for an introductory course, we shall limit our discussions to a relatively simple two-dimensional example. Crank* has written an excellent treatise dealing exclusively with mathematical solutions for more complex diffusion problems.

The classical approach to an exact solution of the Laplace equation is the separation-of-variables technique. We shall illustrate this approach by applying it to a relatively simple two-dimensional physical situation. Consider a thin rectangular passage in a catalyst particle. Component A diffuses into the passage across the top surface and, upon reaching one of the other three surfaces, undergoes an instantaneous reaction to produce B; that is, the concentration of A at the three surfaces of the passage will be zero. Figure 26.6 illustrates these boundary conditions.

The passage is W units wide and L units long. The concentration at the surface,

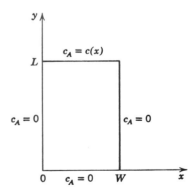

Figure 26.6 Two-dimensional model for a catalyst passage.

---

*J. Crank, *The Mathematics of Diffusion*, Oxford University Press, London, 1957.

$y = L$, is nonuniform, expressed functionally as $c(x)$. If, within the passage, there is no net bulk motion; that is, $N_A = -N_B$, the concentration distribution in the passage is defined by Laplace's equation,

$$\frac{\partial^2 c_A}{\partial x^2} + \frac{\partial^2 c_A}{\partial y^2} = 0 \tag{25-23}$$

Equation (25-23) is a linear, homogeneous partial differential equation. This type of equation can often be integrated by assuming a product solution of the form

$$c_A(x, y) = X(x) Y(y) \tag{26-48}$$

where $X(x)$ is a function of x only and $Y(y)$ is a function of y only. Upon substituting equation (26-48) into equation (25-23), we obtain an expression in which the variables are separated,

$$-\frac{1}{X}\frac{d^2 X}{dx^2} = \frac{1}{Y}\frac{d^2 Y}{dy^2} \tag{26-49}$$

The left-hand side of this equation is a function of x only, while the right-hand side is a function of y only. Since neither side can change as x and y vary, both must be equal to a constant, say λ^2. We have, therefore, two ordinary differential equations,

$$\frac{d^2 X}{dx^2} + \lambda^2 X = 0 \tag{26-50}$$

and

$$\frac{d^2 Y}{dy^2} - \lambda^2 Y = 0 \tag{26-51}$$

The general solution to equation (26-50) is

$$X = A \cos \lambda x + B \sin \lambda x \tag{26-52}$$

and the general solution to equation (26-51) is

$$Y = De^{-\lambda y} + Ee^{\lambda y} \tag{26-53}$$

According the equation (26-48), the concentration is defined in terms of the product XY: consequently,

$$c_A = (A \cos \lambda x + B \sin \lambda x)(De^{-\lambda y} + Ee^{\lambda y}) \tag{26-54}$$

where A, B, D, and E are constants to be evaluated from the four boundary conditions

$$\text{at } x = 0 \qquad c_A = 0$$
$$\text{at } x = W \qquad c_A = 0$$
$$\text{at } y = 0 \qquad c_A = 0$$

and

$$\text{at } y = L \qquad c_A = c(x)$$

The constants in equation (26-54) may be evaluated by the following substitutions: for the first condition at $x = 0$,

$$A(De^{-\lambda y} + Ee^{\lambda y}) = 0$$

for the second condition at $x = W$,

$$(A \cos \lambda W + B \sin \lambda W)(De^{-\lambda y} + Ee^{\lambda y}) = 0$$

and for the third condition,

$$(A \cos \lambda x + B \sin \lambda x)(D + E) = 0$$

The third condition can be satisfied only if $D = -E$, and the first condition only if $A = 0$. Using these results, we find that the second condition simplifies to

$$DB \sin \lambda W(e^{-\lambda y} - e^{\lambda y}) = 2DB \sin \lambda W \sinh \lambda y = 0 \qquad (26\text{-}55)$$

Neither B nor D can be zero if a solution other than the trivial solution $c_A = 0$ throughout the passage is desired. Since this expression is true for all y values, the condition specified by equation (26-55) can only be satisfied if $\sin \lambda W$ is zero; that is, $\lambda = n\pi/W$, where $n = 1, 2, 3, \ldots$. There exists a different solution for each integer n and each solution has a separate integration constant A_n. Summing these solutions, we obtain

$$c_A = \sum_{n=1}^{\infty} A_n \sin \frac{n\pi x}{W} \sinh \frac{n\pi y}{W} \qquad (26\text{-}56)$$

The last boundary condition, at $y = L$, stipulates

$$c_A = c_A(x) = \sum_{n=1}^{\infty} A_n \sin \frac{n\pi x}{W} \sinh \frac{n\pi L}{W} \qquad (26\text{-}57)$$

The constant A_n can be evaluated from equation (26-57), once the profile of $c_A(x)$ is given at the surface, $y = L$. An equation describing the variation of c_A with x and y can be obtained, after substituting the value of A_n into equation (26-56).

The separation-of-variables method can be extended to three-dimensional cases by assuming that c_A is equal to the product $X(x)Y(y)Z(z)$, and substituting this expression for c_A into the differential equation. If the variables can be separated, three second-order ordinary differential equations are obtained which may be integrated by using the given boundary conditions.

GRAPHICAL SOLUTION BY FLUX PLOTTING

An approximate solution of the Laplace equation for a two-dimensional system can be obtained graphically by plotting the potential field. This method was used in Chapter 17 to solve the Laplace equation for a heat-transfer situation by plotting the potential field for isotherms and heat-flow lines. The analogous

Laplace mass-transfer equation (25-23) can be solved by a potential field plot of mass-flow and constant-concentration lines. The steps involved in applying this technique are identical for both of these molecular transport phenomena.

The object of a graphical solution is to construct a network consisting of constant-concentration lines and lines indicating the direction of mass flow. The basic principles of the graphical method, as established for molecular heat transfer in section 17.4, will be illustrated by considering molecular mass transfer through an infinitely long, flat plate as illustrated in Figure 26.7.

When the two surface concentrations, c_{A_1}, and c_{A_2}, are constant over the faces of the plate, the mass-flow lines run perpendicular to the composition lines as shown. The basic principle of flux plotting for a system of any configuration is to draw, often by trial and error, the constant-concentration lines and the mass-flow lines so that they are perpendicular at every point of intersection. They also must satisfy the boundary conditions. After the network is established and the concentration distribution known, the rate of mass transfer can be evaluated. By proper spacing of the mass flow lines, the same amount of mass is transferred in each flow tube formed between adjacent mass flow lines. The total rate of mass transfer equals the mass flow per tube, $N_{A,x} \Delta x = D_{AB} \Delta c_A$, times the number of tubes, irrespective of the size of the squares.

In constructing a flux plot, the general procedure is to divide the body into curvilinear squares by trial and error while satisfying the boundary conditions imposed on the transport process. In terms of the mass flux plot, the general requirements are as follows:

1. Constant-concentration and mass-flow lines intersect each other at right angles while forming a network of curvilinear squares.

2. Diagonals of the curvilinear squares bisect each other at 90° and bisect any corners in the body.

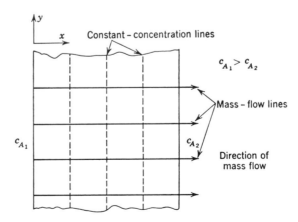

Figure 26.7 Constant-concentration lines and mass-flow lines in an infinitely long, flat plate.

3. Constant-concentration lines are parallel to constant-concentration boundaries at the boundary.

4. Mass-flow lines are perpendicular to constant-concentration boundaries at the boundary.

5. Mass-flow lines leading to a corner of a constant-concentration boundary bisect the angle between the surfaces of the boundary formed at the corner.

A graphical solution, just like an analytical solution, which satisfies the defining differential equation and the appropriate boundary conditions, is an unique solution. By satisfying all of the above requirements, a correct solution to the Laplace mass transfer equation is obtained.

ANALOGICAL SOLUTION

The recognition of similarity between any two or more transport phenomena permits the analysis of each phenomenon by analogous mathematical methods. As shown earlier in section 17.4, the two-dimensional Laplace equation may be used to describe the similar phenomena of electrical potential distribution in an electrical field,

$$\frac{\partial^2 E}{\partial x^2} + \frac{\partial^2 E}{\partial y^2} = 0$$

and of temperature distribution in a temperature field,

$$\frac{\partial^2 T}{\partial x^2} + \frac{\partial^2 T}{\partial y^2} = 0$$

Since the mass-transfer phenomenon is analogous to the heat-transfer phenomenon, as shown by the Laplace mass-transfer equation,

$$\frac{\partial^2 c_A}{\partial x^2} + \frac{\partial^2 c_A}{\partial y^2} = 0 \tag{25-23}$$

we should also expect the electrical potential, E, to be an analog of the concentration potential, c_A. In other words, the constant-voltage lines in an electric field correspond to constant-composition lines in a mass-transfer flux field, and lines of electric current flow correspond to the mass-flow lines. This permits the use of the *analog field plotter* for solving mass-transfer problems as well as for solving heat-transfer problems.

NUMERICAL SOLUTIONS

The solution techniques discussed thus far for multidimensional molecular mass transfer have considerable utility when the geometry and boundary conditions are simple enough to permit their use. Analytical solutions require relatively simple functions and geometries; flux plotting requires equi-potential boundaries. When the situation of interest becomes sufficiently complex or when the boundary conditions preclude the use of simple solution techniques, one may be forced to use a numerical solution.

Again, the recognition of the similarity between molecular mass transfer and the transfer of energy by conduction permits us to predict the resulting equations from the equations developed in section 17.4. The reader is referred to this section where the concepts of numerical problem formulation and solutions were introduced.

The partial differentials encountered in equation (25-23) can be set up as finite differences. We can write the finite difference of $\partial^2 c_A / \partial x^2$ to be

$$\frac{\partial^2 c_A}{\partial x^2} = \frac{\partial(\partial c_A / \partial x)}{\partial x} = \frac{\dfrac{(c_{A,i+1,j} - c_{A,i,j})}{\Delta x} + \dfrac{(c_{A,i-1,j} - c_{A,i,j})}{\Delta x}}{\Delta x}$$

$$= \frac{c_{A,i+1,j} - 2c_{A,i,j} + c_{A,i-1,j}}{(\Delta x)^2} \qquad (26\text{-}58)$$

The finite difference form of $\partial^2 c_A / \partial y^2$ is

$$\frac{\partial^2 c_A}{\partial y^2} = \frac{c_{A,i,j+1} - 2c_{A,i,j} + c_{A,i,j-1}}{(\Delta y)^2} \qquad (26\text{-}59)$$

Upon substitution of equations (26-58) and (26-59) into equation (25-23), we obtain

$$\frac{c_{A,i+1,j} - 2c_{A,i,j} + c_{A,i-1,j}}{(\Delta x)^2} + \frac{c_{A,i,j+1} - 2c_{A,i,j} + c_{A,i,j-1}}{(\Delta y)^2} = 0 \qquad (26\text{-}60)$$

The adjacent node indices are as shown in Figure 17.19. The grid is normally conveniently established with constant node width, Δx, equal to the constant node height, Δy. When this is true equation (26-60) simplifies to

$$c_{A,i+1,j} + c_{A,i-1,j} + c_{A,i,j+1} + c_{A,i,j-1} - 4c_{A,i,j} = 0 \qquad (26\text{-}61)$$

This final equation states that in a square grid pattern, for steady-state mass transfer in the absence of any chemical production, the concentration of the diffusing species at a given point, $c_{A,i,j}$, is equal to the arithmetic average of the concentrations of its adjacent nodes.

Equation (26-61) can be applied to a square grid drawn for any two-dimensional geometry. At the four boundaries of this grid the concentrations are known or fixed. The internal concentrations are unknown. For example, for point (2, 2) equation (26-61) becomes

$$c_{A3,2} + c_{A1,2} + c_{A2,3} + c_{A2,1} - 4c_{A2,2} = 0$$

If this is an internal point all five concentrations will be unknown, but if the point is within Δx or Δy distance of the boundary, one of the concentrations will be known. We can write N equations with N unknowns if there are N internal nodes.

This means that there will be a large number of simultaneous linear algebraic equations which must be solved to determine the concentration profile. Obviously, the use of a digital computer will help simplify the solution of the equations.

26.4 SIMULTANEOUS MOMENTUM, HEAT, AND MASS TRANSFER

In previous sections, we have considered steady-state mass transfer independent of the other transport phenomena. Many physical situations involve the simultaneous transfer of mass and either energy or momentum, and in a few cases, the simultaneous transfer of mass, energy and momentum. The drying of a wet surface by a hot, dry gas is an excellent example in which all three transport phenomena are involved. Energy is transferred to the cooler surface by convection and radiation; mass and its associated enthalpy are transferred back into the moving gas stream. The simultaneous transport processes are more complex, requiring the simultaneous treatment of each transport phenomenon involved.

In this section, we consider two examples involving the simultaneous transfer of mass and a second transport phenomenon.

SIMULTANEOUS HEAT AND MASS TRANSFER

Generally, a diffusion process is accompanied by the transport of energy, even within an isothermal system. Since each diffusing species carries its own individual enthalpy, a heat flux at a given plane as described by

$$\frac{\mathbf{q}_D}{A} = \sum_{i=1}^{n} \mathbf{N}_i \bar{H}_i \tag{26-62}$$

where $\mathbf{q}_D/A$ is the heat flux due to the diffusion of mass past the given plane, and $\bar{H}_i$ is the partial molar enthalpy of species i in the mixture. When a temperature difference exists, energy will also be transported by one of the three heat transfer mechanisms; for example, the equation for total energy transport by conduction and molecular diffusion becomes

$$\frac{\mathbf{q}}{A} = -k\nabla T + \sum_{i=1}^{n} \mathbf{N}_i \bar{H}_i \tag{26-63}$$

If the heat transfer is by convection, the first energy transport term in equation (26-63) would be replaced by the product of the convective heat transfer coefficient and a ΔT driving force.

A process important in many engineering processes as well as in day-to-day events involves the condensation of a vapor upon a cold surface. Examples of this process include the "sweating" on cold water pipes and the condensation of moist vapor on a cold window pane. Figure 26.8 illustrates the process which involves a

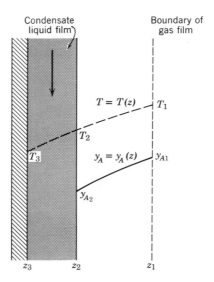

Figure 26.8 Vapor condensation on a cold surface.

film of condensed liquid flowing down a cold surface and a film of gas through which the condensate is transferred by molecular diffusion. This process involves the simultaneous transfer of heat and mass.

The following conditions will be stipulated for this particular steady-state physical situation. Pure component A will condense from a binary gas mixture. By psychrometry, the composition, y_{A_1}, and the temperature, T_1, are known at the plane z_1. The temperature of the condensing surface, T_3, is also known. By heat-transfer considerations, the convective heat-transfer coefficients for the condensate liquid film and the gas film can be calculated from equations given in Chapter 20. For example, in the gas phase, when the carrier gas is air and the vapor content of the diffusing species is relatively low the heat-transfer coefficient for natural convection can be estimated by equation (20-5),

$$\mathrm{Nu}_L = 0.68 + \frac{0.670\,\mathrm{Ra}_L^{1/4}}{[1+(0.492/\mathrm{Pr})^{9/16}]^{4/9}} \qquad (20\text{-}5)$$

Using the general differential equation for mass transfer, equation (25-11), we see that the differential equation which describes the mass transfer in the gas phase is

$$\frac{d}{dz}N_{A,z} = 0 \qquad (26\text{-}64)$$

Equation (26-64) stipulates that the mass flux in the z direction is constant over the diffusion path. To complete the description of the process, the proper form of

Fick's law must be chosen. If component A is diffusing through a stagnant gas, the flux is defined by equation (26-3)

$$N_{A,z} = \frac{-cD_{AB}}{1-y_A} \frac{dy_A}{dz} \tag{26-4}$$

Since a temperature profile exists within the film, and the diffusion coefficient and total gas concentration vary with temperature, this variation with z must often be considered. Needless to say, this complicates the problem and requires additional information before equation (26-3) can be integrated.

When the temperature profile is known or can be approximated, the variation in the diffusion coefficient can be treated. For example, if the temperature profile is of the form

$$\frac{T}{T_1} = \left(\frac{z}{z_1}\right)^n \tag{26-65}$$

the relation between the diffusion coefficient and the length parameter may be determined by using equation (24-41) as follows.

$$D_{AB} = D_{AB}|_{T_1}\left(\frac{T}{T_1}\right)^{3/2} = D_{AB}|_{T_1}\left(\frac{z}{z_1}\right)^{3n/2} \tag{26-66}$$

The variation in the total concentration due to the temperature variation can be evaluated by

$$c = \frac{P}{RT} = \frac{P}{RT_1(z/z_1)^n}$$

The flux equation now becomes

$$N_{A,z} = \frac{-PD_{AB}|_{T_1}}{RT_1(1-y_A)}\left(\frac{z}{z_1}\right)^{n/2}\frac{dy_A}{dz} \tag{26-67}$$

This is the same approach used in example 2, Chapter 15, which discussed heat transfer by conduction when the thermal conductivity was a variable.

Over a small temperature range, an average diffusion coefficient and the total molar concentration may be used. With this assumption, equation (26-3) simplifies to

$$N_{A,z} = -\frac{(cD_{AB})_{avg}}{(1-y_A)}\frac{dy_A}{dz} \tag{26-68}$$

Integrating this equation between the boundary conditions

$$\text{at } z = z_1 \qquad y_A = y_{A_1}$$

and

$$\text{at } z = z_2 \qquad y_A = y_{A_2}$$

we obtain the relation

$$N_{A,z} = \frac{(cD_{AB})_{avg}(y_{A_1} - y_{A_2})}{(z_2 - z_1)y_{B,lm}} \tag{26-69}$$

The temperature, T_2, is needed for evaluating $(cD_{AB})_{avg}$, the temperature difference between the liquid surface and the adjacent vapor, and the vapor pressure of species A at the liquid surface. This temperature may be evaluated from heat-transfer considerations. The total energy flux through the liquid surface also passes through the liquid film. This can be expressed by

$$\frac{q_z}{A} = h_{liquid}(T_2 - T_3) = h_c(T_1 - T_2) + N_{A,z}M_A(H_1 - H_2) \qquad (26\text{-}70)$$

where h_{liquid} is the convective heat transfer coefficient in the liquid film; h_c is the natural convective heat transfer coefficient in the gas film; M_A is the molecular weight of A; and H_1 and H_2 are the enthalpies of the vapor at plane 1 and the liquid at plane 2, respectively, for species A per unit mass. It is important to realize that there are two contributions to the energy flux entering the liquid surface from the gas film, convective heat transfer, and the energy carried by the condensing species.

To solve equation (26-70), a trial-and-error solution is required. If a value for the temperature of the liquid surface is assumed, T_2, h_c, and $(cD_{AB})_{avg}$ may be calculated. The equilibrium composition, y_{A_2}, can be determined from thermodynamic relations. For example, if Raoult's law holds,

$$p_{A_2} = x_A P_A$$

where x_A for a pure liquid is 1.0, and the partial pressure of A above the liquid surface is equal to the vapor pressure P_A. By Dalton's law, the mole fraction of A in the gas immediately above the liquid is

$$y_{A_2} = \frac{p_{A_2}}{P} \quad \text{or} \quad \frac{P_A}{P}$$

where P is the total pressure of the system, and P_A is the vapor pressure of A at the assumed temperature T_2. Knowing $(cD_{AB})_{avg}$ and y_{A_2}, we can evaluate N_{Az} by equation (26-69). The liquid-film heat-transfer coefficients can be evaluated, using equations presented in Chapter 20. A value is now known for each term in equation (26-70). When the left- and right-hand sides of the equation are equal, the correct temperature of the liquid surface has been assumed. If the initially assumed temperature does not yield an equality, additional values must be assumed until equation (26-70) is satisfied.

There are several industrial processes in which heat and mass transfer between a gas and liquid occur simultaneously. The concentration of sulfuric acid in a chamber sulfuric acid plant was one of the earliest of these processes. Heat supplied by hot gases produced the evaporation of water and the desorption of nitrous oxides. A more recent development involves the cooling of rockets during reentry by the sublimation of ablative material. Other processes involving simultaneous heat and mass transfer are water cooling and the humidification or dehumidification of air.

SIMULTANEOUS MOMENTUM AND MASS TRANSFER

In several mass-transfer operations, mass is exchanged between two phases. An important example which we have previously encountered is *absorption*, the selective dissolution of one of the components of a gas mixture by a liquid. A wetted-wall column, as illustrated in Figure 26.9, is commonly used to study the mechanism of this mass-transfer operation, since it provides a well-defined area of contact between the two phases. In this operation a thin liquid film flows along the wall of the column while in contact with a gas mixture. The length of contact between the two phases is relatively short during normal operation. Since only a

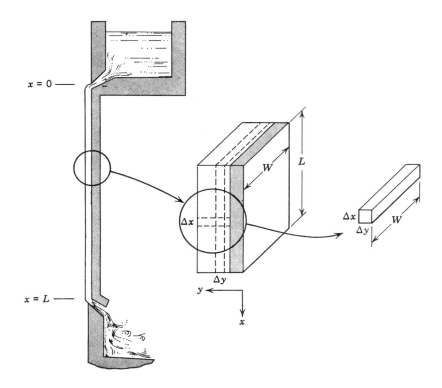

Figure 26.9 Absorption into a falling-liquid film.

small quantity of mass is absorbed, the properties of the liquid are assumed to be unaltered; the velocity of the falling film will thus be virtually unaffected by the diffusion process.

The process involves both momentum and mass transfer. In Chapter 8, the laminar flow of a fluid down an inclined plane was discussed. When the angle of inclination is 90°, the results obtained in section 8.2 can be used to describe the falling-film velocity profile. With this substitution, the differential equation for momentum transfer becomes

$$\frac{d\tau_{yx}}{dy} + \rho g = 0$$

and the boundary conditions which must be satisfied are

$$\text{at } y = 0 \qquad v_x = 0$$

and

$$\text{at } y = \delta \qquad \frac{\partial v_x}{\partial y} = 0$$

The final expression for the velocity profile is given by

$$v_x = \frac{\rho g \delta^2}{\mu}\left[\frac{y}{\delta} - \frac{1}{2}\left(\frac{y}{\delta}\right)^2\right]$$

The maximum velocity will be at the edge of the film where $y = \delta$; its value is

$$v_{max} = \frac{\rho g \delta^2}{2\mu}$$

Substituting this result into the velocity profile, we obtain another form of the expression for v_x,

$$v_x = 2v_{max}\left[\frac{y}{\delta} - \frac{1}{2}\left(\frac{y}{\delta}\right)^2\right] \tag{26-71}$$

The differential equation for mass transfer can be obtained by using the general differential equation of mass transfer and eliminating the irrelevant terms or by the making of a balance over the control volume, $\Delta x\,\Delta y W$, as shown in Figure 26.9. It is important to note that the y component of the mass flux, $N_{A,y}$, is associated with the negative y direction, according to the axes previously established in our fluid-flow considerations. The mass balance over the control volume is

$$N_{A,x}|_{x+\Delta x} W\,\Delta y - N_{A,x}|_x W\,\Delta y + N_{A,y}|_{y+\Delta y} W\,\Delta x - N_{A,y}|_y W\,\Delta x = 0$$

Dividing by $W\,\Delta x\,\Delta y$ and letting Δx and Δy approach zero, we obtain the differential equation

$$\frac{\partial N_{A,x}}{\partial x} + \frac{\partial N_{A,y}}{\partial y} = 0 \tag{26-72}$$

The one-directional molar fluxes are defined by

$$N_{A,x} = -D_{AB}\frac{\partial c_A}{\partial x} + x_A(N_{A,x} + N_{B,x}) \tag{26-73}$$

and

$$N_{A,y} = -D_{AB}\frac{\partial c_A}{\partial y} + x_A(N_{A,y} + N_{B,y}) \tag{26-74}$$

As previously mentioned, the time of contact between the vapor and liquid is relatively short; thus a negligible concentration gradient will develop in the x direction, and equation (26-73) will reduce to

$$N_{A,x} = x_A(N_{A,x} + N_{B,x}) = c_A v_x \tag{26-75}$$

The convective transport term in the negative y direction, $x_A(N_{A,y} + N_{B,y})$, involves multiplying two extremely small values and is negligible; thus equation (26-74) becomes

$$N_{A,y} = -D_{AB}\frac{\partial c_A}{\partial y} \tag{26-76}$$

Substituting equations (26-75) and (26-76) into equation (26-72), we obtain

$$\frac{\partial(c_A v_x)}{\partial x} - D_{AB}\frac{\partial^2 c_A}{\partial y^2} = 0 \tag{26-77}$$

or, since v_x is dependent upon y only,

$$v_x\frac{\partial c_A}{\partial x} - D_{AB}\frac{\partial^2 c_A}{\partial y^2} = 0 \tag{26-78}$$

The velocity profile, as defined by equation (26-71), may be substituted into equation (26-78) yielding

$$2v_{max}\left[\frac{y}{\delta} - \frac{1}{2}\left(\frac{y}{\delta}\right)^2\right]\frac{\partial c_A}{\partial x} = D_{AB}\frac{\partial^2 c_A}{\partial y^2} \tag{26-79}$$

The boundary conditions for mass transfer into the falling film are

$$\text{at } x = 0 \qquad c_A = 0$$

$$\text{at } y = 0 \qquad \frac{\partial c_A}{\partial y} = 0$$

and

$$\text{at } y = \delta \qquad c_A = c_{A_0}$$

Johnstone and Pigford[*] solved equation (26-79) and obtained, for the dimensionless concentration profile, the expression

$$\frac{c_A|_{x=L} - c_A|_{y=\delta}}{c_A|_{x=0} - c_A|_{y=\delta}} = 0.7857e^{-5.1213n} + 0.1001e^{-39.318n}$$

$$+ 0.03500e^{-105.64n}$$

$$+ 0.01811e^{-204.75n}$$

$$+ \cdots \tag{26-80}$$

[*] H. F. Johnstone and R. L. Pigford, *Trans. AIChE*, **38**, 25 (1942).

where $c_A|_{x=L}$ is the concentration of solute at the bottom of the column; $c_A|_{y=\delta}$ is the concentration of the solute at the gas-liquid interface; $c_A|_{x=0}$ is the concentration of the solute at the top of the column; n is the ratio $D_{AB}L/\delta^2 v_{max}$; L is the height of the column; δ is the film thickness; v_{max} is the maximum velocity in the film, located at the film surface; and D_{AB} is the diffusion coefficient of the solute in the liquid.

The specific case in which solute A penetrates only a short distance into the liquid film because of a slow rate of diffusion or a short time of exposure can be treated by the *penetration theory* model developed by Higbie.* As solute A is transferred into the film at $y = \delta$, the effect of the falling film on the diffusing species is such that the fluid may be considered to be flowing at the uniform velocity, v_{max}. Figure 26.10 illustrates the penetration depth. Solute A will not be affected by the presence of the wall; thus the fluid may be considered to be of infinite depth. With these simplifications, equation (26-79) reduces to

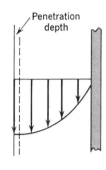

Figure 26.10 Penetration depth in a falling film.

$$v_{max}\frac{\partial c_A}{\partial x} = D_{AB}\frac{\partial^2 c_A}{\partial y^2} \qquad (26\text{-}81)$$

with the boundary conditions

$$\text{at } x = 0 \qquad c_A = 0$$
$$\text{at } y = \delta \qquad c_A = c_{A_0}$$

and

$$\text{at } y = -\infty \qquad c_A = 0$$

Equation (26-81) can be transformed into a form commonly encountered in unsteady-state mass transfer. If ξ is set equal to $\delta - y$, the transformed equation and boundary conditions become

$$v_{max}\frac{\partial c_A}{\partial x} = D_{AB}\frac{\partial^2 c_A}{\partial \xi^2} \qquad (26\text{-}82)$$

and

$$\text{at } x = 0 \qquad c_A = 0$$
$$\text{at } \xi = 0 \qquad c_A = c_{A_0}$$

and

$$\text{at } \xi = \infty \qquad c_A = 0$$

This partial differential equation can be solved by using Laplace transforms. On applying the transforms in the x direction, we obtain an ordinary differential

*R. Higbie, *Trans. AIChE*, **31**, 368–389 (1935).

equation in the s domain

$$v_{max}s\bar{c}_A - 0 = D_{AB}\frac{d^2\bar{c}_A(\xi, s)}{d\xi^2}$$

or

$$\frac{d^2\bar{c}_A}{d\xi^2} - \frac{v_{max}s\bar{c}_A}{D_{AB}} = 0 \tag{26-83}$$

This ordinary equation is readily solved to give

$$\bar{c}_A = A_1 \exp\left(\sqrt{\frac{v_{max}s}{D_{AB}}}\xi\right) + B_1 \exp\left(-\sqrt{\frac{v_{max}s}{D_{AB}}}\xi\right) \tag{26-84}$$

The constants A_1 and B_1 are evaluated, using the two transformed boundary conditions:

$$at\ \xi = 0 \qquad \bar{c}_A(0, s) = \frac{c_{A0}}{s}$$

$$at\ \xi = \infty \qquad \bar{c}_A(\infty, s) = 0$$

yielding the solution

$$c_A = \frac{c_{A0}}{s} \exp\left(-\frac{v_{max}s}{D_{AB}}\xi\right) \tag{26-85}$$

Equation (26-85) can be transformed back to the x domain by taking the inverse Laplacian, yielding

$$c_A(x, \xi) = c_{A0}\left[1 - \text{erf}\left(\frac{\xi}{4\frac{D_{AB}x}{v_{max}}}\right)\right] \tag{26-86}$$

$$c_A(x, \xi) = c_{A0}\left[1 - \text{erf}\left(\frac{\xi}{4D_{AB}t_{exp}}\right)\right]$$

where the time of exposure is defined by $t_{exp} = x/v_{max}$.

The error function, a mathematical form that is commonly encountered in transient problems, was discussed in Chapter 18. Similar to other mathematical functions, tables have been prepared of the error function and one of these tables is presented in Appendix L.

The local mass flux at the surface, where $\xi = 0$ or $y = \delta$, is obtained by differentiating equation (26-86) with respect to ξ and then inserting the derivative into equation (26-76).

$$N_{A,y}|_{\xi=0} = N_{A,y}|_{y=\delta} = -D_{AB}\frac{\partial c_A}{\partial y}\bigg|_{y=\delta}$$

The unidirectional flux becomes

$$N_{A,y}|_{y=\delta} = c_{A0}\sqrt{\frac{D_{AB}v_{max}}{\pi x}} \tag{26-85}$$

or

$$N_{A,y}\big|_{y=\delta} = c_{A_0} \sqrt{\frac{D_{AB}}{\pi t_{exp}}} \qquad (26\text{-}86)$$

As previously shown in section 26.2, when the diffusion was accompanied by a rapid chemical disappearance of the diffusing component, the *penetration model* mass flux varies as the diffusion coefficient to the $\frac{1}{2}$ power.

26.5   CLOSURE

In this chapter we have considered solutions to molecular mass-transfer problems. The defining differential equations were established through the use of a control volume expression for the conservation of mass or by using the general differential equation for mass transfer. It is hoped that this two-pronged attack will provide the student with an insight into the various terms contained in the general differential equation, and thus enable him to decide whether they are relevant or irrelevant to any specific situation.

One-directional systems both with and without chemical production were considered. Two models of mass transfer, film and penetration, were introduced. These models will be used in Chapter 28 to evaluate and explain convective mass-transfer coefficients.

Four types of solutions to mass-transfer problems in more than one direction were discussed. These methods included analytical, graphical, analogical and numerical techniques.

PROBLEMS

26.1 A steady-state Arnold cell is used to determine the diffusivity of methanol in air at 298 K and 1.013×10^5 Pa. If the results agree with the value reported in Appendix J.1, and the cell has a 0.80 cm^2 cross-sectional area and a diffusion path of 16 cm, how much methanol must be supplied to the cell to maintain a constant liquid level. At 298 K, the vapor pressure of methanol is 1.7×10^4 Pa and its specific gravity is 0.7914.

26.2 The Arnold cell, described in problem 26.1, is to be operated as a pseudo-steady-state cell. If the tube was initially filled to within 3 cm of the lip of the tube, how long would it take for the methanol level to fall to the following levels:
(a) 4 cm;
(b) 6 cm;
(c) 12 cm.

26.3 The diffusivity of ethanol through air was determined in a steady-state Arnold evaporating cell. The cell, having a cross-sectional area of 0.82 cm^2, was operated at 297 K and 1.013×10^5 Pa pressure. The length of the diffusion path was 15.0 cm.
(a) If 0.0445 cm^3 of ethanol was evaporated in 10 hr of steady-state operation, what should be the value of the diffusivity of ethanol in air?
(b) Calculate the same diffusivity, using the Hirschfelder equation, and compare the two diffusion coefficient values.

26.4 In the case of diffusion through a nondiffusing medium, the mass flux of component A is described by equation (26-8).
(a) How is the flux of component A influenced if the pressure on the system is doubled?
(b) Since a partial pressure gradient for component A must exist if it is transferred, and since the total pressure for a given system often remains constant, a partial pressure gradient for component B must also exist. If this is so, how can component B be considered a nondiffusing ($N_B = 0$) gas?

26.5 Determine the time for one mole of ethanol to diffuse through a 4-mm thick stagnant film of water if the concentrations at the leading and exiting planes are 0.1 and 0.0 kg moles/m^3, respectively. Consider a mass transfer area of 100 cm^2 and a water film temperature of 283 K. The results will help confirm the concept of penetration depths that were discussed in section 26.4

26.6 The permeability of solids by gases is experimentally determined by steady-state diffusion measurements. The diffusing solute is introduced at one side of a membrane and removed from the other side as a gas.

For a diatomic gas, A_2, which dissociates upon dissolving into a solid, Sievert's law relates the concentration of A atoms in the surface layer of the membrane, c_{A_1}, in equilibrium with the applied pressure, p_1, of the diatomic gas by the relation

$$c_{A_1} = k (p_1)^{1/2}$$

This same equation also holds at the other surface of the membrane for the off-gas pressure, p_2. Sievert's law is a variant of Henry's law for gases that dissociate upon dissolving.
(a) Prove that the rate of diffusion of a diatomic gas from a high-pressure reservoir, p_1, through a membrane of thickness z, into a low-pressure reservoir, p_2, is

$$J_{A_2} = \frac{D_{AB} k (p_1^{1/2} - p_2^{1/2})}{z}$$

where D_{A_2} is the diffusivity of A_2 through the membrane. When standard pressures are employed with a membrane of standard thickness, J_{A_2}, is called the *permeability* of A_2.

(b) A piece of laboratory equipment operating at 700°C had hydrogen gas at 8 atm that was separated from a continuously evacuated space by a 8 cm^2 nickel disk, 2 mm thick. The solubility of hydrogen in nickel at 1 atm pressure and 700°C is approximately 7.0 cm^3/100 g of nickel. The diffusivity of hydrogen through nickel at 700°C is 6×10^{-5} cm^2/s, and the density of nickel at 700°C is 9.0 g/cm^3. Calculate the number of cubic centimeters of hydrogen per hour that diffuses through the nickel.

26.7 In the oxidation of many metals, an oxide film is formed on the surface of the metal. For the oxidation to proceed, oxygen must diffuse through the oxide film to the surface of the metal. The oxide that is produced has a larger volume than the metal that is consumed; accordingly, the diffusion path increases with time. Eventually, the oxidation becomes diffusion controlled and the dissolved oxygen concentration at the oxide-metal interface becomes essentially zero. If a pseudo-steady-state condition may be assumed, develop an expression that tells the depth of the oxide film as a function of time, oxygen concentration of the free surface of the oxide film, and the diffusivity of oxygen through the oxide.

26.8 Ammonia, NH$_3$, is selectively removed from an air-NH$_3$ mixture by absorption into water. In this steady-state process, ammonia is transferred by molecular diffusion through a stagnant gas layer 2 cm thick and then through a stagnant water layer 1 cm thick. The concentration of ammonia

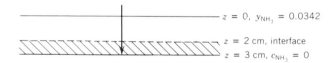

at the outer boundary of the gas layer is 3.42 mole percent and the concentration at the lower boundary of the water layer is essentially zero. The temperature of the system is 15°C and the total pressure on the system is 1 atm. The concentration at the interface between the gas and the liquid phases is given by the following equilibrium data:

Equilibrium Data for Ammonia in Air over Aqueous Solutions

p_{NH_3} (mm Hg)	5	10	15	20	25	30
c_{NH_3} (mol/cm^3)(10^6)	6.1	11.9	20.0	32.1	53.6	84.8

Determine the rate of diffusion of ammonia. At 15°C, the diffusivity of ammonia in air is 0.215 cm^2/s and in liquid water it is 1.77×10^{-5} cm^2/s.

26.9 In the treatment of wastewater, chlorine gas is bubbled through water as a disinfecting agent. Compare the time that is required for 1 mole of chlorine to diffuse through a 5-mm thick film of
(a) air at 289 K;
(b) liquid water at 289 K;
when the chlorine concentration levels are 0.04 mol/m^3 on the one edge of the film and 0.01 mol/m^3 on the other edge. The pressure on the system is 1.013×10^5 Pa.

26.10 In a hot combustion chamber, oxygen diffuses through air to the carbon surface where it reacts to make CO and/or CO_2. The mole fraction of oxygen at $z = 0$ is 0.21. The reaction at the surface may be assumed to be instantaneous. No reaction occurs in the gas film.

Determine the rate of oxygen diffusion per hour through one square meter of area if
(a) only carbon monoxide is produced at the carbon surface;
(b) only carbon dioxide is produced at the carbon surface;
(c) the following instantaneous reaction occurs at the carbon surface:

$$4C + 3O_2 \rightarrow 2CO + 2CO_2$$

26.11 An ethanol/water vapor mixture is being rectified by contact with an alcohol/water liquid solution. The alcohol is transferred from the liquid to the vapor phase and the water is transferred in the opposite direction. Both components are diffusing through a gas film 0.1 mm thck. The temperature is 368 K and the pressure is 1.013×10^5 Pa. At that temperature, the latent heat of vaporization of the alcohol and water are 1.122×10^6 J/kg and 2.244×10^6 J/kg, respectively. The mole fraction of ethanol is 0.80 on one side of the gas film and 0.20 on the other side of the film. Calculate the rate of diffusion of ethanol and of water in kg/s through one square meter of area.

26.12 The following sketch illustrates the gas-phase diffusion in the neighborhood of a catayltic surface. Component A diffuses through a stagnant film to the catalytic surface where it is instantaneously converted to B. The product B diffuses away from the catalytic surface, back through the stagnant film.

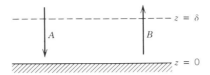

(a) If the reaction on the surface is $A \rightarrow B$, determine an equation that will predict the concentration profile; i.e., y_A at various z levels.
(b) If the reaction on the surface is $A \rightarrow 3B$, determine an equation that will predict the concentration profile.

26.13 A liquid mixture of A and B is being separated by a distillation operation. The more volatile component A is transferred from the liquid phase into the vapor phase while B is transferred in the opposite direction. Both components are diffusing through a gas film 1 in. thick. At the temperature of the system, the difference in the latent heats of vaporization for the two compounds predicts that 2 mol of A will vaporize per mole of B that is condensing. The mole fraction of A is 0.80 at one side of the film and 0.3 on the other side.

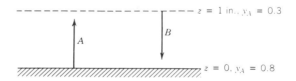

Determine an equation that will predict the concentration in mole fraction of A at various distances, z, from the liquid surface.

26.14 A tank containing water has its top open to the air. The tank is cylindrical with a diameter of 1 m. The liquid level is maintained at a level 1 m below the top of the tank as shown in sketch (a)

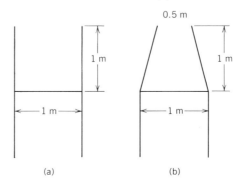

(a) How many moles of water are lost per hour if dry air at 310 K is blown across the top of the tank?

(b) It is proposed to add a tapered top to this water tank as shown in sketch (b). What will be the loss per hour in this case if dry air at 310 K is blown across the top of the tank?

26.15 Two large gas tanks, maintained at the same temperature and pressure, are connected by a circular 0.15-m diameter duct, which is 3 m in length. One tank contains a uniform mixture of 60 mole % ammonia and 40 mole % air and the other tank contains a uniform mixture of 20 mole % ammonia and 80 mole % air. The system is at 273 K and 1.013×10^5 Pa pressure. Determine the rate of ammonia transfer between the two tanks, assuming a steady-state mass transfer.

26.16 If the two tanks that are described in problem 26.15 had been connected by a truncated conical duct, having internal diameters of 0.20 and 0.1 m, respectively, at the duct's larger and smaller ends, determine the rate of ammonia transfer between the two tanks. Assume the ammonia diffuses along the duct in the direction of decreasing diameter.

26.17 An early mass-transfer study of oxygen transport in human tissue won a Nobel prize for August Krough. By considering a tissue cylinder surrounding each blood vessel, he proposed that the diffusion of oxygen away from the blood vessel into the annular tissue was accompanied by a zero-order reaction; that is, $R_A = -m$, where m is a constant. The reaction was necessary to explain the metabolic consumption of the oxygen to produce carbon dioxide. Since the tissue cylinders were believed to be arranged in a hexagonal bundle, the following boundary conditions were suggested:

at $r = R_1$, $p_A = p_{A_1}$,   the average oxygen pressure value between the arterial and venous ends of the blood vessel

at $r = R_2$, $\dfrac{dp_A}{dr} = 0$,   due to the identical oxygen fluxes from the neighboring tissue cylinders

Determine the concentration profile, p_A, as a function of r, and the flux of oxygen that enters the tissue cylinder.

26.18 A pulverized coal particle burns in air at 1145 K. The process is limited by diffusion of the oxygen counterflow to the CO_2 and CO that are formed at the particle surface. Twice as much CO is formed as CO_2. Assume that the coal is pure carbon with a density of 1280 kg/cm^3 and that the particle is spherical with an initial diameter of 0.015 cm. Air (21 mole% oxygen) exists several diameters away from the sphere.

Under the conditions of the combustion process, the diffusivity of oxygen in the gas mixture may be assumed to be 10^{-4} m^2/s. Determine the time that is necessary to reduce the diameter to 0.005 cm.

26.19 Estimate how long it will take to reduce the diameter of a hemispherical drop of water that lies on a flat surface from 0.25 to 0.05 in. if the water evaporates by molecular diffusion through an "effective film" of air, 0.25 in thick, which covers the drop. The main environment beyond the effective film can be assumed to be free of water vapor. The water in the drop will be maintained at a temperature such that the vapor pressure of the water will always be 76 mm Hg. The pressure of the system is 757 mm Hg. At this pressure and at the average temperature of the air film, the diffusion coefficient of water in air may be taken as $0.22 \text{ cm}^2/\text{s}$.

Initially, neglect the effect of the gas phase that is required to replace the evaporated liquid. How would your estimate be affected if you were to consider the bulk motion of the phase that is required to replace the evaporating liquid?

26.20 In the catalytic cracking of hydrocarbon oils, hot gases of heavy hydrocarbons diffuse to the catalytic surface where they decompose by the following reaction:

$$H \rightarrow 2P$$

The product, P, diffuses back into the gas stream.

A kinetic investigation verified that the reaction taking place on a spherical catalyst particle occurs so rapidly that one may presume the rate of diffusion in the stagnant film surrounding the particle controls the overall reaction.

(a) Develop an expression for the cracking reaction rate in terms of the gas-phase properties, the concentration of compound H in the bulk of the gas phase, the diameter of the catalyst particle, and the thickness of the stagnant film surrounding the catalyst.

(b) Evaluate the concentration profile, expressing the mole fraction of H as a function of the distance from the particle's surface.

26.21 If the flux of the heavy hydrocarbon that is described in problem 26.20 is 1.42×10^{-3} lb mol/(hr)(ft^2) at the surface of the spherical catalyst and the atmosphere surrounding the catalyst contains 30 mole% H and 70 mole% inert gas, determine the radius, R, of the catalyst in inches at that specific instant. The gaseous atmosphere is at 1000°C and 1.5 atm and the diffusivity of H in the gas mixture is $0.13 \text{ cm}^2/\text{s}$.

26.22 Carbon dioxide diffuses through a 0.1 in. stagnant air film to a NaOH solution where it instantaneously disappears by a chemical reaction. Estimate the rate of transfer across one square foot of area if the temperature and pressure of the system are 70°F and 1 atm, respectively, and if the concentration of carbon dioxide at the outer edge of the air layer is 4 mole%. Determine the concentration profile for this process.

26.23 Uranium hexafluoride, UF_6, is to be used as a fuel in an experimental nuclear power reactor. The rate of production of UF_6 is required for the reactor design. The UF_6 is to be prepared by exposing pure uranium pellets, spherical in shape, to fluorine gas at 1000 K and 1 atm pressure.

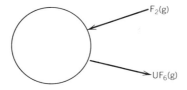

The molecular diffusion of fluorine to the pellet surface is thought to be the controlling step. If the reaction

$$U + 3F_2 \rightarrow UF_6$$

occurs irreversibly and instantaneously on the pellet surface and the diffusivity of fluorine through uranium hexafluoride is $0.273 \text{ cm}^2/\text{s}$ at 1000 K and 1 atm, determine the moles of UF_6 that is produced when the pellet's diameter is 0.4 cm.

26.24 Determine the time that is required to reduce the diameter of the uranium pellet that is described in problem 26.23 from 0.4 to 0.2 cm. The specific gravity of uranium is 19.05.

26.25 A fluidized coal combustion tower has been proposed for a new power plant. Due to the high sulfur content in most coals, sulfur dioxide will often be one of the combustion products. In an effort to help eliminate this hazardous pollutant, it has been proposed to fluidize crushed limestone simultaneously with the coal. At the combustion temperatures, the limestone will decompose to calcium oxide, which will, in turn, adsorb sulfur dioxide from its immediate surrounding environment.

Develop an expression for the sulfur dioxide adsorption rate in terms of the gas-phase properties, the concentration of the sulfur dioxide in the surrounding environment and on the surface of the calcium oxide, and the diameter of the calcium oxide spherical pellets.

26.26 The decomposition of A_n to form A occurs on a catalytic surface according to the reaction

$$A_n \rightarrow nA$$

This reaction is so rapid that diffusion through the stagnant film surrounding the catalyst controls the rate of the decomposition.

Develop an expression for the decomposition rate in terms of the fluid properties, the concentrations of A and A_n in the bulk of the fluid phase, and the thickness of the stagnant film if the catalyst is
(a) spherical radius of R;
(b) a flat plate of length L and width W;
(c) a cylindrical rod of radius R and length L.

26.27 Reconsider problem 26.26 with the reaction involving the isomerization of A to form A_n according to $nA \rightarrow A_n$.

26.28 A 20-cm long, cylindrical graphite (pure carbon) rod is inserted into an oxidizing atmosphere at 1145 K and 1.013×10^5 Pa pressure. The oxidizing process is limited by the diffusion of oxygen counterflow to the carbon monoxide that is formed on the cylindrical surface. Under the conditions of the combustion process, the diffusivity of oxygen in the gas mixture may be assumed to be $1.0 \times 10^{-5} \text{ m}^2/\text{s}$.
(a) Determine the moles of CO that are produced at the surface of the rod per second at the time when the diameter of the rod is 1.0 cm and the oxygen concentration that is 1.0-cm radial distance from the rod is 40 mole%. Assume a steady-state process.
(b) What would be the composition of oxygen 1.0-cm radial distance from the center of the rod.

26.29 Estimate the time that is required to reduce the diameter of a naphthalene sphere from 1.0 to 0.5 cm when the sphere is suspended in an effectively infinite amount of air at 165°F and 1 atm pressure. Naphthalene has a molecular weight of 128 g/mol, a density of $71.1 \text{ lb}_m/\text{ft}^3$, and a vapor pressure of 5 mm Hg at 165°F.

26.30 In a cylindrical nuclear rod that contains fissionable material, the rate of production of neutrons is proportional to the neutron concentration. Assuming that only thermal neutrons are produced, determine the neutron concentration profile in a fuel rod of radius R.

26.31 The following sketch illustrates the gas-phase diffusion in the neighborhood of a catalytic surface. Component A diffuses through a stagnant film containing only A and B. Upon reaching the catalytic surface, it is instantaneously converted into species B by the reaction $A \rightarrow B$. When

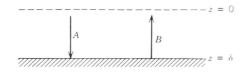

B diffuses back into the stagnant film, it begins to decompose by the first-order reaction

$$B \rightarrow A$$

The rate of formation of component A within the film is equal to $R_A = k_1 y_B$, moles A produced/(time)(volume), where y_B is the concentration of B expressed in mole fraction.

Determine the rate at which A enters the gas film if this is a steady-state process.

26.32 Reconsider problem 26.31 and determine the concentration profile of compound A in the stagnant film if in the film B decomposes to form A and if A reacts to form B, both by first-order reactions

$$A \underset{k_1}{\overset{k_1'}{\rightleftarrows}} B, \qquad R_A = k_1 y_B - k_1' y_A.$$

Simultaneously, A is instantaneously reacting to form B on the catalytic surface.

26.33 Compound A is converted to compound B on the "active sites" within a porous catalytic particle manufactured in the shape of thin disks; the surface area of the edge of the disks are small in comparison with that of the two circular faces. The rate of disappearance of A is related to the concentration of A within the catalyst by

$$R_A = kac_A$$

where a is the available catalytic surface area per unit volume of catalyst.

In terms of an "effective diffusion coefficient" $D_{A,\text{eff}}$, the concentration of A on the surface c_{As}, the reaction rate constant ka, and one half the thickness of the disks L, develop an expression for
(a) the concentration profile within the catalyst;
(b) the molar flux at the surface $z = \pm L$.

26.34 Reconsider problem 26.33 for a spherical catalyst of radius R.

26.35 A cooled metal tube, 8 cm in diameter, is used as a condenser for removing water from a water vapor-air mixture. Under steady-state conditions, a liquid film is formed on the exterior of the tube; its thickness is equivalent to a heat-transfer coefficient of $2500 \, \text{J/m}^2 \cdot \text{s} \cdot \text{K}$. At one point in the condenser, the temperature of the metal surface is 310 K, and the gas mixture is at 370 K and 1.013×10^5 Pa. The gas and liquid phases are in equilibrium at the surface of the water film. The gas contains 65 mole% water vapor. If the water vapor diffuses by molecular diffusion through a path length that is 10% of the tube radius, determine the rate of water condensation as kg/s per meter of tube length.

27
UNSTEADY-STATE MOLECULAR DIFFUSION

Transient processes, in which the concentration at a given point varies with time, are referred to as *unsteady-state processes* or *time-dependent processes*. This variation in concentration is associated with a variation in the mass flux. Many common examples of unsteady-state transfer may be cited. These generally fall into two categories: the process which is in an unsteady state only during its initial startup, and the process which is a batch or closed-system operation throughout its duration.

In Chapter 26, problems involving steady-state molecular mass transfer were considered. However, before steady-state conditions can be reached, some time must elapse after the mass-transfer process is initiated before the transient conditions completely disappear. In our previous solutions, we simply assumed that the period of transition had passed and that steady-state conditions had been established.

In this chapter, we shall consider problems and their solutions which deal with unsteady-state molecular diffusion. The time-dependent differential equations are simple to derive from the general differential equation for mass transfer. The equation of continuity for component A,

$$\nabla \cdot \mathbf{n}_A + \frac{\partial \rho_A}{\partial t} - r_A = 0 \tag{25-5}$$

or

$$\nabla \cdot \mathbf{N}_A + \frac{\partial c_A}{\partial t} - R_A = 0 \tag{25-11}$$

contains the time-dependent or unsteady-state term. The solution to these partial differential equations is generally difficult, involving relatively advanced mathematical techniques. A detailed discussion of the mathematics of diffusion is beyond the scope of this book; an excellent reference on this subject is a treatise by Crank.* In subsequent sections of this chapter we shall consider methods for

---

*J. Crank, *The Mathematics of Diffusion*, Oxford Univ. Press, London, 1958.

571

solving practical problems of unsteady-state mass transfer; however, we shall not go into the mathematical details required to obtain the solutions.

27.1　ANALYTICAL SOLUTIONS

Although the differential equations for unsteady-state diffusion are easy to establish, most solutions to these equations have been limited to situations involving simple geometries and boundary conditions, and a constant diffusion coefficient. Many solutions are for one-directional mass transfer as defined by Fick's second "law" of diffusion,

$$\frac{\partial c_A}{\partial t} = D_{AB}\frac{\partial^2 c_A}{\partial z^2} \tag{27-1}$$

This partial differential equation describes a physical situation in which there is no bulk motion contribution, that is, $\mathbf{v} = 0$, and no chemical reaction, that is, $R_A = 0$. This situation is encountered when the diffusion takes place in solids, in stationary liquids, or in systems having equimolar counterdiffusion. Due to the extremely slow rate of diffusion within liquids, the bulk motion contribution of Fick's first law (i.e., $y_A \Sigma \mathbf{N}_i$) approaches the value of zero for dilute solutions; accordingly, this system also satisfies Fick's second law of diffusion.

It may be advantageous to express equation (27-1) in terms of other concentration units. For example, the mass density of species A, ρ_A, is equal to $M_A c_A$; by multiplying both sides of equation (27-1) by the constant molecular weight of A, we obtain

$$\frac{\partial \rho_A}{\partial t} = D_{AB}\frac{\partial^2 \rho_A}{\partial z^2} \tag{27-2}$$

If the density of the given phase remains essentially constant during the mass-transfer period, the density of species A can be divided by the total density, ρ_A/ρ; this ratio is the weight fraction of A, w_A and our equation becomes

$$\frac{\partial w_A}{\partial t} = D_{AB}\frac{\partial^2 w_A}{\partial z^2} \tag{27-3}$$

However, when the phase loses a considerable amount of the solute the total density, ρ, will not be constant and equation (27-3) cannot be used to explain the transient mass transfer. Under these circumstances, it is preferable to divide by the density of the given phase on a solute-free basis (as an example, during the drying of a piece of wood, the density of the moisture-free solid will be constant). On dividing equation (27-2) by the constant $\rho_{A\text{-free}}$, we obtain

$$\frac{\partial\left(\dfrac{\rho_A}{\rho_{A\text{-free}}}\right)}{\partial t} = D_{AB}\frac{\partial^2\left(\dfrac{\rho_A}{\rho_{A\text{-free}}}\right)}{\partial z^2}$$

or

$$\frac{\partial w_A{}'}{\partial t} = D_{AB} \frac{\partial^2 w_A{}'}{\partial z^2} \qquad (27\text{-}4)$$

where $w_A{}'$ is the weight fraction of A divided by one minus the weight fraction of A.

Equations (27-1) through (27-4) are similar in form to Fourier's second "law" of heat conduction

$$\frac{\partial T}{\partial t} = \alpha \frac{\partial^2 T}{\partial z^2} \qquad (27\text{-}5)$$

thereby establishing an analogy between transient molecular diffusion and heat conduction.

The solution to Fick's second "law" usually has one of two standard forms. It may appear in the form of a trigonometric series which converges for large values of time, or it may involve a series of error functions or related integrals which are most suitable for numerical evaluation at small values of time. These solutions are commonly obtained by using the mathematical techniques of separation of variables or Laplace transforms.

Transient diffusion under conditions of negligible surface resistance. The simplest time-dependent process is encountered when a body is subjected to a sudden change in the surrounding environment which brings its surface concentration to $c_{A,s}$. To illustrate the analytical method of solving this class of transient diffusion problems, consider the drying of a large sheet of wood which has a uniform thickness L. The initial concentration distribution through the sheet will be assumed to be an arbitrary function of z. This transport process is analogous to the heating of a body under conditions of negligible surface resistance as discussed in Chapter 18.

The solution for the concentration history must satisfy equation (27-1) and the initial and boundary conditions:

$$c_A = c_{A_0}(z) \qquad \text{at } t = 0, \quad \text{for } 0 \le z \le L$$
$$c_A = c_{A,s} \qquad \text{at } z = 0, \quad \text{for } t > 0$$
$$c_A = c_{A,s} \qquad \text{at } z = L, \quad \text{for } t > 0$$

Easier applied boundary conditions are obtained if the concentrations are expressed in terms of $Y = (c_A - c_{A,s})/(c_{A_0} - c_{A,s})$, the unaccomplished concentration change. The partial differential equation becomes

$$\frac{\partial Y}{\partial t} = D_{AB} \frac{\partial^2 Y}{\partial z^2} \qquad (27\text{-}6)$$

with the initial and boundary conditions

$$Y = Y_0(z) \quad \text{at } t = 0, \quad \text{for } 0 \le z \le L$$
$$Y = 0 \quad \text{at } z = 0, \quad \text{for } t > 0$$
$$Y = 0 \quad \text{at } z = L, \quad \text{for } t > 0$$

The analogous heat conduction equation (18-10) was solved in Chapter 18. The solution to our transient mass transfer equation will be analogous to equation (18-12)

$$Y = \frac{2}{L} \sum_{n=1}^{\infty} \sin\left(\frac{n\pi z}{L}\right) e^{-(n\pi/2)^2 X_D} \int_0^L Y_0 \sin\left(\frac{n\pi z}{L}\right) dz \qquad (27\text{-}7)$$

where X_D is the relative time ratio, $D_{AB}t/x_1^2$, with x_1 being the characteristic length of $L/2$. If the sheet of wood has a uniform concentration, $Y_0(z) = Y_0$, the concentration distribution will be analogous to equation (18-13).

$$\frac{c_A - c_{A,s}}{c_{A_0} - c_{A,s}} = \frac{4}{\pi} \sum_{n=1}^{\infty} \frac{1}{n} \sin\left(\frac{n\pi z}{L}\right) e^{-(n\pi/2)^2 X_D}, \qquad n = 1, 3, 5, \ldots \quad (27\text{-}8)$$

The mass flux, $N_{A,z}$, at any plane in the sheet of wood may be evaluated by

$$N_{A,z} = -D_{AB} \frac{\partial c_A}{\partial z} \qquad (27\text{-}9)$$

In the case of the infinite flat sheet with an initial uniform concentration distribution of c_{A_0}, the mass flux at any given time t is

$$N_{A,z} = \frac{4 D_{AB}}{L}(c_{A,s} - c_{A_0}) \sum_{n=1}^{\infty} \cos\left(\frac{n\pi z}{L}\right) e^{-(n\pi/2)^2 X_D}, \qquad n = 1, 3, 5, \ldots$$
$$(27\text{-}10)$$

Transient diffusion in a semi-infinite medium. Another important case of transient mass diffusion which is amenable to an analytical solution is the one-directional mass transfer into a semi-infinite stationary medium with a fixed surface concentration. For example, we might like to describe the absorption of oxygen from air in the aeration of a lake or the solid-phase diffusion process involved in the case-hardening of mild steel in a carburizing atmosphere. Figure 27.1 depicts the concentration profiles as the time increases for a semi-infinite medium which had a uniform initial concentration of c_{A_0} and which was subjected to a constant surface concentration of $c_{A,s}$.

The differential equation to be solved is

$$\frac{\partial c_A}{\partial t} = D_{AB} \frac{\partial^2 c_A}{\partial z^2} \qquad (27\text{-}11)$$

Figure 27.1 Transient diffusion in a semi-infinite medium.

and the initial and boundary conditions are

$$c_A = c_{A_0} \quad \text{at } t = 0, \quad \text{for all } z$$

$$c_A = c_{A,s} \quad \text{at } z = 0, \quad \text{for all } t$$

$$c_A = c_{A_0} \quad \text{as } z \to \infty, \quad \text{for all } t$$

The solution to this problem may be accomplished in a variety of ways, among which are the Laplace transformation and the separation of variables. The analogous heat conduction problem, heat transfer to a semi-infinite wall, was described by equation (18-20) in Chapter 18. By analogy, we can immediately write the solution to our diffusion problem

$$\frac{c_A - c_{A_0}}{c_{A,s} - c_{A_0}} = 1 - \text{erf}\left(\frac{z}{2\sqrt{D_{AB}t}}\right) \tag{27-12}$$

or

$$\frac{c_{A,s} - c_A}{c_{A,s} - c_{A_0}} = \text{erf}\left(\frac{z}{2\sqrt{D_{AB}t}}\right)$$

The error function, which appears in the solution of many unsteady-state transport problems, has the properties

$$\text{erf}(-\phi) = -\text{erf}\,\phi$$

$$\text{erf}(0) = 0$$

$$\text{erf}(\infty) = 1.0$$

An abbreviated listing of values of erf ϕ is included in Appendix L. Equation (27-12) will be used to explain the case hardening of steel in the following example.

EXAMPLE 1

A preheated piece of mild steel, having an initial concentration of 0.20% by weight carbon, is exposed to a carburizing atmosphere for 1 hr. Under the process conditions, the surface concentration of carbon is 0.70%. If the diffusivity of carbon through steel is 1.0×10^{-11} m^2/s at the process temperature, determine the carbon composition at 0.01 cm, 0.02 cm, and 0.04 cm below the surface.

Since the overall concentration in the mild steel is very small, its density may be considered constant; accordingly

$$\frac{c_{A,s} - c_A}{c_{A,s} - c_{A0}} = \frac{\omega_{A,s} - \omega_A}{\omega_{A,s} - \omega_{A0}} = \text{erf}\left(\frac{z}{2\sqrt{D_{AB}t}}\right)$$

$$= \frac{0.007 - \omega_A}{0.007 - 0.002} = \text{erf}\left(\frac{z}{2\sqrt{(1 \times 10^{-11}\ \text{m}^2/\text{s})(3600\ \text{s})}}\right)$$

$$= \frac{0.007 - \omega_A}{0.005} = \text{erf}\left(\frac{z}{3.79 \times 10^{-4}}\right)$$

or

$$\omega_A = 0.007 - 0.005\ \text{erf}\left(\frac{z}{3.79 \times 10^{-4}}\right)$$

At the first depth, $z = 0.01$ cm $= 1 \times 10^{-4}$ m

$$\text{erf}\left(\frac{z}{3.79 \times 10^{-4}}\right) = \text{erf}\left(\frac{1 \times 10^{-4}}{3.79 \times 10^{-4}}\right) = \text{erf}(0.264)$$

From Appendix L, the erf (0.264) equals 0.291; then $\omega_A = 0.007 - 0.005(0.291) = 0.0055$ or 0.55% carbon.

At the second depth, $z = 0.02$ cm $= 2 \times 10^{-4}$ m

$$\text{erf}\left(\frac{z}{3.79 \times 10^{-4}}\right) = \text{erf}(0.528) = 0.545$$

$$\omega_A = 0.007 - 0.005(0.545)$$

$$= 0.0043 \text{ or } 0.43\% \text{ carbon}$$

At the third depth, $z = 0.04$ cm $= 4 \times 10^{-4}$ m

$$\text{erf}\left(\frac{z}{3.79 \times 10^{-4}}\right) = \text{erf}(1.055) = 0.866$$

$$\omega_A = 0.007 - 0.005(0.866)$$

$$= 0.0027 \text{ or } 0.27\% \text{ carbon}$$

27.2 CONCENTRATION-TIME CHARTS FOR SIMPLE GEOMETRIC SHAPES

In our analytical solutions, the unaccomplished change, Y, was found to be a function of the relative time, X_D. The mathematical solutions, for the unsteady-state mass transfer in several simple shapes with certain restrictive boundary conditions, have been presented in a wide variety of charts to facilitate their use. Two forms of these charts are available in the Appendix.

The "Gurney-Lurie" charts present solutions for the flat plate, sphere, and long cylinder. Since the defining partial differential equations for heat conduction and molecular diffusion are analogous, these charts may be used to solve either transport phenomenon. For molecular diffusion, the charts are in terms of four dimensionless ratios:

$$Y = \text{unaccomplished concentration change} = \frac{c_{A,s} - c_A}{c_{A,s} - c_{A_0}}$$

$$X_D = \text{relative time} = \frac{D_{AB}t}{x_1^2}$$

$$n = \text{relative position} = \frac{x}{x_1}$$

$$m = \text{relative resistance} = \frac{D_{AB}}{k_c x_1}$$

The characteristic length, x_1, is the distance from the midpoint to the position of interest. The relative resistance, m, is a ratio of the convective mass transfer resistance to the internal molecular resistance to the mass transfer.

These charts may be used to evaluate concentration profiles for cases involving molecular mass transfer into, as well as out of, bodies of the specified shapes if the following conditions are satisfied:

(a) Fick's second law of diffusion is assumed; that is, no fluid motion, $\mathbf{v} = 0$, no production term, $R_A = 0$, and constant mass diffusivity.
(b) The body has an initial uniform concentration, c_{A_0}.
(c) The boundary is subjected to a new condition which remains constant with time.

For shapes where the transport takes place from only one of the faces, the dimensionless ratios are calculated as if the thickness were twice the true value; that is, for a slab of thickness $2a$, the relative time, X_D, is considered to be $D_{AB}t/4a^2$.

Although the charts were drawn for one-dimensional transport, they may be combined to yield solutions for two- and three-dimensional transfer. In two dimensions, Y_a evaluated with the width, $x_1 = a$, and Y_b evaluated with the depth, $x_1 = b$, are combined to give

$$Y = Y_a Y_b \tag{27-13}$$

A summary of these combined solutions follows:

1. For transport from a rectangular bar with sealed ends,

$$Y_{\text{bar}} = Y_a Y_b \tag{27-14}$$

where Y_a is evaluated with width $x_1 = a$, and Y_b is evaluated with thickness $x_1 = b$.

2. For transport from a rectangular parallelepiped,

$$Y_{\text{parallelepiped}} = Y_a Y_b Y_c \qquad (27\text{-}15)$$

where Y_a is evaluated with width $x_1 = a$, Y_b is evaluated with thickness $x_1 = b$, and Y_c is evaluated with depth $x_1 = c$.

3. For transport from a cylinder, including both ends,

$$Y_{\text{cylinder plus ends}} = Y_{\text{cylinder}} Y_a \qquad (27\text{-}16)$$

The use of these charts will be shown in the following example.

EXAMPLE 2

A slab of wood, 12 in. by 12 in. by 1 in., is exposed to relatively dry air. The edges are initially sealed to limit the drying process to the large flat faces of the slab. The internal liquid diffuses to the surface, where it evaporates into the passing air stream. The moisture content on the surface remains constant at 7 wt %. After 10 hr of drying the center moisture content decreases from 15 to 10 wt %. If the convective mass transfer coefficient can be considered sufficiently large in value that the relative resistance, m, is essentially zero, calculate:

(a) The effective diffusion coefficient.
(b) The center moisture content if all six faces are used for the same drying period.
(c) The time necessary to lower the center moisture content of a 1-ft cube made from the same wood, from 15 to 10 wt % if all six faces are used. Assume that the effective diffusion coefficient calculated in (a) is constant throughout the cube.

Since the concentration change is small over a long period of time, the density of the wood will be assumed to be essentially constant over a differential time period; accordingly, Fick's second law will describe the drying operation. Concentrations may be converted from the terms of c_A to those commonly used in drying, that is, the mass of moisture per unit mass of dry solid by multiplying each c_A term by the molecular weight of A and dividing by the weight of dry solid per unit volume. By doing this each concentration term will be expressed in terms of a basis that remains constant throughout the drying process. The unaccomplished concentration ratio, Y, becomes $(w_{A,s}' - w_A')/(w_{A,s}' - w_{A0}')$, where w_A' represents the moisture content that is expressed in mass of water per mass of dry solid. The unaccomplished change for case (a) is evaluated by

$$w_{A,s}' = \frac{0.15}{1 - 0.15} = 0.176 \text{ lb}_m \text{ water/lb}_m \text{ dry wood}$$

$$w_{A0}' = \frac{0.50}{1 - 0.50} = 1.000 \text{ lb}_m \text{ water/lb}_m \text{ dry wood}$$

$$w_A' = \frac{0.32}{1 - 0.32} = 0.471 \text{ lb}_m \text{ water/lb}_m \text{ dry wood}$$

$$Y = \frac{w_{A,s}' - w_A'}{w_{A,s}' - w_{A0}'} = \frac{0.176 - 0.471}{0.176 - 1.00} = 0.358$$

The desired composition, w_A', is the center moisture concentration at $x = 0$. The corresponding relative position, n, is $x/x_1 = 0$. The relative resistance, m, was stipulated to be 0. From Figure F.7 in the appendix when $Y = 0.358$, $n = 0$, and $m = 0$, the relative time, X_D, is 0.51. Since the drying takes place from both faces, $x_1 = \frac{1}{24}$ ft; then

$$D_{AB} = \frac{X_D x_1^2}{t} = \frac{(0.51)(1/24)^2}{10 \text{ hr}} = 8.85 \times 10^{-5} \text{ ft}^2/\text{hr} \qquad (2.29 \times 10^{-9} \text{ m}^2/\text{s})$$

(b) When the other faces are also used for drying, their unaccomplished changes must be evaluated. For both $X_{D,a}$, $X_{D,b}$, the semithickness x_1 is $\frac{1}{2}$ ft; the relative time for both edges is

$$X_D = \frac{D_{AB} t}{x_1^2} = \frac{(8.85 \times 10^{-5} \text{ ft}^2/\text{hr})(10 \text{ hr})}{(\frac{1}{2} \text{ ft})^2} = 0.003$$

For a slab when $X_D = 0.003$, $n = 0$, and $m = 0$, the Y is approximately 1.0; then

$$Y_{\text{parallelpiped}} = Y_a Y_b Y_c = (1.0)(1.0)(0.358) = 0.358$$

and

$$w_A' = 0.471 \text{ lb}_m \text{ water/lb}_m \text{ dry wood}$$

This solution illustrates the relative unimportance of the edges on a large, thin slab.

(c) For the 1-ft cube in which the mass is being transferred from all six faces, the unaccomplished change is

$$Y = Y_a Y_b Y_c = Y_a^3$$

For this case,

$$Y_a = Y^{1/3} = (0.358)^{1/3} = 0.710$$

$$x_1 = 0.5 \text{ ft}$$

$$n = x/x_1 = 0$$

and

$$m = 0$$

and by Figure F.7, $X_D = 0.23$; accordingly,

$$t = \frac{X_D x_1^2}{D_{AB}} = \frac{(0.23)(0.5 \text{ ft})^2}{(8.85 \times 10^{-5} \text{ ft}^2/\text{hr})} = 650 \text{ hr} \qquad (2.34 \times 10^6 \text{ s})$$

27.3 NUMERICAL METHODS FOR TRANSIENT MASS TRANSFER ANALYSIS

In many time-dependent processes, the initial concentration distribution may be nonuniform and/or the two boundaries may have different concentration levels. Both conditions eliminate the use of concentration-time charts for evaluating these complex cases. Instead, one must employ a numerical solution technique.

Let us reconsider Fick's second law of diffusion, equation (27-1)

$$\frac{\partial c_A}{\partial t} = D_{AB} \frac{\partial^2 c_A}{\partial z^2} \tag{27-1}$$

The partial derivative of the concentration with respect to time can be expressed as a finite difference for a given position located at z

$$\frac{\partial c_A}{\partial t} = \frac{c_{A,z}|_{t+\Delta t} - c_{A,z}|_t}{\Delta t} \tag{27-17}$$

where $c_{A,z}|_t$ is the concentration at a fixed z at time t and $c_{A,z}|_{t+\Delta t}$ is the concentration at the same fixed z at time $t+\Delta t$. The second-order partial derivative on the right-hand side of the equation can be expressed in finite form by

$$\frac{\partial^2 c_A}{\partial x^2} = \frac{\dfrac{c_{A,t}|_{z+\Delta z} - c_{A,t}|_z}{\Delta z} - \dfrac{c_{A,t}|_z - c_{A,t}|_{z-\Delta z}}{\Delta z}}{\Delta z}$$

or

$$\frac{\partial^2 c_A}{\partial x^2} = \frac{c_{A,t}|_{z+\Delta z} - 2c_{A,t}|_z + c_{A,t}|_{z-\Delta z}}{(\Delta z)^2} \tag{27-18}$$

where $c_{A,t}|_{z+\Delta z}$ is the concentration at time t at a position to the right or forward from z and $c_{A,t}|_{z-\Delta z}$ is the concentration at time t at a position Δz to the left or backward from z. Upon substitution of equations (27-17) and (27-18) into equation (27-1), we obtain

$$\frac{c_{a,z}|_{t+\Delta t} - c_{A,z}|_t}{\Delta t} = D_{AB} \frac{c_{A,t}|_{z+\Delta z} - 2c_{A,t}|_z + c_{A,t}|_{z-\Delta z}}{(\Delta z)^2}$$

or

$$c_{A,z}|_{t+\Delta t} - c_{A,z}|_t = \frac{D_{AB}\Delta t}{(\Delta z)^2}[c_{A,t}|_{z+\Delta z} - 2c_{A,t}|_z + c_{A,t}|_{z-\Delta z}] \tag{27-19}$$

According to equation (27-19), the concentration at position z and at the new time $t+\Delta t$, $c_{A,z}|_{t+\Delta t}$, can be evaluated from three concentrations that are known at the time t. Use of this equation is often referred to as an explicit numerical method since it permits the concentration at a new time to be evaluated explicitly from concentrations that are known for the previous time. The accuracy of the solution improves as smaller values of the intervals Δt and Δz are used; however, the number of calculations that are required to obtain the solution for a finite amount of time and a finite distance increases as Δt and Δz decreases. The use of a digital computer is ideally suited for such computations.

In equation (27-19) the dimensionless ratio, $D_{AB}\Delta t/(\Delta z)^2$, has resulted quite naturally. This ratio of terms, which resembles the mass-transfer Fourier modulus, X_D, is extremely important in obtaining a solution since it relates the time increment to the node size, Δz. It has been found that equation (27-19) is numerically "stable" when

$$\frac{D_{AB}\Delta t}{(\Delta z)^2} \leq \tfrac{1}{2} \tag{27-20}$$

For the case where $D_{AB}\Delta t/(\Delta z)^2 = \tfrac{1}{2}$, equation (27-19) reduces to

$$c_{A,z}|_{t+\Delta t} = \frac{c_{A,t}|_{z+\Delta z} + c_{A,t}|_{z-\Delta z}}{2} \tag{27-21}$$

This expression forms the basis of a graphical solution that is quite similar to the Schmidt plotting technique, discussed in section 18.3. Our equation indicates that the concentration at node z, after a time interval Δt has elapsed, is equal to the arithmetic mean of the concentrations at the node points that are Δz units away at the start of the time interval.

Consider an infinitely thick wall that has a surface at which the concentration of component A is constant at $c_{A,s}$. The initial concentration within the wall is known; for the sake of our discussions, let us say that it is constant at c_{A0}. Figure 27.2 illustrates the concentration distribution at time $t = 0$ by a heavy line. The wall is divided into layers, each of them Δz thick. Each layer is labeled by an integer, numbering away from the surface. These lines are referred to as concentration reference lines. According to equation (27-1), a straight line ① connecting $c_{A,s}$ and $c_{A,2}$ locates $c_{A,1}'$ at the point where line ① intersects plane 1. $c_{A,1}'$ is the new concentration at this plane after one Δt time interval.

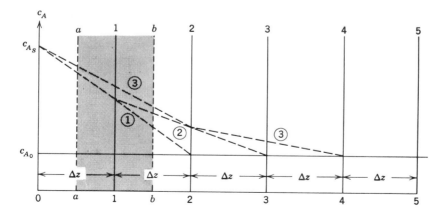

Figure 27.2 Schmidt plot in an infinitely thick wall.

In the same manner, it can be shown that the concentration at any reference plane at a time $(n+1) \Delta t$ is the arithmetic mean of the concentrations of the adjacent planes evaluated at $n \Delta t$, or

$$c_{A,i}{}^{n+1} = \frac{c_{A,i-1}{}^{n} + c_{A,i+1}{}^{n}}{2} \tag{27-22}$$

where the superscript n refers to the number of Δt time intervals and the subscript i refers to the concentration reference plane.

Using equation (27-22), we can continue with our illustration over additional time intervals. Figure 27.2 illustrates the graphical technique. For the second time interval, Δt_2, line ② is drawn between concentration $c_{A_1}{}'$ at reference line 1 and c_{A_0} at reference line 3. This line intersects reference line 2 at concentration $c_{A_2}{}''$. For the third time interval, Δt_3, two lines ③ are drawn, one between $c_{A,s}$ and the new $c_{A_2}{}''$ on reference line 2 and one between $c_{A_2}{}''$ and c_{A_0} on reference line 4. These lines indicate that concentration at lines 1 and 3 will be $c_{A_1}{}'''$ and $c_{A_3}{}'''$ at the end of the third time interval. The same procedure can be continued for additional time intervals. It is important to realize that constant Δz and Δt values are used throughout the solution.

The rate of mass flow per unit area into the wall at any instant, $N_{A,z}{}^{n}$, can be obtained from the slope of the concentration profile between the surface and reference line 1. The algebraic expression is

$$N_{A,z}{}^{n} = \frac{D_{AB}(c_{A,s} - c_{A_1}{}^{n})}{\Delta x} \tag{27-23}$$

This convenient graphical technique is based upon the assumption that the diffusion coefficient is constant and that the body initially has a known concentration profile. The resulting solution is a fair approximation to the more rigorous analytical solutions of Fick's second "law" of diffusion. Its accuracy can be improved by using smaller Δz subdivisions.

The Schmidt method may be applied to any initial condition. As an illustration we still reconsider the slab of wood from example 2.

EXAMPLE 3

A slab of wood, 12 in. by 12 in. by 1 in. is exposed to relatively dry air. The edges of the slab are sealed to limit the drying process to the large flat faces of the slab. The internal liquid diffuses to the surface, where it evaporates into the passing air stream. Initially, the moisture content of the wood is 35 wt %. During the drying operation, the moisture content of the surface remains constant at 7.0 weight percent. If the effective diffusion coefficient is 8.68×10^{-5} ft^2/hr, determine the center moisture content after 4.4 hr.

The corresponding concentrations on a dry basis are

$$w_{A0}{}' = \frac{0.35}{1 - 0.35} = 0.539 \text{ lb}_m \text{ water/lb}_m \text{ dry wood}$$

and

$$w_{A,s}' = \frac{0.07}{1-0.07} = 0.075 \text{ lb}_m \text{ water/lb}_m \text{ dry wood}$$

The incremental length, Δz, is chosen to be 0.1 in. Therefore, by equation (27-19) the incremental time

$$\Delta t = \frac{(\Delta z)^2}{2D_{AB}} = \frac{(0.1 \times \frac{1}{12} \text{ ft})^2}{(2)(8.68 \times 10^{-5} \text{ ft}^2/\text{hr})} = 0.4 \text{ hr}$$

The required number of time intervals is 4.4/0.4 or 11. In Figure 27.3 the Schmidt plot for this problem is illustrated. The center moisture content after 4.4 hr of drying equals 0.4 lb$_m$ water/lb$_m$ dry wood (0.4 kg water/kg dry wood).

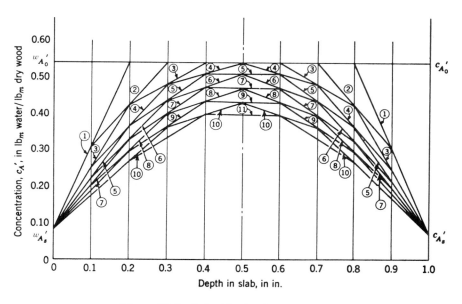

Figure 27.3 Schmidt plot for example 3.

27.4 CLOSURE

In this chapter, we have considered unsteady-state molecular diffusion. The partial differential equations which describe the transient processes were obtained from the general differential equation for mass transfer. However, most of the solutions to these differential equations required advanced mathematics beyond the scope of this book. Two types of solutions to Fick's second law of diffusion were presented. Charts for solving unsteady-state molecular transport problems were introduced, and an explicit numerical method was presented, which can be used for obtaining a solution of an unidirectional transient mass transfer by use of a digital computer or a modified Schmidt plot.

PROBLEMS

27.1 Develop equations that are equivalent to equations (27-6) through (27-10) that would permit the evaluation of the mass flux, $n_{A,z}$, at any time in terms of the concentration unit w_A'.

27.2 The evaluation of diffusion rates within metals is an important metallurgical problem. To measure the diffusion coefficient, a thin layer of the particular solute is applied to the end of a bar and allowed to diffuse into the bar. After an appropriate time, thin sections are removed parallel to the initial interface. These sections, having constant thickness, are analyzed for the concentration of the solute. This technique is now used with a radioactive tracer as the diffusing solute, since the concentration of the tracer can be determined with higher accuracy than is possible by using chemical analysis.

A short length of copper-5% nickel alloy was joined to a bar of pure copper. After 100 000 s, sections were taken parallel to the plane of joining with the aid of a lathe. Each section was analyzed for its nickel content. Arbitrarily setting the last section in which there was no nickel at zero distance, the following *average* compositions were obtained for the particular section:

% Nickel		0.10	0.21	0.41	0.67	1.05	1.31	1.96
z', distance at start of section $\times 10^4$ m		0.000	0.508	1.016	1.524	2.032	2.540	3.048
% Nickel	2.52	3.05	3.69	4.02	4.32	4.63	4.82	4.92
$z' \times 10^4$	3.555	4.064	4.572	5.080	5.588	6.096	6.604	7.112

(a) Show that the following expression satisfies Fick's second law and the boundary conditions of the diffusion process,

$$c_A = \frac{M}{\sqrt{4\pi D_A t}} e^{-z^2/4D_A t}$$

where c_A is the concentration of the diffusing solute, M is the quantity of solute that is applied to the end of the bar, t is time, and z is the distance in the direction normal to the initial solute film.

27.3 Estimate the depth below the surface of a plate of mild steel at which the concentration of carbon may be expected to decrease to 40% of its initial value as a result of exposure to strongly decarburizing conditions at 1700°F for (a) 1 h and (b) for 10 h if the diffusivity of carbon in steel at 1700°F is 1×10^{-7} cm^2/s.

27.4 A porous sphere with a $\frac{1}{2}$-in. diameter is saturated with ethanol. The void space in the solid provides sufficient pores so that molecular diffusion can take place through the liquid in the passage. The sphere is dropped into a large, well-agitated reservoir of pure water. If the concentration of the ethanol at the center of the sphere is lowered to 27% of its initial concentration in 30 h, what would be the concentration at the center of the sphere after 37.4h?

27.5 A slab of clay, 5 cm thick, was dried from both flat surfaces by being exposed to a drying air stream. The initial moisture content was 18 wt% and the four thin edges were sealed to prevent evaporation. Under the specified drying conditions, the drying was controlled by the internal diffusion of liquid water to the surface. The surface moisture was constant throughout the process at 4 wt%. In 20 000 s, the center moisture content had fallen to 10 wt%. If the relative resistance to mass transfer at the surface was essentially zero, calculate:
(a) the effective diffusivity;
(b) under the same drying conditions, the time that is necessary to reduce the center moisture content to 6 wt%;
(c) the time that is necessary to lower the center moisture content of a cylinder 0.1 m long and 0.1 m in diameter from 18 to 10 wt% if the cylinder is made from the same clay as in part (a) and dried under the same drying conditions;
(d) repeat (c) for a cylinder that is 0.5 m long and 0.1 m in diameter.

27.6 Plot the moisture concentration profile in the clay slab that is described in problem 27.5 after 36 000 s of drying.

27.7 Determine the time that is necessary to reduce the center moisture content to 10 wt% if a 5-cm-thick slab of clay is placed on a belt passing through a continuous drier, thus restricting the drying to only one of the flat surfaces. The initial moisture content will be 15 wt% and the surface moisture content under the constant drying conditions will be maintained at 4 wt%. The effective diffusivity of water through clay is 1.3×10^{-4} cm^2/s.

27.8 A large semideep lake, which initially had a uniform oxygen concentration of 2 kg/m^3, has its surface concentration suddenly raised and maintained at a 9 kg/m^3 concentration level. Sketch a concentration profile, c_A, as a function of the depth, z, for the period of
(a) 3600 s;

(b) 36 000 s;
(c) 360 000 s
if the lake is at a temperature of 283 K.

27.9 A slab of white pine, 5 cm thick, has a moisture content of 45 wt% at the start of the drying process. The equilibrium moisture content is 14 wt% for the humidity conditions in the drying air. The ends and edges are covered with a moisture-resistant coating to prevent evaporation. The diffusivity of water through the white pine may be assumed to be $1 \times 10^{-9} \, \text{m}^2/\text{s}$.

Initially, low airflow rates were used. This produced a drying process in which the ratio of the surface resistance to the internal resistance to diffusion was equal to 0.25. It was found that the length of time that is necessary to reduce the moisture content at the centerline to 25 wt% was too long. Accordingly, the drying wind velocity was increased until the ratio of the resistances approached zero.

Determine the time of drying for each of the described processes.

27.10 Plot a curve showing the concentration ratio for hydrogen,

$$\frac{c_A - c_{A,s}}{c_{A0} - c_{A,s}}$$

as a function of distance as it diffuses into a sheet of mild steel that is 6 mm thick. The diffusivity of hydrogen is equal to $1.6 \times 10^{-2} \, e^{-9200/RT} \, \text{cm}^2/\text{s}$, where T is in degrees K and $R = 1.98$. Samples of the sheet of steel are exposed to hydrogen at 1 atm pressure and 500°C for periods of
(a) 10 min;
(b) 1 hr;
(c) 10 h.

27.11 A large tanker truck overturns and spills a herbicide over a field. If the mass diffusivity of the fluid in the soil is $1 \times 10^{-8} \, \text{m}^2/\text{s}$ and the fluid remains on the soil for 1800 s before evaporating into the air, determine the depth at which plant and insect life is likely to be destroyed if a concentration of 0.1% by weight will destroy most life.

27.12 A slab of white pine, 2 in. thick, has the following initial moisture content at the start of the drying process:

z, in	0.0	0.2	0.4	0.6	0.8	1.0	1.2	1.4	1.6	1.8	2.0
wt%	46.0	48.0	49.0	49.5	50.0	50.0	50.0	49.5	49.0	48.0	46.0

where z is the distance from one of the large flat surfaces. The ends and edges will be covered with a sealant to prevent evaporation. If the drying conditions maintain a constant 13 wt% surface moisture at both surfaces and the diffusivity of water through the pine is 4×10^{-5} ft^2/hr, determine the time that is necessary to lower the moisture content at the centerline to 35 wt%.

27.13 The slab of white pine that is described in problem 27.12 is to be dried in another dryer that maintains the moisture content at $z = 0$ in. at 13 wt% and the moisture content at $z = 2.0$ in. at 15 wt%. Determine the time that is necessary to lower the moisture content at the centerline to 30 wt%.

27.14 A large semideep lake, which initially had a uniform oxygen concentration of 1.5 kg/m^3, has its surface concentration suddenly raised to and maintained at a 8 kg/m^3 concentration level. Sketch a concentration profile, c_A, as a function of depth, z, for the period of 3600 s by using
(a) equation (27-12);
(b) a modified Schmidt plot with incremental Δz equal to 3 mm.
The lake is at 283 K.

27.15 A charcoal briquet, approximately spherical in shape with a 2-cm radius, has an initial moisture content of 400 kg/m^3. It is placed in a forced air dryer, which produces a surface moisture concentration of 10 kg/m^3. If the diffusivity of water in the charcoal is 1.3×10^{-6} m^2/s and the surface resistance is negligible, estimate the time that is required to dry the center of the briquet to a moisture concentration of 50 kg/m^3.

27.16 The concentration profile resulting from transient diffusion from a large sheet of wood under conditions of negligible surface resistance is described by equation (27-8). Use this equation to develop an equation for predicting the average concentration, $\overline{c_A}$; evaluate and plot the dimensionless average concentration profile, $(\overline{c_A} - c_{A,s})/(c_{A0} - c_{A,s})$ as a function of the dimensionless relative time ratio, X_D.

27.17 A slab of Douglas fir, 2 in. thick, has a moisture content of 45 wt% at the start of the drying process. The equilibrium moisture content is 14 wt% for the humidity conditions in the drying air. Due to the size of the slab, the drying only takes place from the large flat faces. If the relative resistance to mass transfer at the surface is negligible and the diffusivity of water through the fir is 4×10^{-5} ft^2/hr, determine the drying time that is required to reduce the moisture content at the centerline to 25 wt% by using
(a) the unsteady-state charts;
(b) a modified Schmidt plot.

27.18 A lawn-mower oak roller, having an initial moisture content of 55 wt%, is placed in a drying kiln where its surface moisture is maintained at

20 wt%. If the maximum moisture content of the dried roller is set at 30 wt%, how long must the 4 in. in diameter by 18 in. in length roller be dried when

(a) the ends of the roller are sealed with a vapor barrier;
(b) the cylinder surface is sealed with a vapor barrier;
(c) the drying occurs from the entire surface.

The surface resistance may be assumed as negligible and the diffusivity of moisture through the oak is 4×10^{-5} ft^2/hr.

27.19 For the conditions that are specified in problem 27.17, write a computer program that will explicitly solve the transient mass transfer, using a $\Delta z = 0.1$ in. Plot the concentration profile at the final drying time.

27.20 A porous cylinder, 1 in. in diameter and 3 ft long, is saturated with an alcohol. The void space in the solid provides sufficient pores so that the molecular diffusion can take place through the liquid in the passage. The cylinder is dropped into a large well-agitated reservoir of pure water. The agitation maintains a concentration of 1 wt% alcohol at the surface of the cylinder.

    If the concentration at the center of the cylinder drops from 30 wt% alcohol to 18 wt% in 10 h, determine the concentration wt% at the center after 15 h.

27.21 A slab of white pine, 2 in. thick, was initially stored in a lumber yard where evaporation from the one exposed surface to the surrounding atmosphere produced a moisture content profile of

$$\text{wt\%} = 5z + 30$$

where z is the depth above the bottom surface in inches. The slab is to be dried in a drier in which the drying medium will maintain a constant 13 wt% surface moisture content at both surfaces. The diffusivity of water through pine at the drying temperature is 4×10^{-5} ft^2/hr. Determine the time that is necessary to lower the moisture content at the centerline to 25 wt%.

27.22 If the slab of white pine that is described in problem 27.21 is dried for 40 h, determine the moisture content at the centerline.

27.23 Resolve problem 27.21 with the use of a computer program. Write the program using $\Delta z = 0.1$ in. Present a plot showing the concentrations within the slab for the final drying time.

28
CONVECTIVE
MASS TRANSFER

Mass transfer by convection involves the transport of material between a boundary surface and a moving fluid or between two relatively immiscible, moving fluids. The rate equation for convective mass transfer has been expressed previously in the form

$$N_A = k_c \, \Delta c_A \qquad (24\text{-}52)$$

where the mass flux, N_A, occurs in the direction of a decreasing concentration. This simple equation is the defining relation for k_c, the *convective mass-transfer coefficient*. It is analogous to the defining equation for the convective heat-transfer coefficient,

$$q/A = h \, \Delta T \qquad (15\text{-}11)$$

Recalling the discussions of the heat-transfer coefficient in Chapter 15, we should realize the determination of the mass-transfer coefficient is not a simple undertaking. Both transport coefficients are related to the properties of the fluid, the dynamic characteristics of the flowing fluid, and the geometry of the specific system of interest.

In light of the close similarity between the convective heat- and mass-transfer equations, we may expect that the analytical treatment of the heat-transfer coefficient in Chapter 19 might be applied to the mass-transfer coefficient. This we will do in the analyses to follow; considerable use will be made of the developments and concepts of Chapters 9 through 14.

28.1 FUNDAMENTAL CONSIDERATIONS IN CONVECTIVE MASS TRANSFER

From our early discussions dealing with a fluid flowing past a surface, we may recall that there is postulated a layer, sometimes extremely thin, close to the surface where the flow is laminar. Thus molecular mass transfer will always be present and will play a role in any convection process. If the fluid flow is laminar, then all of the transport between the surface and the moving fluid will be by molecular means. If, on the other hand, the fluid flow is turbulent, there will be a physical movement of packets of material across streamlines, transported by the eddies present in the turbulent flow. As in the case of heat transfer, higher mass-transfer rates are associated with turbulent conditions. The distinction between laminar and turbulent flow will be an important consideration in any convective situation.

The hydrodynamic boundary layer, analyzed in Chapter 12, plays a major role in convective mass transfer. We shall also define and analyze a concentration boundary layer which will be vital to the analysis of the convective mass-transfer process. This layer is similar, but not necessarily equal in thickness to the thermal boundary layer which was discussed in Chapter 19.

When the mass transfer involves a solute dissolving at a steady rate from a solid surface and then diffusing into a moving fluid, the convective mass-transfer coefficient is defined by

$$N_A = k_c(c_{A,s} - c_A) \tag{28-1}$$

In this equation the flux, N_A, represents the moles of solute A leaving the interface per time and unit interfacial area. The composition of the solute in the fluid at the interface, $c_{A,s}$, is the composition of the fluid in equilibrium with the solid at the temperature and pressure of the system. The quantity c_A represents the composition at some point within the fluid phase. When the concentration boundary layer is defined, c_A can be chosen as the concentration of component A at the edge of the boundary layer and expressed as $c_{A\infty}$. If the flow were in a closed conduit, the composition, c_A, could be the bulk concentration or the *mixing-cup concentration*. The mixing-cup composition is the concentration one would measure if the fluid at a plane were collected and thoroughly mixed; i.e., an average composition of the bulk flow.

There are four methods of evaluating convective mass-transfer coefficients which will be discussed in this chapter. These are

1. dimensional analysis coupled with experiment;
2. exact boundary-layer analysis;
3. approximate boundary-layer analysis;
4. analogy between momentum, energy, and mass transfer.

Each of these methods will be considered in the sections to follow.

28.2　SIGNIFICANT PARAMETERS IN CONVECTIVE MASS TRANSFER

Dimensionless parameters are often used to correlate convective transport data. In momentum transfer we encountered the Reynolds and the Euler numbers. In the correlation of convective heat-transfer data, the Prandtl and the

Nusselt numbers were important. Some of the same parameters, along with some newly defined dimensionless ratios, will be useful in the correlation of convective mass-transfer data. In this section, we shall consider the physical interpretation of three such ratios.

The molecular diffusivities of the three transport phenomena have been defined as

momentum diffusivity, $\nu = \mu/\rho$

thermal diffusivity, $\alpha = \dfrac{k}{\rho c_p}$

and

mass diffusivity, D_{AB}

As we have noted earlier, each of the diffusivities has the dimensions L^2/t; thus a ratio of any two of these must be dimensionless. The ratio of the molecular diffusivity of momentum to the molecular diffusivity of mass is designated the *Schmidt number,*

$$\frac{\text{momentum diffusivity}}{\text{mass diffusivity}} = \text{Sc} \equiv \frac{\nu}{D_{AB}} = \frac{\mu}{\rho D_{AB}} \tag{28-2}$$

The Schmidt number plays a role in convective mass transfer analogous to that of the Prandtl number in convective heat transfer. The ratio of the thermal diffusivity to the molecular diffusivity of mass is designated the *Lewis number,*

$$\frac{\text{thermal diffusivity}}{\text{mass diffusivity}} = \text{Le} \equiv \frac{k}{\rho c_p D_{AB}} \tag{28-3}$$

The Lewis number is encountered when a process involves the simultaneous convective transfer of mass and energy. The Schmidt and the Lewis numbers are observed to be combinations of fluid properties; thus each number may be treated as a property of the diffusing system.

Consider the mass transfer of solute A from a solid to a fluid flowing past the surface of the solid. The concentration profile is depicted in Figure 28.1. For such

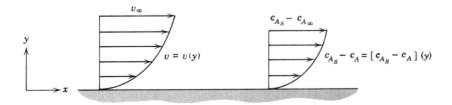

Figure 28.1 Concentration and velocity profiles for a fluid flowing past a solid surface.

a case the mass transfer between the surface and the fluid may be written as

$$N_A = k_c(c_{A,s} - c_{A,\infty}) \qquad (28\text{-}4)$$

Since the mass transfer at the surface is by molecular diffusion, the mass transfer may also be described by

$$N_A = -D_{AB}\frac{dc_A}{dy}\bigg|_{y=0}$$

when the boundary concentration, $c_{A,s}$, is constant, this equation simplifies to

$$N_A = -D_{AB}\frac{d(c_A - c_{A,s})}{dy}\bigg|_{y=0} \qquad (28\text{-}5)$$

Equations (28-4) and (28-5) may be equated, since they define the same flux of component A leaving the surface and entering the fluid. This gives the relation

$$k_c(c_{A,s} - c_{A,\infty}) = -D_{AB}\frac{d}{dy}(c_A - c_{A,s})\bigg|_{y=0}$$

which may be rearranged into the following form:

$$\frac{k_c}{D_{AB}} = \frac{-d(c_A - c_{A,s})/dy|_{y=0}}{(c_{A,s} - c_{A,\infty})} \qquad (28\text{-}6)$$

Multiplying both sides of equation (28-6) by a significant length, L, we obtain the following dimensionless expression:

$$\frac{k_c L}{D_{AB}} = \frac{-d(c_A - c_{A,s})/dy|_{y=0}}{(c_{A,s} - c_{A,\infty})/L} \qquad (28\text{-}7)$$

The right-hand side of equation (28-7) is the ratio of the concentration gradient at the surface to an overall or reference concentration gradient; accordingly, it may be considered a ratio of the molecular mass-transport resistance to the convective mass-transport resistance of the fluid. Since the development of equation (28-7) parallels the development of equation (19-5) for the Nusselt number encountered in convective heat transfer, the ratio $k_c L/D_{AB}$ in this book will be referred to as the *mass-transfer Nusselt number*, Nu_{AB}, thus emphasizing the similarity between the two transfer phenomena. It is generally known as the *Sherwood number*, Sh.

These three parameters, Sc, Nu_{AB} or Sh, and Le will be encountered in the analyses of convective mass transfer in the following sections.

28.3 DIMENSIONAL ANALYSIS OF CONVECTIVE MASS TRANSFER

Dimensional analysis predicts the various dimensionless parameters which are helpful in correlating experimental data. There are two important mass-transfer processes which we shall consider, mass transfer into a stream flowing under forced convection and mass transfer into a phase which is moving under natural-convection conditions.

TRANSFER INTO A STREAM FLOWING UNDER FORCED CONVECTION

Consider the transfer of mass from the walls of a circular conduit to a fluid flowing through the conduit. The transfer is a result of the concentration driving force, $c_{A,s} - c_A$. The important variables, their symbols, and their dimensional representations are listed below:

Variable	Symbol	Dimensions
tube diameter	D	L
fluid density	ρ	M/L^3
fluid viscosity	μ	M/Lt
fluid velocity	v	L/t
fluid diffusivity	D_{AB}	L^2/t
mass-transfer coefficient	k_c	L/t

The above variables include terms descriptive of the system geometry, the flow, the fluid properties, and the quantity which is of primary interest, k_c.

By the Buckingham method of grouping the variables as presented in Chapter 11, we can determine that there will be three dimensionless groups. With D_{AB}, ρ, and D as the core variables, the three pi groups to be formed are

$$\pi_1 = D_{AB}{}^a \rho^b D^c k_c$$

$$\pi_2 = D_{AB}{}^d \rho^e D^f v$$

and

$$\pi_3 = D_{AB}{}^g \rho^h D^i \mu$$

Writing π_1 in dimensional form,

$$\pi_1 = D_{AB}{}^a \rho^b D^c k_c$$

$$1 = \left(\frac{L^2}{t}\right)^a \left(\frac{M}{L^3}\right)^b (L)^c \left(\frac{L}{t}\right)$$

equating the exponents of the fundamental dimensions on both sides of the equation, we have for

$$L: \quad 0 = 2a - 3b + c + 1$$

$$t: \quad 0 = -a - 1$$

and

$$M: \quad 0 = b$$

The solution of these equations for the three unknown exponents yields

$$a = -1$$

$$b = 0$$

and

$$c = 1$$

thus $\pi_1 = k_c D / D_{AB}$, which is the mass-transfer Nusselt or Sherwood number. The other two pi groups could be determined in the same manner, yielding

$$\pi_2 = \frac{Dv}{D_{AB}}$$

and

$$\pi_3 = \frac{\mu}{\rho D_{AB}} \equiv \text{Sc}$$

the Schmidt number. Dividing π_2 by π_3, we obtain

$$\frac{\pi_2}{\pi_3} = \left(\frac{Dv}{D_{AB}}\right)\left(\frac{D_{AB}\rho}{\mu}\right) = \frac{Dv\rho}{\mu} \equiv \text{Re}$$

the Reynolds number. The result of the dimensional analysis of forced-convection mass transfer in a circular conduit indicates that a correlating relation could be of the form,

$$\text{Nu}_{AB} = f(\text{Re}, \text{Sc}) \tag{28-8}$$

which is analogous to the heat-transfer correlation,

$$\text{Nu} = f(\text{Re}, \text{Pr}) \tag{19-7}$$

TRANSFER INTO A PHASE WHOSE MOTION IS DUE TO NATURAL CONVECTION

Natural convection currents will develop if there exists any variation in density within a liquid or gas phase. The density variation may be due to temperature differences or to relatively large concentration differences.

In the case of natural convection involving mass transfer from a vertical plane wall to an adjacent fluid, the variables will differ from those used in the forced-convection analysis. The important variables, their symbols, and dimensional representations are listed below:

Variable	Symbol	Dimensions
characteristic length	L	L
fluid diffusivity	D_{AB}	L^2/t
fluid density	ρ	M/L^3
fluid viscosity	μ	M/LT
buoyant force	$g\,\Delta\rho_A$	$M/L^2 t^2$
mass-transfer coefficient	k_c	L/t

By the Buckingham theorem, there will be three dimensionless groups. With D_{AB}, L, and μ as the core variables, the three pi groups to be formed are

$$\pi_1 = D_{AB}{}^a L^b \mu^c k_c$$

$$\pi_2 = D_{AB}{}^d L^e \mu^f \rho$$

and

$$\pi_3 = D_{AB}{}^g L^h \mu^i g \, \Delta\rho_a$$

Solving for the three pi groups, we obtain

$$\pi_1 = \frac{k_c L}{D_{AB}} \equiv \mathrm{Nu}_{AB}$$

the mass-transfer Nusselt or Sherwood number;

$$\pi_2 = \frac{\rho D_{AB}}{\mu} \equiv \frac{1}{\mathrm{Sc}}$$

the reciprocal of the Schmidt number; and

$$\pi_3 = \frac{L^3 g \, \Delta\rho_A}{\mu D_{AB}}$$

Multiplying π_2 and π_3, we obtain a parameter which is analogous to the Grashof number in natural-convection heat transfer,

$$\pi_2 \pi_3 = \left(\frac{\rho D_{AB}}{\mu}\right)\left(\frac{L^3 g \, \Delta\rho_A}{\mu D_{AB}}\right)$$

$$= \frac{L^3 \rho g \Delta\rho_A}{\mu^2} = \frac{L^3 g \Delta\rho_A}{\rho \nu^2} \equiv \mathrm{Gr}_{AB}$$

The result of the dimensional analysis of natural-convection mass transfer suggests a correlating relation of the form

$$\mathrm{Nu}_{AB} = f(\mathrm{Gr}_{AB}, \mathrm{Sc}) \tag{28-9}$$

For both forced and natural convection, relations have been obtained by dimensional analysis which suggest that a correlation of experimental data may be in terms of three variables instead of the original six. This reduction in variables has aided investigators who have suggested correlations of these forms to provide many of the empirical equations reported in Chapter 30.

28.4 EXACT ANALYSIS OF THE LAMINAR CONCENTRATION BOUNDARY LAYER

Blasius developed an exact solution for the hydrodynamic boundary layer for laminar flow parallel to a flat surface. This solution was discussed in section 12.5.

An extension of the Blasius solution was made in section 19.4 to explain convective heat transfer. In an exactly analogous manner we shall also extend the Blasius solution to include convective mass transfer for the same geometry and laminar flow.

The boundary-layer equations considered in the steady-state momentum transfer included the two-dimensional, incompressible continuity equation,

$$\frac{\partial v_x}{\partial x}+\frac{\partial v_y}{\partial y}=0 \tag{12-11b}$$

and the equation of motion in the x direction, for constant ν and pressure,

$$v_x\frac{\partial v_x}{\partial x}+v_y\frac{\partial v_x}{\partial y}=\nu\frac{\partial^2 v_x}{\partial y^2} \tag{12-11a}$$

For the thermal boundary layer, the equation describing the energy transfer in a steady, incompressible, two-dimensional, isobaric flow with constant thermal diffusivity was

$$v_x\frac{\partial T}{\partial x}+v_y\frac{\partial T}{\partial y}=\alpha\frac{\partial^2 T}{\partial y^2} \tag{19-15}$$

An analogous differential equation applies to mass transfer within a concentration boundary layer if no production of the diffusing component occurs and if the second derivative of c_A with respect to x, $\partial^2 c_A/\partial x^2$, is much smaller in magnitude than the second derivative of c_A with respect to y. This equation written for steady, incompressible two-dimensional flow with constant mass diffusivity is

$$v_x\frac{\partial c_A}{\partial x}+v_y\frac{\partial c_A}{\partial y}=D_{AB}\frac{\partial^2 c_A}{\partial y^2} \tag{28-10}$$

The concentration boundary layer is shown schematically in Figure 28.2. The following are the boundary conditions for the three boundary layers:

momentum: $\dfrac{v_x}{v_\infty}=0$ at $y=0$ and $\dfrac{v_x}{v_\infty}=1$ at $y=\infty$

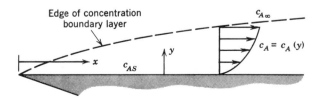

Figure 28.2. The concentration boundary layer for laminar flow past a flat surface.

or, since the velocity in the x direction at the wall, $v_{x,s}$ is zero

$$\frac{v_x - v_{x,s}}{v_\infty - v_{x,s}} = 0 \quad \text{at } y = 0 \qquad \text{and} \qquad \frac{v_x - v_{x,s}}{v_\infty - v_{x,s}} = 1 \quad \text{at } y = \infty$$

thermal: $\quad \dfrac{T - T_s}{T_\infty - T_s} = 0 \quad \text{at } y = 0 \qquad \text{and} \qquad \dfrac{T - T_s}{T_\infty - T_s} = 1 \quad \text{at } y = \infty$

and *concentration:*

$$\frac{c_A - c_{A,s}}{c_{A,\infty} - c_{A,s}} = 0 \quad \text{at } y = 0 \qquad \text{and} \qquad \frac{c_A - c_{A,s}}{c_{A,\infty} - c_{A,s}} = 1 \quad \text{at } y = \infty$$

The similarity in the three differential equations, (12-11a), (19-15), and (28-10), and the boundary conditions suggests that similar solutions should be obtained for the three transfer phenomena. In Chapter 19 the Blasius solution for equation (12-11a) was modified and successfully applied to explain convective heat transfer when the ratio of the momentum to thermal diffusivity, $\nu/\alpha = \text{Pr} = 1$. The same type of solution should also describe convective mass transfer when the ratio of the momentum to mass diffusivity, $\nu/D = \text{Sc} = 1$. Using the nomenclature defined in Chapter 12.

$$f' = 2\frac{v_x}{v_\infty} = 2\frac{v_x - v_{x,s}}{v_\infty - v_{x,s}} = 2\frac{c_A - c_{A,s}}{c_{A,\infty} - c_{A,s}} \tag{28-11}$$

and

$$\eta = \frac{y}{2}\sqrt{\frac{v_\infty}{\nu x}} = \frac{y}{2x}\sqrt{\frac{x v_\infty}{\nu}} = \frac{y}{2x}\sqrt{\text{Re}_x} \tag{28-12}$$

the Blasius solution to the momentum boundary layer

$$\frac{df'}{d\eta} = f''(0) = \frac{d[2(v_x/v_\infty)]}{d[(y/2x)\sqrt{\text{Re}_x}]}\bigg|_{y=0} = 1.328$$

suggests an analogous solution for the concentration boundary layer

$$\frac{df'}{d\eta} = f''(0) = \frac{d[2(c_A - c_{A,s})/(c_{A,\infty} - c_{A,s})]}{d[(y/2x)\sqrt{\text{Re}_x}]}\bigg|_{y=0} = 1.328 \tag{28-13}$$

Equation (28-13) may be rearranged to obtain an expression for the concentration gradient at the surface,

$$\frac{dc_A}{dy}\bigg|_{y=0} = (c_{A,\infty} - c_{A,s})\left[\frac{0.332}{x}\text{Re}_x^{1/2}\right] \tag{28-14}$$

It is important to recall that the Blasius solution for equation (12-11a) did not involve a velocity in the y direction at the surface; accordingly, equation (28-14) involves the important assumption that the rate at which mass enters or leaves the boundary layer at the surface is so small that it does not alter the velocity profile predicted by the Blasius solution.

When the velocity in the y direction at the surface, $v_{y,s}$, is essentially zero, the bulk contribution term in Fick's equation for the mass flux in the y direction is also zero. The mass transfer from the flat surface into the laminar boundary layer is described by

$$N_{A,y} = -D_{AB}\frac{\partial c_A}{\partial y}\bigg|_{y=0} \tag{28-15}$$

Upon substituting equation (28-14) into equation (28-15), we obtain

$$N_{A,y} = -D_{AB}\left[\frac{0.332\,\mathrm{Re}_x^{1/2}}{x}\right](c_{A,\infty} - c_{A,s})$$

or

$$N_{A,y} = D_{AB}\left[\frac{0.332\,\mathrm{Re}_x^{1/2}}{x}\right](c_{A,s} - c_{A,\infty}) \tag{28-16}$$

The mass flux of the diffusing component was defined in terms of the mass-transfer coefficient by

$$N_{A,y} = k_c(c_{A,s} - c_{A,\infty}) \tag{28-4}$$

The right-hand sides of equations (28-16) and (28-4) may be equated to give

$$k_c = \frac{D_{AB}}{x}[0.332\,\mathrm{Re}_x^{1/2}]$$

or

$$\frac{k_c x}{D_{AB}} = \mathrm{Nu}_{AB} = 0.332\,\mathrm{Re}_x^{1/2} \tag{28-17}$$

Equation (28-17) is restricted to systems having a Schmidt number, Sc, of one and low mass transfer rates between the flat plate and the boundary layer.

A graphical presentation of the solution to the concentration boundary layer equation (28-10) by Hartnett and Eckert* is depicted in Figure 28.3. Curves representing positive and negative values of the surface boundary parameter, $(v_{y,s}/v_\infty)(\mathrm{Re}_x)^{1/2}$ are shown. The positive values apply when the mass transfer from the flat plate is into the boundary layer, and the negative values describe mass transfer from the fluid to the plate. As this surface boundary parameter approaches a zero value, the mass-transfer rate diminishes until it is considered to have no effect upon the velocity profile. The slope of the zero line, evaluated at $y = 0$, is 0.332 as predicted by equation (28-13).

In most physical operations involving mass transfer, the surface boundary parameter is negligible, and the low-mass-transfer Blasius type of solution is used to define the transfer into the laminar boundary layer. The vaporization of a volatile material into a gas stream flowing at low pressures is a case in which the low-mass-transfer assumption cannot be made.

---

* J. P. Hartnett and E. R. G. Eckert, *Trans. A.S.M.E.*, **13**, 247 (1957).

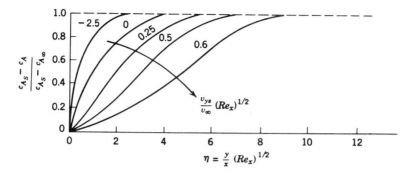

Figure 28.3 Concentration profiles for mass transfer in a laminar boundary layer on a flat plate.

For a fluid with a Schmidt number other than unity, similar curves to those shown in Figure 28.3 can be defined. The similarity in differential equations and boundary conditions suggests a treatment for convective mass transfer analogous to Pohlhausen's solution for convective heat transfer. The concentration boundary layer is related to the hydrodynamic boundary layer by

$$\frac{\delta}{\delta_c} = Sc^{1/3} \qquad (28\text{-}18)$$

where δ is the thickness of the hydrodynamic boundary layer and δ_c is the thickness of the concentration boundary layer; thus the Blasius η term must be multiplied by $Sc^{1/3}$. A plot of the dimensionless concentration versus $\eta Sc^{1/3}$ for $v_{y,s} = 0$ is shown in Figure 28.4. The concentration variation given in this form

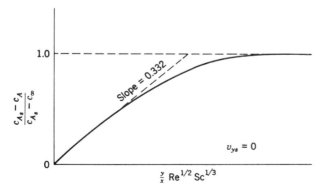

Figure 28.4 Concentration variation for laminar flow over a flat plate.

leads to an expression for the convective mass-transfer coefficient similar to equation (28-17). At $y = 0$, the concentration gradient is

$$\left.\frac{\partial c_A}{\partial y}\right|_{y=0} = (c_{A,\infty} - c_{A,s})\left[\frac{0.332}{x} \mathrm{Re}_x^{1/2}\mathrm{Sc}^{1/3}\right] \tag{28-19}$$

which, when used with equation (28-15), yields

$$\frac{k_c x}{D_{AB}} = \mathrm{Nu}_{x,AB} = 0.332\,\mathrm{Re}_x^{1/2}\,\mathrm{Sc}^{1/3} \tag{28-20}$$

The mean mass-transfer coefficient, which applies over a plate of width W and length L, may be obtained by integration. For a plate of these dimensions, the total mass transfer rate, W_A, may be evaluated by

$$W_A = \overline{k_c}A(c_{A,s} - c_{A,\infty}) = \int_A k_c(c_{A,s} - c_{A,\infty})\,dA$$

$$= \overline{k_c}WL(c_{A,s} - c_{A,\infty})$$

$$= (c_{A,s} - c_{A,\infty})\int_A \frac{0.332 D_{AB}\,\mathrm{Re}_x^{1/2}\mathrm{Sc}^{1/3}\,dA}{x}$$

Accordingly,

$$\overline{k_c}WL = 0.332\,WD_{AB}\mathrm{Sc}^{1/3}\int_0^L \frac{\mathrm{Re}_x^{1/2}}{x}\,dx$$

$$\overline{k_c}L = 0.332 D_{AB}\mathrm{Sc}^{1/3}\left(\frac{v_\infty\rho}{\mu}\right)^{1/2}\int_0^L x^{-1/2}\,dx$$

$$= 0.664 D_{AB}\mathrm{Sc}^{1/3}\left(\frac{v_\infty\rho}{\mu}\right)^{1/2}L^{1/2}$$

and

$$\frac{\overline{k_c}L}{D_{AB}} = \mathrm{Nu}_{L,AB} = 0.664\mathrm{Re}_L^{1/2}\mathrm{Sc}^{1/3} \tag{28-21}$$

The local Nusselt number at a distance x downstream is related to the mean Nusselt number for the plate by the relation

$$\mathrm{Nu}_{L,AB} = 2\mathrm{Nu}_{x,AB}|_{x=L} \tag{28-22}$$

which is analogous to the convective heat-transfer result

$$\mathrm{Nu}_L = 2\mathrm{Nu}_x|_{x=L} \tag{19-27}$$

Equations (28-20) and (28-21) have been experimentally verified.[*] It is interesting to note that this entirely different analysis has produced results of the same

[*] W. J. Christian and S. P. Kezios, *A.I.Ch.E. J.*, **5**, 61 (1959).

form predicted in section 28.3 by dimensional analysis for forced-convective mass transfer,

$$\text{Nu}_{AB} = f(\text{Re, Sc}) \tag{28-8}$$

Reconsidering the dimensionless concentration profiles of Harnett and Eckert as presented in Figure 28.3, we can observe that the slope of each ⁀urve, when evaluated at $y = 0$, decreases as the positive surface boundary parameter, $(v_{ys}/v_\infty)(\text{Re})^{1/2}$, increases. Since the magnitude of the transfer coefficient is directly related to the slope by the relation

$$k_c = D_{AB} \frac{d[(c_{A,s} - c_A)/(c_{A,s} - c_{A,\infty})]}{dy} \Bigg|_{y=0} \tag{28-23}$$

the decrease in slope indicates that the systems having higher values of the surface boundary parameter will have lower mass-transfer coefficients.

When both energy and mass are transferred through the laminar boundary layer, the dimensionless profiles in Figure 28.3 may also represent the dimensionless temperature profiles if the Prandtl and Schmidt numbers for the system are both unity. In the previous paragraph, it was pointed out that the mass-transfer coefficient diminishes in magnitude as mass is transferred into the boundary layer from the surface; accordingly, we should also expect the heat-transfer coefficient to diminish as mass is transferred into the boundary layer. This may be accomplished by forcing a fluid through a porous plate out into the boundary layer or by sublimating the plate material itself. These simultaneous heat and mass transfer processes, often referred to as *transpiration cooling* and *ablation*, respectively, are used to help reduce the large heat effects during the re-entry of a missile into the earth's atmosphere.

EXAMPLE 1

The mass transfer coefficient for a turbulent boundary layer formed over a flat plate has been correlated in terms of a local Nusselt number by

$$\text{Nu}_{x,AB} = 0.0292 \text{Re}_x^{4/5} \text{Sc}^{1/3} \tag{28-24}$$

where x is the distance downstream from the leading edge of the flat plate; the transition from laminar to turbulent flow occurs at $\text{Re}_x = 3 \times 10^5$.

(a) Develop an expression for the mean mass transfer coefficient for a flat plate of length L.

By definition,

$$\bar{k}_c = \frac{\int_0^L k_c \, dx}{\int_0^L dx} = \frac{\int_0^{L_t} k_{c,\text{lam}} \, dx + \int_{L_t}^L k_{c,\text{turb}} \, dx}{L} \tag{28-25}$$

where L_t is the measured distance from leading edge to the transition point. $k_{c,\text{lam}}$ is defined by equation (28-20)

$$k_{c,\text{lam}} = 0.332 \frac{D_{AB}}{x} (\text{Re}_x)^{1/2} (\text{Sc})^{1/3}$$

$k_{c,\text{turb}}$ is defined by equation (28-24)

$$k_{c,\text{turb}} = 0.0292\frac{D_{AB}}{x}(\text{Re}_x)^{4/5}(\text{Sc})^{1/3}$$

Upon substitution of these two equations in our equation for the mean mass-transfer coefficient, we obtain

$$\bar{k}_c = \frac{\displaystyle\int_0^{L_t}\frac{0.332D_{AB}(\text{Re}_x)^{1/2}}{x}(\text{Sc})^{1/3}\,dx + \int_{L_t}^{L}\frac{0.0292D_{AB}(\text{Re}_x)^{4/5}}{x}(\text{Sc})^{1/3}\,dx}{L}$$

where L_t is the distance from the leading edge of the plane to the transition point where the $\text{Re}_x = 3\times10^5$.

$$\bar{k}_c = \frac{0.332D_{AB}\left(\dfrac{v}{\nu}\right)^{1/2}(\text{Sc})^{1/3}\displaystyle\int_0^{L_t}x^{-1/2}\,dx + 0.0292D_{AB}\left(\dfrac{v}{\nu}\right)^{4/5}(\text{Sc})^{1/3}\int_{L_t}^{L}x^{-1/5}\,dx}{L}$$

$$\bar{k}_c = \frac{0.664D_{AB}\left(\dfrac{v}{\nu}\right)^{1/2}(\text{Sc})^{1/3}L_t^{1/2} + 0.0365D_{AB}\left(\dfrac{v}{\nu}\right)^{4/5}(\text{Sc})^{1/3}[(L)^{4/5}-(L_t)^{4/5}]}{L}$$

$$\bar{k}_c = \frac{0.664D_{AB}(\text{Re}_t)^{1/2}(\text{Sc})^{1/3} + 0.0365D_{AB}(\text{Sc})^{1/3}[(\text{Re}_L)^{4/5}-(\text{Re}_t)^{4/5}]}{L} \qquad (28\text{-}26)$$

(b) A beaker of ethyl alcohol was accidently upset, covering the top, smooth surface of a laboratory bench. The exhaust fan in the laboratory hood produced a 6-m/s air flow parallel to the surface, flowing across the 1-m wide bench. The air was maintained at 289 K and 1 atm $(1.013\times10^5\,\text{Pa})$. The vapor pressure of ethyl alcohol at 289 K is 4000 Pa. Determine the amount of alcohol evaporating from one square meter surface area each 60 sec.

At 289 K, the kinematic viscosity is $1.48\times10^{-5}\,\text{m}^2/\text{s}$ and the mass diffusivity of ethanol in air is $1.26\times10^{-5}\,\text{m}^2/\text{s}$. For this system,

$$\text{Sc} = \frac{\nu}{D_{AB}} = \frac{1.48\times10^{-5}\,\text{m}^2/\text{s}}{1.26\times10^{-5}\,\text{m}^2/\text{s}} = 1.17$$

The Reynolds number for the entire 1-m length is evaluated to be

$$\text{Re}_L = \frac{vL}{\nu} = \frac{(6\,\text{m/s})(1\,\text{m})}{1.48\times10^{-5}\,\text{m}^2/\text{s}} = 4.05\times10^5$$

Since this is greater than 3×10^5, we recognize that there will be a transition point where the boundary layer changes from laminar to turbulent flow. This transition point can be evaluated from the transition Reynolds number, $\text{Re}_t = 3\times10^5$.

$$L_t = \frac{\text{Re}_t\nu}{v} = \frac{(3\times10^5)(1.48\times10^{-5}\,\text{m}^2/\text{s})}{6\,\text{m/s}} = 0.74\,\text{m}$$

We can evaluate the mean mass-transfer coefficient by using equation (28-26) derived in part (a)

$$\bar{k}_c = \frac{0.664(1.26 \times 10^{-5}\ \text{m}^2/\text{s})(3 \times 10^5)^{1/2}(1.17)^{1/3}}{1\ \text{m}}$$

$$+ \frac{0.0365(1.26 \times 10^{-5}\ \text{m}^2/\text{s})(1.17)^{1/3}[(4.05 \times 10^5)^{4/5} - (3 \times 10^5)^{4/5}]}{1\ \text{m}}$$

$$= 0.00483 + 0.00313 = 0.00796\ \text{m/s}$$

The concentration of ethyl alcohol in the vapor immediately above the liquid surface can be evaluated by

$$c_{A,s} = \frac{P_A}{RT} = \frac{4000\ \text{Pa}}{\left(8.314\dfrac{\text{Pa} \cdot \text{m}^3}{\text{mol} \cdot \text{K}}\right)(289\ \text{K})}$$

$$= 1.66\ \text{mol/m}^3$$

The amount of alcohol leaving the surface is $\bar{k}_c(c_{A,s} - c_{A,\infty})$ or

$$W_A = (0.00796\ \text{m/s})(1\ \text{m}^2)(1.66\ \text{mol/m}^3)$$

$$= 1.32 \times 10^{-2}\ \text{mol/s} = 0.793\ \text{mol/60 s} \quad (0.104\ \text{lb mole/hr})$$

28.5 APPROXIMATE ANALYSIS OF THE CONCENTRATION BOUNDARY LAYER

When the flow is other than laminar or the configuration is other than a flat plate, few exact solutions presently exist for the transport in a boundary layer. The approximate method developed by von Kármán to describe the hydrodynamic boundary layer can be used for analyzing the concentration boundary layer. The use of this approach was discussed in Chapters 12 and 19.

Consider a control volume which is located in the concentration boundary layer as illustrated in Figure 28.5. This volume, designated by the dashed line, has

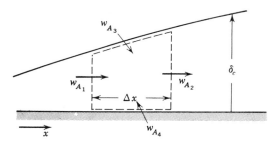

Figure 28.5 The concentration boundary-layer control volume.

a width of Δx, a height equal to the thickness of the concentration boundary layer, δ_c, and a unit depth. A steady-state molar mass balance over the control volume produces the relation

$$W_{A_1} + W_{A_3} + W_{A_4} = W_{A_2} \tag{28-27}$$

where W_A is the molar rate of mass transfer of component A. At each surface, the molar rate is expressed as

$$W_{A_1} = \int_0^{\delta_c} c_A v_x \, dy \bigg|_x$$

$$W_{A_2} = \int_0^{\delta_c} c_A v_x \, dy \bigg|_{x+\Delta x}$$

$$W_{A_3} = c_{A,\infty} \left[\frac{\partial}{\partial x} \int_0^{\delta_c} v_x \, dy \right] \Delta x$$

and

$$W_{A_4} = k_c (c_{A,s} - c_{A,\infty}) \, \Delta x$$

In terms of these molar rates, equation (28-27) may be rewritten as

$$\int_0^{\delta_c} c_A v_x \, dy \bigg|_x + c_{A,\infty} \left[\frac{\partial}{\partial x} \int_0^{\delta_c} v_x \, dy \right] \Delta x + k_c (c_{A,s} - c_{A,\infty}) \, \Delta x = \int_0^{\delta_c} c_A v_x \, dy \bigg|_{x+\Delta x} \tag{28-28}$$

Rearranging, dividing each term by Δx, and evaluating the results in the limit as Δx approaches zero, we obtain

$$\frac{d}{dx} \int_0^{\delta_c} c_A v_x \, dy = c_{A,\infty} \left[\frac{d}{dx} \int_0^{\delta_c} v_x \, dy \right] + k_c (c_{A,s} - c_{A,\infty})$$

or

$$\frac{d}{dx} \int_0^{\delta_c} (c_A - c_{A,\infty}) v_x \, dy = k_c (c_{A,s} - c_{A,\infty}) \tag{28-29}$$

Equation (28-29) is analogous to equations (12-38) and (19-30). In order to solve equation (28-29), the velocity and the concentration profiles must be known; normally these profiles are unknown and must be assumed. Some of the boundary conditions which must be satisfied by the assumed boundary conditions are

(1) $\qquad\qquad\qquad v_x = 0 \qquad$ at $y = 0$

(2) $\qquad\qquad\qquad v_x = v_\infty \qquad$ at $y = \delta$

(3) $\qquad\qquad\qquad \dfrac{\partial v_x}{\partial y} = 0 \qquad$ at $y = \delta$

and, according to equation (12-33)

(4)
$$\frac{\partial^2 v_x}{\partial y^2} = 0 \qquad \text{at } y = 0$$

The assumed concentration profile must satisfy the corresponding boundary conditions in terms of concentrations

(1) $c_A - c_{A,s} = 0$ at $y = 0$ (28-30)

(2) $c_A - c_{A,s} = c_{A,\infty} - c_{A,s}$ at $y = \delta_c$ (28-31)

(3) $\frac{\partial}{\partial y}(c_A - c_{A,s}) = 0$ at $y = \delta_c$ (28-32)

and

(4) $\frac{\partial^2}{\partial y^2}(c_A - c_{A,s}) = 0$ at $y = 0$ (28-33)

If we reconsider the laminar flow parallel to a flat surface, we can use the von Kármán integral equation (28-29) to obtain an approximate solution; the results can be compared to the exact solution, equation (28-20) and thus verify how well we have assumed the velocity and the concentration profiles. As our first approximation, let us consider a power-series expression for the concentration variation with y

$$c_A - c_{A,s} = a + by + cy^2 + dy^3$$

Application of the boundary conditions will result in the following expression:

$$\frac{c_A - c_{A,s}}{c_{A,\infty} - c_{A,s}} = \frac{3}{2}\left(\frac{y}{\delta_c}\right) - \frac{1}{2}\left(\frac{y}{\delta_c}\right)^3 \tag{28-34}$$

If the velocity profile is assumed in the same power-series form, then the resulting expression, as obtained in Chapter 12, is

$$\frac{v_x}{v_\infty} = \frac{3}{2}\left(\frac{y}{\delta}\right) - \frac{1}{2}\left(\frac{y}{\delta}\right)^3 \tag{12-40}$$

Upon substituting equations (28-34) and (12-40) into the integral expression (28-29) and solving, we obtain

$$\text{Nu}_{x,AB} = 0.36\text{Re}_x^{1/2}\text{Sc}^{1/3} \tag{28-35}$$

which is close to the exact solution expressed in equation (28-20).

Although this result is not the correct relation, it is sufficiently close to the exact solution to indicate that the integral method may be used with some degree of confidence in other situations in which an exact solution is unknown. The accuracy of the method depends entirely on the ability to assume good velocity and concentration profiles.

The von Kármán integral equation (28-29) has been used to obtain an approximate solution for the turbulent boundary layer over a flat plate. With the velocity profile approximated by

$$v_x = \alpha + \beta y^{1/7}$$

and the concentration profile approximated by

$$c_A - c_{A,\infty} = \eta + \xi y^{1/7}$$

the local Nusselt number for the turbulent layer is found to be

$$\text{Nu}_{x,AB} = 0.0292 \, \text{Re}_x^{4/5} \tag{28-24}$$

The reader is encouraged to carry out the derivation of equation (28-24) in one of the problems at the end of this chapter.

28.6 MASS, ENERGY, AND MOMENTUM TRANSFER ANALOGIES

In the previous analyses of convective mass transfer, we have recognized the similarities in the differential equations for momentum, energy, and mass transfer and in the boundary conditions when the transport gradients were expressed in terms of dimensionless variables. These similarities have permitted us to predict solutions for the similar transfer processes. In this section, we shall consider several analogies among transfer phenomena which have been proposed because of the similarity in their mechanisms. The analogies are useful in understanding the transfer phenomena and as a satisfactory means for predicting behavior of systems for which limited quantitative data are available.

The similarity among the transfer phenomena and, accordingly, the existence of the analogies, require that the following five conditions exist within the system:

1. The physical properties are constant.

2. There is no energy or mass produced within the system. This, of course, infers that no homogeneous chemical reactions may occur.

3. There is no emission or absorption of radiant energy.

4. There is no viscous dissipation.

5. The velocity profile is not affected by the mass transfer; thus there is a low rate of mass transfer.

REYNOLDS ANALOGY

The first recognition of the analogous behavior of momentum and energy transfer was reported by Reynolds.* Although this analogy is limited in application, it has served as the catalyst for seeking better analogies, and it has been

---

* O. Reynolds, *Proc. Manchester Lit. Phil. Soc.*, **8** (1874).

used successfully in analyzing the complex boundary-layer phenomena of aerodynamics.

Reynolds postulated that the mechanisms for transfer of momentum and energy were identical. We have observed in our earlier discussions on laminar boundary layers that this is true if the Prandtl number, Pr, is unity. From our previous consideration in section 28.4, we can extend the Reynolds postulation to include the mechanism for the transfer of mass if the Schmidt number, Sc, is also unity. For example, if we consider the laminar flow over a flat plate where Sc = 1, the concentration and velocity profiles within the boundary layers are related by

$$\frac{\partial}{\partial y}\left(\frac{c_A - c_{A,s}}{c_{A,\infty} - c_{A,s}}\right)\Bigg|_{y=0} = \frac{\partial}{\partial y}\left(\frac{v_x}{v_\infty}\right)\Bigg|_{y=0} \tag{28-36}$$

Recalling that at the boundary next to the plate, where $y = 0$, we may express the mass flux in terms of either the mass-diffusivity or the mass-transfer coefficient by

$$N_{A,y} = -D_{AB}\frac{\partial}{\partial y}(c_A - c_{A,s})\big|_{y=0} = k_c(c_{A,s} - c_{A,\infty}) \tag{28-37}$$

We can combine equations (28-36) and (28-37) and take advantage that D_{AB} equals μ/ρ when the Schmidt number equals 1 to achieve an expression that relates the mass-transfer coefficient to the velocity gradient at the surface,

$$k_c = \frac{\mu}{\rho v_\infty}\frac{\partial v_x}{\partial y}\bigg|_{y=0} \tag{28-38}$$

The coefficient of skin friction was related in Chapter 12 to this same velocity gradient by

$$C_f = \frac{\tau_0}{\rho v_\infty^2/2} = \frac{2\mu(\partial v_x/\partial y)|_{y=0}}{\rho v_\infty^2} \tag{12-2}$$

Using this definition, we can rearrange equation (28-38) to obtain the mass-transfer Reynolds analogy for systems with a Schmidt number of one,

$$\frac{k_c}{v_\infty} = \frac{C_f}{2} \tag{28-39}$$

Equation (28-39) is analogous to the energy-transfer Reynolds' analogy for systems with a Prandtl number of one. This analogy was discussed in Chapter 19 and may be expressed by

$$\frac{h}{\rho v_\infty c_p} = \frac{C_f}{2} \tag{19-36}$$

Experimental data for mass transfer into gas streams agree approximately with equation (28-39) if the system has a Schmidt number near 1, and if the resistance to flow is that due to skin friction; accordingly, equation (28-39) should not be used to describe situations in which form drag is involved.

TURBULENT-FLOW CONSIDERATIONS

In a majority of practical applications the flow in the main stream is turbulent rather than laminar. Although many investigators have contributed considerably to the understanding of turbulent flow, so far no one has succeeded in predicting convective transfer coefficients or friction factors by direct analysis. This is not too surprising when we recall from our earlier discussions on turbulent flow, in section 13.1, the flow at any point is subject to irregular fluctuations in direction and velocity. Accordingly, any particle of the fluid undergoes a series of random movements, superimposed on the main flow. These eddy movements bring about mixing throughout the turbulent core. This process is often referred to as "eddy diffusion." The value of the eddy mass diffusivity will be very much larger than the molecular diffusivity in the turbulent core.

In an effort to characterize this type of motion, Prandtl proposed the mixing-length hypothesis as discussed in Chapter 13. In this hypothesis, any velocity fluctuation v_x' is due to the y-directional motion of an eddy through a distance equal to the mixing length L. The fluid eddy, possessing a mean velocity, $\bar{v}_x|_y$, is displaced into a stream where the adjacent fluid has a mean velocity, $\bar{v}_x|_{y+L}$. The velocity fluctuation is related to the mean-velocity gradient by

$$v_x' = \bar{v}_x|_{y+L} - \bar{v}_x|_y = \pm L \frac{d\bar{v}_x}{dy} \tag{13-10}$$

The total shear stress in a fluid was defined by the expression

$$\tau = \mu \frac{d\bar{v}_x}{dy} - \rho \overline{v_x' v_y'} \tag{13-8}$$

The substitution of equation (13-10) into (13-8) gives

$$\tau = \rho[\nu + L v_y'] \frac{d\bar{v}_x}{dy} \tag{28-40}$$

or

$$\tau = \rho[\nu + \epsilon_M] \frac{d\bar{v}_x}{dy} \tag{28-41}$$

where $\epsilon_M = L v_y'$ is designated the eddy momentum diffusivity. It is analogous to the molecular momentum diffusivity, ν.

We may now similarly analyze mass transfer in turbulent flow, since this transport mechanism is also due to the presence of the fluctuations or eddies. In Figure 28.6, the curve represents a portion of the turbulent concentration profile with mean flow in the x direction. The instantaneous rate of transfer of component A in the y direction is

$$N_{A,y} = c_A' v_y' \tag{28-42}$$

where $c_A = \bar{c}_A + c_A'$, the temporal average plus the instantaneous fluctuation in the concentration of component A. We can again use the concept of the mixing

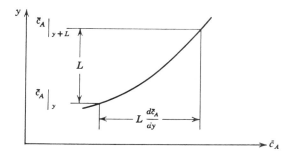

Figure 28.6 Portion of turbulent concentration profile curve, showing the Prandtl mixing length.

length to define the concentration fluctuation by the following relation:

$$c_A{}' = \bar{c}_A|_{y+L} - \bar{c}_A|_y = L\frac{d\bar{c}_A}{dy} \tag{28-43}$$

Inserting equation (28-43) into equation (28-42), we obtain an expression for the turbulent transfer of mass by eddy transport. The total mass transfer normal to the direction of flow is

$$N_{A,y} = -D_{AB}\frac{d\bar{c}_A}{dy} - \overline{v'_yL}\frac{d\bar{c}_A}{dy}$$

or

$$N_{A,y} = -(D_{AB} + \epsilon_D)\frac{d\bar{c}_A}{dy} \tag{28-44}$$

where $\epsilon_D = \overline{Lv_y{}'}$ is designated as the eddy mass diffusivity.

By similar reasoning, an expression was derived in Chapter 19 for convective heat transfer,

$$\frac{q_y}{A} = -\rho c_p(\alpha + \epsilon_H)\frac{d\bar{T}}{dy} \tag{19-49}$$

where α is the molecular thermal diffusivity and ϵ_H is the eddy thermal diffusivity.

Eddy diffusion plays an important role in a number of mass-transfer processes. For instance, there is mass transfer between a fluid flowing past solids in heterogeneous catalytic reactors, blast-furnaces, driers and so on. As a result of the eddy diffusion, transport in the turbulent core is rapid, reducing any gradient in composition. As the wall is approached the turbulence is progressively damped until in the immediate neighborhood of the solid surface, it essentially disappears, and the transport is almost entirely by molecular diffusion. The majority of the resistance to transfer occurs in the boundary layer near the surface where the gradient of the composition is steepest.

THE PRANDTL AND VON KÁRMÁN ANALOGIES

In Chapter 19, the Prandtl analogy for heat and momentum transfer was developed when consideration was given to the effect of both the turbulent core and the laminar sublayer. The same reasoning with regard to mass and momentum transfer can be used to develop a similar analogy. For the laminar sublayer the eddy diffusivities of momentum and mass are negligible, and at the surface the shear stress, τ_s, and the mass flux, $N_{A,y,s}$, are constant. Equation (28-41) may be integrated over the thickness of the sublayer, giving

$$\int_0^{v_x|_\xi} dv_x = \frac{\tau_s}{\rho\nu} \int_0^\xi dy$$

or

$$v_x|_\xi = \frac{\tau_s \xi}{\rho\nu} \tag{28-45}$$

Equation (28-44) may also be integrated over the thickness of the sublayer, yielding

$$\int_{c_{A,s}}^{c_A|_\xi} d\bar{c}_A = -\frac{N_{A,y,s}}{D_{AB}} \int_0^\xi dy$$

or

$$(c_{A,s} - c_A|_\xi) = \frac{N_{A,y,s}}{D_{AB}} \xi \tag{28-46}$$

Eliminating ξ from these two equations, we obtain

$$\frac{\rho\nu v_x|_\xi}{\tau_s} = \frac{D_{AB}}{N_{A,y,s}}(c_{A,s} - c_A|_\xi) \tag{28-47}$$

The Reynolds analogy, $k_c/v_\infty = C_f/2 = \tau_s/\rho v_\infty^2$, may be used in the turbulent core, from $y = \xi$ to y at the bulk conditions. The mass flux in the turbulent core becomes

$$N_{A,y} = k_c(c_A|_\xi - c_{A,\infty}) = \frac{\tau_s}{\rho(v_\infty - v_x|_\xi)}(c_A|_\xi - c_{A,\infty}) \tag{28-48}$$

Eliminating $c_A|_\xi$ between equations (28-47) and (28-48) we obtain

$$\frac{c_{A,s} - c_{A,\infty}}{N_{A,y}} = \frac{\rho}{\tau_s}\left[v_\infty + v_x|_\xi\left(\frac{\nu}{D_{AB}} - 1\right)\right] \tag{28-49}$$

Substituting the defining equations

$$C_f \equiv \frac{\tau_s}{\rho(v_\infty^2/2)}$$

$$k_c \equiv \frac{N_A}{(c_{A,s} - c_{A,\infty})}$$

and

$$Sc \equiv \frac{\nu}{D_{AB}}$$

into equation (28-49), we may simplify the relation to

$$\frac{1}{k_c} = \frac{2}{C_f v_\infty^2} [v_\infty + v_{x|\xi}(Sc - 1)]$$

or in slightly different form,

$$\frac{k_c}{v_\infty} = \frac{C_f/2}{1 + (v_{x|\xi}/v_\infty)(Sc - 1)} \tag{28-50}$$

Note that equation (28-50) simplifies to the Reynolds analogy with the restriction $Sc = 1$. In Chapter 13, the laminar sublayer was defined by $v^+ = y^+ = 5$, where $v^+ = v_{x|\xi}/(v_\infty \sqrt{c_f/2})$; thus

$$v^+ = \frac{v_{x|\xi}}{v_\infty \sqrt{C_f/2}} = 5$$

or

$$\frac{v_{x|\xi}}{v_\infty} = 5\sqrt{\frac{C_f}{2}} \tag{28-51}$$

Substituting for $v_{x|\xi}/v_\infty$ in equation (28-50), we obtain an analogy for convective mass transfer similar to the *Prandtl analogy* for convective heat transfer,

$$\frac{k_c}{v_\infty} = \frac{C_f/2}{1 + 5\sqrt{C_f/2}(Sc - 1)} \tag{28-52}$$

Rearranging and multiplying both sides of equation (28-52) by $v_\infty L/D_{AB}$, where L is a characteristic length, we obtain

$$\frac{k_c}{v_\infty} \frac{v_\infty L}{D_{AB}} = \frac{(C_f/2)(v_\infty L/D_{AB})(\rho\mu/\rho\mu)}{1 + 5\sqrt{C_f/2}(Sc - 1)}$$

or

$$Nu_{L,AB} = \frac{(C_f/2)ReSc}{1 + 5\sqrt{C_f/2}(Sc - 1)} \tag{28-53}$$

Equations (28-52) and (28-53) are analogous to the Prandtl momentum-energy transfer analogy, equation (19-57).

von Kármán extended the Prandtl analogy by considering the so-called "buffer layer" in addition to the laminar sublayer and the turbulent core. This led to the development of the *von Kármán analogy*,

$$Nu = \frac{(C_f/2)RePr}{1 + 5\sqrt{C_f/2}\{Pr - 1 + \ln[(1 + 5Pr)/6]\}} \tag{19-58}$$

for momentum and energy transfer. The von Kármán analysis for mass transfer yields

$$\text{Nu}_{AB} = \frac{(C_f/2)\text{ReSc}}{1 + 5\sqrt{C_f/2}\{\text{Sc} - 1 + \ln\left[(1 + 5\text{Sc})/6\right]\}} \tag{28-54}$$

or

$$\frac{k_c}{v_\infty} = \frac{C_f/2}{1 + 5\sqrt{C_f/2}\{\text{Sc} - 1 + \ln\left[(1 + 5\text{Sc})/6\right]\}} \tag{28-55}$$

The results of most analogies can be put in a general form, as illustrated in equations (28-52) and (28-55) in which the denominator of the right-hand side is a complex group of terms which serve as a correction to the simple Reynolds analogy.

CHILTON-COLBURN ANALOGY

Chilton and Colburn,[*] using experimental data, sought modifications to the Reynold's analogy that would not have the restrictions that Pr and Sc numbers must be equal to 1. They defined the *j factor for mass transfer*,

$$j_D \equiv \frac{k_c}{v_\infty}(\text{Sc})^{2/3}$$

This factor is analogous to the *j* factor for heat transfer that is defined by equation (19-39). Based on data collected in both laminar and turbulent flow regimes, they found

$$j_D \equiv \frac{k_c}{v_\infty}(\text{Sc})^{2/3} = \frac{C_f}{2} \tag{28-56}$$

This analogy is valid for gases and liquids within the range of $0.6 < \text{Sc} < 2500$. Equation (28-56) can be shown to satisfy the exact solution for laminar flow over a flat plate,

$$\text{Nu}_{x,AB} = 0.332\text{Re}_x^{1/2}\text{Sc}^{1/3} \tag{28-20}$$

If both sides of this equation are divided by $\text{Re}_x \text{Sc}^{1/3}$, we obtain

$$\frac{\text{Nu}_{x,AB}}{\text{Re}_x \text{Sc}^{1/3}} = \frac{0.332}{\text{Re}_x^{1/2}} \tag{28-57}$$

This equation reduces to the *Chilton-Colburn analogy* when we substitute into the above expression the Blasius solution for the laminar boundary layer,

$$\frac{\text{Nu}_{x,AB}}{\text{Re}_x \text{Sc}^{1/3}} = \frac{\text{Nu}_{x,AB}}{\text{Re}_x \text{Sc}}\text{Sc}^{2/3} = \frac{C_f}{2}$$

[*] A. P. Colburn, *Trans. A.I.Ch.E.*, **29**, 174–210 (1933); T. H. Chilton and A. P. Colburn, *Ind. Eng. Chem.*, **26**, 1183 (1934).

or

$$\left(\frac{k_c x}{D_{AB}}\right)\left(\frac{\mu}{x v_\infty \rho}\right)\left(\frac{\rho D_{AB}}{\mu}\right)(\text{Sc})^{2/3} = \frac{k_c \text{Sc}^{2/3}}{v_\infty} = \frac{C_f}{2} \tag{28-58}$$

The complete Chilton-Colburn analogy is

$$j_H = j_D = \frac{C_f}{2} \tag{28-59}$$

which relates all three types of transport in one expression. Equation (28-59) is exact for flat plates, and is satisfactory for systems of other geometry provided no form drag is present. For systems where form drag is present, it has been found that neither j_H or j_D equals $C_f/2$; however, when form drag is present

$$j_H = j_D \tag{28-60}$$

or

$$\frac{h}{\rho v_\infty c_p}(\text{Pr})^{2/3} = \frac{k_c}{v_\infty}(\text{Sc})^{2/3} \tag{28-61}$$

Equation (28-61) relates convective heat and mass transfer; it permits the evaluation of one unknown transfer coefficient through information obtained for another transfer phenomenon. It is valid for gases and liquids within the ranges $0.6 < \text{Sc} < 2500$ and $0.6 < \text{Pr} < 100$.

The Chilton-Colburn analogy for heat and mass transfer has been observed to hold for many different geometries; for example; flow over flat plates, flow in pipes, and flow around cylinders. In the following two examples, we will apply the Chilton-Colburn analogy to predict a correlating equation for mass transfer and to derive the important wet-bulb line equation that is used in psychrometric charts.

EXAMPLE 2

Dittus and Boelter proposed the following equation for correlating the heat-transfer coefficient for turbulent flow in a pipe

$$\text{Nu} = \frac{hD}{k} = 0.023 \, \text{Re}^{0.8} \, \text{Pr}^{1/3}$$

What should be the corresponding equation for the mass-transfer coefficient when the transfer is to a turbulent fluid flowing in a pipe?

According to the Chilton-Colburn relationship (28-61)

$$\frac{h}{\rho v_\infty c_p}(\text{Pr})^{2/3} = \frac{k_c}{v_\infty}(\text{Sc})^{2/3}$$

or

$$h = k_c \rho c_p \left(\frac{\text{Sc}}{\text{Pr}}\right)^{2/3}$$

Upon substituting this into the Dittus-Boelter equation, we obtain

$$k_c \rho c_p \left(\frac{Sc}{Pr}\right)^{2/3} \frac{D}{k} = 0.023 \ Re^{0.8} \ Pr^{1/3}$$

or

$$\frac{k_c D}{D_{AB}} \frac{D_{AB} \rho}{\mu} \frac{\mu c_p}{k} \left(\frac{Sc}{Pr}\right)^{2/3} = 0.023 \ Re^{0.8} \ Pr^{1/3}$$

$$\frac{k_c D}{D_{AB}} \cdot \frac{1}{Sc} \cdot Pr \left(\frac{Sc}{Pr}\right)^{2/3} = 0.023 \ Re^{0.8} \ Pr^{1/3}$$

This simplifies to

$$Nu_{AB} = \frac{k_c D}{D_{AB}} = 0.023 \ Re^{0.8} \ Pr^{1/3}$$

Linton and Sherwood,[*] considering mass transfer into turbulent streams flowing through pipes, correlated their data by

$$\frac{k_c D}{D_{AB}} = 0.023 \ Re^{0.83} \ Pr^{1/3}$$

for

$$2000 < Re < 70\,000$$

$$1000 < Sc < 2260$$

EXAMPLE 3

Dry air at atmospheric pressure blows across a thermometer whose bulb has been covered with a dampened wick. This classical "wet-bulb" thermometer indicates a steady-state temperature reached by a small amount of liquid evaporating into a large amount of unsaturated vapor-gas mixture. The thermometer reads 65°F. At this temperature the following properties were evaluated:

vapor pressure of water	0.3056 psi
density of air	0.076 lb_m/ft^3
latent heat of vaporization of water	1057 Btu/lb_m
Prandtl number	0.72
Schmidt number	0.61
specific heat, c_p, of air	0.24 $Btu/lb_m\,°F$

What is the temperature of the dry air?

Equation (28-1) defines the molar flux of water evaporating

$$N_{H_2O} = k_c (c_{H_2O,s} - c_{H_2O,\infty})$$

The energy required to evaporate this water is supplied by convective heat transfer; thus

$$\frac{q}{A} = h(T_\infty - T_s) = \lambda M_{H_2O} N_{H_2O}$$

[*] W. H. Linton and T. K. Sherwood, *Chem. Engr. Prog.*, **46**, 258 (1950).

where λ is the latent heat of vaporization of water at the surface temperature. This equation may be solved for the bulk temperature

$$T_\infty = \frac{\lambda M_{H_2O} N_{H_2O}}{h} + T_s$$

If we substitute equation (28-1) into this equation, we obtain

$$T_\infty = \lambda M_{H_2O} \frac{k_c}{h}(c_{H_2O,s} - c_{H_2O,\infty}) + T_s$$

Chilton-Colburn j factors give us a relationship for the k_c/h ratio

$$j_H = j_D$$

$$\frac{h}{\rho v_\infty c_p}(Pr)^{2/3} = \frac{k_c}{v_\infty}(Sc)^{2/3}$$

$$\frac{k_c}{h} = \frac{1}{\rho c_p}\left(\frac{Pr}{Sc}\right)^{2/3}$$

When this expression is substituted into our equation for the bulk temperature, we obtain

$$T_\infty = \frac{\lambda M_{H_2O}}{\rho c_p}\left(\frac{Pr}{Sc}\right)^{2/3}(c_{H_2O,s} - c_{H_2O,\infty}) + T_s$$

The concentrations are

$$c_{H_2O,s} = \frac{\left(\frac{0.3056}{14.7}\,atm\right)}{\left(0.73\frac{atm \cdot ft^3}{lb\ mole\ °R}\right)(525°R)} = 5.42 \times 10^{-5}\frac{lb\ mole}{ft^3}$$

$$c_{H_2O,\infty} = 0$$

Upon substitution of the known values, we obtain

$$T = \frac{(1057\ Btu/lb_m)(18\ lb_m/lb\ mole)}{(0.076\ lb_m/ft^3)(0.24\ Btu/lb_m\ °F)}\left(\frac{0.72}{0.61}\right)^{2/3}\left(5.42 \times 10^{-5}\frac{lb_m}{ft^3}\right) + 65°F$$

$$T = 128.2°F \quad (326\ K)$$

28.7 MODELS FOR CONVECTIVE MASS-TRANSFER COEFFICIENTS

Convective mass-transfer coefficients have been used in the design of mass-transfer equipment for many years. However, in most cases, they have been empirical coefficients which were determined from experimental investigations. A theoretical explanation of the coefficients will require a better understanding of the mechanism of turbulence, since they are directly related to the dynamic characteristics of the flow. In Chapter 26, two possible models for explaining convective mass transfer were introduced. Both the film theory and the penetration theory have been widely applied.

The *film theory* is based upon the presence of a fictitious film of fluid in laminar flow next to the boundary which offers the same resistance to mass transfer as actually exists in the entire flowing fluid. In other words, all resistance to transfer is assumed to exist in a fictitious film in which the transport is entirely by molecular diffusion. The film thickness, δ, must extend beyond the laminar sublayer to include an equivalent resistance encountered as the concentration changes within the buffer layer and the turbulent core. For diffusion through a nondiffusing layer or stagnant fluid this theory predicts the mass-transfer coefficient to be

$$k_c = \frac{D_{AB}}{\delta} \frac{P}{p_{B,1m}} \tag{26-9}$$

as developed in Chapter 26. For equimolar counterdiffusion, the mass-transfer coefficient was expressed as

$$k_c{}^0 = \frac{D_{AB}}{\delta} \tag{26-24}$$

In both cases, the convective mass-transfer coefficient is directly related to the molecular mass diffusivity. Obviously, the fictitious film thickness, δ, can never be measured, since it does not exist. Because of this and because of its apparent inadequacy in physically explaining convective mass transfer, other theories and models have been postulated to describe this phenomenon.

The *penetration theory* was originally proposed by Higbie[*] to explain the mass transfer in the liquid phase during gas absorption. It has been applied to turbulent flow by Danckwerts[†] and many other investigators when the diffusing component only penetrates a short distance into the phase of interest because of its rapid disappearance through chemical reaction or its relatively short time of contact.

Higbie considered mass to be transferred into the liquid phase by unsteady-state molecular transport. With this concept, the mass flux at the interface between the liquid and the gas phases was expressed as

$$N_A = \sqrt{\frac{D_{AB}}{\pi t_{\exp}}}(c_{A,s} - c_{A,\infty}) \tag{26-86}$$

Danckwerts applied this unsteady-state concept to the absorption of component A in a turbulent liquid stream. His model assumes that the motion of the liquid is constantly bringing fresh liquid eddies from the interior up to the surface, where they displace the liquid elements previously on the surface. While on the surface,

[*] R. Higbie, *Trans. A.I.Ch.E.*, **31**, 368–389 (1935).
[†] P. V. Danckwerts, *Ind. Eng. Chem.*, **43**, 1460–67 (1951).

each element of the liquid becomes exposed to the second phase and mass is transferred into the liquid as though it were stagnant and infinitely deep; the rate of transfer is dependent upon the exposure time. Many different assumptions can be made relative to the surface renewal. For instance, each element of the surface may have the same exposure time before being replaced; this infers that the instantaneous mass transfer will occur according to equation (26-86). The total solute penetrating the eddy in an exposure time, t_{exp}, is

$$\int_0^{t_{exp}} N_A \, dt = (c_{A,s} - c_{A,\infty}) \sqrt{\frac{D_{AB}}{\pi}} \int_0^{t_{exp}} t^{-1/2} \, dt$$

$$= 2(c_{A,s} - c_{A,\infty})\left(\frac{D_{AB}t_{exp}}{\pi}\right)^{1/2}$$

and the average rate of transfer during the exposure is obtained by dividing this equation by the time of exposure

$$N_A = 2(c_{A,s} - c_{A,\infty})\left(\frac{D_{AB}}{\pi t_{exp}}\right)^{1/2} \tag{28-62}$$

Danckwerts modified the assumption of constant exposure period by proposing an "infinite" range of ages for the elements at the surface. Surface age distribution functions were introduced to predict the probability of an element of surface being replaced by a fresh eddy. The rate of surface renewal was believed to be constant for a given degree of turbulence and equal to a surface-renewal factor s. The rate of mass transfer with random surface renewal is

$$N_A = \sqrt{D_{AB}s}(c_{A,s} - c_{A,\infty}) \tag{28-63}$$

The values of s are currently obtained by experimental investigations. The surface-renewal concept has been very successful in the explanation and analysis of convective mass transfer, particularly when the mass transport is accompanied by chemical reactions in the liquid phase;[*][†] considerable development and experimental verification are needed to define this model clearly.

A detailed discussion of mass-transfer coefficients for chemically reacting systems is not treated in this text. In our earlier discussions of molecular mass transfer associated with a chemical reaction, in section 26.2, mass transfer was shown to depend upon the rate constant of the chemical reaction. We should expect a similar dependency for the convective mass-transfer coefficient. Excellent discussions on this subject are available.[*][†]

Toor and Marchello[‡] have pointed out that the penetration concept of Danckwerts is valid only when the surface renewal is relatively rapid, thus

[*] P. V. Danckwerts, *Gas-Liquid Reactions*, McGraw-Hill Book Co., New York, 1970.

[†] G. Astarita, *Mass Transfer with Chemical Reaction*, Elsevier Publishing Co., Amsterdam, 1967.

[‡] H. L. Toor and J. M. Marchello, *A.I.Ch.E. J.*, **1**, 97 (1958).

providing young elements at the surface on a continuous basis. For older elements at the surface, a steady-state concentration gradient is established as predicted by the film theory; accordingly, the rate of mass transfer should be directly proportional to the molecular mass diffusivity. At low Schmidt numbers, a steady concentration gradient is set up very rapidly in any new surface element so that unless the rate of surface renewal is high enough to remove a major fraction of the surface elements before they are penetrated, most of the surface behaves as older elements. As the Schmidt number increases, the time necessary to set up the steady gradient increases rapidly, and accordingly, relatively low surface renewal rates are sufficient to keep most of the elements from being penetrated. When conditions are such that the surface contains both young and older elements, the transfer characteristics are intermediate between the film and penetration models. The convective mass-transfer coefficients will be proportional to a power of the molecular mass diffusivity between 0.5 and 1.0. These conclusions have been supported by experimental data.

In both the film and penetration models, the mass transfer involves an interface between two moving fluids. When one of the phases is a solid, the fluid velocity parallel to the surface at the interface must be zero; accordingly, we should expect the need of a third model, the boundary layer model, for correlating the data involving a solid subliming into a gas or a solid dissolving into a liquid. For diffusion through a laminar boundary layer, the average mass-transfer coefficient was found to be

$$\bar{k}_c = 0.664 \frac{D_{AB}}{L} \text{Re}_L^{1/2} \text{Sc}^{1/3} \tag{28-21}$$

This shows the mass-transfer coefficient varies as $D_{AB}^{2/3}$, which is typical of boundary-layer calculations.

28.8   CLOSURE

In this chapter, we have discussed the principles of mass transfer by forced convection, the significant parameters which help describe convective mass transfer, and the models proposed to explain the mechanism of convective transport. We have seen that the transfer of mass by convection is intimately related to the dynamic characteristics of the flowing fluid, particularly to the fluid in the vicinity of the boundary. Because of the close similarities in the mechanisms of momentum, energy, and mass transfer, we were able to use the same four methods for evaluating convective mass-transfer coefficients which were originally developed to analyze convective heat-transfer coefficients. In all four analyses, the mass-transfer coefficient was correlated by the general equation

$$\text{Nu}_{AB} = f(\text{Re}, \text{Sc})$$

Mass transfer into turbulent streams was discussed, and the eddy mass diffusivity was defined. Analogies were presented for convective mass transfer into turbulent streams.

PROBLEMS

28.1 Mass-transfer data were obtained for the vaporization of naphthalene into a turbulent gas stream flowing in an annulus. Both the inside rod and the outer pipe of the annular conduit were made of naphthalene. To correlate the data, the investigators predicted that the mass-transfer coefficient would depend on the velocity of the flowing stream, v, the diameter of the rod, D_0, the inside diameter of the outer pipe, D_i, the density and viscosity of the gas medium, and the diffusivity of naphthalene in the gas medium. Arrange these variables into dimensionless groups that the investigators might elect to use to correlate their data.

28.2 In section 28.3, the transfer of mass from the walls of a circular conduit to a fluid flowing in a horizontal conduit was discussed. Reconsider this transport phenomenon; however, this time consider that the transfer is from a vertical tube as a result of a concentration difference, Δc_A. Determine the variables that are necessary to describe the mass-transfer coefficient and arrange these variables in dimensionless groups.

28.3 In a mass-transfer spray column, a liquid is sprayed into a gas stream, and mass is exchanged between the liquid and gas phase. The mass of the drops that are formed from a spray nozzle is considered a function of the nozzle diameter; acceleration of gravity; surface tension of the liquid against the gas; fluid density, viscosity, and velocity; and the viscosity and density of the gas medium. Arrange these variables in dimensionless groups. Should any other variables have been included?

28.4 In applying dimensional analysis to explain the mass-transfer coefficient, one must consider the geometry involved, a variable to explain the flow characteristics of the moving stream, and the properties of the moving stream. Predict the variables that are necessary to explain the mass-transfer coefficient for a gas stream flowing over a flat plate and arrange these variables into dimensionless *pi* groups.

28.5 A long cylinder of porous clay, initially having a uniform water concentration of c_{A_0}, is suddenly inserted into an air stream that has a water content of $c_{A,\infty}$. If the radius of the cylinder is r_0 and the average mass-transfer coefficient from the cylinder into the air stream is k_c, show by means of dimensional analysis that the concentration profile within the cylinder can be expressed in terms of the parameters

$$\frac{C_A(r) - c_{A,\infty}}{c_{A_0} - c_{A,\infty}} \frac{r}{r_0} \frac{D_{AB}}{k_c r_0} \quad \text{and} \quad \frac{D_{AB}t}{r_0^2}$$

28.6 A 1-m square, thin plate of solid naphthalene is oriented parallel to a stream of air flowing at 20 m/s. The air is at 1.013×10^5 Pa and 310 K. The naphthalene plate remains at 290 K; at this temperature, the diffusivity of naphthalene in air is 5.61×10^{-6} m^2/s and its vapor pressure is 26 Pa. Determine

(a) the value of the mass-transfer coefficient at a point 0.20 m downstream from the leading edge;

(b) the moles of naphthalene per hour lost from the section of the plate 0.5 to 0.75 m downstream from the leading edge.

28.7 If the local Nusselt number for the laminar boundary layer that is formed over a flat plate is

$$\mathrm{Nu}_{AB,x} = 0.332 \, \mathrm{Re}_x^{1/2} \, \mathrm{Sc}^{1/3}$$

and for the turbulent boundary layer is

$$\mathrm{Nu}_{AB,x} = 0.0292 \, \mathrm{Re}_x^{4/5} \, \mathrm{Sc}^{1/3}$$

obtain an expression for the average film-transfer coefficient, $\bar{k}_c$, when the Reynolds number for the plate is

(a) $\mathrm{Re}_L = 100\,000$;

(b) $\mathrm{Re}_L = 1\,500\,000$.

The transition from laminar to turbulent flow occurs at $\mathrm{Re}_x = 3 \times 10^5$.

28.8 A pan containing water is placed in a wind tunnel, where it is exposed to a 7 m/s wind. The pan contains water at a uniform depth of 1 cm and a length of 4 m in the direction in which the wind is blowing, and it is quite wide. The water in the pan is at a constant temperature of 292 K and the total pressure on the system is 1.013×10^5 Pa. Under these conditions, the vapor pressure of water is 2000 Pa, the mass diffusivity of water in air is 2.5×10^{-5} m^2/s, and the kinematic viscosity of the air is 1.5×10^{-5} m^2/s. If the transition from laminar to turbulent flow occurs at $\mathrm{Re}_x = 3 \times 10^5$, determine the length of time that is required to evaporate all the water.

28.9 The boundary layer solution for a flat plate predicted the following equations:

for laminar flow: $\quad \dfrac{k_c x}{D_{AB}} = 0.332 \, \mathrm{Re}_x^{1/2} \, \mathrm{Sc}^{1/3}$

for turbulent flow: $\quad \dfrac{k_c x}{D_{AB}} = 0.0292 \, \mathrm{Re}_x^{4/5} \, \mathrm{Sc}^{1/3}$

with the transition occurring at $\mathrm{Re}_x = 3 \times 10^5$.

A 20 ft/s wind flows parallel to a pan containing water. The kinematic viscosity of the air is 1.7×10^{-4} ft^2/s and the mass diffusivity of water in air at the temperature and pressure of the system is 2.8×10^{-4} ft^2/s. Determine

(a) the point value of the film transfer coefficient, k_c, at a distance 4 ft from the leading edge of the pan;

(b) the average $\bar{k}_c$ value for the surface between $x = 1$ and 1.5 ft;

(c) the average $\bar{k}_c$ value for the entire surface if the pan is 5 ft in length.

28.10 A thin plate of solid salt, NaCl, measuring 6 in. by 6 in., is to be dragged through seawater (edgewise) at a velocity of 2 ft/s. The 64°F seawater has a salt concentration of 0.0309 g/cm^3; if saturated, the seawater would have a concentration of 35 g/cm^3. The kinematic viscosity of seawater is approximately 1.10×10^{-5} ft^2/s. Estimate the rate at which the salt goes into solution if the edge effects can be ignored.

28.11 Plot the local mass-transfer coefficient, k_c, when dry air at 340 K and 1.013×10^5 Pa flows over a 2.0-m long pan of water with a velocity of 4 m/s. Assume the water is maintained at a constant temperature of 340 K.

28.12 A 1-ft square reservoir of water is inserted into the bottom of a large air duct. The water temperature is 65°F, creating a surface water vapor concentration of 5.15×10^{-5} lb mole/ft^3. If 0.10 lb$_m$ of water evaporates from the pan per hour when a 20 ft/s airstream flows parallel to the surface of the water, how much water would evaporate if a 10 ft/s airstream were flowing parallel to the surface of the water.

28.13 In using the von Kármán approximate solution for solving the laminar concentration-boundary layer, we must assume a concentration profile. Equation (28-35) was obtained with a power series concentration profile:

$$c_A - c_{A,s} = a + by + cy^2 + dy^3$$

Apply the boundary conditions for a laminar concentration boundary layer and evaluate the constants a, b, c, and d.

28.14 Given, for the form of the velocity and concentration profiles in the case of a turbulent boundary layer over a flat plate, the expression

$$v_x = \alpha + \beta y^{1/7}$$

and

$$c_A - c_{A,\infty} = \eta + \xi y^{1/7}$$

where α, β, η, and ξ are constants to be determined from the appropriate boundary conditions, verify the following expressions for a system with Sc = 1:

(a) The boundary-layer thickness

$$\delta_c = 0.371 x \, (\text{Re}_x)^{-1/5}$$

(b) The local mass-transfer coefficient

$$k_c = 0.0289 v_\infty (\text{Re}_x)^{-1/5}$$

In your solution, it will be important to recall from Chapter 13, in the turbulent layer,

$$\frac{\tau_s}{\rho} = 0.0225 v_\infty^2 \left(\frac{\nu}{v_\infty \delta}\right)^{1/4}$$

28.15 The following concentration profile has been proposed for use in the integral expression that was developed for the concentration boundary layer

$$c_A - c_{A,s} = 1 - a \cos (by)$$

(a) What boundary conditions are necessary for the evaluation of constants a and b?
(b) What complete expression for $c_A - c_{A,s}$ results from application of these boundary conditions?
(c) Is this proposed concentration profile a wise selection? Give reasons for your decision?

28.16 By the approximate integral analysis of the concentration boundary layer, derive the appropriate integral equation of mass transfer from an ablative plate under forced convection; that is, a steady, incompressible two-dimensional flow over a flat porous surface through which a fluid is injected with a velocity v_{y0} that is normal to the surface.

28.17 A concentration profile of the form

$$c_A - c_{A,s} = a + bye^{cy}$$

has been proposed for the use of the von Kármán integral expression that was developed for the concentration boundary layer. Is this proposed concentration profile a wise selection? Give reasons for your decision.

28.18 Assuming a linear velocity distribution and a linear concentration profile in the laminar boundary layer over a flat plate:
(a) Derive the velocity and concentration profile equations.
(b) Upon application of the von Kármán momentum integral equation, the shear stress at the wall can be shown to be

$$\frac{\tau_s}{\rho} = \frac{1}{6} v_\infty^2 \frac{d\delta}{dx}$$

Use this relationship and the solution to von Kármán's concentration integral equation to derive a relationship between the hydrodynamic boundary-layer thickness, δ, the concentration boundary-layer thickness, δ_c, and the Schmidt number.

28.19 The boundary layer solution for a flat plate provided the following equations:

$$\text{laminar flow:} \qquad \frac{k_c x}{D_{AB}} = 0.332 \, \text{Re}_x^{1/2} \, \text{Sc}^{1/3}$$

$$\text{turbulent flow:} \qquad \frac{k_c x}{D_{AB}} = 0.0292 \, \text{Re}_x^{4/5} \, \text{Sc}^{1/3}$$

with the transition occurring at $\text{Re}_x = 3 \times 10^5$.

A 15 ft/s wind flows parallel to a pan containing water. The kinematic viscosity of the air is $1.81 \times 10^{-4} \, \text{ft}^2/\text{s}$, the mass diffusivity of water in air at the temperature and pressure of the system is $2.81 \times 10^{-4} \, \text{ft}^2/\text{s}$, and the thermal diffusivity, α, is $2.37 \times 10^{-4} \, \text{ft}^2/\text{s}$. The density of the air is $0.0735 \, \text{lb}_m/\text{ft}^3$ and the heat capacity of the air is $0.24 \, \text{Btu}/(\text{lb}_m)(^\circ\text{F})$.

(a) Determine the point value of the mass-transfer coefficient, k_c, at a distance of 4.5 ft from the leading edge of the pan.

(b) Based on the reported mass-transfer data, predict the point value of the heat-transfer coefficient at the same location.

28.20 Douglas and Churchill* correlated heat transfer from a single cylinder with the following relation:

$$\text{Nu}_D = 0.46 \, \text{Re}_D^{1/2} + 0.00128 \, \text{Re}$$

when the flowing medium was air. In the case where the fluid is a liquid, it was recommended that the Nusselt number in this relationship be multiplied by $1.1 \, \text{Pr}^{1/3}$, yielding

$$\text{Nu}_D = (0.506 \, \text{Re}^{1/2} + 0.00141 \, \text{Re}) \, \text{Pr}^{1/3}$$

This equation, in association with the Chilton-Colburn analogy, can be used to predict the mass-transfer coefficient for a cylinder.

Estimate the mass-transfer coefficient, k_c, for the dissolution of sodium chloride from a cast cylinder, 1.5 cm in diameter, which is placed normal to a flowing water stream. The velocity of the 300 K water stream is 10 m/s.

28.21 Seider and Tate correlated their heat-transfer data for laminar flow in tubes by the equation

$$\text{Nu}_L = 1.86 \left(\text{Pe} \frac{D}{L} \right)^{1/3}$$

where $\text{Pe} = \text{Re} \, \text{Pr}$. This equation satisfied the data when the viscosity variation effects could be neglected. Intuition suggests that a similar

---

* M. L. M. Douglas and S. W. Churchill, *Chem. Engr. Prog. Symp. Ser.*, **51**, 17, 57 (1956).

equation for mass transfer into laminar flow in tubes should be

$$Nu_{L,AB} = 1.86 \left(Re \ Sc \frac{D}{L} \right)^{1/3}$$

Use the Chilton-Colburn analogy, $j_H = j_D$, to develop an equation for $Nu_{L,AB}$ for laminar flow in tubes. Is the preceding equation correct?

28.22 McAdams[*] presented the following equation for heat transfer from a single sphere

$$Nu_D = 0.37 \ Re_D^{0.6} \ Pr^{1/3}$$

when the gases are flowing in the range $20 < Re_D < 150\ 000$.

Use the Chilton-Colburn analogy to predict an equation for correlating mass-transfer data in this same Reynolds number range.

28.23 Dry air, flowing at a velocity of 1.5 m/s, enters a 6-m-long, 0.15-m-diameter tube at 310 K and 1.013×10^5 Pa pressure. The inner surface of the tube is lined with a felt material (diameter to roughness ratio, D/e, of 10 000), which is continuously saturated with water at 290 K. Assuming constant temperature of the air and the pipe, determine the rate that water must be added to keep the felt continuously saturated. It is important to realize that the bulk composition of the gas stream will be continuously increasing with the length.

28.24 A 1-foot square, thin plate of naphthalene is oriented parallel to a stream of air flowing at 100 fps. The air is at 1 atm pressure and 100°F and the plate is at 85°F. Determine the rate of vaporization from the plate. The diffusivity of naphthalene in air at 32°F and 1 atm is $0.199 \ ft^2/hr$ and the vapor pressure of naphthalene at 85°F is 0.188 mm Hg.

28.25 Air at 100°F and 1 atm flows over a naphthalene mothball. Since naphthalene exerts a vapor pressure of 5 mm Hg at 100°F, it will sublime into the passing airstream, which has a negligibly small concentration of naphthalene in the bulk airstream. Determine
(a) the heat-transfer coefficient;
(b) the mass-transfer coefficient;
(c) the molar flux of naphthalene into the airstream;
if a $\frac{3}{4}$-in. mothball is suspended into a 5 ft/s airstream. The physical properties at the film temperature are:

mass diffusivity $= 0.37 \ ft^2/hr$
kinematic viscosity of air $= 0.651 \ ft^2/hr$

[*] W. H. McAdams, *Heat Transmission*, Third Edition, McGraw-Hill Book Company, New York, 1949.

thermal diffusivity of air $= 0.92 \text{ ft}^2/\text{hr}$
density of air $= 0.071 \text{ lb}_m/\text{ft}^3$
thermal conductivity of air $= 0.0156 \text{ Btu}/(\text{hr})(\text{ft})(°\text{F})$

28.26 Several thin sheets of naphthalene, 0.25 cm thick and 10 cm square, are arranged parallel to each other with their centers at 1 cm intervals. Air at 273 K and 1.013×10^5 Pa enter this sandwich arrangement with a bulk velocity of 15 m/s. At 273 K, the molecular diffusivity for naphthalene in air is $5.14 \times 10^{-6} \text{ m}^2/\text{s}$, the Schmidt number is 2.57, and the vapor pressure of naphthalene is 1 Pa. Determine the concentration of naphthalene in the air as it leaves the arrangement, evaluating the mass-transfer coefficient with
(a) Reynolds analogy;
(b) von Kármán analogy;
(c) Chilton-Colburn analogy.
Determine the length of time the sheets must be exposed before half their mass will have sublimed under these conditions.

28.27 A small water drop, falling through air in a spray drying tower, has its diameter reduced as evaporation occurs from its surface. If we assume the temperature of the liquid within the drop remains at 60°F and the drying air is at 100°F, determine the moisture composition of the drying medium. The following properties are available:

kinematic viscosity of air $= 0.181 \times 10^{-3} \text{ ft}^2/\text{s}$
thermal diffusivity of air $= 0.225 \times 10^{-3} \text{ ft}^2/\text{s}$
mass diffusivity of water in air $= 0.297 \times 10^{-3} \text{ ft}^2/\text{s}$
density of air $= 0.072 \text{ lb}_m/\text{ft}^3$
thermal conductivity of air $= 0.014 \text{ Btu}/(\text{hr})(\text{ft})(°\text{F})$
heat capacity of air $= 0.24 \text{ Btu}/(\text{lb}_m)(°\text{F})$
latent heat of vaporization
  at 60°F $= 1060 \text{ Btu}/\text{lb}_m$
  at 100°F $= 1037 \text{ Btu}/\text{lb}_m$
vapor pressure of water
  at 60°F $= 13.2 \text{ mm Hg}$
  at 100°F $= 49.1 \text{ mm Hg}$

28.28 Monrad and Pelton[*] presented the following correlation for heat-transfer coefficients for water and air in an annular space:

$$\frac{h_i}{c_p G} = 0.023 \left(\frac{D_2}{D_1}\right)^{0.5} \left(\frac{D_e G}{\mu}\right)^{-0.2} \left(\frac{c_p \mu}{k}\right)^{-2/3}$$

where D_1 is the inside diameter of the annulus; D_2 is the outside diameter

---

[*] C. C. Monrad and J. F. Pelton, *Trans. A.I.Ch.E.*, **38**, 593 (1942).

of the annulus; D_e is the equivalent diameter of the annulus; and h_i is the heat-transfer coefficient for the inside wall of the annulus.

In studying rates of diffusion of naphthalene into air, an investigator replaced a 1-ft section of the inner pipe with a naphthalene rod. The annulus was composed of a 2-in.-OD brass inner pipe surrounded by a 3-in.-ID brass pipe. While operating at a mass velocity within the annulus of 2.5 lb_m air/sec ft^2 at 1 atm and 32°F, the investigator determined that the partial pressure of the naphthalene in the exiting gas stream was 0.003 mm. Under the conditions of the investigation, the Schmidt number of the air was 2.57, the viscosity of the air was 0.0175 centipoises, and the vapor pressure of naphthalene was 10 mm. Determine the individual gas-film coefficient predicted by these data.

28.29 A "cooling bag," commonly used for storing water in the desert, is made of porous canvas. A small amount of water diffuses through the canvas and evaporates from the surface of the bag. The evaporation of the water cools the surface of the bag and a temperature driving force is established.

Determine the temperature of the ambient air by using mass-transfer considerations, if the following values are found to hold:

surface temperature of the bag	= 68°F
Prandtl number of the ambient air	= 0.72
Schmidt number of the ambient air	= 0.61
density of the air at film temperature	= 0.072 lb_m/ft^3
viscosity of air at film temperature	= 0.018 cp
thermal conductivity of ambient air	= 0.014 Btu/(hr)(ft)(°F)
heat capacity of ambient air	= 0.24 Btu/(lb_m)(°F)
bulk velocity of ambient air	= 0.5 mile/hr
latent heat of vaporization of water at 68°F	= 1055.5 Btu/lb_m
vapor pressure of water at 68°F	= 0.339 lb_f/in^2
partial pressure of water in the ambient air	= 0.193 lb_f/in^2

28.30 Water at 70°F flows through a 2-in.-ID pipe at an average velocity of 10 fps. A 2-ft section of the pipe is replaced with a tube made of solid sodium chloride. Compare the mass-transfer coefficient of the sodium chloride in water using the mass-transfer forms of the
(a) Prandtl analogy;
(b) von Kármán analogy;
(c) Chilton-Colburn analogy.

28.31 If $\epsilon_M = \epsilon_H = \epsilon_D$ in a tube, determine the conditions that are necessary for the velocity, temperature, and concentrations to be completely similar.

28.32 In studying the sublimation of naphthalene into an airstream, an investigator constructed a 10-ft-long annular duct. The inner pipe was made from a 1-in.-OD, solid naphthalene rod; this was surrounded by a 2-in.-ID naphthalene pipe.

Air at 60°F and an average pressure of 1 atm flowed through the annular space at a bulk velocity of 50 ft/s. At 60°F, naphthalene has a vapor pressure of 0.039 mm Hg and a diffusivity in air of 0.206 ft²/hr. Determine the composition of the airstream exiting from the tube.

28.33 Air passes through a naphthalene tube that has an inside diameter of 2.5 cm, flowing at a bulk velocity of 15 m/s. The air is at 283 K and an average pressure of 1.013×10^5 Pa. Assuming that the change in pressure along the tube is negligible and that the naphthalene surface is at 283 K, determine the length of tube that is necessary to produce a naphthalene concentration in the exiting gas stream of 4.75×10^{-4} mol/m³. At 283 K, naphthalene has a vapor pressure of 3 Pa and a diffusivity in air of 5.40×10^{-6} m²/s.

28.34 If the length of the tube that is described in problem 28.34 had been 2 m, find the rate of naphthalene sublimation from the tube in kg/s.

28.35 Sherwood and Woertz* obtained the following water vapor concentration data in a turbulent stream of carbon dioxide, flowing at a Reynolds number of 102 000 in a duct 5.06 cm wide:

Position, in cm from water wall	Partial pressure of water, in mm Hg
0 (water wall)	21.12
0.27	17.14
0.43	17.00
1.08	16.58
1.72	16.08
2.37	15.74
2.69	15.49
3.34	15.12
4.63	14.27
4.79	13.90
5.06 (brine wall)	9.54

Water was transferred at a constant rate of 7.14 g/min from one vertical wall, covered with water, to the opposite wall, covered by a strong film

*T. K. Sherwood and B. B. Woertz, *Ind. Eng. Chem.* **31**, 1034 (1939).

of $CaCl_2$ solution. The total cross-sectional area was 12 630 cm^2, and the temperature of the experiment was 23°C.

(a) Prepare a graph of the partial pressure of water vs. the position in the duct, and calculate the value of the eddy diffusivity for the main central portion, in which the gradient is essentially a straight line.

(b) What fraction of the total resistance to the water-vapor transfer is offered in the main turbulent core?

29
INTERPHASE
MASS TRANSFER

In the previous chapters, we have discussed the transfer of mass within a single phase. Many mass-transfer operations, however, involve the transfer of material between two contacting phases. These phases may be a gas stream contacting a liquid, two liquid streams if they are immiscible, or a fluid flowing past a solid. In this chapter, we shall consider the mechanism of steady-state mass transfer between phases. Chapter 30 presents empirical equations for mass-transfer coefficients for interphase transfer obtained from experimental investigations, and Chapter 31 presents methods of applying interphase concepts to the design of mass transfer equipment.

29.1 EQUILIBRIUM

The transport of mass by either molecular or convective transport mechanisms has been shown to be directly dependent upon the concentration gradient of the diffusing species within a single phase. When equilibrium is established, the concentration gradient and, in turn, the net diffusion rate of the diffusing species become zero within the phase. Transfer between two phases also requires a departure from equilibrium which might exist between the average or bulk concentrations within each phase. Since the deviation from equilibrium provides the concentration gradient within a phase, it is necessary to consider interphase equilibria in order to describe interphase mass transfer.

It is convenient to consider the equilibrium characteristics of a particular system and then generalize the results for other systems. Consider a two-phase system involving a gas contacting a liquid; for the sake of discussion, let the initial system composition include air and ammonia in the gas phase and only water in the liquid phase. When initially brought into contact, ammonia is transferred into the water in which it is soluble, and water is vaporized into the gas phase. If the gas-liquid mixture is contained within an isothermal, isobaric container, a dynamic equilibrium between the two phases will eventually be established. A portion of the molecules entering the liquid phase returns to the gas phase at a rate dependent upon the concentration of ammonia in the liquid phase and the vapor pressure exerted by the ammonia in the aqueous solution. Similarly, a portion of the water vaporizing into the gas phase recondenses into the solution. Dynamic equilibrium is indicated by a constant concentration of ammonia in the liquid phase and a constant concentration or partial pressure of ammonia in the gas phase.

This equilibrium condition can be altered by adding more ammonia to the isothermal, isobaric container. After a period of time a new dynamic equilibrium will be established with a different concentration of ammonia in the liquid and a different partial pressure of ammonia in the gas. Obviously, one could continue to add more ammonia to the system; each time a new equilibrium will be reached. Figure 29.1 illustrates an equilibrium curve which shows the relationship between

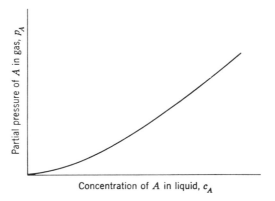

Figure 29.1 Equilibrium distribution of solute A between a gas and a liquid phase at control temperature.

the concentration of the solute in the liquid phase and the partial pressure of the solute in the gas phase. There are many graphical forms of equilibrium data due to the many ways of expressing concentrations in each of the phases. We shall find use for many types of equilibrium plots in Chapter 31.

Equations relating the equilibrium concentrations in the two phases have been developed and are presented in textbooks on thermodynamics. For the case of nonideal gas and liquid phases, the relations are generally complex. However, in cases involving ideal gas and liquid phases, some fairly simple yet useful relations are known. For example, when the liquid phase is ideal, Raoult's law applies,

$$p_A = x_A P_A \tag{29-1}$$

where p_A is the equilibrium partial pressure of component A in the vapor phase above the liquid phase; x_A is the mole fraction of A in the liquid phase; and P_A is the vapor pressure of pure A at the equilibrium temperature. When the gas phase is ideal, Dalton's law is obeyed,

$$p_A = y_A P \qquad (29\text{-}2)$$

where y_A is the mole fraction of A in the gas phase; and P is the total pressure of the system. When both phases are ideal, the two equations may be combined to obtain a relation between the concentration terms, x_A and y_A; at constant pressure and temperature, the combined Raoult-Dalton equilibrium law stipulates

$$y_A P = x_A P_A \qquad (29\text{-}3)$$

Another equilibrium relation which is found to be true for dilute solutions is *Henry's law*. The law is expressed by

$$p_A = Hc_A \qquad (29\text{-}4)$$

where H is the Henry's law constant; and c_A is the equilibrium composition of A in the liquid phase. An equation similar to the Henry's law relation describes the distribution of a solute between two immiscible liquids. This equation, the "distribution-law" equation, is

$$c_{A,\text{liquid},1} = K c_{A,\text{liquid},2} \qquad (29\text{-}5)$$

where c_A is the concentration of solute A in the specified liquid phase; and K is the distribution coefficient.

A complete discussion of equilibria and equilibrium relations must be left to thermodynamics textbooks. However, the following basic concepts common to all systems involving the distribution of a component between two phases are descriptive of interphase mass transfer:

1. At a fixed set of conditions, such as temperature and pressure, Gibbs' phase rule stipulates that a set of equilibrium relations exists which may be shown in the form of an equilibrium distribution curve.

2. When a system is in equilibrium, there is no net mass transfer between the phases.

3. When a system is not in equilibrium, components or a component of the system will be transported in such a manner as to cause the system composition to shift toward equilibrium. If sufficient time is permitted, the system will eventually reach equilibrium.

The following examples illustrate the application of equilibrium relations for determining equilibrium compositions.

EXAMPLE 1

The combined Raoult-Dalton equilibrium relation may be used to determine phase compositions for the binary system, benzene-toluene, at low pressures and temperatures.

Determine the composition of the vapor in equilibrium with a liquid containing 0.6 mole fraction of benzene at 68°F.

The partial pressures of benzene and toluene will be evaluated, using equation (29-1),

$$p_B = x_B P_B \quad \text{and} \quad p_T = x_T P_T$$

At 68°F, the vapor pressures are as follows:

$$P_B = 0.0986 \text{ atm}$$

and

$$P_T = 0.0297 \text{ atm}$$

thus

$$p_B = 0.6(0.0986) = 0.059 \text{ atm}$$

and

$$p_T = 0.4(0.0297) = 0.012 \text{ atm}$$

By Dalton's law, the total pressure is the sum of the partial pressures,

$$P = 0.059 + 0.012 = 0.071 \text{ atm}$$

We can now evaluate the vapor composition, using equation (29-2)

$$y_B = \frac{p_B}{P} = \frac{0.059}{0.071} = 0.83$$

$$y_T = \frac{p_T}{P} = \frac{0.012}{0.071} = 0.17$$

EXAMPLE 2

The Henry's law constant for oxygen dissolved in water is 4.01×10^4 atm/mole fraction at 20°C. Determine the saturation concentration of oxygen in water which is exposed to dry air at 1 atm and 20°C.

Henry's law can be expressed in terms of the mole fraction units by

$$p_A = H' x_A$$

where H' is 4.01×10^4 atm/mole fraction or 4.06×10^9 Pa/mol of O_2 per total mol of solution.

From example 24.1, we recognize that dry air contains 21 mole percent oxygen. By Dalton's law,

$$p_A = y_A P = 0.21(1.013 \times 10^5 \text{ Pa}) = 2.13 \times 10^4 \text{ Pa}$$

The mole fraction of the liquid at the interface is computed using Henry's law

$$x_A = \frac{p_A}{H'} = \frac{2.13 \times 10^4}{4.06 \times 10^9} = 5.25 \times 10^{-6} \frac{\text{mol } O_2}{\text{mol soln}}$$

For one cubic meter of very dilute solution, the moles of water in the solution will be approximately

$$n_{H_2O} = (1 \text{ m}^3)(1 \times 10^3 \text{ kg H}_2\text{O/m}^3)\left(\frac{1}{0.018 \text{ kg/mol}}\right)$$

$$= 5.56 \times 10^4 \text{ mol}$$

The total moles in the solution is essentially the moles of water since the concentration of oxygen is quite low; accordingly the moles of oxygen in one cubic meter of solution is

$$n_{H_2O} = (5.25 \times 10^{-6} \text{ mol O}_2/\text{mol soln})(5.56 \times 10^4 \text{ mol soln})$$

$$= 0.292 \text{ mol of O}_2$$

The saturation concentration is

$$(0.292 \text{ mol/m}^3)(0.032 \text{ kg/mol}) = 9.34 \times 10^{-3} \text{ kg O}_2/\text{m}^3 \qquad (9.34 \text{ mg/liter})$$

29.2 TWO-RESISTANCE THEORY

Interphase mass transfer involves three transfer steps, the transfer of mass from the bulk conditions of one phase to the interfacial surface, transfer across the interface into the second phase, and finally transfer to the bulk conditions of the second phase. A two-resistance theory, initially suggested by Whitman,[*] is often used to explain this process. The theory has two principal assumptions: the rate of mass transfer between the two phases is controlled by the rates of diffusion through the phases on each side of the interface, and no resistance is offered to the transfer of the diffusing component across the interface. The transfer of component A from the gas phase to the liquid phase may be graphically illustrated as in Figure 29.2, with a partial pressure gradient from the bulk gas composition, $p_{A,G}$,

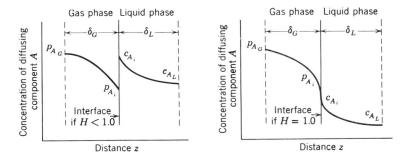

Figure 29.2 Concentration gradients between two contacting phases.

* W. G. Whitman, *Chem. Met. Engr.*, **29**, (4), 147 (1923).

to the interfacial gas composition, $p_{A,i}$, and a concentration gradient in the liquid from, $c_{A,i}$, at the interface to the bulk liquid concentration, $c_{A,L}$. If no resistance to mass transfer exists at the interfacial surface, $p_{A,i}$ and $c_{A,i}$ are equilibrium concentrations; these are the concentration values which would be obtained if the two phases had been in contact for an infinite period of time. The concentrations $p_{A,i}$ and $c_{A,i}$ are related by thermodynamic relations as discussed in section 29.1. The interfacial partial pressure, $p_{A,i}$, can be less than, equal to, or greater than $c_{A,i}$ according to the equilibrium conditions at the temperature and pressure of the system. When the transfer is from the liquid phase to the gas phase, $c_{A,L}$ will be greater than $c_{A,i}$ and $p_{A,i}$ will be greater than $p_{A,G}$.

INDIVIDUAL MASS-TRANSFER COEFFICIENTS

Restricting our discussion to the steady-state transfer of component A, we can describe the rates of diffusion in the z direction on each side of the interface by the equations

$$N_{A,z} = k_G(p_{A,G} - p_{A,i}).$$ (29-6)

and

$$N_{A,z} = k_L(c_{A,i} - c_{A,L})$$ (29-7)

where k_G is the *convective mass-transfer coefficient in the gas phase*, in [moles of A transferred/(time)(interfacial area) (Δp_A units of concentration)]; and k_L is the *convective mass-transfer coefficient in the liquid phase*, in [moles of A transferred/(time)(interfacial area) (Δc_A units of concentration)]. The partial pressure difference, $p_{A,G} - p_{A,i}$, is the driving force necessary to transfer component A from the bulk gas conditions to the interface separating the two phases. The concentration difference, $c_{A,i} - c_{A,L}$, is the driving force necessary to continue the transfer of A into the liquid phase.

Under steady-state conditions, the flux of mass in one phase must equal the flux of mass in the second phase. Combining equations (29-6) and (29-7), we obtain

$$N_{A,z} = k_G(p_{A,G} - p_{A,i}) = -k_L(c_{A,L} - c_{A,i})$$ (29-8)

The ratio of the two convective mass-transfer coefficients may be obtained from equation (29-8) by rearrangement, giving

$$-\frac{k_L}{k_G} = \frac{p_{A,G} - p_{A,i}}{c_{A,L} - c_{A,i}}$$ (29-9)

In Figure 29.3, the application of equation (29-9) for the evaluation of the interfacial compositions for a specific set of bulk compositions as represented by point O is illustrated. The point O represents conditions found at one plane within a mass exchanger. The conditions at another plane could be quite different.

In Table 29.1, the most often encountered individual-phase mass-transfer coefficients are listed and the interrelations between them are noted. A zero

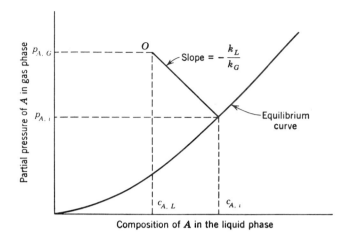

Figure 29.3 Interfacial compositions as predicted by the two-resistance theory.

TABLE 29.1 INDIVIDUAL MASS-TRANSFER COEFFICIENTS

Gas Phase		
Rate equation		Units of coefficient
Diffusion of A through non-diffusing B	Equimolar counterdiffusion	
$N_A = k_G \, \Delta p_A$	$N_A = k_G{}^0 \, \Delta p_A$	$\dfrac{\text{moles of } A \text{ transferred}}{(\text{time})(\text{area})(\text{pressure})}$
$N_A = k_c \, \Delta c_A$	$N_A = k_c{}^0 \, \Delta c_A$	$\dfrac{\text{moles of } A \text{ transferred}}{(\text{time})(\text{area})(\text{mol/volume})}$
$N_A = k_y \, \Delta y_A$	$N_A = k_y{}^0 \, \Delta y_A$	$\dfrac{\text{moles of } A \text{ transferred}}{(\text{time})(\text{area})(\text{mole fraction})}$
$N_A = k_Y \, \Delta Y_A$		$\dfrac{\text{moles of } A \text{ transferred}}{(\text{time})(\text{area})(\text{mol } A/\text{mol } B)}$
$n_A = k_{\mathscr{H}} \, \Delta \mathscr{H}_A$		$\dfrac{\text{mass of } A \text{ transferred}}{(\text{time})(\text{area})(\text{mass } A/\text{mass } B)}$

TABLE 29—(CONT'D.)

Gas Phase

$$k_G = \frac{k_y}{P} = \frac{k_c}{RT} \qquad\qquad k_G{}^0 = \frac{k_G p_{B,lm}}{P}$$

$$k_G{}^0 = \frac{k_y{}^0}{P} = \frac{k_c{}^0}{RT} \qquad\qquad k_y{}^0 = k_y \frac{p_{B,lm}}{P}$$

$$\mathscr{H} = \text{the specific humidity} \qquad k_c{}^0 = k_c \frac{p_{B,lm}}{P} = k_c \frac{c_{B,lm}}{c}.$$

Liquid Phase		
Rate equation		Units of coefficient
Diffusion of A through non-diffusing B	Equimolar counterdiffusion	
$N_A = k_L\,\Delta c_A$	$N_A = k_L{}^0\,\Delta c_A$	$\dfrac{\text{moles of } A \text{ transferred}}{(\text{time})(\text{area})(\text{mol/volume})}$
$N_A = k_x\,\Delta x_A$	$N_A = k_x{}^0\,\Delta x_A$	$\dfrac{\text{moles of } A \text{ transferred}}{(\text{time})(\text{area})(\text{mole fraction})}$
$k_L = \dfrac{k_x}{c}$		$k_L{}^0 = k_L \dfrac{c_{B,lm}}{c} = k_L x_{B,lm}$
$k_L{}^0 = \dfrac{k_x{}^0}{c}$		$k_x{}^0 = k_x x_{B,lm}$

superscript on the mass transfer coefficient for equimolar counterdiffusion is used to designate no net mass transfer into the phase, according to equation (26-24). It is important to realize that there are many other different mass-transfer coefficients for other specific mass-transfer situations; for example, when $N_A = -2N_B$, etc. This table may be helpful in explaining why there are so many different units given for individual coefficients.

OVERALL MASS-TRANSFER COEFFICIENTS

It is quite difficult to measure physically the partial pressure and concentration at the interface. It is therefore convenient to employ overall coefficients based on an overall driving force between the bulk compositions, $p_{A,G}$ and $c_{A,L}$. This treatment is similar to the one used in Chapter 15 when the overall heat transfer coefficient, U, was defined. An *overall mass transfer coefficient* may be defined in terms of a partial pressure driving force. This coefficient, K_G, must account for the entire diffusional resistance in both phases; it is defined by

$$N_A = K_G(p_{A,G} - p_A{}^*) \qquad (29\text{-}10)$$

where $p_{A,G}$ is the bulk composition in the gas phase; $p_A{}^*$ is the partial pressure of A in equilibrium with the bulk composition in the liquid phase, $c_{A,L}$, and K_G is the overall mass-transfer coefficient based on a partial pressure driving force, in moles of A transferred/(time)(interfacial area)(pressure). Since the equilibrium distribution of solute A between the gas and liquid phases is unique at the pressure and temperature of the system, then $p_A{}^*$, in equilibrium with $c_{A,L}$, is as good a measure of $c_{A,L}$ as $c_{A,L}$ itself, and it is on the same basis as $p_{A,G}$. An overall mass transfer coefficient, K_L, including the resistance to diffusion in both phases in terms of liquid phase concentration driving force, is defined by

$$N_A = K_L(c_A{}^* - c_{A,L}) \qquad (29\text{-}11)$$

where $c_A{}^*$ is the concentration of A in equilibrium with $p_{A,G}$ and is accordingly, a good measure of $p_{A,G}$; K_L is the overall mass-transfer coefficient based on a liquid concentration driving force, in [moles of A transferred/(time)(interfacial area) (moles/volume)]. Figure 29.4 illustrates the driving forces associated with

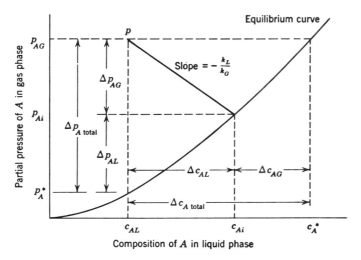

Figure 29.4   Concentration driving forces for two-resistance theory.

each phase and the overall driving forces. The ratio of the resistance in an individual phase to the total resistance may be determined by

$$\frac{\text{resistance in the gas phase}}{\text{total resistance in both phases}} = \frac{\Delta p_{A,\text{gas film}}}{\Delta p_{A,\text{total}}} = \frac{1/k_G}{1/K_G} \tag{29-12}$$

and

$$\frac{\text{resistance in the liquid phase}}{\text{total resistance in both phases}} = \frac{\Delta c_{A,\text{liquid film}}}{\Delta c_{A,\text{total}}} = \frac{1/k_L}{1/K_L} \tag{29-13}$$

A relation between these overall coefficients and the individual phase coefficients can be obtained when the equilibrium relation is linear as expressed by

$$p_{A,i} = m c_{A,i} \tag{29-14}$$

This condition is always encountered at low concentrations, where Henry's law is obeyed; the proportionality constant is then the Henry's law constant, H. Utilizing equation (29-14), we may relate the gas- and liquid-phase concentrations by

$$p_{A,G} = m c_A{}^*$$

$$p_A{}^* = m c_{A,L}$$

and

$$p_{A,i} = m c_{A,i}$$

Rearranging equation (29-10), we obtain

$$\frac{1}{K_G} = \frac{p_{A,G} - p_A{}^*}{N_{A,z}} = \frac{p_{A,G} - p_{A,i}}{N_{A,z}} + \frac{p_{A,i} - p_A{}^*}{N_{A,z}}$$

or in terms of m,

$$\frac{1}{K_G} = \frac{(p_{A,G} - p_{A,i})}{N_{A,z}} + \frac{m(c_{A,i} - c_{A,L})}{N_{A,z}} \tag{29-15}$$

The substitution of equations (29-6) and (29-7) into the above relation relates K_G to the individual phase coefficients by

$$\frac{1}{K_G} = \frac{1}{k_G} + \frac{m}{k_L} \tag{29-16}$$

A similar expression for K_L may be derived as follows:

$$\frac{1}{K_L} = \frac{c_A{}^* - c_{A,L}}{N_{A,z}} = \frac{(p_{A,G} - p_{A,i})}{m N_{A,z}} + \frac{(c_{A,i} - c_{A,L})}{N_{A,z}}$$

or

$$\frac{1}{K_L} = \frac{1}{m k_G} + \frac{1}{k_L} \tag{29-17}$$

Equations (29-16) and (29-17) stipulate that the relative magnitudes of the individual phase resistances depend on the solubility of the gas, as indicated by the magnitude of the proportionality constant. For a system involving a soluble gas, such as ammonia in water, m is very small. From equation (29-16), we may conclude that the gas-phase resistance is essentially equal to the overall resistance in such a system. When this is true, the major resistance to mass transfer lies in the gas phase, and such a system is said to be *gas phase controlled*. Systems involving gases of low solubility, such as carbon dioxide in water, have such a large value of m that equation (29-17) stipulates that the gas-phase resistance may be neglected, and the overall coefficient, K_L, is essentially equal to the individual liquid phase coefficient, k_L. This type of system is designated *liquid-phase controlled*. In many systems, both phase resistances are important and must be considered when evaluating the total resistance.

In Chapter 28, the individual phase convective coefficients, k_L and k_G, were shown to be dependent on the nature of the diffusing component, on the nature of the phase through which the component is diffusing, and also on the flow conditions of the phase. Even when the individual coefficient, k_G, is essentially independent of concentration, the overall coefficient, K_G, may vary with the concentration unless the equilibrium line is straight. This is also true for the overall coefficient, K_L. Accordingly, the overall coefficients should be employed only at conditions similar to those under which they were measured and should not be employed for other concentration ranges unless the equilibrium curve for the system is straight over the entire range of interest.

The two-resistance theory, including the addition of resistances, was proposed by Lewis and Whitman* in 1924 as the two-film theory. Although originally proposed in terms of the film model for convective mass transfer, it is equally applicable to the individual phase coefficients evaluated by either the film or the penetration theory. The assumption of negligible interfacial resistance has not been adequately verified; in fact, many investigators have shown that a resistance does exist if dust particles or other foreign particles are carried by the liquid. Nevertheless, most industrial data have been interpreted in terms of the two-resistance theory.

The application of the two-resistance theory to the absorption of ammonia vapor in water is illustrated in the following example.

EXAMPLE 2

In an experimental study of the absorption of NH_3 by water in a wetted-wall column, the value of K_G was found to be 0.205 lb mole NH_3/hr ft^2 atm. At one point in the column, the gas contained 8 mole percent NH_3 and the liquid phase concentration was 0.004 mole of NH_3 per ft^3 of solution. The temperature was 68°F, and the total pressure was one atmosphere. Eighty-five percent of the total resistance to the mass transfer was found to be in the gas phase. If Henry's constant at 68°F is 0.215 atm/(lb mole NH_3/ft^3 of solution), calculate the individual film coefficients and the interfacial compositions.

*W. K. Lewis and W. G. Whitman, *Ind. Eng. Chem.*, **16**, 1215 (1924).

The total resistance in both phases, according to equation (29-12), is

$$\frac{1}{K_G} = \frac{1}{0.205} = 4.88 \text{ hr ft}^2 \text{ atm/lb mole}$$

Since the resistance in the gas phase, $1/k_G$, is 85% of the total resistance, we may evaluate the individual gas-phase coefficient by

$$\frac{1}{k_G} = 0.85(4.88) = 4.15 \text{ hr ft}^2 \text{ atm/lb mole NH}_3$$

and

$$k_G = \frac{1}{4.15} = 0.241 \text{ lb mole NH}_3/\text{hr ft}^2 \text{ atm}$$

The liquid phase coefficient, k_L, is evaluated as follows:

$$\frac{1}{K_G} = \frac{1}{k_G} + \frac{m}{k_L}$$

or

$$4.88 = 4.15 + \frac{m}{k_L}$$

Thus

$$\frac{m}{k_L} = 0.73 \text{ hr ft}^2 \text{ atm/lb mole}$$

and

$$K_L = \frac{0.215 \text{ atm/(lb mole NH}_3/\text{ft}^3 \text{ solution})}{0.73 \text{ (hr)(ft}^2)(\text{atm})/\text{lb mole}}$$

or

$$k_L = 0.294 \text{ lb mole NH}_3/(\text{hr})(\text{ft}^2)(\text{lb mole NH}^3/\text{ft}^3 \text{ solution})$$

At one point in the column, the bulk concentrations in both phases are

$$p_{A,G} = y_A P = (0.08)(1 \text{ atm}) = 0.08 \text{ atm}$$

and

$$c_{A,L} = 0.004 \text{ lb mole NH}_3/\text{ft}^3 \text{ solution}$$

Upon introducing the Henry's law constant, we find that the partial pressure, p_A^*, in equilibrium with the bulk liquid concentration is

$$p_A^* = Hc_{A,L} = 0.215 \text{ atm/(lb mole NH}_3/\text{ft}^3 \text{ solution})(0.004 \text{ lb mole NH}_3/\text{ft}^3 \text{ solution})$$

or

$$p_A^* = 0.00086 \text{ atm}$$

The mass flux, as expressed by equation (29-10), becomes

$$N_{A,z} = K_G(p_{A,G} - p_A{}^*)$$

$$= 0.205 \frac{\text{lb mole}}{\text{hr ft}^3 \text{ atm}} (0.08 - 0.00086) \text{atm}$$

$$= 1.62 \times 10^{-2} \text{ lb mole/hr ft}^2$$

The interfacial composition can be determined, using equations (29-6) and (29-7), as follows:

$$N_{A,z} = k_G(p_{A,G} - p_{A,i})$$

thus

$$1.62 \times 10^{-2} \frac{\text{lb mole}}{\text{hr ft}^2} = 0.241 \frac{\text{lb mole}}{\text{hr ft}^2 \text{ atm}} (0.08 - p_{A,i}) \text{ atm}$$

$$p_{A,i} = 0.0128 \text{ atm}$$

$$N_{A,z} = k_L(c_{A,i} - c_{A,L})$$

$$1.62 \times 10^{-2} \frac{\text{lb mole}}{\text{hr ft}^2} = 0.294 \frac{\text{lb mole}}{(\text{hr})(\text{ft}^2)(\text{lb mole/ft}^3)} (c_{A,i} - 0.004) \frac{\text{lb mole NH}_3}{\text{ft}^3}$$

and

$$c_{A,i} = 0.117 \frac{\text{lb mole NH}_3}{\text{ft}^3 \text{ solution}} \quad (1.88 \times 10^3 \text{ mole/m}^3)$$

29.3 CLOSURE

In this chapter we have considered the mechanism of steady-state mass transfer between phases. The two-resistance theory was presented. This theory defines the mass transfer in each phase as a function of the concentration driving force and the individual mass-transfer coefficient according to the equations

$$N_{A,z} = k_G(p_{A,G} - p_{A,i})$$

and

$$N_{A,z} = k_L(c_{A,i} - c_{A,L})$$

The overall mass-transfer coefficients were defined by

$$N_{A,z} = K_G(p_{A,G} - p_A{}^*)$$
$$N_{A,z} = K_L(c_A{}^* - c_{A,L})$$

and related to the individual coefficients by the relations

$$\frac{1}{K_G} = \frac{1}{k_G} + \frac{m}{k_L}$$

and

$$\frac{1}{K_L} = \frac{1}{mk_G} + \frac{1}{k_L}$$

PROBLEMS

29.1 Determine the composition of the vapor in equilibrium with a 60 mole percent toluene-40 mole percent benzene liquid mixture that is contained in a closed vessel at 90°C. Predict the pressure at which the equilibrium exists.

Vapor Pressure Data in Millimeters of Mercury

	Temperature, °C						
	60	70	80	90	100	110	120
Benzene	348	540	756	1008	1338	1740	2215
Toluene	150	206	287	404	557	741	990

29.2 Determine the composition of the liquid in equilibrium with a vapor containing 40 mole percent benzene-60 mole percent toluene if the system exists in a vessel under 1 atm pressure. Predict the temperature at which the equilibrium exists.

29.3 Determine the composition of the vapor in equilibrium with a 60 mole percent toluene-40 mole percent benzene liquid solution that is contained in a vessel existing under 1 atm pressure. Predict the temperature at which the equilibrium exists.

29.4 Normal heptane, C_7H_{16}, has a vapor pressure of 1.06×10^5 Pa at 373 K and normal octane, C_8H_{18}, has a vapor pressure of 4.71×10^4 Pa at 373 K.
(a) What would be the composition of a heptane-octane solution that boils at 373 K under a 9.3×10^4 Pa pressure?
(b) What would be the composition of the vapor in equilibrium with the solution that is described in (a)?
(c) What is the total pressure exerted by a vapor in equilibrium with a 373 K solution containing 37 mole percent normal heptane and 63 mole percent normal octane?

29.5 A liquid solution containing 49 mole benzene and 21 mole toluene is slowly heated to a pressure of 1.013×10^5 Pa in a closed container.

(a) Determine the amount of liquid that is present if this mixture is heated to 363 K.

(b) What would be the composition of the liquid phase and the gas phase in equilibrium at 363 K and 1.013×10^5 Pa.

29.6 A solution with oxygen dissolved in water containing 1×10^{-3} g $O_2/100$ g H_2O is in contact with a large volume of ordinary air at 10°C and a total pressure of 1 atm. The Henry's law constant for the oxygen–water system at 10°C, H', equals 3.27×10^4 atm/mole fraction of oxygen in the liquid.

(a) Will the solution gain or lose oxygen?

(b) What will be the concentration of oxygen in the final equilibrium solution?

(c) Will the solution gain or lose oxygen if it is heated to 40°C where the Henry's law constant for the oxygen–water system, H', equals 5.35×10^4 atm/mole fraction of oxygen in the liquid.

29.7 Determine the value of Henry's law constant, in atm/mole fraction of ammonia (as $x_{NH_3,i} \to 0$) for ammonia, NH_3, in water. The following equilibrium data at 298 K were reported in the *Chemical Engineering Handbook*[*]:

partial pressure NH₃, mm Hg	3.4	7.4	9.1	12.0	15.3	19.4	23.5
wt of NH₃/100 wt of water	0.5	1.0	1.2	1.6	2.0	2.5	3.0

29.8 Determine the value of Henry's law constant, in mm Hg/(g mole/liter) of chlorine for the chlorine–water system. The following equilibrium data at 293 K were reported in the *Chemical Engineering Handbook*:[*]

partial pressure of Cl₂, in mm Hg	5	10	30	50	100	150
solubility, in g of Cl₂/liter	0.438	0.575	0.937	1.210	1.773	2.27

29.9 Twenty pounds of dry gaseous ammonia, NH_3, and 500 ft³ of dry air measured at 20°C and 760 mm pressure are mixed and then brought into contact with 100 lb of water in a closed container. There is 1200 ft³ of gas space over the water. After a long period of time, the system reaches equilibrium. Assuming that the gas-space volume remains constant and that the temperature of the system is 20°C, determine

(a) the concentration of ammonia in the liquid in lb_m $NH_3/100$ lb_m H_2O;

[*] J. H. Perry, *Chemical Engineering Handbook*, Fifth Edition, McGraw-Hill Book Company, New York, 1973, Chapt. 3.

(b) the partial pressure of water in the gas space;
(c) the total pressure in the gas space.

At 20°C, the partial pressure of ammonia, in mm Hg, over aqueous solutions is as follows:

lb_m NH$_3$/100 lb_m H$_2$O	7.5	10	15	20	25
partial pressure NH$_3$, in mm Hg	50.0	69.6	114.0	166.0	227.0

29.10 In the absorption of component A from an airstream into an aqueous stream, the bulk compositions of the two adjacent streams were analyzed to be $p_{A,G}$ equals 0.10 atm and $c_{A,L}$ eqauals 0.25 lb mole/ft^3. The Henry's constant for this system is 0.265 atm/(lb mole A/ft^3 of solution). The overall gas coefficient, K_G, was equal to 0.055 lb mole A/(hr)(ft^2)(atm).

If 57% of the total resistance of mass transfer is encountered in the gas film, determine
(a) the gas-film coefficient, k_G;
(b) the concentration on the liquid side of the interface, $c_{A,i}$;
(c) the liquid-film coefficient, k_L;
(d) the mass flux of A.

29.11 An absorption tower, operating at 20°C and 1 atm pressure, was used to absorb SO$_2$ from an air mixture into water. At one point in the equipment the partial pressure of the SO$_2$ in the gas stream was 30 mm, and the concentration of the contacting liquid stream was 0.0344 mole SO$_2$/ft^3 of solution. The individual film mass-transfer coefficients at 20°C and 1 atm, were $k_L = 1.3$ lb mole/(hr)(ft^2)(lb mole/ft^3 solution) and $k_G = 0.295$ lb mole/hr ft^2 atm. Equilibrium data at 20°C are as follows:

partial pressure SO$_2$ in mm Hg	0.5	3.2	8.5	26	59
concentration in lb mole SO$_2$/ft^3 soln	0.0191	0.0911	0.1738	0.388	0.681

(a) Evaluate the interfacial concentrations, $c_{A,i}$ and $p_{A,i}$;
(b) Fill in the values for the following table, giving the various coefficients and associated driving forces:

Coefficient	Driving force
$k_G =$ _____	$p_{A,G} - p_{A,i} =$ _____
$k_L =$ _____	$c_{A,i} - c_{A,L} =$ _____
$K_G =$ _____	$p_{A,G} - p_A^* =$ _____
$K_L =$ _____	$c_A^* - c_{A,L} =$ _____

(c) What percentage of the overall mass-transfer resistance is in the gas film?

29.12 In a wetted-wall tower where ammonia, NH_3, was stripped from an ammonia–water solution into an airstream, the overall liquid coefficient, K_L, was 0.05 lb mole/(hr)(ft^2)(lb mole ft^3). At a plane in the tower, the bulk concentration of the falling liquid stream was 0.25 lb mole NH_3/ft^3 of solution and the partial pressure of ammonia in the gas stream was 0.03 atm. For dilute solutions of ammonia in water at the operating temperature, the equilibrium partial pressure may be evaluated by

$$p_{A,i} = 0.215 c_{A,i}$$

where $p_{A,i}$ is the partial pressure in atmospheres and $c_{A,i}$ is the equilibrium concentration of ammonia in water in lb moles of ammonia per cubic feet of solution. If the gas phase offered 75% of the total resistance to mass transfer, calculate
(a) the individual gas-film coefficient, k_G;
(b) the overall gas-transfer coefficient, K_G;
(c) the interfacial concentrations, $p_{A,i}$ and $c_{A,i}$.

29.13 In a textbook on unit operations, the following statement was made:

> For a given mass transfer operation, K_L was experimentally measured to be 30 lb moles/(hr)(ft^2)(lb moles/ft^3) and k_L was evaluated to be 10 lb moles/(hr)(ft^2)(lb moles/ft^3). Since this is a 1 to 3 ratio, there must have been $\frac{2}{3}$ of the resistance to mass transfer in the gas phase.

Criticize or verify this statement.

29.14 For a system in which component A is transferring from the gas to the liquid phase, the equilibrium relation is given by

$$p_{A,i} = 0.75 c_{A,i}$$

where $p_{A,i}$ is the equilibrium partial pressure in atm and $c_{A,i}$ is the equilibrium concentration in moles per liter. At one point in the atmospheric mass-transfer equipment, the liquid stream contains 0.045 mol/liter and the gas stream contains 9.0 mole % A. The individual gas-film coefficient, k_G, at this point in the apparatus has a value of 2.7 mol/(s)(m^2)(atm). Thirty percent of the overall resistance to mass transfer is known to be encountered in the liquid phase. Evaluate
(a) the molar flux of A;
(b) the liquid interfacial concentration of A;
(c) the overall mass-transfer coefficient, K_G.

29.15 In the absorption of ammonia, NH_3, into water from an air–ammonia mixture in an absorption tower operated at 60°F and 1 atm, the individual film coefficients were estimated to be $k_L = 0.205$ lb mole $NH_3/(hr)(ft^2)$ $\times$(lb mole NH_3/ft^2) and $k_G = 0.240$ lb mole $NH_3/(hr)(ft^2)(atm)$. The equilibrium partial pressure of ammonia over dilute solutions of ammonia in water at 60°F is given by

$$p_{NH_3,i} = 0.215c_{NH_3,i}$$

with $p_{NH_3,i}$ in atm and $c_{A,i}$ in lb mole NH_3/ft^3 of solution. Determine the following mass-transfer coefficients:

(a) k_y;
(b) k_c for the gas film;
(c) K_G;
(d) K_y;
(e) K_L.

29.16 Reconsider problem 29.15. Draw a plot to scale of the partial pressure vs. the concentration in the liquid, indicating the significant driving forces that go with the various individual and overall mass-transfer coefficients when the liquid stream contains 5×10^{-3} lb mole NH_3/ft^3 and the gas stream contains 1 mole % NH_3.

What percent of the total resistance to mass transfer is encountered in the liquid film?

29.17 Chlorine water for pulp bleaching is being prepared by absorbing chlorine gas in water within a packed tower operating at 293 K and 1.013×10^5 Pa pressure. At one point in the tower, the chlorine pressure in the gas is 4×10^4 Pa and the concentration in the liquid is 1 kg/m^3. Data on the solubility of chlorine in water at 293 K are given in problem 29.8. If 80% of the resistance to mass transfer lies in the liquid phase, what are the interfacial concentrations?

29.18 A packed tower has been designed to strip component A from an aqueous stream into a countercurrent flowing airstream. At a given plane in the tower, the compositions of the two adjacent streams are $p_{A,G} = 30$ mm Hg and $c_{A,L} = 0.25$ lb mole of A/ft^3 of solution. Under the given flow conditions, the overall mass-transfer liquid coefficient, K_L, is equal to 0.4 lb mole/(hr)(ft^2)(lb mole of A/ft^3 of solution) and 40% of the resistance to mass transfer is encountered in the liquid phase. At the tower's operating conditions of 60°F and 1 atm pressure, the system satisfies Henry's law with a Henry's constant of 167 mm/(lb mole of A/ft^3 of solution). Determine

(a) k_G, lb moles of $A/(hr)(ft^2)(atm)$;
(b) k_L, lb moles of $A/(hr)(ft^2)(lb$ mole/ft$^3)$;
(c) $p_{A,i}$, mm;
(d) K_y, lb moles of $A/(hr)(ft^2)(\Delta y_A)$.

29.19 In a wetted-wall tower that is used for the absorption of ammonia, NH_3, into water from an air–ammonia mixture, the overall gas mass-transfer coefficient, K_G, was 0.191 lb mole/(hr)(ft^2)(atm). The tower was operated at 60°F and 2 atm pressure. At the top of the tower, the exiting gas contained 0.5% by volume NH$_3$ and the entering liquid that this gas contracted was pure water. If the gas phase offered 80% of the resistance to mass transfer, calculate

(a) the gas-film coefficient, k_G;
(b) the liquid-film coefficient, k_L;
(c) the overall liquid mass-transfer coefficient, K_L;
(d) the interfacial concentrations, $p_{A,i}$ and $c_{A,i}$.

For dilute solutions of ammonia in water at 60°F, the equilibrium partial pressure is given by

$$p_{NH_3,i} = 0.215c_{NH_3,i}$$

where $p_{NH_3,i}$ is in atm and c_{NH_3} is in lb mole NH$_3$/ft^3 of solution.

29.20 An absorption tower operates at a temperature of 293 K and an average pressure of 1.013×10^5 Pa. Water flows down through the column and absorbs sulfur dioxide from a gas stream flowing countercurrent to the liquid stream. At the top of the tower, the water entering is essentially free of sulfur dioxide and the gas stream at this plane contains 4 mole percent SO$_2$. If the liquid-phase mass-transfer coefficient, k_x, equals 4.0 lb mole/(hr)(ft^2)(Δx_{SO_2}) and the gas-phase mass-transfer coefficient, k_y, equals 0.3 lb mole/(hr)(ft^2)(Δy_{SO_2}), determine the overall mass-transfer coefficients K_x and K_y at the top of the tower. What percent of the total resistance to mass transfer is encountered in the liquid film? Equilibrium data for this system are available in problem 29.11.

29.21 A packed, absorption tower was used to absorb compound A from a gas mixture into solvent B. At one point in the tower, the partial pressure of A in the gas stream was 0.15 atm and the concentration of A in the contracting liquid stream was 6.24×10^{-5} lb mole/ft^3. The mass transfer between the gas stream and the liquid stream at that point in the tower was 0.0295 lb mole/(hr)(ft^2). The individual gas-film transfer coefficient, k_G, was found to be 0.295 lb mole/(hr)(ft^2)(atm). A laboratory experiment verified that the system satisfied Henry's law and that the liquid composition, 6.24×10^{-5} lb mole/ft^3, was in equilibrium with a partial pressure of 0.03 atm.

(a) Fill in the values for the following table:

Coefficient		Driving Force	
$k_G =$	_____	$p_{A,G} - p_{A,i} =$	_____
$k_L =$	_____	$c_{A,i} - c_{A,L} =$	_____
$K_G =$	_____	$p_{A,G} - p_A^* =$	_____
$K_L =$	_____	$c_A^* - c_{A,L} =$	_____

(b) What percent of the overall mass-transfer resistance was in the gas film?

29.22 Benzene and toluene are being separated in a packed distillation tower. At a point in the tower, the gas phase is 70% benzene, while the adjacent contacting liquid is 60 mole percent benzene. The temperature at this point is 362.6 K and the pressure is 1.013×10^5 Pa. If 60% of the resistance to mass transfer is encountered in the gas phase, determine the interfacial composition.

30
CONVECTIVE
MASS-TRANSFER
CORRELATIONS

Thus far, we have considered convective mass transfer from an analytical viewpoint, and from relations developed from the analogous transport of momentum or convective heat transfer. Although these considerations have given an insight into the process of convective transport, the validity of the analysis must be proven by comparing it with experimental data. In this chapter, we shall present equations based upon experimental results. There will be no attempt to review all of the mass-transfer investigations; reviews have been presented in several excellent references.* However, correlations will be presented to show that the form of the equations is as predicted by the analytical expressions derived in Chapter 28. Additional correlations will be given for those situations which have not been successfully treated analytically. The application of these correlations in design calculations will be discussed in Chapter 31.

30.1 MASS TRANSFER TO PLATES, SPHERES, AND CYLINDERS

Extensive data have been obtained for the transfer of mass between a moving fluid and certain shapes, such as flat plates, spheres, and cylinders. The techniques employed include sublimation of a solid, vaporization of a liquid into air, and the dissolution of a solid into water.

---

* W. S. Norman, *Absorption, Distillation and Cooling Towers*, John Wiley & Sons, New York, 1961; C. J. Geankoplis, *Mass Transfer Phenomena*, Holt, Rinehart and Winston, New York, 1972; A. H. P. Skelland, *Diffusional Mass transfer*, John Wiley & Sons, New York, 1974; T. K. Sherwood, R. L. Pigford, and C. R. Wilke, *Mass Transfer*, McGraw-Hill Book Company, New York, 1975; R. E. Treybal, *Mass Transfer Operations*, McGraw-Hill Book Company, New York, 1980.

By correlating the data in terms of dimensionless parameters, these empirical equations can then be extended to other moving fluids and geometrically similar surfaces.

FLAT PLATE

Several investigators have measured the evaporation from a free liquid surface or the sublimation from a flat, volatile-solid surface into a controlled air stream. These data have been found to satisfy favorably the theoretical equations for laminar and turbulent boundary layers,

$$\mathrm{Nu}_{AB,L} = 0.664\ \mathrm{Re}_L^{1/2} \mathrm{Sc}^{1/3} \ \text{(laminar)} \qquad \mathrm{Re}_L < 3 \times 10^5 \qquad (28\text{-}21)$$

$$\mathrm{Nu}_{AB,L} = 0.036\ \mathrm{Re}_L^{0.8} \mathrm{Sc}^{1/3} \ \text{(turbulent)} \qquad \mathrm{Re}_x > 3 \times 10^5 \qquad (28\text{-}26)$$

These equations may be expressed in terms of the j factor by recalling that

$$j_D = \frac{k_c}{v_\infty}\mathrm{Sc}^{2/3} = \frac{k_c L}{D_{AB}} \cdot \frac{\mu}{L v_\infty \rho} \cdot \frac{D_{AB}\rho}{\mu} \cdot \left(\frac{\mu}{\rho D_{AB}}\right)^{2/3}$$

$$= \frac{\mathrm{Nu}_{AB,L}}{\mathrm{Re}_L\,\mathrm{Sc}^{1/3}} \qquad (30\text{-}1)$$

Upon rearranging equations (28-21) and (28-26) into the form of equation (30-1), we obtain

$$j_D = 0.664\ \mathrm{Re}_L^{-1/2} \ \text{(laminar)} \qquad \mathrm{Re}_L < 3 \times 10^5 \qquad (30\text{-}2)$$

and

$$j_D = 0.037\ \mathrm{Re}_L^{-0.2} \ \text{(turbulent)} \qquad \mathrm{Re}_x > 3 \times 10^5 \qquad (30\text{-}3)$$

These equations may be used if the Schmidt number is in the range $0.6 < \mathrm{Sc} < 2500$. The j factor for mass transfer is also equal to the j factor for heat transfer in the Prandtl number range of $0.6 < \mathrm{Pr} < 100$ and is also equal to $C_f/2$. At a distance x from the leading edge of the flat plate, the exact solution to the laminar boundary layer,

$$\mathrm{Nu}_{AB,x} = \frac{k_c x}{D_{AB}} = 0.332\ \mathrm{Re}_x^{1/2}\ \mathrm{Sc}^{1/3} \qquad (28\text{-}20)$$

agrees with experimental data.

SINGLE SPHERE

Investigators have studied the mass transfer from single spheres and have correlated the mass-transfer Nusselt number by direct addition of terms representing transfer by purely molecular diffusion and transfer by forced convection, in the form

$$\mathrm{Nu}_{AB} = \mathrm{Nu}_{AB_0} + C\,\mathrm{Re}^m\,\mathrm{Sc}^{1/3}$$

where C and m are correlating constants. For very low Reynolds number, the Nusselt number should approach a value of 2.0. This value can be derived theoretically by considering the molecular diffusion from a sphere into a large volume of stagnant fluid. Accordingly, the generalized equation becomes

$$Nu_{AB} = 2.0 + C\, Re^m\, Sc^{1/3}$$

For transfer into liquid streams, the equation of Brian and Hales*

$$Nu_{AB} = (4.0 + 1.21\, Pe_{AB}^{2/3})^{1/2} \tag{30-4}$$

correlates data that are obtained when the mass-transfer Peclet number, Pe_{AB}, is less than 10 000. This Peclet number is equal to the product of the Reynolds and Schmidt numbers; i.e., $Pe_{AB} = Re\, Sc$. For Peclet numbers greater than 10 000, Levich† recommended the simpler relationship

$$Nu_{AB} = 1.01\, Pe_{AB}^{1/3} \tag{30-5}$$

The Fröessling equation‡

$$Nu_{AB} = 2.0 + 0.552\, Re^{1/2}\, Sc^{1/3} \tag{30-6}$$

correlates the data for transfer into gases at Reynolds numbers ranging from 2 to 800 and a Schmidt number range of 0.6 to 2.7. Data of Evnochides and Thodos‖ have extended the Fröessling equation to a Reynolds number range of 1500 to 12 000, with a Schmidt number range of 0.6 to 1.85. Eqautions (30-4), (30-5), and (30-6) can be used to describe forced convection mass-transfer coefficients only when the effects of free or natural convections are negligible; that is, when

$$Re \geq 0.4\, Gr_{AB}^{1/2}\, Sc^{-1/6} \tag{30-7}$$

The following equations of Steinberger and Treybal§ are recommended when the transfer occurs in the presence of natural convection

$$Nu_{AB} = Nu_{AB,nc} + 0.347\, (Re\, Sc^{1/2})^{0.62} \tag{30.8}$$

for

$$1 \leq Re \leq 3 \times 10^4$$

$$0.6 \leq Sc \leq 3200$$

where

$$Nu_{AB,nc} = 2.0 + 0.569\, (Gr_{AB}Sc)^{0.25} \qquad Gr_{AB}Sc < 10^8 \tag{30-9}$$

or

$$Nu_{AB,nc} = 2.0 + 0.0254\, (Gr_{AB}Sc)^{1/3}\, Sc^{0.244} \qquad Gr_{AB}Sc < 10^8 \tag{30-10}$$

The correlating equations for a single sphere are used in the following example.

* P. L. T. Brian and H. B. Hales, *A.I.Ch.E. J.*, **15**, 419 (1969).
† V. G. Levich, *Physicochemical Hydrodynamics*, Prentice-Hall, Englewood Cliffs, N.J. 1962.
‡ N. Fröessling, *Gerlands Beitr. Geophys.*, **52**, 170 (1939).
‖ S. Evnochides and G. Thodos, *A.I.Ch.E. J.*, **5**, 178 (1959).
§ R. L. Steinberger and R. E. Treybal, *A.I.Ch.E. J.*, **6**, 227 (1960).

EXAMPLE 1

Estimate the distance a spherical drop of water, originally 1.0 mm in diameter, must fall in quiet, dry air at 50°C in order to reduce its volume by 50%. Assume that the velocity of the drop is its terminal velocity evaluated at its mean diameter and that the water remains at 20°C. Evaluate all properties of the gas film at 35°C.

By considering a force balance on a spherical particle falling in a fluid medium, we can show that the terminal velocity of the particle is

$$v_0 = \sqrt{\frac{4d_p(\rho_P - \rho)g}{3C_D\rho}}$$

where d_p is the diameter of the particle, ρ_P is the density of the particle, ρ is the density of the fluid, g is the acceleration due to gravity, and C_D is the drag coefficient. C_D is a function of the Reynolds number of the particle as illustrated in Figure 12.4.

The arithmetic mean diameter is evaluated by

$$\bar{d}_p = \frac{d_{p,|t_1} + d_{p,|t_2}}{2} = \frac{d_{p,|t_1} + (1/2)^{1/3}d_{p,|t_1}}{2}$$

$$= 0.897\, d_{p,|t_1} = (0.897)(1 \times 10^{-3}\text{m}) = 8.97 \times 10^{-4}\,\text{m}$$

and the arithmetic mean radius, $\bar{r}$, is 4.48×10^{-4} m.

At 20°C or 293 K, the density of the water droplet, ρ_P is 9.95×10^2 kg/m³. At 35°C or 308 K, the density of the air, ρ, is 1.14 kg/m³ and the viscosity of air is 1.91×10^{-5} Pa · s. Upon substitution of these values into the terminal velocity equation, we obtain

$$v_0 = \sqrt{\frac{(4)(8.97 \times 10^{-4}\,\text{m})(9.95 \times 10^2\,\text{kg/m}^3 - 1.14\,\text{kg/m}^3)(9.8\,\text{m/s}^2)}{(3)(1.14\,\text{kg/m}^3)C_D}}$$

$$= \sqrt{\frac{10.22}{C_D}}$$

By trial and error, we can guess a v_0; calculate a Reynolds number; read C_D from Figure 12.4; and check our guessed value of v_0 by the above equation. Try $v_0 = 3.62$ m/s.

$$\text{Re} = \frac{d_p v_0 \rho}{\mu} = \frac{(8.97 \times 10^{-4}\text{m})(3.62\,\text{m/s})(1.14\,\text{kg/m}^3)}{(1.91 \times 10^{-5}\,\text{Pa} \cdot \text{s})\left(\dfrac{\text{kg/m} \cdot \text{s}}{\text{Pa} \cdot \text{s}}\right)}$$

$$= 194$$

from Figure 12.4, $C_D = 0.78$, then $v_0 = \sqrt{\dfrac{10.22}{0.78}} = 3.62$ m/s.

Equation (30.7) will be evaluated to verify whether natural convection effects are important. The mass transfer Grashof number is

$$\text{Gr}_{AB} = \frac{d_p^3 \rho g \, \Delta\rho_A}{\mu^2}$$

where $\Delta\rho_A$ is the difference in density of the saturated gas at the temperature of the water droplet, 293 K, and the dry gas at 308 K. For our system, $\Delta\rho_A = 0.027\ 2\ \mathrm{kg/m^3}$; the density of the saturated gas was evaluated using the moles of water per mole of dry air value obtained from a humidity chart.

$$\mathrm{Gr}_{AB} = \frac{(8.97 \times 10^{-4}\,\mathrm{m})^3 (1.14\ \mathrm{kg/m^3})(9.8\ \mathrm{m/s^2})(0.027\ 2\ \mathrm{kg/m^3})}{(1.91 \times 10^{-5}\ \mathrm{Pa \cdot s})^2 \left(\dfrac{\mathrm{kg/m \cdot s}}{\mathrm{Pa \cdot s}}\right)^2}$$

$$= 0.60$$

From Appendix J.1, the gas diffusivity, D_{AB}, for water vapor in air at 298 K is $0.260 \times 10^{-4}\ \mathrm{m^2/s}$; this value can be corrected to the desired temperature by

$$D_{AB} = 0.260 \times 10^{-4}\ \mathrm{m/s^2} \left(\frac{308\ \mathrm{K}}{298\ \mathrm{K}}\right)^{3/2} = 0.273 \times 10^{-4}\ \mathrm{m^2/s}$$

The Schmidt number is

$$\mathrm{Sc} = \frac{\mu}{\rho D_{AB}} = \frac{(1.91 \times 10^{-5}\ \mathrm{Pa \cdot s})\left(\dfrac{\mathrm{kg/m \cdot s}}{\mathrm{Pa \cdot s}}\right)}{(1.14\ \mathrm{kg/m^3})(0.273 \times 10^{-4}\ \mathrm{m^2/s})} = 0.61$$

In equation (30-6)

$$0.4\ \mathrm{Gr}_{AB}{}^{1/2}\mathrm{Sc}^{-1/6} = 0.4(0.6)^{1/2}(0.61)^{-1/6} = 0.336$$

This is smaller than the Reynolds number, indicating that natural convection effects are negligible. The Fröessling equation (30-6) can be used to evaluate the mass-transfer coefficient

$$\frac{k_c d_p}{D_{AB}} = 2.0 + 0.552\ \mathrm{Re}^{1/2}\ \mathrm{Sc}^{1/3}$$

$$k_c = \frac{D_{AB}}{d_p}[2.0 + 0.552\ \mathrm{Re}^{1/2}\ \mathrm{Sc}^{1/3}]$$

$$= \frac{(0.273 \times 10^{-4}\ \mathrm{m^2/s})}{(8.97 \times 10^{-4}\ \mathrm{m})}[2.0 + 0.552(194)^{1/2}(0.61)^{1/3}]$$

$$= 0.26\ \mathrm{m/s}$$

The average rate of evaporation is

$$W_A = 4\pi \bar{r}^2 N_A = 4\pi \bar{r}^2 k_c (c_{A,s} - c_{A,\infty})$$

The dry-air concentration, $c_{A,\infty}$, is zero. The surface concentration is evaluated from the vapor pressure of water at 293 K.

$$c_{A,s} = \frac{P_A}{RT} = \frac{2.33 \times 10^3\ \mathrm{Pa}}{(8.314\ \mathrm{Pa \cdot m^3/mol \cdot K})(293\ \mathrm{K})}$$

$$= 0.956\ \mathrm{mol/m^3}$$

When we substitute the known values into the rate of evaporation equation, we obtain

$$W_A = 4\pi(4.48 \times 10^{-4} \text{ m})^2(0.26 \text{ m/s})(0.956 \text{ mol/m}^3)$$
$$= 6.28 \times 10^{-7} \text{ mol/s}$$
$$= 1.13 \times 10^{-8} \text{ kg/s}$$

The amount of water evaporated, m, is

$$m = \rho \, \Delta V = \rho(V_{t_1} - V_{t_2}) = \rho(V_{t_1} - 0.5 V_{t_1}) = \frac{\rho V_{t_1}}{2}$$
$$= \frac{\rho}{2}\frac{4\pi}{3}r_{t_1}^3 = \frac{4}{6}\pi(9.95 \times 10^2 \text{ kg/m}^3)(4.48 \times 10^{-4} \text{ m})^3$$
$$= 1.87 \times 10^{-7} \text{ kg} = 1.87 \times 10^{-4} \text{ g}$$

The time necessary to reduce the volume by 50% is

$$t = \frac{m}{w_A} = \frac{1.87 \times 10^{-4} \text{ g}}{1.13 \times 10^{-5} \text{ g/s}} = 16.5 \text{ s}$$

The distance of fall is equal to $v_0 t$ or $(3.62 \text{ m/s})(16.5 \text{ s}) = 60 \text{ m} (196 \text{ ft})$.

SINGLE CYLINDER

Several investigators have studied the sublimation from a solid cylinder into air flowing normal to its axis. Additional results on the dissolution of solid cylinders into a turbulent water stream have been reported. Bedingfield and Drew* correlated the available data by

$$\frac{k_G P \text{Sc}^{0.56}}{G_M} = 0.281(\text{Re}')^{-0.4} \tag{30-11}$$

for

$$400 < \text{Re}' < 25\,000$$
$$0.6 < \text{Sc} < 2.6$$

where Re' is the Reynolds number in terms of the diameter of the cylinder, G_M is the molar gas-mass velocity, and P is the total pressure.

The full analogy among momentum, heat, and mass transfer breaks down when the flow is around bluff bodies, such as spheres and cylinders. The total drag force includes the form drag in addition to the skin friction; accordingly, the j factor will not equal $C_f/2$. The analogy between heat and mass transfer, $j_H = j_D$, still holds. Accordingly, the mass-transfer coefficient for a single cylinder that does not satisfy the specified ranges for equation (30-11) can be evaluated by using the Chilton-Colburn analogy and one of the heat-transfer relations discussed in section 20.3.

---

* C. H. Bedingfield and T. B. Drew, *Ind. Eng. Chem.*, **42**, 1164 (1950).

EXAMPLE 2

In a humidification apparatus, water flows in a thin film down the outside of a vertical, circular cylinder. Dry air at 100°F and 1 atm flows at right angles to the 3-in.-diameter, 4-ft-long cylinder at a velocity of 15 ft/s. The liquid temperature is 60°F. Calculate the rate at which liquid must be supplied to the top of the cylinder if the entire surface of the cylinder is to be used for the evaporating process, but restricting any water from going beyond the bottom of the cylinder.

The properties of the airstream will be evaluated at the film temperature, $t_f = (100 + 60)/2 = 80°F$. The density of the air may be determined by

$$\rho = \frac{nM_{air}}{V} = \frac{PM_{air}}{RT} = \frac{(1 \text{ atm})(29 \text{ lb}_m/\text{lb mole})}{\left(\dfrac{0.73 \text{ atm ft}^3}{\text{lb mole °R}}\right)(540°R)}$$

$$= 0.0735 \text{ lb}_m/\text{ft}^3$$

Other variables that are required to evaluate the Reynolds number are

$$v_\infty = 15 \text{ ft/s}$$

$$\mu = 1.24 \times 10^{-5} \text{ lb}_m/(\text{ft})(\text{s})$$

$$d_c = \frac{3 \text{ in.}}{12 \text{ in./ft}} = 0.25 \text{ ft}$$

giving

$$\text{Re}' = \frac{d_c v_\infty \rho}{\mu} = \frac{(0.25 \text{ ft})(15 \text{ ft/s})(0.0735 \text{ lb}_m/\text{ft}^3)}{1.24 \times 10^{-5} \text{ lb}_m/(\text{ft})(\text{s})} = 17\,784$$

From Appendix Table J.1, the diffusivity of water in air at 298 K is 0.260 cm^2/s, which corrected for temperature and corrected to engineering units becomes

$$D_{AB} = (0.260 \text{ cm}^2/\text{s})\left(\frac{300}{298}\right)^{3/2}\left(3.87\frac{\text{ft}^2/\text{hr}}{\text{cm}^2/\text{s}}\right) = 1.016 \text{ ft}^2/\text{hr}$$

The Schmidt number may now be evaluated,

$$\text{Sc} = \frac{\mu}{\rho D_{AB}} = \frac{[1.24 \times 10^{-5} \text{ lb}_m/(\text{ft})(\text{s})](3600 \text{ s/hr})}{(0.0735 \text{ lb}_m/\text{ft}^3)(1.016 \text{ ft}^2/\text{hr})} = 0.60$$

The molar mass velocity of the air normal to the cylinder is

$$G_M = (15 \text{ ft/s})(3600 \text{ s/hr})(0.0735 \text{ lb}_m/\text{ft}^3)(\text{lb mole}/29 \text{ lb})$$

$$= 136.9 \text{ lb mole}/(\text{hr})(\text{ft}^2)$$

Upon substitution of the known values into equation (30-11),

$$\frac{k_G P \text{Sc}^{0.56}}{G_M} = 0.281 \, (\text{Re}')^{-0.4}$$

$$\frac{k_G (760 \text{ mm})}{(136.9 \text{ lb mole}/(\text{hr})(\text{ft}^2))}(0.60)^{0.56} = 0.281(17\,784)^{-0.4}$$

$$k_G = 1.34 \times 10^{-3} \text{ lb mole}/(\text{hr})(\text{ft}^2)(\text{mm})$$

The vapor pressure of water at 60°F is 13 mm. The mass flux is defined by equation (29-6) as

$$N_A = k_G(p_{H_2O,i} - p_{H_2O,\infty})$$
$$= [0.00134 \text{ lb mole}/(\text{hr})(\text{ft}^2)(\text{mm})](13 \text{ mm})$$
$$= 0.0174 \text{ lb mole}/(\text{hr})(\text{ft}^2)$$

The feed rate of water may now be determined for the cylinder that has a surface area of $\pi d_c L = \pi(0.25 \text{ ft})(4 \text{ ft}) = 3.14 \text{ ft}^2$

$$\text{feed rate} = [0.0174 \text{ lb mole}/(\text{hr})(\text{ft}^2)](18 \text{ lb}_m/\text{lb mole})(3.14 \text{ ft}^2)$$
$$= 0.984 \text{ lb/hr} \quad (1.24 \times 10^{-4} \text{ kg/s})$$

30.2 MASS TRANSFER INVOLVING TURBULENT FLOW THROUGH PIPES

Mass transfer from the inner wall of a tube to a moving fluid has been studied extensively. Most of the data have been obtained for vaporization of liquids into air; data have also been measured for mass transfer into a moving liquid.

Gilliland and Sherwood* studied the vaporization of nine different liquids into air. Their correlation is

$$\frac{k_c D}{D_{AB}} \frac{p_{B,lm}}{P} = 0.023 \text{ Re}^{0.83}\text{Sc}^{0.44} \tag{30-12}$$

where D is the inner diameter of the pipe; $p_{B,lm}$ is the log mean composition of the carrier gas, evaluated between the surface and the bulk stream composition; P is the total pressure; D_{AB} is the mass diffusivity of the diffusing component A in the flowing carrier gas B; and Re and Sc are the dimensionless parameters evaluated at the bulk conditions of the flowing stream; this expression has been found to be reliable over the range

$$2000 < \text{Re} < 35\,000$$
$$0.6 < \text{Sc} < 2.5$$

In a subsequent study, Linton and Sherwood† extended the range of Schmidt number when they investigated the dissolution of benzoic acid, cinnamic acid, and β-naphthol. The combined results of Gilliland and Sherwood and Linton and Sherwood were correlated by the relation

$$\frac{k_L D}{D_{AB}} = 0.023 \text{ Re}^{0.83}\text{Sc}^{1/3} \tag{30-13}$$

---

* E. R. Gilliland and T. K. Sherwood, *Ind. Eng. Chem.*, **26**, 516 (1934).
† W. H. Linton and T. K. Sherwood, *Chem. Eng. Prog.*, **46**, 258 (1950).

for $2000 < \text{Re} < 70\,000$ and $1000 < \text{Sc} < 2260$. Since $k_L D/D_{AB}$ is the mass-transfer Nusselt number or the Sherwood number, the similarity between equation (30-13) and the Dittus-Boelter equation for energy transfer (20-26) becomes apparent. This once again verifies the analogous behavior of these two transport phenomena.

30.3 MASS TRANSFER IN WETTED-WALL COLUMNS

Most data on mass transfer between a pipe surface and a flowing fluid have been obtained by the use of wetted-wall columns, as described in Chapter 26. The principal reason for the use of these columns for mass-transfer investigations is that the contacting area between the two phases can be accurately measured.

The convective mass-transfer coefficient for the gas stream is defined by the correlating equation (30-12). The convective mass-transfer coefficient for the falling liquid film was correlated by Vivian and Peaceman* by the relation

$$\frac{k_L z}{D_{AB}} = 0.433(\text{Sc})^{1/2}\left(\frac{\rho^2 g z^3}{\mu^2}\right)^{1/6}(\text{Re}_L)^{0.4} \qquad (30\text{-}14)$$

where z is the length of contact; D_{AB} is the mass diffusivity of the diffusing component A into liquid B; ρ is the density of liquid B; μ is the viscosity of liquid B; g is the acceleration due to gravity; Sc is the Schmidt number evaluated at the liquid film temperature; and Re_L is the Reynolds number of the liquid flowing down the tube; that is, $4\Gamma/\mu$, where Γ is the mass flow rate of liquid per unit wetted perimeter. The liquid film coefficients were found to be 10 to 20% lower than the theoretical equation for absorption in laminar films. This may have been due to ripples along the liquid surface or to disturbances in the liquid flow at the two ends of the wetted-wall column. These discrepancies between the theoretical and measured rates of mass transfer have often led to the suggestion that a resistance to the mass transfer exists at the gas-liquid interface. Investigations by Scriven and Pigford,† Raimondi and Toor,‡ and Chiang and Toor§ have substantiated that the interfacial resistance is negligible in normal mass-transfer operations. The correlation of Gilliland and Sherwood, equation (30-12), will be used to determine the gas mass-transfer coefficient in the following example.

EXAMPLE 3

A 2-in.-ID wetted-wall column is being used to strip CO_2 from an aqueous solution by an air stream flowing at 2.5 fps. At one point in the column, the concentration of the CO_2 in

* J. E. Vivian and D. W. Peaceman, *A.I.Ch.E. J.*, **2**, 437 (1956).
† L. E. Scriven and R. L. Pigford, *A.I.Ch.E. J.*, **4**, 439 (1958).
‡ P. Raimondi and H. L. Toor, *A.I.Ch.E. J.*, **5**, 86 (1959).
§ S. H. Chiang and H. L. Toor, *A.I.Ch.E. J.*, **5**, 165 (1959).

the air stream is 1 mole percent. At the same point in the column, the concentration of CO_2 in the water is 0.5 mole percent. Determine the gas mass-transfer coefficient and the mass flux at the point in the column. The column is operated at 10 atm and 25°C.

The gas convective mass-transfer coefficient is evaluated, using equation (30-12), as

$$\frac{k_c D}{D_{AB}} \frac{p_{B,lm}}{P} = 0.023 Re^{0.83} Sc^{0.44}$$

where A is the diffusing species, CO_2, and the dimensionless parameters and the mass diffusivity are evaluated by using bulk conditions in the air stream. The mass diffusivity of CO_2 in air at 273 K and 1 atmosphere is equal to 0.136 cm²/sec, according to Table J.1. This value may be corrected to the bulk conditions as follows:

$$D_{AB} \text{ at 298 K and 10 atm} = (0.136 \text{ cm}^2/\text{s})\left(\frac{1 \text{ atm}}{10 \text{ atm}}\right)\left(\frac{298 \text{ K}}{273 \text{ K}}\right)^{3/2} = 0.016 \frac{\text{cm}^2}{\text{s}}$$

or

$$D_{AB} = (0.016 \text{ cm}^2/\text{s})\left(3.87 \frac{\text{ft}^2/\text{hr}}{\text{cm}^2/\text{s}}\right) = 0.06 \text{ ft}^2/\text{hr}$$

Since the bulk composition of the gas is 99% air, we shall assume that the viscosity and density of air represent the bulk conditions. These values are

$$\mu_{air} = 0.018 cp = (0.018)(2.42) lb_m/\text{ft hr}$$

or

$$\mu_{air} = (0.018)(6.72 \times 10^{-4}) \text{ lb}_m/\text{ft s}$$

and

$$\rho_{air} = \frac{P}{RT} M_{air} = \frac{(10 \text{ atm})(29 \text{ lb}_m/\text{lb mole})}{(1.315 \text{ atm ft}^3/\text{lb mole K})(298 \text{ K})} = 0.74 \text{ lb}_m/\text{ft}^3$$

The Schmidt and Reynolds numbers are calculated to be

$$Sc = \frac{\mu}{\rho D_{AB}} = \frac{(0.018 \times 2.42 \text{ lb}_m/\text{ft hr})}{(0.74 \text{ lb}_m/\text{ft}^3)(0.06 \text{ ft}^2/\text{hr})} = 0.98$$

and

$$Re = \frac{D v_\infty \rho}{\mu} = \frac{(2/12 \text{ ft})(2.5 \text{ fps})(0.74 \text{ lb}_m/\text{ft}^3)}{(0.018 \times 6.72 \times 10^{-4} \text{ lb}_m/\text{ft s})} = 25\,500$$

At 25°C, the Henry's law constant for CO_2 in water is 1.64×10^3 atm/mole fraction CO_2 in solution. With a concentration of 0.005 mole fraction CO_2, the partial pressure of CO_2 at the interface is 8.2 atm. The interfacial and bulk concentrations in terms of pressure are as follows:

at the interface,

$$p_A = 8.2 \text{ atm}$$

and

$$p_B = 1.8 \text{ atm}$$

and in the bulk gas phase,

$$p_A = 0.1 \text{ atm}$$

and

$$p_B = 9.9 \text{ atm}$$

The log mean partial pressure of constituent B is evaluated to be

$$p_{B,lm} = \frac{9.9 \text{ atm} - 1.8 \text{ atm}}{\ln(9.9 \text{ atm}/1.8 \text{ atm})} = 4.75 \text{ atm}$$

Substituting these values into equation (30-12), we can evaluate the mass-transfer coefficient as

$$k_c = 0.023 \, \text{Re}^{0.83} \text{Sc}^{0.44} \left(\frac{P}{p_{B,lm}}\right)\left(\frac{D_{AB}}{D}\right)$$

$$k_c = (0.023)(25\,500)^{0.83}(0.98)^{0.44} \frac{(10 \text{ atm})}{4.75 \text{ atm}} \frac{(0.060 \text{ ft}^2/\text{hr})}{(2/12 \text{ ft})}$$

$$= 78.5 \text{ ft/hr} = 78.5 \frac{\text{lb mole}}{\text{hr ft}^2 \text{ lb mole/ft}^3}$$

The mass flux is

$$N_A = k_c(c_{A,i} - c_{A,\infty}) = \frac{k_c}{RT}(p_{A,i} - p_{A,\infty})$$

or

$$N_a = \frac{(78.5 \text{ lb mole/hr ft}^2 \text{ lb mole/ft}^3)}{(1.315 \text{ atm ft}^3/\text{lb mole K})(298 \text{ K})}(8.2 \text{ atm} - 0.1 \text{ atm})$$

$$= 1.62 \text{ lb mole/hr ft}^2 \qquad (2.20 \text{ mol/s} \cdot \text{m}^2)$$

30.4 MASS TRANSFER IN PACKED AND FLUIDIZED BEDS

Packed and fluidized beds are commonly used in industrial mass-transfer operations, including adsorption, ion exchange, chromatography, and gaseous reactions that are catalyzed by solid surfaces. Numerous investigations have been conducted for measuring mass-transfer coefficients in packed beds and correlating the results. In general, the agreement among the investigators is poor, which is to be expected when one realizes the experimental difficulties. Sherwood, Pigford, and Wilke* presented a graphical representation of most of the data for mass

---

* T. K. Sherwood, R. L. Pigford, and C. R. Wilke, *Mass Transfer*, McGraw-Hill Book Company, New York, 1975.

transfer in packed beds with single-phase fluid and gas flows. They found that a single straight line through the experimental points did a fair job of representing all the data; this line is represented by a fairly simple equation

$$j_D = 1.17\, \text{Re}^{-0.415} \qquad 10 < \text{Re} < 2500 \qquad (30\text{-}15)$$

where

$$\text{Re} = \frac{d_p u_{\text{ave}} \rho}{\mu}$$

u_{ave} = superficial fluid velocity

d_p = diameter of sphere having the same

surface or volume as the particle

This equation may be employed for engineering estimates.

Most of the earlier correlations for packed beds failed to account for variations in the void fraction of the beds, ϵ, which in beds of spheres and pellets can range from 0.3 to 0.5. Mass transfer between liquids and beds of spheres was investigated by Wilson and Geankoplis* who correlated their data by

$$\epsilon j_D = \frac{1.09}{\text{Re}'''} \qquad (30\text{-}16)$$

for $0.0016 < \text{Re}''' < 55$, $165 < \text{Sc} < 70\,600$ and $0.35 < \epsilon < 0.75$, and by

$$\epsilon j_D = \frac{0.25}{(\text{Re}''')^{0.31}} \qquad (30\text{-}17)$$

for $55 < \text{Re}''' < 1500$ and $165 < \text{Sc} < 10\,690$. The Reynolds number, Re''', is defined in terms of the diameter of the spheres, d_p, and the superficial mass velocity of the fluid, G, in mass per unit time per unit cross section of the tower without packing. The void fraction in the packed bed is designated as ϵ, the volume void space between the solid particles divided by the total volume of void space plus the solid particles. These values range from about 0.30 to 0.50 in most packed beds. The correlation of Gupta and Thodos,†

$$\epsilon j_D = \frac{2.06}{(\text{Re}''')^{0.575}} \qquad (30\text{-}18)$$

is recommended for mass transfer between gases and beds of spheres in the Reynolds number range $90 < \text{Re}''' < 4000$. Data above this range indicate a transitional behavior and are reported in a graphical form by Gupta and Thodos.‡

---

* E. J. Wilson and C. J. Geankoplis, *Ind. Eng. Chem. Fund.*, **5**, 9 (1966).
† A. S. Gupta and G. Thodos, *A.I.Ch.E. J.*, **9**, 751 (1963).
‡ A. S. Gupta and G. Thodos, *Ind. Eng. Chem. Fund.*, **3**, 218 (1964).

Mass transfer in both gas and liquid fluidized beds of spheres has been correlated by Gupta and Thodos[*] with the equation

$$\epsilon j_D = 0.010 + \frac{0.863}{(Re''')^{0.58} - 0.483} \tag{30-19}$$

A detailed discussion of heat and mass transfer in fluidized beds is provided in the book by Kunii and Levenspiel.[†]

30.5 MASS TRANSFER WITH CHEMICAL REACTION

Many applications of mass transfer are found in conjunction with chemical reactions. Consider the absorption of carbon dioxide from a gas phase into a liquid phase. The maximum amount of absorbed carbon dioxide is limited by the equilibrium concentration. If a dilute caustic solution is used as the liquid phase, the carbon dioxide will be absorbed and then will react to form a nonvolatile carbonate; the equilibrium concentration is lowered to zero. As a direct consequence of the chemical reaction, the quantity of fluid required in the absorption process is reduced.

Either the mass-transfer or the chemical-reaction rate can be the controlling factor in this simultaneous process; more frequently, neither rate completely dominates. The overall mass-transfer rate is expressed in terms of the resistance to both rates. In general, treatment of simultaneous mass transfer and chemical reaction is extremely complicated. The reader is referred to two references on the subject.[‡][§]

30.6 CAPACITY COEFFICIENTS FOR INDUSTRIAL TOWERS

Although the wetted-wall column, as described in Chapter 26, has a definite interfacial surface area, the corresponding area in other types of equipment, which will be described in Chapter 31, is virtually impossible to measure. For this reason, an engineering factor a must be introduced to represent the interfacial surface area per unit volume of the mass-transfer equipment. Both a and the mass-transfer coefficient depend on the physical geometry of the equipment and on the flow rates of the two contacting, immiscible streams; accordingly, they are normally correlated together as the *capacity coefficient*, $k_c a$. The units of $k_c a$ are

[*] A. S. Gupta and G. Thodos, *A.I.Ch.E. J.*, **8**, 608 (1962).
[†] D. Kunii and O. Levenspiel, *Fluidization Engineering*, Wiley, New York (1969).
[‡] G. Astarita, *Mass Transfer With Chemical Reaction*, Elsevier, Amsterdam (1967).
[§] P. V. Danckwerts, *Gas-Liquid Reactions*, McGraw-Hill Book Company, New York (1970).

moles of A transferred/(hr)(volume)(moles of A/volume). The capacity coefficient will be encountered in the basic design equations of Chapter 31.

Empirical equations for capacity coefficients must be experimentally obtained for each type of mass-transfer operation. Such a correlation was obtained by Sherwood and Holloway[*] in the first comprehensive investigation of liquid-film mass-transfer coefficients in packed absorption towers. The experimental results for a variety of packings were represented by

$$\frac{k_L a}{D_{AB}} = \alpha \left(\frac{L}{\mu}\right)^{1-n} \left(\frac{\mu}{\rho D_{AB}}\right)^{0.5} \qquad (30\text{-}20)$$

where $k_L a$ is the mass-transfer capacity coefficient, in lb mole/hr ft^3(lb mole/ft^3); L is the liquid rate, in lb/hr ft^2; μ is the viscosity of the liquid, in lb/hr ft; ρ is the density of the liquid, in lb/ft^3; and D_{AB} is the liquid mass diffusivity of component A in liquid B, in ft^2/hr. The values of the constant α and the exponent n for various packing are given in Table 30.1.

TABLE 30.1   PACKING COEFFICIENTS
FOR (30-20)

Packing	α	n
2-in. rings	80	0.22
$1\frac{1}{2}$-in. rings	90	0.22
1-in. rings	100	0.22
$\frac{1}{2}$-in. rings	280	0.35
$\frac{3}{8}$-in rings	550	0.46
$1\frac{1}{2}$-in. saddles	160	0.28
1-in. saddles	170	0.28
$\frac{1}{2}$-in. saddles	150	0.28
3-in. spiral tiles	110	0.28

Further correlations for capacity coefficients can be found in treatises on mass-transfer operations in the discussion of each specific operation and each specific type of tower.[†]

30.7   CLOSURE

In this chapter, we have presented correlating equations for convective mass transfer coefficients obtained from experimental investigations. These correla-

[*] T. K. Sherwood and F. A. Holloway, *Trans. A.I.Ch.E.*, **36**, 21, 39 (1940).

[†] T. K. Sherwood, R. L. Pigford and C. R. Wilke, *Mass Transfer*, McGraw-Hill Book Company, New York, 1975; R. E. Treybal, *Mass Transfer Operations*, McGraw-Hill Book Company, New York, 1980; C. J. King, *Separation Processes*, McGraw-Hill Book Company, New York, 1971; W. S. Norman, *Absorption, Distillation and Cooling Towers*, Wiley, 1961; A. H. P. Skelland, *Diffusional Mass Transfer*, Wiley, New York, 1974.

tions have verified the validity of the analysis of convective transport as presented in Chapter 28. In Chapter 31, methods will be developed for applying the capacity coefficient correlations to the design of mass-transfer equipment.

PROBLEMS

30.1 Benzene, C_6H_6, flows in a thin film down the outside surface of a vertical plate, 1.5 m wide and 3 m long. The liquid temperature is 289 K. Benzene-free air at 303 K and 1.013×10^5 Pa pressure flows across the width of the plate parallel to the surface. At the average temperature of the gas film, the diffusivity of benzene in air is 9.51×10^{-6} m^2/s. At 289 K, the vapor pressure of benzene is 8000 Pa.
(a) Calculate the rate at which the liquid should be supplied to the top of the plate so that evaporation will just prevent it from reaching the bottom of the plate if the wind has a velocity of 5 m/s.
(b) What rate should the liquid be supplied if the velocity is 3 m/s.

30.2 A 1-ft square reservoir is inserted into the bottom of a large air duct to help humidify the heating airstream. The surface of the water is kept at a level even with the bottom of the duct, and its temperature is 65°F. Air at 100°F and 1 atm flows parallel to the surface with a bulk velocity of 20 ft/s. If 0.10 lb$_m$/hr of water evaporates from the pan, determine the moisture content of the airstream.

30.3 A 1-m square, thin plate of solid naphthalene is oriented parallel to a stream of air flowing at 20 m/s. The air is at 310 K and 1.013×10^5 Pa. The naphthalene remains at 290 K; at this temperature, the diffusivity of naphthalene in air is 5.61×10^{-6} m^2/s and its vapor pressure is 26 Pa. Determine
(a) the value of the mass-transfer coefficient at a point 0.3 m downstream from the leading edge;
(b) the moles of naphthalene per hour lost from the section of the plate 0.5 to 0.75 m downstream from the leading edge.

30.4 A thin plate of solid salt, NaCl, measuring 15 by 15 cm, is to be dragged through seawater (edgewise) at a velocity of 0.6 m/s. The 251 K seawater has a salt concentration of 0.0309 g/cm^3; if saturated, the seawater would have a concentration of 35 g/cm^3. The kinematic viscosity of seawater is approximately 1.10×10^{-5} ft^2/s. Estimate the rate at which the salt goes into solution if the edge effects can be ignored.

30.5 If the local Nusselt number for the laminar boundary layer formed over a flat plate is

$$\mathrm{Nu}_{AB,x} = 0.332\,\mathrm{Re}_x^{1/2}\,\mathrm{Sc}^{1/3}$$

and for the turbulent boundary layer is

$$\mathrm{Nu}_{AB,x} = 0.0292\,\mathrm{Re}_x^{4/5}\,\mathrm{Sc}^{1/3}$$

obtain an expression for the mass-transfer j factor, j_D, for each flow region and then evaluate an expression for the average film-transfer coefficient when the Reynolds number for the entire plate is
(a) $\mathrm{Re}_L = 100\,000$;
(b) $\mathrm{Re}_L = 1\,500\,000$.

30.6 Air at 50°F and 1 atm pressure flows over a naphthalene mothball that is suspended in an airstream moving with a velocity of 3 miles/hr. Since the naphthalene exerts a vapor pressure of 0.021 mm Hg at 50°F, it will sublime into the passing airstream. The transfer is so slow that the sublimation from the surface at any instant can be treated as though it were occurring from a sphere of constant diameter. Determine the molar flux of naphthalene from a $\frac{1}{2}$-in.-diameter mothball.

30.7 Find the length of time that is required to sublime the naphthalene mothall that is described in problem 30.6 if the diameter is reduced to half its original value. Assume that the mothball remains spherical. Is this a good assumption?

30.8 Investigators studying the mass transfer from single spheres, correlated the mass-transfer Nusselt number by

$$\mathrm{Nu}_{AB} = 2.0 + C\,\mathrm{Re}^m\,\mathrm{Sc}^{1/3}$$

The value of 2.0, which represents the contribution by molecular diffusion into a large volume of stagnant air, has been designated Nu_{AB_0}. By considering the diffusion from a fixed diameter sphere, derive Nu_{AB_0} for molecular diffusion and evaluate what assumptions must be made for it to be equal to 2.0.

30.9 A 2-cm naphthalene mothball is suspended in an air duct. Estimate the instantaneous mass-transfer coefficient
(a) if the air is quiescent about the mothball and is at 290 K and 1.013×10^5 Pa. The vapor pressure of naphthalene at 290 K is 26 Pa;
(b) if the air is flowing at a velocity of 1.75 m/s and is at 290 K and 1.013×10^5 Pa.

30.10 Estimate the distance a spherical drop of water, originally 2 mm in diameter, must fall in quiet air at 295 K and 1.013×10^5 Pa pressure in order to reduce its volume by 50%. The free-fall terminal velocity of water drops in air was reported by Sherwood and Pigford*

Diameter, mm	0.05	0.2	0.5,	1.0	2.0	3.0
Velocity, ft/s	0.18	2.3	7.0	12.7	19.2	23.8

Assume the drop remains spherical and that the liquid temperature remains constant at 290 K.

30.11 Evaluate the distance that the spherical drop of water described in problem 30.10 falls if its velocity is always equal to the terminal velocity that is associated with its initial diameter.

30.12 Estimate the mass-transfer coefficient, k_G, for the vaporization of naphthalene from a 4.5-in.-diameter naphthalene cylinder. The cylinder has been inserted into a wind tunnel normal to an airstream flowing at 3 ft/s. Under the specified flow conditions, naphthalene exerts a vapor pressure equal to 5 mm Hg and the physical properties of the airstream at the film temperature are

mass diffusivity	$0.37 \text{ ft}^2/\text{hr}$
kinematic viscosity of air	$0.651 \text{ ft}^2/\text{hr}$
thermal diffusivity of air	$0.92 \text{ ft}^2/\text{hr}$
density of air	$0.071 \text{ lb}_m/\text{ft}^3$
heat capacity of air	$0.24 \text{ Btu}/(\text{lb}_m)(°F)$
thermal conductivity of air	$0.0156 \text{ Btu}/(\text{hr})(\text{ft})(°F)$

30.13 Predict the heat-transfer coefficient for a cylinder when the flow conditions are as specified in problem 30.12.

30.14 Douglas and Churchill† correlated heat transfer from a single cylinder with the following relation:

$$Nu_D = 0.46 \, Re_D^{1/2} + 0.001\,28 \, Re$$

when the flowing medium was air. In the case where the fluid is a liquid, it was recommended that the Nusselt number in this relationship be multiplied by $1.1 \, Pr^{1/3}$, yielding

$$Nu_D = (0.506 \, Re^{1/2} + 0.001\,41 \, Re) \, Pr^{1/3}$$

This equation, in assocaition with the Chilton-Colburn analogy, can be used to predict the mass-transfer coefficient for a cylinder.

---

* T. K. Sherwood and R. L. Pigford, *Absorption and Extraction*, McGraw-Hill Book Company, New York, 1952.
† M. J. M. Douglas and S. W. Churchill, *Chem. Engr. Prog. Symp. Ser.*, *51*, **17**, 57 (1956).

Estimate the mass-transfer coefficient, k_L, for the dissolution of sodium chloride from a cast cylinder, 1.5 cm in diameter, which is placed normal to a flowing water stream. The velocity of the 300 K water stream is 10 m/s.

30.15 Estimate the individual mass-transfer coefficient, k_L, for the dissolution of sodium chloride from a cast sphere, 1.5 cm in diameter, if placed in a flowing water stream. The velocity of the 300 K water stream is 10 m/s.

30.16 Four kilograms of benzene per hour flows in a thin film down the outside surface of a vertical, circular cylinder. Dry air at 317 K and 1.013×10^5 Pa flows at right angles to the 8-cm-diameter cylinder at a velocity of 8 m/s. The liquid is at a temperature of 289 K where it exerts a vapor pressure of 8000 Pa. Determine the length of the cylinder if the entire surface of the cylinder is used for the evaporating process, but restricting any benzene from going beyond the bottom of the cylinder.

30.17 Since no correlations were presented in this chapter that would permit you to predict the mass-transfer coefficient with laminar flow within tubes, use the Chilton-Colburn analogy and the Sieder-Tate, equation (20-25), to develop an equation for the mass-transfer Nusselt number, $Nu_{L,AB}$, for laminar flow in tubes.

30.18 Water at 70°F flows through a 2-in.-ID pipe at an average velocity of 10 ft/s. A 2-ft section of the pipe is replaced with a tube made from a block of solid sodium chloride. Compare the mass-transfer coefficient of the sodium chloride in water evaluated by using equation (30-13) with that evaluated by using the Chilton-Colburn analogy.

30.19 Dry air, flowing at a velocity of 1.5 m/s, enters a 6-m-long, 0.15-m-diameter tube at 310 K and 1.013×10^5 Pa pressure. The inner surface of the tube is lined with a felt material (diameter to roughness ratio, D/e, of 10 000), which is continuously saturated with water at 290 K. Assuming constant temperature of the air and the pipe, determine the rate water must be added to keep the felt continuously saturated.

It is important to realize that the bulk composition of the gas stream will be continuously increasing with the length of the tube.

30.20 Toluene, supplied to the top of a circular cylinder, 3 in. in diameter, flows in a thin film down the outside surface of the vertical cylinder. Dry air at 100°F and 1 atm pressure flows at right angles to the wetted cylinder at a velocity of 25 ft/s. The temperature of the toluene is 65°F; at this temperature, its vapor pressure is 20 mm Hg. At the average temperature of the gas film, the diffusivity of toluene in air is 0.333 ft²/hr. If all the toluene is completely evaporated just before reaching the bottom of the

8-ft-tall cylinder, determine the rate at which toluene must be supplied to the top of the cylinder.

30.21 An air–water vapor gas stream is flowing countercurrent to a falling liquid film in a 3-cm-ID wetted-wall column. The gas and liquid streams are at 298 K and the total pressure within the column is 1.013×10^5 Pa. At one point in the column, the airflow rate is 9.5×10^{-4} m^3/s and the average partial pressure of the water vapor in the gas stream is 665 Pa. The kinematic viscosity of the gas stream is 1.7×10^{-5} m^2/s. Determine the mass-transfer coefficient, k_G.

30.22 If a film model for the mass-transfer coefficient evaluated in problem 30.21 is assumed, evaluate the "effective film" thickness.

30.23 In studying the sublimation of naphthalene into an airstream, an investigator constructed a short section of an annular duct; the inner pipe was made from a 1-in.-OD solid naphthalene rod and the outer wall was made by drilling a 2-in.-ID hole in a long block of naphthalene. He proposed using equation (30-12) for correlating the data, using an equivalent diameter in the mass-transfer Nusselt number and in the Reynolds number.

Air at 60°F and an average pressure of 1 atm flowed through the annular space at a bulk velocity of 50 ft/s. At 60°F, naphthalene has a vapor pressure of 0.039 mm Hg and a diffusivity in air of 0.206 ft^2/hr. Evaluate the mass-transfer coefficient for the annulus.

30.24 Data for the absorption of ammonia, NH_3, from an air-NH_3 gas stream by a falling liquid film were obtained in a 1-in.-ID wetted-wall column operated at 75°F and 1 atm pressure. The gas mixture had the following properties:

molecular weight	28.9
viscosity of the gas	0.0431 lb$_m$/(hr)(ft)
diffusivity of NH_3 in air	0.766 ft^2/hr
density of the gas	0.081 lb$_m$/ft^3
average NH_3 composition	1.0%

(a) Evaluate the individual mass-transfer coefficient, k_G, for a mass velocity of 5250 lb$_m$ gas/(hr)(ft^2);

(b) From the design equations to be discussed in Chapter 31, the overall mass-transfer coefficient, K_G, was evaluated to be 0.80 lb mole NH_3/(hr)(ft^2)(atm). Evaluate the percent of the total resistance to transfer that was encountered in the gas film.

30.25 Wilke and Hougan[*] reported the mass transfer in beds of granular solid. Air was blown through a bed of porous celite pellets wetted with water, and by evaporating this water under adiabatic conditions, they reported gas-film coefficients for packed beds. In one run, the following data were reported:

effective particle diameter	0.01872 ft
gas stream mass velocity	601 $lb_m/(hr)(ft^2)$
temperature at the surface	100°F
pressure	733 mm

They reported a value for k_G of 3.250 lb mole/(hr)(ft²)(atm).

With the assumption that the properties of the gas mixture are the same as air, calulate the gas-film mass-transfer coefficient and compare your value with the reported value by using

(a) the approximate correlation, equation (30-15);

(b) the Gupta and Thodus correlation (30-18) if the packed bed had a void fraction of 0.75.

30.26 R. P. Whitney and J. E. Vivian[†] measured rates of absorption of SO_2 in water and found the following expression for 1-in. Raschig rings at 70°F

$$k_L a = 0.044 L^{0.82}$$

They reported a $k_L a$ value of 224 lb mole/(hr)(ft³)(lb mole/ft³) for the experimental run that involved a liquid flow rate of 1500 $lb_m/(hr)(ft^2)$.

(a) What is the reliability of their correlating equation?

(b) Compare the experimental value with a value evaluated by using the Sherwood and Holloway equation (30-20). The diffusivity of SO_2 in water at 70°F is 6.58×10^{-5} ft²/hr.

[*] C. R. Wilke and O. A. Hougan, *Trans. A.I.Ch.E.*, **41**, 445 (1945).
[†] R. P. Whitney and J. E. Vivian, *Chem. Eng. Progr.*, **45**, 323 (1949).

31
MASS-TRANSFER
EQUIPMENT

In Chapter 29, the theory currently used to explain the mechanism of mass transfer between phases was introduced. In Chapter 30, correlations for the convective mass-transfer coefficients used in the interphase transfer equations were listed. This chapter will include the development of methods for applying the transport equations to the design of *continuous-contacting mass transfer equipment*. It is important to realize that design procedures are not restricted to the design of new equipment, for they may also be applied in analyzing existing equipment for possible improvement in performance.

The presentation or development of mass transfer from the defining equations to the final design equations has been completely analogous to our earlier treatment of energy transfer. Convective mass-transfer coefficients were defined and the related theories presented in Chapter 28; the definitions and methods of analysis were similar to those presented in Chapter 19 for convective heat-transfer coefficients. An overall driving force and an overall transfer coefficient expressed in terms of individual convective coefficients have been developed to explain the transfer mechanism of both transport processes. In Chapter 22, by integrating the appropriate energy transfer relation, we were able to evaluate the area of a heat exchanger. Accordingly, we should expect to find similar mass-transfer relations which can be integrated to yield the total contact area within a mass exchanger.

31.1 TYPES OF MASS-TRANSFER EQUIPMENT

A substantial number of industrial operations are concerned with the problem of changing the compositions of solutions and mixtures, using interphase mass-transfer principles. Typical examples of such operations could include: (1) the transfer of a solute from the gas phase into the liquid phase, as encountered in absorption, dehumidification, and distillation; (2) the transfer of a solute from the liquid phase into the gas phase as encountered in desorption and humidification; (3) the transfer of a solute from one liquid phase into a second immiscible liquid phase as encountered in liquid-liquid extraction; (4) the transfer of a solute from a solid into a fluid phase as encountered in drying and leaching; and (5) the transfer of a solute from a fluid onto the surface of a solid as encountered in adsorption and ion exchange.

Mass-transfer operations are commonly conducted in towers which are designed to provide intimate contact of the two phases. This equipment may be classified into one of four general types according to the method used to produce the interphase contact. Many varieties and combinations of these types exist or are possible; we will restrict our discussion to the major classifications.

Spray towers consist of large open chambers through which the gas phase flows and into which the liquid is introduced by spray nozzles or other atomizing devices. Figure 31.1 illustrates the direction of phase flow in a spray tower. The

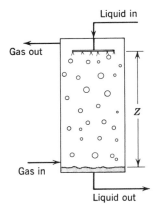

Figure 31.1 Spray tower.

liquid is introduced as a spray and falls because of gravity, countercurrent to the rising gas stream. The spray nozzle is designed to subdivide the liquid into a large number of small drops; for a given liquid flow rate smaller drops provide a greater interphase contact area across which mass is transferred. However, care in design is exercised to avoid producing drops so fine that they become entrained in the exiting gas stream. The distance which the drop falls determines the contact time and, in turn, influences the amount of mass transferred between the two phases in continuous contact. Resistance to transfer within the gas phase is reduced by the swirling motion of the falling liquid droplets. Spray towers are used for mass transfer of highly soluble gases where the gas-phase resistance normally controls the rate of mass transfer.

Exactly opposite in principle to the spray tower is the *bubble tower,* in which the gas is dispersed into the liquid phase in the form of fine bubbles. The small gas bubbles provide the desired large interphase contact area. Mass transfer takes place both during the bubble formation and as the bubbles rise up through the liquid. The moving bubbles reduce the liquid-phase resistance. Bubble towers are used with systems in which the liquid phase controls the rate of mass transfer; that is, the absorption of relatively insoluble gases. Figure 31.2 illustrates the contact length and the direction of phase flow in a bubble tower. The basic mass-transfer mechanism involved in bubble towers is also encountered in batch bubble tanks or ponds where the gas is dispersed at the bottom of the tanks. Such equipment is commonly encountered in the reaeration of wastewater.

Packed towers is the third general type of mass-transfer equipment which involves a continuous countercurrent contact of two immiscible phases. These towers are vertical columns which have been filled with packing as illustrated in Figure 31.3. A variety of packing materials are used, ranging from specially

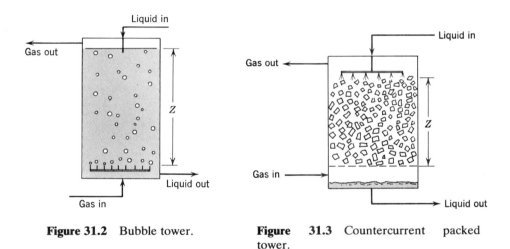

Figure 31.2 Bubble tower.   **Figure 31.3** Countercurrent packed tower.

designed ceramic packing to crushed rock. The liquid is distributed over the packing and flows down the packing surface as thin films or subdivided streams. The gas generally flows upward, countercurrent to the falling liquid. Both phases are well agitated. Thus this type of equipment may be used for gas-liquid systems in which either of the phase resistances controls or in which both resistances are important.

Special types of packed towers are used to cool water so that it can be recirculated as a heat-transfer medium. These structures are made of wood-slat decks, having louver construction so that air can flow across each deck. The water is sprayed above the top deck and then trickles down through the various decks to a bottom collection basin. Cooling towers may be classified as natural draft when sufficient natural wind is available to carry away the humid air or as forced or induced draft when a fan is used. In the forced-draft towers, air is pulled into louvers at the bottom of the structure and then flows up through the decks countercurrent to the water flow.

Bubble-plate and *sieve-plate towers* are commonly used in industry. They represent the combined transfer mechanisms observed in the spray and the bubble towers. At each plate, bubbles of gas are formed at the bottom of a liquid pool by forcing the gas through small holes drilled in the plate or under slotted caps immersed in the liquid. Interphase mass transfer occurs during the bubble formation, and as the bubbles rise up through the agitated liquid pool. Additional mass transfer takes place above the liquid pool because of spray carry-over produced by the active mixing of the liquid and gas on the plate. Such plates are arranged one above the other in a cylindrical shell as schematically illustrated in Figure 31.4. The liquid flows downward, crossing first the upper plate and then the

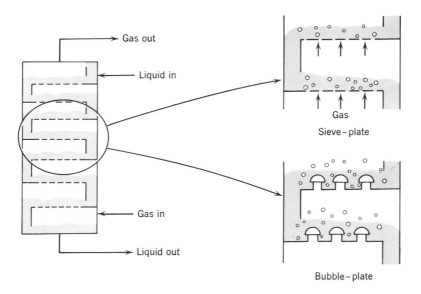

Figure 31.4 Plate towers.

plate below. The vapor rises through each plate. As Figure 31.4 illustrates, the contact of the two phases is stepwise. Such towers cannot be designed by equations which are obtained by integrating over a continuous area of interphase contact. Instead, they are designed by stagewise calculations which are developed and used in design courses of stagewise operations. We shall not consider the design of plate towers in this book; our discussions will be limited to continuous-contact equipment.

31.2 BATCH MASS-TRANSFER TANKS OR PONDS

In the treatment of waste water, undesirable gases are frequently stripped or desorbed from the water and oxygen is absorbed into the water when bubbles of air are dispersed near the bottom of aeration tanks or ponds. The introduction of compressed gas into small-orifice dispersers, such as perforated pipes, porous sparger tubes, and porous plates, produces small bubbles of the gas which rise through the liquid.

As the bubbles rise, solute can be transferred from the gas to the liquid or from the liquid to the gas depending upon the concentration driving force. Since the gases involved are normally only slightly soluble in the liquid, the mass transfer is calculated using the overall liquid mass transfer coefficient,

$$N_A = K_L(c_A{}^* - c_{A,L}) \tag{29-11}$$

The rate of solute added to the liquid contained within the tank is

$$W_A = K_L A(c_A{}^* - c_{A,L})$$

where A is the total interfacial area of contact between the gas bubbles and the liquid solution. This equation can be expressed in differential form by

$$\frac{d \text{ (moles of } A)}{dt} = K_L A(c_A{}^* - c_{A,L})$$

In a constant volume tank, this equation can be expressed in terms of the accumulation of the solute

$$\frac{dc_{A,L}}{dt} = K_L \frac{A}{V}(c_A{}^* - c_{A,L}) \tag{31-1}$$

As bubbles rise, the equilibrium concentration, $c_A{}^*$, remains essentially constant; accordingly, the variables can be separated

$$\frac{dc_{A,L}}{c_A{}^* - c_{A,L}} = K_L \frac{A}{V} dt$$

and integrated between the time limits zero and t and the corresponding concentration limits c_{A,L_0} and c_{A,L_t} to obtain

$$\ln\left[\frac{c_A{}^* - c_{A,L_0}}{c_A{}^* - c_{A,L_t}}\right] = K_L \frac{A}{V} t \tag{31-2}$$

The ratio, A/V, represents the total interfacial area of the bubbles per unit volume of solution in the tank. Since small bubbles produce a larger interfacial area per volume of gas fed to the disperser, thus increasing the magnitude of this ratio, and larger bubbles increase the mixing within the tank, thus increasing the magnitude of the overall mass transfer coefficient, K_L, the gas disperser must be designed to optimize the combined $K_L(A/V)$ transfer factor.

Eckenfelder* developed a general correlation for transfer of oxygen from air bubbles rising in a column of still water,

$$K_L \frac{A}{V} = \frac{\theta_g Q_g^{1+n} h^{0.78}}{V} \qquad (31\text{-}3)$$

where θ_g is a correlating constant dependent on the type of disperser, Q_g is the gas flow rate in standard cubic feet per minute, n is a correlating constant that is dependent on the size of the small orifices in the disperser and h is the depth below the liquid surface at which the air is introduced to the aeration tank. Typical data for a sparger aeration unit, correlated according to equation (31-3) are presented in Figure 31.5.

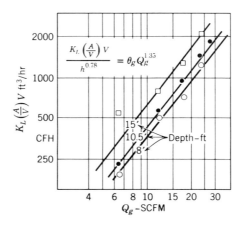

Figure 31.5   Oxygen transfer factor of a single sparger unit in an aeration tank.

In the following example, the data of Eckenfelder will be used to determine the time required to increase the oxygen level in an aeration pond.

EXAMPLE 1
A 20 000 ft³ aeration pond is aerated with 15 spargers, each using compressed air at a rate of 15 standard ft³/min. The spargers will be located 15 ft below the surface of the pond. Find the time required to raise the dissolved oxygen from 2 mg/liter to 5 mg/liter if the water temperature is 20°C.
From Figure 31.5, the transfer factor, $K_L(A/V)V$, for a single sparger is 1200 ft³/hr, and for the system

$$K_L \frac{A}{V} = \frac{(1200 \text{ ft}^3/\text{hr/sparger})(15 \text{ spargers})}{(20\ 000 \text{ ft}^3)} = 0.90 \text{ hr}^{-1}$$

---

* W. W. Eckenfelder, Jr., *J. Sanit. Engr. Div.*, Amer. Soc. Civ. Engr., **85**, SA4, 89 (1959).

The average hydrostatic pressure of the rising air bubble is equal to the arithmetic mean of the pressure at the top and bottom of the pond.

$$P_{bottom} = 1 \text{ atm} + (15 \text{ ft } H_2O)(0.0295 \text{ atm/ft } H_2O) = 1.44 \text{ atm}$$

$$P_{mean} = \frac{1 \text{ atm} + 1.44 \text{ atm}}{2} = 1.22 \text{ atm}$$

Since the mole fraction of oxygen in air is 0.21, the partial pressure of oxygen within the bubbles will be $0.21(1.22) = 0.256$ atm. The equilibrium concentration of a slightly soluble gas is related to its partial pressure by Henry's law. At 20°C, this law stipulates for oxygen

$$p_{O_2}(\text{atm}) = (4.01 \times 10^4 \text{ atm/mole fraction}) \, x_{O_2}$$

Accordingly,

$$x_{O_2} = \frac{0.256}{4.01 \times 10^4} = 6.38 \times 10^{-6}$$

For one liter of solution, which is essentially pure water, the equilibrium concentration in milligrams per liter can be calculated:

$$\text{moles of water} = \frac{(1000 \text{ cc of water})(1 \text{ g/cc water})}{(18 \text{ g water/mol})}$$

$$= 55.6 \text{ mol}$$

moles of oxygen
in the liter of water $= (x_{O_2})(\text{moles solution})$

$$= (6.38 \times 10^{-6})(55.6) = 3.55 \times 10^{-4} \text{ mol}$$

grams of oxygen/liter $= (3.55 \times 10^{-4} \text{ mol})(32 \text{ g/mol})$

$$\doteq 1.135 \times 10^{-2} \text{ g/liter}$$

$$c_A{}^* = 11.35 \text{ mg/liter}$$

Using equation (31-2), we can solve for the required time

$$t = \ln\left(\frac{c_A{}^* - c_{A_0}}{c_A{}^* - c_{A,L_t}}\right)\left(\frac{1}{K_L(A/V)}\right) = \ln\left[\frac{(11.35-2)}{(11.35-5)}\right]\left(\frac{1}{0.90/\text{hr}}\right)$$

$$= 0.43 \text{ hr} = 25.8 \text{ min}$$

31.3 MASS BALANCES FOR CONTINUOUS CONTACT TOWERS: OPERATING LINE EQUATIONS

There are four important fundamentals which constitute the basis for continuous-contact equipment design:

1. Material and enthalpy balances, involving the equations of conservation of mass and energy;

2. Interphase equilibrium;
3. Mass-transfer equations;
4. Momentum-transfer equations.

Interphase equilibrium relations are defined by laws of thermodynamics as discussed in section 29.1. Momentum-transfer equations, as presented in section 9.3, are used to define the pressure drop within the equipment. We shall not treat this subject in this chapter, since it was previously discussed. *The material and enthalpy balances are important, since they provide expressions for evaluating the bulk compositions of the two contacting phases at any plane in the tower as well as the change in bulk compositions between two planes in the tower.* The mass-transfer equations will be developed in differential form, combined with a differential material balance, and then integrated over the area of interfacial contact to provide the length of contact required in the mass exchanger.

COUNTERCURRENT FLOW

Consider any steady-state mass-transfer operation which involves the countercurrent contact of two insoluble phases as schematically shown in Figure 31.6. The two insoluble phases will be identified as phase G and phase L.

At the bottom of the mass-transfer tower, the flow rates and concentrations are defined as follows:

G_1 is the total moles of phase G entering the tower per hour per cross-sectional area of the tower;

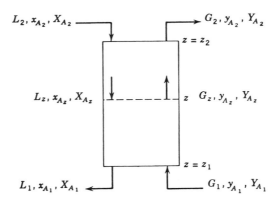

Figure 31.6 Steady-state countercurrent process.

L_1 is the total moles of phase L leaving the tower per hour per cross-sectional area of the tower;

y_{A_1} is the mole fraction of component A in G_1, expressed as moles of A per total moles in phase G;

x_{A_1} is the mole fraction of component A in L_1, expressed as moles of A per total moles in phase L.

Similarly at the top of the tower, or plane z_2, the total moles of each phase will be G_2 and L_2, and the compositions of each stream will be y_{A_2} and x_{A_2}. An overall, macroscopic mass balance for component A around the steady-state mass exchanger, in which there is no chemical production or disappearance of A, requires

$$\begin{bmatrix} \text{moles of } A \text{ entering} \\ \text{the tower} \end{bmatrix} = \begin{bmatrix} \text{moles of } A \text{ leaving} \\ \text{the tower} \end{bmatrix}$$

or

$$G_1 y_{A_1} + L_2 x_{A_2} = G_2 y_{A_2} + L_1 x_{A_1} \tag{31-4}$$

A mass balance for component A around plane $z = z_1$ and the arbitrary plane z stipulates

$$G_1 y_{A_1} + L_z x_{A_z} = G_z y_{A_z} + L_1 x_{A_1} \tag{31-5}$$

Simpler relations, and certainly easier equations to use, may be expressed in terms of *solute-free concentration units*. The concentration of each phase will be defined as follows:

Y_A is the moles of A in G per mole of A-free G; that is,

$$Y_A = \frac{y_A}{1 - y_A} \tag{31-6}$$

and X_A is the moles of A in L per mole of A-free L; that is,

$$X_A = \frac{x_A}{1 - x_A} \tag{31-7}$$

The flow rates now become L_S and G_S, where L_S is the *moles of phase L on a solute-free basis*; that is, moles of the carrier solvent in phase L per hour per cross-sectional area of the tower; and G_S is the *moles of phase G on a solute-free basis*; that is, moles of the carrier solvent in phase G per hour per cross-sectional area of the tower. The overall balance on component A may be written, using the solute-free terms as

$$G_S Y_{A_1} + L_S X_{A_2} = G_S Y_{A_2} + L_S X_{A_1}$$

or

$$G_S(Y_{A_1} - Y_{A_2}) = L_S(X_{A_1} - X_{A_2}) \tag{31-8}$$

Rearranging, we obtain

$$\frac{L_S}{G_S} = \frac{Y_{A_1} - Y_{A_2}}{X_{A_1} - X_{A_2}}$$

Equation (31-8) is an equation of a straight line which passes through the points (X_{A_1}, Y_{A_1}) and (X_{A_2}, Y_{A_2}) with a slope of L_S/G_S. A mass balance on component A around plane z_1 and the arbitrary plane $z = z$ in solute-free terms is

$$G_S Y_{A_1} + L_S X_{A,z} = G_S Y_{A,z} + L_S X_{A_1}$$

or

$$G_S(Y_{A_1} - Y_{A,z}) = L_S(X_{A_1} - X_{A,z}) \tag{31-9}$$

Rearranging, we obtain

$$\frac{L_S}{G_S} = \frac{Y_{A_1} - Y_{A,z}}{X_{A_1} - X_{A,z}}$$

As before, equation (31-9) is an equation of a straight line, one which passes through the points (X_{A_1}, Y_{A_1}) and $(X_{A,z}, Y_{A,z})$ with a slope of L_S/G_S. Two straight lines having the same slope and a point in common lie on the same straight line. Equation (31-9) is, therefore, a general expression relating the bulk compositions of the two phases at any plane in the mass exchanger. Since it defines operating conditions within the equipment, it is designated the *operating-line equation for countercurrent operations*. In our discussions on interphase transfer in section 29.2, point O of Figure 29.4 is one of many points which lie on the operating line. Figures 31.7 and 31.8 illustrate the location of the operating line relative to the equilibrium line when the transfer is from phase G to phase L and from phase L to phase G.

A mass balance for component A over the differential length, dz, is easily obtained by differentiating equation (31-9). This differential equation,

$$L_S \, dX_A = G_S \, dY_A \tag{31-10}$$

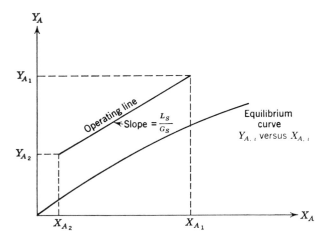

Figure 31.7 Steady-state countercurrent process, transfer from phase G to phase L.

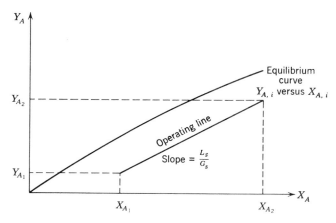

Figure 31.8 Steady-state countercurrent process, transfer from phase L to phase G.

relates the moles transferred from one phase to the second phase per hour per cross-sectional area available in the length dz.

It is important that the student recognize the difference between equations (31-5) and (31-9). Although both equations describe the mass balance for component A, only equation (31-9) is an equation of a straight line. When written in the solute-free units, X and Y, the operating line is straight because the mole-ratio concentrations are based on the constant quantities, L_S and G_S. When written in mole-fraction units; x and y, the total moles in a phase, L or G, change as the solute is transferred into or out of the phase; this produces a curved operating line on x-y coordinates.

In the design of mass-transfer equipment, the flow rate of at least one phase and three of the four entering and exiting compositions must be fixed by the process requirements. The necessary flow rate of the second phase is often a design variable. For example, consider the case in which phase G, with a known G_S, changes in composition from Y_{A_1} to Y_{A_2} by transferring solute to a second phase which enters the tower with composition X_{A_2}. According to equation (31-8), the operating line must pass through point (X_{A_2}, Y_{A_2}) and must end at the ordinate Y_{A_1}. Three possible operating lines are shown in Figure 31.9. Each line has a different slope, L_S/G_S, and since G_S is fixed by the process requirement, each line represents a different quantity, L_S, of the second phase. In fact, as the slope decreases, L_S decreases. The *minimum* L_S which may be used corresponds to the operating line ending at point P_3. This quantity of the second phase corresponds to an operating line which touches the equilibrium line. If we recall from Chapter 29 the definition of driving forces, we should immediately recognize that the closer the operating line is to the equilibrium curve, the smaller will be the driving force for overcoming any mass-transfer resistance. At the point of tangency, the diffusional driving force is zero; thus mass transfer between the two phases cannot occur. This then represents a limiting condition, the *minimum*

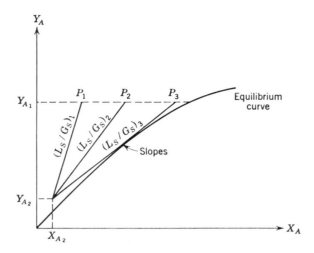

Figure 31.9   Operating-line locations.

L_S/G_S *ratio for mass transfer.* In the case of equilibrium curves which are concave upward, the minimum L_S/G_S ratio corresponds to the second phase L_1 leaving in equilibrium with the entering phase G_1; that is, point (X_1, Y_1) lies on the equilibrium curve.

EXAMPLE 2

Ammonia is to be absorbed from air at 68°F and atmospheric pressure in a countercurrent packed tower, using water at 68°F as the absorbent. An inlet gas rate of 1540 ft^3/hr and an ammonia-free water rate of 75 lb$_m$/hr will be used. If the ammonia, NH_3, concentration is reduced from 3.52 to 1.29% by volume, determine the ratio $(L_S/G_S)_{actual}/(L_S/G_S)_{minimum}$. Equilibrium data for the system at 68°F and 1 atm are as follows:

X, lb mole NH_3/lb mole H_2O	0.0164	0.0252	0.0349	0.0455	0.0722
Y, lb mole NH_3/lb mole air	0.021	0.032	0.042	0.053	0.08

The total moles of gas entering the tower per hour may be evaluated, using the ideal gas law

$$GA = \text{moles entering gas/hr} = \frac{\dot{V}P}{RT} = \frac{(1540 \text{ ft}^3/\text{hr})(1 \text{ atm})}{(0.73 \text{ ft}^3 \text{ atm/lb mole } °R)/(528 \text{ } °R)}$$

$$= \frac{4 \text{ lb mole}}{\text{hr}}$$

The gas enters the tower with a mole fraction of ammonia equal to 0.0352. Choosing a cross-sectional area for the tower of A ft^2, we may evaluate the moles of phase G on a solute-free basis as

$$G_S = G(1 - y_{NH_3}) = \left(\frac{4 \text{ lb moles}}{\text{hr}}\right)\left(\frac{0.9648}{A \text{ ft}^2}\right) = \frac{3.85}{A}\frac{\text{lb moles}}{\text{hr ft}^2}$$

The moles of phase L on a solute-free basis are

$$L_S = 75\frac{\text{lb}_m}{\text{hr}}\frac{\text{lb mole}}{18 \text{ lb}_m}\frac{1}{A \text{ ft}^2} = \frac{4.17}{A}\frac{\text{lb mole}}{\text{hr ft}^2}$$

The ratio of the actual L_S to G_S values is evaluated as

$$\left(\frac{L_S}{G_S}\right)_{\text{actual}} = \frac{4.17}{A}\times\frac{A}{3.85} = 1.08\frac{\text{mole NH}_3\text{-free } L \text{ phase}}{\text{mole NH}_3\text{-free } G \text{ phase}}$$

The compositions of the known streams, G_1, G_2, and L_2, on a solute-free basis, are evaluated from the known mole fractions as

$$Y_{NH_3}|_1 = \frac{y_{NH_3}|_1}{1 - y_{NH_3}|_1} = \frac{0.0352}{0.9648} = 0.0365$$

$$Y_{NH_3}|_2 = \frac{y_{NH_3}|_2}{1 - y_{NH_3}|_2} = \frac{0.0129}{0.9871} = 0.0131$$

and

$$X_{NH_3}|_2 = \frac{x_{NH_3}|_2}{1 - x_{NH_3}|_2} = 0.0$$

The exiting composition, $X_{NH_3}|_1$, can be evaluated by

$$G_S(Y_{A_1} - Y_{A_2}) = L_S(X_{A_1} - X_{A_2})$$

$$\frac{3.85}{A}(0.0365 - 0.0131) = \frac{4.17}{A}(X_{A_1} - 0)$$

or

$$X_{NH_3}|_1 = 0.0216 \tag{31-8}$$

The composition of the solution in equilibrium with $Y_{NH_3}|_1 = 0.0365$ is obtained from the equilibrium curves as $X_{NH_3}|_{\text{equil}} = 0.0296$.

The actual and minimum operating lines are shown in Figure 31.10. The slope of the minimum operating line is

$$\left(\frac{L_S}{G_S}\right)_{\text{minimum}} = \frac{\Delta Y}{\Delta X} = \frac{0.0365 - 0.0131}{0.0296 - 0} = 0.79\frac{\text{mole NH}_3\text{-free } L \text{ phase}}{\text{mole NH}_3\text{-free } G \text{ phase}}$$

The desired ratio, $(L_S/G_S)_{\text{actual}}/(L_S/G_S)_{\text{minimum}}$, is then a ratio of the two values, $1.08/0.79$ or 1.37.

Cocurrent flow. For steady-state mass-transfer operations involving cocurrent contact of two insoluble phases, as shown in Figure 31.11, the overall mass balance for component A is

$$L_S X_{A_2} + G_S Y_{A_2} = L_S X_{A_1} + G_S Y_{A_1}$$

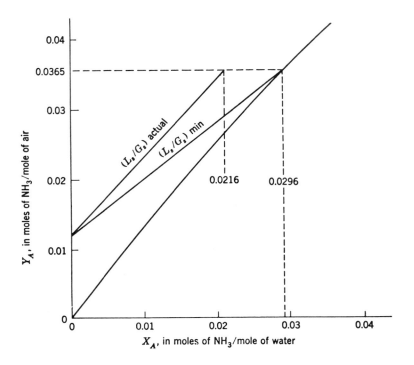

Figure 31.10   Solution to example 2.

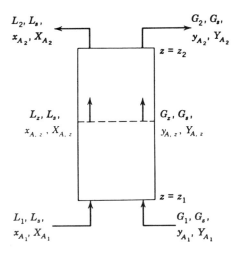

Figure 31.11   Steady-state cocurrent process.

or

$$L_S(X_{A_2} - X_{A_1}) = G_S(Y_{A_1} - Y_{A_2}) \tag{31-11}$$

The mass balance on component A around planes z_1 and an arbitrary plane z stipulates that

$$L_S X_{A,z} + G_S Y_{A,z} = L_S X_{A_1} + G_S Y_{A_1}$$

or

$$L_S(X_{A,z} - X_{A_1}) = G_S(Y_{A_1} - Y_{A,z}) \tag{31-12}$$

Both (31-11) and (31-12) are equations of straight lines which pass through the common point, (X_{A_1}, Y_{A_1}) and have the same slope, $-L_S/G_S$. Equation (31-12) is the general expression which relates the compositions of the two contacting phases at any plane in the equipment. It is designated the *operating-line equation for cocurrent operations.* Figures 31.12 and 31.13 illustrate the location of the

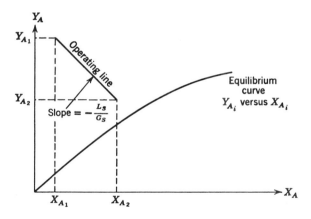

Figure 31.12 Steady-state cocurrent process, transfer from phase G to phase L.

operating line relative to the equilibrium line. A mass balance for component A over the differential length, dz, for cocurrent flow is

$$L_S \, dX_A = -G_S \, dY_A \tag{31-13}$$

As in the case of countercurrent flow, there is a *minimum L_S/G_S ratio* for cocurrent mass transfer operations established from the fixed process variables, G_S, Y_{A_1}, Y_{A_2}, and X_{A_1}. Its evaluation involves the same procedure as discussed for countercurrent flow.

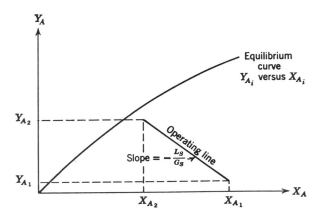

Figure 31.13 Steady-state cocurrent process, transfer from phase L to phase G.

EXAMPLE 3

The ammonia-air feed stream described in example 2 is fed cocurrently with an ammonia-free water stream. The ammonia concentration is to be reduced from 3.52 to 1.29% by volume, using a water stream 1.37 times the minimum. Determine (a) the minimum L_S/G_S ratio, (b) the actual water rate, and (c) the concentration in the exiting aqueous stream.

In example 2, the following compositions were evaluated:

$$\text{Entering } Y_{\text{NH}_3}|_1 = 0.0365$$

$$\text{Exiting } Y_{\text{NH}_3}|_2 = 0.0131$$

and

$$\text{Entering } X_{\text{NH}_3}|_1 = 0.0$$

The moles of G on a solute-free basis were evaluated to be $3.85/A$ lb mole/hr ft^2. In Figure 31.14 the minimum and the actual operating lines are shown. For these operating lines,

$$\left(\frac{L_S}{G_S}\right)_{\text{minimum}} = \frac{Y_{\text{NH}_3}|_1 - Y_{\text{NH}_3}|_2}{X_{\text{NH}_3}|_2 - X_{\text{NH}_3}|_1} = \frac{0.0365 - 0.0131}{0.01}$$

$$= 2.34 \frac{\text{lb mole NH}_3\text{-free } L \text{ phase}}{\text{lb mole NH}_3\text{-free } G \text{ phase}}$$

and

$$\left(\frac{L_S}{G_S}\right)_{\text{actual}} = 1.37\left(\frac{L_S}{G_S}\right)_{\text{minimum}} = (1.37)(2.34)$$

$$= 3.21 \frac{\text{lb mole NH}_3\text{-free } L \text{ phase}}{\text{lb mole NH}_3\text{-free } G \text{ phase}}$$

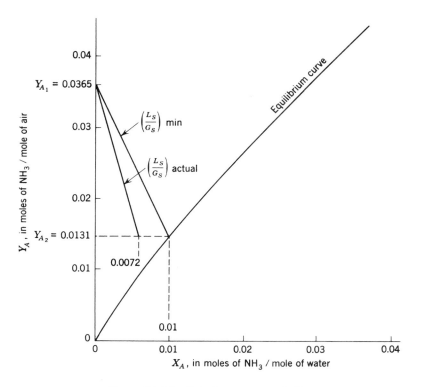

Figure 31.14 Solution to example 3.

The composition of the exiting liquid stream can be evaluated with the slope of the actual operating line by

$$\left(\frac{L_S}{G_S}\right)_{\text{actual}} = 3.21 = \frac{Y_{\text{NH}_3}|_1 - Y_{\text{NH}_3}|_2}{X_{\text{NH}_3}|_2 - X_{\text{NH}_3}|_1} = \frac{0.0365 - 0.0131}{X_{\text{NH}_3}|_2}$$

or

$$X_{\text{NH}_3}|_2 = \frac{0.0232}{3.21} = 0.0072 \frac{\text{mole NH}_3}{\text{mole NH}_3\text{-free water}}$$

The moles of NH_3-free water fed to the tower, L_S, is also evaluated, using the value of $(L_S/G_S)_{\text{actual}}$

$$\left(\frac{L_S}{G_S}\right)_{\text{actual}} = 3.21 \frac{\text{lb mole NH}_3\text{-free } L \text{ phase}}{\text{lb mole NH}_3\text{-free } G \text{ phase}}$$

then

$$L_S = 3.21 G_S = 3.21\left(\frac{3.85}{A}\right) = \frac{12.4}{A} \frac{\text{lb mole}}{\text{hr ft}^2} \left(\frac{0.303}{A} \frac{\text{kg}}{\text{s} \cdot \text{m}^2}\right)$$

31.4 ENTHALPY BALANCES FOR CONTINUOUS CONTACT TOWERS

Many mass-transfer operations are isothermal. This is especially true when we are dealing with dilute mixtures. However, when large quantities of solute are transferred, the heat of mixing can produce a temperature rise in the receiving phase. If the temperature of the phase changes the equilibrium solubility of the solute will be altered, and in turn, the diffusion driving forces will be altered.

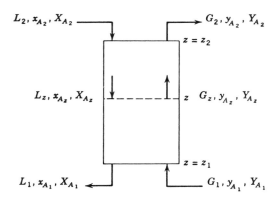

Figure 31.15 Steady-state countercurrent process.

Consider the steady-state countercurrent process illustrated in Figure 31.15. An enthalpy balance around the planes $z = z_2$ and z is

$$L_2 H_{L_2} + G H_G = G_2 H_{G_2} + L H_L \tag{31-14}$$

where H is the molal enthalpy of the stream at its particular temperature, pressure, and concentration. The enthalpies are normally based upon a reference of pure solute-free carrier solvent and pure solute at a chosen base temperature, T_0. The normal enthalpy of a liquid mixture is evaluated above this base temperature by the relation

$$H_L = c_{pL}(T_L - T_0)M_{\text{avg}} + \Delta H_S \tag{31-15}$$

where H_L is the enthalpy of the liquid stream, in Btu/lb mole or kJ/mol of L; $c_{p,L}$ is the heat capacity of the mixture on a mass basis, Btu/lb$_m$ °F or kJ/kg · K; T_L is the temperature of the mixture in °F or K; M_{avg} is the average molecular weight of the mixture; and ΔH_s is the integral heat of solution, evaluated at the base temperature, T_0, and at the concentration of the mixture in Btu/lb mole or kJ/mol.

The molal enthalpy of a gas mixture, with the same base temperature and standard state of the solute is expressed as

$$H_G = [y_{\text{solute}} c_{p,G \text{ solute}} M_{\text{solute}} + (1 - y_{\text{solute}})(c_{p,G \text{ solute-free } G\text{-phase}})$$

$$(M_{\text{solute-free } G\text{-phase}})](T_G - T_0) + y_{\text{solute}} h_{f,g \text{ solute}} M_{\text{solute}} \qquad (31\text{-}16)$$

where H_G is the enthalpy of the gas stream, in Btu/lb mole or kJ/mol of G; $c_{p,G}$ is the heat capacity in the gas phase in Btu/lb$_m$ °F or kJ/kg·K; T_G is the temperature of the gas mixture in °F or K; M is the molecular weight; and $h_{f,g \text{ solute}}$ is the heat of vaporization of the solute in Btu/lb$_m$ or kJ/kg. The integral heat of solution, ΔH_s, is zero for ideal solutions and essentially zero for gas mixtures. For nonideal solutions, it is a negative quantity if heat is evolved on mixing and a positive quantity if heat is absorbed on mixing.

Equation (31-14) may be used to compute the temperature of a given phase at any plane within the mass-transfer equipment. The calculations involve the simultaneous application of the mass balance in order to know the flow rate of the stream associated with the particular enthalpy term.

31.5 MASS-TRANSFER CAPACITY COEFFICIENTS

The individual mass-transfer coefficient, k_G, was defined by the expression

$$N_A = k_G(p_{A,G} - p_{A,i}) \qquad (29\text{-}6)$$

and the overall mass-transfer coefficient was defined by a similar equation in terms of the overall driving force in partial pressure units,

$$N_A = K_G(p_{A,G} - p_A{}^*) \qquad (29\text{-}10)$$

In both expressions, the interphase mass transfer was expressed as moles of A transferred per unit time per unit area per unit driving force in terms of partial pressure. In order to use these equations in the design of mass exchangers, the interphase contact area must be known. Although the wetted-wall column, as described in Chapter 26, has a definite interfacial surface area, the corresponding area in other types of equipment is virtually impossible to measure. For this reason the factor a must be introduced to represent the interfacial surface area per unit volume of the mass-transfer equipment. The mass transfer within a differential height, dz, per unit cross-sectional area of the mass exchanger is

$$N_A \left[\frac{\text{moles of } A \text{ transferred}}{(\text{hr})(\text{interfacial area})} \right] \left[a \left(\frac{\text{interfacial area}}{\text{ft}^3} \right) \right] dz \text{ (ft)}$$

$$= \frac{\text{moles of } A \text{ transferred}}{(\text{hr})(\text{cross-sectional area})}$$

or, in terms of the mass-transfer coefficients,

$$N_A a\, dz = k_G a (p_{A,G} - p_{A,i})\, dz \qquad (31\text{-}17)$$

and

$$N_A a\, dz = K_G a (p_{A,G} - p_A{}^*)\, dz \qquad (31\text{-}18)$$

Since both the factor a and the mass-transfer coefficients depend on the geometry of the mass-transfer equipment and on the flow rates of the two contacting, immiscible streams, they are commonly combined as a product. The *individual capacity coefficient*, $k_G a$, and the *overall capacity coefficient*, $K_G a$, are each experimentally evaluated as a combined process variable. The units of the gas phase capacity coefficient are

$$k_G a \left[\frac{\text{moles of } A \text{ transferred}}{(\text{hr})(\text{interfacial area})(\text{pressure})} \right] \left[\frac{\text{interfacial area}}{\text{volume}} \right]$$

$$= \frac{\text{moles of } A \text{ transferred}}{(\text{hr})(\text{volume})(\text{pressure})}$$

the most often encountered units are g moles of $A/\text{s} \cdot \text{m}^3 \cdot \text{Pa}$ or lb moles of $A/(\text{hr})(\text{ft}^3)(\text{atm})$. The capacity coefficients in terms of liquid concentration driving forces are similarly defined by

$$N_A a\, dz = k_L a (c_{A,i} - c_{A,L})\, dz \qquad (31\text{-}19)$$

and

$$N_A a\, dz = K_L a (c_A{}^* - c_{A,L})\, dz \qquad (31\text{-}20)$$

The most common units for the liquid phase capacity coefficients are g moles of $A/\text{s} \cdot \text{m}^3 \cdot \text{g moles of } A/\text{m}^3$ of solution or lb moles of $A/(\text{hr})(\text{ft}^3)$ (lb moles of A/ft^3 of solution). Mass-transfer capacity equations in terms of $k_y a$, $k_x a$, $k_Y a$ and $k_X a$ are similarly defined.

31.6 CONTINUOUS-CONTACT EQUIPMENT ANALYSIS

The moles of the diffusing component A transferred per time per cross-sectional area have been defined by two entirely different concepts, the material balance and the mass-transfer equations. For equipment involving the continuous-contact between the two immiscible phases, these two equations may be combined and the resulting expression integrated to provide a defining relation for the unknown height of the mass exchanger.

CONSTANT OVERALL CAPACITY COEFFICIENT

Consider an isothermal, countercurrent mass exchanger used to achieve a separation in a system which has a constant overall mass transfer coefficient $K_Y a$

through the concentration range involved in the mass-transfer operations. The mass balance for component A over the differential length dz described by

$$\frac{\text{moles of } A \text{ transferred}}{(\text{hr})(\text{cross-sectional area})} = L_S \, dX_A = G_S \, dY_A \qquad (31\text{-}10)$$

The mass transfer of component A in the differential length, dz, is defined by

$$\frac{\text{moles of } A \text{ transferred}}{(\text{hr})(\text{cross-sectional area})} = N_A a \, dz = K_Y a (Y_{AG} - Y_A^*) \, dz \qquad (31\text{-}21)$$

Combining these two equations and rearranging, we obtain

$$dz = \frac{G_S \, dY_A}{K_Y a (Y_A - Y_A^*)} \qquad (31\text{-}22)$$

or, for the length of the mass exchanger,

$$z = \frac{G_S}{K_Y a} \int_{Y_{A_1}}^{Y_{A_2}} \frac{dY_A}{Y_A - Y_A^*} \qquad (31\text{-}23)$$

The evaluation of the right-hand-side of this equation requires graphical integration. As discussed in section 31.3, we may evaluate $Y_{A,G} - Y_A^*$ from the plot of Y_A versus X_A, as illustrated in Figure 31.16. This vertical distance between the

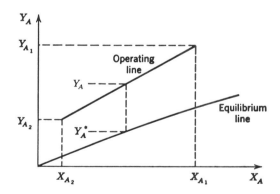

Figure 31.16 Evaluation of $Y_A - Y_A^*$, the overall driving force.

operating line and the equilibrium line represents the overall driving force in Y units. The driving force may be determined for each value of $Y_{A,G}$, and its reciprocal may then be plotted versus $Y_{A,G}$, as illustrated in Figure 31.17. Having the area under the curve in Figure 31.17, we may evaluate the length of the mass exchanger by equation (31-23).

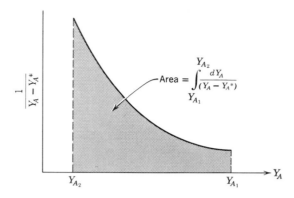

Figure 31.17 Graphical evaluation of the integral $\int_{Y_{A_1}}^{Y_{A_2}} dY_A/(Y_A - Y_A^*)$.

A similar but more complex equation could be derived in terms of the overall capacity coefficient $K_G a$,

$$z = \frac{G_S}{K_G a} \int_{Y_{A_1}}^{Y_{A_2}} \frac{dY_A}{(p_{A,G} - p_A^*)} \tag{31-24}$$

Since two different units for gas concentrations are involved, the equation is a little more difficult to evaluate. The length of the mass exchanger can also be determined by an equation written in terms of the overall liquid capacity coefficient, $K_X a$, if the coefficient is constant over the concentration range involved in the mass transfer operation,

$$z = \frac{L_S}{K_X a} \int_{X_{A_2}}^{X_{A_1}} \frac{dX_A}{X_A^* - X_A} \tag{31-25}$$

The overall driving force, $X_A^* - X_{A,L}$, is the horizontal difference between the operating line and the equilibrium line values on a plot similar to Figure 31.16.

VARIABLE OVERALL CAPACITY COEFFICIENT-ALLOWANCE FOR RESISTANCE IN BOTH GAS AND LIQUID PHASE

In Chapter 29, the overall coefficient was found to vary with concentration unless the equilibrium line was straight; accordingly, we should expect that the overall capacity coefficient will also vary when the slope of the equilibrium line varies within the region which includes the bulk and interfacial concentrations. With slightly curved equilibrium lines, one may safely use the design equations (31-23), (31-24), and (31-25); however, in the case of equilibrium lines with more pronounced curvature, the exact calculations should be based on one of the individual capacity coefficients.

The mass balance for components A over the differential length dz is

$$L_S \, dX_A = G_S \, dY_A \tag{31-10}$$

Differentiating equation (31-6), we obtain

$$dY_A = \frac{dy_A}{(1-y_A)^2}$$

This relation may be substituted into equation (31-10) to give

$$L_S \, dX_A = G_S \frac{dy_A}{(1-y_A)^2} \tag{31-26}$$

The mass transfer of component A in the differential length, dz, is defined in terms of the individual gas-phase capacity coefficient by

$$N_A a \, dz = k_G a (p_{A,G} - p_{A,i}) \, dz \tag{31-27}$$

Combining equations (31-26) and (31-27) and rearranging, we obtain

$$dz = G_S \frac{dy_A}{k_G a (p_{A,G} - p_{A,i})(1-y_A)^2}$$

or

$$dz = \frac{G_S \, dy_A}{k_G a P (y_A - y_{A,i})(1-y_A)^2} \tag{31-28}$$

As discussed in Chapter 29, the interfacial compositions $y_{A,i}$ and $x_{A,i}$ may be found for each point on the operating line by drawing a line from the point toward the equilibrium line. The slope of this line is $-k_L/k_G$ on a p_A versus c_A plot, or is $-ck_L/k_G P$ on a y_A versus x_A plot, where c is the molar concentration in the liquid phase. In Figure 31.18, the location of the interfacial compositions on both plots is

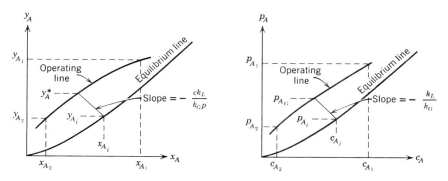

Figure 31.18 Determination of interfacial composition for transfer from phase G to phase L.

illustrated. It is important to recall from the discussion in section 31.3 that the operating line is not straight on plots of y_A versus x_A and p_A versus c_A, except when we are dealing with relatively dilute gas and liquid mixtures. Knowing the interfacial composition $y_{A,i}$ for every bulk composition y_A in the gas stream, we may graphically integrate equation (31-28) to obtain the length of the mass exchanger.

LOGARITHMIC-MEAN DRIVING FORCE

Although the graphical integration procedure must be employed in most practical design calculations, it is sometimes possible to use a much simpler equation based upon a logarithmic-mean driving force. When the two contacting streams are relatively dilute, the equilibrium curve and the operating line may both be linear in terms of the mole fractions over the range of concentration involved in the mass-transfer operation. Under these conditions, $G_1 \approx G_2 \approx G$ and $L_1 \approx L_2 \approx L$. The mass balance for component A may be approximated by

$$L(x_{A_1} - x_A) = G(y_{A_1} - y_A) \tag{31-29}$$

or

$$L \, dx_A = G \, dy_A \tag{31-30}$$

The rate of interphase transfer may be expressed in terms of the overall gas-phase capacity coefficient by

$$N_A a \, dz = K_G a(p_{A,G} - p_A{}^*) \, dz$$

or

$$N_A a \, dz = K_G a P(y_A - y_A{}^*) \, dz \tag{31-31}$$

Since the operating and equilibrium lines are straight, the difference in the ordinates of the two lines must vary linearly in composition. Designating the difference $y_A - y_A{}^*$ by Δ, we see that this linearity stipulates

$$\frac{d\Delta}{dy_A} = \frac{\Delta_{\text{end}_1} - \Delta_{\text{end}_2}}{y_{A_1} - y_{A_2}} = \frac{\Delta_1 - \Delta_2}{y_{A_1} - y_{A_2}} \tag{31-32}$$

Combining equations (31-30) and (31-31) and substituting equation (31-32) into the resulting expression, we obtain

$$dz = \frac{G}{K_G a P} \frac{dy_A}{y_A - y_A{}^*} = \frac{G}{K_G a P} \frac{dy_A}{\Delta}$$

or

$$dz = \frac{G}{K_G a P} \frac{y_{A_1} - y_{A_2}}{\Delta_1 - \Delta_2} \frac{d\Delta}{\Delta} \tag{31-33}$$

Integrating over the length of the mass exchanger, we obtain

$$z = \frac{G}{K_G a P} \frac{y_{A_1} - y_{A_2}}{\Delta_1 - \Delta_2} \ln \frac{\Delta_1}{\Delta_2}$$

or

$$z = \frac{G}{K_G a P} \frac{y_{A_1} - y_{A_2}}{(y_A - y_A{}^*)_{lm}} \tag{31-34}$$

where

$$\Delta_{lm} = \frac{\Delta_1 - \Delta_2}{\ln \Delta_1/\Delta_2} = (y_A - y_A{}^*)_{lm}$$

$$= \frac{(y_A - y_A{}^*)_{end_1} - (y_A - y_A{}^*)_{end_2}}{\ln[(y_A - y_A{}^*)_{end_1}/(y_A - y_A{}^*)_{end_2}]} \tag{31-35}$$

A similar expression in terms of the overall liquid-phase capacity coefficient is

$$z = \frac{L(x_{A_1} - x_{A_2})}{K_L a c (x_A{}^* - x_A)_{lm}} \tag{31-36}$$

where

$$(x_A{}^* - x_A)_{lm} = \frac{(x_A{}^* - x_A)_{end_1} - (x_A{}^* - x_A)_{end_2}}{\ln[(x_A{}^* - x_A)_{end_1}/(x_A{}^* - x_A)_{end_2}]} \tag{31-37}$$

EXAMPLE 4

Ammonia is to be absorbed from air at 68°F and atmospheric pressure in a countercurrent packed tower, 6.07 in. in diameter, using ammonia-free water as the absorbent. The inlet gas rate will be 390 ft^3/min and the inlet water rate will be 24.6 lb$_m$/min. Under these conditions, the overall capacity coefficient, $K_Y a$, may be assumed to be 4.60 lb mole/hr ft^3 ΔY_{NH_3}. The ammonia concentration will be reduced from 0.0825 mole fraction to 0.003 mole fraction. The tower will be cooled, the operation thus taking place essentially at 68°F; the equilibrium data of example 1 may be used. Determine the length of the mass exchanger.

The concentrations of three of the streams were given; these may be expressed on a NH$_3$-free basis by

$$Y_{NH_3}|_1 = \frac{y_{NH_3}|_1}{1 - y_{NH_3}|_1} = \frac{0.0825}{0.9175} = 0.09$$

$$Y_{NH_3}|_2 = \frac{y_{NH_3}|_2}{1 - y_{NH_3}|_2} = \frac{0.003}{0.997} = 0.003$$

and

$$X_{NH_3}|_2 = 0.0$$

The area of the tower is equal to $\pi D^2/4$ or $(\pi/4)[(6.07/12)^2] = 0.201$ ft^2. The gas enters the exchanger at plane 1, and its flow rate is

$$G = \frac{\dot{V}P}{RT} \frac{1}{A} = \frac{(390 \text{ ft}^3/\text{min})(1 \text{ atm})}{(0.73 \text{ ft}^3 \text{ atm/lb mole } °R)(528°R)} \frac{1}{0.201 \text{ ft}^2}$$

$$= 5.03 \text{ lb mole gas/(min)(ft}^2)$$

The NH_3-free gas flow rate is

$$G_S = (5.03)\frac{\text{lb mole}}{(\text{min})(\text{ft}^2)}\left(\frac{0.9175 \text{ mole air}}{\text{mole gas}}\right) = 4.61 \text{ lb mole air/min ft}^2$$

The NH_3-free water flow rate is evaluated using the flow into the exchanger at plane 2,

$$L_S = (24.6 \text{ lb}_m/\text{min})\frac{\text{lb mole}}{18 \text{ lb}_m}\frac{1}{0.201 \text{ ft}^2} = 6.80 \text{ lb moles } H_2O/\text{min ft}^2$$

The liquid leaves the exchanger at plane 1; its concentration is evaluated, using the countercurrent material balance equation,

$$G_S(Y_{NH_3}|_1 - Y_{NH_3}|_2) = L_S(X_{NH_3}|_1 - X_{NH_3}|_2)$$

or

$$4.61(0.09 - 0.003) = 6.80(X_{NH_3}|_1 - 0.00)$$

The exiting concentration, $X_{NH_3}|_1$, is 0.059. The operating and equilibrium lines are shown in Figure 31.19.

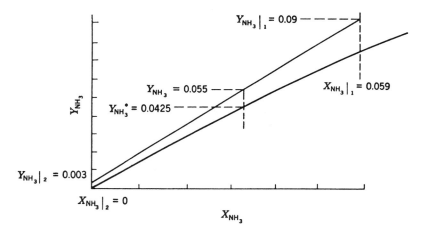

Figure 31.19   The operating line for example 4.

Since this is not a case in which the gas and liquid concentrations are dilute enough for us to assume straight equilibrium and operating lines on the plot of y versus x, the height of the tower must be evaluated by the graphical integration procedure, using

$$z = \frac{G_S}{K_Y a}\int_{Y_{A_1}}^{Y_{A_2}}\frac{dY_A}{Y_A - Y_A{}^*} \tag{31-23}$$

or

$$z = \frac{4.61 \text{ lb mole of } NH_3\text{-free gas/hr ft}^2}{4.6 \text{ lb mole of } NH_3/\text{hr ft}^3 \, \Delta Y \text{ units}} \cdot \int_{NH_3|_1}^{NH_3|_2}\frac{dY_{NH_3}}{Y_{NH_3} - Y_{NH_3}{}^*}$$

TABLE 31.1 EXAMPLE 3 GAS COMPOSITIONS

Y_A	Y_A^*	$Y_A - Y_A^*$	$1/(Y_A - Y_A^*)$
0.003	0	0.003	333.3
0.01	0.0065	0.0035	296
0.02	0.0153	0.0047	212.5
0.035	0.0275	0.0075	133.3
0.055	0.0425	0.0125	80.0
0.065	0.0503	0.0147	68.0
0.075	0.508	0.017	58.9
0.09	0.0683	0.0217	47.6

In Table 31.1, Y_{NH_3} is the composition evaluated at a point on the operating line and $Y_{NH_3}^*$ is that composition on the equilibrium line directly below the Y_{NH_3} value. An example of these compositions is illustrated in Figure 31.19. The integral

$$\int_{Y_{NH_3|1}}^{Y_{NH_3|2}} \frac{dY_{NH_3}}{Y_{NH_3} - Y_{NH_3}^*}$$

is graphically evaluated in Figure 31.20.

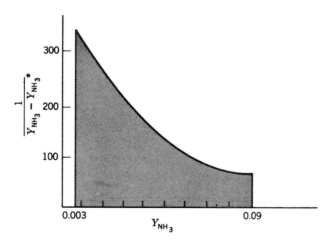

Figure 31.20 Graphical evaluation of the integral.

The area under the curve is the numerical value of the integral. The resulting length of the mass exchanger is

$$z = \frac{4.61}{4.60} 10.95 = 11.0 \text{ ft}$$

EXAMPLE 5

Recently enacted legislation requires that the effectiveness of the absorption system, currently employed to reduce the level of a pollutant in a process off-gas, must be improved to meet new state requirements. Under current operating conditions, $13.6 \text{ mol/s} \cdot \text{m}^2$ of off-gas flows countercurrent to $27.2 \text{ mol/s} \cdot \text{m}^2$ of water in the 12-m tall absorber. The concentration of A is reduced from 2 to 0.5 mole percent.

It has been suggested that the required concentration level in the off-gas may be obtained by doubling the water rate in the existing tower. Its diameter is large enough to permit this higher flow rate. If it is assumed that the process is gas-film controlled and that the overall gas capacity coefficient, $K_G a$, is proportional to the mass velocity of the solvent raised to the 0.4 power, determine the concentration of A in the effluent gas when the same 12-m tower and the same gas mass velocity are used with the doubled liquid flow rate.

Equilibrium for the system is defined by a modified Henry's law; $y_A{}^* = 1.5x_A$.

Since the concentrations are in the dilute range where the equilibrium curve is a straight line, the height of the tower may be evaluated by equation (31-34)

$$z = \frac{G(y_{A_1} - y_{A_2})}{K_G a P(y_A - y_A{}^*)_{lm}}$$

An overall material balance on the existing system establishes the composition of its exiting liquid stream.

$$G(y_{A_1} - y_{A_2}) = L(x_{A_1} - x_{A_2})$$

$$(13.6 \text{ mol/s} \cdot \text{m}^2)(0.02 - 0.005) = (27.2 \text{ mol/s} \cdot \text{m}^2)(x_{A_1} - 0)$$

$$x_{A_1} = 0.0075$$

The compositions at each end of the tower for the existing system are:

$$\text{Bottom: } y_{A_1} = 0.02$$

$$x_{A_1} = 0.0075$$

$$y_{A_1}{}^* = 1.5x_{A_1} = 1.5(0.0075) = 0.0113$$

$$\text{Top: } y_{A_2} = 0.005$$

$$x_{A_2} = 0.0$$

$$y_{A_2}{}^* = 1.5x_{A_2} = 0.0$$

For the existing tower, $(y_A - y_A{}^*)_{lm}$ is evaluated by equation (31-35)

$$(y_A - y_A{}^*)_{lm} = \frac{(y_A - y_A{}^*)_{end_1} - (y_A - y_A{}^*)_{end_2}}{\ln\left[\dfrac{(y_A - y_A{}^*)_{end_1}}{(y_A - y_A{}^*)_{end_2}}\right]}$$

$$= \frac{(0.02 - 0.0113) - (0.005 - 0)}{\ln\dfrac{(0.02 - 0.0113)}{(0.005 - 0.0)}} = 0.0067$$

Upon substituting known values into equation (31-34), we obtain

$$12 = \frac{(13.6 \text{ mol/s} \cdot \text{m}^2)(0.02 - 0.005)}{K_G a_1 P (0.0067)}$$

$$12 = \frac{(13.6 \text{ mol/s} \cdot \text{m}^2)(2.24)}{K_G a_1 P} \tag{31-38}$$

When we consider the proposed system, we obtain the following relationship with equation (31-34)

$$12 = \frac{(13.6 \text{ mol/s} \cdot \text{m}^2)(0.02 - y_{A_2})}{K_G a_{II} P (y_A - y_A{}^*)_{lm}} \tag{31-39}$$

Upon equating and simplifying equations (31-38) and (31-39) we obtain

$$\frac{2.24}{K_G a_1} = \left. \frac{(0.02 - y_{A_2})}{K_G a_{II}(y_A - y_A{}^*)_{lm}} \right|_{II}$$

or

$$\frac{K_G a_{II}(2.24)}{K_G a_1} = \left. \frac{(0.02 - y_{A_2})}{(y_A - y_A{}^*)_{lm}} \right|_{II} \tag{31-40}$$

Since the capacity coefficient is proportional to the mass velocity of the solvent raised to the 0.4 power, the ratio of the capacity coefficients is

$$\frac{K_G a_{II}}{K_G a_1} = \left[\frac{L_{II}}{L_I}\right]^{0.4} = \left[\frac{(54.4 \text{ mol/s} \cdot \text{m}^2)(0.018 \text{ kg/mol})}{(27.2 \text{ mol/s} \cdot \text{m}^2)(0.018 \text{ kg/mol})}\right]^{0.4} = 1.32$$

Accordingly, equation (31-40) becomes

$$\frac{0.02 - y_{A_2}}{(y_A - y_A{}^*)_{lm}} = (1.32)(2.24) = 2.95 \tag{31-41}$$

This equation requires a trial-and-error solution. With a guess of $y_{A_2} = 0.0021$, we can make an overall balance to establish x_{A_1} for the proposed system.

$$G(y_{A_1} - y_{A_2}) = L(x_{A_1} - x_{A_2})$$

$$(13.6 \text{ mol/s} \cdot \text{m}^2)(0.02 - 0.0021) = (54.4 \text{ mol/s} \cdot \text{m}^2)(x_{A_1} - 0.0)$$

$$x_{A_1} = 0.0045$$

The compositions at each end of the proposed tower are

Bottom: $y_{A_1} = 0.02$

$$x_{A_1} = 0.0045$$

$$y_{A_1}{}^* = 1.5 x_{A_1} = 1.5(0.0045) = 0.0067$$

Top: $y_{A_2} = 0.0021$ (estimated value)

$$x_{A_2} = 0.0$$

$$y_{A_2}{}^* = 1.5 x_{A_2} = 1.5(0.0) = 0.0$$

The $(y_A - y_A^*)_{lm}$ for the proposed tower is evaluated by equation (31-35)

$$(y_A - y_A^*)_{lm} = \frac{(y_A - y_A^*)_{end_1} - (y_A - y_A^*)_{end_2}}{\ln\left[\dfrac{(y_A - y_A^*)_{end_1}}{(y_A - y_A^*)_{end_2}}\right]}$$

$$(y_A - y_A^*)_{lm} = \frac{(0.02 - 0.0067) - (0.0021 - 0.0)}{\ln\left[\dfrac{(0.02 - 0.0067)}{(0.0021 - 0.0)}\right]} = 0.00607$$

$$\frac{0.02 - y_{A_2}}{(y_A - y_A^*)_{lm}} = \frac{0.02 - 0.0021}{0.00607} = 2.95$$

This satisfies equation (31-41); as a result of the trial-and-error solution the concentration of A in the effluent gas is 0.21%.

PACKED TOWER DIAMETER

The packed tower is the most commonly encountered continuous-contacting equipment in gas-liquid operations. A variety of packing materials are used, ranging from specially designed ceramic or plastic packing to crushed rock. The packing is chosen to promote a large area of contact between the phases, with a minimum resistance to the flow of the two phases. Table 31.2 list some of the properties of packing frequently used in industry.

We have previously established that the height of a continuous-contact tower is determined by the rate of mass transfer. The diameter of the tower is established to handle the flow rates of the two phases to be treated.

As illustrated in Figure 31.21, the pressure drop suffered by the gas phase as it flows through the packing is influenced by the flow rates of both phases. This is to be expected since both phases will be competing for the free cross section that is available for the streams to flow through. Let us consider a tower operating with a fixed liquid flow rate L'; below the region marked A, the quantity of liquid retained in the packed bed will remain reasonably constant with changing gas velocities. As the gas flow rate increases, the interphase friction increases and a greater quantity of liquid is held up in the packing. This is known as *loading*. Finally, at a certain value of the gas flow rate, G', the holdup is so high that the tower starts to fill with liquid. The tower cannot be operated above this *flooding velocity*, which is a function of the liquid velocity, the fluid properties, and the characteristics of the packing.

In Figure 31.22, a correlation is given for the flooding velocity in a random packed tower. Absorbers and desorbers are designed to operate well below the pressure drop that is associated with flooding; typically, they are designed for gas pressure drops of 200 to 400 N/m^2 per meter of packed depth. The abscissa on this figure involves a ratio of the superficial liquid and gas-mass flow rates and the densities of the gas and liquid phases. The ordinate involves the superficial gas-mass flow rate, the liquid-phase viscosity, the liquid and gas densities, a packing characteristic, C_f, which can be obtained from Table 31.2, and two

TABLE 31.2   TOWER PACKING CHARACTERISTICS*

Packing	Nominal size, in. (mm)					
	$\frac{1}{4}$ (6)	$\frac{1}{2}$ (13)	$\frac{3}{4}$ (19)	1 (25)	$1\frac{1}{2}$ (38)	2 (50)
Raschig rings						
Ceramic						
ϵ	0.73	0.63	0.73	0.73	0.71	0.74
c_f	1600	909	255	155	95	65
a_p ft^2/ft^3	240	111	80	58	38	28
Metal						
ϵ	0.69	0.84	0.88	0.92		
c_f	700	300	155	115		
a_p ft^2/ft^3	236	128	83.5	62.7		
Berl saddles						
Ceramic						
ϵ	0.60	0.63	0.66	0.69	0.75	0.72
c_f	900	240	170	110	65	45
a_p ft^2/ft^3	274	142	82	76	44	32
Intralox saddles						
Ceramic						
ϵ	0.75	0.78	0.77	0.775	0.81	0.79
c_f	725	200	145	98	52	40
a_p ft^2/ft^3	300	190	102	78	59.5	36
Plastic						
ϵ				0.91		0.93
c_f				33		56.5
a_p ft^2/ft^3				63		33
Pall rings						
Plastic						
ϵ				0.90	0.91	0.92
c_f				52	40	25
a_p ft^2/ft^3				63	39	31
Metal						
ϵ				0.94	0.95	0.96
c_f				48	28	20
a_p ft^2/ft^3				63	39	31

* R. E. Treybal, *Mass-Transfer Operations*, McGraw-Hill Book Company, New York, 1980.

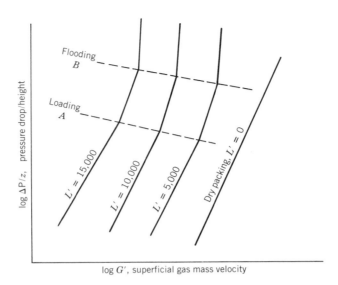

Figure 31.21   Typical gas pressure drop for countercurrent, packed tower.

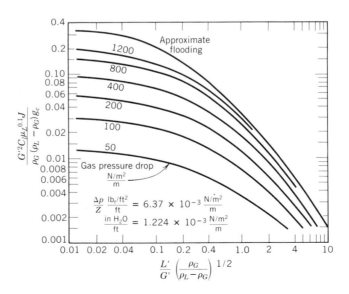

Figure 31.22   Flooding and pressure drop in random-packed towers.

constants. For SI units, g_c equals 1, and J equals 1; for U.S.–English units, μ_L is in centipoise, densities are in lb_m/ft^3, mass flow rates are in $lb_m/(ft^2)(hr)$, g_c equals 4.18×10^8, and J equals 1.502.

Example 6 will illustrate how we may evaluate the diameter of a packed tower by using Figure 31.22.

EXAMPLE 6

A packed tower is to be used to reduce the ammonia concentration in the gas stream from 4.0 to 0.3% by volume. A water stream is fed to the top of the tower at a rate of 30.5 lb_m/min, and the gas is fed countercurrently at a rate of 390 ft^3/min. The tower, which is packed with 1-in. Raschig rings, is operated isothermally at 68°F and 1 atm.

Determine
(a) the composition of the exiting liquid stream;
(b) the diameter of the absorption tower if the gas pressure drop is limited to 200 N/m^2 per meter of packing.

The concentration of three of the streams may be expressed on a NH_3-free basis

$$Y_{NH_3}|_1 = \frac{y_{NH_3}|_1}{1 - y_{NH_3}|_1} = \frac{0.04}{0.96} = 0.0417$$

$$Y_{NH_3}|_2 = \frac{y_{NH_3}|_2}{1 - y_{NH_3}|_2} = \frac{0.003}{0.997} = 0.003$$

$$X_{NH_3}|_2 = 0$$

The gas enters the tower at plane 1 and its molar flow rate is

$$G_1 = \frac{\dot{V}P}{RT} \cdot \frac{1}{A} = \frac{(390 \text{ ft}^3/\text{min})(1 \text{ atm})}{(0.73 \text{ ft}^3 \cdot \text{atm/lb mole} \cdot {}^\circ\text{R})(528{}^\circ\text{R})} \cdot \frac{1}{A}$$

$$= \frac{1.01}{A} \frac{\text{lb mole}}{(\text{min})(\text{ft}^2)}$$

The NH_3-free gas flow rate is

$$G_s = G_1(1 - y_{NH_3}|_1) = \frac{1.01}{A} \frac{\text{lb mole}}{(\text{min})(\text{ft}^2)}(1.0 - 0.04) = \frac{0.97}{A} \frac{\text{lb mole}}{(\text{min})(\text{ft}^2)}$$

The NH_3-free water flow rate is

$$L_s = L_2(1 - x_{NH_3}|_2) = \frac{30.5}{A} \frac{\text{lb}_m/\text{min}}{18 \text{ lb/lb mole}} = \frac{1.694}{A} \frac{\text{lb mole}}{(\text{min})(\text{ft}^2)}$$

By an overall balance for ammonia, the concentration of the exiting liquid stream can be established.

$$L_s(X_{NH_3}|_1 - X_{NH_3}|_2) = G_s(Y_{NH_3}|_1 - Y_{NH_3}|_2)$$

$$\frac{1.694}{A} \frac{\text{lb mole}}{(\text{min})(\text{ft}^2)}(X_{NH_3}|_1 - 0) = \frac{0.97}{A} \frac{\text{lb mole}}{(\text{min})(\text{ft}^2)}(0.0417 - 0.003)$$

$$X_{NH_3}|_1 = 0.022$$

or

$$x_{NH_3}|_1 = \frac{X_{NH_3}|_1}{1 + X_{NH_3}|_1} = \frac{0.022}{1.022} = 0.0217$$

The liquid flow rate at end 1 is

$$L_1 = \frac{L_s}{1 - x_{NH_3}|_1} = \frac{1.694}{A} \frac{\text{lb mole}/(\text{min})(\text{ft}^2)}{(1 - 0.0215)} = \frac{1.732}{A} \frac{\text{lb mole}}{(\text{min})(\text{ft}^2)}$$

The maximum mass flow rates for both phases will occur at end 1; accordingly, we will use these flow rates to determine the diameter of the tower.

On a unit area basis, the liquid stream at end 1 will contain 1.694 lb mole H_2O/min and $(1.732 - 1.694) = 0.038$ lb mole NH_3/min. The total liquid-mass flow rate is

$$\frac{1.694}{A} \frac{\text{lb mole } H_2O}{(\text{min})(\text{ft}^2)} \cdot \frac{18 \text{ lb}_m}{\text{lb mole}} + \frac{0.038}{A} \frac{\text{lb mole } NH_3}{(\text{min})(\text{ft}^2)} \cdot \frac{17 \text{ lb}_m}{\text{lb mole}}$$

or

$$\frac{31.14}{A} \frac{\text{lb}_m}{(\text{min})(\text{ft}^2)}$$

The average molecular weight of the gas mixture is $(0.04)(17) + (0.96)(29) = 28.5 \text{ lb}_m/\text{lb mole}$. The total gas mass flow rate is

$$\frac{1.01}{A} \frac{\text{lb mole}}{(\text{min})(\text{ft}^2)} \cdot \frac{28.5 \text{ lb}_m}{\text{lb mole}} = \frac{28.8}{A} \frac{\text{lb}_m}{(\text{min})(\text{ft}^2)}$$

The ratio of the two mass flow rates

$$\frac{L'}{G'} = \frac{(31.14/A)[\text{lb}_m/(\text{min})(\text{ft}^2)]}{(28.8/A)[\text{lb}_m/(\text{min})(\text{ft}^2)]} = 1.081$$

Notice this ratio can be evaluated without knowing either the diameter or the cross-sectional area. The density of the gas stream entering the tower is

$$\rho_G = \frac{n}{V} \cdot MW = \frac{P}{RT} \cdot MW = \frac{1 \text{ atm}}{(0.73 \text{ ft}^3 \text{ atm}/\text{lb mole }°R)(528°R)} \cdot \frac{28.5 \text{ lb}_m}{\text{lb mole}}$$

$$= 0.0739 \text{ lb}_m/\text{ft}^3$$

The density of the dilute aqueous stream will be essentially that of water. The abscissa for Figure 31.22 becomes

$$\frac{L'}{G'}\left(\frac{\rho_G}{\rho_L - \rho_G}\right)^{1/2} = 1.081\left(\frac{0.0739 \text{ lb}_m/\text{ft}^3}{62.2 - 0.0739 \text{ lb}_m/\text{ft}^3}\right)^{1/2} = 0.037$$

At a pressure drop of 200 N/m^2 per meter of packing, this abscissa value indicates an ordinate value of 0.049.

$$0.049 = \frac{G'^2 C_f \mu_L^{0.1} J}{\rho_G(\rho_L - \rho_G)g_c}$$

Upon rearrangement,

$$G'^2 = \frac{0.049\rho_G(\rho_L - \rho_G)g_c}{C_f \mu_L^{0.1} J}$$

From Table 31.2, we can evaluate the C_f for 1-in. Raschig rings to be 160 and the viscosity of the aqueous stream can be determined to be 1.3 cp; accordingly,

$$G'^2 = \frac{0.049(0.0739 \text{ lb}_m/\text{ft}^3)(62.2 - 0.0739 \text{ lb}_m/\text{ft}^3)4.18 \times 10^8}{(160)(1.3 \text{ cp})^{0.1}(1.502)}$$

$$= 381\,159$$

$$G' = 617.4 \text{ lb}_m/(\text{hr})(\text{ft}^2)$$

Since the gas feed rate to the tower is $(28.8/A)[\text{lb}_m/(\text{min})(\text{ft}^2)]$ or $(1728/A)$ $[\text{lb}_m/(\text{hr})(\text{ft}^2)]$, the cross-sectional area of the tower is

$$A = \frac{1728 \text{ lb/hr}}{617.4 \text{ lb}_m/(\text{hr})(\text{ft}^2)} = 2.80 \text{ ft}^2$$

The area is $\pi D^2/4$; accordingly, the diameter is $[(2.8)(4)/\pi]^{1/2} = 1.89$ ft.

31.7 CLOSURE

Continuous-contact mass exchangers are designed by integrating an equation which relates the mass balance and the mass-transfer relations for a differential area of interfacial contact. In this chapter, we have described the four major types of mass-transfer equipment. The fundamental equations for the design of continuous-contact equipment have been developed. A mass balance for the diffusing component A in terms of solute-free concentration units produced the following important operating line equations:

steady-state countercurrent operations,

$$G_S(Y_{A_1} - Y_{A_2}) = L_S(X_{A_1} - X_{A_2})$$

and

$$G_S \, dY_A = L_S \, dX_A$$

steady-state cocurrent operations,

$$G_S(Y_{A_1} - Y_{A_2}) = L_S(X_{A_2} - X_{A_1})$$

and

$$G_S \, dY_A = -L_S \, dX_A$$

Because of the difficulty in measuring the interphase contact area within most mass-transfer equipment, the factor a, the interphase surface area per unit volume of the mass exchanger, was introduced. The product of the mass-transfer convective coefficient and the factor a was designated the mass-transfer capacity coefficient. The mass transferred in a differential length per cross-sectional area was expressed in empirical equations, one of which was

$$N_A a \, dz = K_G a (p_{A,G} - p_A{}^*) \, dz$$

For equipment involving the continuous-contact between two immiscible phases, the differential mass balance and the differential mass transfer equations were combined to produce the following design equations:

Constant overall capacity coefficient $K_Y a$,

$$z = \frac{G_S}{K_Y a} \int_{Y_{A_1}}^{Y_{A_2}} \frac{dY_A}{(Y_A - Y_A{}^*)}$$

The integral evaluation must be accomplished graphically.

Variable overall capacity coefficient-allowance for resistance in both gas and liquid phases,

$$z = \int_{Y_{A_1}}^{Y_{A_2}} \frac{dY_A}{k_G a P (y_A - y_{A,i})(1 - y_A)^2}$$

This integral is also evaluated graphically.

Straight equilibrium and operating lines on the x-y coordinates—the log mean driving force,

$$z = \frac{G(y_{A_1} - y_{A_2})}{K_G a P (y_A - y_A{}^*)_{lm}}$$

where

$$(y_A - y_A{}^*)_{lm} = \frac{(y_A - y_A{}^*)_1 - (y_A - y_A{}^*)_2}{\ln \left[(y_A - y_A{}^*)_1 / (y_A - y_A{}^*)_2 \right]}$$

The similarity between mass and energy transfer was further emphasized in this chapter. Using a combined term representing the total resistance, $K_G a$ as compared to UA, and a total resistance, $(p_{A,G} - p_A{}^*)$ as compared to $\Delta T_{overall}$, we have obtained design equations for mass exchangers by integrating over the area of contact.

PROBLEMS

31.1 A $10\,000\text{-ft}^3$ basin is to be aerated with 5 spargers, each sparger using air at a rate of 15 SCFM. Find the time that is necessary to raise the dissolved oxygen level in the wastewater from 8×10^{-5} to 3×10^{-4} mol/liter if the temperature of the water is 10°C and the depth of the water is 10.5 ft. The dissolved solids content will be low enough so that Henry's law will be obeyed, with a Henry's law constant of 3.27×10^4 atm/mole fraction.

31.2 In the previous problem, wastewater at 10°C in a 10.5-ft-deep basin was treated by flowing 15 SCFM of air to each of the 5 spargers. Predict the time if

(a) 10 spargers, each using 10 SCFM of air, had been used;

(b) 15 SCFM of pure oxygen had been fed to the 5 spargers, assuming that $K_L a$ for oxygen is essentially equal to $K_L a$ for air;

(c) the wastewater in the 10.5-ft basin had been 20°C, where the Henry's law constant is a 4.01×10^4 atm/mole fraction.

31.3 A 15 000-ft^3 basin is to be aerated with 6 spargers, each sparger using air at the rate of 15 SCFM. If 10°C wastewater, initially containing a dissolved oxygen level of 5×10^{-5} mole/liter, and filling the basin to a depth of 15 ft, is aerated for 2 hr, determine the final dissolved oxygen level. The dissolved solid content will be low enough so that Henry's law is obeyed, with a Henry's law constant of 3.27×10^4 atm/mole fraction.

31.4 A 10 000 ft^3 basin, currently equipped with 5 spargers, is being used to raise the dissolved oxygen level in 10°C of wastewater from 6×10^{-5} to 3×10^{-4} mole/liter in 4 hr. The spargers are located 15 ft below the wastewater surface. The dissolved solid content will be low enough so that Henry's law for oxygen is obeyed, with a Henry's law constant of 3.27×10^4 atm/mole fraction.

It is desired to accomplish the desired oxygen level in one half the time. Due to a restriction in the sparger design, the airflow can only be increased by 40%. How many spargers will be required to be added to the system if the maximum airflow is used?

31.5 A 10 000-ft^3 basin, equipped with 10 gas dispersers of a new design, is to be designed to strip H_2S from wastewater. The dispersers, each using air at a rate of 15 SCFM, are to be located 15 ft below the liquid surface. An analysis of the 10°C wastewater indicates that the initial H_2S concentration is 3×10^{-4} mol/liter.

In a pilot investigation of the proposed new disperser, 15 SCFM of air was discharged from a single dispenser located 15 ft below the surface of the water contained in a 1000-ft^3 basin. This water was also at 10°C. After 4 hr of aeration, the dissolved oxygen concentration level increased from 4×10^{-5} to 3×10^{-4} mol/liter. Determine

(a) the $K_L a$ for oxygen for the disperser when operated 15 ft below the liquid surface with 15 SCFM of air;

(b) the $K_L a$ for H_2S for the new disperser when operated 15 ft below the liquid surface with 15 SCFM of air, assuming the penetration model for convective mass-transfer coefficient holds. The mass diffusivity of oxygen in water is 2.14×10^{-5} cm^2/s and of H$_2$S in water is 1.4×10^{-5} cm^2/s;

(c) the concentration of the H$_2$S in the 10 000-ft^3 basin after 3 hr of operation. The Henry's law constant for H$_2$S at 10°C is 0.0367×10^4 atm/mole fraction.

31.6 A scheme for the removal of H$_2$S from a gas by scrubbing with water at 293 K and 10 atm is being considered. The initial composition of the feed gas is 2.5 mole percent H$_2$S. A final gas stream containing only 0.1 mole percent H$_2$S is desired. The absorbing water will enter the absorption tower free of any H$_2$S. At the given temperature and pressure, the system will follow Henry's law, according to the following relationship;

$$Y_{H_2S} = 48.3 X_{H_2S}$$

(a) For a countercurrent absorber, determine the moles of water that is required per mole H$_2$S-free carrier air if 1.5 times the minimum ratio is used.

(b) Determine the composition of the exiting liquid.

31.7 The absorber for problem 31.6 would have been excessively tall; accordingly, various schemes are being considered for using two shorter absor-

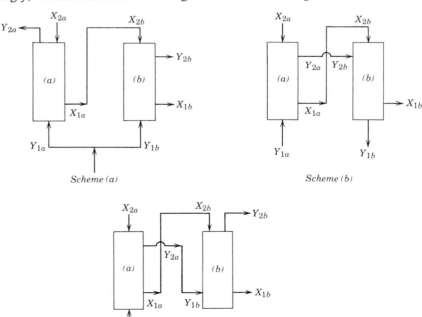

Scheme (a)

Scheme (b)

Scheme (c)

bers side by side. Make freehand sketches of operating diagrams, one for each scheme, showing the relation between the operating lines for the two absorbers and the equilibrium curve. Indicate, in general, where the concentrations of each stream will be in respect to each other but do not compute the actual concentrations.

31.8 Another designer who was asked to propse a design for the desorption of a volatile component from a solution, decided to use two towers. With the proposed flow schemes given in problem 31.7, make freehand sketches for each scheme, showing the relation between the operating lines for the two desorbers and the equilibrium curve. Indicate where the concentrations of each stream will be in respect to each other but do not compute the actual concentrations.

31.9 Benzene that has been free from coal gas in a scrubber is recovered by passing the benzene-wash oil solution through a tower in contact with steam. The entering liquid stream contains 8 mole percent benzene and the stream is benzene free. It is desired to recover 80% of the benzene by using a steam rate 1.4 times the minimum steam rate. A wash oil (liquid solvent-benzene free) flow rate of 6.94 mol/s will be used. Determine the required moles of steam per second if
(a) a countercurrent tower is used;
(b) a cocurrent tower is used.
Equilibrium data for the benzene-wash oil-steam system are as follows:

$X \dfrac{\text{mol benzene}}{\text{mol wash oil}}$	0.00	0.02	0.04	0.06	0.08	0.10	0.12	0.14
$Y \dfrac{\text{mol benzene}}{\text{mol steam}}$	0.00	0.07	0.14	0.22	0.31	0.405	0.515	0.65

31.10 Air containing 4% ammonia, NH_3, passes countercurrent to a water stream in a packed absorber. Twenty-five cubic meters per minute of gas and 28.6 kg/min of water enter the tower. The tower is operated at 295 K and 1.013×10^5 Pa. For dilute concentrations at 295 K, the equilibrium concentrations are related by $y/x = 1.08$. If the exit gas contains only 0.5% NH_3, determine
(a) the ratio $(L_s/G_s)_{\text{actual}}/(L_s/G_s)_{\text{minimum}}$;
(b) the concentration of the exiting liquid stream.

31.11 It is desired to strip component A from a liquid mixture that intially contains 4 mole percent of A. For each 104.2 lb mole of liquid fed to the tower, 50 lb mole of steam will be used for the stripping operation. The solubility of A in the solvent, at the temperature and pressure of the proposed mass-transfer operation, is given in the following table:

X_A	0.000	0.005	0.0150	0.0300	0.0600
Y_A	0.000	0.0101	0.0204	0.0304	0.0417

Calculate the percentage of A that is stripped from the liquid mixture if
(a) the liquid mixture and the streams flow countercurrent in an infinitely tall tower;
(b) the liquid mixture and steam are brought into contact with each other and allowed to remain in contact for a long, long period of time.

31.12 Two gas streams are to be treated in the same packed absorption tower. One hundred and five pound moles per hour, containing 4% component A, is fed to the bottom of the tower. A second gas stream, containing 2% A, is fed at a rate of 52 lb mol/hr at a point in the column where the rising gas stream's composition within the column is also 2% A. The exiting gas stream is to contain only 0.5% A. An A-free water stream is fed at the top of the tower at the rate of 116 lb mole/hr. At the temperature and pressure of the tower, the equilibrium data are as follows:

$X\dfrac{\text{lb mole } A}{\text{lb mole H}_2\text{O}}$	0.00	0.02	0.04	0.06	0.08	0.10
$Y\dfrac{\text{lb mole } A}{\text{lb mole air}}$	0.00	0.055	0.102	0.155	0.205	0.260

Determine
(a) the composition of the liquid stream within the tower at the level where the second gas stream enters the tower;
(b) the composition of the liquid stream leaving the bottom of the tower.

31.13 An absorber, packed to the height of 35 ft, is employed by an industrial company as a gas scrubber. A dilute air-ammonia mixture containing 4.93% NH_3 is drawn through the tower that is countercurrent to an initially ammonia-free water stream. The inlet gas rate is 100 lb mole/(hr)(ft^2) and the inlet NH_3-free water rate provides an actual L_s/G_s ratio that is 1.4 times the minimum L_s/G_s ratio. At the temperature and pressure of the absorber, the equilibrium concentrations are related by $y/x = 1.075$. If the exiting gas stream contains 0.2% ammonia, determine
(a) the pounds of ammonia absorbed per hour per square feet of the cross-sectional area;
(b) the composition of the exiting liquid stream;
(c) the overall mass-transfer coefficient, $K_Y a$, lb mole/(hr)(ft^3)(ΔY_A).

31.14 If the absorber described in problem 31.13 had been packed with 2-in. plastic Pall rings, determine what percentage the actual gas mass flow rate is of the flooding rate. The tower is to be operated at 68°F and 1 atm pressure.

31.15 Pulp bleaching chlorine water containing 4 g Cl_2/liter is being prepared by the absorption of chlorine into water within a packed tower operating at 20°C and 1 atm pressure. The gas enters the tower with a chlorine partial pressure of 0.9 atm; the exiting gas has a partial pressure of 0.4 atm. Pure water is fed to the top of the tower. For the dilute solutions encountered in this tower, the liters of solution may be considered equal to the liters of water.

  If L' equals the liters of water divided by the molecular weight of chlorine and C equals the solubility of chlorine, grams per liter, $L'dC$ equals the moles of chlorine that are absorbed by the liquid stream.
  (a) Determine the ratio L'/G_s;
  (b) Plot the operating line on the curve for chlorine pressure vs. the solubility of chlorine;
  (c) Plot the operating line on the curve Y vs. the solubility of chlorine.

31.16 An absorber, packed to the height of 4.5 m, has been designed to reduce the concentration level of a mercaptan pollutant in an exhaust gas stream from 5 to 0.3%. A nonvolatile, mercaptan-free solvent stream is fed to the top of the tower, thus flowing countercurrent to the 0.236 m^3/s of the exhaust gas stream. The solvent stream leaves the bottom of the tower containing 3.05 mole percent mercaptan. At the pressure and temperature of the tower, 293 K and 1.013×10^5 Pa, equilibrium data for the mercaptan-solvent system are as follows:

$X \dfrac{\text{mol mercaptan}}{\text{mol mercaptan-free solvent}}$	0.00	0.01	0.02	0.03	0.04
$Y \dfrac{\text{mol mercaptan}}{\text{mol mercaptan-free air}}$	0.00	0.0045	0.0145	0.0310	0.0545

Determine the overall capacity coefficient, K_Ya, by
(a) use of graphical integration;
(b) use of Equation (31-34);
if the cross-sectional area of the tower is 0.2 m^2.

31.17 Determine the diameter of the absorber that is described in problem 31.16 if the tower is packed with 1-in. Intralox saddles and the gas pressure drop within the packing is limited to 300 N/m^3 of packing. The gas enters the tower with an average molecular weight of 30.1 and the liquid stream leaving the bottom of the tower has an average molecular weight of 180 and a viscosity of 3 cp. and a specific gravity of 0.81.

31.18 An absorber, packed to the height of 2.4 m, is employed to reduce the concentration of a gas stream from 6.5 to 1.0% A. An aqueous stream, initially containing 1.0% A, is fed to the top of the tower. Twenty thousand cubic feet per hour of the gas mixture at 60°F and 1 atm is fed to the tower. The equilibrium data for this system are

X_A	0.00	0.01	0.02	0.03	0.04	0.05	0.06	0.07
Y_A	0.00	0.002	0.005	0.010	0.021	0.036	0.055	0.079

Determine
(a) the amount of liquid stream that is required per hour if the tower is operated countercurrently with a L_s/G_s ratio that is 1.5 times the minimum L_s/G_s;
(b) the amount of liquid stream that is required per hour if the tower is operated cocurrently with a L_s/G_s ratio that is 1.5 times the minimum L_s/G_s.

31.19 A mass-transfer tower is to be designed for reducing the organic concentration from 0.0394 to 0.0131 mg mol/liter. Five thousand gallons per hour of wastewater will be fed to the 2-ft-diameter tower, countercurrent to a stripping airstream that will enter the tower completely free of the organic material. If a volumetric flow rate, Q_L in liters/hr, to the gas flow rate, G in mg mole/hr, ratio of 1.4 is used, the overall capacity coefficient, $K_L a$, is equal to 36 lb mole/(hr)(ft^3)(lb mole/ft^3). The equilibrium data are

$c_A \times 10^3$ mole organic/liter	0.014	0.0240	0.0349	0.0495
y_A mole fraction organic	0.018	0.030	0.042	0.053

Determine
(a) the minimum gas flow, G, which might be used;
(b) the height of the tower.

31.20 An absorber, packed to the height of 15 ft, is currently being used to remove a pollutant from an exhaust gas stream. Thirty thousand cubic feet per hour of gas at 60°F and 1 atm, containing 5% mercaptan is fed to the bottom of the absorption tower. By feeding a nonvolatile, mercaptan-free solvent stream to the top of the tower, the mercaptan concentration in the exiting stream is reduced to 0.3%. The solvent stream leaves the bottom of the tower containing 3.65 mole percent mercaptan. At the pressure and temperature of the tower, the equilibrium for the mercaptan-solvent system may be represented by $Y_A = 0.8X_A$. The cross-sectional area of the tower is 2 ft^2. Determine
(a) the molar composition of the liquid stream flowing countercurrent to the gas stream at the point in the tower where the bulk gas composition is 2% mercaptan;

(b) the overall mass-transfer capacity coefficient, $K_Y a$, lb mole/(hr)(ft^3) × (ΔY_A);

(c) the composition at the interface, y_{Ai}, at the point in the tower where the gas composition is 2% if 70% of the resistance to mass-transfer is in the gas phase.

31.21 A tower, 15 cm in diameter, is to be used to reduce the ammonia, NH$_3$, concentration in a gas stream from 3.6 to 0.3%. The water is fed at a rate of 14.5 mol/s and the gas is fed countercurrently at the rate of 8 mol/s. The tower operates isothermally at 293 K and 1.013×10^5 Pa pressure. The overall capacity coefficient, $K_Y a$, may be assumed to be 71 mol/s · m^3 · ΔY_{NH_3}. If the equilibrium data at 293 K are as follows:

$X \dfrac{\text{mol NH}_3}{\text{mol H}_2\text{O}}$	0.00	0.0164	0.0252	0.0349	0.0445	0.0722
$Y \dfrac{\text{mol NH}_3}{\text{mol air}}$	0.00	0.021	0.032	0.042	0.053	0.080

evaluate

(a) the height of the tower required;

(b) the ratio $(L_s/G_S)_{\text{actual}}$ to $(L_s/G_s)_{\text{minimum}}$.

31.22 The absorption of water in sulfuric acid is an exothermic process. Describe the effect this would have on the required height of a mass-transfer column as compared to the height that is evaluated if isothermal conditions are assumed.

31.23 Air containing 4.5% ammonia, NH$_3$, passes countercurrent to a water stream in a packed absorber. The inlet gas rate is 135 mol/s · m^2; this provides an actual L_s/G_s ratio that is 1.4 times the minimum L_s/G_s ratio. The tower operates at a constant temperature of 293 K and 1.013×10^5 Pa pressure. For dilute concentrations at 293 K, a modified Henry's law applies, $y/x = 1.075$. If the exit gas contains 0.4% ammonia, determine

(a) the kg/s · m^2 of ammonia absorbed;

(b) the height of the absorber if the overall gas-capacity coefficient, $K_G a$, is 7.9×10^{-4} mol/s · m^3 · Pa.

31.24 In the sulfite pulping process,* the cooking liquor is prepared by absorbing sulfur dioxide in two packed towers (Jenssen towers) arranged for counter-current flow. The effluent gas stream leaving the first tower is passed to the bottom of the second tower in which it contacts a countercurrently

---

* This problem was presented in the Technical Association of the Pulp and Paper Industry publication, "Chemical Engineering Problems in the Pulp and Paper Industry."

moving stream of water, entering at the top. The following conditions describe the operation of a particular tower:

water flow entering	270 gallons/min
total volume of gas entering	838 ft^3/min
composition of gas entering	14.8 volume % SO$_2$
	85.2 volume % inerts
SO$_2$ content of exhaust gas	1.0%
pressure throughout the tower	1 atm
isothermal tower temperature	30°C
maximum liquor flow rate	75 lb$_m$/ft^2 min
maximum gas flow rate	1.5 fps
liquor specific gravity	1.0
overall capacity	
mass-transfer coefficient, $K_L a$	0.217 lb$_m$ SO$_2$/min ft^3(lb$_m$ SO$_2$/ft^3)

The equilibrium data at 30°C are as follows:

y', $\dfrac{\text{lb}_m \text{ SO}_2}{\text{ft}^3 \text{ inert}}$	0.026	0.025	0.020	0.015	0.010	0.005	0.0016
c, $\dfrac{\text{lb}_m \text{ SO}_2}{\text{ft}^3 \text{ liquor}}$	0.780	0.749	0.624	0.480	0.330	0.168	0.050

(a) Calculate the tower diameter and height of packed section that are necessary to satisfy the given conditions.
(b) What error in tower height would result if a log-mean driving force were assumed?

31.25 Estimate the height of an absorber that is packed with 1-in. Raschig rings if the tower absorbs 80% of component A from an inlet gas stream that initially contains 5 mole percent A. Solvent S, having a composition of 0.01 mole fraction A, is fed at a 20 lb mole/(hr)(ft^2) rate that is counter-current to the gas stream that enters at a rate of 10 lb mole/(hr)(ft^2). The overall capacity coefficient, $K_G a$, may be assumed to be constant over the entire column and equal to 0.9 lb mole/(hr)(ft^3)(atm). The solubility of A in the solvent at the temperature and pressure of the tower is given in the following table:

x_A	0.000	0.010	0.020	0.030	0.040	0.050
y_A	0.000	0.0025	0.0085	0.0205	0.0430	0.0800

31.26 Gas from a Mannheim furnace contains 12.5% HCl by volume and 87.5% inerts. This gas is scrubbed by HCl-free water at atmospheric pressure and a temperature of 20°C. In order to recover 97% of the HCl in the furnace gas, it has been decided to operate the tower at 1.64 times the minimum L_s/G_s ratio. The equilibrium relation for solutions of HCl in water is given by the data in the following table:

x	0.210	0.243	0.287	0.330	0.353	0.375	0.400	0.425
y	0.0023	0.0095	0.0215	0.0523	0.0852	0.135	0.203	0.322

(a) What would be the exit concentration of the acid stream?
(b) The tower is to be of stoneware, 2 ft in diameter. Ten thousand cubic feet per hour of Mannheim gas, measured at 20°C and 1 atm pressure, will be blown into the tower. The overall capacity coefficient, $K_G a$, may be taken as 2.0 lb mole/(hr)(ft³)(atm). Determine the height of the tower.

31.27 A packed tower is to be designed for the absorption of 80% of the SO_2 in a burner gas by a countercurrent water stream. The burner gas, whose analysis on a dry basis is

$$SO_2 \quad 10 \text{ volume \%}$$
$$O_2 \quad 12 \text{ volume \%}$$
$$N_2 \quad 78 \text{ volume \%}$$

will enter the column at 50°C, saturated with water vapor. The wet volume of the entering gas will be 250 ft³/min. On the basis of an empty tower, this volume of gas will have a velocity of 0.50 ft/s. The entering water stream will initially contain 0.026 wt % SO_2 and will leave with 0.375 wt % SO_2. Under these operating conditions, the capacity coefficients are predicted to have values of $k_G a = 2.55$ lb mole/(hr)(ft³)(atm) and $k_L a = 53.1$ lb mole/(hr)(ft³)(lb mole/ft³). Smoothed equilibrium data at 50°C are listed in the table below:

partial pressure SO_2, mm Hg	12.0	20.8	29.0	46.0	82.0
concentration, $g\,SO_2/100\,g\,H_2O$	0.10	0.15	0.20	0.30	0.50
density of solution, lb_m/ft^3	61.71	61.73	61.74	61.77	61.83

Determine the diameter of the tower and the height of packing required.

NOMENCLATURE

$\mathbf{a}$,   acceleration; ft/s^2, m/s^2.

A,   area; ft^2, m^2.

A_p,   projected area of surface; ft^2, m^2; eq. (12-3).

c,   total molar concentration; lb mole/ft^3; mol/m^3.

$c_A{}^*$,   concentration of A in equilibrium with the bulk composition of gas phase, $p_{A,G}$: lb mole/ft^3; mol/m^3; eq. (29-11).

c_{Ao},   concentration of A at time $t = 0$; lb mole/ft^3; mol/m^3; chapter 27.

$c_{A,i}$,   liquid molar concentration of A at the interface; lb mole/ft^3, mol/m^3; section 29.2.

$c_{A,L}$,   liquid molar concentration of A in the bulk stream; lb mole/ft^3, mol/m^3; section 29.2.

$c_{A,s}$,   concentration of A at the surface; lb mole/ft^3, mol/m^3; section 28.4.

$c_{A,\infty}$,   concentration of A in the bulk stream; lb mole/ft^3, mol/m^3; section 28.4.

c_i,   molar concentration of species i; lb mole/ft^3, mol/m^3; eq. (24-4).

c_p,   heat capacity; Btu/lb °F, J/kg K.

$\bar{C}$,   average random molecular velocity; m/s; sections 7.3, 15.1, and 24.2.

C_C,   capacity rate of cold fluid stream; Btu/hr °F, kW/K; eq. (22-1).

C_d,   correlating parameter for nucleate boiling; dimensionless; eq. (21-4).

C_D,   drag coefficient; dimensionless; eq. (12-3).

C_f,   coefficient of friction; dimensionless; eq. (12-2).

C_H,   capacity rate of hot fluid stream; Btu/hr °F, kW/K; eq. (22-1).

C_{sf},   correlating coefficient for nucleate boiling; dimensionless; Table 21.1.

d_c,   diameter of cylinder; ft, m; Table 30.1.

d_p,   diameter of spherical particle; ft, m; Table 30.1.

D,   tube diameter; ft, m.

D_{AB},   mass diffusivity or diffusion coefficient for component A diffusing through component B; ft^2/hr, m^2/s; eq. (24-15).

$D_{A,\text{eff}}$,   catalyst "effective" diffusion coefficient; ft^2/hr, m^2/s; eq. (24-48).

D_{eq},   equivalent diameter; ft, m; eq. (14-18).

$D_{K,\text{eff}}$,   Knudsen diffusion coefficient, ft^2/hr, m^2/s; eq. (24-51).

e, pipe roughness; in., mm; eq. (14-2).

e, specific energy or energy per unit mass; Btu/lb_m, J/kg; section 6.1.

E, total energy of system; Btu, J; section 6.1.

E, total emissive power; $Btu/hr\ ft^2$, W/m^2; eq. (23-2).

E, electrical potential; volt; sections 17.4, 26.3.

E_b, black body emissive power; $Btu/hr\ ft^2$, W/m^2; eq. (23-8).

f, dependent variable used in Blasius solution of boundary layer; dimensionless; eq. (12-13).

f', similarity parameter for convective analysis of boundary layer, prime denotes derivative with respect to η; dimensionless; eq. (19-16).

f_D, Darcy friction factor; dimensionless; eq. (14-4).

f_f, Fanning friction factor; dimensionless; eq. (14-3).

F, force; lb_f, N; section 1.2.

F, correction factor for compact heat exchangers configurations; dimensionless; eq. (22-14).

F_{ii}, view factor for radiant heat transfer; dimensionless; section 23.7.

$\bar{F}_{ii}$, reradiating view factor; dimensionless; section 23.8.

$\mathscr{F}_{ii}$, gray body view factor; dimensionless; section 23.8.

$\mathfrak{F}$, Faraday constant; 9.652×10^4 abs. coulombs/g equivalent; eq. (24-47).

g, acceleration due to gravity; ft/s^2, m/s^2.

g_c, dimensional conversion factor; $32.2\ ft\ lb_m/lb_f\ s^2$, $1\ kg \cdot m/s^2 \cdot N$.

G, irradiation; $Btu/hr\ ft^2$, W/m^2; section 23.10.

G, mass velocity; $lb_m/ft^2\ hr$, $g/m^2 \cdot s$; Table 30.1.

G, total moles of the gas phase per time per cross-sectional area; lb moles/ft^2hr, g moles/$m^2 \cdot s$; eq. (31-4).

G', superficial gas-mass flow rate; $lb_m/(hr)(ft^2)$; section 31.6.

G_b, mass velocity of bubbles; lb_m/ft^2s, $kg/m^2 \cdot s$; eq. (21-5).

G_M, molar velocity; lb moles/ft^2 hr, g moles/$m^2 \cdot s$; Table 30.1.

G_s, moles of gas stream on a solute-free basis per time per cross-sectional area; lb moles/ft^2 hr, g moles/$m^2 \cdot s$; eq. (31-8).

h, convective heat transfer coefficient; $Btu/hr\ ft^2\ °F$, $W/m^2 \cdot K$; eq. (15-11).

h, depth of pool in aeration tank; ft; eq. (31-3).

h_L, head loss, $\Delta P/\rho$; $ft\ lb_f/lb_m$, $Pa/kg/m^3$; eq. (14-3).

h_L, heat-transfer coefficient for conduit flow applying over distance L from the entrance with $L < 60D$; $Btu/hr\ ft^2\ °F$, $W/m^2 \cdot K$; eq. (20-33).

h_r, radiation heat-transfer coefficient; $Btu/hr\ ft^2\ °F$, $W/m^2 \cdot K$; section 23.12.

h_∞, heat-transfer coefficient for internal flow for $L > 60D$; $Btu/hr\ ft^2\ °F$, $W/m^2 \cdot K$; eq. (20-33).

H, Henry's law constant; concentration of gas phase/concentration of liquid phase; eq. (29-4).

H, moment of momentum; $lb_m\ ft^2/s$, $kg \cdot m^2/s$; eq. (5-7).

H_i, enthalpy of given species i; Btu, J.

$\bar{H}_i$, partial molar enthalpy of species i; Btu/lb mole, J/mol; eq. (26-62).

$\mathscr{H}$, humidity, mass H_2O/mass dry air; lb_m/lb_m, kg/kg; Table 29.1.

I, intensity of radiation; Btu/hr ft^2, W/m^2; section 23.3.

j', j factor for heat transfer with tube bundles; dimensionless; Figures 20.13, 20.14.

j_D, j factor for mass transfer, Chilton-Colburn analogy; dimensionless; eq. (28-56).

j_H, j factor for heat transfer, Colburn analogy; dimensionless; eq. (19-38).

$\mathbf{j}_i$, mass flux relative to the mass-average velocity; lb$_m$/ft^2 hr, kg/m^2 · s; eq. (25-17).

J, radiosity; Btu/hr ft^2, W/m^2; section 23.10.

$\mathbf{J}_i$, molar flux relative to the molar-average velocity; lb mole/hr ft^2, mol/m^2 · s; eq. (24-16).

k, thermal conductivity; Btu/hr ft °F, W/m · K; eq. (15-1).

k, rate constant for chemical reaction, used to define r_A and R_A; section 25.3.

k^0, mass-transfer coefficient with no net mass transfer into film; lb mole/ft^2 s Δc_A; mol/m^2 · s · Δc_A; eq. (26-24).

k_c, convective mass-transfer coefficient; lb mole/ft^2 hr Δc_A, mol/m^2 · s · mol/m^3; eq. (24-52).

$\bar{k}_c$, mean convective mass-transfer coefficient; lb mole/ft^2 hr Δc_A, mol/m^2 · s · mol/m^3; eq. (28-25).

k_G, convective mass-transfer coefficient in the gas phase; lb mole/ft^2 hr atm, mol/m^2 · s · Pa; eq. (29-6).

k_L, convective mass-transfer coefficient in the liquid phase; lb mole/ft^2 hr lb mole/ft^3, mol/m^2 · s · mol/m^3; eq. (29-7).

$k_G a$, individual gas-capacity coefficient; lb mole/hr ft^3 atm, mole/s · m^3 · Pa; eq. (39-17).

$k_L a$, individual liquid-capacity coefficient; lb mole/hr ft^3 Δc_A, mol/s · m^3 mol/m^3; eq. (31-19).

K_G, overall mass-transfer coefficient in the gas phase; lb mole/hr ft^2 atm, mol/s · m^2 · Pa; eq. (29-10).

K_L, overall mass-transfer coefficient in the liquid phase; lb mole/hr ft^2 Δc_A, mol/s · m^2 · mol/m^3; eq. (29-11).

$K_G a$, overall gas-capacity coefficient; lb mole/hr ft^3 atm, mol/s · m^3 · Pa; eq. (31-18).

$K_L a$, overall liquid-capacity coefficient; lb mole/hr ft^3 Δc_A, mol/s · m^3 · mol/m^3; eq. (31-20).

$K_X a$, overall liquid capacity coefficient based on ΔX_A driving force; lb mole/hr ft^3 ΔX_A, mol/s · m^3 · ΔX_A; eq. (31-22).

$K_Y a$, overall gas capacity coefficient based on ΔY_A driving force; lb mole/hr ft^3 ΔY_A, mol/s · m^3 · ΔY_A; eq. (31-21).

L, mixing length; eqs. (13-10), (19-40), and (28-45).

L, characteristic length; ft, m.

L, total moles of liquid phase per time per cross-sectional area; lb mole/hr ft^2, mol/s · m^2; eq. (31-4).

L', superficial liquid-mass flow rates; lb$_m$/(hr)(ft^2); section 31.6

L_{eq}, equivalent length; ft, m; eq. (14-17).

L_m, molar liquid mass velocity; lb mole/hr ft^2, mol/s $\cdot$ m^2; Table 30.1.

L_s, moles of liquid phase on a solute-free basis per time per cross-sectional area; lb mole/hr ft^2, mol/s $\cdot$ m^2; eq. (31-8).

m, mass of molecule; section 7.3.

m, relative resistance $= D_{AB}/k_c x_1$; dimensionless; section 27.2.

m, slope of the equilibrium line; units gas concentration per units of liquid concentration; eq. (29-14).

M, moment; lb$_m$ ft^2/s^2, kg $\cdot$ m^2/s^2.

M_i, molecular weight of species; lb/lb mole, kg/mol.

n, packed bed constant; dimensionless; eq. (30-17).

n, number of species in a mixture; eqs. (24-1), (24-3), and (24-6).

n, relative position $= x/x_1$; dimensionless; section 27.2.

N, molecules per unit volume; section 7.3.

n, outward directed unit normal vector; eq. (4-1).

n_i, number moles of species i.

n$_i$, mass flux relative to a set of stationary axes; lb$_m$/hr ft^2, kg/s $\cdot$ m^2; eq. (24-22).

N$_i$, molar flux relative to a set of stationary axes; lb mole/hr ft^2, mol/s $\cdot$ m^2; eq. (24-21).

NTU, number of transfer units; dimensionless; section 22.4.

$p_A{}^*$, partial pressure of A in equilibrium with bulk composition in liquid phase, $c_{A,L}$; atm, Pa; eq. (29-10).

$p_{A,G}$, partial pressure of component A in the bulk gas stream; atm, Pa; section 29.2.

$p_{A,i}$, partial pressure of component A at the interface; atm, Pa; section 29.2.

p_i, partial pressure of species i; atm, Pa.

$p_{B,lm}$, log mean of partial pressure of the non-diffusing gas; atm, Pa; eq. (26-8).

P, total pressure; atm, Pa.

P, total linear momentum of system; lb$_m$ ft/s, kg $\cdot$ m/s; eq. (5-1).

P_c, critical pressure; atm, Pa.

P_i, vapor pressure species i; atm, Pa.

q, heat flow rate; Btu/hr, W; eq. (15-1).

$\dot{q}$, volumetric energy generation rate; Btu/hr ft^3, W/m^3; eq. (16-1).

Q, heat transfer; Btu, J; section 6.1.

Q_g, gas flow rate, ft^3/min; eq. (31-3).

r, radial distance in both cylindrical and spherical coordinates; ft, m.

r, radius; ft, m.

r_{crit}, critical radius of insulation; ft, m; eq. (17-13).

R, radius of sphere, ft, m; eq. (26-35).

R, gas constant; atm ft^3/lb mole °R, Pa $\cdot$ m^3/mol $\cdot$ K; example 1, section 24.1.

R_t, thermal resistance; hr °F/Btu, K/W; eq. (15-16).

r_A, rate of the production of mass A within the control volume; lb$_m$/ft^3 hr, kg/m^3 $\cdot$ s; eq. (25-5).

R_A,	rate of production of moles A within control volume; lb mole/ft^3 hr, mol/m$^3 \cdot$ s; eq. (25-11).
s,	surface renewal factor; eq. (28-63).
S,	shape factor; ft or m; eq. (15-19).
S,	cross-sectional area; ft^2, m^2; section 26.1.
$\mathbf{S}$,	force intensity on control volume; lb$_f$/s^2 ft, kg/s$^2 \cdot$ m; section 1.2.
t,	time; hr, s.
T,	absolute temperature; °R, K.
T_b,	normal boiling temperature; K; eq. (24-38).
T_c,	critical temperature; K; eq. (24-36).
T_f,	film temperature; °F, K; eq. (19-28).
T_{sat},	temperature of saturated liquid-vapor mixtures; °F, K; Figure 21.1.
U,	overall heat transfer coefficient; Btu/hr ft^2 °F, W/m$^2 \cdot$ K; eq. (15-17).
$\mathbf{v}_0$,	terminal velocity; ft/s, m/s.
v_x,	x-component of velocity, $\mathbf{v}$: ft/s, m/s.
v_y,	y-component of velocity, $\mathbf{v}$; ft/s, m/s.
v_z,	z component of velocity, $\mathbf{v}$; ft/s, m/s.
v^+,	dimensionless velocity.
$\mathbf{v}$,	velocity, ft/s, m/s.
$\mathbf{v}$,	mass-average velocity for multicomponent mixture; ft/s, m/s; eq. (25-13).
$\mathbf{v}_i$,	velocity of given species i; ft/s, m/s.
$\mathbf{v}_i - \mathbf{v}$,	diffusion velocity of species i relative to mass-average velocity; ft/s, m/s; section 24.1.
$\mathbf{v}_i - \mathbf{V}$,	diffusion velocity of species i relative to the molar-average velocity; ft/s, m/s; section 24.1.
V,	gas volume; ft^3, m^3; eq. (24-5).
V_b,	molecular volume at the normal boiling point, cm^3/g mole; eq. (24-34).
V_c,	critical molecular volume; cm^3/g mole; eq. (24-35).
$\mathbf{V}$,	molar-average velocity; ft/sec, m/s; eq. (24-14).
W,	work done; Btu, J; section 6.1.
w_A,	mass rate of flow of species A; lb$_m$/hr, g/s.
w_A',	moisture content; lb moisture/lb dry solid, kg moisture/kg dry solid; section 27.2.
W_s,	shaft work; Btu, J; section 6.1.
W_δ,	normal stress work; Btu, J; section 6.1.
W_τ,	shear work; Btu, J; section 6.1.
x,	rectangular coordinate.
x_1,	characteristic length; ft, m; section 27.2.
x_A,	mole fraction in either liquid or solid phase; dimensionless; eq. (24-7).
X_A,	mole of A/mole A-free liquid; eq. (31-7).
X_D,	relative time, $D_{AB}t/x_1^2$; dimensionless; section 27.2.
y,	rectangular coordinate.
y^+,	dimensionless distance; eq. (13-18).

y_A,   mole fraction in the gas phase; eqs. (24-7), (24-8).

$y_{B,lm}$,   log mean mole fraction of the carrier gas; eq. (26-6).

$y_n{}'$,   log fraction of component n in a gas mixture on species i-free basis; eq. (24-42).

Y,   parameter in heat exchanger analysis; dimensionless; eq. (22-12).

Y,   unaccomplished change; dimensionless; section 27.2.

Y_A,   mole of A/mole A-free gas; eq. (31-6).

z,   distance in the z direction; ft, m.

z,   rectangular coordinate.

Z,   wall collision frequency; eq. (7-8).

Z,   parameter in heat exchanger analysis; dimensionless; eq. (22-13).

α,   absorptivity; dimensionless; section 23.2.

α,   thermal diffusivity; ft^2/hr, m^2/s; eq. (16-17).

α,   packed bed constant; eq. (30-17).

β,   geometric factor for stagnation point heat transfer; Table 20.3.

β,   coefficient of thermal expansion; $1/°F$, $1/K$; eq. (19-10).

δ,   boundary layer thickness; ft, m; eq. (12-28).

δ,   thickness of stagnant or laminar layer; ft, m; eq. (26-9).

δ_c,   concentration boundary layer thickness; ft, m; eq. (28-18).

δ_t,   thermal boundary layer thickness; ft, m; eq. (19-22).

ϵ,   emissivity; dimensionless; eq. (23-2).

ϵ,   volume void; eq. (30-16).

ϵ_{AB},   a Lennard-Jones parameter; ergs; eq. (24-40).

ϵ_D,   eddy mass diffusivity, ft^2/hr, m^2/s; eq. (28-44).

ϵ_H,   eddy thermal diffusivity; ft^2/hr, m^2/s; section 19.7.

ϵ_i,   a Lennard-Jones parameter; ergs; eqs. (24-37), (24-38).

ϵ_M,   eddy momentum diffusivity or eddy viscosity; ft^2/hr, m^2/s; eq. (13-13).

η,   dependent variable used by Blasius in solution of boundary layer; dimensionless; eq. (12-12).

η,   similarity parameter for convection analysis; dimensionless; eq. (19-7).

η_F,   fin efficiency; dimensionless; section 17.3, Figure 17.11.

θ,   temperature parameter $= T - T_\infty$; $°F$, K; section 17.3.

θ,   fractional void space of a catalyst; eq. (24-49).

θ,   angle in cylindrical or spherical coordinates; radians.

θ_g,   correlating constant; eq. (31-3).

κ,   Boltzmann constant; ergs/K; sections (23.4), (24.2).

λ,   molecular mean free path; sections 7.3, 15.2, 24.2.

λ,   wave length of thermal radiation; microns; section 23.4.

λ,   ionic conductance; (amp/cm^2) $(volt/cm)$ $(g\ equivalent/cm^3)$; eq. (24-47).

μ,   viscosity; $lb_m/ft\ s$, $Pa \cdot s$; eq. (7-4).

μ_B,   viscosity of solvent B; cp; eq. (24-45).

μ_c,   chemical potential of given species; Btu/lb mole, J/mol; eq. (24-23).

ν,   frequency; hz; section 23.1.

ν,   kinematic viscosity, μ/ρ; ft^2/s, m^2/s.

π,   Pi groups in dimensional analysis; sections 14.1, 19.3, 28.3.

ρ,   density of a fluid; lb_m/ft^3, kg/m^3; section 1.2.

ρ,   mass density of mixture; lb_m/ft^3, kg/m^3; eq. (24-1).

ρ,   reflectivity; dimensionless; section 23.2.

ρ_i,   mass concentration of species i, lb_m/ft^3, kg/m^3; eq. (24-13).

σ,   surface tension; lb_f/ft, N/m; Chapter 21.

σ,   Stefan-Boltzmann constant; 0.1714×10^{-8} Btu/hr ft^2 °R^4, 5.672×10^{-8} W/m$^2 \cdot$ K^4; eq. (15-13).

σ_{AB},   Lennard-Jones parameter; Å; eq. (24-39).

σ_i,   a Lennard-Jones parameter; Å; eqs. (24-34), (24-35), (24-36).

σ_{ii},   normal stress; lb_f/in^2, N/m^2; section 1.2.

τ,   tortuosity of a catalyst; eq. (24-49).

τ,   transmissivity; dimensionless; section 23.2.

τ_{ij},   shear stress; lb_f/in^2, N/m^2; section 1.2.

τ_0,   shear stress at the surface; lb_f/in^2, N/m^2; eq. (12-30).

ϕ,   velocity potential; section 10.4.

ϕ,   angle in spherical coordinates; radians.

ω,   angular velocity; 1/s.

ω_i,   mass fraction of species i; dimensionless; eq. (24-2).

$\omega/2$,   vorticity; eq. (10-4).

Γ,   flow rate of condensate film per unit width; lb_m/ft s, $kg/m \cdot s$; eq. (21-13).

Δ, $y_A - y_A{}^*$; dimensionless; eq. (31-32).

Δa,   equivalent to $a_2 - a_1$, where a is any variable and 1 and 2 refer to two control surfaces.

ΔT_{lm},   logarithmic mean temperature difference; °F, K; eq. (22-9).

$\mathscr{E}$,   heat exchanger effectiveness; dimensionless; eq. (22-17).

Φ_B,   association parameter; eq. (24-45).

Ψ,   stream function; section 10.2.

Ω,   solid angle; radians; section 23.3.

Ω_D,   collision integral: eq. (24-33) and Appendix K.

Ω_k,   Lennard-Jones collision integral; eq. (15-7) and Appendix K.

Ω_μ,   Lennard-Jones collision integral; eq. (7-10) and Appendix K.

DIMENSIONLESS PARAMETERS

Bi,   Biot number, $(hV/A)/k$; eq. (18-7).

Eu,   Euler number, $P/\rho v^2$; eq. (11-2).

Fo,   Fourier number, $\alpha t/(V/A)^2$; eq. (18-8).

Fr,   Froude number, v^2/gL; eq. (11-1).

Gr,   Grashof number, $\beta g \rho^2 L^3 \Delta T/\mu^2$; eq. (19-12).

Gr$_{AB}$,   mass-transfer Grashof number, $L^3 g \Delta \rho_A/\rho \nu^3$; eq. (28-9).

Gz,   Graetz number, $(\pi/4)(D/x)$ Re Pr; section 20.2.

Le,   Lewis number, $k/\rho c_p D_{AB}$; eq. (28-3).

Nu,   Nusselt number, hL/k; eq. (19-6).

Nu$_{AB}$, mass-transfer Nusselt number, $k_c L/D_{AB}$: eq. (28-7).

Pe, Peclet number, $Dv\rho c_p/k$ = Re Pr; section 20.2.

Pe$_{AB}$, mass-transfer Peclet number, Dv/D_{AB} = Re Sc; eq. (30-4).

Pr, Prandtl number, $\nu/\alpha = \mu c_p/k$; eq. (19-1).

Re, Reynolds number, $lv\rho/\mu = Lv/\nu$; eq. (11-3).

Sc, Schmidt number, $\mu/\rho D_{AB}$; eq. (28-2).

Sh, Sherwood number, $k_c L/D_{AB}$; eq. (28-7).

St, Stanton number, $h/\rho v c_p$; eq. (19-8).

˙MATHEMATICAL OPERATIONS

D/Dt, substantial derivative; eq. (9-4).

div **A**, or $\nabla \cdot \mathbf{A}$, divergence of a vector.

erf ϕ, the error function of ϕ; Appendix L.

exp x, or e^x, exponential function of x.

ln x, logarithm of x to the base e.

log$_{10}$ x, logarithm of x to base 10.

$$\nabla = \frac{\partial}{\partial x}\mathbf{e}_x + \frac{\partial}{\partial y}\mathbf{e}_y + \frac{\partial}{\partial z}\mathbf{e}_z.$$

APPENDIX A
TRANSFORMATIONS OF THE OPERATORS ∇ AND ∇² TO CYLINDRICAL COORDINATES

THE OPERATOR ∇ IN CYLINDRICAL COORDINATES

In cartesian coordinates ∇ is written:

$$\nabla = e_x \frac{\partial}{\partial x} + \mathbf{e}_y \frac{\partial}{\partial y} + \mathbf{e}_z \frac{\partial}{\partial z} \qquad (A\text{-}1)$$

When transforming this operator into cylindrical coordinates, both the unit vectors and the partial derivatives must be transformed.

A cylindrical coordinate system and a cartesian coordinate system are shown in Figure A.1. The following relations are observed to exist between the cartesian and cylindrical coordinates:

$$z = z \qquad x^2 + y^2 = r^2 \qquad \tan\theta = \frac{y}{x} \qquad (A\text{-}2)$$

Thus

$$\left(\frac{\partial}{\partial z}\right)_{\text{cyl}} = \left(\frac{\partial}{\partial z}\right)_{\text{cart}} \qquad (A\text{-}3)$$

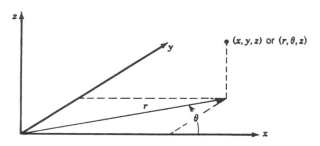

Figure A.1 Cylindrical and cartesian coordinates.

722

while from the chain rule,

$$\left(\frac{\partial}{\partial x}\right) = \frac{\partial}{\partial r}\frac{\partial r}{\partial x} + \frac{\partial}{\partial \theta}\frac{\partial \theta}{\partial x}$$

As

$$\frac{\partial r}{\partial x} = \frac{x}{r} = \cos\theta$$

$$\frac{\partial \theta}{\partial x} = -\frac{y}{x^2 \sec^2\theta} = -\frac{y}{r^2} = -\frac{\sin\theta}{r}$$

thus

$$\left(\frac{\partial}{\partial x}\right) = \cos\theta\left(\frac{\partial}{\partial r}\right) - \frac{\sin\theta}{r}\left(\frac{\partial}{\partial \theta}\right) \tag{A-4}$$

In a similar manner,

$$\frac{\partial}{\partial r} = \frac{\partial}{\partial r}\frac{\partial r}{\partial y} + \frac{\partial}{\partial \theta}\frac{\partial \theta}{\partial y}$$

where

$$\frac{\partial r}{\partial y} = \frac{y}{r} = \sin\theta \quad \text{and} \quad \frac{\partial \theta}{\partial y} = \frac{1}{x^2 \sec^2\theta} = \frac{\cos\theta}{r}$$

Thus $(\partial/\partial y)$ becomes

$$\left(\frac{\partial}{\partial y}\right) = \sin\theta\left(\frac{\partial}{\partial r}\right) + \frac{\cos\theta}{r}\left(\frac{\partial}{\partial \theta}\right) \tag{A-5}$$

The unit vectors must also be transformed. Resolving the unit vectors into their x, y, and z-direction components, we obtain

$$\mathbf{e}_z = \mathbf{e}_z \tag{A-6}$$

$$\mathbf{e}_x = \mathbf{e}_r\cos\theta - \mathbf{e}_\theta\sin\theta \tag{A-7}$$

$$\mathbf{e}_y = \mathbf{e}_r\sin\theta + \mathbf{e}_\theta\cos\theta \tag{A-8}$$

Substituting the above relations into equation (A-1), we obtain

$$\mathbf{e}_x\frac{\partial}{\partial x} = \mathbf{e}_r\cos^2\theta\frac{\partial}{\partial r} - \mathbf{e}_r\frac{\sin\theta\cos\theta}{r}\frac{\partial}{\partial \theta} - \mathbf{e}_\theta\sin\theta\cos\theta\frac{\partial}{\partial r} + \mathbf{e}_\theta\frac{\sin^2\theta}{r}\frac{\partial}{\partial \theta}$$

$$\mathbf{e}_y\frac{\partial}{\partial y} = \mathbf{e}_r\sin^2\theta\frac{\partial}{\partial r} + \mathbf{e}_r\frac{\sin\theta\cos\theta}{r}\frac{\partial}{\partial \theta} + \mathbf{e}_\theta\sin\theta\cos\theta\frac{\partial}{\partial r} + \mathbf{e}_\theta\frac{\cos^2\theta}{r}\frac{\partial}{\partial \theta}$$

and

$$\mathbf{e}_z\frac{\partial}{\partial z} = \mathbf{e}_z\frac{\partial}{\partial z}$$

Adding the above relations, we obtain, after noting that $\sin^2\theta + \cos^2\theta = 1$,

$$\nabla = \mathbf{e}_r\left(\frac{\partial}{\partial r}\right) + \frac{\mathbf{e}_\theta}{r}\left(\frac{\partial}{\partial \theta}\right) + \mathbf{e}_z\left(\frac{\partial}{\partial z}\right) \tag{A-9}$$

THE OPERATOR ∇^2 IN CYLINDRICAL COORDINATES

A unit vector may not change magnitude; however, its direction may change. Cartesian unit vectors do not change their absolute directions, but in cylindrical coordinates both $\mathbf{e}_r$ and $\mathbf{e}_\theta$ depend upon the angle θ. Since these vectors change direction, they have derivatives with respect to θ. As $\mathbf{e}_r = \mathbf{e}_x \cos \theta + \mathbf{e}_r \sin \theta$, and $\mathbf{e}_\theta = -\mathbf{e}_x \sin \theta + \mathbf{e}_r \cos \theta$, it may be seen that

$$\frac{\partial}{\partial r}\mathbf{e}_r = 0 \qquad \frac{\partial}{\partial r}\mathbf{e}_\theta = 0$$

while

$$\frac{\partial}{\partial \theta}\mathbf{e}_r = \mathbf{e}_\theta \tag{A-10}$$

and

$$\frac{\partial}{\partial \theta}\mathbf{e}_\theta = -\mathbf{e}_r \tag{A-11}$$

Now the operator $\nabla^2 = \nabla \cdot \nabla$ and thus

$$\nabla \cdot \nabla = \nabla^2 = \left(\mathbf{e}_r\frac{\partial}{\partial r} + \frac{\mathbf{e}_\theta}{r}\frac{\partial}{\partial \theta} + \mathbf{e}_z\frac{\partial}{\partial z}\right) \cdot \left(\mathbf{e}_r\frac{\partial}{\partial r} + \frac{\mathbf{e}_\theta}{r}\frac{\partial}{\partial \theta} + \mathbf{e}_z\frac{\partial}{\partial z}\right)$$

Performing the indicated operations, we obtain

$$\mathbf{e}_r\frac{\partial}{\partial r} \cdot \nabla = \frac{\partial^2}{\partial r^2}$$

$$\frac{\mathbf{e}_\theta}{r}\frac{\partial}{\partial \theta} \cdot \nabla = \frac{\mathbf{e}_\theta}{r} \cdot \frac{\partial}{\partial \theta}\left(\mathbf{e}_r\frac{\partial}{\partial r}\right) + \frac{\mathbf{e}_\theta}{r} \cdot \frac{\partial}{\partial \theta}\left(\frac{\mathbf{e}_\theta}{r}\frac{\partial}{\partial \theta}\right)$$

or

$$\frac{\mathbf{e}_\theta}{r}\frac{\partial}{\partial \theta} \cdot \nabla = \frac{1}{r}\frac{\partial}{\partial r} + \frac{1}{r^2}\frac{\partial^2}{\partial \theta^2}$$

and

$$\mathbf{e}_z\frac{\partial}{\partial z} \cdot \nabla = \frac{\partial^2}{\partial z^2}$$

Thus the operator ∇^2 becomes

$$\nabla^2 = \frac{\partial^2}{\partial r^2} + \frac{1}{r}\frac{\partial}{\partial r} + \frac{1}{r^2}\frac{\partial^2}{\partial \theta^2} + \frac{\partial^2}{\partial z^2} \tag{A-12}$$

or

$$\nabla^2 = \frac{1}{r}\frac{\partial}{\partial r}\left(r\frac{\partial}{\partial r}\right) + \frac{1}{r^2}\frac{\partial^2}{\partial \theta^2} + \frac{\partial^2}{\partial z^2} \tag{A-13}$$

APPENDIX B
SUMMARY OF DIFFERENTIAL VECTOR OPERATIONS IN VARIOUS COORDINATE SYSTEMS

CARTESIAN COORDINATES

Coordinate system

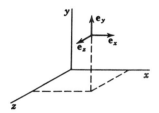

Figure B.1 Unit vectors at the point (x, y, z).

Gradient

$$\boldsymbol{\nabla} P = \frac{\partial P}{\partial x}\mathbf{e}_x + \frac{\partial P}{\partial y}\mathbf{e}_y + \frac{\partial P}{\partial z}\mathbf{e}_z \tag{B-1}$$

Divergence

$$\boldsymbol{\nabla} \cdot \mathbf{v} = \frac{\partial v_x}{\partial x} + \frac{\partial v_y}{\partial y} + \frac{\partial v_z}{\partial z} \tag{B-2}$$

Curl

$$\boldsymbol{\nabla} \times \mathbf{v} = \begin{cases} \left(\dfrac{\partial v_z}{\partial y} - \dfrac{\partial v_y}{\partial z}\right)\mathbf{e}_x \\[2mm] \left(\dfrac{\partial v_x}{\partial z} - \dfrac{\partial v_z}{\partial x}\right)\mathbf{e}_y \\[2mm] \left(\dfrac{\partial v_y}{\partial x} - \dfrac{\partial v_x}{\partial y}\right)\mathbf{e}_z \end{cases} \tag{B-3}$$

725

Laplacian of a scalar

$$\nabla^2 T = \frac{\partial^2 T}{\partial x^2} + \frac{\partial^2 T}{\partial y^2} + \frac{\partial^2 T}{\partial z^2} \tag{B-4}$$

CYLINDRICAL COORDINATES

Coordinate system

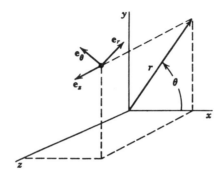

Figure B.2 Unit vectors at the point (r, θ, z).

Gradient

$$\nabla P = \frac{\partial P}{\partial r}\mathbf{e}_r + \frac{1}{r}\frac{\partial P}{\partial \theta}\mathbf{e}_\theta + \frac{\partial P}{\partial z}\mathbf{e}_z \tag{B-5}$$

Divergence

$$\nabla \cdot \mathbf{v} = \frac{1}{r}\frac{\partial}{\partial r}(rv_r) + \frac{1}{r}\frac{\partial v_\theta}{\partial \theta} + \frac{\partial v_z}{\partial z} \tag{B-6}$$

Curl

$$\nabla \times \mathbf{v} = \left\{ \begin{array}{l} \left(\dfrac{1}{r}\dfrac{\partial v_z}{\partial \theta} - \dfrac{\partial v_\theta}{\partial z}\right)\mathbf{e}_r \\[2ex] \left(\dfrac{\partial v_r}{\partial z} - \dfrac{\partial v_z}{\partial r}\right)\mathbf{e}_\theta \\[2ex] \left\{\dfrac{1}{r}\left[\dfrac{\partial}{\partial r}(rv_\theta) - \dfrac{\partial v_r}{\partial \theta}\right]\right\}\mathbf{e}_z \end{array} \right\} \tag{B-7}$$

Laplacian of a scalar

$$\nabla^2 T = \frac{1}{r}\frac{\partial}{\partial r}\left(r\frac{\partial T}{\partial r}\right) + \frac{1}{r^2}\frac{\partial^2 T}{\partial \theta^2} + \frac{\partial^2 T}{\partial z^2} \tag{B-8}$$

SPHERICAL COORDINATES

Coordinate system

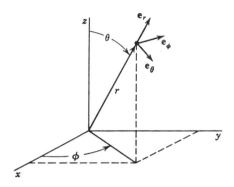

Figure B.3 Unit vectors at the point (r, θ, ϕ).

Gradient

$$\nabla P = \frac{\partial P}{\partial r}\mathbf{e}_r + \frac{1}{r}\frac{\partial P}{\partial \theta}\mathbf{e}_\theta + \frac{1}{r \sin \theta}\frac{\partial P}{\partial \phi}\mathbf{e}_\phi \tag{B-9}$$

Divergence

$$\nabla \cdot \mathbf{v} = \frac{1}{r^2}\frac{\partial}{\partial r}(r^2 v_r) + \frac{1}{r \sin \theta}\frac{\partial}{\partial \theta}(v_\theta \sin \theta) + \frac{1}{r \sin \theta}\frac{\partial v_\phi}{\partial \phi} \tag{B-10}$$

Curl

$$\nabla \times \mathbf{v} = \begin{Bmatrix} \dfrac{1}{r \sin \theta}\left[\dfrac{\partial}{\partial \theta}(v_\phi \sin \theta) - \dfrac{\partial v_\theta}{\partial \phi}\right]\mathbf{e}_r \\[2ex] \left[\dfrac{1}{r \sin \theta}\dfrac{\partial v_r}{\partial \phi} - \dfrac{1}{r}\dfrac{\partial}{\partial r}(rv_\phi)\right]\mathbf{e}_\theta \\[2ex] \dfrac{1}{r}\left[\dfrac{\partial}{\partial r}(rv_\theta) - \dfrac{\partial v_r}{\partial \theta}\right]\mathbf{e}_\phi \end{Bmatrix} \tag{B-11}$$

Laplacian of a scalar

$$\nabla^2 T = \frac{1}{r^2}\frac{\partial}{\partial r}\left(r^2\frac{\partial T}{\partial r}\right) + \frac{1}{r^2 \sin \theta}\frac{\partial}{\partial \theta}\left(\sin \theta\frac{\partial T}{\partial \theta}\right) + \frac{1}{r^2 \sin^2 \theta}\frac{\partial^2 T}{\partial \phi^2} \tag{B-12}$$

APPENDIX C
SYMMETRY OF
THE STRESS TENSOR

The shear stress $\tau_{i,j}$ can be shown to be equal to $\tau_{j,i}$ by the following simple argument. Consider the element of fluid shown in Figure C.1. The sum of the

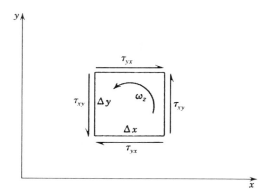

Figure C.1   Free body of element.

moments on the element will be related to the angular acceleration by

$$\sum M = I\dot{\omega} \tag{C-1}$$

where I is the mass moment of inertia of the element. Substituting into equation (C-1),

$$-(\tau_{y,x}\,\Delta x\,\Delta z)\,\Delta y + (\tau_{x,y}\,\Delta y\,\Delta z)\,\Delta x = \rho\,\Delta x\,\Delta y\,\Delta z\frac{(\Delta x^2 + \Delta y^2)}{12}\dot{\omega}_z$$

where the moment of inertia of a rectangular prism has been used for the element.
The volume of the element $\Delta x\,\Delta y\,\Delta z$ may be canceled to yield

$$\rho\left(\frac{\Delta x^2 + \Delta y^2}{12}\right)\dot{\omega}_z = \tau_{x,y} - \tau_{y,x} \tag{C-2}$$

Now the difference in shear stress is seen to depend upon the size of the element. As the element shrinks to a point, Δx and Δy approach zero independently, and we obtain, in the limit,

$$\tau_{x,y} = \tau_{y,x}$$

or, as this can be done about any axis,

$$\tau_{i,j} = \tau_{j,i}$$

Another way to look at equation (C-2) is to determine the angular acceleration ω_z as the element shrinks to a point. The angular acceleration at a point must be finite, hence $\tau_{y,x}$ and $\tau_{x,y}$ must be equal.

APPENDIX D
THE VISCOUS CONTRIBUTION TO THE NORMAL STRESS

The normal stress, σ, may be divided into two parts, the pressure contribution, $-P$, and a viscous contribution, σ_v. The viscous contribution to the normal stress is obtained by analogy with Hooke's law for an elastic solid. In Hooke's law for three-dimensional stress, the normal stress, $\sigma_{x,x}$, in the x direction is related to the strains in the x, y, z directions by*

$$\sigma_{x,x} = 2G\epsilon_x + \frac{2G\eta}{1-2\eta}(\epsilon_x + \epsilon_y + \epsilon_z) \qquad \text{(D-1)}$$

where G is the shear modulus; η is Poisson's ratio; and ϵ is the axial strain.

When Newton's viscosity relation was discussed, the shear strain in a solid was seen to be analogous to the rate of shear strain in a fluid. Accordingly, the axial strain in a solid, ϵ_x, is taken to be analogous to the axial strain rate in a fluid, $\partial v_x/\partial x$.

When the velocity derivatives are substituted for the strains in equation (D-1), and the viscosity is used in place of the shear modulus, we obtain

$$(\sigma_{x,x})_{\text{viscous}} = 2\mu\frac{\partial v_x}{\partial x} + \lambda \boldsymbol{\nabla} \cdot \mathbf{v} \qquad \text{(D-2)}$$

Here the sum of the strain-rate derivatives is observed to be equal to $\boldsymbol{\nabla} \cdot \mathbf{v}$, and the second coefficient has been designated λ and is called the bulk viscosity or second viscosity coefficient. The total normal stress in the x-direction becomes

$$\sigma_{x,x} = -P + 2\mu\frac{\partial v_x}{\partial x} + \lambda \boldsymbol{\nabla} \cdot \mathbf{v} \qquad \text{(D-3)}$$

---

* A more familiar form is

$$\sigma_{x,x} = \frac{E}{(1+\eta)(1-2\eta)}[(1-\eta)\epsilon_x + \eta(\epsilon_y + \epsilon_z)]$$

The shear modulus G has been replaced with its equivalent, $E/2(1+\eta)$.

If the corresponding normal stress components in the y and z directions are added together, we obtain

$$\sigma_{x,x} + \sigma_{y,y} + \sigma_{z,z} = -3P + (2\mu + 3\lambda)\nabla \cdot \mathbf{v}$$

so that the average normal stress $\bar{\sigma}$ is given by

$$\bar{\sigma} = -P + \left(\frac{2\mu + 3\lambda}{3}\right)\nabla \cdot \mathbf{v}$$

Thus, unless $\lambda = -\frac{2}{3}\mu$, the average stress will depend upon the flow properties rather than the fluid property, P. Stokes assumed that $\lambda = -\frac{2}{3}\mu$, and experiments have indicated that λ is of the same order of magnitude as μ for air. As $\nabla \cdot \mathbf{v} = 0$ in an incompressible flow, the value of λ is of no concern except for compressible fluids.

The resulting expressions for normal stress in a newtonian fluid are

$$\sigma_{x,x} = -P + 2\mu\frac{\partial v_x}{\partial x} - \frac{2}{3}\mu\nabla \cdot \mathbf{v} \tag{D-4}$$

$$\sigma_{y,y} = -P + 2\mu\frac{\partial v_y}{\partial y} - \frac{2}{3}\mu\nabla \cdot \mathbf{v} \tag{D-5}$$

$$\sigma_{z,z} = -P + 2\mu\frac{\partial v_z}{\partial z} - \frac{2}{3}\mu\nabla \cdot \mathbf{v} \tag{D-6}$$

APPENDIX E
THE NAVIER-STOKES EQUATIONS FOR CONSTANT ρ AND μ IN CARTESIAN, CYLINDRICAL, AND SPHERICAL COORDINATES

CARTESIAN COORDINATES

x direction

$$\rho\left(\frac{\partial v_x}{\partial t}+v_x\frac{\partial v_x}{\partial x}+v_y\frac{\partial v_x}{\partial y}+v_z\frac{\partial v_x}{\partial z}\right)=-\frac{\partial P}{\partial x}+\rho g_x+\mu\left(\frac{\partial^2 v_x}{\partial x^2}+\frac{\partial^2 v_x}{\partial y^2}+\frac{\partial^2 v_x}{\partial z^2}\right) \qquad \text{(E-1)}$$

y direction

$$\rho\left(\frac{\partial v_y}{\partial t}+v_x\frac{\partial v_y}{\partial x}+v_y\frac{\partial v_y}{\partial y}+v_z\frac{\partial v_y}{\partial z}\right)=-\frac{\partial P}{\partial y}+\rho g_y+\mu\left(\frac{\partial^2 v_y}{\partial x^2}+\frac{\partial^2 v_y}{\partial y^2}+\frac{\partial^2 v_y}{\partial z^2}\right) \qquad \text{(E-2)}$$

z direction

$$\rho\left(\frac{\partial v_z}{\partial t}+v_x\frac{\partial v_z}{\partial x}+v_y\frac{\partial v_z}{\partial y}+v_z\frac{\partial v_z}{\partial z}\right)=-\frac{\partial P}{\partial z}+\rho g_z+\mu\left(\frac{\partial^2 v_z}{\partial x^2}+\frac{\partial^2 v_z}{\partial y^2}+\frac{\partial^2 v_z}{\partial z^2}\right) \qquad \text{(E-3)}$$

CYLINDRICAL COORDINATES

r direction

$$\rho\left(\frac{\partial v_r}{\partial t}+v_r\frac{\partial v_r}{\partial r}+\frac{v_\theta}{r}\frac{\partial v_r}{\partial \theta}-\frac{v_\theta^2}{r}+v_z\frac{\partial v_r}{\partial z}\right)$$

$$=-\frac{\partial P}{\partial r}+\rho g_r+\mu\left[\frac{\partial}{\partial r}\left(\frac{1}{r}\frac{\partial}{\partial r}(rv_r)\right)+\frac{1}{r^2}\frac{\partial^2 v_r}{\partial \theta^2}-\frac{2}{r^2}\frac{\partial v_\theta}{\partial \theta}+\frac{\partial^2 v_r}{\partial z^2}\right] \qquad \text{(E-4)}$$

θ direction

$$\rho\left(\frac{\partial v_\theta}{\partial t}+v_r\frac{\partial v_\theta}{\partial r}+\frac{v_\theta}{r}\frac{\partial v_\theta}{\partial \theta}+\frac{v_r v_\theta}{r}+v_z\frac{\partial v_\theta}{\partial z}\right)$$

$$=-\frac{1}{r}\frac{\partial P}{\partial \theta}+\rho g_\theta+\mu\left[\frac{\partial}{\partial r}\left(\frac{1}{r}\frac{\partial}{\partial r}(rv_\theta)\right)+\frac{1}{r^2}\frac{\partial^2 v_\theta}{\partial \theta^2}+\frac{2}{r^2}\frac{\partial v_r}{\partial \theta}+\frac{\partial^2 v_\theta}{\partial z^2}\right] \tag{E-5}$$

z direction

$$\rho\left(\frac{\partial v_z}{\partial t}+v_r\frac{\partial v_z}{\partial r}+\frac{v_\theta}{r}\frac{\partial v_z}{\partial \theta}+v_z\frac{\partial v_z}{\partial z}\right)$$

$$=-\frac{\partial P}{\partial z}+\rho g_z+\mu\left[\frac{1}{r}\frac{\partial}{\partial r}\left(r\frac{\partial v_z}{\partial r}\right)+\frac{1}{r^2}\frac{\partial^2 v_z}{\partial \theta^2}+\frac{\partial^2 v_z}{\partial z^2}\right] \tag{E-6}$$

SPHERICAL COORDINATES*

r direction

$$\rho\left(\frac{\partial v_r}{\partial t}+v_r\frac{\partial v_r}{\partial r}+\frac{v_\theta}{r}\frac{\partial v_r}{\partial \theta}+\frac{v_\phi}{r\sin\theta}\frac{\partial v_r}{\partial \phi}-\frac{v_\phi^2}{r}-\frac{v_\theta^2}{r}\right)$$

$$=-\frac{\partial P}{\partial r}+\rho g_r+\mu\left[\nabla^2 v_r-\frac{2}{r^2}v_r-\frac{2}{r^2}\frac{\partial v_\theta}{\partial \theta}-\frac{2}{r^2}v_\theta\cot\theta-\frac{2}{r^2\sin\theta}\frac{\partial v_\phi}{\partial \phi}\right] \tag{E-7}$$

θ direction

$$\rho\left[\frac{\partial v_\theta}{\partial t}+v_r\frac{\partial v_\theta}{\partial r}+\frac{v_\theta}{r}\frac{\partial v_\theta}{\partial \theta}+\frac{v_\phi}{r\sin\theta}\frac{\partial v_\theta}{\partial \phi}+\frac{v_r v_\theta}{r}-\frac{v_\phi^2\cot\theta}{r}\right]$$

$$=-\frac{1}{r}\frac{\partial P}{\partial \theta}+\rho g_\theta+\mu\left[\nabla^2 v_\theta+\frac{2}{r^2}\frac{\partial v_r}{\partial \theta}-\frac{v_\theta}{r^2\sin^2\theta}-\frac{2\cos\theta}{r^2\sin^2\theta}\frac{\partial v_\phi}{\partial \phi}\right] \tag{E-8}$$

φ direction

$$\rho\left(\frac{\partial v_\phi}{\partial t}+v_r\frac{\partial v_\phi}{\partial r}+\frac{v_\theta}{r}\frac{\partial v_\phi}{\partial \theta}+\frac{v_\phi}{r\sin\theta}\frac{\partial v_\phi}{\partial \phi}+\frac{v_\phi v_r}{r}+\frac{v_\theta v_\phi}{r}\cot\theta\right)$$

$$=-\frac{1}{r\sin\theta}\frac{\partial P}{\partial \phi}+\rho g_\phi+\mu\left[\nabla^2 v_\phi-\frac{v_\phi}{r^2\sin^2\theta}+\frac{2}{r^2\sin\theta}\frac{\partial v_r}{\partial \phi}+\frac{2\cos\theta}{r^2\sin^2\theta}\frac{\partial v_\theta}{\partial \phi}\right] \tag{E-9}$$

* In the above equations,

$$\nabla^2=\frac{1}{r^2}\frac{\partial}{\partial r}\left(r^2\frac{\partial}{\partial r}\right)+\frac{1}{r^2\sin\theta}\frac{\partial}{\partial \theta}\left(\sin\theta\frac{\partial}{\partial \theta}\right)+\frac{1}{r^2\sin^2\theta}\frac{\partial^2}{\partial \phi^2}$$

APPENDIX F
CHARTS FOR SOLUTION OF UNSTEADY TRANSPORT PROBLEMS

TABLE F.1 SYMBOLS FOR UNSTEADY-STATE CHARTS

	Parameter symbol	Molecular mass transfer	Heat conduction
Unaccomplished change, a dimensionless ratio	Y	$\dfrac{c_{A_1} - c_A}{c_{A_1} - c_{A_0}}$	$\dfrac{T_\infty - T}{T_\infty - T_0}$
Relative time	X	$\dfrac{D_{AB}t}{x_1^{\,2}}$	$\dfrac{\alpha t}{x_1^{\,2}}$
Relative position	n	$\dfrac{x}{x_1}$	$\dfrac{x}{x_1}$
Relative resistance	m	$\dfrac{D_{AB}}{k_c x_1}$	$\dfrac{k}{h x_1}$

T = temperature
c_A = concentration of component A
x = distance from center to any point
t = time
k = thermal conductivity
h, k_c = convective transfer coefficients
α = thermal diffusivity
D_{AB} = mass diffusivity

Subscripts:
  0 = initial condition at time $t = 0$
  1 = boundary
  A = component A
  ∞ = reference condition for temperature

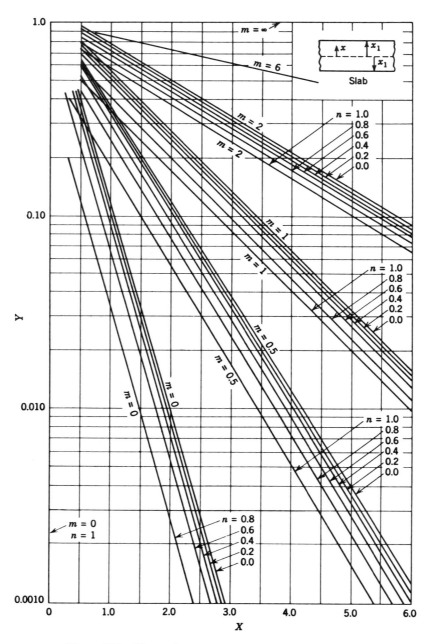

Figure F.1   Unsteady-state transport in a large flat slab.

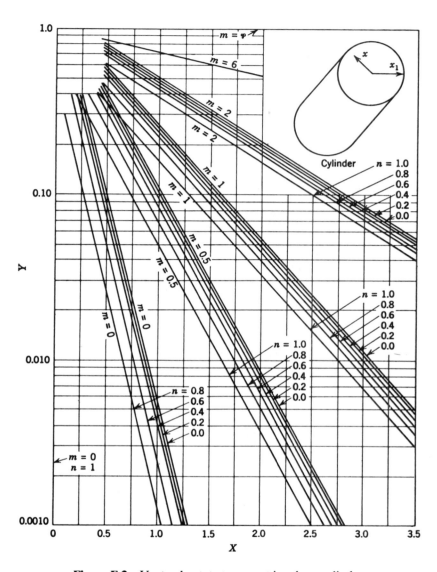

Figure F.2 Unsteady-state transport in a long cylinder.

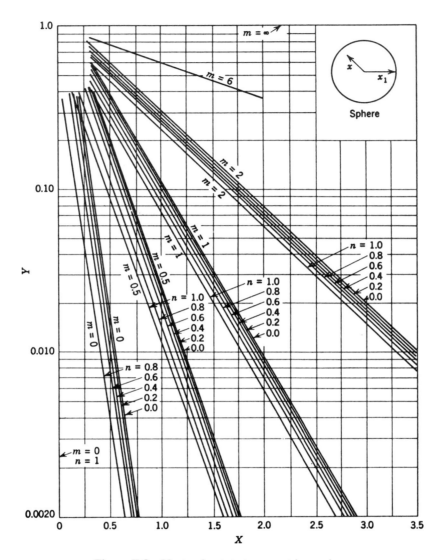

Figure F.3   Unsteady-state transport in a sphere.

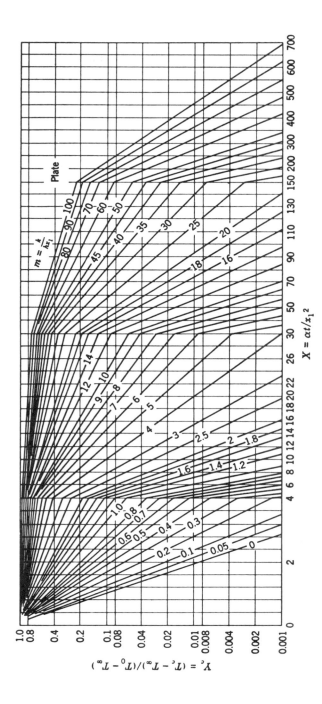

Figure F.4 Center temperature history for an infinite plate.

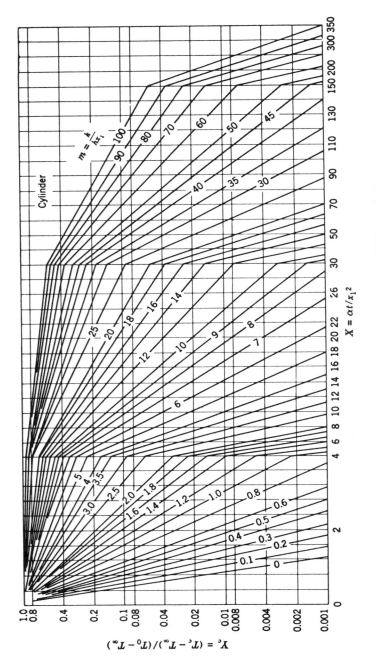

Figure F.5 Center temperature history for an infinite cylinder.

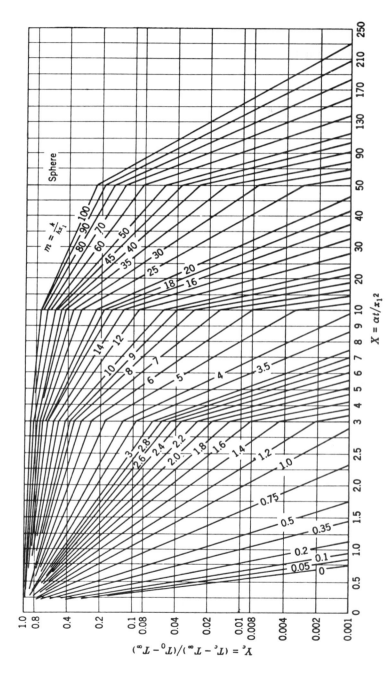

Figure F.6 Center temperature history for a sphere.

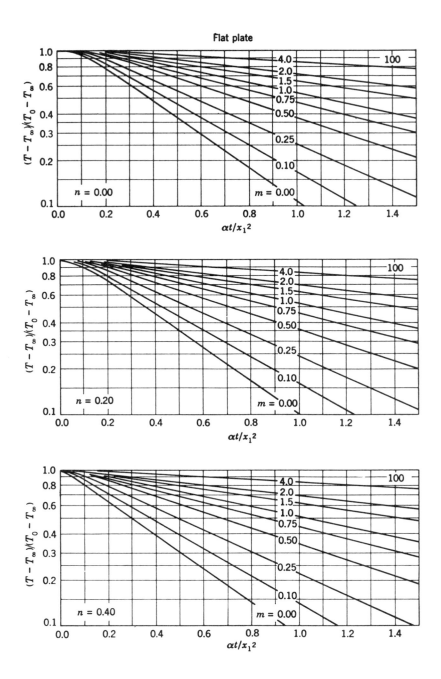

Figure F.7 Charts for solution of unsteady transport problems: flat plate.

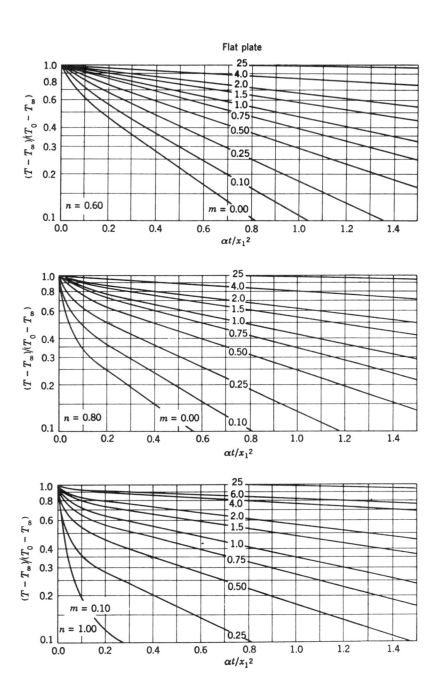

Figure F.7 Continued.

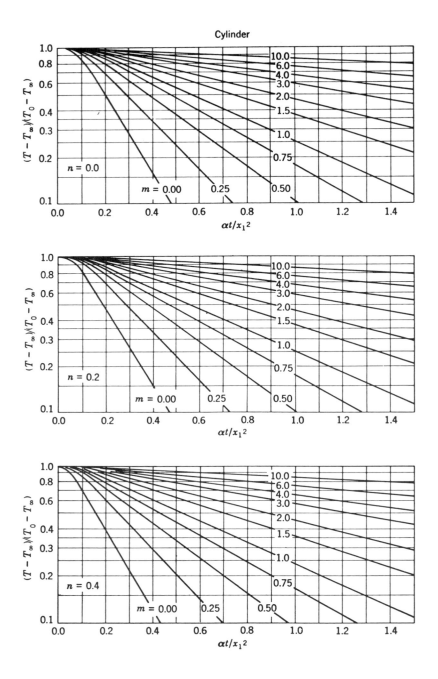

Figure F.8   Charts for solution of unsteady transport problems: cylinder.

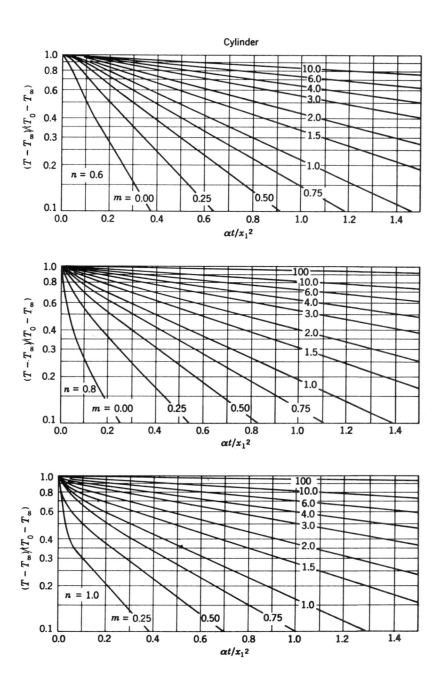

Figure F.8 Continued.

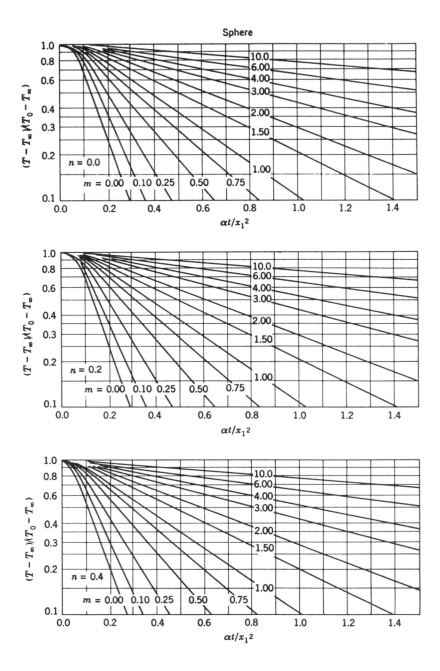

Figure F.9 Charts for solution of unsteady transport problems: sphere.

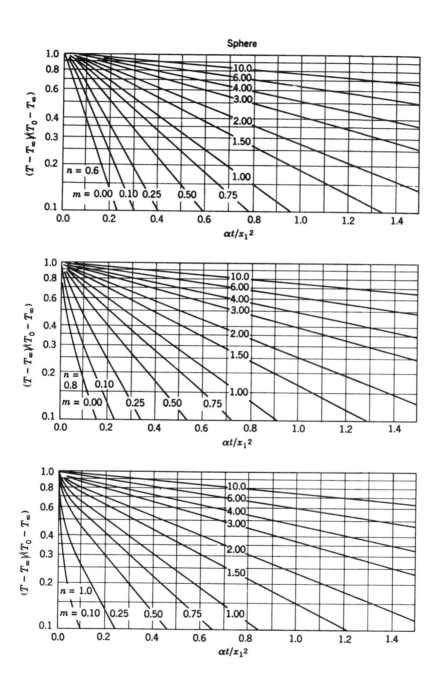

Figure F.9   Continued.

APPENDIX G
PROPERTIES OF THE
STANDARD ATMOSPHERE*

TABLE 6.1 ENGLISH UNITS

h (ft)	T (°F)	a (fps)	P lb/ft²)	ρ (slug/ft³)	$\mu \times 10^7$ (slug/ft sec)
0	59.00	1117	2116.2	0.002378	3.719
1,000	57.44	1113	2040.9	0.002310	3.699
2,000	51.87	1109	1967.7	0.002242	3.679
3,000	48.31	1105	1896.7	0.002177	3.659
4,000	44.74	1102	1827.7	0.002112	3.639
5,000	41.18	1098	1760.8	0.002049	3.618
6,000	37.62	1094	1696.0	0.001988	3.598
7,000	34.05	1090	1633.0	0.001928	3.577
8,000	30.49	1086	1571.9	0.001869	3.557
9,000	26.92	1082	1512.8	0.001812	3.536
10,000	23.36	1078	1455.4	0.001756	3.515
11,000	19.80	1074	1399.8	0.001702	3.495
12,000	16.23	1070	1345.9	0.001649	3.474
13,000	12.67	1066	1293.7	0.001597	3.453
14,000	9.10	1062	1243.2	0.001546	3.432
15,000	5.54	1058	1194.3	0.001497	3.411
16,000	1.98	1054	1147.0	0.001448	3.390
17,000	−1.59	1050	1101.1	0.001401	3.369
18,000	−5.15	1046	1056.9	0.001355	3.347
19,000	−8.72	1041	1014.0	0.001311	3.326
20,000	−12.28	1037	972.6	0.001267	3.305
21,000	−15.84	1033	932.5	0.001225	3.283
22,000	−19.41	1029	893.8	0.001183	3.262
23,000	−22.97	1025	856.4	0.001143	3.240

* Data taken from NACA TN 1428.

h (ft)	T (°F)	a (fps)	P (lb/ft^2)	ρ (slug/ft^3)	$\mu \times 10^7$ (slug/ft sec)
24,000	−26.54	1021	820.3	0.001104	3.218
25,000	−30.10	1017	785.3	0.001066	3.196
26,000	−33.66	1012	751.7	0.001029	3.174
27,000	−37.23	1008	719.2	0.000993	3.153
28,000	−40.79	1004	687.9	0.000957	3.130
29,000	−44.36	999	657.6	0.000923	3.108
30,000	−47.92	995	628.5	0.000890	3.086
31,000	−51.48	991	600.4	0.000858	3.064
32,000	−55.05	987	573.3	0.000826	3.041
33,000	−58.61	982	547.3	0.000796	3.019
34,000	−62.18	978	522.2	0.000766	2.997
35,000	−65.74	973	498.0	0.000737	2.974
40,000	−67.6	971	391.8	0.0005857	2.961
45,000	−67.6	971	308.0	0.0004605	2.961
50,000	−67.6	971	242.2	0.0003622	2.961
60,000	−67.6	971	150.9	0.0002240	2.961
70,000	−67.6	971	93.5	0.0001389	2.961
80,000	−67.6	971	58.0	0.0000861	2.961
90,000	−67.6	971	36.0	0.0000535	2.961
100,000	−67.6	971	22.4	0.0000331	2.961
150,000	113.5	1174	3.003	0.00000305	4.032
200,000	159.4	1220	0.6645	0.00000062	4.277
250,000	−8.2	1042	0.1139	0.00000015	3.333

TABLE 6.2   SI UNITS—PROPERTIES OF THE STANDARD ATMOSPHERE

h (m)	T (K)	a (m/s)	P (Pa)	ρ (kg/m^3)	$\mu \times 10^5$ (Pa·s)
0	288.2	340.3	1.0133×10^5	1.225	1.789
500	284.9	338.4	0.95461	1.167	1.774
1 000	281.7	336.4	0.89876	1.111	1.758
1 500	278.4	334.5	0.84560	1.058	1.742
2 000	275.2	332.5	0.79501	1.007	1.726
2 500	271.9	330.6	0.74692	0.9570	1.710
3 000	268.7	328.6	0.70121	0.9093	1.694
3 500	265.4	326.6	0.65780	0.8634	1.678
4 000	262.2	324.6	0.61660	0.8194	1.661
4 500	258.9	322.6	0.57753	0.7770	1.645
5 000	255.7	320.5	0.54048	0.7364	1.628
5 500	252.4	318.5	0.50539	0.6975	1.612
6 000	249.2	316.5	0.47218	0.6601	1.595
6 500	245.9	314.4	0.44075	0.6243	1.578
7 000	242.7	312.3	0.41105	0.5900	1.561
7 500	239.5	310.2	0.38300	0.5572	1.544
8 000	236.2	308.1	0.35652	0.5258	1.527
8 500	233.0	306.0	0.33154	0.4958	1.510
9 000	229.7	303.8	0.30801	0.4671	1.493
9 500	226.5	301.7	0.28585	0.4397	1.475
10 000	223.3	299.5	0.26500	0.4135	1.458
11 000	216.8	295.2	0.22700	0.3648	1.422
12 000	216.7	295.1	0.19399	0.3119	1.422
13 000	216.7	295.1	0.16580	0.2666	1.422
14 000	216.7	295.1	0.14170	0.2279	1.422
15 000	216.7	295.1	0.12112	0.1948	1.422
16 000	216.7	295.1	0.10353	0.1665	1.422
17 000	216.7	295.1	8.8497×10^3	0.1423	1.422
18 000	216.7	295.1	7.5652	0.1217	1.422
19 000	216.7	295.1	6.4675	0.1040	1.422
20 000	216.7	295.1	5.5293	0.08891	1.422
25 000	221.5	298.4	2.5492	0.04008	1.448
30 000	226.5	301.7	1.1970	0.01841	1.475
35 000	236.5	308.3	0.57459	0.008463	1.529
40 000	250.4	317.2	0.28714	0.003996	1.601
45 000	264.2	325.8	0.14910	0.001966	1.671
50 000	270.7	329.8	7.9779×10^1	0.001027	1.704
55 000	265.6	326.7	4.27516	0.0005608	1.678
60 000	255.8	320.6	2.2461	0.0003059	1.629
65 000	239.3	310.1	1.1446	0.0001667	1.543
70 000	219.7	297.1	5.5205×10^0	0.00008754	1.438
75 000	200.2	283.6	2.4904	0.00004335	1.329
80 000	180.7	269.4	1.0366	0.00001999	1.216

APPENDIX H
PHYSICAL PROPERTIES
OF SOLIDS

Material	ρ (lb$_m$/ft³) (68°F)	ρ (kg/m³) (293 K)	c_p (Btu/lb$_m$ °F) (68°F)	c_p (J/kg·K ×10⁻²) (293 K)	α (ft²/hr) (68°F)	α (m²/s ×10⁵) (293 K)	k (Btu/hr ft °F) °F (68)	k (Btu/hr ft °F) °F (212)	k (Btu/hr ft °F) °F (572)	k (W/m·K) K (293)	k (W/m·K) K (373)	k (W/m·K) K (573)
Metals												
Aluminum	168.6	2 701.1	0.224	9.383	3.55	9.16	132	133	133	229	229	230
Copper	555	8 890	0.092	3.854	3.98	10.27	223	219	213	386	379	369
Gold	1 206	19 320	0.031	1.299	4.52	11.66	169	170	172	293	294	298
Iron	492	7 880	0.122	5.110	0.83	2.14	42.3	39	31.6	73.2	68	54
Lead	708	11 300	0.030	1.257	0.80	2.06	20.3	19.3	17.2	35.1	33.4	29.8
Magnesium	109	1 750	0.248	10.39	3.68	9.50	99.5	96.8	91.4	172	168	158
Nickel	556	8 910	0.111	4.560	0.87	2.24	53.7	47.7	36.9	93.0	82.6	63.9
Platinum	1 340	21 500	0.032	1.340	0.09	0.23	40.5	41.9	43.5	70.1	72.5	75.3
Silver	656	10 500	0.057	2.388	6.42	16.57	240	237	209	415	410	362
Tin	450	7 210	0.051	2.136	1.57	4.05	36	34	—	62	59	—
Tungsten	1 206	19 320	0.032	1.340	2.44	6.30	94	87	77	160	150	130
Uranium α	1 167	18 700	0.027	1.131	0.53	1.37	16.9	17.2	19.6	29.3	29.8	33.9
Zinc	446	7 150	0.094	3.937	1.55	4.00	65	63	58	110	110	100
Alloys												
Aluminum 2024	173	2 770	0.230	9.634	1.76	4.54	70.2			122		
Brass (70% Cu. 30% Ni)	532	8 520	0.091	3.812	1.27	3.28	61.8	73.9	85.3	107	128	148
Constantan (60% Cu. 40% Ni)	557	8 920	0.098	4.105	0.24	0.62	13.1	15.4		22.7	26.7	
Iron, cast	455	7 920	0.100	4.189	0.65	1.68	29.6	26.8		51.2	46.4	
Nichrome V	530	8 490	0.106	4.440	0.12	0.31	7.06	7.99	9.94	12.2	13.8	17.2
Stainless steel	488	7 820	0.110	4.608	0.17	0.44	9.4	10.0	13	16	17.3	23
Steel, mild (1% C)	488	7 820	0.113	4.733	0.45	1.16	24.8	24.8	22.9	42.9	42.9	39.0

Material	ρ (lb$_m$/ft³) (68°F)	ρ (kg/m³) (293 K)	c_p (Btu/lb$_m$°F) (68°F)	c_p (J/kg·K) ×10⁻² (293 K)	α (ft²/hr) (68°F)	α (m²/s) ×10⁵ (293 K)	k (Btu/hr ft °F) (68)	k (212)	k (572)	k (W/m·K) (293)	k (373)	k (573)
Nonmetals												
Asbestos	36	580	0.25	10.5			0.092	0.11	0.125	0.159	0.190	0.21
Brick (fire clay)	144	2 310	0.22	9.22				0.65			1.13	
Brick (masonry)	106	1 670	0.20	8.38			0.38			0.66		
Brick (chrome)	188	3 010	0.20	8.38				0.67			1.16	
Concrete	144	2 310	0.21	8.80			0.70			1.21		
Corkboard	10	160	0.4	17			0.025			0.043		
Diatomaceous earth, powdered	14	*220	0.2	8.4			0.03			0.05		
Glass, window	170	2 720	0.2	8.4			0.45			0.78		
Glass, Pyrex	140	2 240	0.2	8.4			0.63	0.67	0.84	1.09	1.16	1.45
Kaolin firebrick	19	300							0.052			0.09
85% Magnesia	17	270					0.038	0.041		0.066	0.071	
Sandy loam, 4% H$_2$O	104	1 670	0.4	17			0.54			0.94		
Sandy loam, 10% H$_2$O	121	1 940					1.08			1.87		
Rock wool	10	160	0.2	8.4			0.023	0.033		0.040	0.057	
Wood, oak ⊥ to grain	51	820	0.57	23.9			0.12			0.21		
Wood, oak ‖ to grain	51	820	0.57	23.9			0.23			0.40		

APPENDIX I
PHYSICAL PROPERTIES OF GASES AND LIQUIDS*

* All gas properties are for atmospheric pressure.

Gases

Air

T (°F)	ρ (lb_m/ft^3)	c_p (Btu/lb_m °F)	$\mu \times 10^5$ (lb_m/ft sec)	$\nu \times 10^3$ (ft^2/sec)	k (Btu/hr ft °F)	α (ft^2/hr)	Pr	$\beta \times 10^3$ (1/°F)	$g\beta\rho^2/\mu^2$ (1/°F · ft^3)
0	0.0862	0.240	1.09	0.126	0.0132	0.639	0.721	2.18	4.39×10^6
30	0.0810	0.240	1.15	0.142	0.0139	0.714	0.716	2.04	3.28
60	0.0764	0.240	1.21	0.159	0.0146	0.798	0.711	1.92	2.48
80	0.0735	0.240	1.24	0.169	0.0152	0.855	0.708	1.85	2.09
100	0.0710	0.240	1.28	0.181	0.0156	0.919	0.703	1.79	1.76
150	0.0651	0.241	1.36	0.209	0.0167	1.06	0.698	1.64	1.22
200	0.0602	0.241	1.45	0.241	0.0179	1.24	0.694	1.52	0.840
250	0.0559	0.242	1.53	0.274	0.0191	1.42	0.690	1.41	0.607
300	0.0523	0.243	1.60	0.306	0.0203	1.60	0.686	1.32	0.454
400	0.0462	0.245	1.74	0.377	0.0225	2.00	0.681	1.16	0.264
500	0.0413	0.247	1.87	0.453	0.0246	2.41	0.680	1.04	0.163
600	0.0374	0.251	2.00	0.535	0.0270	2.88	0.680	0.944	79.4×10^3
800	0.0315	0.257	2.24	0.711	0.0303	3.75	0.684	0.794	50.6
1000	0.0272	0.263	2.46	0.906	0.0337	4.72	0.689	0.685	27.0
1500	0.0203	0.277	2.92	1.44	0.0408	7.27	0.705	0.510	7.96

Air

T (K)	ρ (kg/m³)	$c_p \times 10^{-3}$ (J/kg · K)	$\mu \times 10^5$ (Pa · s)	$\nu \times 10^5$ (m²/s)	$k \times 10^2$ (W/m · K)	$\alpha \times 10^5$ (m²/s)	Pr	$g\beta\rho^2/\mu^2$ (1/K · m³)
250	1.4133	1.0054	1.5991	1.1315	2.2269	1.5672	0.722	4.638×10^8
260	1.3587	1.0054	1.6503	1.2146	2.3080	1.6896	0.719	2.573
280	1.2614	1.0057	1.7503	1.3876	2.4671	1.9448	0.713	1.815
300	1.1769	1.0063	1.8464	1.5689	2.6240	2.2156	0.708	1.327
320	1.1032	1.0073	1.9391	1.7577	2.7785	2.5003	0.703	0.9942
340	1.0382	1.0085	2.0300	1.9553	2.9282	2.7967	0.699	0.7502
360	0.9805	1.0100	2.1175	2.1596	3.0779	3.1080	0.695	0.5828
400	0.8822	1.0142	2.2857	2.5909	3.3651	3.7610	0.689	0.3656
440	0.8021	1.0197	2.4453	3.0486	3.6427	4.4537	0.684	0.2394
480	0.7351	1.0263	2.5963	3.5319	3.9107	5.1836	0.681	0.1627
520	0.6786	1.0339	2.7422	4.0410	4.1690	5.9421	0.680	0.1156
580	0.6084	1.0468	2.9515	4.8512	4.5407	7.1297	0.680	7.193×10^6
700	0.5040	1.0751	3.3325	6.6121	5.2360	9.6632	0.684	3.210
800	0.4411	1.0988	3.6242	8.2163	5.7743	11.9136	0.689	1.804
1000	0.3529	1.1421	4.1527	11.1767	6.7544	16.7583	0.702	0.803

Steam

T (°F)	ρ (lb$_m$/ft^3)	c_p (Btu/lb$_m$ °F)	$\mu \times 10^5$ (lb$_m$/ft sec)	$\nu \times 10^3$ (ft^2/sec)	k (Btu/hr ft °F)	α (ft^2/hr)	Pr	$\beta \times 10^3$ (1/°F)	$g\beta\rho^2/\mu^2$ (1/°F · ft^3)
212	0.0372	0.493	0.870	0.234	0.0145	0.794	1.06	1.49	0.873×10^6
250	0.0350	0.483	0.890	0.254	0.0155	0.920	0.994	1.41	0.698
300	0.0327	0.476	0.960	0.294	0.0171	1.10	0.963	1.32	0.493
400	0.0289	0.472	1.09	0.377	0.0200	1.47	0.924	1.16	0.262
500	0.0259	0.477	1.23	0.474	0.0228	1.85	0.922	1.04	0.148
600	0.0234	0.483	1.37	0.585	0.0258	2.29	0.920	0.944	88.9×10^3
800	0.0197	0.498	1.63	0.828	0.0321	3.27	0.912	0.794	37.8
1000	0.0170	0.517	1.90	1.12	0.0390	4.44	0.911	0.685	17.2
1500	0.0126	0.564	2.57	2.05	0.0580	8.17	0.906	0.510	3.97

T (K)	ρ (kg/m³)	$c_p \times 10^{-3}$ (J/kg · K)	$\mu \times 10^5$ (Pa · s)	$\nu \times 10^5$ (m²/s)	$k \times 10^2$ (W/m · K)	$\alpha \times 10^5$ (m²/s)	Pr	$g\beta\rho^2/\mu^2$ (1/K · m³)
				Steam				
380	0.5860	2.0592	12.70	2.1672	2.4520	2.0320	1.067	5.5210×10^7
400	0.5549	2.0098	13.42	2.4185	2.6010	2.3322	1.037	4.1951
450	0.4911	1.9771	15.23	3.1012	2.9877	3.0771	1.008	2.2558
500	0.4410	1.9817	17.03	3.8617	3.3903	3.8794	0.995	1.3139
550	0.4004	2.0006	18.84	4.7053	3.8008	4.7448	0.992	0.8069
600	0.3667	2.0264	20.64	5.6286	4.2161	5.6738	0.992	0.5154
650	0.3383	2.0555	22.45	6.6361	4.6361	6.6670	0.995	0.3415
700	0.3140	2.0869	24.25	7.7229	5.0593	7.7207	1.000	0.2277
750	0.2930	2.1192	26.06	8.8942	5.4841	8.8321	1.007	0.1651
800	0.2746	2.1529	27.86	10.1457	5.9089	9.9950	1.015	0.1183

Nitrogen

T ($^\circ$F)	ρ (lb$_m$/ft^3)	c_p (Btu/lb$_m$ $^\circ$F)	$\mu \times 10^5$ (lb$_m$/ft sec)	$\nu \times 10^3$ (ft^2/sec)	k (Btu/hr ft $^\circ$F)	α (ft^2/hr)	Pr	$\beta \times 10^3$ (1/$^\circ$F)	$g\beta\rho^2/\mu^2$ (1/$^\circ$F · ft^3)
0	0.0837	0.249	1.06	0.127	0.0132	0.633	0.719	2.18	4.38×10^6
30	0.0786	0.249	1.12	0.142	0.0139	0.710	0.719	2.04	3.29
60	0.0740	0.249	1.17	0.158	0.0146	0.800	0.716	1.92	2.51
80	0.0711	0.249	1.20	0.169	0.0151	0.853	0.712	1.85	2.10
100	0.0685	0.249	1.23	0.180	0.0154	0.915	0.708	1.79	1.79
150	0.0630	0.249	1.32	0.209	0.0168	1.07	0.702	1.64	1.22
200	0.0580	0.249	1.39	0.240	0.0174	1.25	0.690	1.52	0.854
250	0.0540	0.249	1.47	0.271	0.0192	1.42	0.687	1.41	0.616
300	0.0502	0.250	1.53	0.305	0.0202	1.62	0.685	1.32	0.457
400	0.0443	0.250	1.67	0.377	0.0212	2.02	0.684	1.16	0.263
500	0.0397	0.253	1.80	0.453	0.0244	2.43	0.683	1.04	0.163
600	0.0363	0.256	1.93	0.532	0.0252	2.81	0.686	0.944	0.108
800	0.0304	0.262	2.16	0.710	0.0291	3.71	0.691	0.794	0.0507
1000	0.0263	0.269	2.37	0.901	0.0336	4.64	0.700	0.685	0.0272
1500	0.0195	0.283	2.82	1.45	0.0423	7.14	0.732	0.510	0.00785

T (K)	ρ (kg/m³)	$c_p \times 10^{-3}$ (J/kg·K)	$\mu \times 10^5$ (Pa·s)	$\nu \times 10^5$ (m²/s)	$k \times 10^2$ (W/m·K)	$\alpha \times 10^5$ (m²/s)	Pr	$g\beta\rho^2/\mu^2$ (1/K·m³)
				Nitrogen				
250	1.3668	1.0415	1.5528	1.1361	2.2268	1.5643	0.729	3.0362×10^8
300	1.1383	1.0412	1.7855	1.5686	2.6052	2.1981	0.713	1.3273
350	0.9754	1.0421	2.0000	2.0504	2.9691	2.9210	0.701	0.6655
400	0.8533	1.0449	2.1995	2.5776	3.3186	3.7220	0.691	0.3697
450	0.7584	1.0495	2.3890	3.1501	3.6463	4.5811	0.688	0.2187
500	0.6826	1.0564	2.5702	3.7653	3.9645	5.4979	0.684	0.1382
600	0.5688	1.0751	2.9127	5.1208	4.4549	7.4485	0.686	6.237×10^6
700	0.4875	1.0980	3.2120	6.5887	5.0947	9.5179	0.691	3.233
800	0.4266	1.1222	3.4896	8.1800	5.5864	11.6692	0.700	1.820
1000	0.3413	1.1672	4.0000	11.7199	6.4419	16.1708	0.724	0.810

Oxygen

T (°F)	ρ (lb$_m$/ft^3)	c_p (Btu/lb$_m$ °F)	$\mu \times 10^5$ (lb$_m$/ft sec)	$\nu \times 10^3$ (ft^2/sec)	k (Btu/hr ft °F)	α (ft^2/hr)	Pr	$\beta \times 10^3$ (1/°F)	$g\beta\rho^2/\mu^2$ (1/°F · ft^3)
0	0.0955	0.219	1.22	0.128	0.0134	0.641	0.718	2.18	4.29×10^6
30	0.0897	0.219	1.28	0.143	0.0141	0.718	0.716	2.04	3.22
60	0.0845	0.219	1.35	0.160	0.0149	0.806	0.713	1.92	2.43
80	0.0814	0.220	1.40	0.172	0.0155	0.866	0.713	1.85	2.02
100	0.0785	0.220	1.43	0.182	0.0160	0.925	0.708	1.79	1.74
150	0.0720	0.221	1.52	0.211	0.0172	1.08	0.703	1.64	1.19
200	0.0665	0.223	1.62	0.244	0.0185	1.25	0.703	1.52	0.825
250	0.0168	0.225	1.70	0.276	0.0197	1.42	0.700	1.41	0.600
300	0.0578	0.227	1.79	0.310	0.0209	1.60	0.700	1.32	0.442
400	0.0511	0.230	1.95	0.381	0.0233	1.97	0.698	1.16	0.257
500	0.0458	0.234	2.10	0.458	0.0254	2.37	0.696	1.04	0.160
600	0.0414	0.239	2.25	0.543	0.0281	2.84	0.688	0.944	0.103
800	0.0349	0.246	2.52	0.723	0.0324	3.77	0.680	0.794	49.4×10^3
1000	0.0300	0.252	2.79	0.930	0.0366	4.85	0.691	0.685	25.6
1500	0.0224	0.264	3.39	1.52	0.0465	7.86	0.696	0.510	7.22

Oxygen

T (K)	ρ (kg/m³)	$c_p \times 10^{-3}$ (J/kg·K)	$\mu \times 10^5$ (Pa·s)	$\nu \times 10^5$ (m²/s)	$k \times 10^2$ (W/m·K)	$\alpha \times 10^5$ (m²/s)	Pr	$g\beta\rho^2/\mu^2$ (1/K·m³)
250	1.5620	0.9150	1.7887	1.1451	2.2586	1.5803	0.725	2.9885×10^8
300	1.3007	0.9199	2.0633	1.5863	2.6760	2.2365	0.709	1.2978
350	1.1144	0.9291	2.3176	2.0797	3.0688	2.9639	0.702	0.6469
400	0.9749	0.9417	2.5556	2.6214	3.4616	3.7705	0.695	0.3571
450	0.8665	0.9564	2.7798	3.2081	3.8298	4.6216	0.694	0.2108
500	0.7798	0.9721	2.9930	3.8382	4.1735	5.5056	0.697	0.1330
550	0.7089	0.9879	3.1966	4.5092	4.5172	6.4502	0.700	8.786×10^6
600	0.6498	1.0032	3.3931	5.2218	4.8364	7.4192	0.704	5.988

T (°F)	ρ (lb$_m$/ft^3)	c_p (Btu/lb$_m$ °F)	$\mu \times 10^5$ (lb$_m$/ft sec)	$\nu \times 10^3$ (ft^2/sec)	k (Btu/hr ft °F)	α (ft^2/hr)	Pr	$\beta \times 10^3$ (1/°F)	$g\beta\rho^2/\mu^2$ (1/°F · ft^3)
				Carbon dioxide					
0	0.132	0.193	0.865	0.0655	0.00760	0.298	0.792	2.18	16.3×10^6
30	0.124	0.198	0.915	0.0739	0.00830	0.339	0.787	2.04	12.0
60	0.117	0.202	0.965	0.0829	0.00910	0.387	0.773	1.92	9.00
80	0.112	0.204	1.00	0.0891	0.00960	0.421	0.760	1.85	7.45
100	0.108	0.207	1.03	0.0953	0.0102	0.455	0.758	1.79	6.33
150	0.100	0.213	1.12	0.113	0.0115	0.539	0.755	1.64	4.16
200	0.092	0.219	1.20	0.131	0.0130	0.646	0.730	1.52	2.86
250	0.0850	0.225	1.32	0.155	0.0148	0.777	0.717	1.41	2.04
300	0.0800	0.230	1.36	0.171	0.0160	0.878	0.704	1.32	1.45
400	0.0740	0.239	1.45	0.196	0.0180	1.02	0.695	1.16	1.11
500	0.0630	0.248	1.65	0.263	0.0210	1.36	0.700	1.04	0.485
600	0.0570	0.256	1.78	0.312	0.0235	1.61	0.700	0.944	0.310
800	0.0480	0.269	2.02	0.420	0.0278	2.15	0.702	0.794	0.143
1000	0.0416	0.280	2.25	0.540	0.0324	2.78	0.703	0.685	75.3×10^3
1500	0.0306	0.301	2.80	0.913	0.0340	4.67	0.704	0.510	19.6

Carbon dioxide

T (K)	ρ (kg/m^3)	$c_p \times 10^{-3}$ (J/kg $\cdot$ K)	$\mu \times 10^5$ (Pa $\cdot$ s)	$\nu \times 10^5$ (m^2/s)	$k \times 10^2$ (W/m $\cdot$ K)	$\alpha \times 10^5$ (m^2/s)	Pr	$g\beta\rho^2/\mu^2$ (1/K $\cdot$ m^3)
250	2.1652	0.8052	1.2590	0.5815	1.2891	0.7394	0.793	1.1591×10^9
300	1.7967	0.8526	1.4948	0.8320	1.6572	1.0818	0.770	0.4178
350	1.5369	0.8989	1.7208	1.1197	2.0457	1.4808	0.755	0.2232
400	1.3432	0.9416	1.9318	1.4382	2.4604	1.9454	0.738	0.1186
450	1.1931	0.9803	2.1332	1.7879	2.8955	2.4756	0.721	6.786×10^7
500	1.0733	1.0153	2.3251	2.1663	3.3523	3.0763	0.702	4.176
550	0.9756	1.0470	2.5073	2.5700	3.8208	3.7406	0.685	2.705
600	0.8941	1.0761	2.6827	3.0004	4.3097	4.4793	0.668	1.814

T (°F)	ρ (lb$_m$/ft³)	c_p (Btu/lb$_m$ °F)	$\mu \times 10^5$ (lb$_m$/ft sec)	$\nu \times 10^3$ (ft²/sec)	k (Btu/hr ft °F)	α (ft²/hr)	Pr	$\beta \times 10^3$ (1/°F)	$g\beta\rho^2/\mu^2$ (1/°F · ft³)
					Hydrogen				
0	0.00597	3.37	0.537	0.900	0.092	4.59	0.713	2.18	87 000
30	0.00562	3.39	0.562	1.00	0.097	5.09	0.709	2.04	65 700
60	0.00530	3.41	0.587	1.11	0.102	5.65	0.707	1.92	50 500
80	0.00510	3.42	0.602	1.18	0.105	6.04	0.705	1.85	42 700
100	0.00492	3.42	0.617	1.25	0.108	6.42	0.700	1.79	36 700
150	0.00450	3.44	0.653	1.45	0.116	7.50	0.696	1.64	25 000
200	0.00412	3.45	0.688	1.67	0.123	8.64	0.696	1.52	17 500
250	0.00382	3.46	0.723	1.89	0.130	9.85	0.690	1.41	12 700
300	0.00357	3.46	0.756	2.12	0.137	11.1	0.687	1.32	9 440
400	0.00315	3.47	0.822	2.61	0.151	13.8	0.681	1.16	5 470
500	0.00285	3.47	0.890	3.12	0.165	16.7	0.675	1.04	3 430
600	0.00260	3.47	0.952	3.66	0.179	19.8	0.667	0.944	2 270
800	0.00219	3.49	1.07	4.87	0.205	26.8	0.654	0.794	1 080
1000	0.00189	3.52	1.18	6.21	0.224	33.7	0.664	0.685	571
1500	0.00141	3.62	1.44	10.2	0.265	51.9	0.708	0.510	158

Hydrogen

T (K)	ρ (kg/m³)	c_p (J/kg·K)	$\mu \times 10^6$ (Pa·s)	$\nu \times 10^6$ (m²/s)	k (W/m·K)	$\alpha \times 10^4$ (m²/s)	Pr	$g\beta\rho^2/\mu^2 \times 10^{-6}$ (1/K·m³)
50	0.5095	10.501	2.516	4.938	0.0362	0.0633	0.78	
100	0.2457	11.229	4.212	17.143	0.0665	0.2410	0.711	333.8
150	0.1637	12.602	5.595	34.178	0.0981	0.4755	0.719	55.99
200	0.1227	13.504	6.813	55.526	0.1282	0.7717	0.719	15.90
250	0.0982	14.059	7.919	80.641	0.1561	1.131	0.713	6.03
300	0.0818	14.314	8.963	109.57	0.182	1.554	0.705	2.72
350	0.0702	14.436	9.954	141.79	0.206	2.033	0.697	1.39
400	0.0613	14.491	10.864	177.23	0.228	2.567	0.690	0.782
450	0.0546	14.499	11.779	215.73	0.251	3.171	0.680	0.468
500	0.0492	14.507	12.636	256.83	0.272	3.811	0.674	0.297
600	0.0408	14.537	14.285	350.12	0.315	5.311	0.659	0.134
700	0.0349	14.574	15.890	455.30	0.351	6.901	0.660	0.0677
800	0.0306	14.675	17.40	568.63	0.384	8.551	0.665	0.0379
1000	0.0245	14.968	20.160	822.86	0.440	11.998	0.686	00145
1200	0.0205	15.366	22.75	1109.80	0.488	15.492	0.716	0.00667

T ($°F$)	ρ (lb_m/ft^3)	c_p ($Btu/lb_m\,°F$)	$\mu \times 10^5$ ($lb_m/ft\,sec$)	$\nu \times 10^3$ (ft^2/sec)	k ($Btu/hr\,ft\,°F$)	α (ft^2/hr)	Pr	$\beta \times 10^3$ ($1/°F$)	$g\beta\rho^2/\mu^2$ ($1/°F \cdot ft^3$)
					Carbon monoxide				
0	0.0832	0.249	1.05	0.126	0.0128	0.620	0.749	2.18	4.40×10^6
30	0.0780	0.249	1.11	0.142	0.0134	0.691	0.744	2.04	3.32
60	0.0736	0.249	1.16	0.157	0.0142	0.775	0.740	1.92	2.48
80	0.0709	0.249	1.20	0.169	0.0146	0.828	0.737	1.85	2.09
100	0.0684	0.249	1.23	0.180	0.0150	0.884	0.735	1.79	1.79
150	0.0628	0.249	1.32	0.210	0.0163	1.04	0.730	1.64	1.19
200	0.0580	0.250	1.40	0.241	0.0174	1.20	0.726	1.52	0.842
250	0.0539	0.250	1.48	0.275	0.0183	1.36	0.722	1.41	0.604
300	0.0503	0.251	1.56	0.310	0.0196	1.56	0.720	1.32	0.442
400	0.0445	0.253	1.73	0.389	0.0217	1.92	0.718	1.16	0.248
500	0.0399	0.256	1.85	0.463	0.0234	2.30	0.725	1.04	0.156
600	0.0361	0.259	1.97	0.545	0.0253	2.71	0.723	0.944	0.101
800	0.0304	0.266	2.21	0.728	0.0288	3.57	0.730	0.794	48.2×10^3
1000	0.0262	0.273	2.43	0.929	0.0324	4.54	0.740	0.685	25.6
1500	0.0195	0.286	3.00	1.54	0.0410	7.35	0.756	0.510	6.93

T (K)	ρ (kg/m³)	$c_p \times 10^{-3}$ (J/kg·K)	$\mu \times 10^5$ (Pa·s)	$\nu \times 10^5$ (m²/s)	$k \times 10^2$ (W/m·K)	$\alpha \times 10^5$ (m²/s)	Pr	$g\beta\rho^2/\mu^2$ (1/K·m³)
				Carbon monoxide				
250	1.3669	1.0425	1.5408	1.1272	2.1432	1.5040	0.749	3.0841×10^8
300	1.1382	1.0422	1.7854	1.5686	2.5240	2.1277	0.737	1.3273
350	0.9753	1.0440	2.0097	2.0606	2.8839	2.8323	0.727	0.6590
400	0.8532	1.0484	2.2201	2.6021	3.2253	3.6057	0.722	0.3623
450	0.7583	1.0550	2.4189	3.1899	3.5527	4.4408	0.718	0.2133
500	0.6824	1.0642	2.6078	3.8215	3.8638	5.3205	0.718	0.1342
550	0.6204	1.0751	2.7884	4.4945	4.1587	6.2350	0.721	8.843×10^6
600	0.5687	1.0870	2.9607	5.2061	4.4443	7.1894	0.724	6.025

Chlorine

T (°F)	ρ (lb$_m$/ft^3)	c_p (Btu/lb$_m$ °F)	$\mu \times 10^6$ (lb$_m$/ft sec)	$\nu \times 10^3$ (ft^2/sec)	k (Btu/hr ft °F)	α (ft^2/hr)	Pr	$\beta \times 10^3$ (1/°F)	$g\beta\rho^2/\mu^2 \times 10^{-6}$ (1/°F · ft^3)
0	0.211	0.113	8.06	0.0381	0.00418	0.175	0.785	2.18	48.3
30	0.197	0.114	8.40	0.0426	0.00450	0.201	0.769	2.04	36.6
60	0.187	0.114	8.80	0.0470	0.00480	0.225	0.753	1.92	28.1
80	0.180	0.115	9.07	0.0504	0.00500	0.242	0.753	1.85	24.3
100	0.173	0.115	9.34	0.0540	0.00520	0.261	0.748	1.79	19.9
150	0.159	0.117	10.0	0.0629	0.00570	0.306	0.739	1.64	13.4

Helium

T (°F)	ρ (lb$_m$/ft³)	c_p (Btu/lb$_m$ °F)	$\mu \times 10^7$ (lb$_m$/ft sec)	$\nu \times 10^3$ (ft²/sec)	k (Btu/hr ft °F)	α (ft²/hr)	Pr	$\beta \times 10^3$ (1/°F)	$g\beta\rho^2/\mu^2$ (1/°F · ft³)
0	0.0119	1.24	122	1.03	0.0784	5.30	0.698	2.18	66 800
30	0.0112	1.24	127	1.14	0.0818	5.89	0.699	2.04	51 100
60	0.0106	1.24	132	1.25	0.0852	6.46	0.700	1.92	40 000
80	0.0102	1.24	135	1.32	0.0872	6.88	0.701	1.85	33 900
100	0.00980	1.24	138	1.41	0.0892	7.37	0.701	1.79	29 000
150	0.00900	1.24	146	1.63	0.0937	8.36	0.703	1.64	20 100
200	0.00829	1.24	155	1.87	0.0977	9.48	0.705	1.52	14 000
250	0.00772	1.24	162	2.09	0.102	10.7	0.707	1.41	10 400
300	0.00722	1.24	170	2.36	0.106	11.8	0.709	1.32	7 650
400	0.00637	1.24	185	2.91	0.114	14.4	0.714	1.16	4 410
500	0.00572	1.24	198	3.46	0.122	17.1	0.719	1.04	2 800
600	0.00517	1.24	209	4.04	0.130	20.6	0.720	0.994	1 850
800	0.00439	1.24	232	5.28	0.145	27.6	0.722	0.794	915
1000	0.00376	1.24	255	6.78	0.159	35.5	0.725	0.685	480
1500	0.00280	1.24	309	11.1	0.189	59.7	0.730	0.510	135

T (°F)	ρ (lb$_m$/ft³)	c_p (Btu/lb$_m$ °F)	$\mu \times 10^5$ (lb$_m$/ft sec)	$\nu \times 10^3$ (ft²/sec)	k (Btu/hr ft °F)	α (ft²/hr)	Pr	$\beta \times 10^3$ (1/°F)	$g\beta\rho^2/\mu^2$ (1/°F · ft³)
				Sulfur dioxide					
0	0.195	0.142	0.700	3.59	0.00460	0.166	0.778	2.03	50.6×10^6
100	0.161	0.149	0.890	5.52	0.00560	0.233	0.854	1.79	19.0
200	0.136	0.157	1.05	7.74	0.00670	0.313	0.883	1.52	8.25
300	0.118	0.164	1.20	10.2	0.00790	0.407	0.898	1.32	4.12
400	0.104	0.170	1.35	13.0	0.00920	0.520	0.898	1.16	2.24
500	0.0935	0.176	1.50	16.0	0.00990	0.601	0.958	1.04	1.30
600	0.0846	0.180	1.65	19.5	0.0108	0.711	0.987	0.994	0.795

Liquids

Water

T (°F)	ρ (lb$_m$/ft³)	c_p (Btu/lb$_m$ °F)	$\mu \times 10^3$ (lb$_m$/ft sec)	$\nu \times 10^5$ (ft²/sec)	k (Btu/hr ft °F)	$\alpha \times 10^3$ (ft²/hr)	Pr	$\beta \times 10^4$ (1/°F)	$g\beta\rho^2/\mu^2 \times 10^{-6}$ (1/°F·ft³)
32	62.4	1.01	1.20	1.93	0.319	5.06	13.7	-0.350	
60	62.3	1.00	0.760	1.22	0.340	5.45	8.07	0.800	17.2
80	62.2	0.999	0.578	0.929	0.353	5.67	5.89	1.30	48.3
100	62.1	0.999	0.458	0.736	0.364	5.87	4.51	1.80	107
150	61.3	1.00	0.290	0.474	0.383	6.26	2.72	2.80	403
200	60.1	1.01	0.206	0.342	0.392	6.46	1.91	3.70	1 010
250	58.9	1.02	0.160	0.272	0.395	6.60	1.49	4.70	2 045
300	57.3	1.03	0.130	0.227	0.395	6.70	1.22	5.60	3 510
400	53.6	1.08	0.0930	0.174	0.382	6.58	0.950	7.80	8 350
500	49.0	1.19	0.0700	0.143	0.349	5.98	0.859	11.0	17 350
600	42.4	1.51	0.0579	0.137	0.293	4.58	1.07	17.5	30 300

Water

T (K)	ρ (kg/m³)	c_p (J/kg·K)	$\mu \times 10^6$ (Pa·s)	$\nu \times 10^6$ (m²/s)	k (W/m·K)	$\alpha \times 10^6$ (m²/s)	Pr	$g\beta\rho^2/\mu^2 \times 10^{-9}$ (1/K·m³)
273	999.3	4226	1794	1.795	0.558	0.132	13.6	
293	998.2	4182	993	0.995	0.597	0.143	6.96	2.035
313	992.2	4175	658	0.663	0.633	0.153	4.33	8.833
333	983.2	4181	472	0.480	0.658	0.160	3.00	22.75
353	971.8	4194	352	0.362	0.673	0.165	2.57	46.68
373	958.4	4211	278	0.290	0.682	0.169	1.72	85.09
473	862.8	4501	139	0.161	0.665	0.171	0.94	517.2
573	712.5	5694	92.2	0.129	0.564	0.139	0.93	1766.0

Aniline

T (°F)	ρ (lb$_m$/ft³)	c_p (Btu/lb$_m$°F)	$\mu \times 10^5$ (lb$_m$/ft sec)	$\nu \times 10^5$ (ft²/sec)	k (Btu/hr ft°F)	$\alpha \times 10^3$ (ft²/hr)	Pr	$\beta \times 10^3$ (1/°F)	$g\beta\rho^2/\mu^2 \times 10^{-6}$ (1/°F·ft³)
60	64.0	0.480	305	4.77	0.101	3.29	52.3		
80	63.5	0.485	240	3.78	0.100	3.25	41.8		
100	63.0	0.490	180	2.86	0.100	3.24	31.8	0.45	17.7
150	61.6	0.503	100	1.62	0.0980	3.16	18.4		
200	60.2	0.515	62	1.03	0.0962	3.10	12.0		
250	58.9	0.527	42	0.714	0.0947	3.05	8.44		
300	57.5	0.540	30	0.522	0.0931	2.99	6.28		

Ammonia

T (°F)	ρ (lb$_m$/ft³)	c_p (Btu/lb$_m$°F)	$\mu \times 10^5$ (lb$_m$/ft sec)	$\nu \times 10^5$ (ft²/sec)	k (Btu/hr ft°F)	$\alpha \times 10^3$ (ft²/hr)	Pr	$\beta \times 10^3$ (1/°F)	$g\beta\rho^2/\mu^2 \times 10^{-7}$ (1/°F·ft³)
−60	43.9	1.07	20.6	0.471	0.316	6.74	2.52	0.94	132
−30	42.7	1.07	18.2	0.426	0.317	6.93	2.22	1.02	265
0	41.3	1.08	16.9	0.409	0.315	7.06	2.08	1.1	467
30	40.0	1.11	16.2	0.402	0.312	7.05	2.05	1.19	757
60	38.5	1.14	15.0	0.391	0.304	6.92	2.03	1.3	1130
80	37.5	1.16	14.2	0.379	0.296	6.79	2.01	1.4	1650
100	36.4	1.19	13.5	0.368	0.287	6.62	2.00	1.5	2200
120	35.3	1.22	12.6	0.356	0.275	6.43	2.00	1.68	3180

Freon-12

T (°F)	ρ (lb$_m$/ft³)	c_p (Btu/lb$_m$ °F)	$\mu \times 10^5$ (lb$_m$/ft sec)	$\nu \times 10^5$ (ft²/sec)	k (Btu/hr ft °F)	$\alpha \times 10^3$ (ft²/hr)	Pr	$\beta \times 10^4$ (1/°F)	$g\beta\rho^2/\mu^2 \times 10^{-6}$ (1/°F·ft³)
-40	94.5	0.202	125	1.32	0.0650	3.40	14.0	9.10	168
-30	93.5	0.204	123	1.32	0.0640	3.35	14.1	9.60	179
0	90.9	0.212	116	1.28	0.0578	3.00	15.4	11.4	225
30	87.4	0.221	108	1.24	0.0564	2.92	15.3	13.1	277
60	84.0	0.230	99.6	1.19	0.0528	2.74	15.6	14.9	341
80	81.3	0.238	94.0	1.16	0.0504	2.60	16.0	16.0	384
100	78.7	0.246	88.4	1.12	0.0480	2.48	16.3	17.2	439
150	71.0	0.271	74.8	1.05	0.0420	2.18	17.4	19.5	625

n-Butyl Alcohol

T (°F)	ρ (lb$_m$/ft³)	c_p (Btu/lb$_m$ °F)	$\mu \times 10^5$ (lb$_m$/ft sec)	$\nu \times 10^5$ (ft²/sec)	k (Btu/hr ft °F)	$\alpha \times 10^3$ (ft²/hr)	Pr	$\beta \times 10^3$ (1/°F)	$g\beta\rho^2/\mu^2 \times 10^{-6}$ (1/°F·ft³)
60	50.5	0.55	225	4.46	0.100	3.59	44.6		
80	50.0	0.58	180	3.60	0.099	3.41	38.0	0.25	6.23
100	49.6	0.61	130	2.62	0.098	3.25	29.1	0.43	2.02
150	48.5	0.68	68	1.41	0.098	2.97	17.1		

Benzene

T (°F)	ρ (lb$_m$/ft³)	c_p (Btu/lb$_m$ °F)	$\mu \times 10^5$ (lb$_m$/ft sec)	$\nu \times 10^5$ (ft²/sec)	k (Btu/hr ft °F)	$\alpha \times 10^3$ (ft²/hr)	$Pr \times 10^{-2}$	$\beta \times 10^4$ (1/°F)	$g\beta\rho^2/\mu^2 \times 10^{-6}$ (1/°F · ft³)
60	55.2	0.395	44.5	0.806	0.0856	3.93	7.39		
80	54.6	0.410	38	0.695	0.0836	3.73	6.70		
100	53.6	0.420	33	0.615	0.0814	3.61	6.13	7.5	498
150	51.8	0.450	24.5	0.473	0.0762	3.27	5.21	7.2	609
200	49.9	0.480	19.4	0.390	0.0711	2.97	4.73	6.8	980

Hydraulic fluid (MIL-M-5606)

T (°F)	ρ (lb$_m$/ft³)	c_p (Btu/lb$_m$ °F)	$\mu \times 10^5$ (lb$_m$/ft sec)	$\nu \times 10^5$ (ft²/sec)	k (Btu/hr ft °F)	$\alpha \times 10^3$ (ft²/hr)	Pr	$\beta \times 10^3$ (1/°F)	$g\beta\rho^2/\mu^2 \times 10^{-4}$ (1/°F · ft³)
0	55.0	0.400	5550	101	0.0780	3.54	1030	0.76	2.39
30	54.0	0.420	2220	41.1	0.0755	3.32	446	0.68	13.0
60	53.0	0.439	1110	20.9	0.0732	3.14	239	0.60	44.1
80	52.5	0.453	695	13.3	0.0710	3.07	155	0.52	95.7
100	52.0	0.467	556	10.7	0.0690	2.84	136	0.47	132
150	51.0	0.499	278	5.45	0.0645	2.44	80.5	0.32	346
200	50.0	0.530	250	5.00	0.0600	2.27	79.4	0.20	258

Glycerin

T (°F)	ρ (lb$_m$/ft³)	c_p (Btu/lb$_m$ °F)	μ (lb$_m$/ft sec)	$\nu \times 10^2$ (ft²/sec)	k (Btu/hr ft °F)	$\alpha \times 10^3$ (ft²/hr)	$Pr \times 10^{-2}$	$\beta \times 10^3$ (1/°F)	$g\beta\rho^2/\mu^2$ (1/°F·ft³)
30	79.7	0.540	7.2	9.03	0.168	3.91	832		
60	79.1	0.563	1.4	1.77	0.167	3.75	170		
80	78.7	0.580	0.6	0.762	0.166	3.64	75.3	0.30	166
100	78.2	0.598	0.1	0.128	0.165	3.53	13.1		

Kerosene

T (°F)	ρ (lb$_m$/ft³)	c_p (Btu/lb$_m$ °F)	$\mu \times 10^5$ (lb$_m$/ft sec)	$\nu \times 10^5$ (ft²/sec)	k (Btu/hr ft °F)	$\alpha \times 10^3$ (ft²/hr)	Pr	$\beta \times 10^3$ (1/°F)	$g\beta\rho^2/\mu^2 \times 10^{-4}$ (1/°F·ft³)
30	48.8	0.456	800	16.4	0.0809	3.63	163		
60	48.1	0.474	600	12.5	0.0805	3.53	127		
80	47.6	0.491	490	10.3	0.0800	3.42	108	0.58	120
100	47.2	0.505	420	8.90	0.0797	3.35	95.7	0.48	146
150	46.1	0.540	320	6.83	0.0788	3.16	77.9	0.47	192

T ($°F$)	ρ (lb_m/ft^3)	c_p ($Btu/lb_m\,°F$)	$\mu \times 10^5$ ($lb_m/ft\,sec$)	$\nu \times 10^5$ (ft^2/sec)	k ($Btu/hr\,ft\,°F$)	$\alpha \times 10^3$ (ft^2/hr)	Pr	$\beta \times 10^3$ ($1/°F$)	$g\beta\rho^2/\mu^2 \times 10^{-4}$ ($1/°F \cdot ft^3$)
				Liquid hydrogen					
−435	4.84	1.69	1.63	0.337	0.0595	7.28	1.67		
−433	4.77	1.78	1.52	0.319	0.0610	7.20	1.59		
−431	4.71	1.87	1.40	0.297	0.0625	7.09	1.51	7.1	2.59
−429	4.64	1.96	1.28	0.276	0.0640	7.03	1.41		
−427	4.58	2.05	1.17	0.256	0.0655	6.97	1.32		
−425	4.51	2.15	1.05	0.233	0.0670	6.90	1.21		

T ($°F$)	ρ (lb_m/ft^3)	c_p ($Btu/lb_m\,°F$)	$\mu \times 10^5$ ($lb_m/ft\,sec$)	$\nu \times 10^5$ (ft^2/sec)	$k \times 10^3$ ($Btu/hr\,ft\,°F$)	$\alpha \times 10^5$ (ft^2/hr)	Pr	$\beta \times 10^3$ ($1/°F$)	$g\beta\rho^2/\mu^2 \times 10^{-8}$ ($1/°F \cdot ft^3$)
				Liquid oxygen					
−350	80.1	0.400	38.0	0.474	3.1	9.67	172		
−340	78.5	0.401	28.0	0.356	3.4	10.8	109		
−330	76.8	0.402	21.8	0.284	3.7	12.0	85.0		
−320	75.1	0.404	17.4	0.232	4.0	12.2	63.5	3.19	186
−310	73.4	0.405	14.8	0.202	4.3	14.5	50.1		
−300	71.7	0.406	13.0	0.181	4.6	15.8	41.2		

Bismuth

T (°F)	ρ (lb$_m$/ft^3)	c_p (Btu/lb$_m$ °F)	$\mu \times 10^3$ (lb$_m$/ft sec)	$\nu \times 10^6$ (ft^2/sec)	k (Btu/hr ft °F)	α (ft^2/hr)	Pr	$\beta \times 10^3$ (1/°F)	$g\beta\rho^2/\mu^2 \times 10^{-9}$ (1/°F·ft^3)
600	625	0.0345	1.09	1.75	8.58	0.397	0.0159		
700	622	0.0353	0.990	1.59	8.87	0.405	0.0141	0.062	0.786
800	618	0.0361	0.900	1.46	9.16	0.408	0.0129	0.065	0.985
900	613	0.0368	0.830	1.35	9.44	0.418	0.0116	0.068	1.19
1000	608	0.0375	0.765	1.26	9.74	0.427	0.0106	0.071	1.45
1100	604	0.0381	0.710	1.17	10.0	0.435	0.00970	0.074	1.72
1200	599	0.0386	0.660	1.10	10.3	0.446	0.00895	0.077	2.04
1300	595	0.0391	0.620	1.04	10.6	0.456	0.00820		

Mercury

T (°F)	ρ (lb$_m$/ft^3)	c_p (Btu/lb$_m$°F)	$\mu \times 10^3$ (lb$_m$/ft sec)	$\nu \times 10^6$ (ft^2/sec)	k (Btu/hr ft °F)	α (ft^2/hr)	Pr	$\beta \times 10^3$ (1/°F)	$g\beta\rho^2/\mu^2 \times 10^{-9}$ (1/°F·ft^3)
40	848	0.0334	1.11	1.31	4.55	0.161	0.0292		1.57
60	847	0.0333	1.05	1.24	4.64	0.165	0.0270		1.76
80	845	0.0332	1.00	1.18	4.72	0.169	0.0252		1.94
100	843	0.0331	0.960	1.14	4.80	0.172	0.0239		2.09
150	839	0.0330	0.893	1.06	5.03	0.182	0.0210		2.38
200	835	0.0328	0.850	1.02	5.25	0.192	0.0191		2.62
250	831	0.0328	0.806	0.970	5.45	0.200	0.0175		2.87
300	827	0.0328	0.766	0.928	5.65	0.209	0.0160		3.16
400	819	0.0328	0.700	0.856	6.05	0.225	0.0137	0.084	3.70
500	811	0.0328	0.650	0.803	6.43	0.243	0.0119		4.12
600	804	0.0328	0.606	0.754	6.80	0.259	0.0105		4.80
800	789	0.0329	0.550	0.698	7.45	0.289	0.0087		5.54

Sodium

T ($°F$)	ρ (lb_m/ft^3)	c_p (Btu/lb_m $°F$)	$\mu \times 10^3$ ($lb_m/ft\ sec$)	$\nu \times 10^6$ (ft^2/sec)	k ($Btu/hr\ ft\ °F$)	α (ft^2/hr)	Pr	$\beta \times 10^3$ ($1/°F$)	$g\beta\rho^2/\mu^2 \times 10^{-6}$ ($1/°F \cdot ft^3$)
200	58.1	0.332	0.489	8.43	49.8	2.58	0.0118		68.0
250	57.6	0.328	0.428	7.43	49.3	2.60	0.0103		87.4
300	57.2	0.324	0.378	6.61	48.8	2.64	0.00903		110
400	56.3	0.317	0.302	5.36	47.3	2.66	0.00725		168
500	55.5	0.309	0.258	4.64	45.5	2.64	0.00633	0.15	224
600	54.6	0.305	0.224	4.11	43.1	2.58	0.00574		287
800	52.9	0.304	0.180	3.40	38.8	2.41	0.00510		418
1000	51.2	0.304	0.152	2.97	36.0	2.31	0.00463		548
1300	48.7	0.305	0.120	2.47	34.2	2.31	0.00385		795

APPENDIX J
MASS-TRANSFER DIFFUSION COEFFICIENTS IN BINARY SYSTEMS

TABLE J.1   BINARY MASS DIFFUSIVITIES IN GASES*

System	T, K	$D_{AB}P$, cm^2 atm/s	$D_{AB}P$, m^2 Pa/s
Air			
Ammonia	273	0.198	2.006
Aniline	298	0.0726	0.735
Benzene	298	0.0962	0.974
Bromine	293	0.091	0.923
Carbon dioxide	273	0.136	1.378
Carbon disulfide	273	0.0883	0.894
Chlorine	273	0.124	1.256
Diphenyl	491	0.160	1.621
Ethyl acetate	273	0.0709	0.718
Ethanol	298	0.132	1.337
Ethyl ether	293	0.0896	0.908
Iodine	298	0.0834	0.845
Methanol	298	0.162	1.641
Mercury	614	0.473	4.791
Naphthalene	298	0.0611	0.619
Nitrobenzene	298	0.0868	0.879
n-Octane	298	0.0602	0.610
Oxygen	273	0.175	1.773
Propyl acetate	315	0.092	0.932
Sulfur dioxide	273	0.122	1.236
Toluene	298	0.0844	0.855
Water	298	0.260	2.634
Ammonia			
Ethylene	293	0.177	1.793

* R. C. Reid and T. K. Sherwood, *The Properties of Gases and Liquids*, McGraw-Hill Book Company, New York, 1958, Chap. 8.

System	T, K	$D_{AB}P$, cm^2 atm/s	$D_{AB}P$, m^2 Pa/s
Argon			
Neon	293	0.329	3.333
Carbon dioxide			
Benzene	318	0.0715	0.724
Carbon disulfide	318	0.0715	0.724
Ethyl acetate	319	0.0666	0.675
Ethanol	273	0.0693	0.702
Ethyl ether	273	0.0541	0.548
Hydrogen	273	0.550	5.572
Methane	273	0.153	1.550
Methanol	298.6	0.105	1.064
Nitrogen	298	0.165	1.672
Nitrous oxide	298	0.117	1.185
Propane	298	0.0863	0.874
Water	298	0.164	1.661
Carbon monoxide			
Ethylene	273	0.151	1.530
Hydrogen	273	0.651	6.595
Nitrogen	288	0.192	1.945
Oxygen	273	0.185	1.874
Helium			
Argon	273	0.641	6.493
Benzene	298	0.384	3.890
Ethanol	298	0.494	5.004
Hydrogen	293	1.64	16.613
Neon	293	1.23	12.460
Water	298	0.908	9.198
Hydrogen			
Ammonia	293	0.849	8.600
Argon	293	0.770	7.800
Benzene	273	0.317	3.211
Ethane	273	0.439	4.447
Methane	273	0.625	6.331
Oxygen	273	0.697	7.061
Water	293	0.850	8.611
Nitrogen			
Ammonia	293	0.241	2.441
Ethylene	298	0.163	1.651
Hydrogen	288	0.743	7.527
Iodine	273	0.070	0.709
Oxygen	273	0.181	1.834

System	T, K	$D_{AB}P$, cm²atm/s	$D_{AB}P$, m² Pa/s
Oxygen			
Ammonia	293	0.253	2.563
Benzene	296	0.0939	0.951
Ethylene	293	0.182	1.844

TABLE J.2 BINARY MASS DIFFUSIVITIES IN LIQUIDS*

Solute A	Solvent B	Temperature, in K	Solute Concentration, in g mole/liter or kg mole/m³	Diffusivity cm²/s × 10⁵ or m²/s × 10⁹
Chlorine	Water	289	0.12	1.26
Hydrogen	Water	273	9	2.7
chloride			2	1.8
		283	9	3.3
			2.5	2.5
		289	0.5	2.44
Ammonia	Water	278	3.5	1.24
		288	1.0	1.77
Carbon dioxide	Water	283	0	1.46
		293	0	1.77
Sodium	Water	291	0.05	1.26
chloride			0.2	1.21
			1.0	1.24
			3.0	1.36
			5.4	1.54
Methanol	Water	288	0	1.28
Acetic acid	Water	285.5	1.0	0.82
			0.01	0.91
		291	1.0	0.96
Ethanol	Water	283	3.75	0.50
			0.05	0.83
		289	2.0	0.90
n-Butanol	Water	288	0	0.77
Carbon dioxide	Ethanol	290	0	3.2
Chloroform	Ethanol	293	2.0	1.25

* R. E. Treybal, *Mass Transfer Operations*, McGraw-Hill Book Company, New York, 1955, p. 25.

TABLE J.3 BINARY DIFFUSIVITIES IN SOLIDS*

Solute	Solid	K	Diffusivity, cm^2/s or $m^2/s \times 10^4$	Diffusivity, ft^2/hr
Helium	Pyrex	293	4.49×10^{-11}	1.74×10^{-10}
		773	2.00×10^{-8}	7.76×10^{-8}
Hydrogen	Nickel	358	1.16×10^{-8}	4.5×10^{-8}
		438	1.05×10^{-7}	4.07×10^{-7}
Bismuth	Lead	293	1.10×10^{-16}	4.27×10^{-16}
Mercury	Lead	293	2.50×10^{-15}	9.7×10^{-15}
Antimony	Silver	293	3.51×10^{-21}	1.36×10^{-20}
Aluminum	Copper	293	1.30×10^{-30}	5.04×10^{-30}
Cadmium	Copper	293	2.71×10^{-15}	1.05×10^{-14}

* R. M. Barrer, *Diffusion In and Through Solids*, The Macmillan Company, New York, 1941.

APPENDIX K
LENNARD-JONES CONSTANTS

TABLE K.1   THE COLLISION INTEGRALS, Ω_μ AND Ω_D, BASED ON THE LENNARD-JONES POTENTIAL*

$\kappa T/\epsilon$	$\Omega_\mu = \Omega_k$ (For viscosity and thermal conductivity)	Ω_D (For mass diffusivity)	$\kappa T/\epsilon$	$\Omega_\mu = \Omega_k$ (For viscosity and thermal conductivity)	Ω_D (For mass diffusivity)
			1.25	1.424	1.296
0.30	2.785	2.662	1.30	1.399	1.273
0.35	2.628	2.476	1.35	1.375	1.253
0.40	2.492	2.318	1.40	1.353	1.233
0.45	2.368	2.184	1.45	1.333	1.215
0.50	2.257	2.066	1.50	1.314	1.198
0.55	2.156	1.966	1.55	1.296	1.182
0.60	2.065	1.877	1.60	1.279	1.167
0.65	1.982	1.798	1.65	1.264	1.153
0.70	1.908	1.729	1.70	1.248	1.140
0.75	1.841	1.667	1.75	1.234	1.128
0.80	1.780	1.612	1.80	1.221	1.116
0.85	1.725	1.562	1.85	1.209	1.105
0.90	1.675	1.517	1.90	1.197	1.094
0.95	1.629	1.476	1.95	1.186	1.084
1.00	1.587	1.439	2.00	1.175	1.075
1.05	1.549	1.406	2.10	1.156	1.057
1.10	1.514	1.375	2.20	1.138	1.041
1.15	1.482	1.346	2.30	1.122	1.026
1.20	1.452	1.320	2.40	1.107	1.012

* Taken from J. O. Hirschfelder, R. B. Bird, and E. L. Spotz, *Chem. Revs.*, **44**, 205 (1949).

$\kappa T/\epsilon$	$\Omega_\mu = \Omega_k$ (For viscosity and thermal conductivity)	Ω_D (For mass diffusivity)	$\kappa T/\epsilon$	$\Omega_\mu = \Omega_k$ (For viscosity and thermal conductivity)	Ω_D (For mass diffusivity)
2.50	1.093	0.9996	4.50	0.9464	0.8610
2.60	1.081	0.9878	4.60	0.9422	0.8568
2.70	1.069	0.9770	4.70	0.9382	0.8530
2.80	1.058	0.9672	4.80	0.9343	0.8492
2.90	1.048	0.9576	4.90	0.9305	0.8456
3.00	1.039	0.9490	5.0	0.9269	0.8422
3.10	1.030	0.9406	6.0	0.8963	0.8124
3.20	1.022	0.9328	7.0	0.8727	0.7896
3.30	1.014	0.9256	8.0	0.8538	0.7712
3.40	1.007	0.9186	9.0	0.8379	0.7556
3.50	0.9999	0.9120	10.0	0.8242	0.7424
3.60	0.9932	0.9058	20.0	0.7432	0.6640
3.70	0.9870	0.8998	30.0	0.7005	0.6232
3.80	0.9811	0.8942	40.0	0.6718	0.5960
3.90	0.9755	0.8888	50.0	0.6504	0.5756
4.00	0.9700	0.8836	60.0	0.6335	0.5596
4.10	0.9649	0.8788	70.0	0.6194	0.5464
4.20	0.9600	0.8740	80.0	0.6076	0.5352
4.30	0.9553	0.8694	90.0	0.5973	0.5256
4.40	0.9507	0.8652	100.0	0.5882	0.5170

TABLE K.2 LENNARD-JONES FORCE CONSTANTS CALCULATED
FROM VISCOSITY DATA*

Compound	Formula	ϵ_A/κ, in K	σ, in Å
Acetylene	C_2H_2	185	4.221
Air		97	3.617
Argon	A	124	3.418
Arsine	AsH_3	281	4.06
Benzene	C_6H_6	440	5.270
Bromine	Br_2	520	4.268
i-Butane	C_4H_{10}	313	5.341
n-Butane	C_4H_{10}	410	4.997
Carbon dioxide	CO_2	190	3.996

* R. C. Reid and T. K. Sherwood, *The Properties of Gases and Liquids*, McGraw-Hill Book Company, New York, 1958.

Compound	Formula	ϵ_A/κ, in K	σ, in Å
Carbon disulfide	CS_2	488	4.438
Carbon monoxide	CO	110	3.590
Carbon tetrachloride	CCl_4	327	5.881
Carbonyl sulfide	COS	335	4.13
Chlorine	Cl_2	357	4.115
Chloroform	$CHCl_3$	327	5.430
Cyanogen	C_2N_2	339	4.38
Cyclohexane	C_6H_{12}	324	6.093
Ethane	C_2H_6	230	4.418
Ethanol	C_2H_5OH	391	4.455
Ethylene	C_2H_6	205	4.232
Fluorine	F_2	112	3.653
Helium	He	10.22	2.576
n-Heptane	C_7H_{16}	282†	8.88†
n-Hexane	C_6H_{14}	413	5.909
Hydrogen	H_2	33.3	2.968
Hydrogen chloride	HCl	360	3.305
Hydrogen iodide	HI	324	4.123
Iodine	I_2	550	4.982
Krypton	Kr	190	3.60
Methane	CH_4	136.5	3.822
Methanol	CH_3OH	507	3.585
Methylene chloride	CH_2Cl_2	406	4.759
Methyl chloride	CH_3Cl	855	3.375
Mercuric iodide	HgI_2	691	5.625
Mercury	Hg	851	2.898
Neon	Ne	35.7	2.789
Nitric Oxide	NO	119	3.470
Nitrogen	N_2	91.5	3.681
Nitrous Oxide	N_2O	220	3.879
n-Nonane	C_9H_{20}	240	8.448
n-Octane	C_8H_{18}	320	7.451
Oxygen	O_2	113	3.433
n-Pentane	C_5H_{12}	345	5.769
Propane	C_3H_8	254	5.061
Sulfur dioxide	SO_2	252	4.290
Water	H_2O	356	2.649
Xenon	Xe	229	4.055

† Calculated from virial coefficients.

APPENDIX L
THE ERROR FUNCTION*

ϕ	erf ϕ	ϕ	erf ϕ
0	0.0	0.85	0.7707
0.025	0.0282	0.90	0.7970
0.05	0.0564	0.95	0.8209
0.10	0.1125	1.0	0.8427
0.15	0.1680	1.1	0.8802
0.20	0.2227	1.2	0.9103
0.25	0.2763	1.3	0.9340
0.30	0.3286	1.4	0.9523
0.35	0.3794	1.5	0.9661
0.40	0.4284	1.6	0.9763
0.45	0.4755	1.7	0.9838
0.50	0.5205	1.8	0.9891
0.55	0.5633	1.9	0.9928
0.60	0.6039	2.0	0.9953
0.65	0.6420	2.2	0.9981
0.70	0.6778	2.4	0.9993
0.75	0.7112	2.6	0.9998
0.80	0.7421	2.8	0.9999

* J. Crank, *The Mathematics of Diffusion*, Oxford Univ. Press, London, 1958.

APPENDIX M
STANDARD PIPE SIZES

Nominal pipe size, in.	Outside diam., in.	Schedule No.	Wall thickness, in.	Inside diam., in.	Cross sectional area metal, in.^{2}	Inside sectional area, ft^2
$\frac{1}{8}$	0.405	40	0.068	0.269	0.072	0.00040
		80	0.095	0.215	0.093	0.00025
$\frac{1}{4}$	0.540	40	0.088	0.364	0.125	0.00072
		80	0.119	0.302	0.157	0.00050
$\frac{3}{8}$	0.675	40	0.091	0.493	0.167	0.00133
		80	0.126	0.423	0.217	0.00098
$\frac{1}{2}$	0.840	40	0.109	0.622	0.250	0.00211
		80	0.147	0.546	0.320	0.00163
		160	0.187	0.466	0.384	0.00118
$\frac{3}{4}$	1.050	40	0.113	0.824	0.333	0.00371
		80	0.154	0.742	0.433	0.00300
		160	0.218	0.614	0.570	0.00206
1	1.315	40	0.133	1.049	0.494	0.00600
		80	0.179	0.957	0.639	0.00499
		160	0.250	0.815	0.837	0.00362
$1\frac{1}{2}$	1.900	40	0.145	1.610	0.799	0.01414
		80	0.200	1.500	1.068	0.01225
		160	0.281	1.338	1.429	0.00976
2	2.375	40	0.154	2.067	1.075	0.02330
		80	0.218	1.939	1.477	0.02050
		160	0.343	1.689	2.190	0.01556
$2\frac{1}{2}$	2.875	40	0.203	2.469	1.704	0.03322
		80	0.276	2.323	2.254	0.02942
		160	0.375	2.125	2.945	0.02463
3	3.500	40	0.216	3.068	2.228	0.05130
		80	0.300	2.900	3.016	0.04587
		160	0.437	2.626	4.205	0.03761

Nominal pipe size, in.	Outside diam., in.	Schedule No.	Wall thickness, in.	Inside diam., in.	Cross sectional area metal, in.^{2}	Inside sectional area, ft^2
4	4.500	40	0.237	4.026	3.173	0.08840
		80	0.337	3.826	4.407	0.07986
		120	0.437	3.626	5.578	0.07170
		160	0.531	3.438	6.621	0.06447
5	5.563	40	0.258	5.047	4.304	0.1390
		80	0.375	4.813	6.112	0.1263
		120	0.500	4.563	7.963	0.1136
		160	0.625	4.313	9.696	0.1015
6	6.625	40	0.280	6.065	5.584	0.2006
		80	0.432	5.761	8.405	0.1810
		120	0.562	5.501	10.71	0.1650
		160	0.718	5.189	13.32	0.1469
8	8.625	20	0.250	8.125	6.570	0.3601
		30	0.277	8.071	7.260	0.3553
		40	0.322	7.981	8.396	0.3474
		60	0.406	7.813	10.48	0.3329
		80	0.500	7.625	12.76	0.3171
		100	0.593	7.439	14.96	0.3018
		120	0.718	7.189	17.84	0.2819
		140	0.812	7.001	19.93	0.2673
		160	0.906	6.813	21.97	0.2532
10	10.75	20	0.250	10.250	8.24	0.5731
		30	0.307	10.136	10.07	0.5603
		40	0.365	10.020	11.90	0.5475
		60	0.500	9.750	16.10	0.5158
		80	0.593	9.564	18.92	0.4989
		100	0.718	9.314	22.63	0.4732
		120	0.843	9.064	26.34	0.4481
		140	1.000	8.750	30.63	0.4176
		160	1.125	8.500	34.02	0.3941
12	12.75	20	0.250	12.250	9.82	0.8185
		30	0.330	12.090	12.87	0.7972
		40	0.406	11.938	15.77	0.7773
		60	0.562	11.626	21.52	0.7372
		80	0.687	11.376	26.03	0.7058
		100	0.843	11.064	31.53	0.6677
		120	1.000	10.750	36.91	0.6303
		140	1.125	10.500	41.08	0.6013
		160	1.312	10.126	47.14	0.5592

Outside diam., in.	Wall thickness		Inside diam., in.	Cross-sectional area, ft^2	Inside sectional area, ft^2
	B.W.G. and Stubs' gage	in.			
$\frac{1}{2}$	12	0.109	0.282	0.1338	0.000433
	14	0.083	0.334	0.1087	0.000608
	16	0.065	0.370	0.0888	0.000747
	18	0.049	0.402	0.0694	0.000882
	20	0.035	0.430	0.0511	0.001009
$\frac{3}{4}$	12	0.109	0.532	0.2195	0.00154
	13	0.095	0.560	0.1955	0.00171
	14	0.083	0.584	0.1739	0.00186
	15	0.072	0.606	0.1534	0.00200
	16	0.065	0.620	0.1398	0.00210
	17	0.058	0.634	0.1261	0.00219
	18	0.049	0.652	0.1079	0.00232
1	12	0.109	0.782	0.3051	0.00334
	13	0.095	0.810	0.2701	0.00358
	14	0.083	0.834	0.2391	0.00379
	15	0.072	0.856	0.2099	0.00400
	16	0.065	0.870	0.1909	0.00413
	17	0.058	0.884	0.1716	0.00426
	18	0.049	0.902	0.1463	0.00444
$1\frac{1}{4}$	12	0.109	1.032	0.3907	0.00581
	13	0.095	1.060	0.3447	0.00613
	14	0.083	1.084	0.3042	0.00641
	15	0.072	1.106	0.2665	0.00677
	16	0.065	1.120	0.2419	0.00684
	17	0.058	1.134	0.2172	0.00701
	18	0.049	1.152	0.1848	0.00724

Outside diam., in.	Wall thickness		Inside diam., in.	Cross-sectional area, ft²	Inside sectional area, ft²
	B.W.G. and Stubs' gage	in.			
$1\frac{1}{2}$	12	0.109	1.282	0.4763	0.00896
	13	0.095	1.310	0.4193	0.00936
	14	0.083	1.334	0.3694	0.00971
	15	0.072	1.358	0.3187	0.0100
	16	0.065	1.370	0.2930	0.0102
	17	0.058	1.384	0.2627	0.0107
	18	0.049	1.402	0.2234	0.0109
$1\frac{3}{4}$	10	0.134	1.482	0.6803	0.0120
	11	0.120	1.510	0.6145	0.0124
	12	0.109	1.532	0.5620	0.0128
	13	0.095	1.560	0.4939	0.0133
	14	0.083	1.584	0.4346	0.0137
	15	0.072	1.606	0.3796	0.0141
	16	0.065	1.620	0.3441	0.0143
2	10	0.134	1.732	0.7855	0.0164
	11	0.120	1.760	0.7084	0.0169
	12	0.109	1.782	0.6475	0.0173
	13	0.095	1.810	0.5686	0.0179
	14	0.083	1.834	0.4998	0.0183
	15	0.072	1.856	0.4359	0.0188
	16	0.065	1.870	0.3951	0.0191

AUTHOR INDEX

SUBJECT INDEX